Safety of Biologics Therapy

Brian A. Baldo

Safety of Biologics Therapy

Monoclonal Antibodies, Cytokines, Fusion Proteins, Hormones, Enzymes, Coagulation Proteins, Vaccines, Botulinum Toxins

Springer

Brian A. Baldo (Retired)
AUA(Pharmacy); BSc(Hons); PhD
Formerly:
Royal North Shore Hospital of Sydney
Head, Molecular Immunology Unit
Kolling Institute of Medical Research and
University of Sydney
Sydney, NSW, Australia

ISBN 978-3-319-30470-0 ISBN 978-3-319-30472-4 (eBook)
DOI 10.1007/978-3-319-30472-4

Library of Congress Control Number: 2016943849

Printed on acid-free paper

This Springer imprint is published by Springer Nature
The registered company is Springer International Publishing AG Switzerland

Dedicated to the memories of my mother and father and to Gail MacDiarmid for the cherished years of support, partnership, and mutual devotion

Preface

In writing this book, the author's primary intention was to produce an up-to-date text book on approved biologic therapies as far as that is possible in this time of rapidly evolving and seemingly ever-expanding developments in biotherapeutic research and the introduction of new and novel biopharmaceuticals. Emergence of the disciplines of genomics and proteomics, together with molecular biological approaches to elucidate the functions of single genes, continues to reveal the complexities and multifaceted nature of diseases such as cancer, autoimmunity, and metabolic disorders and to identify potential targets for the development of new drug therapies. Targeted approaches, long practiced in relation to peptide hormones and enzymes, now so often drive the extraordinary interest in, and development of, monoclonal antibody, fusion protein, and cytokine therapies. Added stimulus has been provided by regulatory authorities in efforts to encourage the development of diagnostic agents and treatments for rare diseases previously neglected because of inadequate financial returns from very small markets. In particular, The US Food and Drug Authority (FDA) Office of Orphan Products Development provides incentives for the study and development of products for so-called orphan diseases, that is, diseases with fewer than 200,000 patients in the USA. This initiative has, for example, transformed the extent and nature of the research and development of enzymes as replacement therapies for lysosomal storage diseases and led to the introduction of monoclonal antibody therapy for the rare paroxymal nocturnal hemoglobinuria and atypical hemolytic uremic syndrome by targeting complement C5. These advances, among many others presented in this monograph, would almost certainly not have been made in the absence of recombinant DNA technology, today's sequencing methods, application of modern bioinformatics, and parallel proteome analyses by application of techniques such as mass spectroscopy.

An attempt has been made to cover those biologics that are currently the main product classes with regulatory approval in the USA and/or European Union and which show every indication of remaining important therapies over at least the next decade and beyond. Due to considerations of established therapeutic relevance and space constraints, coverage has been almost totally restricted to products given regulatory approval. This is reflected in the three chapters devoted to monoclonal antibodies, the dominant biotherapeutic agents in terms of diversity of target recognition and approved indications, and the products with the highest global sales revenue in today's biopharmaceutical market. Although there are well over 300 monoclonal antibodies in development, coverage here at June 2016 is restricted to the 50 (counting alemtuzumab [MabCampath® and Lemtrada®] and Denosumab [Prolia® and Xgeva®] as two not four antibodies) currently approved by the FDA or European Medicines Agency or both. Unsurprisingly, recombinant preparations dominate the different categories of approved biologics, and because of their inherent advantages including production of large quantities of pure human materials without the need to purify crude extracts, their ease of genetic and chemical manipulation to reduce side effects and accentuate or reduce selected properties, consistency of supply, minimal batch-to-batch variation, reduced cost of production, and their safety of manufacture, this is certain to continue. Coverage is extended to the relatively small number of cytokines approved for therapy out of more than 130 of these known pleiotropic immune modulators of immune and inflammatory responses and to the growing list of approved fusion proteins most of which are made up of an effector peptide (such as a cytokine, growth factor, etc.) linked to an antibody Fc fusion partner or human albumin. Known, studied, and used as therapies for many years, peptide and glycoprotein hormones, now mainly as recombinant products, are examined in some detail together with other related and/or modified hormone products produced to effect therapeutic improvements, alter pharmacokinetic and pharmacodynamic properties, or reduce adverse effects. In addition to enzymes as replacement therapies for lysosomal storage diseases, a number of other enzymes indicated for disorders as diverse as cystic fibrosis, Dupuytren's contracture, vitreomacular adhesion, myocardial infarction, and acute lymphocytic leukemia are examined. Descriptions, approved indications, and usage, of 22 approved coagulation or clotting factor preparations, essential for maintaining homeostasis, are reviewed together with the clotting cascade, an emphasis on safety aspects, and new product developments. Vaccination, an indispensable public health measure described as "the greatest triumph of modern immunology and the most successful exploitation of our knowledge of the workings of the immune response," has nevertheless not always been afforded the respect it deserves in modern medical practice. While vaccines are not free of associated adverse events, those that have been recorded together with the known and suspected effects induced by additives and possible contaminants are examined for all 46 approved vaccine preparations presented. Botulinum neurotoxins, surprising in their sheer number of clinical applications (a few approved, many not), which now include muscular, neurologic, gastrointestinal, urologic, ophthalmic, and oropharyngeal disorders, have exceeded the most optimistic early estimates of their usage. This already large list of approved and potential indications is currently being further enlarged by evaluations in off-label treatments outside controlled clinical trials. Such relatively uncontrolled activity is a reminder of the need to remain aware of the potentially extreme toxicity of the botulinum neurotoxins and to record and report adverse events when they occur. In the light of the seemingly exorbitant costs associated with many of today's biologic therapies, follow-on biologics or biosimilars offer the promise of fostering competition, allowing the treatment of more patients at lower cost, and helping to lower ever-increasing government health costs. Difficulties in achieving the required comparability or similarity, safety evaluations, and eventual regulatory approval are considered in the book's final chapter.

While efforts have been made to unify the text by cross-referencing and interconnecting common or related subjects, no comprehensive and scrupulous referencing to the original literature that is standard for scientific papers has been undertaken since this would have considerably increased the size of the book and been at odds with its intended organization and textbook style. As a reasonable compromise and with the aim of assisting the reader to locate original sources and extend understanding, carefully selected suggestions for Further Reading have been included at the end of each chapter. The chapters on cytokines, fusion proteins, and enzymes are based on the author's previous publications for Springer. These publications, quoted in the Further Reading lists, provide a comprehensive reference list for Chaps. 5, 6, and 9. Further Reading selections for these and other chapters have been selected to guide the interested reader to the most significant studies in the original literature and preview potentially important future developments.

It is with sincere appreciation and thanks that the author acknowledges the skills, dedication, cooperation, and help, given by long-standing collaborator Dr. Nghia H. Pham in the joint preparation of numerous tables and a widely diverse range of figures, many of the latter demonstrating individuality and produced with some artistry as well as relevance, scientific accuracy, and dedication to detail.

In conclusion, with biologic therapies continuing to demonstrate extraordinary growth in the origin and nature of the agents employed, the introduction of new disease targets, the enlarging range of approved indications, and the increasing understanding of each agent's spectrum of associated risks and adverse events, the continued development and innovation seen in biologic therapies over the last few years seem certain to be sustained. Given this ongoing expansion of new knowledge, research, and development in therapeutic biologics, the author remains open and ready to consider all comments in an ongoing effort to remain abreast of new developments, improve the book, and correct any errors.

Sydney, Australia Brian A. Baldo

Contents

Chapter 1
Approved Biologics Used for Therapy and Their Adverse Effects

Biologics

"Biologics" in this book refer to therapies that are prepared from materials made or expressed in living organisms. They may simply be isolated proteins such as enzymes and blood products or, as is increasingly the case, preparations produced by recombinant DNA technology. Biologics, sometimes also referred to as "biotherapeutics" or "biopharmaceuticals," are covered by a number of different definitions depending on the perspective of the interested party, with researchers from different disciplines, biotechnologists, chemists, clinicians, legislators, and regulatory agencies, to name only a few, requiring or excluding aspects that reflect their interest and involvement. Whereas a biologist, chemist, or clinician may see a biologic used for therapy as material derived from, or related to, a living organism, for example, cells, cell extracts, or molecules composed of protein, peptide, complex carbohydrate, lipid, or nucleic acid, a regulatory authority will also consider how such agents are to be classified and assessed for characterization, manufacturing, and control; product development; identity, purity, and potency; and so on. In other words, in a regulatory context, "biologics" does not necessarily correspond to common usage or usage in everyday medical research and likewise for the terms "biotechnology medicine/drug," "biological medical product/drug," and "biopharmaceuticals," used outside the regulatory environment.

US Guidelines

In 1902, in the Biologics Control Act passed by the US Congress, biologics and biologic products were defined as "any virus, therapeutic serum, toxin, antitoxin or analogous product applicable to the prevention, treatment or cure of diseases or injuries of man." Although this definition has changed over time, uncertainties, difficulties, and impreciseness have remained for agencies, many groups, and individuals

B.A. Baldo, *Safety of Biologics Therapy*, DOI 10.1007/978-3-319-30472-4_1

concerned and/or working with biologic therapeutics. In particular, the term "analogous" has remained undefined and open to varying degrees of relatedness. Since 1902, the Congress has expanded the list of biologics to include other products including vaccines, blood products, and some other proteins, but polypeptides prepared by chemical synthesis remained excluded. In the 1938 Federal Food, Drug, and Cosmetic Act, "drug" was defined as a substance for the investigation, prevention, or cure of disease, but no guidance was forthcoming to distinguish biologic and non-biologic drugs. However, in 1944, the Congress did declare that a requirement for a new drug application (NDA) did not apply to biologics. The latter are now marketed under the provisions of the Public Health Service Act requiring a Biologics License Application (BLA) showing the agent is "safe, pure, and potent." By 1947, hormones had been excluded from the list of biologics, and with the arrival of the new age of biotechnology in the mid- to late 1980s, US Food and Drug Administration (FDA) issued a policy statement saying that agents would be regarded as biologics "based on the intended use of each product on a case-by-case basis." Following an Intercenter Agreement, in June 2003, the FDA transferred some of the therapeutic biologic products that had formerly been reviewed and regulated by the Center for Biologics Evaluation and Research (CBER) to the Center for Drug Evaluation and Research (CDER). Therapeutic biological products transferred to the CDER include monoclonal antibodies (mAbs), cytokines, fusion proteins, some enzymes, growth factors, non-vaccine immunomodulators, and therapeutic proteins derived from animals, plants, and microorganisms and recombinant versions of these products. Remaining with the CBER are cellular products of human, animal, or bacterial origin; gene therapy products such as nucleic acids, viruses and genetically engineered microorganisms; vaccines; allergenic extracts; antitoxins, antivenins, and venoms; and blood, blood components, and plasma-derived products. The approved biologics covered in this monograph are comprised of therapeutic biological products from both the CDER and CBER lists. In 2012, the FDA issued a draft guidance addressing and distinguishing the long-standing proposed differences between proteins, peptides, and chemically synthesized polypeptides. A protein was defined as any alfa amino acid polymer with a specific defined sequence that is greater than 40 amino acids in size. From this definition, it followed that peptides have fewer than 40 amino acids and are therefore not proteins. A chemically synthesized polypeptide was defined as an alfa amino acid polymer that is made entirely by chemical synthesis and has fewer than 100 amino acids. Until the draft guidance is finalized, these definitions can be seen as proposals, but regardless of the definitions finally declared and adopted and for the coverage of biologics in this volume, peptides with less than 40 amino acids and chemically synthesized polypeptides (whether they contain fewer or more than 100 amino acids) with regulatory approval for therapeutic use in humans will be considered first and foremost as biologics regardless of size or method of preparation. In keeping with this approach, peptide hormones (Chap. 7) and glycoprotein hormones (Chap. 8) are logically included in the coverage of biologics licensed for marketing as approved therapeutic agents. Note that regardless of the method of manufacture, hormones require a NDA. In summary, in the US, distinguishing a product from other drugs and classifying it as a biologic on the basis of existing

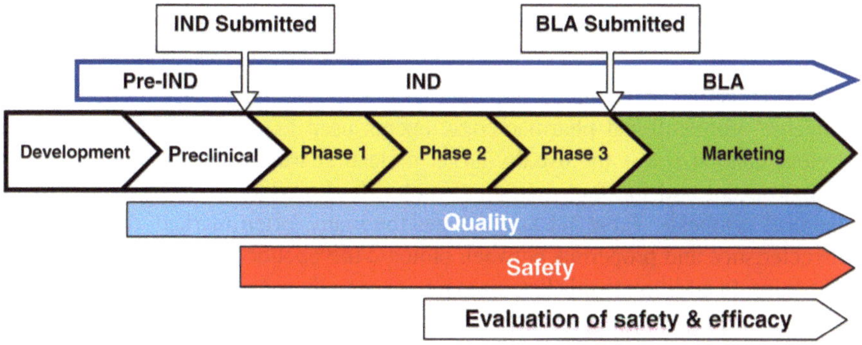

Fig. 1.1 Phases in biological product development under an investigational new drug (ING) application leading to clinical trials with evaluations for safety and efficacy, a Biologics License Application (BLA) and ultimately licensing and marketing. Reproduced from Vatsan RS, Bross PF, Liu K, et al. J Immunother Cancer 2013;1:5. http://www.immunotherapyofcancer.org/content/1/1/5, an open access article distributed under the terms of the Creative Commons Attribution License

definitions is not straightforward since detailed legislated guidance has never been provided. Recently, the FDA has mentioned the application of a so-called bright-line rule for distinguishing proteins suggesting a much-needed shift from ad hoc to jurisdictional decision-making.

From the early preliminary specifications for product characterization, Fig. 1.1 summarizes the phases in biological product development under an investigational new drug (IND) application leading on to clinical trials with evaluations for safety and efficacy, a BLA, and ultimately licensing and marketing.

European Guidelines

From a 2001 Directive of the European Parliament and the Council of Six on the community code relating to medicinal products for human use, "biological medicines" are defined as "products, the active substance of which is a biological substance." In turn, a "biological substance" is defined as "a substance that is produced by or extracted from a biological source and that needs for its characterization and the determination of its quality a combination of physico-chemical-biological testing, together with the production process and its control." In the European Union, the European Medicines Agency (EMA) defines a "biological medicinal product" as "a protein or nucleic acid-based pharmaceutical substance used for therapeutic or in vivo diagnostic purposes, which is produced by means other than direct extraction from a native (nonengineered) biological source." This definition essentially appears to restrict "biological medicinal products" to recombinant preparations including mAbs, cytokines, fusion proteins, some hormones (such as insulin, glucagon, growth hormone), enzymes (e.g., alteplase and enzymes used for enzyme replacement therapy), and coagulation proteins (factors VIIa, VIII, IX, XIII, and antithrombin).

Biologics and Small Molecule Drugs

In distinguishing a purified, homogeneous biologic agent from other "small molecule" conventional pharmacologic agents used as drugs, an immediate and obvious comparison is the usually larger molecular weight and complexity of the biologic agent and the fact that small molecule drugs are generally prepared by chemical synthesis. Table 1.1 summarizes the many, often marked, differences in characteristics and properties between biologic therapeutic agents and small molecule drugs. In addition to this list, biologics, especially mAbs and other recombinant proteins, may have a number of potential advantages as therapeutic agents. To begin with, they generally have a shorter development cycle which may be as little as 3–5 years compared to, typically, 7–8 years for most small molecule drugs. An often

Table 1.1 Comparison of properties of biologic products[a] and small molecule drugs

Biologics	Small molecule drugs
Generally high molecular weight[b]	Generally low molecular weight[c]
Often heterogeneous mixtures; may include variants	Homogeneous
Structure may not be well defined or known	Well-defined structure
Complex physicochemical characteristics	Physicochemically far less complex
May be synthesized (e.g., short peptides) but usually made with the aid of or from live cells and organisms	Usually organic molecules prepared by chemical synthesis
Usually many critical process steps in preparation and manufacture	Fewer critical steps involved
Usually not easily characterized	Not difficult to characterize and therefore usually well characterized
Often not stable; usually heat sensitive	Usually relatively stable
Controlled storage conditions often required to overcome instability	Often stable for storage at room temperature for long periods
Usually administered parenterally	Often administered orally
May have long half-life (days to weeks) allowing daily to monthly dosing	Relatively short half-life with dosing required every few hours
High selectivity and specificity for target	High potential for off-target effects
Catabolized to amino acids, sugars, lipids etc. Limits toxicity	Metabolism by liver enzymes such as cytochrome P450 may lead to toxicity
Do not readily penetrate cells and cross the blood-brain barrier	Smaller size often allows cell entry and passage across blood-brain barrier (especially if lipid soluble)
Often immunogenic	Without linkage to a carrier, usually not immunogenic
High cost of development	Cost high but often less than a biologic
Cost of treatments for patients and health budgets often very high	Treatment cost usually lower, often considerably so

[a] Biologics as defined in this monograph
[b] Generally > 2–5 kDa but may be significantly greater
[c] Generally < 0.5 kDa

unique difference shown by biologics is their pleiotropism in biological effects, well demonstrated, for example, by many cytokines (Chap. 5). Biologics by virtue of recombinant methodologies, precise targeting, exquisite specificity, and high binding affinity can sometimes offer new treatments previously seen as a long, difficult, uncertain, and costly process because of the difficulties involved in finding and developing small molecule agonists and antagonists to interact with a particular metabolite or with receptors, some newly suspected or identified. This can be seen, for example, with mAbs (Chaps. 2, 3, and 4) and fusion proteins (Chap. 6), developed on the back of research identifying the mechanisms underlying particular diseases, e.g., the specific targeting of complement component C5 by the mAb eculizumab (Soliris®) for the treatment of paroxysmal nocturnal hemoglobinuria, the development of the fusion protein etanercept (Enbrel®) to treat rheumatoid arthritis by linking the tumor necrosis factor receptor to human IgG Fc, and the treatment of arthritic diseases by the mAb tocilizumab specifically targeted to the interleukin-6 receptor, IL-6R (Chap. 4). The capacity to quickly devise and prepare sufficient quantities of highly purified material, often too difficult/costly to extract from natural sources or to synthesize, for example, recombinant enzymes prepared and brought to market under Orphan Drug Designation programs for treating lysosomal storage diseases (see section "Protein Therapeutics" below and Chap. 9), is a further example of the efficacy, efficiency, and speed of the modern successful application of recombinant technologies to treat diseases formerly regarded as beyond immediate or even longer-term reach of conventional small molecule pharmacologic therapies. Biologics may also have an inherent versatility allowing manipulations and changes to overcome serious defects/side effects that cannot be achieved with most small molecule drugs. A good demonstration of this is the marked reduction of immunogenicity achieved in the evolution of mouse monoclonal antibodies through the progressive development of less immunogenic and therefore safer, human-mouse chimeras, humanized, and finally fully human, therapeutic mAbs (Chap. 2).

Note that some biologics have both a biological component (usually a protein) and an attached small molecule, both of which are necessary for the therapeutic action of the conjugate. Antibody-drug conjugates (Chaps. 2 and 3) are an example of this product form.

Protein Therapeutics

Given their ubiquitous presence, diverse roles, and importance in the body, it should not be surprising that proteins dominate the growing list of the more than 200 approved biotherapeutic agents used in medicine today. From the abundant albumin, important for the osmolarity and volume of the blood, to vaccines; myriad enzymes; targeted antibodies, fusion proteins, and receptors; and so-called factors involved in blood clotting, homeostasis, and thrombosis; to the extremely potent botulinum neurotoxins and tiny concentrations of hormones and cytokines that act as signaling and immunomodulating molecules, respectively, proteins, often in recombinant form, comprise the majority of FDA- and EMA-approved biologics. Over the last decade, it is probably true to say that no group of therapeutic agents has had such a successful history of use and wide disease application as the steadily growing collection of 50 mAbs currently approved by the FDA. From 25 different antibodies with approved indications covering blood, solid tumor, and skin cancers to a similar sized range of others specifically developed for the management of a variety of diseases including chronic asthma, cryopyrin-associated periodic syndrome, macular degeneration, paroxysmal nocturnal hemoglobinuria, autoimmune disorders such as Crohn's disease and rheumatoid arthritis, bone loss, psoriasis, systemic lupus erythematosus, and hypercholesterolemia and prevention of organ rejection, mAb development continues to be extended and refined. The biotechnical advances, pharmacokinetics, clinical applications, and adverse effects of these antibodies continue to be well studied, and together with the expanding list of fusion proteins often utilizing the IgG Fc fragment to achieve FcRn receptor-mediated recycling, precise, high-affinity, approved targeted therapies are certain to continue to expand. Such close attention is yet to be directed at a relatively more slowly expanding list of approved cytokines, a number of which, like enzymes used in enzyme replacement therapies, have been developed as orphan drugs available in highly purified, well-characterized recombinant form. These biologics have been developed for clinical use via programs administered by the FDA Office of Orphan Products Development (OOPD). The OOPD provides incentives for the development of products (drugs, biologics, devices, medicinal foods) for the diagnosis and/or treatment of rare conditions, that is, diseases or disorders that affect fewer than 200,000 people in the US or where developers/manufacturers are not expected to cover the total costs of developing and marketing the agents. Since 1983, more than 400 drugs and biologic products for rare diseases have been brought to market under the Orphan Drug Designation programs, a marked increase over the ten industry-supported products developed and marketed in the decade prior to 1983.

Some Complexities of Protein Therapeutics: Perceived Advantages and Some Problems

Protein therapeutics prepared by recombinant DNA technology and produced in bacterial, yeast, mammalian, and insect cells or transgenic animals and plants are often assumed to generally provide a more efficient and less expensive means of production and, importantly, to also remove the risk of transmissible disease that may be present in material isolated from human and other animal tissues. Perceived advantages of protein therapeutics over small MW drugs include specificity of action and potent therapeutic efficiency, more predictable behavior after administration, fewer side effects including the expected lower immunogenicity due to their human origin, faster regulatory approval time, and their uniqueness allowing better patent protection for developers and marketers. Not surprisingly, clinical experience may sometimes reveal significant departures from the expected outcomes. As mentioned above, unlike small drug molecules, protein therapeutics are typically more complex, and there is the possibility of heterogeneity due to changes in amino acid sequence, the presence and degree of glycosylation, folding, and protein-protein interactions. Even small differences which are often difficult to control can affect a protein's purity, specificity, potency, and safety. Some proteins may have a short half-life in plasma requiring frequent parenteral administration and ultimately poor patient compliance. This may be offset to at least some extent by the degree of glycosylation and that, in turn, may affect the protein's activity, potency, and immunogenicity. In relation to side effects, it is sometimes assumed that structural identity or near identity with the natural human protein automatically guarantees a more restricted adverse events profile, but as will be revealed in later chapters, this is not necessarily true.

The Evolving Biologics Market

The first biologic recombinant preparation, human insulin (Humulin®), was approved for US market in 1982. By 2009, sales of biologics had reached ~$93 billion; by 2013–2014, more than 200 biologics were in the market; and about one third of the biopharmaceutical research and development pipeline was made up of biologics. Approximately one quarter of all new drugs approved by the FDA in 2014 were biologics. It is predicted that by 2016, biologics will make up half of the world's 20 top-selling drugs, and by 2018, biologic medicine sales will account for almost half of the world's 100 biggest sellers. IMS Health, US-based healthcare information provider, forecast that the biologic share of the $1.2 trillion of global spending on medicines in 2017 will be $221 billion or approximately 20 % of the total market. This growth is primarily being driven by mAbs; in 2012, four of the top five biologics were mAbs and that dominance is continuing. In 2012, the fusion protein etanercept generated worldwide sales of $8.37 billion, second to the one of the biggest selling drugs of all time, adalimumab (Humira®). Forecasts for 2013–2017 estimate revenue of $45 billion for etanercept and $58 billion for adalimumab.

Given their high research and development costs, and potential, often realized, to improve on existing treatments or provide approved efficacious treatments for diseases previously lacking any effective therapies, it is not too surprising that most biologic therapies are so expensive that they can be a major financial burden for both patients and government health budgets. The average monthly cost of treating rheumatoid arthritis with adalimumab, for example, has been calculated to be ~$1800, about 30 times the cost of treatment with methotrexate. Inevitably, the price premium of many biologics and the increasing activities of counties with poor protection laws for intellectual property has seen the appearance and growth of non-original biologic products and so-called biosimilars developed to be marketed after expiration of the original patent. Biosimilars (Chap. 13) offer the potential of treatments at least the equal of the original product but at lower cost, thus making expensive approved therapies more accessible to more patients while at the same time stimulating lower cost competition. However, despite the estimated $600 million per year being invested in their development, biosimilars currently account for less than 0.5 % of biologic products' revenue, and this is unlikely to show a dramatic increase until the largest market, the USA, gets into full stride with a suitable and supportive regulatory framework (Chap. 13). In the meantime, large and small biopharmaceutical companies are pursuing the next-generation biologics, sometimes referred to as "biobetters" or "biosuperiors." These are not copies of existing marketed biologics but drugs with a different primary structure and/or other differences such as in glycosylation or lipid content.

Adverse Drug Reactions

Definitions

Over 40 years ago, the World Health Organization (WHO) defined an adverse drug reaction as "a response to a drug that is noxious and unintended and occurs at doses normally used in man for the prophylaxis, diagnosis or therapy or disease, or for the modification of physiological function." Following questioning of this definition, for example, its vagueness and adequacy in covering minor reactions, its inclusion of "noxious," and use of the term "drug," Edwards and Aronson attempted to improve existing definitions and remove existing ambiguities. Their proposed definition "an appreciably harmful or unpleasant reaction, resulting from an intervention related to the use of a medicinal product, which predicts hazard from future administration and warrants prevention or specific treatment, or alteration of the dosage regimen, or withdrawal of the product" provides a succinct and satisfying coverage and is often referred to, but, for many, the directness and simplicity of the FDA definition "any undesirable experience associated with the use of a medicinal product in a patient" is adequate. The event is said to be serious when the patient outcome is death, life-threatening, hospitalization (initial or prolonged), disability or permanent damage, congenital anomaly/birth defect, required intervention to prevent permanent impairment or damage (devices), and other serious important medical events (e.g., allergic bronchospasm, serious blood dyscrasias, or seizures or convulsions that do not result in hospitalization).

Terminology: Adverse Reactions and Adverse Events

The US Department of Health and Human Services (with the National Institutes of Health and National Cancer Institute [NCI]) in its Common Terminology Criteria for Adverse Events (CTCAE) version 4.03, June 2010 (CTCAE v. 5.0, 2015, is in draft review stage), defines CTCAE as "a descriptive terminology which can be utilized for adverse events reporting." For such reporting, a grading system, or severity scale, has been issued. The use of the term "events" in the issued criteria draws attention to the appropriateness or otherwise of the terms "adverse reactions" and "adverse events," whether these terms are interchangeable or, in at least some cases, different, and whether or not there is some, or frequent, misuse of these terms. An essential difference between the terms is that while an adverse reaction is a reaction directly related to use of the drug, an adverse event in a patient may not necessarily be directly drug related during the treatment period. In the CTCAE v. 4, an adverse event is defined as "any unfavourable and unintended sign (including an abnormal laboratory finding), symptom, or disease temporally associated with the use of a medical treatment or procedure that may or may *not* be considered related to the medical treatment or procedure." In the same way, an "adverse effect" should refer to the adverse or harmful consequences experienced during or after the treatment period, whether or not the observed effects were a direct consequence of the administered agent. Meaning of the term "side effect" should be the least ambiguous of the terms associated with drug administration. A side effect is *any* event caused by the administered agent other than the desired therapeutic effect, whether the effect is beneficial or detrimental. It must be said that it is not uncommon to see some or all of the above terms used as synonyms. In the absence of widely agreed definitions, and bearing in mind the frequent difficulty of clearly and unequivocally establishing a direct connection between an observed adverse effect and an administered agent, the use of the term "adverse event" is perhaps the one that is most often appropriate and, of the four terms discussed (adverse reaction, adverse event, adverse effect, side effect), the one least likely to be used incorrectly.

CTCAE guidelines refer to the severity of an adverse event by assigning a grade 1–5 together with a clinical description of severity for each adverse event. Descriptions of the grades are shown in Table 1.2. Grades 1 and 2 relate to severity that is mild to moderate, grade 3 is seen as severe but not life-threatening, grade 4 is life-threatening, and grade 5 (not relevant to some adverse events) represents death.

There is mounting evidence-derived opinion that physicians underestimate and under report adverse events and that patient reporting of adverse events leads to improved identification, accuracy, and completeness of data in clinical trials. A patient-centered approach to complement physician reporting, the so-called Patient-Reported Outcomes of the CTCAE (PRO-CTCAE) measurement system, has been developed by the NCI with the aim of ultimately improving safety. Note that PROs have already established their value in related areas such as treatment preferences and quality of life assessments.

Table 1.2 Common Terminology Criteria for Adverse Events (CTCAE)[a]

Grade[b]	Clinical description of severity
1	Mild; asymptomatic or mild symptoms; clinical or diagnostic observations only; intervention not indicated
2	Moderate; minimal, local, or noninvasive intervention indicated; limiting age-appropriate instrumental activities of daily living
3	Severe or medically significant but not immediately life-threatening; hospitalization or prolongation of hospitalization indicated; disabling; limiting self-care activities of daily living
4	Life-threatening consequences; urgent intervention indicated
5	Death related to adverse event

[a]Version 4.03, June 2010, US Department of Health and Human Services; National Institutes of Health; National Cancer Institute. CTCAE v5.0 in draft review 2015
[b]Not all grades are appropriate for all adverse events

Classification of Adverse Drug Reactions

Within the disciplines of pharmacology and pharmacovigilance, adverse drug reactions were originally distinguished on the basis of dose-related and non-dose-related reactions, also known as type A and type B reactions, respectively (Table 1.3). Type A reactions, making up about 80 % or more of reactions, can be anticipated from the drug's pharmacological actions, are predictable and dose dependent, and resolve when the dose is reduced or therapy is withdrawn. Type B reactions are generally unrelated to the drug's pharmacological actions, are unpredictable and independent of dose, and generally resolve when treatment is stopped, but reactions sometimes continue and may even progress. These unpredictable reactions, sometimes called bizarre reactions, are made up of different responses that may be broadly distinguished as immune-mediated hypersensitivities and non-immune sensitivities. The latter, which may involve mechanisms such as the activation of bradykinin and drug-induced redirection of arachidonic acid metabolism from the cyclooxygenase to the lipoxygenase pathway, are further divided into pseudoallergic reactions, idiosyncratic reactions, and intolerances (Table 1.3, Fig. 1.2). Examples of pseudoallergy are most reactions to nonsteroidal anti-inflammatory drugs which interfere in the biosynthetic pathways for the synthesis of prostanoids from arachidonic acid and direct mast cell degranulation seen with neuromuscular blocking drugs, some opioids, contrast media, and vancomycin. Halothane hepatitis and malignant hypothermia are examples of idiosyncratic reactions, while nonsteroidal anti-inflammatory drugs and contrast media may also induce nonimmune intolerances in some patients.

Hypersensitivities

On the basis of the Gell and Coombs classification of immune-mediated reactions, hypersensitivities are subdivided into four categories, types I, II, III, and IV. Type I reactions, often referred to as the "true" allergies, are generally immediate in onset (seconds to ~1 h) and mediated by IgE antibodies; type II hypersensitivities, or cytotoxic reactions, are antibody IgG or IgM mediated: type III or immune complex reactions again are mediated by IgG or IgM antibodies; and type IV hypersensitivities, also called delayed hypersensitivities, are T-cell-mediated reactions. Table 1.4 sets out the main features of the four hypersensitivity states with respect to reaction onset times, immune reactants involved, effector mechanisms, cutaneous/intradermal responses to antigen, examples of diseases in each category, and the small molecule drugs and biologic products known to be implicated in each of the different reaction types. Although the Gell and Coombs classification has been a valuable guide for over 50 years, some adverse responses to drugs do not fit neatly into the four hypersensitivity types. Examples include some reactions to contrast media and nonsteroidal anti-inflammatory drugs and skin reactions alopecia, folliculitis, and hyperpigmentation.

Table 1.3 Classification of adverse drug reactions

Reaction type	Examples of reaction	Main features of reaction
A. Augmented pharmacologic effects[a]	– Toxic (intolerant) reactions—e.g., serotonin syndrome to opioids, antidepressants; digoxin toxicity – Side effects—e.g., bronchospasm to β-blocker in hypertensive patient; dry mouth to antidepressants	– Majority of ADRs – Common – Predictable – Usually dose dependent – Related to pharmacologic reaction of drug – Low mortality
B. Bizarre[b] (see Fig. 1.1)	– Immunologic reactions – Idiosyncratic reactions – Pseudoallergy – Intolerance	– Relatively uncommon – Unpredictable – Rarely dose dependent[c] – Unrelated to drug's pharmacologic action – High mortality
C. Chronic (continuous) effects	– Corticosteroid-induced suppression of hypothalamic-pituitary-adrenal axis – Renal papillary necrosis caused by phenacetin	– Uncommon – Cumulative dose and long-term exposure required
D. Delayed effects	– Carcinogenesis – Teratogenesis—e.g., vaginal adenocarcinoma induced by diethylstilbestrol	– Time-related—apparent some time after drug exposure – Uncommon – Usually dose related
E. End-of-treatment effects (withdrawal effects)	– Opiate withdrawal syndrome – β-Blocker withdrawal	– Occurs with little or no delay after withdrawal of drug – Uncommon
F. Failure of therapy	– Resistance to drug action—e.g., resistant bacteria to antibiotic or tumor to chemotherapy – Oral contraceptive dose too low	– Common – Usually dose related – May be caused by drug interactions
G. Genetic reactions[b,d]	– Abnormal drug metabolism due to inherited factors (alleles of P450 (CYP), *N*-acetyltransferase, pseudocholinesterase) – HLA-drug hypersensitivity associations (e.g., abacavir, carbamazepine, allopurinol) – Succinylcholine sensitivity – Porphyria	– Abnormal drug metabolism appears to be uncommon – Pharmacogenomic studies still in early stages – Ethnicity seems to matter for some drugs, e.g., carbamazepine

From Baldo BA, Pham NH. Drug allergy. Clinical aspects, diagnosis, mechanisms, structure-activity relationships. New York: Springer; 2013. With permission from Springer Science + Business Media

[a]Said to account for ~80 % of ADRs

[b]There is evidence that some type B reactions are under genetic control

[c]Exceptions exist, e.g., with type IV hypersensitivity skin reactions, responses to vaccines, desensitization with increasing dosages of drug

[d]ADRs likely to be multigenetic phenomena

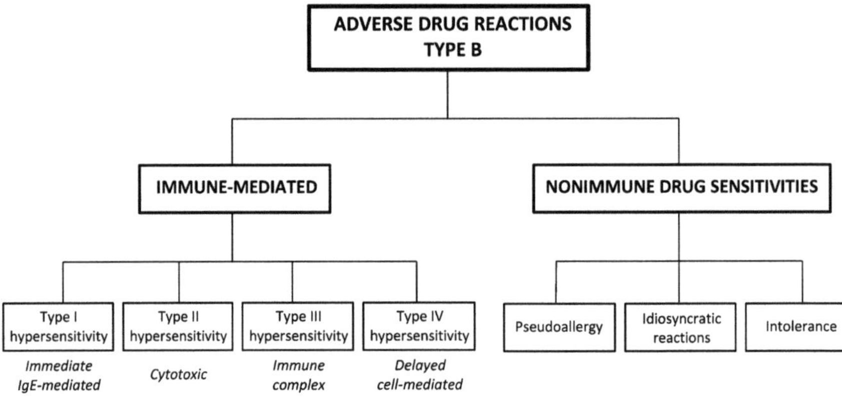

Fig. 1.2 Classification of immune and nonimmune sensitivities to drugs including biologics used for therapy. The four types of immune reactions, types I, II, II, and IV, are referred to here as hypersensitivities and nonimmune sensitivities as intolerances. A type I hypersensitivity is often called an "allergy." Examples of biologics in each of the four hypersensitivity types are mentioned in the text. Adapted from Baldo BA, Pham NH. Drug allergy. Clinical aspects, diagnosis, mechanisms, structure-activity relationships. New York: Springer; 2013. With kind permission from Springer Science+Business Media

Type I (immediate) hypersensitivities, anaphylaxis and urticaria, are rarely seen following biotherapy, but both reactions are known to occur with many different biologic agents. Because of the immunogenic potential of mAbs, especially murine antibodies ibritumomab tiuxetan and blinatumomab, the rat-mouse chimera catumaxomab, and human-mouse chimeras abciximab, basiliximab, brentuximab, cetuximab, infliximab, rituximab, siltuximab, and dinutuximab, mAbs commonly carry warnings of possible hypersensitivity, including anaphylaxis. This small risk is lessened with the so-called humanized mAbs which are human antibodies with complementarity-determining regions (usually murine) grafted into human framework V_H and V_L regions. The risk is further reduced with fully humanized mAbs, but the possibilities of generating anti-idiotype antibodies and the presence of antibodies to some drugs, including mAbs, in normal sera, and sera of pretreated patients, remain (see Chap. 2, Sections "Monoclonal Antibodies for Therapy" and "Evolution of Therapeutic Monoclonal Antibodies: From Mouse to Man"). Anaphylaxis has been recorded for the mAbs cetuximab, rituximab, trastuzumab, pertuzumab, ibritumomab tiuxetan, brentuximab, adalimumab, infliximab, and belimumab, and other type I hypersensitivities, urticaria and angioedema, have been reported following mAb administration. Along with the mAbs, enzymes used for therapy, especially those given as enzyme replacement therapies for some lysosomal storage disorders, are well known possible, but rare, causes of anaphylaxis, a fact demonstrated by FDA boxed warnings issued for alglucosidase, elosulfase alfa, idursulfase, and laronidase and warnings/precautions for the β-glucocerebrosidases taliglucerase alfa and velaglucerase. There have also been reports of anaphylaxis to agalsidase beta, galsulfase, glucarpidase, and imiglucerase. Of approved enzymes used for conditions other than enzyme replacement therapy, L-asparaginase,

Table 1.4 Hypersensitivities to drugs according to the Gell and Coombs classification[a]

Type of hypersensitivity	I	II	III	IV
Other designations	Immediate; anaphylactic	Cytotoxic	Immune complex	Delayed; cell mediated; T-cell mediated
Time for reaction to reach maximum	Seconds to 30 minutes[b]	Hours (~1 day)	3–10 h	24–72 h
Immune reactant(s)	IgE antibody	IgG (and IgM) antibodies	IgG antibody (and/or IgM)	Th1, Th2, and/or Th17 cells Cytotoxic lymphocytes
Effector mechanism	Mast cell and basophil activation	Complement fixation, phagocytes, NK cells (Fc receptor cells)	Complement Phagocytes	Macrophage activation Cytotoxic lymphocytes Eosinophil activation
Intradermal response to antigen	Wheal and flare	Lysis and necrosis	Erythema and edema	Erythema and induration
Histology	Degranulated mast cells; cellular infiltrates including neutrophils[b]	Immunofluorescence shows antibody, complement, neutrophils	Acute inflammatory reaction Mainly neutrophils	Perivascular inflammation Mainly mononuclear cells
Sensitivity transferred by	Serum IgE antibody	Serum antibody	Serum antibody	Lymphoid cells
Examples of disease states	Erythema; urticaria; angioedema; respiratory symptoms; GI symptoms; anaphylaxis	Drug-induced hemolytic anemia, thrombocytopenia, agranulocytosis (immune form)	Serum sickness Drug-induced vasculitis Hypersensitivity pneumonitis (combined types III and IV)	Allergic contact dermatitis; psoriasis; maculopapular exanthema; AGEP; FDE; DRESS; SJS; TEN; EM

(continued)

Table 1.4 (continued)

Type of hypersensitivity	I	II	III	IV
Drugs implicated	mAbs[c]; enzymes[d]; fusion proteins[e]; cytokines[f]; β-lactams; other antibacterials; NMBDs; some NSAIDs	mAbs[g]; β-lactams; quinine; quinidine; sulfonamides; NSAIDs; procainamide; gold; carbamazepine; propylthiouracil; ticlopidine	mAbs[h]; β-lactams; ciprofloxacin; sulfonamides; tetracycline; NSAIDs; carbamazepine; allopurinol; gold; methyldopa	mAbs[i]; NSAIDs; β-lactams; other antibiotics; anticonvulsants; antimalarials; local anesthetics; barbiturates; quinolones; dapsone

AGEP acute generalized exanthematous pustulosis, *DRESS* drug reaction with eosinophilia and systemic symptoms, *EM* erythema multiforme, *FDE* fixed drug eruption, *mAbs* monoclonal antibodies, *NMBDs* neuromuscular blocking drugs, *NSAIDs* nonsteroidal anti-inflammatory drugs, *SJS* Stevens-Johnson syndrome, *TEN* toxic epidermal necrolysis

Adapted from Baldo BA, Pham NH. Drug allergy. Clinical aspects, diagnosis, mechanisms, structure-activity relationships. New York: Springer; 2013. With permission from Springer Science+Business Media

[a]Coombs RRA, Gell PGH. In: Gell PGH et al. (eds). Clinical Aspects of Immunology. Oxford: Blackwells. 1975, p. 761–81

[b]Late reaction may occur ~3–4 h after immediate reaction, peak at ~12 h and subside by ~24 h

[c]Cetuximab, rituximab, trastuzumab, pertuzumab, ibritumomab tiuxetan, brentuximab, adalimumab, infliximab, and belimumab

[d]For example, alglucosidase, elosulfase alfa, idursulfase, laronidase, taliglucerase alfa, velaglucerase, agalsidase beta, galsulfase, glucarpidase, and imiglucerase. Approved enzymes not used for enzyme replacement therapy: L-asparaginase, reteplase, collagenase, pegloticase, and rasburicase

[e]For example, etanercept

[f]For example, interferon beta-1a, oprelvekin, anakinra, darbepoetin alfa

[g]For example, rituximab, alemtuzumab, bevacizumab

[h]For example, rituximab, adalimumab, infliximab, certolizumab pegol, alirocumab, omalizumab

[i]For example, Brentuximab, adalimumab, ibritumomab tiuxetan, rituximab, and infliximab

reteplase, collagenase, pegloticase, and rasburicase have provoked anaphylaxis, and, for the latter two enzymes, this is emphasized by an FDA black box warning. For most proteins, the potential for inducing a hypersensitivity reaction, including anaphylaxis, is real and needs to be anticipated, but the incidences of type I hypersensitivities to approved cytokines, fusion proteins (in particular etanercept), coagulation proteins, and vaccines are significantly lower than seen with mAbs and enzymes. Cytokines implicated in type I reactions, including anaphylaxis, are interferon beta-1a, oprelvekin, anakinra, and darbepoetin alfa.

While adverse drug reactions manifesting as anaphylaxis, urticaria, and angioedema are usually readily identifiable as type I hypersensitivities with IgE antibody involvement, the underlying mechanism(s) of many **infusion reactions** with signs and symptoms of one or more of cytokine release syndrome (section "Cytokine Release Syndrome"), an anaphylactic/anaphylactoid reaction, and direct toxicity, are often less easy to define with confidence and precision. Infusion of many biologics, particularly mAbs, provokes a characteristic infusion syndrome, usually within one or a few hours during/after the first administration. Whereas most reactions are mild to moderate with symptoms often described as "flu"-like with fever, chills, rigors, headache, nausea, asthenia, rash, and pruritus, a small number of patients, mostly at the first or second infusion, show potentially fatal symptoms resembling an IgE antibody-mediated reaction with hypotension, cardiac arrest, bronchospasm, and urticaria. Trastuzumab, rituximab, and cetuximab induce the highest incidence of infusion reactions; approximately 80 % of fatal reactions to rituximab occurred after the first infusion, and, even after the fourth and eighth infusions, reactions persisted in 30 % and 14 % of patients, respectively (Chap. 3, section "Infusion Reactions and Cytokine Release Syndrome"). Reaction incidences and severity to humanized and fully human mAbs are significantly lower. Mechanisms of the reactions are incompletely understood. Release of tumor necrosis factor (TNF) and interleukin-6 (IL-6) has been suggested since these cytokines produce some symptoms similar to those seen in type I allergic responses and regression of symptoms in some cancer patients given chemotherapy has been found to coincide with decreases in serum cytokine concentrations. It has been suggested that the severity of infusion reactions is related to the number of circulating lymphocytes with severe reactions seen when counts exceed 50×10^9/L.

The efficacy, safety, and costs of desensitization to mAbs following 41 mAb-induced hypersensitivity reactions were recently assessed in a 16-step desensitization protocol over a 6.7 h period in patients with cancer, vasculitis, and hematological and connective tissue diseases. MAbs implicated were cetuximab, trastuzumab, bevacizumab, tocilizumab, rituximab, and infliximab. Rapid desensitization was found to be cost effective and safe for hypersensitive patients to remain on their first line therapy.

Many cutaneous reactions, for example, allergic contact dermatitis, psoriasis, fixed drug eruption, erythema multiforme (EM), and the life-threatening toxidermias Stevens-Johnson syndrome (SJS) and toxic epidermal necrolysis (TEN), are also not difficult to diagnose as immune cell-mediated **delayed type IV reactions**. Monoclonal antibodies most commonly implicated in type IV hypersensitivities are brentuximab, adalimumab, ibritumomab tiuxetan, rituximab, and infliximab; the latter three have provoked cases of EM, SJS, and TEN.

Some cytopenias, tumor lysis syndrome (section "Tumor Lysis Syndrome"), and kidney, embryo-fetal, neurological, some pulmonary, liver, and heart toxicities appear to be due to direct cytotoxic actions and/or other nonimmune mechanisms. However, some seemingly drug-induced adverse events such as cytopenias (thrombocytopenia, neutropenia, and anemia); some pulmonary, liver, and heart toxicities; and a number of syndromes (capillary leak [section "Capillary Leak Syndrome"], cytokine release [section "Cytokine Release Syndrome"], systemic inflammatory response [section "Systemic Inflammatory Response Syndrome"], immune reconstitution inflammatory [section "Immune Reconstitution Inflammatory Syndrome"], and macrophage activation [section "Macrophage Activation Syndrome"] syndromes) may have at least an indirect connection to immune processes. Rituximab may cause **type II cytotoxic hypersensitivities**, and autoimmune forms of thrombocytopenia and hemolytic anemia (AIHA), both type II hypersensitivities, occur predominately in patients with lymphocytic leukemias, often treated with mAbs. As well as thrombocytopenia and anemia, late onset neutropenia (LON) occurs in some cancer patients treated with mAbs, especially rituximab. While the mechanism of LON is not well understood, direct cytotoxicity is unlikely, autoantibodies may be involved, and the condition may be a true type II hypersensitivity response. Severe anemia has been reported in 1.1–5.2 % of patients given rituximab monotherapy, and cases of severe AIHA, intravascular hemolysis, and multi-organ ischemia due to an autoimmune anti-Pr cold agglutinin have occurred following rituximab therapy. Alemtuzumab has also been implicated in AIHA and bevacizumab has been linked to cases of renal thrombotic microangiopathy. It should be noted that "hypersensitivity" in the hypersensitivity type I–IV Gell and Coombs definition is a much misused term, and in oncology and some other medical specialties, its strict and correct usage along with efforts to investigate underlying mechanisms of possible hypersensitivity responses is not always observed.

Type III hypersensitivities, in particular serum sickness (or perhaps more correctly, a serum sickness-like reaction) and drug-induced vasculitis, occur in response to biologics with mAbs again being the agents most often involved. The frequency of occurrence of these disorders is said to be underestimated, an unsurprising conclusion given the absence of mechanistic studies of many of the observed cytopenias. Rituximab-induced vasculitis is the subject of a number of reports; cutaneous vasculitis after adalimumab, infliximab, and certolizumab pegol is known; and alirocumab carries an FDA warning for hypersensitivity vasculitis. Chimeric mAbs have the potential to cause serum sickness, and this has been seen with rituximab and infliximab but also with humanized antibodies such as omalizumab. Some drug-induced pulmonary events are hypersensitivity reactions resulting from drug interaction with the immune system and involvement of drug-reactive antibodies or T cells. Hypersensitivity pneumonitis, for example, is a combined type III and type IV hypersensitivity reaction, but overall, pulmonary adverse events provoked by biologic agents, as well as small molecule drugs, comprise a heterogeneous group of lung diseases. Again, mechanisms have generally not always been elucidated. Monoclonal antibodies most commonly associated with pulmonary adverse events are cetuximab, rituximab, alemtuzumab, bevacizumab, trastuzumab, and panitumumab.

Other Adverse Drug Reaction Categories

In addition to the type A (augmented pharmacologic effect) and type B (bizarre) reactions in the classification of adverse drug reactions (Table 1.3), five other reaction types are distinguished. From the early 1980s, three of these categories, C, D, and E, were included in the classification. Reaction type C, demonstrated by corticosteroid-induced suppression and designated chronic or continuous effects, is seen uncommonly and is related to cumulative dose and time with long-term exposure required. Reaction type D, called delayed effects and illustrated by stilbestrol-induced vaginal carcinoma, is uncommon and time and dose related. Category E, end-of-treatment (or withdrawal) effects, can be seen with little or no delay as, for example, in opiate withdrawal syndrome. More recently, a sixth category, F, failure of therapy, has been added. This commonly seen reaction type may be caused by drug interactions and is usually dose related. Resistance of a tumor to immunotherapy is an example of a type F reaction. In some classifications of adverse drug reactions, a seventh category G, genetic reactions, is included. Sensitivity to succinylcholine and some HLA-drug hypersensitivity associations are examples of these reactions that appear, at this early stage, to be uncommon.

For the extensive range of diverse biologics already in use, the next few years will see increasing attention paid to the adverse drug reaction categories A–G as adverse events are examined ever more closely and safety data continues to enlarge. The lists of reactions that fit into categories A and B are already extensive for the approved biologic therapies, and effects related to cumulative dose, time and dose, drug withdrawal, resistance to drug action, and genetics will be progressively enlarged, better defined, interpreted, and understood.

Syndromes That May Be Associated with Biologic Therapies

A number of systemic potentially life-threatening syndromes may occur with low frequency during or following the administration of a variety of biologic agents including mAbs, fusion proteins, growth factors, interleukins, and a few enzymes such as L-asparaginase. These syndromes, some discussed in later chapters, are not exclusively associated with biotherapy and may occur with some small molecule drugs or even in cases unrelated to drug therapy.

Capillary Leak Syndrome

Capillary leak syndrome, also known as systemic capillary leak syndrome, vascular leak syndrome, or Clarkson's disease, manifests as an increase in body weight, malaise, weakness, and sometimes abdominal pain, myalgia, pyrexia, vomiting, and diarrhea. Symptoms are variable and causes not well understood. An example of the condition may be seen following infusion of interleukin (IL)-2 for metastatic cancer. Within 24 h there is an increase in vascular permeability accompanied by extravasation of fluids and proteins resulting in peripheral and interstitial edema, pleural and pericardial effusions, ascites, and, in severe form, pulmonary and cardiovascular failure. Complications such as renal failure, stroke, ischemia, deep vein thrombosis, and rhabdomyolysis may occur. Erythematous cutaneous eruptions that often accompany the syndrome may be induced by cytokine activation of endothelial cells, and it has been suggested that these cells may have a role in the events underlying the syndrome. Biologic agents that have been associated with capillary leak syndrome include mAbs (alemtuzumab, basiliximab, bevacizumab, catumaxomab, dinutuximab), IL-1, IL-2, oprelvekin (recombinant IL-11), interferons alfa and beta-1b, filgrastim, and the fusion protein denileukin diftitox.

Cytokine Release Syndrome

Cytokine release syndrome (CRS), also called acute infusion reaction or cytokine storm, has been described by S. A. Grupp and coworkers, Children's Hospital of Philadelphia, as "a constellation of inflammatory symptoms resulting from cytokine elevations associated with T cell engagement and proliferation." Although reactions are usually short term (resolving within 24 h) and mild with fever, flu-like symptoms, and myalgia, some patients experience an acute inflammatory response with hypotension, pulmonary edema, vascular leak, coagulopathy, and even multi-organ failure. CRS has been associated with some mAbs and with antigens CD3, CD20, and CD52 on T cells, CD20 on B cells, CD52 on monocytes, and CD28, the receptor for B7.1 (CD80) and B7.2 (CD86). CRS is the most common toxicity associated with the bispecific T-cell-engaging fusion protein blinatumomab (Chap. 3, section "Blinatumomab") targeted to CD3 and CD19 and was recently implicated in a death induced by this agent. Type I hypersensitivity and CRS share some signs and symptoms such as nausea, cough, dyspnea, bronchospasm, hypotension, rash, itching, and urticaria. Symptoms are thought to be caused by cytokines such as TNF and interleukins following cell destruction.

It has been stated that in some patients with severe CRS, symptoms associated with macrophage activation syndrome (MAS) or hemophagocytic lymphohistiocytosis (HLH) (see below) occur, and the cytokine profile resembles that seen in HLH. These symptoms include high fever accompanied by hepatosplenomegaly, liver dysfunction, coagulopathy, hypofibrinogenemia, and hyperferritinemia, and it has been suggested that CRS may represent a spectrum of symptoms with some individuals developing HLH and/or MAS. Besides blinatumomab, approved biologics implicated in the induction of CRS include IL-1, IL-2, IL-6, TNF, and the mAbs alemtuzumab, catumaxomab, and rituximab.

Hemophagocytic Lymphohistiocytosis

HLH is a rare disorder of immune regulation in which a highly or overstimulated immune response leads to pathologic immune activation and excessive inflammation, an outcome resembling life-threatening cytokine storm. Two forms of the syndrome are distinguished: primary HLH, also known as familial HLH or familial erythrophagocytic lymphohistiocytosis, an autosomal recessive disorder, and secondary (or acquired) HLH occurring after infection, malignancy, or immunodeficiency. Diagnostic criteria required to establish a diagnosis for HLH are fever for seven or more days, splenomegaly, cytopenia, hemophagocytosis, and hypofibrinogenemia or hypertriglyceridemia. Hyperferritinemia may also occur (ferritin levels may be used as a marker for the disease), and cutaneous involvement of erythroderma, purpuric macules and papules, and morbilliform eruptions are seen in more than half of patients. The mechanism of HLH remains unclear although the highly activated immune reaction involves proliferating and activated T cells, macrophage activation, restricted apoptosis of immune cells, the involvement of the pore-forming protein perforin, and natural killer (NK) cells. Large amounts of cytokines, particularly IFN gamma, TNF, and granulocyte-macrophage colony-stimulating factor (GM-CSF), are released following decreased NK-cell activity and increased T-cell activation and expansion. Resultant activation of macrophages and production of IL-1 and IL-6 provoke an inflammatory response, tissue damage, and the associated symptoms. HLH has been reported following treatment with the T-cell-engaging therapy blinatumomab.

Macrophage Activation Syndrome

MAS (see also Chap. 4, section "Approved Indications and Safety of Tocilizumab") is a serious complication of rheumatic disease, seen most often in systemic juvenile idiopathic arthritis (SJIA) (occurring in at least 10 % of patients) and adult-onset Still's disease although it has also been reported in juvenile systemic lupus erythematosus, Kawasaki disease, and periodic fever syndromes. Patients with MAS present with high fever, hepatosplenomegaly, pancytopenia, lymphadenopathy, disseminated intravascular coagulation, hypertriglyceridemia, and elevated ferritin levels. The disorder may occur in a number of different ways and situations: spontaneously, as a complication of another disease, during or after an infection, following a change in drug therapy, or as a result of biologic therapy. MAS can be rapidly fatal, so early diagnosis is essential, but distinguishing it from lupus flare and some infections can be difficult. MAS has been classified as an acquired form of HLH, and as with that disorder, it is characterized by an explosive inflammatory reaction mediated by the uncontrolled proliferation of T cells and macrophages, decreased NK-cell and cytotoxic T-cell functions, macrophages exhibiting hemophagocytic activity, and abnormal expression of perforin. It seems that for most, if not all, of the syndromes provoked by biologics, release of a cascade of inflammatory cytokines is a common feature. Biologics implicated in the induction of MAS include anakinra, the recombinant antagonist of the IL-1 receptor, the fusion proteins etanercept and rilonacept, and the mAbs alemtuzumab, canakinumab, and tocilizumab.

Systemic Inflammatory Response Syndrome

Systemic inflammatory response syndrome (SIRS) is a serious systemic inflammatory disorder related to sepsis with the potential to cause organ dysfunction and failure. It may be caused by an infection or a variety of noninfectious stimuli such as ischemia, trauma, pancreatitis, hemorrhage, adrenal insufficiency, and anaphylaxis or therapy, including biologic agents. SIRS may be diagnosed on the basis of two or more manifestations related to body temperature, heart rate, tachypnea, and white blood cell count. SIRS without infection may involve renal failure, deep vein thrombosis, disseminated intravascular coagulation, gastrointestinal bleeding, anemia, and hyperglycemia. SIRS with infection may lead to sepsis, septic shock, and multiple organ dysfunction syndrome (MODS). In the early events in SIRS following trauma, infection, or other relevant stimuli, an inflammatory cascade is activated producing a multitude of cytokines, in particular the proinflammatory cytokines IL-1, TNF, IL-6, IL-8, and IFN gamma. Infections stimulate the release of more TNF which in turn leads to more IL-6 and IL-8 and a higher rate of fever than is seen with trauma-induced SIRS. SIRS has been reported following infusion of catumaxomab and eculizumab.

Tumor Lysis Syndrome

Unlike some cases of CRS, tumor lysis syndrome (TLS) which may occur 48–72 h after the start of anticancer therapy is not difficult to distinguish from type I immediate hypersensitivity. TLS is seen most often in patients with leukemias and high-grade lymphomas, especially those with a high tumor load, and only rarely in association with solid tumors. The syndrome is the consequence of the induced death of large numbers of cancer cells which results in hyperkalemia, hypercalcemia, hyperphosphatemia, and hyperuricemia, all of which produce a marked ionic imbalance and the possibility of acute renal failure, cardiac arrhythmias, seizures, and death. Biologics most commonly associated with TLS are the mAbs brentuximab vedotin, alemtuzumab, ipilimumab, obinutuzumab, blinatumomab, and rituximab.

Posterior Reversible Encephalopathy Syndrome

Posterior reversible encephalopathy syndrome (PRES), also called reversible posterior leukoencephalopathy syndrome, is an increasingly recognized disorder that shows variable clinical features, in particular, hypertension (but not always), headache, seizures, visual disturbances, and altered consciousness. Magnetic resonance imaging (MRI) reveals edematous changes in the brain of patients with PRES but with the belated demonstration that cortical gray matter was involved as well as subcortical white matter, the more commonly used name and abbreviation, reversible posterior leukoencephalopathy syndrome (RPLS), has now been modified to PRES to reflect this. The underlying pathophysiology of the disorder is thought to involve vasogenic edema caused by hypertension-induced hyperperfusion and damage to cerebral vessels leading to interstitial extravasation of fluid and protein. However, it remains unclear why PRES sometimes occurs in the absence of hypertension and the extent of edema is not always directly related to the blood pressure. An alternative explanation is that PRES results from systemic inflammation which causes endothelial dysfunction, hypoxia, and vasogenic edema, but some cases of PRES appear to occur in the absence of any observed inflammation. Associated clinical disorders include cancer, eclampsia, and autoimmunity. Associated medications are immunosuppressive agents, chemotherapeutics, antiangiogenic agents, and biologics including the interferon alfa, IL-2, granulocyte-colony-stimulating factor (G-CSF), erythropoietin, the fusion protein etanercept, and the mAbs bevacizumab, certolizumab, infliximab, ramucirumab, rituximab, and ustekinumab. Recently, PRES was reported in neuroblastoma patients receiving anti-disialoganglioside (GD2) chimeric mAb therapy. Despite the apparent severity of the symptomatology seen in many cases of PRES, prompt diagnosis and correct treatment including immediate withdrawal of the offending agent, administration of antihypertensives and anticonvulsants, and treatment of associated disorders can often prevent brain damage and restore patients to their pre-PRES condition.

Immune Reconstitution Inflammatory Syndrome

Immune reconstitution inflammatory syndrome (IRIS), also called immune recovery syndrome (see also Chap. 4, section "FDA Warnings and Precautions for Natalizumab Including Immune Reconstitution Inflammatory Syndrome"), is a paradoxical clinical deterioration of a known or new condition after the restoration of immunity. A good example of the syndrome is seen in AIDS, in tuberculosis, or in some cases of immune suppression when recovery of the immune system begins, only to be overwhelmed by an opportunistic infection together with a dysregulated and damaging inflammatory response. The pathogenesis of IRIS is poorly understood. Progression of the disorder, at least in HIV patients, seems to depend on increases in CD4+ T helper cells and CD8+ T suppressor cells, a reduction in T regulatory cells, and a marked release, and activity increase, of inflammatory cytokines. Many different opportunistic pathogens may be involved in IRIS including *Mycobacterium tuberculosis*, cryptococci, hepatitis viruses B and C, and JC polyomavirus. The dysregulated inflammatory response seen to infections in IRIS can also occur in response to noninfectious agents during recovery of immunity after induced immunosuppression. Biologics implicated in cases of IRIS are natalizumab in a localized form in the central nervous system and infliximab and adalimumab in patients infected with tuberculosis.

Progressive Multifocal Leukoencephalopathy

Progressive multifocal leukoencephalopathy (PML) is a rare and usually fatal demyelinating disease characterized by inflammation and progressive damage of the white matter of the brain. It is caused by infection with the JC virus, a common, widely distributed, and normally harmless virus that can become lethal only in patients with severe immune deficiency, for example, in AIDS and immunosuppressed patients but also in autoimmune diseases and patients receiving chemotherapy. PML has been reported after therapy with a number of currently approved biologics including the fusion protein belatacept and mAbs belimumab, infliximab, eculizumab, ofatumumab, natalizumab, and rituximab. In 2009, 57 cases of PML were recorded following rituximab administration in HIV-negative patients. Labeling of the mAb had earlier been amended to indicate the risk of infections including from JC virus. For an extended discussion of PML, see Chap. 4, section "Posterior Multifocal Leukoencephalopathy and Natalizumab."

Summary

- Whereas a biologist, chemist, or clinician may see a biologic used for therapy as material derived from, or related to, a living organism, a regulatory authority will also consider how such agents are to be classified and assessed for characterization, manufacturing, and control, product development, and identity, purity, and potency.
- In 2012, the FDA issued a draft guidance taking up and distinguishing the long-standing proposed differences between proteins, peptides, and chemically synthesized polypeptides. A protein was defined as any alfa amino acid polymer with a specific defined sequence that is greater than 40 amino acids in size. From this definition, it followed that peptides have fewer than 40 amino acids and are therefore not proteins. A chemically synthesized polypeptide was defined as an alfa amino acid polymer that is made entirely by chemical synthesis and has fewer than 100 amino acids. For the coverage of biologics, here, peptides with less than 40 amino acids and chemically synthesized polypeptides with regulatory approval for therapeutic use in humans are considered as biologics regardless of size or method of preparation.
- In the European Union, the European Medicines Agency defines a "biological medicinal product" as "a protein or nucleic acid-based pharmaceutical substance used for therapeutic or in vivo diagnostic purposes, which is produced by means other than direct extraction from a native (nonengineered) biological source." This definition essentially appears to restrict "biological medicinal products" to recombinant preparations.
- In distinguishing a purified, homogeneous biologic agent from other "small molecule" conventional pharmacologic agents used as drugs, an immediate and obvious comparison is the usually larger molecular weight and complexity of the biologic agent and the fact that small molecule drugs are prepared by chemical synthesis.

- Biologics, especially mAbs and other recombinant proteins, may have a number of potential advantages as therapeutic agents. As well as a shorter development cycle, biologics by virtue of recombinant methodologies, precise targeting, exquisite specificity, and high binding affinity can sometimes offer new treatments previously seen as a long, difficult, uncertain, and costly process because of the difficulties involved in finding and developing small molecule agonists and antagonists.
- Recombinant preparations brought to market under Orphan Drug Designation programs for treating rare diseases are an example of the efficacy, efficiency, and speed of the modern successful application of recombinant technologies to treat some diseases formerly regarded as too small a market for an appropriate financial return and beyond immediate or even longer-term reach of conventional small molecule pharmacologic therapies.
- Some biologics have both a biological component (usually a protein) and an attached small molecule, both of which are necessary for the therapeutic action of the conjugate. Antibody-drug conjugates are an example of this product form.
- No group of therapeutic agents has had such a successful history of use and wide disease application as the steadily growing collection of the now 50 mAbs currently approved by the FDA. Half of these are approved for indications covering blood, solid tumor, and skin cancers. Others have been specifically developed for the management of a variety of diseases including chronic asthma, cryopyrin-associated periodic syndrome, macular degeneration, paroxysmal nocturnal hemoglobinuria, autoimmune disorders such as Crohn's disease and rheumatoid arthritis, bone loss, psoriasis, systemic lupus erythematosus, and hypercholesterolemia and prevention of organ rejection.
- Biologics may exhibit heterogeneity due to changes in amino acid sequence, the presence and degree of glycosylation, folding, and protein-protein interactions. Even small differences that are often difficult to control can affect a protein's purity, specificity, potency, and safety.
- By 2013–2014, more than 200 biologics were in the market and about one third of the biopharmaceutical research and development pipeline was made up of biologics. Approximately one quarter of all new drugs approved by the FDA in 2014 were biologics. It is predicted that by 2016, biologics will make up half of the world's 20 top-selling drugs, and by 2018, biologic medicine sales will account for almost half of the world's 100 biggest sellers. It is forecast that the biologics share of the $1.2 trillion of global spending on medicines in 2017 will be $221 billion or approximately 20 % of the total market.
- Biosimilars, developed to be marketed after expiration of the original patent, offer the potential of treatments at least the equal of the original product but at lower cost, thus making expensive approved therapies more accessible to more patients while at the same time stimulating lower cost competition. Despite the estimated $600 million per year being invested in their development, biosimilars currently account for less than 0.5 % of biologic products' revenue.
- An adverse drug reaction may be defined as "an appreciably harmful or unpleasant reaction, resulting from an intervention related to the use of a medicinal product, which predicts hazard from future administration and warrants prevention or specific treatment, or alteration of the dosage regimen, or withdrawal of

the product." For many, the directness and simplicity of the FDA definition, "any undesirable experience associated with the use of a medicinal product in a patient," is adequate.

- The US Department of Health and Human Services in its Common Terminology Criteria for Adverse Events (CTCAE) defines CTCAE as "a descriptive terminology which can be utilized for adverse events reporting." For such reporting, a grading system, or severity scale, has been issued. The CTCAE defines an adverse event as "any unfavourable and unintended sign (including an abnormal laboratory finding), symptom, or disease temporally associated with the use of a medical treatment or procedure that may or may *not* be considered related to the medical treatment or procedure." The severity of an adverse event is graded 1–5 together with a clinical description of severity for each adverse event. Grades 1 and 2 relate to severity that is mild to moderate, grade 3 is seen as severe but not life-threatening, grade 4 is life-threatening, and grade 5 represents death.
- There is evidence that physicians underestimate and under report adverse events and that patient reporting of events leads to improved identification, accuracy, and completeness of data in clinical trials. A patient-centered approach to the assessment of adverse events to complement physician reporting has been developed by the NCI with the aim of better identifying adverse events, understanding toxicity and tolerability, assessing risks, and ultimately improving safety.
- Adverse drug reactions were originally distinguished on the basis of dose-related and non-dose-related reactions, also known as type A and type B reactions, respectively. Type A reactions, making up about 80 % or more of reactions, can be anticipated from the drug's pharmacological actions, are predictable and dose dependent, and resolve when the dose is reduced or therapy is withdrawn. Type B reactions are generally unrelated to the drug's pharmacological actions, are unpredictable and independent of dose, and generally resolve when treatment is stopped, but reactions sometimes continue and may even progress.
- Type B reactions, sometimes called bizarre reactions, are made up of different responses that may be broadly distinguished as immune-mediated hypersensitivities and nonimmune sensitivities. The latter are further divided into pseudoallergic reactions, idiosyncratic reactions, and intolerances.
- On the basis of the Gell and Coombs classification of immune-mediated reactions, hypersensitivities are subdivided into four categories, types I, II, III, and IV. Type I reactions, often referred to as the "true" allergies, are generally immediate in onset (seconds to ~1 h) and mediated by IgE antibodies; type II hypersensitivities, or cytotoxic reactions, are IgG or IgM mediated; type III or immune complex reactions again are mediated by IgG or IgM antibodies; and type IV hypersensitivities, also called delayed hypersensitivities, are T-cell-mediated reactions.
- Type I hypersensitivities, anaphylaxis and urticaria, are rarely seen following biotherapy, but both reactions are known to occur with many different biologic agents. Because of the immunogenic potential of mAbs, especially murine antibodies ibritumomab tiuxetan and blinatumomab, the rat-mouse chimera catu-

maxomab, and human-mouse chimeras abciximab, basiliximab, brentuximab, cetuximab, infliximab, rituximab, siltuximab, and dinutuximab, mAbs commonly carry warnings of possible hypersensitivity, including anaphylaxis. The risk is lessened with the so-called humanized mAbs which are human antibodies with complementarity-determining regions (usually murine) grafted into human framework V_H and V_L regions and further reduced with fully humanized mAbs, but the possibility of generating anti-idiotype antibodies remains.

- Anaphylaxis has been recorded for the mAbs cetuximab, rituximab, trastuzumab, pertuzumab, ibritumomab tiuxetan, brentuximab, adalimumab, infliximab, and belimumab.
- Enzymes given as replacement therapies for some lysosomal storage disorders are well known possible, but rare, causes of anaphylaxis, a fact demonstrated by issued FDA boxed warnings for alglucosidase, elosulfase alfa, idursulfase, and laronidase and warnings/precautions for the β-glucocerebrosidases taliglucerase alfa and velaglucerase. Other approved enzymes, L-asparaginase, reteplase, collagenase, pegloticase, and rasburicase, have provoked anaphylaxis. This is emphasized for the latter two enzymes by an FDA black box warning.
- Incidences of type I hypersensitivities to approved cytokines, fusion proteins (in particular etanercept), coagulation proteins, and vaccines are significantly lower than seen with mAbs and enzymes. Cytokines implicated in type I reactions, including anaphylaxis, are interferon beta-1a, oprelvekin, anakinra, and darbepoetin alfa.
- Infusion reactions with signs and symptoms of one or more of cytokine release syndrome, an anaphylactic/anaphylactoid reaction, and direct toxicity are not always easy to distinguish with confidence and precision. Infusion of many biologics, particularly mAbs, provokes a characteristic infusion syndrome, usually within one or a few hours and during/after the first administration. Whereas most reactions are mild to moderate with symptoms often described as "flu"-like with fever, chills, rigors, headache, nausea, asthenia, rash, and pruritus, a small number of patients, mostly at the first or second infusion, show potentially fatal symptoms resembling an IgE antibody-mediated reaction with hypotension, cardiac arrest, bronchospasm, and urticaria.
- Trastuzumab, rituximab, and cetuximab induce the highest incidences of infusion reactions; approximately 80 % of fatal reactions to rituximab occurred after the first infusion, and even after the fourth and eighth infusions, reactions persisted in 30 % and 14 % of patients, respectively. Reactions to humanized and fully human mAbs are significantly lower. Mechanisms of the reactions are incompletely understood. Release of TNF and interleukin-6 (IL-6) has been suggested since these cytokines produce some symptoms similar to those seen in type I allergic responses. Infusion reactions may be related to the number of circulating lymphocytes with severe reactions seen when counts exceed 50×10^9/L.
- Monoclonal antibodies most commonly implicated in type IV hypersensitivities are brentuximab, adalimumab, ibritumomab tiuxetan, rituximab, and infliximab; the latter three have provoked cases of EM, SJS, and TEN.

- Rituximab may cause type II cytotoxic hypersensitivities and autoimmune forms of thrombocytopenia and hemolytic anemia. Late onset neutropenia occurs in some cancer patients treated with mAbs, especially rituximab. Autoantibodies may be involved, and the condition may be a true type II hypersensitivity response.
- Type III hypersensitivities, in particular serum sickness and drug-induced vasculitis, occur in response to biologics with mAbs again being the agents most often involved. Rituximab-induced vasculitis is the subject of a number of reports. Cutaneous vasculitis after adalimumab, infliximab, and certolizumab pegol is known and alirocumab carries an FDA warning for hypersensitivity vasculitis. Chimeric mAbs have the potential to cause serum sickness as seen with rituximab, infliximab, and omalizumab.
- Some drug-induced pulmonary events are hypersensitivity reactions resulting from interaction with the immune system and involvement of drug-reactive antibodies or T cells. Hypersensitivity pneumonitis, for example, is a combined type III and type IV hypersensitivity reaction. Monoclonal antibodies most commonly associated with pulmonary adverse events are cetuximab, rituximab, alemtuzumab, bevacizumab, trastuzumab, and panitumumab.
- In addition to the type A and type B reactions in the classification of adverse drug reactions, five other reaction types are distinguished. Reaction type C, demonstrated by corticosteroid-induced suppression and designated chronic or continuous effects, is seen uncommonly; reaction type D, called delayed effects and illustrated by stilbestrol-induced vaginal carcinoma, is time and dose related; category E, end-of-treatment (or withdrawal) effects, can be seen with little or no delay as, for example, in opiate withdrawal syndrome; category F, failure of therapy, may be caused by drug interactions and is usually dose related; and category G, genetic reactions, are seen as sensitivity to succinylcholine and some HLA-drug hypersensitivity associations.
- A number of systemic potentially life-threatening syndromes may occur with low frequency during or following the administration of a variety of biologic agents. These are capillary leak syndrome, cytokine release syndrome, hemophagocytic lymphohistiocytosis, macrophage activation syndrome, systemic inflammatory response syndrome, tumor lysis syndrome, posterior reversible encephalopathy syndrome, immune reconstitution inflammatory syndrome, and progressive multifocal leukoencephalopathy.

Further Reading

Baldo BA. Adverse events to monoclonal antibodies used for cancer therapy. Focus on hypersensitivity responses. OncoImmunology. 2013;2(10):e26333. doi:10.4161/onci.26333.

Baldo BA, Pham NH. Drug allergy. Clinical aspects, diagnosis, mechanisms, structure-activity relationships. New York: Springer; 2013. p. 1–13.

Basch E, Reeve BB, Mitchell SA, et al. Development of the National Cancer Institute's patient-reported outcomes version of the Common Terminology Criteria for adverse events (PRO-CTCAE). J Natl Cancer Inst. 2014;106:dju244. doi:10.1093/jnci/dju244.

Druey KM, Greipp PR. Narrative review: the systemic capillary leak syndrome. Ann Intern Med. 2010;153:90–8.

Edwards IR, Aronson JK. Adverse drug reactions: definitions, diagnosis, and management. Lancet. 2000;356:1255–9.

Feske SK. Posterior reversible encephalopathy syndrome: a review. Semin Neurol. 2011;31:202–15.

Giezen TJ, Mantel-Teeuwisse AK, Straus SMJM, et al. Safety-related regulatory actions for biologicals approved in the United States and the European Union. JAMA. 2008;300:1887–96.

Ho RJY. Biotechnology and biopharmaceuticals. Transforming proteins and genes into drugs. 2nd ed. Hoboken, New Jersey. John Wiley and Sons; 2013.

Howard SC, Jones DP, Pui C-H. The tumor lysis syndrome. N Engl J Med. 2002;364:1844–54.

Kinch MS. An overview of FDA-approved biologics medicines. Drug Discov Today. 2015;20: 393–8.

Leader B, Baca QJ, Golan DE. Protein therapeutics: a summary and pharmacological classification. Nat Rev Drug Discov. 2008;7:21–39.

Maude SL, Barrett D, Teachey DT, et al. Managing cytokine release syndrome associated with novel T cell-engaging therapies. Cancer J. 2014;20:119–22.

Rare diseases: common issues in drug development. Guidance for industry. U.S. Department of Health and Human Services; Food and Drug Administration Center for Drug Evaluation and Research; Center for Biologics Evaluation and Research, August 2015. http://www.fda.gov/downloads/Drugs/GuidanceComplianceRegulatoryInformation/Guidances/UCM458485.pdf. Accessed 30 Sept 2015.

Rawlins MD, Thomas SHL. Mechanisms of adverse drug reactions. In: Davies DM, Ferner ER, de Glanville H, editors. Davies' textbook of adverse drug reactions. 5th ed. London: Chapman and Hall Medical; 1998. p. 40–64.

Talbot J, Aronson JK, editors. Stephen's detection and evaluation of adverse drug reactions. 6th ed. Chichester: Wiley-Blackwell; 2012.

Wang W, Singh M, editors. Biological drug products: development and strategies. Hoboken: Wiley; 2014.

Further Reading

Chapter 2
Monoclonal Antibodies: Introduction

Monoclonal Antibodies for Therapy

A basic feature of the normal humoral immune response is its heterogeneity, seen at the levels of antibody class or isotype and subclass and combining site differences reflected in different idiotypic determinants. Antibodies also demonstrate allotypic differences representing genetically determined differences between individuals of the same species. Such antibody heterogeneity is itself a reflection of multiple different B-cell clonal lineages which ultimately differentiate to form plasma cells secreting a net polyclonal population of antibodies. For the purpose of antibody therapy, a pooled polyclonal antibody response directed to multiple antigens may be effective, e.g., for mounting a response to pathogens, but for many forms of therapy where precise targeting, a "perpetual" supply of antibody, greater precision, and reproducibility are needed, uniform and well-defined antibodies from a single long-lived clonal cell line are required. Two properties of polyclonal B cells are impediments to their use for the preparation of targeted therapeutic antibodies, namely, the relatively short life span of plasma cells and the difficulties associated with isolating a single plasma cell of the desired specificity from a culture of antigen-challenged B cells producing a polyclonal population of antibodies. These two fundamental difficulties were essentially overcome 40 years ago by Köhler and Milstein with their production of mouse hybridoma cells prepared by fusing spleen cells from an immunized mouse with mouse myeloma cells. The resultant hybridoma cells retain the capacity to make specific invariant antibody while at the same time the immortalized cells are able to grow indefinitely in culture, continuously secreting monoclonal antibody (mAb). In the continuing efforts to target and modulate individual diseases, mAbs often provide the best means of establishing a proof of concept and at a lower safety risk than many other biologic and small molecule drugs.

B.A. Baldo, *Safety of Biologics Therapy*, DOI 10.1007/978-3-319-30472-4_2

Evolution of Therapeutic Monoclonal Antibodies: From Mouse to Man

Just over a decade after the publication of Köhler and Milstein's strategy and methods for hybridoma technology, the first therapeutic mAb, muromonab-CD3 (Orthoclone OKT3), received regulatory approval. This murine IgG2a mAb directed against the CD3 (T3) receptor on the surface of human T lymphocytes was used as an immunosuppressant to combat steroid-resistant acute allograft rejection in patients with renal, hepatic, and cardiac transplants and for acute graft-versus-host disease. Not unexpectedly, it soon became apparent that the human immune response to mouse proteins and the poor pharmacokinetics of the mAbs were a significant problem for repeated and long-term therapy. The manufacture of muromonab-CD3 was discontinued in 2010 due to declining sales and the emergence of other treatments with similar efficacy and fewer adverse effects. Adverse events associated with the mouse antibody included cytokine release syndrome (Chap. 1), increased risk of infections and malignancies, encephalopathy, pulmonary edema, cardiac arrest, seizures, and chest pain, and as anticipated, some warnings and safety issues were a direct result of the foreign murine antigens, such as the induction of anaphylactoid/anaphylactic reactions. Besides muromonab-CD3, other murine mAbs approved for human therapy include ibritumomab tiuxetan and tositumomab-^{131}I, both targeted to CD20 and indicated for non-Hodgkin lymphoma (Table 2.1). A mouse/rat hybrid hybridoma, catumaxomab, targeted to EpCAM/CD3 and used for malignant ascites, also received regulatory approval. However, it was soon clear that more "humanized" mAbs were needed. The first step in what was to be an ongoing iterative process began with the production of so-called chimeric antibodies in which the variable (antigen-binding) regions of mouse antibodies were incorporated into the constant regions of human immunoglobulins (Fig. 2.1). Chimeric constructs so formed generally proved to have comparable binding affinities to the parent murine mAb and were at least as active in mediating complement-dependent and cell-mediated cytotoxicities (see below, section "Antibody-Dependent Cell-Mediated and Complement-Dependent Cytotoxicities"). Monoclonal antibodies in this category are abciximab, basiliximab, brentuximab vedotin, cetuximab, rituximab, dinutuximab, infliximab, and siltuximab (see Table 2.1 for targets and indications). Despite the important technological advancement in achieving an approximate two-thirds reduction of murine proteins, the risk of a human anti-mouse immune response was not eliminated, and adverse reactions, especially potentially serious hypersensitivities, still occasionally occurred with a frequency considered to be unacceptable. This need to eliminate species recognition differences led to the next iteration designed to reduce and ultimately eliminate potential murine antigens. A further step forward was the production of the so-called humanized mAbs achieved by substituting murine hypervariable or complementarity-determining regions (CDRs) in place of human sequences while retaining the remainder of the antibody as human (Fig. 2.1). This produced antibodies with only ~5–10 % murine proteins and the improved tolerance of these agents by patients and noticeably fewer side

Table 2.1 Therapeutic monoclonal antibodies currently marketed with regulatory approval from the US FDA or EMA or both as at June 2016

INN and trade name	Type of mAb	Cell line	Target[a]	Mechanism of action	Approved indications
Rat-mouse chimera (-axomab)					
Catumaxomab[b] (Removab®)	Rat IgG2b/ Mouse IgG2a bispecfic	Hybrid hybridoma	EpCAM/ CD3	Binds EpCAM, CD3 and FcγRs (via Fc) activating T and accessory cells at tumor site and leading to cell killing	Malignant ascites
Mouse (-omab)					
Blinatumomab[c] (Blincyto®)	Mouse scFvκ–H[d] bispecific	CHO	CD19/CD3[e] epsilon	Links CD19 on malignant B cells with CD3 in T cell receptor destroying tumor cell via perforin and granzymes	Philadelphia chromosome-negative relapsed or refractory B cell precursor acute lymphoblastic leukemia
Ibritumomab tiuxetan[f] (Zevalin®)	Mouse IgG1κ	CHO	CD20[g]	Binds malignant B cells; β emissions induce cell damage	Non-HL
Human-mouse chimeric (-ximab)					
Abciximab (ReoPro®)	Chimeric IgG1κ Fab	Sp2/0	Glycoprotein IIb/IIIa	A receptor antagonist. Binds gp IIb/ IIIa inhibiting platelet aggregation by preventing binding of fibrinogen and von Willebrand factor	Adjunct therapy for prevention of cardiac ischemic complications
Basiliximab (Simulect®)	Chimeric IgG1κ	Sp2/0	α chain IL-2 receptor (CD25)	Competitively inhibits IL-2-mediated activation of T cells involved in allograft rejection	Prevent organ transplant rejection
Brentuximab vedotin[h] (Adcetris®)	Chimeric IgG1κ	CHO	CD30[i]	Binds to CD30 cells; released MMAE binds to and disrupts microtubules causing cell cycle arrest and apoptotic cell death	HL after failure of stem cell transplant or chemotherapy; sALCL after failure of chemotherapy; post auto-HSCT consolidation treatment for HL
Cetuximab (Erbitux®)	Chimeric IgG1κ	Sp2/0	EGFR	Binds to EGFR blocking phosphorylation and activation of kinases inhibiting growth and survival of tumor cells	Colorectal and head and neck cancers

(continued)

Table 2.1 (continued)

INN and trade name	Type of mAb	Cell line	Target[a]	Mechanism of action	Approved indications
Dinutuximab (Unituxin®)	Chimeric IgG1κ	Sp2/0	GD2	Binds cell surface GD2 on neuroblastoma cells and induces cell lysis by ADCC and CDC	Pediatric patients with high risk neuroblastoma[j]
Infliximab (Remicade®)	Chimeric IgG1κ	Sp2/0	TNF	Inhibits binding of TNF to its receptor neutralizing its many proinflammatory actions	Crohn's disease; ulcerative colitis; rheumatoid arthritis; ankylosing spondylitis; psoriatic arthritis; plaque psoriasis
Obiltoxaximab (Anthim®)	Chimeric IgG1κ		*Bacillus anthracis* PA	Binds free PA (K_d 0.33 nM) inhibiting its binding to its receptors and preventing cell entry of anthrax edema and lethal toxins	Inhalational anthrax to *Bacillus anthracis and* prophylaxis in absence of alternative therapies
Rituximab (Rituxan®; MabThera®)	Chimeric IgG1κ	CHO	CD20	Binds CD20 on B cells[k]; Fc recruits effector functions CDC and ADCC	Non-HL; CLL; rheumatoid arthritis; Wegener's granulomatosis; microscopic polyangiitis[l]
Siltuximab (Sylvant®)	Chimeric IgG1κ	CHO	IL-6	Binds to IL-6 preventing its overproduction and lessening symptoms of disease	Multicentric Castelman's disease[m] in patients negative for HIV and HHV-8
Humanized (-zumab)					
Alemtuzumab (Campath®; MabCampath®; Lemtrada®)	Humanized IgG1κ	CHO	CD52[n]	Binds CD52 on leukemic cells and induces antibody-dependent cell-mediated lysis	Campath, MabCampath: B cell CLL. Lemtrada: Multiple sclerosis[o]
Bevacizumab (Avastin®)	Humanized IgG1κ	CHO	VEGF-A	Inhibits angiogenesis and metastatic disease progression by binding VEGF preventing interaction with its receptors	Metastatic colorectal cancer; non-squamous NSCLC; metastatic breast cancer; glioblastoma; EMA: ovarian, fallopian tube, peritoneal cancers
Certolizumab pegol[p] (Cimzia®)	Humanized IgG1κ Fab', pegylated[p]	*E. coli*	TNF	Binds TNF neutralizing its proinflammatory actions. No Fc therefore no CDC or ADCC	Crohn's disease; rheumatoid arthritis

Eculizumab (Soliris®)	Humanized IgG2/4κ	NSO	C5	Inhibits cleavage of C5 and complement-mediated intravascular hemolysis and thrombotic microangiopathy	Paroxysmal nocturnal hemoglobinuria; atypical hemolytic uremic syndrome
Elotuzumab (Empliciti®)	Humanized IgG1κ	NSO	SLAMF7	Targets SLAMF7 on myeloma cells and natural killer (NK) cells, facilitating the latter to kill myeloma cells through ADCC	Given in combination with lenalidomide and dexamethasone to multiple myeloma patients who have received one to three prior therapies
Idarucizumab (Praxbind®)	Humanized IgG1κ Fab	CHO	Dabigatran	Binds to dabigatran and its acylglucuronide metabolites neutralizing the anticoagulent effect	Reversal of anticoagulant effects of dabigatran for surgical/urgent procedures; life-threatening or uncontrolled bleeding
Ixekizumab (Taltz®)	Humanized IgG4κ		IL-17A	Inhibits interaction of IL-17A with its receptor inhibiting release of other inflammatory cytokines and chemokines	Moderate to severe plaque psoriasis
Mepolizumab (Nucala®)	Humanized IgG1κ	CHO	IL-5	Binds IL-5 and blocks receptor, reducing the production and survival of eosinophils and reducing inflammation in asthma	Add-on maintenance treatment of patients 12 years and older with severe asthma and an eosinophilic phenotype
Natalizumab (Tysabri®)	Humanized IgG4κ	NSO	α4 integrin[q]	Binds α4 subunit of α4β1 and α4β7 on leukocytes inhibiting adhesion to counter receptors[r] and prevents leukocyte migration to inflamed tissue	Multiple sclerosis; Crohn's disease[s]
Obinutuzumab (Gazyva®; Gazyvaro®)	Humanized IgG1κ	CHO	CD20	Mediates B cell lysis by CDC, ADCC, ADCP and by activating death signaling	In combination with chlorambucil for previously untreated CLL; in combination with bendamustine or alone as maintenance for follicular lymphoma
Omalizumab (Xolair®)	Humanized IgG1κ	CHO	IgE	Inhibits binding of IgE to FcγRI on mast cells and basophils limiting mediator release and cell-bound IgE	Persistent asthma[t]; chronic idiopathic urticaria
Palivizumab (Synagis®)	Humanized IgG1κ	NSO	RSVF	Reduces pulmonary RSV replication and inhibits virus entering cell	Prevention of lower respiratory tract disease RSV in children

(continued)

Table 2.1 (continued)

INN and trade name	Type of mAb	Cell line	Target[a]	Mechanism of action	Approved indications
Pembrolizumab (Keytruda®)	Humanized IgG4κ	CHO	PD-1	Blocks receptor interaction of PD-L1 and PD-L2 preventing PD-1-mediated inhibition of anti-tumor immune response	Unresectable or metastatic melanoma[u]; refractory metastatic NSCLC tumors that express PD-L1
Pertuzumab (Perjeta®)	Humanized IgG1κ	CHO	HER2	Prevents heterodimer formation and subsequent MAPK and PI3K signaling resulting in cell growth arrest and apoptosis[v]	Combination with trastuzumab and docetaxel for HER2-positive metastatic breast cancer[w]
Ranibizumab (Lucentis®)	Humanized IgG1κ Fab	*E. coli*	VEGF-A	Binds VEGF-A preventing its interaction with receptors and reducing angiogenesis and vascular leakage	Neovascular (wet) age-related macular degeneration; macular edema following retinal vein occlusion; diabetic macular edema
Reslizumab (Cinqair®)	Humanized IgG4κ	NSO	IL-5	Binds IL-5 and blocks receptor, reducing the production and survival of eosinophils and reducing inflammation in asthma	Add-on maintenance of patients with severe asthma aged 18 and older and with an eosinophilic phenotype
Tocilizumab (Actemra®; RoActemra®)	Humanized IgG1κ	CHO	IL-6R	Binds IL-6 receptors inhibiting signaling and proinflammatory effects of IL-6 in joints affected by inflammation	Rheumatoid arthritis; PJIA; SJIA
Trastuzumab (Herceptin®)	Humanized IgG1κ	CHO	HER2	Blocks homodimerization of HER2 thus inhibiting HER2 receptor signaling. Mediates ADCC	Breast cancer overexpressing HER2, metastatic gastric or GE junction adenocarcinoma overexpressing HER2
Ado-trastuzumab emtansine[x] (Kadcyla®)	Humanized IgG1κ	CHO	HER2	See trastuzumab. Also: Binds to tubulin after internalization disrupting microtubules and causing apoptotic cell death	HER2-positive breast cancer in patients who previously received trastuzumab or a taxane
Vedolizumab (Entyvio®)	Humanized IgG1κ	CHO	α4β7 integrin	Blocks interaction of α4β7 integrin with addressin,[y] inhibiting T cell migration into inflamed GI tissue	Adult ulcerative colitis; adult Crohn's disease
Fully human (-umab)					
Adalimumab (Humira®)	Human IgG1κ	CHO	TNF	Binds to TNF preventing receptor activation and articular inflammation[z] characteristic of these disease	Rheumatoid arthritis; psoriatic arthritis; ankylosing spondylitis; plaque psoriasis; Crohn's disease

Alirocumab (Praluent®)	Human IgG1κ	CHO	PCSK9	Inhibits binding PCSK9 to LDLR, LDLR expression ↑, LDL clearance ↑ and LDL-C ↓ in circulation	Heterozygous FH; atherosclerotic CV disease requiring additional ↓ of LDL-C
Belimumab (Benlysta®)	Human IgG1λ	NSO	BlyS	Blocks binding of soluble BlyS[aa] to receptors on B cells inhibiting B cell survival	Systemic lupus erythematosus
Canakinumab (Ilaris®)	Human IgG1κ	Sp2/0	IL-1β	Mutations in NLRP-3 gene leads to excess IL-1 β and inflammation. Antibody binds to IL-1 β neutralizing its action	Cryopyrin-associated periodic syndromes (CAPS) including familial cold autoinflammatory and Muckle-Wells syndromes; SJIA with body weight ≥7.5 kg; NOMAD/CINCA; FCAS/FCU; gouty arthritis
Daratumumab (Darzalex®)	Human IgG1κ	CHO	CD38[ab]	Inhibits growth of tumor cells expressing CD38, leading to apoptosis by Fc-mediated cross-linking and cell lysis induced via CDC, ADCC, and ADCP	Multiple myeloma
Denosumab (Prolia®; Xgeva®)	Human IgG2κ	CHO	RANKL	Binds RANKL preventing receptor activation, osteoclast formation, bone resorption, osteolysis and tumor growth	Bone loss. Prolia: For osteoporosis and to increase bone mass.[ac] Xgeva: For bone metastases from solid tumors and giant cell tumor of bone[ad]
Evolocumab (Repatha®)	Human IgG2λ	CHO	PCSK9	Inhibits binding PCSK9 to LDLR, LDLR expression↑, LDL clearance ↑ and LDL-C↓ in circulation	Primary hyperlipidemia and mixed dyslipidemia; homozygous FH to reduce LDL-C and other lipids[ae]
Golimumab (Simponi®)	Human IgG1κ	Sp2/0	TNF	Binds to TNF preventing receptor activation and articular inflammation[af] characteristic of these diseases	Rheumatoid arthritis; psoriatic arthritis (both in combination with methotrexate); ankylosing spondylitis
Ipilimumab (Yervoy®)	Human IgG1κ	CHO	CTLA-4	Blocks interaction of CTLA-4 with its ligands[ag] augmenting T cell activation and anti-tumor response	Unresectable or metastatic melanoma; adjuvant treatment of cutaneous melanoma with lymph nodes >1 mm after resection

(continued)

Table 2.1 (continued)

INN and trade name	Type of mAb	Cell line	Target[a]	Mechanism of action	Approved indications
Necitumumab (Portrazza®)	Human IgG1κ	NSO	EGFR	Binds human EGFR blocking binding to its ligands. Activation of EGFR has been correlated with angiogenesis, inhibition of apoptosis and malignant progression	NSCLC[ah]
Nivolumab (Opdivo®)	Human IgG4κ	CHO	PD-1	Blocks receptor interaction of PD-L1 and PD-L2 preventing PD-1-mediated inhibition of anti-tumor immune response	Unresectable or metastatic melanoma and disease progression following ipilimumab and, if BRAF V600 positive, a BRAF inhibitor; NSCLC[ai]; MRCC[ai]
Ofatumumab (Arzerra®)	Human IgG1κ	NSO	CD20	Binds CD20 on B cells; Fc recruits effector functions CDC and ADCC	Chronic lymphocytic leukemia refractory to fludarabine and alemtuzumab
Panitumumab (Vectibix®)	Human IgG2κ	CHO	EGFR	Binds to EGFR blocking phosphorylation and activation of kinases inhibiting growth and survival of tumor cells	Metastatic colorectal cancer[aj]
Ramucirumab (Cyramza®)	Human IgG1κ	NSO	VEGFR2	Inhibits activation of VEGFR2 and hence ligand-induced proliferation and migration of endothelial cells	Gastric or GE junction adeno-carcinoma; metastatic NSCLC with docetaxel after platinum therapy; with FOLFIRI for metastatic colorectal cancer[ak]
Raxibacumab (ABthrax®)	Human IgG1λ[al]	NSO	*Bacillus anthracis* PA	Binds free PA inhibiting its binding to its receptors and preventing cell entry of anthrax edema and lethal toxins[am]	Inhalational anthrax to *Bacillus anthracis and* prophylaxis in absence of alternative therapies
Secukinumab (Cosentyx®)	Human IgG1κ	CHO	IL-17A	Inhibits interaction of IL-17A with its receptor inhibiting release of other inflammatory cytokines and chemokines	Moderate to severe plaque psoriasis
Ustekinumab (Stelara®)	Human IgG1κ	Sp2/0	IL-12, IL-23	Binds to p40 protein subunit of IL-12 and IL-23 disrupting their signaling and immune and inflammatory responses	Plaque psoriasis; psoriatic arthritis in adults alone or in combination with methotrexate

During the production stage of this book, the humanized Fc-engineered IgG1 kappa mAb **atezolizumab** (Tecentriq®) binding PD-L1 and approved for urothelial carcinoma, was approved by the FDA (see Chap. 3, "Recent Approval: Atezolizumab").

ADCC antibody-dependent cell-mediated cytotoxicity, *ADCP* antibody-dependent cellular phagocytosis, *auto-HSCT* autologous hematopoietic stem cell transplantation, *BlyS* B lymphocyte stimulator, also known as B cell activating factor, BAFF *BRAF* proto-oncogene B-Raf, *C5* complement component 5, *CDC* complement-dependent cytotoxicity, *CHO* Chinese hamster ovary cells, *CINCA* chronic infantile neurological, cutaneous, articular syndrome, *CLL* chronic lymphocytic leukemia, *CTLA-4* cytotoxic T lymphocyte-associated antigen 4 or CD152, *CV* cardiovascular, *EGFR* epidermal growth factor receptor, *EMA* European Medicines Agency, *EpCAM* epithelial cell adhesion molecule, *FCAS* familial cold autoinflammatory syndrome, *FCU* familial cold urticaria, *FDA* US Food and Drug Administration, *FH* familial hypercholesterolemia, *FOLFIRI* Folinic acid (leukovorin), fluoropyrimidine and irinotecan, *GD2* glycolipid disialoganglioside on neuroblastoma, central nervous system and peripheral nerve cells, *GE* gastroesophageal, *GI* gastrointestinal, *HER2* human epidermal growth factor receptor 2, also known as HER2/neu, ErbB2, CD340, p185 or EGFR2, *HIV* human immunodeficiency virus, *HHV-8* human herpesvirus-8, *HL* Hodgkin lymphoma, *INN* International Nonproprietary Name, *LDL* low-density lipoprotein, *LDL-C* LDL-cholesterol, *LDLR* LDL receptor, *MAPK* mitogen-activated protein kinase, *MMAE* cytotoxic agent monomethyl auristatin E, *MRCC* metastatic renal cell carcinoma, NLRP-3 gene cryopyrin or nucleotide-binding domain, leucine rich family, pyrin domain containing 3 gene, NOMID neonatal-onset multisystem inflammatory disease, *NSCLC* non-small cell lung cancer, *NSO* non-Ig-secreting, non-L chain-synthesizing, 8-azaguanine-resistant and HAT-sensitive mouse myeloma cell line, *PA* protective antigen of *B. anthracis* toxin, *PCSK9* proprotein convertase subtilisin/kexin type 9, *PD-1* programmed cell death protein 1 or CD279, PD-L1 programmed cell death protein ligand 1, *PI3K* phosphoinositide 3-kinase, *PJIA*, polyarticular juvenile idiopathic arthritis, *RANKL* receptor activator of nuclear factor kappa-B ligand (CD254), a member of the TNF cytokine family, *RSV* human respiratory syncytial virus (*F* [fusion] viral protein coat antigen), *sALCL* systemic anaplastic large cell lymphoma, *SJIA* active systemic juvenile idiopathic arthritis, *SLAMF7* signaling lymphocytic activation molecule receptor family member 7, also known as CS1, CD2 subunit 1, CD319, *Sp2/0* BALB/c mouse spleen cells fused with P3 myeloma. Cells do not secrete Ig, are resistant to 8-azaguanine and HAT-sensitive, *TNF* tumor necrosis factor, *VEGF* vascular endothelial growth factor (a subfamily of growth factors; includes VEGF-A), *VEGFR2* vascular endothelial growth factor receptor 2; also known as KDR (kinase insert domain-containing receptor), FLK1 (fetal liver kinase 1) or CD309

[a]Target specificity of mAb

[b]Registered by the EMA, Health Canada, and Ministry of Health, Israel but not the FDA

[c]A BiTE (bispecific T cell-engaging)fusion protein

[d]Two single chain variable fragments (scFv) each from an H and L chain to give four peptide chains linked to a single protein, MW ~54 kDa[e]

[e]CD19, a B cell antigen; CD3, part of the T cell receptor

[f]Conjugated to the chelator tiuxetan which links the radioisotope Yttrium-90 or Indium-111 *Yttrium-90 or Indium-111*

[g]Expressed on B lymphocytes where it aids optimal B cell response to T-independent antigens. Encoded by the membrane-spanning MS4A1gene

[h]Conjugated to the cytotoxic agent monomethyl auristatin E (MMAE)

[i]CD30 is a cell membrane protein of the tumor necrosis receptor family expression on activated T and B lymphocytes

[j]In combination with granulocyte-macrophage colony-stimulating factor (GM-CSF), IL-2 and 13-cis-retinoic acid

[k]CD20 is expressed on >90 % of non-hodgkin lymphoma cells

[l]In adults in combination with glucocorticoids

[m]Also called giant lymph node hyperplasia and angiofollicular lymph node hyperplasia

[n]CAMPATH-1 antigen. Present on mature lymphocytes, monocytes, and dendritic cells and is associated with some lymphomas

[o]Withdrawn from the US and Europe in 2012 and relaunched for multiple sclerosis (as Lemtrada®)

(continued)

Table 2.1 (continued)

[p]Attached to PEG (polyethylene glycol)
[q]α-4 integrin (CD49d; VLA4). Natalizumab acts on both $\alpha 4\beta 1$and $\alpha 4\beta 7$ integrins
[r]Receptors for the $\alpha 4$ integrin family include vascular cell adhesion molecule-1 (VCAM-1) expressed on vascular endothelium and mucosal addressin cell adhesion molecule-1 (MAdCAM-1) on vascular endothelial cells of the GI tract. The precise mechanisms of the alleviating effects of natalizumab in multiple sclerosis and Crohn's disease have not been elucidated
[s]In Crohn's disease, natalizumab should not be used in combination with immunosuppressants or TNF inhibitors
[t]Adults and adolescents with a positive skin test or in vitro test to aeroallergens and with symptoms not controlled with inhaled corticosteroids
[u]Also in disease progression following ipilimumab and, if BRAF V600 mutation-positive, a BRAF inhibitor. Note: The FDA states: "An improvement in survival or disease-related symptoms has not yet been established." In June 2015, the FDA accepted for review the supplemental Biologics License Application for pembrolizumab for the treatment of advanced non-small cell lung cancer whose disease has progressed on or after platinum and EGFR therapy
[v]Also mediates ADCC
[w]In patients who have not yet received prior anti-HER2 therapy or chemotherapy for metastatic disease
[x]An antibody-drug conjugate of trastuzumab linked to the maytansinoid cytotoxin DM1 or mertansine (*N*2′-diacetyl-*N*2′-(3-mercapto-1-oxopropyl) maytansine. Also known as trastuzumab emtansine and T-DM1
[y]Mucosal addressin cell adhesion molecule-1 (MAdCAM-1)
[z]Adalimumab also modulates biological responses induced by TNF including adhesion molecules VCAM-1 and intercellular adhesion molecule 1 (ICAM-1)
[aa]Also known as CD257 or tumor necrosis factor ligand superfamily member 13B. Acts as a B cell survival factor
[ab]Cyclic ADP ribose hydrolase, a surface antigen expressed by multiple myeloma cells and found on many immune cells including CD4+, CD8+, and B lymphocytes and natural killer (NK) cells
[ac]Treatment of menopausal women at high risk of fracture, men receiving androgen deprivation for prostate cancer, and women at high risk of fracture receiving aromatase inhibitor therapy for breast cancer
[ad]Not indicated for skeletal-related events in multiple myeloma
[ae]May be given in combination with a statin or with a statin and other lipid-lowering therapies (see Chap. 4, Sect. Evolocumab: Indications and Safety)
[af]Elevated levels of TNF in blood, synovium, and joints are implicated in these chronic inflammatory diseases
[ag]Ligands for CTLA-4: B7-1/B7-2 (CD80/CD86)
[ah]For first-line treatment of patients with metastatic squamous non-small cell lung cancer in combination with gemcitabine and cisplatin
[ai]In March 2015, nivolumab was approved by the FDA for the treatment of patients with metastatic squamous non-small-cell-lung cancer with progression on or after platinum-based chemotherapy; approval was extended to non-squamous non-small-cell lung cancer in October 2015. In November 2015, approval was extended to metastatic renal cell carcinoma in patients who have progressed on anti-angiogenic therapy
[aj]Panitumumab is not recommended for treatment of tumors with KRAS mutations in codon 12 or 13
[ak]Gastric or GE junction adenocarcinoma: As single agent or in combination with paclitaxel after prior fluoropyrimidine or platinum chemotherapy. Colorectal cancer: For patients with disease progression on or after bevacizumab, oxaliplatin, and fluoropyrimidine therapy. Also approved by the EMA as monotherapy and for use with paclitaxel for advanced gastric cancer or GE junction cancer
[al]Note λ light chain
[am]Inhibits angiogenesis in an animal model

effects were soon obvious. Examples of approved humanized mAbs include alemtuzumab, bevacizumab, natalizumab, omalizumab, ranibizumab, trastuzumab, and 12 others (Table 2.1). Note, however, both chimeric and humanized mAbs may still have two other undesirable features in addition to unwanted immunogenicity. In the first place, changes to a small number, or even a single amino acid, especially in the variable region causing slight alterations in the CDRs, may be enough to perturb antibody conformational integrity and, in turn, binding affinity and avidity. Secondly, defects in, or the absence of, glycosylation in expression of the humanized antibodies can markedly affect efficacy and pharmacokinetic properties such as solubility and clearance rates. The ultimate goal though was always to develop fully humanized antibodies and this became possible with the development of phage display and transgenic mouse technologies. The first fully human mAb to reach the US market, adalimumab, was developed with phage display technology. Other examples of fully human mAbs include belimumab, denosumab, golimumab, panitumumab, and ustekinumab. Seventeen of the 50 currently approved mAbs (June 2016) are fully human preparations (Table 2.1). Note, however, that the possibility of the formation of anti-idiotype antibodies always remains so the potential for immunogenicity of even a fully human mAb persists to at least some degree. In addition, antibodies to some drugs, including mAbs, are sometimes found in normal sera and sera of pre-treated patients. These antibodies include rheumatoid factors, anti-glycan, anti-hinge, and anti-allotype antibodies, and although they do not usually interfere with treatments, they may occasionally have clinical consequences.

Technological Advances in the Production of Monoclonal Antibodies

Hybridoma Technology and Immortalization of Human B Cells

The evolution of mAbs for human therapy began with hybridoma technology when human B lymphocytes producing antibody were fused with mouse or human myeloma or lymphoblastoid cells. A lack of suitable human myeloma cell lines proved a drawback for the hybridoma approach, but an alternative means of "immortalizing" human antigen-specific antibody-secreting B cells with Epstein-Barr virus (EBV) was utilized. Unfortunately, EBV-transformed cells have some disadvantages being difficult to clone, producing small amounts of antibody, and being non-malignant, the cells do not grow indefinitely. This led to a combination of the two methods which largely overcomes the pitfalls of the hybridoma and EBV-transformed cell methods alone, and despite the undoubled advances arising from the innovative approaches discussed below, improved methods for cell fusion and B-cell immortalization may see EBV-hybridoma technology not only retained but with its use expanded.

Phage Display

First developed in 1985 and used to produce many different peptides and proteins, fragments of antibody are "displayed" on the surface of bacteriophage following the insertion of the corresponding gene(s) into the phage protein coat gene. By attaching target proteins (or DNA sequences) to a solid phase (e.g., a microtiter plate), displayed proteins complementary to the target can be isolated by washing and elution. Eluted phage can be reproduced in bacteria to generate a large library of target-specific antibody fragments. After repeated amplification in bacteria, the antigen-specific antibody fragments identified by bio-panning can be identified by sequencing the DNA of the interacting phage. To produce human antibodies, reverse transcription of mRNA from human B cells and PCR amplification are used to prepare a library of immunoglobulin heavy and light chain variable (V_H and V_L, respectively) gene segments. Using the polymerase chain reaction (PCR), a random combination of V_H and V_L gene segments is employed to create the gene for a single-chain variable fragment (scFv). This can be displayed on the phage by combinatorial infection and recombination to generate a large antibody repertoire. Those antibody fragments with good affinities are then identified by bio-panning. It is now possible to prepare human mAbs to any desired antigen.

Transgenic (Knockout) Mice

By inactivating or "knocking out" a gene from stem cells of mouse embryos and replacing it with a segment of the desired DNA, the altered stem cells are then grown in vivo producing an animal model with an altered genomic profile and a

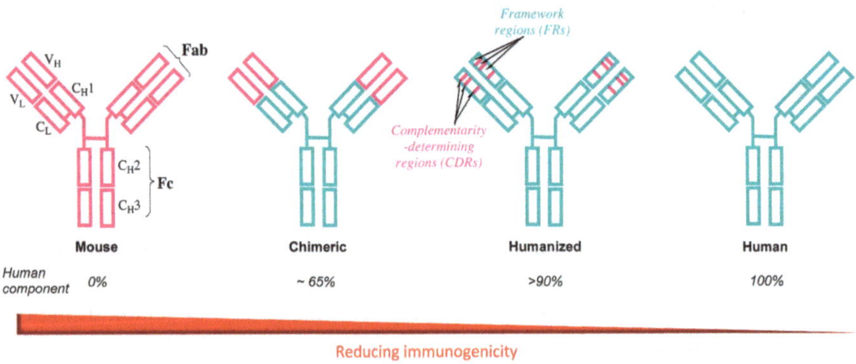

Fig. 2.1 Evolution of the development of therapeutic monoclonal antibodies from murine to fully human proteins to avoid unwanted immunogenicity. Development proceeded stepwise through chimeric constructs incorporating mouse immunoglobulin variable regions into constant regions of human immunoglobulins and via humanized antibodies by substituting mouse complementarity-determining regions (CDRs) in place of human sequences. Fully human antibodies have been developed with the application of phage display and transgenic mice technologies

sometimes changed phenotype. Such genetically modified mice may be used to express human antibody repertoires for the generation of human mAbs. Disruption of mouse immunoglobulin H and L chain loci, followed by the introduction of the equivalent human H and L genes, has generated knockout mice producing human B cells and enabled the expression of additional V gene segments and an impressive expansion of the potential repertoire of antibodies with tailored effector functions. Already a number of human mAbs produced in transgenic mice have been approved for marketing, and increasing numbers of the mAbs are being brought forward into clinical trials. In addition to the almost complete absence of hypersensitivity responses to these antibodies, they generally show similar affinities and pharmacokinetics to antibodies made in humans.

Monoclonal Antibodies from Single Human B Cells by Gene Cloning

This approach to produce mAbs from single human B cells is based on the analysis of the immunoglobulin gene repertoire and reactivity at the single-cell level by the application of reverse transcription-polymerase chain reaction (RT-PCR) and expression vector cloning. By recognition of selected cell surface markers, individual mouse or human B cells are isolated (e.g., by fluorescence-activated cell sorting), and genes coding for V_L and V_H fragments are separately amplified by RT-PCR and combined by PCR. For the final production of human mAbs in vitro, H and L chain gene transcripts from each cell are amplified by RT-PCR before cloning and expression in a mammalian system. This method has the virtue of being able to produce many specific human mAbs in a short period. So-called immunoglobulin gene expression cassettes for the high-throughput isolation of immunoglobulin genes from single human B cells without a cloning step have been developed.

IgG Antibody Subclasses

Although there are five classes of immunoglobulins, IgG is generally favored for the preparation of therapeutic antibodies. The IgG class is further subdivided into four subclasses IgG1, IgG2, IgG3, and IgG4, each with similar intrachain H and L domain disulfides, but the interchain disulfide bridges linking the H and L chains and the two H chains are different (Fig. 2.2). In IgG1, the H and L chains are connected by the first H chain cysteine in the hinge region and the C-terminal cysteine of the L chain. For IgG2, IgG3, and IgG4, the H–L bridge is between the L chain terminal cysteine and the *N*-terminal cysteine of the C_{H1} domain. Some marked amino acid differences occur in the hinge region, especially for the cysteines in IgG2 and IgG3. IgG1 and IgG4 both have two disulfides linking their H chains, whereas IgG2 has 4 and IgG3 has 11 (Fig. 2.2). To date, all of the mAbs, or derived

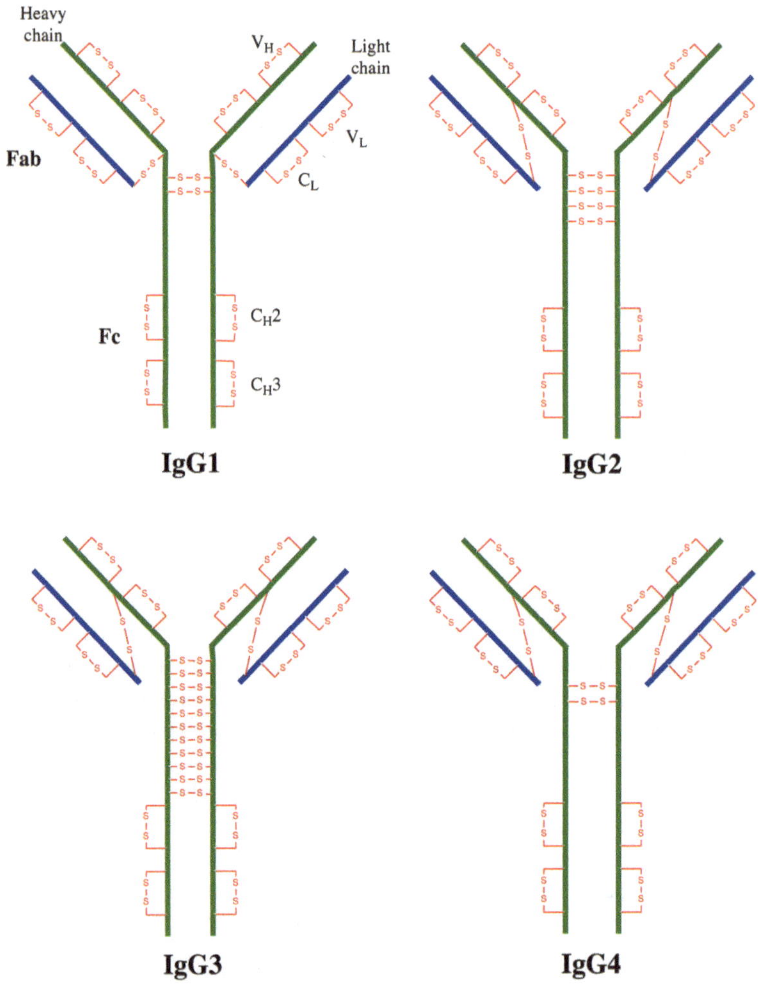

Fig. 2.2 Diagrammatic representation of the four human IgG subclasses, IgG1, IgG2, IgG3, and IgG4, showing the number and locations of the interchain and intrachain disulfide bonds

products (such as abciximab, certolizumab, and blinatumomab), approved for therapy are of the IgG isotype. The main reason for this is the long elimination half-lives of the subclasses which are ~20–21 days for IgG1, IgG2, and IgG4 and ~7 days for IgG3. In addition to the long clearance rate, some biological properties of the IgG subclasses are useful for enhancing or manipulating some effector functions. Table 2.2 summarizes the properties of the four subclasses taken into account when selecting the desired properties for a particular therapeutic antibody. A glance at the list of the 46 currently approved mAbs (Table 2.1) shows that IgG1 is easily the subclass most often selected, while IgG2 and IgG4 are occasionally used. Kappa L chains are overwhelmingly represented over lambda L chains. Apart from its long half-life, IgG1 is the choice for antibody-dependent cell-mediated cytotoxicity

Table 2.2 Properties of IgG subclasses relevant to their use as therapeutic monoclonal antibodies

IgG subclass	Percentage in serum	Half-life (days)	Binding affinity for FcγRIIIa[a]	Complement activation[b]
IgG1	66	~21	+++	++
IgG2	23	~21	+/–	+
IgG3	7	~7	+++	+++
IgG4	4	~21	+ to –	–

+++ high affinity, ++ intermediate affinity, + low affinity, – no affinity

[a]FcγRIIIa (CD16a) is present on adult natural killer (NK) cells, macrophages, and neutrophils. Receptor involved in antibody-dependent cell-mediated cytotoxicity (ADCC). For ADCC: IgG1 ≥ IgG3 > IgG4 >> IgG2

[b]Mediated by binding of complement component C1q to Fc to produce complement-dependent cytotoxicity (CDC). For CDC: IgG3 > IgG1 > IgG2

(ADCC) which is important for cell killing in the treatment of cancer cells, whereas IgG2 and IgG4, which do not mediate cytotoxicity, are choices when cell death is to be avoided. The IgG1 Fc fragment initiates effector functions by binding to FcγRs, C1q, and the FcRn; however, since IgG4 has little or no affinity for C1q, it can be employed when involvement of this host effector function is not wanted, for example, with natalizumab, indicated for multiple sclerosis and Crohn's disease. Unwanted exchange of Fabs between endogenous IgG4 molecules and between IgG4 antibodies given as therapy may occur. Natalizumab, for example, has been shown to exchange Fab arms with endogenous IgG4 in patients given the mAb. Such exchange has important implications for the pharmacokinetic and pharmacodynamic properties of IgG4 therapeutic mAbs as well as for adverse events provoked by these antibodies. Like IgG4, IgG2 may be chosen when soluble antigens are targeted, and neutralization without, or with reduced, effector functions is desired. These circumstances exist, for example, with denosumab used for osteoporosis. The same thinking is followed in the construction of Fc fusion proteins where, for example, with dulaglutide (Chap. 7, section "GLP-1 Receptor Agonists"), the GLP-1 analog receptor agonist is fused to the IgG4 Fc fragment to avoid antibody-dependent cell cytotoxicity (ADCC) and complement-dependent cytotoxicity (CDC). Despite its high affinity for interaction with the FcγRIIIa and high potency for complement activation, the relatively short half-life of IgG3, its susceptibility to proteolysis due to its extended hinge region, and its allotypic polymorphism ensure that it is not used therapeutically in monoclonal form. This is reflected in the current list of therapeutic mAbs with regulatory approval (Table 2.1).

Glycosylation of Monoclonal Antibodies

The efficacy of a mAb is largely dependent on its specificity and avidity for the target but also on posttranslational modifications that help determine antibody stability, immunogenic potential, and effector functions. Glycosylation is a common posttranslational modification that has a critical role in antibody effector functions. This is borne out by manufacturing specifications defining a human-type glycoform pattern produced by mammalian cells, in particular, Chinese hamster ovary (CHO) cells or mouse monoclonal NSO or Sp2/0 cells. Human IgG molecules produced in these systems have a single *N*-linked biantennary glycan at the conserved glycosylation site, asparagine 297 (Asn297), in each of the two C_H2 domains.

Extensively sialylated *N*-linked biantennary glycans containing a number of D-galactose units, plus L-fucose, occur in about 15–20 % of polyclonal human IgG Fab regions. The function of IgG Fab glycosylation is not clear with claims of no effect, a positive effect, or a neutral influence on antigen binding. In one study of the antigen-binding affinities of three mAbs specific for the same antigen but differing in potential *N*-glycosylation sites in the V_H region, two glycosylated antibodies had a 10–50-fold higher affinity for antigen compared to the third, aglycosylated form.

Fc glycosylation is composed of a biantennary heptasaccharide core formed from two β-D-*N*-acetyl-D-glucosamine branches (1-2)-linked to D-mannose residues which are, in turn, α(1-6)- and α(1-3)-linked to a branch-point D-mannose, itself linked to diacetylchitobiose attached to Asn297 (Fig. 2.3). Additional sugars, namely, L-fucose, D-galactose, *N*-acetylneuraminic acid, and *N*-acetyl-D-glucosamine, may be attached to the core. More than 92 % of human IgG Fc glycans are highly fucosylated and less than 10 % of IgG Fc oligosaccharides are sialylated with less than 1 % having two sialic acids. When present, *N*-acetylneuraminic acid is terminally α(2-6)-linked. Further modifications, including the addition of D-galactose linked to the bisecting *N*-acetyl-D-glucosamine and different glycosylation at the two C_H2 Asn297 sites, contribute to the microheterogeneity of Fc glycan structures. All of this allows for a possible 36 different oligosaccharides and potentially more than 400 glycoforms, and because of the paucity of sialylation, 12 neutral oligosaccharides with the potential of 72 glycoforms are commonly seen.

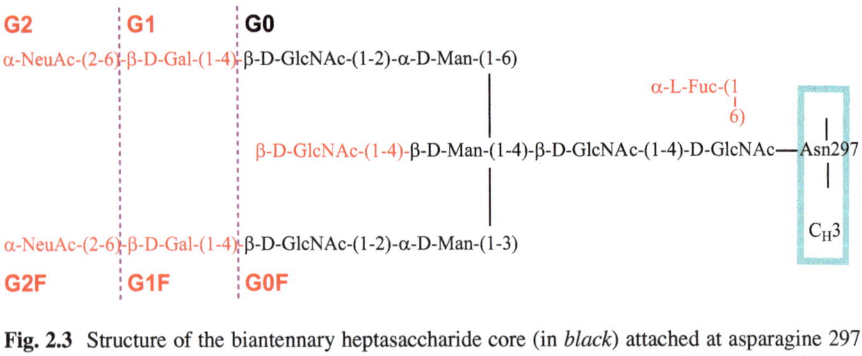

Fig. 2.3 Structure of the biantennary heptasaccharide core (in *black*) attached at asparagine 297 of the IgG Fc. This is termed the G0 glycoform. Additional sugars (in *red*), namely, L-fucose, D-galactose, *N*-acetylneuraminic acid, and *N*-acetyl-D-glucosamine, may be attached to the core structure producing the glycoforms G1 (G0 plus D-galactose on each branch) and G2 (G1 plus *N*-acetylneuraminic acid on each branch). Glycoforms G0F, G1F, and G2F are corresponding structures with the addition of L-fucose

The IgG Fc piece is a homodimer held together by disulfide bonds at the hinge region, a non-covalent pairing of the C_H3 domains and non-covalent interactions between the C_H2-linked oligosaccharides and the C_H2 protein surfaces. This, in turn, influences the conformation of the oligosaccharides in the cavity formed by the two C_H2 domains, and this helps to maintain the H chains of the Fc in an open conformation. Removal of the oligosaccharides leads to collapse of the cavity as seen in deglycosylated preparations and aglycosylated forms, and importantly, the open Fc conformation of human IgG is necessary for normal effector functions mediated via Fc receptors, FcγRI, FcγRII, and FcγRIII, the neonatal Fc receptor (FcRn), and the C1q component of the complement. Glycosylation of the IgG Fc piece is necessary for the biological activities mediated by the Fcγ receptors and C1q (see section "Antibody-Dependent Cell-Mediated and Complement-Dependent Cytotoxicities" below) but not the neonatal receptor, the bacterial Fc-binding proteins SpA and SpG, and the rheumatoid factor. Fc sites that interact with these last four sites involve residues at the junction of the C_H2 and C_H3 domain on both H chains.

Although CHO, NSO, and Sp2/0 cells can produce the desired human glycoforms for therapeutic mAbs, mainly the heptasaccharide core plus L-fucose (G0F) and this structure plus β-D-galactose (1-4)-linked to *N*-acetyl-D-glucosamine on each branch (G1F) (Fig. 2.3), depending on the culture conditions and other factors during production, the glycoforms may not be consistent from clone to clone. Glycosylated products with undesirable properties, e.g., lack of specificity and potency and excessive immunogenicity, may be produced. In relation to immunogenicity, a particular danger is the incorporation of sugars that do not occur naturally in humans, namely, the disaccharide α-D-galactose-(1-3)-D-galactose and the sialic acid and glycolylneuraminic acid, which humans cannot synthesize. These potential problems have led to regulatory authorities setting out requirements for strict glycoform profiles for mAbs.

Concerns have been expressed that the restricted glycoform profiles (mainly G0F and G1F) of mAbs compared to polyclonal IgG may negatively impact the Fc piece and therefore some antibody, effector functions. Some evidence for this was seen in a 15–20-fold increase in ADCC when a bisecting *N*-acetyl-D-glucosamine was added to the mAb oligosaccharide and when increased killing by rituximab resulted from the added bisecting sugar. The absence of L-fucose has also been implicated in effector function performance, demonstrated, for example, by improved ADCC by a non-fucosylated mAb to the human IL-5 receptor and a 40–50-fold increase in ADCC by a non-fucosylated glycoform of trastuzumab (Herceptin®).

Antibody-Dependent Cell-Mediated and Complement-Dependent Cytotoxicities

In the development of a therapeutically effective mAb, target binding and specificity are naturally given a high priority, and in ongoing efforts to enhance clinical efficacy, effector functions are being increasingly studied. This has led to a focus on Fcγ receptors, particularly FcγRIIIa (CD16a), which is known to induce release of cytokines by macrophages and, importantly, for antibody therapy to induce ADCC. Figure 2.4 summarizes in diagrammatic form natural killer (NK) ADCC of antibody-coated target cells affected by antibody Fc interaction with FcγRIIIa on the killer cells. In vitro, mAb-induced ADCC is more effective by several orders of magnitude than the in vivo serum concentrations needed for cell killing in cancer patients. This discrepancy between in vitro and in vivo cytotoxicities is thought to be due to competition in vivo between serum IgG and the mAb, the former inhibiting binding of the mAb to FcγRIIIa on effector cells such

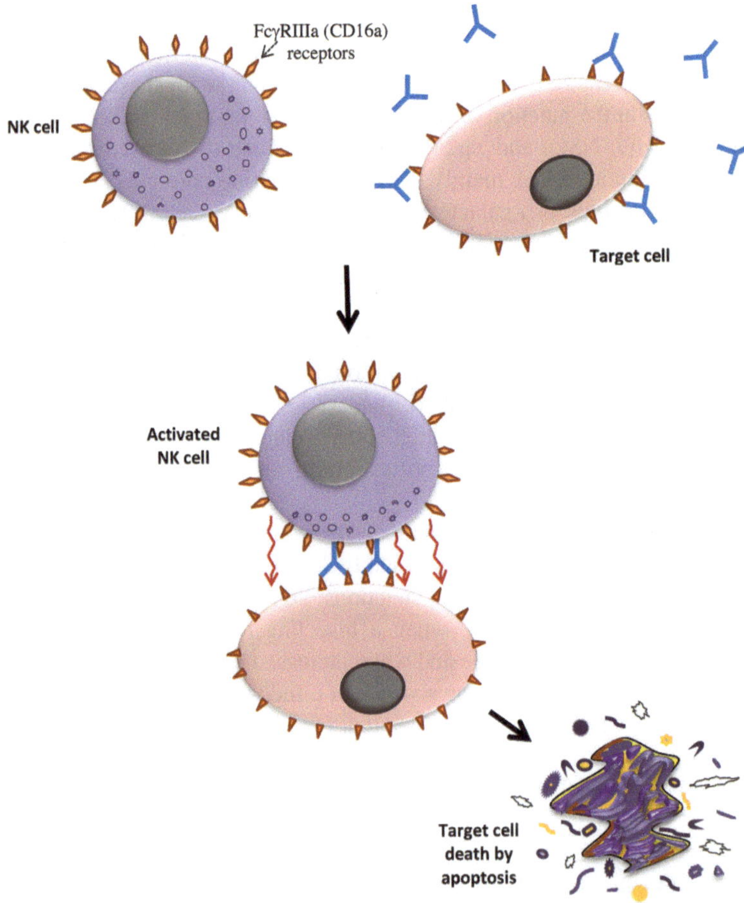

Fig. 2.4 Diagrammatic representation of antibody-dependent cell-mediated cytotoxicity (ADCC) showing the interactions between a natural killer (NK) cell with surface FcγRIIIa (CD16a), antibody molecules (blue), and the target cell. Antibodies bound to their complementary antigens on the surface of target cells interact via their Fc piece with the FcγRIIIa receptors on NK cells. This cross-linking signals and activates the NK cells to kill the target cell which subsequently dies by apoptosis

as NK cells. The importance of ADCC for clinical efficacy, although sometimes queried, is now widely recognized and supported, especially by findings with several mAbs including infliximab (anti-TNF), rituximab (anti-CD20), and trastuzumab (anti-HER2) as well as genetic analyses of FcγR polymorphisms correlated with clinical responses. For example, in cancer patients treated with rituximab, patients carrying the high-affinity FcγRIIIa allotype FcγRIIIa-Val158 show a better clinical response than patients with the low-affinity allotype FcγRIIIa-Phe158. Techniques to boost ADCC and enhance clinical efficacy of mAbs are now being routinely pursued.

One method to enhance ADCC is by making amino acid Fc substitutions that improve both binding to FcγRIIIa and the resultant ADCC. Another successful method is based on knowledge that glycosylation has an important function in ADCC by modulating IgG Fc binding to the Fcγ receptor. Removal of L-fucose from the biantennary glycan core structure (Fig. 2.3) improves FcγRIIIa binding and enhances ADCC without interfering with antigen binding or CDC, and conversely, highly fucosylated IgG1 reduces ADCC. These findings are now well established, but despite the benefits of using non-fucosylated mAbs, there is, as yet, no regulatory requirements that the removal of the sugar from Fc-linked oligosaccharides should be undertaken. Of course, procedures to do this are not easily effected and standardized, but the use of cells with α-1,6-fucosyltransferase gene FUT8 knockout shows promise, and some expression systems such as a CHO fucosyltransferase-deficient cell line and a mutant *Nicotiana* species have been utilized to produce afucosylated recombinant human IgG. A few non-fucosylated products are already in some clinical trials. The presence of bisecting *N*-acetyl-D-glucosamine in the Fc oligosaccharides also increases binding to both FcγRIIIa and ADCC. However, the increases are small compared to the enhancements seen with afucosylated glycoforms. The presence of terminal *N*-acetylneuraminic acid residues, either α-2,3- or α-2,6-linked, on IgG Fc oligosaccharides appears to mediate anti-inflammatory responses and reduce ADCC. The most likely explanation is the decreased binding sialylated IgGs show for Fcγ receptors.

CDC is another important antibody effector function especially in antitumor therapy. Like ADCC, the capacity for CDC (which is retained even after chemotherapy) is important for optimal efficacy of antibodies used for cell killing. CDC is initiated by the binding of C1q to the constant region of targeted cell-bound antibodies. C1q binding sets in motion a complex cascade with successive cleavages of C1r, C1s, C4, and C2 leading to the C3 convertase of the classical pathway and on to C3b which is deposited on the target surface and amplified by the alternative pathway. The cascade proceeds via complement components C5, C6, C7, C8, and C9, finally resulting in the membrane attack complex and membrane perforation. CDC may be enhanced by facilitating the binding of the antibody constant region to C1q. This has been achieved by inserting amino acid mutations into the antibody Fc or hinge regions. In another method for CDC enhancement, described as "Complegent® technology," an IgG1 and IgG3 mixed

H chain variant showed stronger C1q binding and CDC than the parent isotypes. Complegent®-type antibody is claimed to have enhanced alemtuzumab-induced CDC, and its developers suggest that it may be applicable to a wide range of target molecules.

Efforts to improve the efficacy of the next generation of mAbs have made progress, and already it seems that, where desirable, antibodies will be designed with the aim of increasing selected effector functions such as ADCC and CDC (Table 2.2) and perhaps functions such as antibody-dependent cellular phagocytosis (ADCP). It already seems that improved FcγRIIIa binding can be achieved by altering amino acids in the Fc piece, removal of L-fucose from the Fc glycan can improve ADCC, and amino acid mutations in the Fc or hinge regions can increase C1q binding and CDC. Fc engineering to improve effector functions, however, does not stop there. FcRn is the main homeostatic regulator of the human IgG1 isotype (section "IgG Antibody Subclasses"), and as such, it is a focus of attempts to increase antibody half-life and improve antibody transport. The binding site for FcRn is located in the C_H2 – CH3 elbow region distinct from the sites that bind FcγRs and C1q in the lower hinge – but altered binding to FcRn appears to also influence binding to the FcγRs and C1q, ultimately altering cellular mechanisms such as ADCC and CDC. In a recent study from Oslo, a panel of human IgG1 variants, Fc engineered for altered binding to FcRn, was assessed for binding to the FcγRs and C1q, and although binding was generally found to be reduced, one variant showed improved C1q binding. It seems possible therefore that CDC might be modulated by mutations in the C_H3 domain, structurally removed from the C1q site in the hinge and upper C_H2 domain, and this has implications for the ongoing efforts to improve mAb therapeutic efficacy.

Nomenclature for Monoclonal Antibodies

The nomenclature used for mAbs has been adopted by both US Adopted Names (USAN) and WHO's International Nonproprietary Names (INN) for pharmaceuticals. All mAb names end with the stem *-mab* and use a preceding substem (also called a prestem) to distinguish the animal origin of the mAb. A list of these substems is shown in Table 2.3. In addition, a second prestem, preceding the designation for the mAb origin, is usually included in the name as a one-, two-, or three-letter identifier to indicate the mAb target or disease state, for example, "-tu-" for a mAb directed against a tumor(s) or "-ci-" for mAbs with a cardiovascular/circulatory action (Table 2.3). More mAbs indicated for tumors have been given regulatory approval than mAbs for any other diseases or general disorder, and this initially led to recommendations for a further substem divisions to identify individual tumors such as "-co-" for colon, "-me-" for melanoma, "-pr-" for prostate, and so on. This practice has not always been followed and is reflected in the current list of approved mAbs (Table 2.1). Lastly, each mAb has a unique prefix, a few letters not necessarily with any special meaning but selected to identify the mAb as an individual product (Table 2.3).

Table 2.3 Nomenclature for monoclonal antibodies[a]

INN substem (prestem)[b]		Target[c]	
-o-	Mouse	**-ba/b/bac-**	Bacterial
-a-	Rat	**-ci/c-**	Cardiovascular/circulatory
-axo-	Rat-mouse chimera	**-fu/f-**	Antifungal
-e-	Hamster	**-ki/k-**	Interleukin
-i-	Primate	**-le/les-**	Inflammatory lesion
-xi-	Chimeric	**-li/l-**	Immunomodulatory
-xizu-	Chimeric-humanized	**-ne/n-**	Nervous system
-zu-	Humanized	**-so/os/s-**	Bone
-u-	Fully humanized/human	**-tox/toxa-**	Toxin
		-tu/t[d]-	Tumor
		-vi/v-	Viral

INN International Nonproprietary Name

[a]All monoclonal antibodies have the stem (suffix) *-mab* and a unique prefix with no special meaning used to identify the individual product

[b]To identify animal species

[c]To identify target or disease. Placed before substem

[d]Further substem subdivision for tumors: *co/col* colon, *go/got* testis, *go/gov* ovary *ma/mar* mammary, *me/mel* melanoma, *pr/pro* prostate, *tu/tum* miscellaneous tumor

Examples illustrating the application of these nomenclature rules:

Palivizumab. *Pali-*, unique prefix identifier; *-vi-*, targeted to a virus (respiratory syncytial virus); *-zu-*, INN substem for humanized; *-mab*, stem for all monoclonal antibodies

Canakinumab. *Cana-*, unique prefix identifier; *-ki-*, targeted to interleukin Il-1β; *-u-*, INN substem for fully human monoclonal antibody; *-mab*, stem for all monoclonal antibodies

Breakdown of Antibody Type and Approved Indications for the Currently Approved Monoclonal Antibodies

With nine mAb approvals by the FDA in 2015 for secukinumab in January, dinutuximab in March, alirocumab in July, evolocumab in August, idarucizumab in October, and mepolizumab, necitumumab, daratumumab, and elotuzumab in November, plus approvals for obiltoxaximab, ixekizumab, and reslizumab in March 2016, and atezolizumab in May 2016, 50 mAbs are currently (June 2016) approved in the USA and/or Europe. Data on 49 of the mAbs are presented in Table 2.1 along with the antibody type, cell cultures used for production, antibody targets, mechanisms of action, and indications. During the late production stage of this book, the 50th mAb, humanized atezolizumab (Tecentriq®) which binds PD-L1, was approved by the FDA for treatment of urothelial carcinoma (Chap. 3, "Recent Approval: Atezolizumab"). Note that at variance with some classifications, Fc fusion proteins containing the antibody constant region fused to a nonantibody effector protein domain are not considered here as mAbs (see Chap. 6, Fusion Proteins). Also, alemtuzumab and denosumab, each formulated and marketed as two separate products (Campath® and Lemtrada® for alemtuzumab, Prolia® and Xgeva® for denosumab) (Table 2.1), are considered as two different mAbs, not four, and mAb biosimilars (Chap. 13) are not viewed as separate from the originally approved mAb.

Seventeen fully human (-umab) and 21 humanized (-zumab) mAbs comprise the largest groups of the 50 currently approved mAbs. The remaining approved products comprise nine human-mouse chimeric (-ximab), two mouse (-omab), and one rat-mouse chimeric (-axomab) antibodies. All 50 of the antibodies belong to the IgG isotype with the overwhelming number being subclass IgG1. Forty seven have kappa light chains and only three (belimumab, raxibacumab, evolocumab) have lambda light chains. Four of the mAbs are IgG1κ Fab fragments (abciximab, certolizumab, ranibizumab, and idarucizumab), four are subclass IgG2 (catumaxomab rat IgG2b-mouse IgG2a, denosumab, panitumumab, evolocumab), five are IgG4 (natalizumab, pembrolizumab, nivolumab, ixekizumab, reslizumab), while eculizumab is a IgG2/4κ hybrid containing regions from human IgG2 and IgG4 sequences and murine complementarity-determining regions grafted onto the human variable regions. Blinatumomab is the sole BiTE bispecific fusion protein composed of two single-chain variable fragments (scFv) each from an H and L chain (Table 2.1). Atezolizumab is Fc-engineered and non-glycosylated.

Twenty-five of the approved mAbs, including siltuximab for multicentric Castleman's disease, are indicated for oncological therapies, 12 are indicated for inflammatory and/or autoimmune disorders, and 15 individual mAbs are indicated for different disorders including transplant rejection, paroxysmal nocturnal hemoglobinuria, asthma, CAPS, macular degeneration, osteoporosis, hyperlipidemia, and anthrax and RSV infections (Tables 2.1 and 2.4). Alemtuzumab and denosumab have each been indicated for more than one disorder: B-cell chronic lymphocytic leukemia and multiple sclerosis for the former and osteoporosis and bone metastases for the latter. In terms of the numbers of mAbs indicated for cancer therapies, chronic leukemia and breast cancer head the list: alemtuzumab, blinatumomab, obinutuzumab, ofatumumab, and rituximab as treatments for the leukemias and bevacizumab, pertuzumab, trastuzumab, and trastuzumab emtansine for breast cancer. Other tumor indications include lymphomas (brentuximab, ibritumomab tiuxetan, rituximab), colorectal cancer (bevacizumab, cetuximab, panitumumab), melanoma (ipilimumab, nivolumab, pembrolizumab), non-small cell lung cancer (ramucirumab, nivolumab, necitumumab), gastric cancer (ramucirumab, trastuzumab), multiple myeloma (daratumumab and elotuzumab), lung cancer (bevacizumab), metastatic renal cell carcinoma (nivolumab), bone metastases (denosumab), head and neck cancers (cetuximab), glioblastoma (bevacizumab), neuroblastoma in pediatric patients (dinutuximab) and urothelial carcinoma (atezolizumab). The 25 mAbs with specific indications in oncology (Tables 2.1 and 2.4) are discussed in detail in Chap. 3.

A category of mAbs classified according to their approved indications is a group of 12 agents targeted to predominately inflammatory conditions, in particular, rheumatoid arthritis (adalimumab, certolizumab, golimumab, infliximab, tocilizumab), Crohn's disease (adalimumab, certolizumab, infliximab, natalizumab, vedolizumab), psoriasis/psoriatic arthritis (adalimumab, golimumab, infliximab, secukinumab, ixekizumab, ustekinumab), ulcerative colitis (infliximab, vedolizumab), multiple sclerosis (alemtuzumab [also in the antitumor group], natalizumab), and lupus erythematosus (belimumab). The remaining 15 mAbs (including denosumab which is also in the antitumor group) are indicated for diseases/disorders other than cancers and inflammatory conditions (Tables 2.1 and 2.4). Two of the most recently approved

Table 2.4 Therapeutic applications of approved monoclonal antibodies as at June 2016

Oncology[a]	Inflammation and autoimmune disorders[e]	Other disorders[e]
Alemtuzumab[b]	Adalimumab	Abciximab—*cardiac ischemic complications*
Atezolizumab		
Bevacizumab	Alemtuzumab[f]	Alirocumab—*familial hypercholesterolemia*
Blinatumomab	Belimumab	Basiliximab—*organ transplant rejection*
Brentuximab vedotin	Certolizumab pegol	Canakinumab—*CAPS*
Catumaxomab	Golimumab	Denosumab[g]—*osteoporosis and to increase bone mass*
Cetuximab	Infliximab	Ecolizumab—*PNH, AHUS*
Daratumumab	Ixekizumab	Evolocumb—*primary hyperlipidemia, mixed dyslipidemia*
Denosumab[c]	Natalizumab	Idarucizumab—*reverses action of anti-coagulant dabigatran*
Dinutuximab	Secukinumab	Mepolizumab—*eosinophilic asthma*
Elotuzumab	Tocilizumab	Obiltoxaximab—*anthrax*
Ibrutumomab tiuxetan	Ustekinumab	Omalizumab—*asthma, urticaria*
Ipilimumab	Vedolizumab	Palivizumab—*RSV infection*
Necitumumab		Ranibizumab—*macular degeneration*
Nivolumab		Raxibacumab—*anthrax*
Obinutuzumab		Reslizumab—*eosinophilic asthma*
Ofatumumab		
Panitumumab		
Pembrolizumab		
Pertuzumab		
Ramucirumab		
Rituximab		
Siltuximab[d]		
Trastuzumab		
Trastuzumab-emtansine		

AHUS atypical hemolytic uremic syndrome, *CAPS* cryopyrin-associated periodic syndromes, *PNH* paroxymal nocturnal hemoglobinuria, *RSV* human respiratory syncytial virus

[a]For mAb targets, mechanisms of action and approved indications see Table 2.1 and Chap. 3

[b]Approved by FDA in 2001 for lymphocytic leukemia; withdrawn from the US and Europe in 2012 and relaunched for multiple sclerosis as Lemtrada®

[c]Approved by FDA as Xgeva® for bone metastases from solid tumors and giant cell tumor of bone

[d]Indicated for multicentric Castelman's disease, a lymphoproliferative disorder similar in many ways to lymphomas but not considered a true cancer. Castelman's disease is included in the American Cancer Society's cancer information

[e]For mAb targets, mechanisms of action and approved indications see Table 2.1 and Chap. 4

[f]Approved as Lemtrada® for multiple sclerosis by FDA, November 2014

[g]Approved by FDA as Prolia®

mAbs (alirocumab and evolocumab) in this group are aimed at a new target, PCSK9 (proprotein convertase subtilisin/kexin type 9). New approved indications are familial hypercholesterolemia (alirocumab), hyperlipidemia and mixed dyslipidemia (evolocumab), reversal of specific anticoagulant therapy (idarucizumab), pediatric neuroblastoma (dinutuximab), multiple myeloma (daratumumab and elotuzumab), and treatment of asthma with an anti-eosinophilic mAb (mepolizumab, reslizumab). All the mAbs used for non-cancer disorders are discussed in detail in Chap. 4.

Note that in practice, mAb off-label therapies sometimes cover a wider and more diverse range of disorders, and this is reflected in any examination of adverse events provoked by these agents.

Antibody-Drug Conjugates

For mAbs used to treat cancers, antibody-drug (or toxin) conjugates (ADCs) have long been seen as an obvious and simple concept to specifically target and selectively kill diseased cells without unwanted off-target effects. Such a "magic bullet" strategy promising superior therapeutic efficacy is not necessarily restricted to tumor cells and might be potentially employed in other diseases where elimination, restriction, inhibition, or change of certain cell populations is desired. The first ADC to receive regulatory approval was gemtuzumab ozogamicin, an anti-CD33 antibody derivatized with calicheamicin, an enediyne bacterium-derived antitumor antibiotic, approved in 2000 by the FDA under the US orphan drug program for the treatment of acute myelogenous leukemia. At present, only two mAb ADCs, that is, a mAb attached to a "drug" or chemotherapeutic agent, brentuximab vedotin and trastuzumab emtansine, are FDA and EMA approved, as anticancer therapies (Table 2.1). Another "magic bullet with payload" approach is represented by the approved CD20-targeted mAb ibritumomab, used for radioimmunotherapy. In this case, the mAb is conjugated to the chelator tiuxetan which complexes a radioactive isotope (yttrium-90 or indium-111) that kills target, and some nearby CD20-bearing normal, cells by radioactive beta emissions. Trastuzumab emtansine is an ADC formed by conjugating the mAb trastuzumab to the antimitotic agent maytansinoid mertansine (see Chap. 3, section "Ado-trastuzumab Emtansine", and Fig. 3.6), while brentuximab vedotin (Chap. 3, section "Brentuximab Vedotin") is a conjugate of the mAb brentuximab with monomethyl auristatin, a synthetic antimitotic, antineoplastic agent. In October 2015, the FDA granted Breakthrough Therapy Designation to the humanized ADC inotuzumab ozogamicin, a mAb targeted to CD22 and attached to the cytotoxic agent calicheamicin. When inotuzumab reacts with CD22 on malignant B cells of patients with acute lymphoblastic leukemia, the ADC is internalized, the attached toxin is released, and cell death ensues.

Although the concept, strategy, and basic principles of ADCs appear simple, in practice there are many challenges in developing effective and safe therapeutic products. With four factors to consider, namely, the target, the mAb, the linking structure, and the drug/toxin payload, the manufacturing processes are difficult, complex, and specialized. Early problems encountered included linkers that were too stable or not stable enough to avoid systemic toxicity, insufficient internalization and potency of the conjugate, and changes to the physicochemical and/or therapeutic properties of the final antibody-drug/toxin conjugate. For many ADCs, including trastuzumab emtansine, toxin is attached to accessible lysine residues on the antibody, and since there are usually more than 80 of these with about half available for conjugation, highly heterogeneous mixtures may result. Another approach for coupling is to undertake cysteine-based conjugation utilizing reduction of antibody interchain disulfide bonds to yield up to eight thiols that may be targeted for coupling. Site-specific coupling to a selected number of reduced cysteine residues is used in the case of brentuximab vedotin. Although progress has been made in developing ADCs for human therapy, the area must be viewed as still-emerging technology. Nevertheless, the ADC market in 2013 was assessed as $454 million and by 2018 sales are estimated to be ~$5 billion per year.

Future Prospects of Monoclonal Antibody Therapy

With a past regulatory approval rate of about four per year, there will be ~65 mAbs on the market by 2020, but a predicted ongoing market growth rate of 8 %, worldwide sales of ~$125 billion by 2020, and more than 300 mAbs in development could ultimately prove this figure to be an underestimate. For example, in 2014, six mAbs, ramucirumab, siltuximab, vedolizumab, pembrolizumab, blinatumomab, and nivolumab, were granted first approval by the FDA/EMA but this has already been surpassed in 2015 with nine more, secukinumab (first approved in Japan in December 2014), dinutuximab, alirocumab, evolocumab, daratumumab, necitumumab, idarucizumab, mepolizumab, and elotuzumab approved up until the end of November 2015. In the first three months of 2016, three more mAbs, ixekizumab, obiltoxaximab, and reslizumab were granted FDA approval. The ADC inotuzumab ozogamicin has been granted FDA Breakthrough Therapy Designation, and marketing applications have been lodged, or are imminent, for a number of other mAbs. While 50 % (25) of the currently approved mAbs are indicated for therapies in oncology and more than one of the 50 approved antibodies are specific for TNF, CD20, EGFR, HER2, VEGF, *B. anthrac*is PA, PD-1, IL-17A, IL-5, or PCSK9, mAbs aimed at different targets are being introduced steadily. For example, antigens IL-6, GD2, CD38, and SLAMF7 and the anticoagulant dabigatran are already being targeted, and new diseases/disorders Castleman's disease, neuroblastoma, multiple myeloma, eosinophilic asthma, familial hypercholesterolemia, primary hyperlipidemia, and mixed dyslipidemia have been added to the growing list of disorders now amenable to mAb therapy. In comparison to most drugs, mAbs are relatively well tolerated, and this tends to be reflected in a lower risk of safety issues in clinical trials. A consequence of this is the often relatively faster inclusion of mAbs in trials providing opportunities to test proof of concept and a possible more rapid development toward registration and marketing.

Viewing the progress and efforts being made with mAbs and ADCs against the background of a worldwide aging population and an increasing standard of living, including in emerging markets, effective targeted treatment options seem certain to increase significantly within at least the next 3–5 years. This prediction seems to be supported by reports of more than 30 ADCs in current clinical testing with many more in preclinical development. Ongoing examination of new antibody targets and an expanding list of indications also points to a continuing and expanding demand for mAb treatments, more cost-effective products, and a wider market penetration.

Summary

- For humoral immune therapy where precise targeting, a "perpetual" supply of antibody, greater precision, and reproducibility are needed, uniform and well-defined antibodies from a single long-lived clonal cell line are required.
- In 1986, the first therapeutic mAb, muromonab-CD3 (Orthoclone OKT3), a murine IgG2a mAb directed against the CD3 (T3) receptor on the surface of human T lymphocytes, received regulatory approval. It soon became apparent that the human immune response to mouse proteins was a significant problem and more "humanized" mAbs were needed.
- The first step in what was to be an ongoing iterative process was the production of so-called chimeric antibodies in which the variable (antigen-binding) regions of mouse antibodies were incorporated with the constant regions of human immunoglobulins. Despite this important technological advancement, the risk of a human anti-mouse immune response was not eliminated, and adverse reactions, especially potentially serious hypersensitivities, still occasionally occurred.
- The need to eliminate species recognition differences led to the production of so-called humanized mAbs achieved by substituting murine hypervariable or complementarity-determining regions (CDRs) in place of human sequences while retaining the remainder of the antibody as human.
- Although the improved tolerance of these agents by patients and noticeably fewer side effects were soon obvious, the ultimate goal was always to develop fully humanized antibodies. This became possible with the development of phage display and transgenic mouse technologies. The first fully human mAb to reach the US market, adalimumab, was developed with phage display technology.
- In the evolution of mAbs for human therapy beginning with hybridoma technology, technological advances in the production of mAbs can be followed through the immortalization of human B cells with Epstein-Barr virus, phage display, the employment of transgenic (knockout) mice, and the production of mAbs from single human B cells by gene cloning.
- To date, all of the mAbs, and derived products approved for therapy, are of the IgG isotype. Elimination half-lives of the subclasses are ~20–21 days for IgG1, IgG2, and IgG4 and ~7 days for IgG3. IgG1 is easily the subclass most often selected, while IgG2 and IgG4 are occasionally used. Apart from its long half-life, IgG1 is the choice for antibody-dependent cell-mediated cytotoxicity (ADCC) which is important for cell killing in the treatment of cancer cells; IgG2 and IgG4, which do not mediate cytotoxicity, are choices when cell death is to be avoided.

- IgG1 Fc fragment initiates effector functions by binding to FcγRs, C1q, and the FcRn. IgG4 has little or no affinity for C1q and can be employed when involvement of this effector function is not wanted.
- Glycosylation is a common posttranslational modification that has a critical role in antibody effector functions. Human IgG molecules produced in mammalian cells such as Chinese hamster ovary cells have a single *N*-linked biantennary glycan at the conserved glycosylation site, asparagine 297 (Asn297), in each of the two C_H2 domains. Extensively sialylated *N*-linked biantennary glycans containing a number of D-galactose units, plus L-fucose, occur in about 15–20 % of polyclonal human IgG Fab regions.
- Fc glycosylation is composed of a biantennary heptasaccharide core formed from two β-D-galactose-β-D-*N*-acetyl-D-glucosamine branches α(1-6)- and α(1-3)-linked via D-mannose to diacetylchitobiose which, in turn, is linked to Asn297. More than 92 % of human IgG Fc glycans are highly fucosylated and less than 10 % of IgG Fc oligosaccharides are sialylated. Glycosylation of the IgG Fc piece is necessary for the biological activities mediated by the Fcγ receptors and C1q.
- The absence of L-fucose has been implicated in effector function performance, demonstrated by improved ADCC by non-fucosylated mAbs. CDC, initiated by the binding of C1q to the constant region of targeted cell-bound antibodies, is important for optimal efficacy of antibodies used for cell killing.
- FcRn is the main homeostatic regulator of human IgG1 isotype and a focus of attempts to increase antibody half-life and improve antibody transport. Altered binding to FcRn appears to also influence binding to the FcγRs and C1q, ultimately altering cellular mechanisms such as ADCC and CDC.
- Fifty mAbs are currently approved in the USA and/or Europe. Twenty-one humanized and 17 fully human mAbs comprise the largest groups followed by nine human-mouse chimeric, two mouse, and one rat-mouse chimeric mAbs.
- The antibodies belong to the IgG isotype with the overwhelming number being subclass IgG1. Of these, 47 have kappa light chains and only three (belimumab, raxibacumab, evolocumab) have lambda light chains. Four of the mAbs are IgG1κ Fab fragments (abciximab, certolizumab, ranibizumab, idarucizumab), four are subclass IgG2 (catumaxomab rat IgG2b-mouse IgG2a, denosumab, panitumumab, evolocumab), and five are IgG4 (natalizumab, pembrolizumab, nivolumab, ixekizumab, reslizumab), while eculizumab is a IgG2/4κ hybrid.
- Blinatumomab is the sole BiTE bispecific fusion protein composed of two single-chain variable fragments (scFv) each from an H and L chain.
- Twenty-five of the approved mAbs, including siltuximab for multicentric Castleman's disease, are indicated for oncological therapies, 12 for inflammatory and/or autoimmune disorders, and 15 for other different disorders. Two mAbs are approved for more than one indication: alemtuzumab for lymphocytic leukemia (as Campath®) and multiple sclerosis (as Lemtrada®) and denosumab for bone loss (as Prolia®) and bone metastases (as Xgeva®).
- More than one of the 50 approved antibodies are specific for TNF, CD20, α4β7 integrin, EGFR, HER2, VEGF, *B. anthracis* PA, PD-1, IL-17A, IL-5 or PCSK9, but mAbs aimed at different targets are being introduced. Antigens IL-6, GD2, CD38, SLAMF7, and the anticoagulant dabigatran are already being targeted, and

new diseases Castleman's disease, neuroblastoma, hypercholesterolemia, hyperlipidemia, multiple myeloma, urothelial carcinoma, and eosinophilic asthma have been added to the growing list of disorders now amenable to mAb therapy.

- Currently, two monoclonal antibody-drug conjugates (ADCs), brentuximab vedotin and trastuzumab emtansine, are FDA and EMA approved, each as anticancer therapies. Ibritumomab has an attached radioactive label for cell killing. Although the concept, strategy, and basic principles of ADCs appear simple, in practice there are four factors to consider: the target, the mAb, the linking structure, and the drug/toxin payload. This presents many difficult and complex developmental and manufacturing challenges in producing effective and safe targeted therapeutic products.

Further Reading

Chernajovsky Y, Nissim A, editors. Therapeutic antibodies. Handbook of experimental pharmacology, vol. 181. Berlin: Springer; 2008.

Hammers CM, Stanley JR. Antibody phage display: technique and applications. J Invest Dermatol. 2013;134:e17. doi:10.1038/jid.2013.521.

Hwang WYK, Foote J. Immunogenicity of engineered antibodies. Methods. 2005;36:3–10.

Irani V, Guy AJ, Andrew D, et al. Molecular properties of human IgG subclasses and their implications for designing therapeutic monoclonal antibodies against infectious diseases. Mol Immunol. 2015;67:171–82.

Jefferis R. Glycosylation of recombinant antibody therapeutics. Biotechnol Prog. 2005;21:11–6.

Jones TD, Carter PJ, Plückthun A, et al. The INNs and outs of antibody nonproprietary names. mAbs. 2016;8:1–9.

Liu L. Antibody glycosylation and its impact on the pharmacokinetics and pharmacodynamics of monoclonal antibodies and Fc fusion proteins. J Pharm Sci. 2015;104:1866–84.

Lonberg N. Fully human antibodies from transgenic mouse and phage display platforms. Curr Opin Immunol. 2008;20:450–9.

Meyer-Tamaki KB. Preclinical development of monoclonal antibodies. In: Faqi AS, editor. A comprehensive guide to toxicology in preclinical drug development. London: Academic; 2013. p. 489–516.

Nelson AL, Dhimolea E, Reichert JM. Development trends for human monoclonal antibody therapeutics. Nat Rev Drug Discov. 2010;9:767–74.

Ossipow V, Fischer N, editors. Monoclonal antibodies: methods and protocols. Methods in molecular biology, vol. 1131, 2nd ed. New York: Humana Press; 2014.

Reichert JM. Antibodies to watch in 2015. mAbs. 2015;7:1–8.

Reichert JM. Antibodies to watch in 2016. mAbs. 2016;8:197–204.

Shade K-TC, Anthony RM. Antibody glycosylation and inflammation. Antibodies. 2013;2:392–414. doi:10.3390/antib2030392.

Strohl WR, Strohl LM. Therapeutic antibody engineering: Current and future advances driving the strongest growth area in the pharmaceutical industry. Woodhead publishing series in biomedicine no. 11. Cambridge: Woodhead Publishing Ltd; 2012.

Vidarsson G, Dekkers G, Rispens T. IgG subclasses and allotypes: from structure to effector functions. Front Immunol. 2014;5:520. doi:10.3389/fimmu.2014.00520.eCollection2014.

Waterman S. Technology advances enable creation of better ADCs. BioProcess Int. 2014;12 Suppl 6:37–40.

Zhang P, Woen S, Wang T, et al. Challenges of glycosylation analysis and control: an integrated approach to producing optimal and consistent therapeutic drugs. Drug Discov Today. 2016;21:740-65. doi:10.1016/j.drudis.2016.01.006.

Chapter 3
Monoclonal Antibodies Approved for Cancer Therapy

For many years, the mainstay of cancer therapy was the range of so-called chemotherapeutic drugs, small cytotoxic molecules demonstrating nonspecific toxicity, used together with radiation therapy in a rather crude and nondiscriminatory attempt to destroy rapidly dividing malignant cells. This approach often carried with it therapeutic as well as safety limitations since the broad strategy of killing rapidly dividing cells also adversely affected some other normal, healthy cells such as mucosal lining cells and those in the bone marrow and hair follicles. The consequences of this nonspecific therapeutic approach producing an array of toxic effects for patients were often poor tolerance of chemo- and radiation therapies, poor patient compliance, delays and interruptions to therapy, discontinuation of therapy, and ultimately poor survival outcomes. In addition, tumors can become resistant to chemotherapy and radiation treatments, and these resistances may extend to drugs not yet administered to the patient. Effective targeted therapies for cancers without concurrent toxicities have long been desired by oncologists, and from their earliest examples, mAbs specifically directed to selected antigens on many different tumors appeared to offer great promise for both clinicians and patients. Here we cover those mAbs developed as targeted antineoplastic agents that have gone on to receive regulatory approval for specific cancer indications. In keeping with the aims of this monograph, emphasis is placed on the nature of the antibodies, the strategies underlying their use, their mechanisms of action, and their safety issues.

Of the 50 monoclonal antibodies (mAbs) currently approved by the FDA and/or EMA (as at June 2016), half are indicated for the treatment of hematologic, cutaneous, or solid tumor malignancies. This is a reflection of the always-pressing need to make headway against the many different cancers that affect and kill humans of all ages; the knowledge that both innate and adaptive immunities can contribute to the recognition and elimination of malignant cells; the potential for success promised by the relatively nontoxic, specific, and targeted approach

B.A. Baldo, *Safety of Biologics Therapy*, DOI 10.1007/978-3-319-30472-4_3

offered by mAbs; and the potentially large commercial rewards that might follow development of a successful therapy. A combination of different modes of action has been demonstrated or proposed for the approved mAbs used in cancer treatments. Molecular mechanisms involved are predominately a direct cytotoxic action against cancer cells, an inhibitory effect on promitogenic signaling pathways, and immunomodulatory effects leading to the indirect destruction of cancer cells. For a small number of approved mAbs and a larger number in the development pipeline, a direct cytotoxic action is effected by a so-called antibody-drug conjugate (ADC) (Chap. 2, section "Antibody-Drug Conjugates"), whereby cell killing is imparted by an attached bioactive payload of a potentially lethal toxin, drug, cytokine, or radionuclide. Examples are ibritumomab tiuxetan (section "Ibritumomab Tiuxetan"), brentuximab vedotin (section "Brentuximab Vedotin"), and trastuzumab emtansine (section "Ado-trastuzumab Emtansine"). Examples of cell-destructive immunomodulatory effects include antibody-dependent cell cytotoxicity (ADCC) and modulation of immune checkpoints by the targeting of inhibitory pathways regulating signaling between T cells and antigen-presenting cells. Antibodies that induce ADCC and those that modulate immune checkpoints, namely, mAbs that target cytotoxic T lymphocyte antigen 4 (CTLA-4) and programmed cell death protein 1 (PD-1), are discussed later in this chapter.

As well as the importance of understanding the mechanisms of action of mAbs for comprehending tumor recognition and the processes involved in cell destruction, the identification of biomarkers that predict patient's responses is important to help select the patients most likely to benefit from treatment with a particular mAb. Cancer patients are therefore often tested for specific biomarkers known to be predictive of a beneficial or poor response to a targeted agent. Examples are the selection of patients for trastuzumab therapy by testing for HER2 gene amplification and identification of patients unlikely to respond to cetuximab by detection of a mutation in codon 12 of the *KRAS* (Kirsten rat sarcoma 2 viral oncogene homolog) gene. Unfortunately, not all patients predicted to respond on the basis of a biomarker result actually do so. For example, only 25–30 % of breast cancer patients who are HER2 amplification positive respond to trastuzumab. This highlights the need for continued efforts to identify additional biomarkers that select those cancer patients most likely to benefit from therapy with a particular mAb.

Approved Monoclonal Antibodies for Cancer Therapy

Of the 50 mAbs currently approved by the FDA and/or EMA (Table 2.1), Table 3.1 lists 24 different antibodies with regulatory approval for cancer therapy indications together with their targets, warnings, precautions, risks, and safety concerns associated with their use and their recorded common and serious adverse events. Approval of atezolizumab was too recent for inclusion in the table. Extra detail on the safety of each mAb is set out in the following summaries.

Table 3.1 Adverse events associated with approved monoclonal antibodies used for cancer therapy (as at June 2016)

Monoclonal antibody [a] INN and trade names[b]	Target[c]	Warnings, precautions, risks, and safety concerns	Other adverse events,[d] serious and common
Catumaxomab[e] (Removab®)	EpCAM/CD3[f]	Monitor and evaluate for: CRS, SIRS, HAMA/HARA, GI hemorrhage, hepatic disorders, abdominal infection, ileus/intestinal perforation, decreased lymphocyte count	*Systemic*: cytopenias, hepatotoxicity, abdominal disorders, pyrexia, chills, nausea, vomiting, infections, immunogenicity, dyspnea. *Cutaneous*: rash, erythema, allergic dermatitis, hyperhidrosis, pruritus
Blinatumomab[g] (Blincyto®)	CD19/CD3[h] epsilon	*Boxed warning*: CRS, neurological toxicities. *Others*: infections, neutropenia and febrile neutropenia, TLS, elevated liver enzymes, leukoencephalopathy	*Systemic*: HLH, pyrexia, lymphopenia, leukopenia, chills, headache, CNS symptoms (disorientation, confusion, tremor, speech disorders), hypokalemia, pneumonia, sepsis, constipation, peripheral edema. *Cutaneous*: rash
Ibritumomab tiuxetan[i] (Zevalin®)	CD20[j]	*Boxed warning*: serious IR, severe cytopenias,[k] severe mucocutaneous and cutaneous reactions. *Others*: MDS and AML, extravasation, immunization	*Systemic*: infections, asthenia, musculoskeletal symptoms, GI, hemorrhage, hypersensitivity. *Cutaneous*: exfoliative dermatitis, bullous dermatitis, EM, SJS, TEN
Obinutuzumab[l] (Gazyva®, Gazyvaro®)	CD20	*Boxed warning*: hepatitis B virus reactivation, PML. *Others*: IR, TLS, neutropenia, thrombocytopenia, infections, immunization	*Systemic*: anemia, pyrexia, febrile neutropenia, neutropenia, sepsis, pneumonia, thrombocytopenia, arthralgia, respiratory and urinary infections, GI, decreased appetite, sinusitis musculoskeletal disorders, headache, cough
Ofatumumab (Arzerra®)	CD20	IR, hepatitis B virus reactivation, PML, cytopenias, intestinal obstruction, immunization	*Systemic*: infections, pneumonia, neutropenia, pyrexia, dyspnea, cough, diarrhea, URTI, nausea, fatigue, bronchitis. *Cutaneous*: rash, urticaria, hyperhidrosis
Rituximab (MabThera®, Rituxan®)	CD20	*Boxed warning*: fatal IRs, TLS, potentially fatal PML, and severe mucocutaneous reactions. *Others*: hepatitis B virus reactivation, infections, cardiac arrhythmias, bowel obstruction and perforation	*Systemic*: pulmonary events, renal toxicity, neutropenias, serum sickness, anaphylaxis, fever, lymphopenia, chills, asthenia. *Cutaneous*: paraneoplastic pemphigus, lichenoid dermatitis, vesiculobullous dermatitis, SJS, TEN

(continued)

Table 3.1 (continued)

Monoclonal antibody [a] INN and trade names[b]	Target[c]	Warnings, precautions, risks, and safety concerns	Other adverse events,[d] serious and common
Brentuximab vedotin[m] (Adcetris®)	CD30[n]	*Boxed warning*: PML. *Others*: peripheral neuropathy, IR and anaphylaxis, neutropenia, infections, fetal harm, hepatotoxicity, TLS, SJS	*Systemic*: cytopenias, immunogenicity, URTI, pyrexia, nausea, vomiting, fatigue, cough, anaphylaxis. *Cutaneous*: rash, pruritus, SJS, alopecia
Alemtuzumab[o] (Campath®, MabCampath®)	CD52[p]	*Boxed warning*: cytopenias, IR, immunosuppression/ infections[q]. *Others*: immunization	*Systemic*: pulmonary events, immunogenicity, cardiac events, diarrhea, nausea, emesis, insomnia. *Cutaneous*: rash, urticaria, erythema, pruritus
Cetuximab (Erbitux®)	EGFR	*Boxed warning*: serious IR and cardiopulmonary arrest. *Others*: pulmonary toxicity, dermatologic toxicity, hypomagnesemia	*Systemic*: electrolyte imbalance, infection, GI, anaphylaxis, headache, diarrhea. *Cutaneous*: acneiform rash, nail changes, xeroderma, paronychial inflammation, pruritus
Panitumumab[r] (Vectibix®)	EGFR	*Boxed warning*: dermatologic toxicity, IR. *Others*: increased toxicity with bevacizumab and chemotherapy, pulmonary toxicities, electrolyte depletion, ocular events	*Systemic*: pulmonary events,[s] pulmonary embolism, GI, fatigue, abdominal pain, hypomagnesemia. *Cutaneous*[t]: rash, dermatitis "acneiform," erythema, exfoliation, paronychia, skin fissures, photosensitivity, xerosis, pruritus
Necitumumab (Portrazza®)	EGFR	*Boxed warning*: cardiopulmonary arrest, hypomagnesemia. *Others*: venous and arterial thromboembolic events, infusion reactions, dermatologic toxicities, ↑ toxicity and mortality in patients with non-squamous NSCLC, embryofetal toxicity	*Systemic*: vomiting, diarrhea. *Cutaneous*: rash, dermatitis acneiform
Bevacizumab (Avastin®)	VEGF	*Boxed warning*: GI perforation, surgery/wound healing, hemorrhage. *Others*: non-GI fistula, RPLS, IR, CHF, hypertension, arterial/venous thromboembolism, eye disorders, proteinuria, neutropenia/infections, ONJ	*Systemic*: pulmonary events, epistaxis, headache, rectal hemorrhage, dry skin, necrotizing fasciitis, taste alteration, lacrimation disorder, ovarian failure *Cutaneous*: exfoliative dermatitis, alopecia

Ramucirumab (Cyramza®)	VEGFR-2	*Boxed warning*: hemorrhage, GI perforation, impaired wound healing. *Others*: arterial thromboembolic events, IR, RPLS, hypertension, deterioration in patients with cirrhosis, proteinuria including nephrotic syndrome, thyroid dysfunction, embryofetal risk	*Systemic*: hypertension, diarrhea, headache, hyponatremia, neutropenia, epistaxis, stomatitis, immunogenicity[u]
Pertuzumab (Perjeta®)	HER2	*Boxed warning*: cardiomyopathy, embryofetal toxicity. *Others*: IR, hypersensitivity/anaphylaxis	*Systemic*: neutropenias, LVD, peripheral neuropathy, fatigue, GI, asthenia. *Cutaneous*: rash, paronychia, pruritus, alopecia, PPE (in combination therapy)
Trastuzumab (Herceptin®)	HER2	*Boxed warning*: cardiomyopathy,[v] IR, pulmonary toxicity. *Others*: exacerbation of chemotherapy-induced neutropenia, embryofetal toxicity	*Systemic*: neutropenia,[w] anemia, thrombocytopenia, pulmonary events, LVD[v], GI, chills, fever, URTI, anaphylaxis/angioedema, headache, cough, stomatitis, mucosal inflammation. *Cutaneous*: rash, nail disorders, pruritus
Ado-trastuzumab emtansine[x] (Kadcyla®)	HER2	*Boxed warning*: hepatotoxicity, cardiotoxicity, embryofetal toxicity. *Others*: IR, pulmonary toxicity, extravasation, hemorrhage, thrombocytopenia, neurotoxicity	*Systemic*: pulmonary events, fetal harm, LVD, hypersensitivity/IR, nausea, fatigue, anemia, headache, musculoskeletal pain, increased transaminases, constipation. *Cutaneous*: rash, pruritus
Denosumab (Xgeva®, Prolia®)	RANKL	Hypocalcemia, osteonecrosis of the jaw, embryofetal toxicity	*Systemic*: osteomyelitis, hypophosphatemia, dyspnea, fatigue/asthenia, back pain, nausea, extremity pain. *Cutaneous*: rash, pruritus, dermatitis, eczema
Ipilimumab (Yervoy®)	CTLA-4[y]	*Boxed warning*: immune-mediated adverse reactions[z]	*Systemic*: diarrhea, fatigue, colitis. *Cutaneous*: rash, pruritus, dermatitis
Siltuximab (Sylvant®)	IL-6	Not for patients with severe infections or live vaccines, IR, cautionary use in patients with GI perforation risk	*Systemic*: hyperuricemia, URTI, increased weight. *Cutaneous*: rash, pruritus

(continued)

Table 3.1 (continued)

Monoclonal antibody [a] INN and trade names[b]	Target[c]	Warnings, precautions, risks, and safety concerns	Other adverse events,[d] serious and common
Nivolumab (Opdivo®)	PD-1	Immune-mediated adverse reactions,[aa] embryofetal toxicity	*Systemic*: increased ALT, AST, and AP; hyponatremia; hyper- and hypokalemia; hyper- and hypocalcemia; lymphopenia; fatigue; asthenia; musculoskeletal and abdominal pain; dyspnea; cough; GI. *Cutaneous*: rash, pruritus
Pembrolizumab (Keytruda®)	PD-1	Immune-mediated adverse reactions,[ab] embryofetal toxicity	*Systemic*: fatigue, peripheral edema, chills, pyrexia, renal failure, cellulitis, decreased appetite, dyspnea, arthralgia, nausea, diarrhea, cough. *Cutaneous*: rash, pruritus, vitiligo
Dinutuximab (Unituxin®)	GD2	*Boxed warning*: serious IR, neuropathy. *Others*: CLS and hypotension, infection, neurological disorders of the eye, BMS, electrolyte abnormalities, AHUS, embryofetal toxicity	*Systemic*: hypokalemia, pain, fever, hypocalcemia, hyponatremia, anemia, thrombocytopenia, lymphopenia, neutropenia, increased AST and ALT, GI. *Cutaneous*: urticaria
Daratumumab (Darzalex®)	CD38	IR, interference with serological testing,[ac] interference with determination of patient's response and disease progression[ad]	IR, infections, thrombocytopenia, pyrexia, fatigue, nausea, back pain, cough[ae]
Elotuzumab (Empliciti®)	SLAMF7[af]	IR,[ag] infections, second primary malignancies, hepatotoxicity, interference in monitoring M-protein impacting determination of complete response in patients with IgGκ myeloma protein	Fatigue, diarrhea, pyrexia, constipation, cough, peripheral neuropathy, URTI, nasopharyngitis, decreased appetite, pneumonia, headache, pain in the extremity, vomiting, lymphopenia, neutropenia, muscle spasms

Atezolizumab was approved in May 2016, too late to be included in this table. See later this chapter, "Recent Approval: Atezolizumab"

ADCC antibody-dependent cell-mediated cytotoxicity, *AHUS* atypical hemolytic uremic syndrome, *ALT* alanine transaminase, *AML* acute myelogenous leukemia, *AP* alkaline phosphatase, *AST* aspartate transaminase, *BMS* bone marrow suppression, *CHF* congestive heart failure, *CLS* capillary leak syndrome, *CNS* central nervous system, *CRS* cytokine release syndrome, *CTLA-4* cytotoxic T lymphocyte-associated antigen 4, *EGFR* epidermal growth factor receptor (HER1, ErbB1), *EM* erythema multiforme, *EpCAM* epithelial cell adhesion molecule, *GD2* disialoganglioside expressed on tumors of neuroectodermal origin, *GI* gastrointestine/gastrointestinal symptoms, e.g., nausea, diarrhea, vomiting, constipation, etc., *HAMA* human anti-mouse antibody, *HARA* human anti-rat antibody, *HER2* human epidermal growth factor 2, also known as neu, ErbB2, CD340, or p185, *HLH* hemophagocytic lymphohistiocytosis, *IR* infusion reaction, *LVD* left ventricular dysfunction, *MDS* myelodysplastic syndrome, NSCLC non-small cell lung cancer, *ONJ* osteonecrosis of the jaw, *PD-1* programmed cell death protein 1, *PML* progressive multifocal leukoencephalopathy, *PPE* palmar-plantar erythrodysesthesia, *RANKL* receptor activator of nuclear factor kappa-B ligand (CD254), *RPLS* reversible posterior leukoencephalopathy syndrome, *SIRS* systemic inflammatory response syndrome, *SJS* Stevens-Johnson syndrome, *TEN* toxic epidermal necrolysis, *TLS* tumor lysis syndrome, *URTI* upper respiratory tract infection, *VEGF* vascular endothelial growth factor, *VEGFR-2* vascular endothelial growth factor receptor 2

[a]Nomenclature: mAbs of murine origin are given the suffix, or stem, *-omab*; chimeric antibodies in which the V region is spliced into human C region are given the *-ximab* stem; humanized antibodies with murine hypervariable region spliced into human antibody have the *-zumab* stem; and antibodies with complete human sequence are given the *-umab* stem

[b]Approved by FDA or EMA or both

[c]Specificity of antibody

[d]Adverse events in addition to those mentioned as occurring, or potentially likely to occur, and shown in column 3

[e]Registered by EMA, Health Canada, and Ministry of Health, Israel, but not FDA. Catumaxomab is a bispecific mouse—rat hybrid (given suffix *-axomab*) recognizing both EpCAM and CD3

[f]EpCAM (CD326), expressed on epithelial and epithelial-derived neoplasms; CD3—part of TCR complex on T lymphocytes

[g]A BiTE (bispecific T-cell-engaging) fusion protein

[h]CD19, a B-cell antigen; CD3, part of the T-cell receptor

[i]With yttrium-90 or indium-111. Tiuxetan is a chelator

[j]Expressed on B lymphocytes where it aids optimum B-cell response to T-independent antigens

[k]Severe neutropenia, thrombocytopenia, anemia, and lymphopenia. Incidences of thrombocytopenia grades III and IV in ibritumomab tiuxetan-treated non-Hodgkin lymphoma patients were 87 % and 13 %, respectively

[l]Glycoengineered to enrich Fc carbohydrate with non-fucosylated sugars and higher binding to FcγRIII with consequent enhanced ADCC

[m]Conjugated to cytotoxic monomethyl auristatin E (MMAE)

[n]CD30—a cell membrane protein of the tumor necrosis receptor family. Expressed on activated T and B lymphocytes

[o]Withdrawn from the USA and Europe in 2012 to be relaunched for multiple sclerosis

[p]CD52—present on the surface of mature lymphocytes and associated with some lymphomas

[q]In particular *Pneumocystis jirovecii*, CMV, EBV, and herpesvirus

[r]Not indicated for use in combination with chemotherapy due to increased toxicity

[s]Should be discontinued in patients developing interstitial lung disease, pneumonitis, and lung infiltrates

[t]Most common drug-induced reactions following this mAb are skin toxicities

[u]Neutralizing antibodies detected in 1 of 33 patients

(continued)

Table 3.1 (continued)

[v]Greatest risk (LVD) when administered with anthracyclines
[w]Highest risk with myelosuppressive therapy
[x]Called ado-trastuzumab emtansine in the USA to distinguish it from trastuzumab. Trastuzumab linked to the cytotoxin mertansine (DMI), a tubulin inhibitor. Also known as trastuzumab emtansine and T-DM1
[y]Binds CD80/CD86 on antigen-presenting cells
[z]Immune-mediated reactions due to T-cell activation and proliferation—enterocolitis, hepatitis, dermatitis, neuropathies, endocrinopathies, and other immune-mediated reactions including cutaneous and ocular manifestations
[aa]Immune-mediated pneumonitis, colitis, hepatitis, nephritis and renal dysfunction, hypothyroidism, and hyperthyroidism
[ab]Immune-mediated colitis, hepatitis, nephritis, hypothyroidism, and hyperthyroidism
[ac]Daratumumab binds to CD38 on red cells producing a positive Coombs test masking detection of minor antigens in patient's serum
[ad]Daratumumab is a human IgG kappa mAb detected by assays used to monitor endogenous M-protein. In patients with IgG kappa myeloma, this interference may influence the determination of the patient's response and the disease progression
[ae]Postmarketing usage is likely to reveal the occurrence of cytopenias
[af]Signaling lymphocytic activation molecule receptor family member 7, also known as CS1, CD2 subunit 1, and CD319
[ag]Premedicate with dexamethasone, H1 and H2 antihistamines, and acetaminophen

Catumaxomab

The dual antigen recognition specificity of this mouse-rat hybrid mAb (Removab®) (Tables 2.1 and 3.1) is effected by a mouse kappa light chain and IgG2a heavy chain and a rat lambda light chain and IgG2b heavy chain. Binding of the Fc region with Fcγ receptors provides a third functional binding site. The mouse Fab binds to EpCAM, the rat Fab binds to CD3, and the hybrid Fc fragment binds to FcγRI (CD64), FcγRIIa (CD32), and FcγRIIIa (CD16a) on macrophages, NK cells, dendritic cells, and mononuclear cells (Fig. 3.1). EpCAM (CD326), a transmembrane glycoprotein, is shielded by tight junctions in normal tissue but expressed over the whole surface of tumor cells. It promotes tumor growth and metastasis and is overexpressed on epithelial tumors of the gastrointestinal tract, esophagus, head, neck, lung, liver, kidney, ovary, pancreas, and prostate and a number of other organs and tissues. Being so widely and strongly expressed, EpCAM has become a potentially useful target for antibody therapy of various carcinomas. Overexpression of EpCAM tends to be seen in advanced cases of cancer and those with a poor outcome. In relation to its approved indication for malignant ascites, EpCAM is expressed on most epithelial cancers of this type as well as the tumor cells in malignant effusions. The trifunctional action of

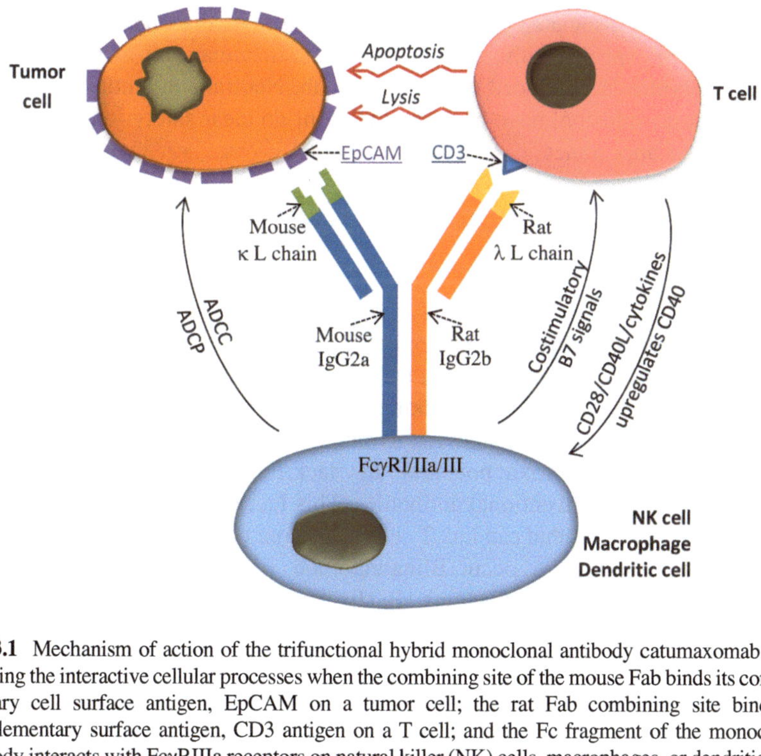

Fig. 3.1 Mechanism of action of the trifunctional hybrid monoclonal antibody catumaxomab summarizing the interactive cellular processes when the combining site of the mouse Fab binds its complementary cell surface antigen, EpCAM on a tumor cell; the rat Fab combining site binds its complementary surface antigen, CD3 antigen on a T cell; and the Fc fragment of the monoclonal antibody interacts with FcγRIIIa receptors on natural killer (NK) cells, macrophages, or dendritic cells

catumaxomab involving antibody recognition of cancer cells expressing EpCAM, CD3 on activated lymphocytes, and effector functions mediated via Fc binding results in T-cell-induced killing, ADCC, complement-dependent cytotoxicity (CDC), and antibody-dependent cellular phagocytosis (ADCP) through interaction with FcγRs on effector cells (Fig. 3.1).

When used to treat malignant ascites, the commonly seen side effects of catumaxomab of fever, nausea, vomiting, and abdominal pain are the result of cytokine release syndrome (CRS) (Chap. 1, section "Cytokine Release Syndrome"; Table 3.1). The cytokines TNF and IL-6, for example, are elevated above normal levels in 60 and 80 %, respectively, of treated patients. Cytokine release is therefore a reflection of both T-cell activation and the mode of action of catumaxomab. Pyrexia, nausea, and vomiting tend to be limited to the duration of therapy and manageable with standard symptomatic treatment. Abdominal pain is generally managed by standard pain medication, while antibody-induced grade 4 adverse events such as ileus are isolated occurrences often related to the malignancy. Lymphopenia, reported in up to 14 % of patients, is reversible, usually within a week. Cutaneous reactions of rash, erythema, pruritus, and catheter-related reactions such as erythema and infection also occur and can be serious. Up to the end of 2012, postmarketing reports from Europe listed eight cases of respiratory disorders, 11 cases of cutaneous reactions, 11 reports of infections, and eight cases of systemic inflammatory response syndrome (SIRS) (Chap. 1, section "Systemic Inflammatory Response Syndrome") with associated pyrexia, tachypnea, tachycardia, and leukocytosis.

As might be expected with rodent antibodies, both human anti-mouse and human anti-rat antibodies are induced by catumaxomab. Neutralizing antibodies, usually after the fourth infusion, have been reported, but no clear safety issues have been linked to their appearance.

Blinatumomab

There is substantial evidence that cytotoxic T lymphocytes can have a major role in controlling the growth of tumors, but therapies aimed at utilizing this observation, for example, the use of CTLA-4-blocking antibodies, have shown only limited success, presumably because of immune avoidance by the cancer cells. Although T cells can show high cytotoxic potential, they lack Fcγ receptors and therefore cannot be recruited by conventional antibodies. This fact led to an approach based on single-chain antibodies that can link T cells and tumor cells bringing them into close proximity for cell killing to occur. Blinatumomab (Blincyto®) (Tables 2.1 and 3.1), a so-called bispecific T-cell-engaging (BiTE) fusion protein composed of two antibody single-chain variable fragments each from an H and L chain, is an ~55 kDa protein derived from the linkage of the four peptide chains from four different genes. One of the two binding specificities is directed to the B-cell antigen CD19, while the other targets CD3, part of the T-cell receptor. Reaction with both antigens is exploited to link malignant B cells of patients with acute lymphoblastic leukemia to cytotoxic T cells, activating them to destroy the tumor cells via production of perforin and granzymes that induce apoptosis. Blinatumomab induces tumor regression at very low doses, for example, partial and complete tumor regressions have been first observed at doses as low as 15 ng/sq m/day. FDA approval for blinatumomab in December 2014 was based on results of a phase II multicenter open-label study of 185 patients with Philadelphia chromosome-negative relapsed or refractory B-cell precursor acute lymphoblastic leukemia. The study revealed that 42 % of participating patients achieved complete or partial remission within two cycles of the drug, the majority improving within the first treatment cycle. Thirty-two percent of patients achieved complete remission for about 6.7 months.

Blinatumomab approval by the FDA was issued with a Risk Evaluation and Mitigation Strategy (REMS). The most common adverse events seen with blinatumomab (Table 3.1) in trials included pyrexia (62 %); neurological toxicity (50 %); headache (36 %); peripheral edema, febrile neutropenia, and nausea (each 25 %); hypokalemia (23 %); rash (21 %); and constipation (20 %). Neurological toxicities, ranging from confusion to tremors, convulsions, and speech disorders and all apparently reversible, frequently interrupted therapy. The majority of adverse events occurred within the first week of treatment and usually faded to grade 1 or less upon further treatment. CRS, sometimes life-threatening or even fatal, was reported in 11 % of patients (see boxed warning, Table 3.1). It has been suggested that CRS, in severe form at least, may be due to activation of macrophages triggered by cytokines released from T cells activated by blinatumomab. This is thought to lead to hemophagocytic lymphohistiocytosis (HLH) (Chap. 1, section "Hemophagocytic Lymphohistiocytosis"), a condition seen in a patient with CRS 36 h after infusion with blinatumomab. Symptoms of fever, circulatory collapse, and respiratory failure accompanied cytopenias, hypofibrinogenemia, and hyperferritinemia. HLH continued after withdrawal of blinatumomab but improved rapidly upon treatment with the IL-6 and inflammation blocker tocilizumab. It has subsequently been suggested that patients given blinatumomab therapy should be monitored for HLH. Unlike antibodies that block CTLA-4, no autoimmune reactions have so far been observed following blinatumomab.

Seen as an "ultra-orphan drug," the pool of potential patients for the mAb is estimated to be ~1000 acute lymphoblastic leukemia patients. This fact is being reflected in the cost of treatment, recently estimated to be approximately $178,000 for two courses of the drug.

Monoclonal Antibodies Targeting CD20: Rituximab, Ibritumomab, Ofatumumab, and Obinutuzumab

CD20 (human B lymphocyte-restricted differentiation antigen Bp35), a 33–35 kDa transmembrane glycosylated phosphoprotein that is part of the MS4A family of proteins, is expressed on the surface of B cells, except for plasmablasts, at all stages of their development until the memory B-cell stage. CD20 also occurs on

B-cell lymphomas, B-cell chronic lymphocytic leukemia, hairy cell leukemia, and melanoma cancer stem cells but is absent from mature plasma cells and other tissues. The antigen may have a functional role in one or more of B-cell growth, differentiation, activation, and proliferation via intracellular signaling or as a calcium channel in association with the B-cell receptor, but, as yet, no CD20 ligand has been identified. In addition to the effector processes initiated when bound to antibody, CD20 may be involved in transmembrane signaling controlling growth and cell death in some tumors. Due to its non-Hodgkin lymphoma (NHL) B-cell expression and the fact that it is not normally shed from cells and is internalized after binding to antibody, CD20 has been seen as an exploitable target for mAbs for the treatment of lymphomas. Four mAbs directed to CD20 are currently approved by the FDA and EMA as antitumor agents. In the order of regulatory approval, these are rituximab, ibritumomab, ofatumumab, and obinutuzumab.

Rituximab

Rituximab (MabThera®, Rituxan®) (Tables 2.1 and 3.1), a human-mouse chimeric IgG1κ antibody, was the first mAb approved by the FDA to treat relapsed or refractory NHL and in 1997 was, in fact, the first mAb approved specifically for cancer therapy. Given in combination with some selected small molecule chemotherapeutic drugs, it is now a first-line therapy for several NHLs, including follicular lymphoma and diffuse large B-cell lymphoma. In 2002, the FDA authorized the use of rituximab as a component of ibritumomab therapy. The mechanisms underlying rituximab's cytotoxicity in cancer therapy are not completely understood. Tumor cell death is attributed to ADCC, CDC, and the induction of apoptosis, but other mechanisms, such as a vaccine-like action of increasing the cytotoxic T-cell response to idiotype antigens on the malignant cells, may also be operative. Rituximab is relatively successful as a treatment for B-cell malignancies, but the mAb is not effective in all patients where variable degrees of tumor resistance occur. This resistance remains poorly understood although many suggested mechanisms have been advanced including loss of CD20 expression, blockade of ADCC, expression of proteins that inhibit CDC, and expression of anti-apoptotic proteins. Several strategies have been devised to overcome tumor cell resistance to rituximab but also to improve anti-CD20 mAb efficacy. In relation to the latter, mAb has been conjugated to a radionuclide as seen with the development of ^{90}Y-ibritumomab tiuxetan (section "Ibritumomab Tiuxetan") and ^{131}I-tositumomab.

By 2012, FDA and EMA official safety data reported for rituximab was based on a total of nearly 6000 exposed patients. A long list of side effects has been reported for the antibody including the serious events detailed in an FDA black box warning, namely, infusion reactions, progressive multifocal leukoencephalopathy (PML) (Chap. 1, section "Progressive Multifocal Leukoencephalopathy"), tumor lysis syndrome (TLS) (Chap. 1, section "Tumor Lysis Syndrome"), skin and mucocutaneous reactions (these four all potentially fatal), infections, hepatitis B reactivation, cardiac arrhythmias, and bowel problems as well as a host of often less serious systemic and cutaneous events and toxicities (Table 3.1). Often seen as CRS, infusion reactions occur on the first infusion in up to 77 % of malignant patients, and this decreases to approximately 10 % after the second infusion. Potentially serious symptoms including hypotension and bronchospasm may complicate the reaction in about 10 % of cases. An early relationship between rituximab administration, the lymphocyte count, and CRS suggested that the appearance of the syndrome correlated with lymphocyte counts higher than 50×10^9/L. Patients with counts exceeding this figure experienced more reactions than patients with lesser numbers of tumor cells. A survey of hypersensitivity reactions to rituximab in patients at the Massachusetts General Hospital between 2006 and 2010 found immediate hypersensitivity in 79 of 901 patients (8.8 %). Approximately three-quarters of the patients showed symptoms after the initial infusion, and 46 % experienced moderate or severe reactions after subsequent infusions. An increased risk of hypersensitivity reactions occurred in patients with advanced disease, and obscurely, Waldenström's macroglobulinemia patients accounted for 10 % of the reactions despite making up only 1 % of the rituximab-treated patients. Other signs of hypersensitivity following rituximab administration are urticaria, cardiovascular and respiratory distress typical of an anaphylactoid/anaphylactic reaction, serum sickness, vasculitis, interstitial pneumonitis, some cutaneous manifestations, and acute respiratory distress syndrome (ARDS). Cough, dyspnea, and bronchospasm are fairly common but pulmonary events can be serious, for example, pulmonary fibrosis which has proved fatal. A review of 62 cases of severe respiratory reactions thought to be caused by rituximab implicated interstitial pneumonitis in three-quarters of the subjects together with other respiratory disorders, bronchiolitis obliterans, organizing pneumonia, pulmonary fibrosis, hypersensitivity pneumonitis, and ARDS. It has been suggested that rituximab should not be administered to patients with lung diseases such as pneumonia, pleural effusion, and collapsed lung.

Infections, bacterial, viral, and fungal, occurring with an incidence of up to 17 %, may be seen during monotherapy with rituximab. Infections during or after rituximab may be rare as well as opportunistic and they are always a concern since they can be serious and even fatal. Cytomegalovirus (CMV) encephalitis is mainly seen in HIV-positive individuals, but it also occurs in occasional malignant melanoma patients treated with rituximab. Another example of a rare organism in a patient receiving rituximab is infection due to *Capnocytophaga bacteraemia*, a gram-negative bacillus found in dog saliva. Fulminant myocarditis due to enterovirus is a further rare pathogenic example associated with the use of the mAb. Pneumocystis pneumonia, an opportunistic fungal infection in rituximab-treated patients caused by *Pneumocystis jirovecii*, is the subject of a number of reports. Mayo Clinic, Rochester, records for the period 1998–2011 revealed 30 patients had the infection, most developed acute hypoxemic respiratory failure, half required admission to intensive care, and 30 % of the patients died. Other reports of *P. jirovecii* pneumonia are numerous and alarming enough for the suggestion to be made that *P. jirovecii* pneumonia prophylaxis should be considered following rituximab therapy.

An incidence of 19 % has been reported for cardiovascular disorders during rituximab monotherapy, and cytopenias, usually mild, occur with only low incidences of severe neutropenia, anemia, and thrombocytopenia. TLS may manifest as renal failure and a 2010 examination of the WHO Collaborating Centre for International Drug Monitoring Adverse Event Data Bank revealed 114 reports associated with rituximab out of 182 case reports of PML. Table 3.1 lists the serious and sometimes fatal mucocutaneous reactions reported for rituximab, most of which appear within approximately the first three months of treatment. Combining rituximab with chemotherapy has produced adverse responses in up to 86 % of malignant patients with 57 % showing severe or serious signs. Hematologic events showed the highest incidence (67 %, 48 % severe or serious), followed by dermatologic reactions (44 %, 2 %), respiratory symptoms (38 %, 4 %), gastrointestinal symptoms (37 %, 2 %), musculoskeletal effects (26 %, 3 %), and cardiovascular disorders (25 %, 3 %). Overall, the incidences of a range of adverse events were higher in patients receiving rituximab plus chemotherapy than in patients receiving either therapy alone. Postmarketing surveillance of rituximab administration has confirmed the importance of infections and respiratory and hematologic events in the safety profile of this widely and heavily used mAb. Infections, with an incidence of 6.6 %, led the number of reports of adverse events in the FDA Adverse Event Reporting System (FAERS) database containing more than 16,700 reports. Hematologic (5.7 %), respiratory (3.7 %), and gastrointestinal symptoms (2.8 %) followed in that order. Infections were again the most commonly reported event (13 %) in the European Pharmacovigilance EudraVigilance Database Management System followed by respiratory events (11 %), hematologic events (10 %), nervous system disorders (7 %), gastrointestinal disorders (6 %), and infusion reactions (4 %). Interesting and noteworthy totals for individual disorders were CRS (44 reports), TLS (145), PML (423), JC virus infections (37), leukoencephalopathy (33), and bone marrow failure (137). Recorded fatalities totaled about 6 %. Other events reported by the FDA under the heading of postmarketing experience are late onset neutropenia; hyperviscosity syndrome in Waldenström's macroglobulinemia; fatal cardiac failure; viral infections; immune/autoimmune events including uveitis, optic neuritis, vasculitis, lupus-like syndrome, serum sickness, pleuritis, and polyarticular arthritis; disease progression of Kaposi's syndrome; bowel obstruction and perforation; fatal bronchiolitis obliterans and interstitial lung disease; and the nervous system disorder, posterior reversible encephalopathy syndrome (PRES) (Chap. 1, section "Posterior Reversible Encephalopathy Syndrome").

Antichimeric antibodies were detected in 4 of 356 (1.1 %) patients with low-grade or follicular NHL receiving rituximab alone. In patients with Wegener's granulomatosis and microscopic polyangiitis treated with rituximab, 23 of 99 (23 %) tested positive for antichimeric antibodies at 18 months. The clinical relevance of the immunogenicity of antichimeric antibodies generally remains unclear although of 11 patients given rituximab for severe pemphigus, two patients who developed antibodies to the mAb experienced an increase in disease activity over an extended time. Of the nine patients who did not develop antibodies, the lesions healed in five, and partial remission was achieved in the other four patients. In rheumatoid arthritis patients receiving rituximab, 273 of 2578 (10.6 %) developed antichimeric antibodies, but there appeared to be no association between the appearance of antibodies and infusion or other adverse reactions.

Ibritumomab Tiuxetan

Ibritumomab tiuxetan (Zevalin®) (Tables 2.1 and 3.1), a murine IgG1κ mAb covalently linked to the chelator tiuxetan by a stable thiourea bond, is radiolabeled with yttrium-90 for therapy or indium-111 for imaging. In the initial dosing schedule set out by the FDA, premedication with acetaminophen and diphenhydramine preceded infusion of rituximab on day 1 and days 7, 8, or 9. Within 4 h of completing the first rituximab infusion, ibritumomab labeled with indium-111 was given as an intravenous injection over 10 min prior to a bioscan for checking biodistribution. The requirement for the bioscan was discontinued in November 2011. Four hours after the second rituximab infusion, yttrium-labeled ibritumomab was given by intravenous injection. Rituximab infusion should be immediately ceased if serious infusion reactions occur and temporarily slowed or interrupted for less severe reactions. Patients should be monitored closely for extravasation during ibritumomab-yttrium-90 therapy, and infusion should be immediately stopped and restarted in another limb if signs of extravasation occur. The two-day half-life of ibritumomab is shorter than that of rituximab (7 days), but this is not necessarily a drawback since prolonged exposure to the beta emissions of the yttrium isotope is not desirable.

Results from clinical trials and an FDA black box warning highlight severe, and potentially fatal, infusion reactions and severe cytopenias as the most serious adverse events experienced with ibritumomab therapy (Table 3.1). Clinical trial results showed almost 60 % of patients experienced cytopenias with severe neutropenia and thrombocytopenia occurring with the highest incidences followed by anemia and hemorrhage. Other serious events are infections, mainly bacterial but some fungal and viral, and potentially fatal myeloid malignancies or dysplasias. Hypersensitivity reactions, generally manifesting as bronchospasm or angioedema, are another potentially dangerous adverse effect. Of the non-hematologic events, gastrointestinal symptoms are commonly seen as well as rare toxidermias such as bullous dermatitis, erythema multiforme, Stevens-Johnson syndrome, and toxic epidermal necrolysis. Postmarketing surveillance has revealed fatal cases of infection, infusion-related cardiac arrest, and cerebral hemorrhage; reports of the severe cutaneous and mucocutaneous reactions were already mentioned, as well as patients with exfoliative dermatitis; and 19 cases of acute myelogenous leukemia or myelodysplastic syndrome were reported in 746 NHL patients (an incidence of 2.5 %). For the term "progressive multifocal leukoencephalopathy," a search in 2010 of the WHO Collaborating Centre for International Drug Monitoring Adverse Event Data Bank retrieved 182 reports, five of which related to ibritumomab tiuxetan. As a comparison, note that the retrieved figure for rituximab was 114.

Being a mouse mAb, it is not surprising that immunization is included in the FDA warnings and precautions for ibritumomab. In fact, human anti-mouse and human antichimeric antibodies have been observed in ~4 % of patients treated with the mAb, but antibody titers do not seem to increase with time and the antibodies generally do not mediate hypersensitivities.

Ofatumumab

Ofatumumab (Arzerra®) (Tables 2.1 and 3.1) is a fully human IgG1κ mAb originally granted orphan drug status by the EMA (2008) and FDA (2009) for the treatment of B-cell CLL. Subsequent accelerated approvals by the FDA, EMA, Australian Therapeutics Goods Administration (TGA), and Health Canada were restricted to CLL patients refractory to fludarabine and alemtuzumab. The reasoning behind this indication is based on the poor survival outcomes of patients who become refractory to fludarabine (normally the cornerstone of treatment of CLL) and alemtuzumab treatments. Although CD20 is sometimes described as an ideal target for mAbs in allowing effector recruitment for ADCC and CDC, different CD20-reactive mAbs can differ in their capacities to trigger programmed cell death. CD20 lodged in the cell membrane is thought to have two extracellular loops, a large one composed of about 44 amino acids and a smaller one which is thought to remain within the plasma membrane. Ofatumumab binds an epitope composed of both the large and small loops, which is a different region recognized by rituximab, and it is this different recognition that helps to increase ofatumumab-induced complement activation and complement-mediated lysis. Ofatumumab's more potent CDC action than that of rituximab is thought to result from the close proximity of the CD20 small loop-binding site to the cell surface which aids the deposition of complement on the cell surface. Ofatumumab induces lysis in cells with high or low CD20 expression (such as freshly isolated CLL cells) and even in complement-resistant B-cell lines and cells expressing complement inhibitory molecules. In addition to its efficient binding of C1q and activation of the complement pathway, ofatumumab induces cell death through ADCC and there is some evidence that binding of the mAb also recruits NK cells.

From clinical trials of ofatumumab, an average of nine adverse events per patient was recorded; serious events occurred in 30–54 % of patients, fatalities in 16 % of patients, and drug-related discontinuations in 14 % of participants. Most adverse events, particularly severe ones, occurred in 27 patients given the highest dose of antibody. Although there are no boxed warnings for ofatumumab, infusion reactions, the possibility of hepatitis B virus reactivation, cytopenias, intestinal obstruction, and PML comprise the warnings, precautions, and risks issued for this mAb (Table 3.1). Infusion reactions following ofatumumab may manifest as bronchospasm, dyspnea, laryngeal and pulmonary edema, flushing, hypotension, hypertension, syncope, cardiac ischemia/infarction, back and abdominal pain, angioedema, pyrexia, rash, and urticaria. Premedication with acetaminophen, an antihistamine, and a corticosteroid may be necessary, and interruption of infusion is recommended for reactions of any severity. Prolonged severe neutropenia and thrombocytopenia may occur, making necessary the regular monitoring of blood and platelet counts.

Carriers of hepatitis B should be monitored for hepatitis B virus infection during treatment with ofatumumab and for 6–12 months following the last infusion.

In an ofatumumab-as-single-agent study of 138 fludarabine-refractory CLL patients (59 fludarabine and alemtuzumab refractory and 79 fludarabine refractory with bulky lymphadenopathy), infusion reactions, nearly all grades 1 or 2, were seen in ~60 % of patients, predominantly during the first and second infusions. The most common adverse events (≥10 % of patients) during treatment were infections (67 %); cough (18 %); anemia and diarrhea (each 16 %); fatigue, fever, and neutropenia (each 15 %); dyspnea (13 %); nausea (11 %); and rash (10 %). As in other trials with ofatumumab, infections were prominent with 189 events in 92 patients; 139 (74 %) were grade 1 or 2 and 37 grade 3 or 4 with pneumonia and other respiratory tract infections the most common. Thirteen infections (sepsis 6, pneumonia 5, *Fusarium* infection 1, and PML 1) led to death.

In 2013, the FAERS database of 1056 reports listed infections (11 %) and white blood cell (10 %), respiratory (5 %), gastrointestinal (3 %), and neurological (3 %) disorders as the most common events provoked by ofatumumab. There were 157 reports of febrile neutropenia, 60 of pneumonia, 27 of sepsis, 37 of thrombocytopenia, 109 of anaphylaxis, 16 of TLS, and 12 of PML. Mucocutaneous events included 14 cases of paraneoplastic pemphigus and 19 of urticaria and 23 patients with rash.

As anticipated from a fully human antibody, ofatumumab has so far shown a relative lack of immunogenicity. In clinical trial investigations, no antibodies to the agent were found in 46 patients after the eighth infusion or in 33 patients after the 12th infusion.

Obinutuzumab

Obinutuzumab (Gazyva®, Gazyvaro®; also known as GA101 and afutuzumab) (Tables 2.1 and 3.1) was approved by the FDA in November 2013 for the treatment of CLL in combination with chlorambucil after earlier receiving Breakthrough Therapy Designation. In combination with bendamustine, approval for the mAb was extended to the treatment of follicular lymphoma in February 2016. Obinutuzumab is a humanized anti-CD20 mAb of the IgG1κ class, MW ~150 kDa, glycoengineered and prepared in Chinese hamster ovary cells. “Glycoengineered” in this case means the presence of non-fucosylated sugars that impart higher binding affinity for FcγIII receptors which, in turn, enhances ADCC and caspase-independent apoptosis (Chap. 2, sections “Glycosylation of Monoclonal Antibodies” and “Antibody-Dependent Cell-Mediated and Complement-Dependent Cytotoxicities”). Modification of elbow hinge sequences in the variable region may also contribute to the apoptotic activity of obinutuzumab. Like its fellow mAbs targeting CD20, obinutuzumab mediates B-cell lysis by engaging immune effector cells, directly activating intracellular pathways signaling cell death, activating the complement cascade, and effecting ADCC and antibody-dependent cellular phagocytosis (ADCP). Although exhibiting less CDC than rituximab, obinutuzumab has demonstrated superior efficacy. Patients with CLL and coexisting illnesses unable to tolerate combined intravenous chemotherapy are likely to be more tolerant of obinutuzumab in its approved indication of the mAb's combination with chlorambucil.

Early observational studies in a phase I investigation of obinutuzumab induction followed by 2 years of maintenance in 22 patients with relapsed B-cell malignancies showed the main adverse events to be infusion reactions, infections, neutropenia, pyrexia, headache, and nausea. These findings were essentially supported by a comparison of obinutuzumab and rituximab, each combined with chlorambucil, in a study of 781 randomly assigned patients with previously untreated CLL. Adverse events were seen more frequently with the obinutuzumab-chlorambucil treatment. Grade 3–4 neutropenia, infusion reactions, anemia, thrombocytopenia, pyrexia, and musculoskeletal pain were the most common events recorded. In addition to a black box warning for hepatitis B virus reactivation and PML, warnings and precautions issued by the FDA cover severe and life-threatening infusion reactions, TLS, the risk of infections, and cytopenias (Table 3.1). Two-thirds of patients experienced infusion reactions by the time of the infusion of the first gram of obinutuzumab. Symptoms included hypotension, tachycardia, dyspnea, and other respiratory problems such as bronchospasm, wheezing, and throat irritation. Late reactions (up to 24 h) have occurred. Premedication with acetaminophen, an antihistamine, and a corticosteroid may be necessary. As is well known for TLS, patients with high tumor burden are at greater risk of acute renal failure, hyperkalemia, hyperuricemia, and hyperphosphatemia which can occur within 12–24 h of the first infusion. Infections, sometimes serious, may be bacterial, fungal, or viral in origin. Grade 3 or 4 cytopenias may occur in patients given obinutuzumab in combination with chlorambucil; incidences of 34 % for neutropenia and 12 % for thrombocytopenia, respectively, were found in trials. Neutropenia may be of late onset and acute thrombocytopenia, occurring within 24 h of infusion, has been seen in up to 5 % of patients. Treatment-related neutropenia also appears to be more common than with other anti-CD20 mAbs. Apart from a few other common, and less serious, adverse events (Table 3.1), more recent clinical trial experience has revealed the possibility of worsening of preexisting cardiac conditions with some fatal cardiac events occurring in patients treated with obinutuzumab. In a recently published nonrandomized, parallel-cohort, phase Ib, multicenter study of the safety and efficacy of obinutuzumab plus fludarabine and cyclophosphamide and obinutuzumab plus the nitrogen mustard bendamustine (SDX-105) used in the treatment of CLL and lymphomas, obinutuzumab in both combinations showed manageable toxicity. In the obinutuzumab-fludarabine/cyclophosphamide cohort, 6 of 21 patients (29 %) experienced serious adverse events, 86 % experienced a grade 3–4 adverse event, 43 % experienced at least one grade 3–4 hematologic event (neutropenia, febrile neutropenia, anemia, or thrombocytopenia), and 52 % had at least one infection. An infusion-related reaction occurred in 91 % of patients. In the obinutuzumab-bendamustine cohort, 9 of 20 patients (45 %) experienced serious adverse events, 85 % had a grade 3–4 event, and 60 % experienced at least one grade 3–4 hematologic event. Ninety percent of patients had an infusion-related reaction, four of which were serious and most reactions occurred with the first dose. Although cutaneous side effects of rituximab are well known and cover a wide spectrum of effects from mild sweating and pruritus to severe toxic epidermal necrolysis, paraneoplastic pemphigus, and lichenoid eruptions, reports of dermatologic

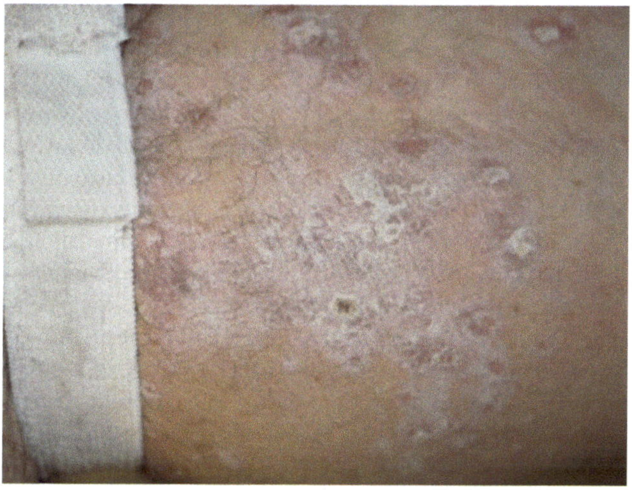

Fig. 3.2 A lichenoid eruption in a 62-year-old man treated with obinutuzumab for follicular non-Hodgkin lymphoma. The reaction, of psoriasiform appearance, was a widespread violaceous lichenoid maculopapular eruption on the trunk, back, arms, and legs. Reproduced from Bakkour W and Coulson IH. GA101 (a novel anti-CD20 monoclonal antibody)-induced lichenoid eruption. Dermatol Ther 2012;2:3. doi:10.1007/s13555-012-0003-9, an open-access article distributed under the terms of the Creative Commons Attribution License

reactions to other anti-CD20 mAbs are surprisingly rare. A fairly recent case of lichenoid eruption (Fig. 3.2) ascribed to obinutuzumab was claimed to be the first case report of a cutaneous side effect to the mAb.

For the recently FDA-approved indication of follicular lymphoma, obinutuzumab with bendamustine is followed by obinutuzumab maintenance in patients refractory to, or who have relapsed after, treatment with a rituximab-containing regimen. The most common adverse reactions seen in treated patients were neutropenia, pyrexia, anemia, thrombocytopenia, infusion reactions, respiratory and urinary infections, arthralgia, asthenia, gastrointestinal symptoms, fatigue, and sinusitis. The most common grade 3/4 adverse reactions were febrile neutropenia, neutropenia, pneumonia, infusion reactions, sepsis, and pyrexia (Table 3.1).

Up to 1 year after obinutuzumab therapy, 9 of 70 treated patients (12.9 %) tested positive for antibodies to the mAb. Neutralizing activity of the antibodies was not investigated.

Brentuximab Vedotin

Brentuximab vedotin (Adcetris®) (Tables 2.1 and 3.1) is a chimeric IgG1κ mAb conjugated to the cytotoxic agent monomethyl auristatin E (MMAE) and targeted to CD30 or TNFRSF8, a 120 kDa cell membrane glycoprotein of the tumor necrosis receptor family expressed on activated T and B lymphocytes. Recognized in 2007 as an orphan drug for Hodgkin lymphoma, later for systemic anaplastic large cell lymphoma (sALCL), and recently for mycosis fungoides, brentuximab vedotin received accelerated approval by the FDA in 2011 for Hodgkin lymphoma after failure of autologous stem cell transplant (ASCT) or after failure of at least two prior multiagent chemotherapy regimens in patients who are not ASCT candidates and for sALCL after failure of at least one prior multiagent chemotherapy regimen. In September 2015, the FDA expanded its approval of brentuximab vedotin to post-autologous hematopoietic stem cell transplantation consolidation treatment of patients with Hodgkin lymphoma who are at risk of relapse or progression.

CD30 is overexpressed in Hodgkin lymphoma, ALCL, cutaneous T-cell lymphoma, and mediastinal B-cell lymphoma. This relatively restricted expression and its lack of expression on most healthy cells make CD30 an attractive target for immunotherapy. CD30 limits the proliferative potential of autoreactive CD8 T cells protecting against autoimmunity and regulates apoptosis through nuclear factor kappa-light-chain-enhancer of activated B-cell (NF-κB) activation. TNF receptor-associated factors 2 and 5 (TRAF2 and TRAF5) interact with the CD30 receptor expressed by activated T and B cells, mediating the signaling that leads to activation of NF-κB. After binding to cells expressing CD30, the antibody-drug conjugate brentuximab linked to its toxic payload MMAE is internalized before MMAE reaches the lysosomes where it is released, disrupting microtubules and ultimately inducing apoptosis. Free MMAE is also capable of exerting a toxic effect on bystander cells. Note that the half-life of the antibody-drug conjugate is 4–6 days compared to 3–4 days for MMAE. Brentuximab vedotin does not exert its action via CDC or ADCC but it does induce phagocytosis.

The list of warnings and precautions issued by the FDA for brentuximab vedotin covers 11 concerns related to neuropathies, infusion-related reactions, hematologic toxicities, serious infections, TLS, hepatotoxicity, PML, embryofetal toxicity, and serious dermatologic reactions. In clinical trials with Hodgkin lymphoma and sALCL patients, 54 % of subjects experienced neuropathy, predominately peripheral neuropathy which is cumulative. Of these patients, only about half had complete resolution of the condition with 20 % showing no improvement. Although infusion reactions to brentuximab vedotin are mostly mild-moderate, a small number of cases of anaphylaxis have been reported. Severe neutropenia and grade 3 or 4 thrombocytopenia or anemia as well as febrile neutropenia occur. In clinical trials, myelosuppression in severe form showed an incidence of 9–21 %. Serious bacterial, fungal, and viral infections and opportunistic infections such as pneumonia and sepsis occur, and in patients with high tumor burden or a rapidly proliferating tumor, TLS may result. In clinical trials, the incidences of severe reactions and death were higher in patients with severe renal impairment; this may be due to higher exposure to MMAE compared to patients with normal renal function. It is a similar situation for patients with moderate to severe hepatic impairment, but in any case, it should be recognized that serious cases of hepatotoxicity, including some fatalities, have resulted from brentuximab vedotin therapy. Cases have occurred after the first dose or after rechallenge. A black box warning has been issued for PML after 13 post-marketing surveillance reports and three cases among 2000 brentuximab vedotin-treated clinical trial patients. PML and the mucocutaneous toxidermias, Stevens-Johnson syndrome and toxic epidermal necrolysis, constitute three serious adverse events that, although rare, are potentially lethal, therefore requiring awareness and vigilance in assessing new onset signs and symptoms of central nervous symptom and mucocutaneous abnormalities. Common adverse reactions (incidence ≥20 % and regardless of causality) recorded in clinical trials with Hodgkin lymphoma patients given brentuximab vedotin were neutropenia, peripheral sensory neuropathy, URTI, pyrexia, anemia, thrombocytopenia, fatigue, cough, nausea, vomiting, diarrhea, abdominal pain, and rash (Table 3.1). When brentuximab vedotin was given with bleomycin, pulmonary toxicity manifesting as interstitial infiltration and/or inflammation occurred indicating that this combination of drugs is contraindicated. The most common serious events experienced by Hodgkin lymphoma patients given brentuximab vedotin include peripheral motor neuropathy (4 %), abdominal pain (3 %), and pulmonary embolism, pneumonitis, pneumothorax, pyelonephritis, and pyrexia, all with an incidence of 2 %. Experience with sALCL patients given brentuximab vedotin showed that the most commonly occurring adverse reactions were neutropenia, peripheral sensory neuropathy, pyrexia, fatigue, nausea, diarrhea, pain, and rash. The most common serious events experienced by sALCL patients were septic shock, supraventricular arrhythmia, pain in extremity, and urinary tract infection, all with an incidence of 3 %. Other serious adverse events, but with a much lower incidence, were PML, TLS, and Stevens-Johnson syndrome. The safety of brentuximab vedotin was evaluated in 25 Hodgkin lymphoma patients with relapsing disease after allogeneic stem cell transplantation. The most frequent adverse events were cough, fatigue, and pyrexia (each 52 %), nausea and peripheral sensory neuropathy (each 48 %), and dyspnea (40 %). The most common adverse events of severity ≥grade 3 were neutropenia, anemia, thrombocytopenia, and hyperglycemia in that order. CMV was detected in five patients. Overall, the safety data for brentuximab vedotin shows that neutropenia, peripheral neuropathy, the risk of infections, and the worrying occasional case of PML are perhaps the most important adverse events provoked by the antibody-drug conjugate. Reports in the FAERS database at the end of 2012 included 40 cases of peripheral neuropathy (65 neurological disorders overall), 131 cases of infections, and 13 reports of PML. Other adverse events identified during postapproval use include febrile neutropenia, hepatotoxicity, serious and opportunistic infections, hyperglycemia, pancreatitis (including fatal cases), and toxic epidermal necrolysis (including fatal outcomes).

In phase II trials on patients with Hodgkin lymphoma and sALCL, ~7 % of subjects were found to develop serum antibodies to brentuximab vedotin that persisted, whereas 30 % of patients developed antibodies of a transient nature. In all cases the antibodies were directed to the mAb and not the attached MMAE. Patients who developed persistently positive antibodies experienced a higher incidence of infusion reactions; two such patients had to discontinue treatment. Thirty-six (62 %) of 58 patients with anti-brentuximab vedotin antibodies had at least one serum sample with neutralizing antibodies. Whether or not such antibodies affect the efficacy and safety of the antibody-drug conjugate is not yet known.

Alemtuzumab

Alemtuzumab (Campath®, MabCampath®), a humanized IgG1κ mAb with complementarity-determining regions from a rat mAb, is targeted to CD52 (Campath-1 antigen) (Tables 2.1 and 3.1), an abundantly expressed 21–28 kDa 12-amino acid glycoprotein of unknown function with a single *N*-linked oligosaccharide. The early name for alemtuzumab, Campath-1, came from the original rat antibodies utilized in the Pathology Department, University of Cambridge. The name Campath-1H was adopted after the rat antibody hypervariable sequences were grafted onto a human antibody framework. CD52 is expressed on mature normal and malignant B and T lymphocytes, monocytes, macrophages, NK cells, a subpopulation of granulocytes, and dendritic cells as well as lymphoid and male sexual organs but not on erythrocytes, platelets, and hematopoietic stem cells. A figure of 5×10^5 CD52 molecules/cells has been reported for human lymphocytes. Approved by both the FDA and EMA in 2001 for the treatment of B-cell CLL resistant to alkylating agents and later as a single agent for the disease, alemtuzumab as Campath® and MabCampath® was widely used in cancer therapy until 2012 when it was withdrawn from US and European markets although a number of CLL patients could still receive it through specific access programs and some off-label usage in cancer therapy remains. Since 2012, alemtuzumab has been relaunched as Lemtrada® for multiple sclerosis (see Chap. 2, Table 2.1, and Chap. 4, section "Alemtuzumab"). After binding to CD52 on leukemic cells, alemtuzumab exerts its antitumor action primarily by ADCC with a contribution from CDC and induction of apoptosis.

The list of generally nonserious adverse effects occurring in patients given alemtuzumab, many similar to other mAbs, is summarized in Table 3.1. Serious adverse effects do, however, occur. In addition to the immunosuppression that accompanies CLL, cytopenia, resulting from alemtuzumab's mode of action in destroying white blood cells, is also the major adverse event resulting from the mAb's use in treating CLL, hence the FDA boxed warning (Table 3.1). As well as a high incidence of neutropenia (75–85 %), febrile neutropenia and thrombocytopenia (serious in 57 % of cases) may occur, the latter causing hemorrhagic problems and purpura. Infections related to alemtuzumab therapy have been reported with incidences of up to 80 % and serious events have shown incidences as high as 50 %. Bacterial and viral infections are common and are a significant cause of death; in fact, fatalities among patients on alemtuzumab therapy are mostly connected to infections. Protozoal infections are occasionally seen, and opportunistic infections, including pneumocystis pneumonia, herpesvirus and CML infections, candidiasis, JC virus activation, aspergillus, and mucormycosis, are relatively frequent with incidences up to ~40 %. *Mycobacterium tuberculosis*, another possible risk organism for opportunistic infections, has been implicated in cases of tuberculosis in alemtuzumab-treated patients receiving a renal transplant from deceased patients. A retrospective study of nonbacterial infections in 182 Asian patients treated with alemtuzumab showed that the most common infections were reactivation of CML (36.3 %), fungal infections (17 %), and varicella zoster virus (13.7 %).

The authors recommended routine prophylaxis with antiviral drugs, especially for patients receiving allogeneic stem cell transplants. The possibility of infection and a life-threatening outcome during alemtuzumab therapy was emphasized.

The destruction of T cells by alemtuzumab with resultant lymphopenia and consequent release of cytokines can lead to CRS, and potentially nephrotoxic TLS may occur as a result of the rapid and massive destruction of neoplastic cells. Infusion reactions to alemtuzumab, manifesting as pyrexia, chills, nausea, emesis, dyspnea, rash, and urticaria, predictably occur most commonly during the first week of treatment. Serious reactions that can be fatal are known. Signs and symptoms include bronchospasm, ARDS, pulmonary infiltrates, cardiac disorders, angioedema, and anaphylaxis. In neoplastic off-label applications of alemtuzumab therapy, for example, in T-cell lymphoma patients, severe pancytopenia associated with hemophagocytosis and viral reactivations has been noted. Of 4000 reports to the FAERS database during postmarketing surveillance, the largest group covered various infections with 40 % due to viruses, particularly CML, and with fungal infections also prominent. Other frequently reported events were white blood cell abnormalities and immune disorders (Goodpasture's syndrome, Graves' disease, aplastic anemia, Guillain-Barré syndrome, serum sickness, chronic inflammatory demyelinating polyradiculoneuropathy). Idiopathic thrombocytopenic purpura, TLS, and PML each made up from 0.7 to 1.1 % of the reports. Other reported adverse events were cardiomyopathy, decreased ejection fraction, and optic neuropathy.

Antihuman antibodies have been detected in 11 of 133 (8.3 %) patients previously untreated with alemtuzumab and in 4 of 211 (1.9 %) previously treated patients. Two patients in the former group proved positive for neutralizing antibodies to the mAb. The limited available data suggest that human antibodies to alemtuzumab do not adversely affect the mAb's antitumor action. In a novel approach to reduce the immunogenicity of alemtuzumab, an attempt was made to tolerize patients by pretreatment with a high dose of a noncell-binding variant of the mAb differing by a single point mutation in the H2 loop of the H chain. After two cycles of alemtuzumab treatment with and without the variant antibody, anti-alemtuzumab antibodies were found in 145 of 197 patients (74 %) who did not receive the variant protein and in 4 of 19 patients (21 %) in which tolerization was attempted. See also alemtuzumab (Lemtrada®), Chap. 4, section "Alemtuzumab."

Monoclonal Antibodies Targeting Epidermal Growth Factor Receptor: Cetuximab, Panitumumab, and Necitumumab

Tyrosine kinases are important targets for cancer therapy because of their central role in growth factor signaling leading to cell proliferation, differentiation, and survival. One of the four related growth factor receptors, epidermal growth factor receptor (EGFR, HER1, ErbB1), is a member of the ErbB family of receptors, a subfamily of closely related receptor tyrosine kinases. EGFR is a 170 kDa transmembrane glycoprotein cell surface receptor activated by the binding of its specific ligands, epidermal growth factor (EGF), transforming growth factor alfa (TGFα), heparin-binding EGF-like growth factor (HB-EGF), amphiregulin, betacellulin, epigen, and epiregulin. After ligand binding, EGFR undergoes homodimerization although EGFR may also form a heterodimer with another member of the ErbB receptor family, for example, ErbB2 (HER2). After dimerization, the receptor-ligand complex is internalized, the autophosphorylation occurs, and the tyrosine kinase signal transduction pathways lead to regulation of gene transcription involved with cell growth and survival, motility, and proliferation. Signal transduction, resulting from natural ligand-induced activation of the EGFR, leads to activation of the wild-type KRAS protein. In tumors with *KRAS* mutations, the mutant KRAS protein is continually active and appears to be independent of regulation by EGFR. In addition to the importance of EGFR in the processes of normal cellular functions and survival, expression of EGFR may contribute to the development of cancerous cells through effects on angiogenesis, cell cycle progression, inhibition of apoptosis, and metastasis. In a number of different tumors, EGFR and its ligands are associated with growth of the cells, and elevated EGFR tyrosine kinase activity is found in many, if not most, solid tumors including breast, renal, head and neck, colon, non-small cell lung cancer, pancreatic, ovarian, prostate, glioma, and bladder carcinomas. The expression of EGFRs on malignant cells is often greater than the numbers expressed on normal cells. Table 3.2 lists the percentages of different types of tumors overexpressing the EGFR. For cancers such as head and neck, non-small cell lung cancer, and renal carcinoma, most of the tumors express the receptor. The EGFR-binding mAbs cetuximab and panitumumab bind the receptor on both normal and tumor cells, competitively inhibiting binding of the normal ligands. Ligand-induced

Table 3.2 EGFR overexpression of some of the most common human solid tumors

Tumor	Percentage of tumors overexpressing EGFR
Breast	15–37[a]
Renal carcinoma	50–90
Head and neck	80–100
Colorectal	25–100[b]
Non-small cell lung carcinoma	40–80
Pancreatic	30–50
Ovarian	35–70
Glioma	40–92[c]
Bladder	31–48
Gastric	33–81
Prostate	40–90

EGFR epidermal growth factor receptor
[a]Other percentages quoted, 14–91 %
[b]25–77 % also quoted for colon cancer
[c]Other percentages quoted, 40–63 %

autophosphorylation of the receptor and activation of receptor-associated kinases are thus prevented, resulting in inhibition of cell growth and decreases in proinflammatory cytokines and the production of vascular growth factor.

Resistance to EGFR-Targeted Monoclonal Antibodies

Anti-EGFR mAbs such as cetuximab bind to the EGFR with a 5–10 times higher affinity than the natural ligands, preventing both binding of these ligands and subsequent activation of tyrosine kinase-mediated signal transduction pathways. Monoclonal antibodies targeting the EGFR were therefore seen as a new approach for treating a range of solid tumors, one that could potentially avoid some of the limitations and safety issues associated with conventional cytotoxic chemotherapy and radiation therapy. As it turned out, toxicity profiles of the mAbs are different to the toxicities of the conventional therapies (see below), but, disappointingly, mAbs along with other targeted cancer therapies have often proved efficacious in only a minority of patients with beneficial response rates varying from 10 % to about 90 %. This situation is clearly apparent with cetuximab and panitumumab both of which bind to, and inhibit, EGFR signaling. The absence of a significant association between EGFR levels and a clinically improved response to cetuximab in colorectal cancer patients led to the findings that patients with mutations in codons 12 and 13 of the *KRAS* gene almost never benefit from cetuximab and panitumumab treatments, but 10–30 % of patients with the wild-type gene respond to the mAbs. As a result of this understanding, regulatory agencies restricted the use of cetuximab and panitumumab to colorectal cancer patients expressing the wild-type *KRAS* gene, and the American Society of Clinical Oncology issued a provisional clinical opinion stating: "patients with metastatic colorectal cancer who are candidates for anti-EGFR antibody therapy should have their tumor tested for *KRAS* mutations. If a *KRAS* mutation in codon 12 or 13 is detected, then patients with metastatic colorectal carcinoma should not receive anti-EGFR antibody therapy as part of their treatment." Note, however, the codon 12 and 13 mutations of *KRAS* account for only ~40 % of the unresponsive patients, so it must be assumed that other mutations might be conferring resistance, and this appears to be the case with mutations in the *BRAF* (B-RAF proto-oncogene serine/threonine kinase), *PI3KCA* (phosphatidylinositol-4-5-biphosphate 3-kinase, catalytic subunit alfa), *NRAS* (neuroblastoma RAS viral oncogene homolog), and *PTEN* (phosphatase and tensin homolog) genes being significantly associated with low response rates. Mutational analysis of *KRAS* and *PI3KCA* and evaluation of the PTEN protein status in patients with metastatic colorectal cancer treated with cetuximab and panitumumab showed 29 % *KRAS* and 13.6 % *PI3KCA* mutations. The latter were significantly associated with clinical resistance to the mAbs and patients had a worse clinical outcome for survival. The analysis revealed that up to 70 % of patients with metastatic colorectal cancer are unlikely to benefit from treatment with anti-EGFR antibodies.

Cetuximab

Cetuximab (Erbitux®) (Tables 2.1 and 3.1) is a recombinant chimeric human-mouse IgG1κ mAb, MW 152 kDa, that binds the extracellular domain of human EGFR. It is prepared by incorporating the Fv regions of a mouse anti-EGFR antibody with the constant regions of human IgG1 H and L immunoglobulin chains. The structural basis of cetuximab's potent inhibition of the EGFR has been identified and shown to be due to two effects: X-ray crystal structure of the Fab antigen-binding fragment of cetuximab in complex with the soluble extracellular region of EGFR (sEGFR) revealed that the mAb interacts with domain III of sEGFR, interfering with access to the ligand-binding region and sterically preventing the receptor from undergoing dimerization.

Cetuximab is approved by the FDA and EMA for the treatment of metastatic colorectal cancer and head and neck cancer. Specifically, cetuximab's approval for colorectal cancer is as a single agent for treating EGFR-expressing tumors after failure of irinotecan- and oxaliplatin-based regimens or in patients intolerant to irinotecan. Additionally, cetuximab is approved for use in combination with irinotecan for patients' refractory to irinotecan therapy. In head and neck cancer, approval is specified for the treatment of squamous cell carcinoma in combination with radiotherapy and for recurrent or metastatic squamous cell carcinoma in combination with platinum- and 5-fluorouracil-based therapy or carcinoma progressing after platinum therapy. The addition of cetuximab in these combination therapies produces an increase in antitumor effects compared to the nonantibody therapies alone. Binding of cetuximab prevents natural ligand-induced receptor autophosphorylation and the activation of receptor kinases, resulting in inhibition of cell growth, apoptosis, and decreases in matrix metalloproteinases, proinflammatory cytokines, the chemokine IL-8 (CXCL8), and vascular endothelial growth factor production. Lowering of VEGF serum levels by cetuximab indicates that antibody-induced VEGF inhibition may be associated with some antitumor activity.

Serious infusion reactions and the possibility of cardiopulmonary arrest make up an FDA boxed warning for cetuximab. The severity and risk of many of the infusion reactions are underlined by the rapid onset of symptoms of airway obstruction, hypotension, shock, cardiac arrest, myocardial infarction, and loss of consciousness, severe reactions (grades 3 and 4) that occurred in 2–5 % of 1373 patients, and the fact that 90 % of the severe cases occurred with the first infusion despite premedication. In relation to anaphylaxis to cetuximab, IgE antibodies to α-D-galactose-(1-3)-β-D-galactose were found in an alarming number of patients, especially in Southern USA, who experienced severe immediate hypersensitivity reactions after receiving the mAb. The disaccharide recognized by the IgE antibodies was shown to be present on the Fab portion of the chimeric antibody at asparagine 88 of the heavy chain, and, interestingly, most of the allergic patients already had the antibodies in their serum before receiving the mAb. The discovery of delayed-onset anaphylaxis, angioedema, and urticaria after consuming red meat and being linked to tick bites and the presence of the alfa-linked disaccharide in red meat provide a possible explanation for this intriguing phenomenon. At least one protocol for the successful desensitization of cetuximab-induced immediate reactions has been published. Premedication was commenced with prednisolone 12 h and 1 h before and diphenhydramine 30 min before the start of the procedure. The initial infusion dose of cetuximab was 1 μg, doses were doubled every 15 min until a total of 64 mg was achieved, and then a final dose of 325 mg was administered to give a total cumulative dose of 844 mg. The appearance of cutaneous reactions was managed with diphenhydramine, dose and infusion reductions, and a 30 min pause in the ongoing procedure.

Cardiopulmonary arrest and/or sudden death, the second subject of the boxed warning, occurred in 2–3 % of patients treated with cetuximab and radiation or cetuximab and platinum therapy. Most of the fatal cases were in patients with a history of cardiac disease. Other warnings and precautions issued by regulatory agencies are for the possibility of interstitial lung disease which was recorded in 4 of 1570 patients (0.25 %); a variety of dermatologic events already known to occur, for example, "acneiform" rash (with an incidence of 76–88 %, up to 17 % serious), xerosis and fissuring, paronychial inflammation, hypertrichosis, and infectious sequelae such as cellulitis and conjunctivitis; and hypomagnesemia and electrolyte abnormalities. Hypomagnesemia occurred in 55 % of 365 clinical trial patients being given cetuximab, and the condition proved severe (grades 3 and 4) in 6–17 % of participants. Of the more commonly reported adverse events (Table 3.1), gastrointestinal symptoms, infections, electrolyte imbalance, and respiratory and cutaneous disorders rank amongst the highest incidences in both clinical trial results and postmarketing surveillance reports. As with small molecule tyrosine kinase inhibitors such as erlotinib and gefitinib, anti-EGFR antibodies like cetuximab and panitumumab frequently provoke so-called "acneiform" rashes (more correctly, papulopustular eruptions) in a large proportion of patients. These reactions are generally more severe with the mAbs than with the small molecule drugs and tend to be confined to seborrheic regions of the face, scalp, neck, shoulders, and upper trunk (Fig. 3.3a–c). EGFR is expressed by cells in these regions and it is thought that its inhibition leads to defects in the epithelial barrier, allowing the entry of bacteria and ultimately the development of the characteristic rash. Addition of EGFR inhibition to radiotherapy may lead to radiation dermatitis enhancement, producing wet or dry desquamation, necrosis, or cutaneous ulceration (Fig. 3.3d, e). Other far less frequent mucocutaneous adverse effects include pruritus, palmar-plantar rash, telangiectasia, trichomegaly, alopecia, hyperkeratosis, pyrogenic granuloma, skin hyperpigmentation, and mucositis. Severe mucositis is uncommon with EGFR inhibition therapy alone, but it is more likely to occur in combination with cytotoxic chemotherapy or radiotherapy (Fig. 3.3f, g). EGFR inhibition can also affect differentiation of keratinocytes, leading to a decrease in loricrin and the development of xerosis and skin fissures (Fig. 3.3h). Paronychia, which may become superinfected, is a risk for all patients receiving anti-EGFR mAb therapy. As well as paronychia, other nail changes such as periungual pyogenic granuloma may occur (Fig. 3.3i).

Of about 47,000 adverse events recorded in the FAERS database, pneumonia is the most frequently reported event followed by febrile neutropenia. Hypersensitivities include 66 reports of anaphylaxis, and there are 88 cases of

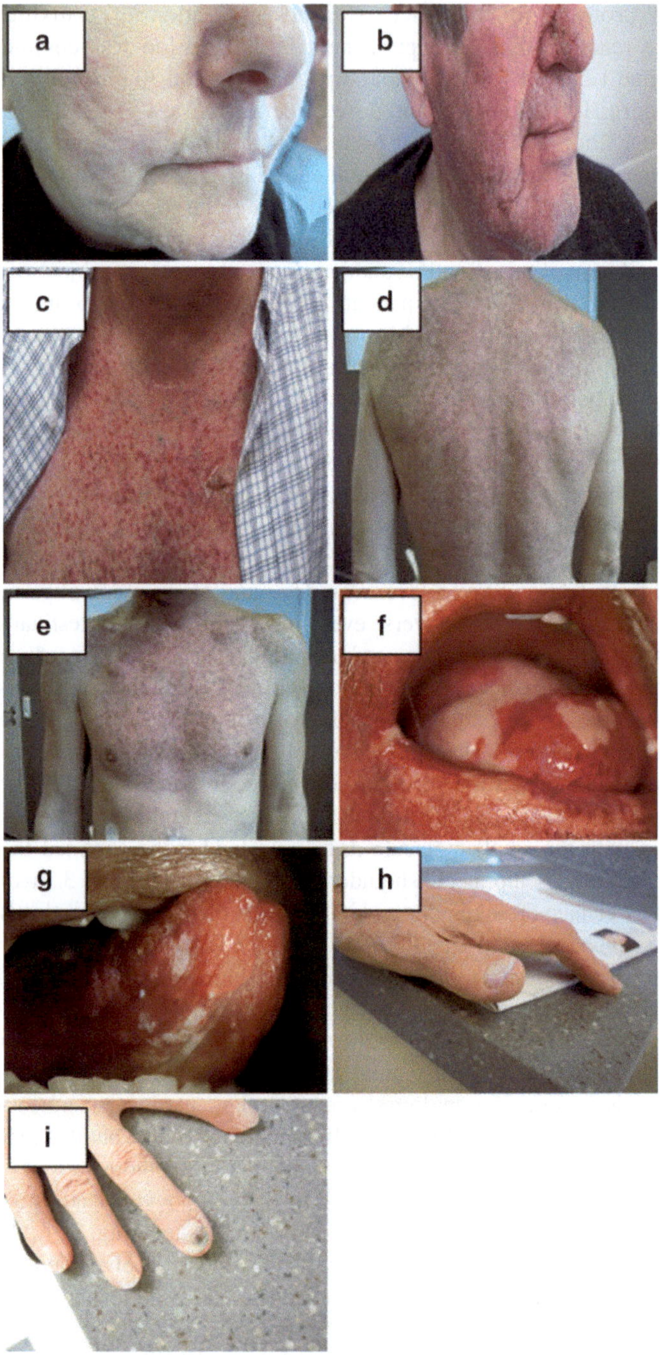

Fig. 3.3 The typical acneiform rash or papulopustular eruption caused by tyrosine kinase inhibitors that bind EGFR manifests as erythematous pruritic papules and pustules. The rash tends to be

papulopustular eruptions, three cases of aseptic meningitis, and two cases of PML. Some obvious discrepancies from these results are found in the European pharmacovigilance database covering nearly 6000 reports and 13,656 adverse events, in particular, records of 997 infusion-related reactions, 585 hypersensitivity responses, 354 anaphylactic reactions, and 38 anaphylactoid reactions. The surprisingly high number of cases of anaphylaxis calls into question the criteria employed to classify the reactions as true, type I IgE antibody-mediated hypersensitivities. Without skin test and rituximab-reactive IgE antibody data on each individual, a diagnosis of anaphylaxis could not be firmly established. Aseptic meningitis is mentioned in 19 reports and PML in two.

Although cetuximab has been reported to trigger a relatively large number of hypersensitivity responses, especially during first infusion sessions, the mAb appears to exhibit rather low immunogenicity. Non-neutralizing human antichimeric antibodies appear to have an incidence of ~4–5 % in treated patients. At least some of these antibodies may be preexisting, in particular so-called "natural" antibodies, mostly of the IgG class specific for the α-1,3-linked D-galactose disaccharide (see above) and antibodies to murine-derived *N*-glycolylneuraminic acid, a structure present in cetuximab but absent in humans.

Panitumumab

Panitumumab (Vectibix®) (Tables 2.1 and 3.1) is a recombinant, fully human IgG2κ mAb that binds the EGFR with high affinity. Like cetuximab, panitumumab is approved for the treatment of colorectal cancer but not yet for head and neck cancer, and it is currently under investigation for the treatment of malignant glioma. For colorectal cancer, both mAbs are equally effective. The antigenic structures recognized by cetuximab have been identified as a large surface conformation on domain III of the EGFR (see above, section "Cetuximab"), and although the complementary structures to the panitumumab combining sites remain poorly defined, it is already clear that both epitopes are not identical. This conclusion is supported by instances of effective treatment with panitumumab in patients with disease progression under cetuximab and the development of resistance to treatment with cetuximab in a colorectal cancer patient who acquired a point mutation in the EGFR domain (Arg

Fig. 3.3 (continued) confined to seborrheic regions of the face, scalp, neck, shoulders, and upper trunk (**a**–**c**). Reactions may occur in 50–100 % of treated patients and tend to be more severe and widespread with EGFR-targeted monoclonal antibodies than small molecule tyrosine kinase inhibitors. When EFGR inhibitors are administered with radiotherapy, a high incidence of radiation dermatitis showing erythema, desquamation, skin necrosis, or ulceration with bleeding may occur (**d**, **e**). Oral complications, most commonly mucositis presenting as stomatitis, are infrequent (**f**, **g**). Skin fissures and cracks (rhagades) may result from xerosis (**h**). Nail changes in the form of paronychia (nail fold inflammation) and periungual pyrogenic granuloma-like lesions may also be seen (**i**). Reproduced from Lacouture ME et al. Clinical practice guidelines for the prevention and treatment of EGFR inhibitor-associated dermatologic toxicities. Support Care Cancer 2011;19:1079–95. Reprinted with permission from Springer Science+Business Media

for Ser at position 468), while panitumumab binding and efficacy remained. It has been suggested that characterization of the binding sites of the two anti-EGFR mAbs could help predict the response to the mAbs in patients with mutations leading to resistance. In an effort to achieve this for panitumumab, epitope recognition was assessed by screening phage display peptide libraries. This approach identified a discontinuous epitope that overlapped with the cetuximab epitope. The overlapping epitopes were shown to consist of 17 amino acids, four of which are targeted by cetuximab and four others by panitumumab. The authors believe that these results have the potential to improve treatments by using the recognition findings to help select patients for EGFR-targeted therapies.

Granted accelerated approval by the FDA in 2006 as a single agent "for the treatment of metastatic colorectal carcinoma with disease progression on or following fluoropyrimidine, oxaliplatin, and irinotecan chemotherapy regimens," panitumumab was subsequently approved by the EMA in 2007 for treatment of wild-type KRAS metastatic colorectal cancer as monotherapy after failure of fluoropyrimidine-, oxaliplatin-, and irinotecan-containing chemotherapy regimens. In 2009 the FDA restricted the indication to patients with wild-type KRAS tumors, and in 2011 the EMA extended the indication to first-line treatment in combination with FOLFOX (folinic acid [leucovorin], fluorouracil, oxaliplatin) or FOLFIRI (folinic acid, fluorouracil, plus the topoisomerase inhibitor irinotecan) and as second-line treatment in combination with FOLFIRI for patients who have received first-line fluoropyrimidine-based chemotherapy, excluding irinotecan. In 2014, the FDA revised its indications for panitumumab to include a combination of the mAb with FOLFOX for first-line treatment.

At first approval of panitumumab in 2006, the FDA issued a boxed warning for dermatologic toxicity and infusion reactions stating that dermatologic toxicities occurred in 89 % of patients with 12 % being CTC grade 3 or higher severe reactions (Chap. 1, section "Terminology: Adverse Reactions and Adverse Events"). Severe infusion reactions were said to occur in ~1 % of patients. In the revised prescribing information issued by the FDA in August 2014, reference to infusion reactions in the boxed warning was removed and dermatologic toxicities were stated to be severe in 15 % of patients. Mucocutaneous diseases provoked by panitumumab contribute to an extensive range of clinical manifestations including erythema, rash, pruritus, skin exfoliation, acneiform dermatitis, xerosis, paronychia, and skin fissures as well as life-threatening infectious complications such as necrotizing fasciitis and abscesses. Life-threatening bullous mucocutaneous diseases with erosions, blisters, and skin sloughing following panitumumab are known, but it is not always easy to ascribe these to antibody-induced inhibition of EGFR or to immune-related drug-induced toxidermias, for example, Stevens-Johnson syndrome or toxic epidermal necrolysis. Other warnings and precautions associated with the use of panitumumab (Table 3.1) relate to the possibilities of severe infusion reactions (grade 3–4) with the reminder that fatal reactions have occurred; severe hypomagnesemia (which showed a 7 % incidence in clinical trials) as well as hypocalcemia and hypokalemia; acute renal failure resulting from severe diarrhea and dehydration when panitumumab is used in combination with chemotherapy; cases, some fatal, of interstitial lung disease and pulmonary fibrosis; ocular toxicities such as keratitis and ulcerative keratitis; dermatologic toxicities caused by sunlight exposure as a result of panitumumab-induced photosensitivity; and the possibility of increased toxicity and mortality when panitumumab is administered in combination with bevacizumab and chemotherapy. In the EMA summary of the safety profile of panitumumab, adverse reactions occurring in ≥20 % of patients are made up of cutaneous disorders, gastrointestinal disorders (diarrhea, nausea, vomiting, constipation, abdominal pain), general disorders (fatigue, pyrexia), infections and infestations, and anorexia. Importantly, when panitumumab was used in combination with chemotherapy, the safety profile was assessed as the adverse events seen with the mAb as monotherapy plus the toxicities of the chemotherapy regimen. When this was done, no new toxicities or worsening of previously recognized toxicities beyond the expected additive effects were reported by the EMA. The most common adverse reactions (≥20 %) listed by the FDA are skin rashes of variable presentation, paronychia, fatigue, nausea, and diarrhea. Other common adverse reactions showing a ≥5 % difference compared to the reactions seen in patients given the best supportive care alone were stomatitis, mucosal inflammation, dyspnea, cough, and an extensive range of skin and subcutaneous tissue disorders.

An interesting recent case report describes the safe treatment with panitumumab of a patient in need of anti-EGFR therapy for rectal carcinoma but who was also allergically sensitized to the disaccharide α-D-galactose-(1-3)-β-D-galactose (see section "Cetuximab"). After allergological testing showed that the patient had α-D-galactose-associated red meat and gelatin allergy and was skin test negative to panitumumab, intraoperative use of gelatin-derived colloids and treatment with cetuximab, both known to carry α-D-galactose residues, were prohibited and panitumumab as anti-EGFR therapy was successfully initiated.

Postmarketing surveillance reports on panitumumab to the FAERS database at the end of 2012 totaled 3987 with 13,830 events. Dermatologic events showed the highest incidence (8.1 %), followed by infections (7.2 %), electrolyte problems (4.7 %), and gastrointestinal disorders (4.2 %). In order of the most frequent events, diarrhea headed the list with 605 reports followed by acneiform dermatitis (595), interstitial lung disease (310), febrile neutropenia (277), hypomagnesemia (276), rash (193), and sepsis (108). Six cases of anaphylaxis, 12 of maculopapular rash, and six of toxic epidermal necrolysis were recorded and angioedema appears to have been reported for the first time. Eye disorders included 25 cases of conjunctivitis and six of keratitis. Postmarketing surveillance reports to the European pharmacovigilance database showed a similar spectrum of adverse events, but incidences of cutaneous reactions (25.8 %), gastrointestinal disorders (9.8 %), and respiratory disorders (8.4 %) were higher.

Immunogenicity of panitumumab appears to be low, a conclusion in keeping with the fully human nature of the antibody. Detection of antibodies to the mAb in 1123 patients receiving panitumumab monotherapy showed an incidence of 0.4 % (five patients) when measured by an ELISA and 3.2 % (36 patients) in a Biacore® assay. Nine patients (0.8 %) had anti-panitumumab neutralizing antibodies. In combination with chemotherapy, the ELISA assay detected an incidence of anti-panitumumab antibodies of 0.9 % (12 of 1297), while the incidence found with the Biacore® assay was 0.7 % (9 of 1296). Only 0.15 % of patients (2 of 1297) had neutralizing antibodies, and, overall, no safety concerns were found in the patients who developed antibodies. In four clinical trials assessing panitumumab plus chemotherapy, antibodies to the mAb, including neutralizing antibodies, were detected with two different immunoassays and a bioassay. Twenty patients (1.8 %) treated with panitumumab in combination with oxaliplatin or irinotecan therapy developed antibodies and two (0.2 %) developed neutralizing antibodies. Patients with tumors expressing wild-type or mutant *KRAS* and patients receiving oxaliplatin or irinotecan therapies showed similar incidences of anti-panitumumab antibodies. The presence of anti-panitumumab antibodies had no noticeable effect on safety or pharmacokinetic profiles.

Necitumumab

Necitumumab (Portrazza®) (Tables 2.1 and 3.1), approved by the FDA in November 2015, is a recombinant human IgG1κ mAb, MW~144.8 kDa, targeted to human EGFR. Indicated for first-line treatment of patients with metastatic squamous non-small cell lung cancer in combination with gemcitabine and cisplatin, necitumumab carries FDA black box warnings for cardiopulmonary arrest and/or sudden death and for hypomagnesemia (Table 3.1). The former has been seen in ~3 % of patients treated with the mAb plus gemcitabine and cisplatin; the latter in more than 80 % of patients treated with the mAb-small molecule drug combination. Cardiopulmonary arrest or sudden death occurred in 15 of 538 patients (2.8 %) treated with necitumumab-gemcitabine-cisplatin compared to three of 541 (0.6 %) of patients given gemcitabine and cisplatin alone. Hypomagnesemia occurred in 83 % of 461 patients treated with necitumumab (20 % severe) compared to 70 % treated with the small MW drugs alone (7 % severe). Patients should be monitored for hypomagnesemia, hypocalcemia, and hypokalemia. Five other warnings/precautions issued for necitumumab are the possible occurrence of venous and arterial thromboembolic events; some fatal, infusion-related reactions; embryofetal toxicity; increased toxicity and mortality in patients with non-squamous non-small cell lung cancer treated with the mAb plus pemetrexed and cisplatin; and dermatologic toxicities. Skin toxicity, generally developing within the first 2 weeks and severe in 8 % of patients, manifests as generalized rash, dermatitis acneiform, xerosis, pruritus, maculopapular rash, and erythema in up to ~79 % of patients receiving necitumumab. Other adverse reactions to necitumumab recorded so far (Table 3.1) are essentially those seen in clinical trials; postmarketing experience will undoubtedly add to this so far small list.

Bevacizumab

Bevacizumab (Avastin®) (Tables 2.1 and 3.1) is a recombinant humanized IgG1κ mAb, MW~149 kDa, that binds to, and inhibits, the biological action of human vascular endothelial growth factor-A (VEGF-A) (Fig. 3.4). VEGFs are a family of secreted proteins with a highly conserved receptor-binding cystine-knot structure comprising five members. Discovered in 1983 and often described as the "founding member" of the family, VEGF-A (also called VEGF and vascular permeability factor [VPF]) is regarded as the most important regulator of the formation of blood vessels in health and disease. The five members of the VEGF family are distinguished as VEGF-A, VEGF-B, VEGF-C, VEGF-D, and placenta growth factor (PlGF, PGF). Before the discovery of the different family members, VEGF-A was simply called VEGF. VEGF-A binds to receptors VEGFR-1 (also called Flt-1) and VEGFR-2 (KDR/Flk-1) although it is the latter receptor that seems to mediate most, if not all, of the cellular responses to VEGF (Fig. 3.4). In promoting angiogenesis, VEGF-A (and other VEGFs) acts in endothelial cells through a family of cognate receptor tyrosine kinases. As well as its function of promoting blood vessel formation in healthy subjects, VEGF-A-induced angiogenesis has a major role in the pathogenesis of a wide range of human diseases, for example, cancers, rheumatoid arthritis, and eye diseases (see also ranibizumab, Chap. 4, section "Ranibizumab," and aflibercept Chap. 6, sections "Aflibercept" and "Aflibercept"). It was this understanding of the functions of VEGF-A that led to the development of bevacizumab and its subsequent application to an impressive range of approved cancer indications and off-label treatments.

VEGFR-1 recruits hematopoietic stem cells, while VEGFR-2 regulates vascular endothelial function. During VEGFR-2 intracellular signaling and according to MJ Cross, L Claesson-Welsh, and others, binding of VEGF-A to the receptor extracellular domain induces dimerization and autophosphorylation of specific intracellular tyrosine residues. Several intracellular proteins, for example, VEGFR-associated protein (VRAP), Sck, and phospholipase C (PLC-γ), bind to specific phosphorylated tyrosine residues in the receptor via their Src homology-2 (SH2) domains leading to phosphorylation and activation of these proteins. PLC-γ activation leads to hydrolysis of membrane phosphatidylinositol 4,5-biphosphate (PIP_2), a generation of second messenger diacylglycerol (DAG) an activator of protein kinase C, and inositol 1,4,5-triphosphate (IP_3) which binds to a receptor on the endoplasmic reticulum releasing intracellular stored Ca^{2+}. Many other proteins are activated including Src, phosphoinositide 3-kinase (PI3K), and p38 mitogen-activated protein kinase (p38 MAPK), and signal transduction downstream produces several physiologic and pathologic effects including proliferation, migration, permeability, and survival. To grow and proliferate, tumors need a matching blood supply. Increased expression of VEGF has been found in many human solid tumors, presumably fueling angiogenesis and assisting tumors to grow aggressively. By binding to VEGF-A, bevacizumab prevents activation of VEGFR-2 (Fig. 3.4) and the generation of new tumor vasculature. It also appears that bevacizumab enhances cytotoxic effects of concurrent chemotherapy.

Bevacizumab, in combination with fluoropyrimidine-based chemotherapy, was approved for metastatic colorectal cancer first by the FDA in 2004 and then by the EMA the following year. In subsequent years, approvals by both agencies were extended to metastatic breast cancer (with paclitaxel or capecitabine); non-squamous, non-small cell lung cancer (with a platinum agent); metastatic renal cell carcinoma (with interferon alfa-2a); and cervical cancer (with paclitaxel and cisplatin or pacli-

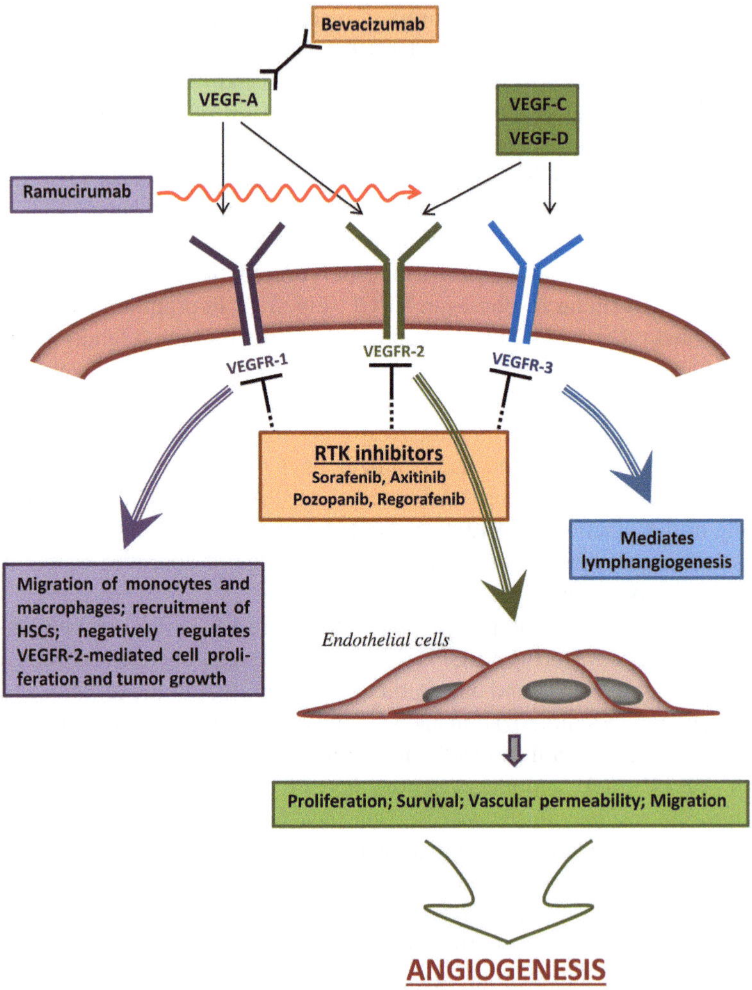

Fig. 3.4 Binding of VEGFR-2 (Flk1) to its ligand initiates receptor dimerization followed by autophosphorylation; activation of multiple downstream pathways including PLC-γ, MAPK, Akt, and Src; proliferation of endothelial cells; and ultimately, via PI3K, increased cell survival, cell migration, and vascular permeability mediated via endothelial nitric oxide synthase. VEGFR-1 (Flt1) is structurally similar to VEGFR-2 but unlike VEGFR-2 is kinase impaired. It may influence angiogenesis by acting as a decoy receptor, binding VEGF-A and preventing it from binding to VEGFR-2, or by heterodimerization with VEGFR-2. VEGFR-1 also binds exclusively VEGF-B and placental growth factor (PlGF). VEGFR-3 (Flt4) mediates lymphangiogenesis in response to VEGF-C and VEGF-D. Targets for bevacizumab, ramucirumab, and small molecule receptor tyrosine kinase inhibitors are shown. *HSCs* hematopoietic stem cells, *RTK* receptor tyrosine kinase

taxel and topotecan). The indication for metastatic breast cancer was withdrawn in 2010 by the FDA due to safety and efficacy concerns. Four adverse events, proteinuria, hypertension, left ventricular dysfunction, and hemorrhagic events, were found to show a statistically significant bevacizumab-associated risk in breast cancer therapy. Glioblastoma was added by the FDA (but not by the EMA) as an indication for bevacizumab monotherapy in 2009, and in 2012 the EMA, with certain requirements, approved bevacizumab in combination with chemotherapy for the treatment of epithelial ovarian, fallopian tube, and primary peritoneal cancers.

Although the lists of approved indications for bevacizumab vary between different regulatory agencies, the relatively wide variety of tumors (about 30, including investigative and preliminary studies) treated by this mAb and its combination with chemotherapy more or less ensures a large number and variety of consequent adverse events (Table 3.1). An extensive list of warnings and precautions is headed by an FDA black box warning for gastrointestinal perforation (which has an incidence of up to 3.2 % in treated patients), complications of surgery and wound healing, and severe or fatal hemorrhage. The incidence of bevacizumab-induced gastrointestinal perforation is estimated to be 0.7–1.2 % with a mortality rate of 11.5–37 %. Because bevacizumab has a half-life of 11–50 days, a delay of 5–8 weeks between bevacizumab treatment and surgery is recommended, and for postoperative initiation of bevacizumab therapy, an interval of 28 days after surgery with a fully healed incision is recommended. Although these delays were observed, a patient treated with bevacizumab before surgery and a second cycle of bevacizumab treatment developed a diaphragmatic rupture, a previously unreported serious adverse event. The case highlights the need for careful monitoring after postoperative administration of bevacizumab. Non-gastrointestinal fistulae, some fatal, involving vaginal, vesical, female genital tract, tracheoesophageal, biliary, renal, and bladder sites, also occur but are uncommon. Hemorrhage may be minor such as epistaxis or severe and sometimes fatal in the form of hemoptysis, gastrointestinal and vaginal bleeding, hematemesis, epistaxis, and CNS hemorrhage. Serious and fatal cases of pulmonary hemorrhage have been reported. Both arterial and venous thromboembolic events, some fatal, have been seen in patients receiving bevacizumab. Recorded arterial events include cerebral infarction, ischemic attacks, myocardial infarction, and angina. Cases of grade 3 or 4 hypertension have an incidence of 5–18 %, the incidence of proteinuria (frequency 21–63 %, serious 3 %) is increased in patients receiving the mAb, PRES has been reported with an incidence of <0.5 %, and although infusion reactions are not a major problem, some results indicate that anaphylactic and anaphylactoid reactions to bevacizumab occur more frequently in patients given the protein in combination with chemotherapy. Frequencies of severe neutropenia and serious infections are 21–26 % and 4–5 %, respectively, with pneumonia and wound and catheter infections prominent. Ovarian failure has been seen in premenopausal women receiving bevacizumab in combination with FOLFOX, emphasizing the need to inform females of reproductive age of the fertility risk prior to commencing treatment with the antibody. Table 3.1 lists some of the most common adverse events caused by bevacizumab administration at a rate of >10 % and at least twice the control rate; additions to this list include rhinitis, dysgeusia, and back pain. Other commonly seen adverse events (any grade) associated with bevacizumab include asthenia, pain, headache, diarrhea, nausea, vomiting, constipation, and stomatitis.

Considering postmarketing surveillance, an analysis of all 351 serious cases associated with bevacizumab and recorded in the French pharmacovigilance database up to the end of 2010 revealed that reactions of the gastrointestinal tract (21.9 %), thromboembolic events (4 %), pulmonary embolism (3.2 %), hypertension (2.7 %), gastrointestinal hemorrhage (2.7 %), and cerebral hemorrhage or vascular accident (2.6 %) were the most frequently reported adverse events. Whereas adverse reactions occurred within a median duration of four cycles, nine of 18 deaths due to an adverse reaction occurred after only one cycle. Reactions causing disability were mainly neurologic in origin (frequency 40 %), especially neuropathy, paralysis, and paresis. A search of the FAERS database for the period 2004–2009 for novel adverse events to bevacizumab revealed the highest number of reports for electrolyte abnormalities followed by cardiovascular events and pneumonitis. Clinically important but unlabeled disorders included necrotizing fasciitis, vessel wall disorders, arrhythmia and conduction disorder, and autoimmune thrombocytopenia. Of 37,000 reports on bevacizumab in the FAERS database at the end of 2012, infections (6.2 %), gastrointestinal (5.5 %), hematologic (3.8 %), and respiratory (3.5 %) events showed the highest incidences. More specifically, the postmarketing period has seen reports for polyserositis, PRES, venous occlusion, gallbladder perforation, nasal septum perforation, arterial thromboembolic events, hemorrhage, and numerous eye disorders (from unapproved intravitreal use), for example, permanent vision loss, endophthalmitis, intraocular inflammation, retinal detachment, increased intraocular pressure, hemorrhage, vitreous floaters, and ocular hyperemia. In the European pharmacovigilance database at the end of 2012, gastrointestinal (incidence 20.4 %), neurologic (8 %), infectious (5 %), hematologic and malignant (4 %), cutaneous (3 %), and renal (2.5 %) disorders made up the most frequent adverse events of the 17,672 mostly serious reports. Hypertension contributed 597 cases, deep vein thrombosis 367, gastrointestinal perforation 333, proteinuria 243, sepsis 172, pneumonia 138, and acute renal failure 133, and there were 86 reported cases of nephrotic syndrome. Hypersensitivity or hypersensitivity-like responses showing the highest frequencies were anaphylaxis with 89 cases and anaphylactoid reactions with 39 cases. For mucocutaneous reactions, palmar-plantar erythrodysesthesia was the most common adverse event with 136 cases. Other less commonly seen adverse events reported during the postmarketing period include various ocular disorders, mesenteric venous occlusion, gastrointestinal ulcer, intestinal necrosis, anastomotic ulceration, gallbladder perforation, osteonecrosis of the jaw, nasal cavity lesions including septum perforation, and dysphonia.

There appears to be a dearth of data on the immunogenic activity of bevacizumab in clinical trials and over the postmarketing years. It has been stated that high titers of antibodies to the mAb were not found in 500 treated patients and the FDA has referred to the detection of anti-bevacizumab serum antibodies in 14 of 2233 (0.63 %) colon cancer patients. Three of the patients had neutralizing antibodies for the mAb, but their clinical significance was not determined.

Ramucirumab

Ramucirumab (Cyramza®) (Tables 2.1 and 3.1), a recombinant fully human IgG1κ mAb, MW ~147 kDa, prepared using a phage display library and affinity maturation selection, has binding specificity for VEGFR-2 [CD309, also known as kinase insert domain-containing receptor (KDR)], the receptor-mediating angiogenesis. VEGF-A and VEGFR-2 are often upregulated in a number of human diseases including cancers, and since uncontrolled and sustained angiogenesis is a major contributor, if not promoter, of tumor growth, targeted inhibition of the development of blood vessels is an accepted antitumor strategy. This strategy can be pursued in at least four ways: by administering small molecule, usually multitargeted, tyrosine kinase receptor inhibitors with anti-angiogenic specificity (e.g., sorafenib, sunitinib, axitinib, pazopanib, and regorafenib) (Fig. 3.4); by the selective VEGF antagonist, the pegylated oligonucleotide pegaptanib (Macugen®; indicated for wet age-related macular degeneration and the only aptamer currently approved for clinical use by the FDA), which binds to extracellular VEGF thereby inhibiting its binding to its receptor; by mAbs binding VEGF as practiced with bevacizumab binding VEGF-A (section "Bevacizumab") or the fusion protein aflibercept which targets VEGF-A, VEGF-B, and PlGF; or by targeting the VEGFR-2 receptor, blocking not only the VEGF-A but also other growth factors including VEGF-C, VEGF-D, and VEGF-E. Ramucirumab is such an antibody, targeting VEGFR-2 with high affinity and thereby inhibiting an array of biological activities including receptor activation and signaling, intracellular Ca^{2+} mobilization, and proliferation and migration of endothelial cells (Fig. 3.4). Tumor angiogenesis, a highly complex process, is the ultimate downstream result of VEGFR-2 activation and signaling. It begins with binding of VEGFR-2 to its ligand which initiates receptor dimerization and leads to intracellular autophosphorylation; activation of multiple downstream pathways including PLC-γ, MAPK, Akt, and Src; proliferation of endothelial cells; and, ultimately, via PI3K, increased cell survival, cell migration, and vascular permeability mediated via endothelial nitric oxide synthase.

Approved by the FDA in April 2014, ramucirumab has had a history of rapid additions to its original indication as a single agent for advanced gastric cancer or gastroesophageal junction adenocarcinoma after fluoropyrimidine- or platinum-containing chemotherapy. In November 2014, the FDA added approval for the option of the addition of paclitaxel to the monotherapy; in December 2014 approval was granted for combination therapy with docetaxel for treatment of metastatic non-small cell lung cancer (on or after platinum-based chemotherapy); and, most recently, in April 2015, approval was forthcoming for the use of ramucirumab with FOLFIRI in second-line treatment of colorectal cancer after prior treatment with bevacizumab, oxaliplatin, and fluoropyrimidine. Granted orphan drug status for gastric cancer by the EMA in 2012, ramucirumab in monotherapy and combination therapy with paclitaxel was, as with the FDA, approved by the agency in December 2014 for advanced gastric and gastroesophageal junction cancers. Ramucirumab is currently undergoing evaluation for hepatocellular carcinoma and breast cancer in phase III studies, so its list of approved indications and usage may be further expanded.

Safety data accumulated so far for ramucirumab is headlined by an FDA boxed warning for an increased risk of hemorrhage (including gastrointestinal hemorrhage that may be severe and sometimes fatal), gastrointestinal perforation, and impaired wound healing. Risk of the latter two adverse events is known to sometimes increase during anti-angiogenic therapy. Both the FDA and EMA have issued a number of other warnings. Serious and sometimes fatal arterial thrombotic events, including myocardial infarction, cardiac arrest, cerebrovascular accident, and cerebral ischemia, were seen in clinical trials, for example, in 1.7 % of 236 patients who were given ramucirumab monotherapy for gastric cancer. On the basis of results obtained with ramucirumab showing an increase in patients with severe hypertension after monotherapy or when combined with paclitaxel or docetaxel, it is recommended that blood pressure should be monitored at least fortnightly. Infusion-related reactions may occur, usually during or following the first or second infusion. An initial incidence of reactions of 16 % in a small number of patients prompted the institution of premedication protocols across clinical trials of the mAb. In severe cases, symptoms include bronchospasm, tachycardia, and hypotension. FDA warnings/precautions also apply for PRES (incidence so far <0.5 %), proteinuria including nephrotic syndrome (incidence 3 % in patients given ramucirumab/FOLFIRI), and thyroid dysfunction (incidence of hypothyroidism in patients given ramucirumab/FOLFIRI, 2.6 %). There have been some reports of clinical deterioration in patients with Child-Pugh scores B or C for cirrhosis following ramucirumab monotherapy, leading to the recommendation that ramucirumab should only be used in such patients if the potential benefits are judged to outweigh the risks.

With ramucirumab now approved by the FDA/EMA as monotherapy or combination therapy for a range of different cancers, safety assessments need to take the different treatment regimens into account. When given as a single agent for gastric cancer, the most commonly seen adverse reactions to ramucirumab were hypertension, diarrhea, headache, and hyponatremia in that order; the most common serious events proved to be anemia and intestinal obstruction. Reactions reported as clinically relevant were neutropenia, epistaxis, intestinal obstruction, arterial thrombotic events, and rash. Combination of the mAb with paclitaxel commonly induced fatigue, diarrhea, and neutropenia with the latter event and febrile neutropenia, constituting the most serious adverse events. Events resulting in discontinuation of therapies were neutropenia and thrombocytopenia, while sepsis and gastrointestinal perforations were assessed as clinically relevant adverse reactions. Safety assessments of the administration of ramucirumab in combination with docetaxel to non-small cell lung cancer patients revealed neutropenia, fatigue/asthenia, and stomatitis/mucosal inflammation as the most common adverse responses. Events most commonly implicated in treatment discontinuations were infusion reactions and epistaxis. The incidences of pulmonary hemorrhage in mAb plus docetaxel-treated patients compared to placebo plus docetaxel patients were not significantly different. The most common serious adverse events with the mAb-docetaxel therapy were febrile neutropenia, pneumonia, and neutropenia. The most common adverse reactions in patients with colorectal cancer given ramucirumab plus FOLFIRI and compared to placebo plus FOLFIRI were diarrhea, neutropenia, decreased appetite, epistaxis, and stomatitis. Neutropenia and thrombocytopenia were the most common causes of treatment discontinuation in the ramucirumab-FOLFIRI group; the main adverse reactions leading to treatment discontinuation due to ramucirumab alone were proteinuria and gastrointestinal perforation. The most serious adverse events following ramucirumab plus FOLFIRI were diarrhea, intestinal obstruction, and febrile neutropenia.

The EMA and FDA refer to the immunogenicity data on ramucirumab assembled so far from clinical trials. As summarized by the EMA, in 527 ramucirumab-treated patients, 11 (2.2 %) developed anti-ramucirumab antibodies compared to 2 (0.5 %) of control patients. No neutralizing antibodies were detected and no infusion-related reactions occurred. The FDA has reported a 3 % incidence (86 of 2890 patients) of anti-ramucirumab antibodies in 23 clinical trials. Fourteen of the 86 antibody-positive patients had neutralizing antibodies.

While anti-VEGFR-2 therapy has shown some encouraging signs of efficacy, the often-seen modest survival benefits are a reminder that our understanding of angiogenic resistance mechanisms remains poor. This may be because although VEGFR-2 is undoubtedly a major, if not *the* major, signaling pathway for the biological processes and events of angiogenesis and VEGF-C and VEGF-D signaling through VEGFR-2 is blocked by ramucirumab (Fig. 3.4), the VEGFR-2 pathway is not the only one—a number of others, for example, P1GF, platelet-derived growth factor (PDGF), angiopoietin, and ephrin pathways, may contribute to the formation of tumor blood vessels. New approaches are clearly needed to overcome tumor resistance to anti-angiogenesis therapies, and given ramucirumab's long half-life (up to ~15 days), it has been suggested that combination anti-angiogenesis strategies should be sought. Investigations to identify suitable angiogenic biomarkers should also be pursued to optimize patient selection, treatment efficacy, and toxicity minimization.

Monoclonal Antibodies Targeting Human Epidermal Growth Factor 2 (HER2): Pertuzumab, Trastuzumab, and Ado-trastuzumab Emtansine

Human epidermal growth factor receptor 2 or HER2 (also known as HER2/neu, ErbB2, CD340, and p185) is a member of the erythroblastic leukemia viral oncogene (gene ErbB) family of four members, the others being HER1 (EGFR, ErbB1), HER3 (ErbB3), and HER4 (ErbB4). The ErbB proteins are each receptor tyrosine kinases related to EGFR (section "Monoclonal Antibodies Targeting Epidermal Growth Factor Receptor: Cetuximab, Panitumumab, and Necitumumab"). The structure of the HER2 receptor, a 185 kDa protein, consists of an extracellular ligand-binding domain, a transmembrane spanning section, and an intracellular protein tyrosine kinase domain with a regulatory carboxy-terminal. An intracellular tyrosine kinase domain also exists for HER1 and HER4 but not HER3. HER2 is inactive in the monomeric state and needs to be in the dimeric or oligomeric state for activation. Although there are 11 growth factors that activate the ErbB receptors (epidermal growth factor [EGF], transforming growth factor alfa [TGFα], heparin-binding EGF-like growth factor [HB-EGF], amphiregulin, betacellulin, epigen, epiregulin, and neuregulins 1, 2, 3, and 4), there is no known natural ligand for HER2. Activation of receptor kinase function proceeds mainly via ligand-mediated hetero- or homodimerization. In addition to the ligand-dependent pathway for activation, ligand-independent receptor activation can occur with activation being triggered by overexpression of HER2 and a high concentration of cell surface receptors that results in the formation of HER2/HER2 homodimers. Overexpression of HER2 leads to constitutive activation of the growth factor signaling pathways with a consequent favorable environment for breast cancer cell growth. Heterodimers with EGFR, HER3, and HER4 also form. For receptor dimerization, HER2 is not only the preferred but also the most important partner. HER3 has a high affinity for HER2 and the HER2/HER3 heterodimer appears to be the most potent in promoting the signal transduction process and tumor promotion. Autophosphorylation of the tyrosine residues on the intracellular domain of HER2 activates the MAPK, PI3K/Akt, PLC-γ, PKC, and signal transducer and activator of transcription (STAT) pathways leading to cell survival and proliferation.

Overexpression of HER2 as a result of gene amplification is found in ~15–25 to 30 % of human breast cancers, and this overexpression tends to correlate with tumors that are more aggressive and with poorer prognosis. Three mAbs targeting HER2 are currently approved by the FDA and EMA, pertuzumab, trastuzumab, and the ADC prepared by conjugating trastuzumab to the cytotoxin, mertansine.

Pertuzumab

Pertuzumab (Perjeta®, 2C4) (Tables 2.1 and 3.1) is a recombinant humanized IgG1κ mAb, MW~148 kDa, targeting the extracellular dimerization domain (subdomain II) of the HER2 protein thereby blocking heterodimer formation and ligand-dependent signaling via two major pathways, MAPK and PI3K. Inhibiting downstream signaling arrests cell growth and leads to apoptosis, while ADCC further contributes to the mode of action of pertuzumab. The action of pertuzumab is independent of the level of expression of HER2. Although pertuzumab and trastuzumab target the same receptor, they recognize different binding sites. Trastuzumab (section "Trastuzumab," below) binds to subdomain IV of the HER2 extracellular (ligand-binding) domain inhibiting ligand-independent signaling and complementing the mechanism of action of pertuzumab which inhibits ligand-dependent signaling, particularly between HER2 and HER3. This combination of actions potently activates cell survival and proliferation (Fig. 3.5b). The clinical efficacy of pertuzumab alone is not impressive, but together with trastuzumab, the combination produces a more complete and effective, if not synergistic, blockade of the HER2-driven signaling pathways.

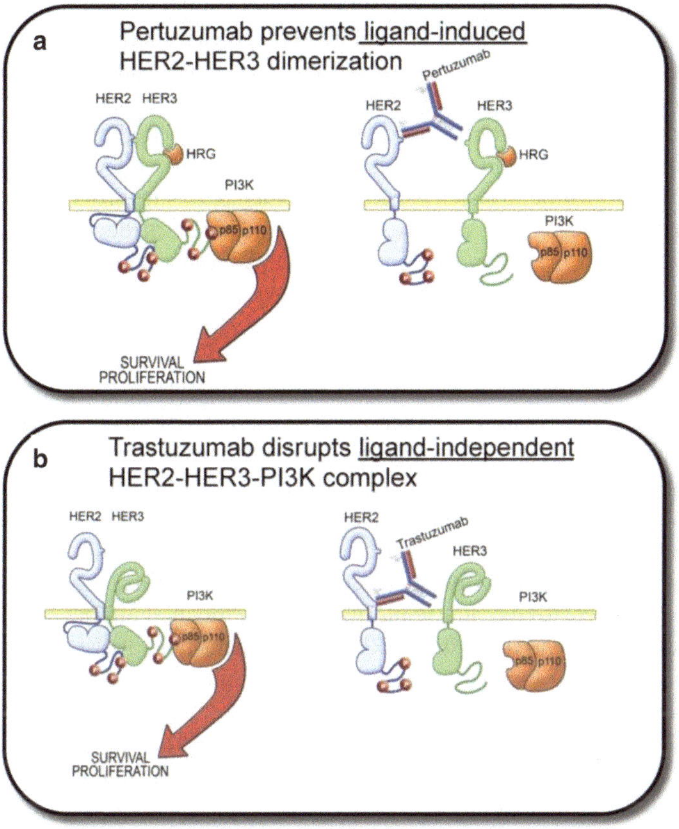

Fig. 3.5 Diagrammatic representation of the general mechanisms of action of pertuzumab and trastuzumab, respectively, in disrupting ligand-independent and ligand-dependent HER2/HER3 interactions that result in antiproliferative effects in HER2-positive tumor cells. (**a**) Pertuzumab blocks the ligand-induced HER2–HER3 dimerization by binding to HER2 extracellular subdomain II. (**b**) Ligand-independent HER2/HER3 interaction is followed by HER3 phosphorylation activating the PI3K signaling pathway leading to cell survival and proliferation. Trastuzumab disrupts the HER2–HER3 interaction by binding to extracellular subdomain IV of HER2. Reproduced from Junttila TT et al. Ligand-independent HER2/HER3/PI3K complex is disrupted by trastuzumab and is effectively inhibited by the PI3K inhibitor GDC-0941. Cancer Cell 2009;15:429–40. Modified and reprinted with permission from Elsevier Limited

Pertuzumab was granted approval by the FDA in June 2012 for use in combination with trastuzumab and docetaxel in patients with HER2-positive metastatic breast cancer who have not received prior anti- HER2 therapy or chemotherapy. The approved indications and usage also cover the same combination as neoadjuvant therapy of patients with HER2-positive, locally advanced, inflammatory, or early stage breast cancer as part of a complete treatment regimen for early breast cancer. The EMA gave final approval for pertuzumab in the same combination therapy in March 2013.

The FDA has issued a boxed warning for pertuzumab covering cardiomyopathy manifesting as congestive heart failure and decreased left ventricular ejection fraction (LVEF) and for embryofetal toxicity that may result in fetal death and birth defects. It is believed that patients who have received prior treatment with anthracyclines or prior radiotherapy to the chest may be at higher risk of decreased LVEF. The incidence of left ventricular systolic dysfunction (LVSD) was found to be higher in pertuzumab-treated patients compared to trastuzumab- and docetaxel-treated patients, and there was an increased incidence of declines in LVEF in those treated with the mAb-drug combination. Administration of pertuzumab to pregnant women can harm the fetus, and the need for this warning is backed up by the observation of oligohydramnios, delayed development of fetal kidneys, and embryofetal death in pregnant monkeys. Infusion-related reactions and hypersensitivity responses including anaphylaxis are the subjects of the remaining FDA warnings for pertuzumab. Described as "hypersensitivity, anaphylactic reaction, acute infusion reaction, or cytokine release syndrome occurring during an infusion or on the same day as the infusion," the frequency of reactions to pertuzumab alone when the mAb was given on day one without trastuzumab and docetaxel was 13 % compared to 9.8 % in the placebo group. Grade 3 and 4 reactions made up less than 1 % of the infusion reactions. Pyrexia, chills, fatigue, headache, asthenia, hypersensitivity, and vomiting were the main symptoms. In a clinical trial involving 804 patients with HER2-positive metastatic breast cancer, patients were randomized to receive pertuzumab in combination with trastuzumab and docetaxel or placebo in combination with trastuzumab and docetaxel. After a median study treatment time of 18.1 months for the pertuzumab group, the most common adverse reactions (>30 %) were diarrhea, neutropenia, alopecia, nausea, fatigue, rash, and peripheral neuropathy. The most common grade 3–4 adverse reactions (>2 %) were neutropenia, febrile neutropenia, leukopenia, diarrhea, peripheral neuropathy, anemia, asthenia, and fatigue. Reactions showing a clear higher incidence in the pertuzumab-treated group compared with the placebo-treated group were leukopenia, diarrhea, rash, pruritus, dry skin, and febrile neutropenia. There was an increased incidence of febrile neutropenia in Asian patients in both treatment arms although it was significantly higher in the pertuzumab-treated group (26 %) than the placebo-treated group (12 %). Reactions judged to be clinically relevant in the pertuzumab-treated group and reported in <10 % of patients were LVSD, pleural effusion, hypersensitivity, and paronychia. Results of trials of the pertuzumab drug combination in neoadjuvant treatment of breast cancer showed a similar spectrum of adverse reactions with upper respiratory tract infections as the main addition.

Because of the known trastuzumab-related cardiotoxicity and black box warnings of cardiomyopathy for both pertuzumab and trastuzumab, particular attention was paid to possible cardiac effects when the two mAbs were used together. In fact, cardiotoxicity to pertuzumab has been documented in several studies. In a phase I clinical study, two of 21 patients showed a reduction in LVEF and one developed congestive heart failure after two cycles of pertuzumab. Decreases in LVEF have also been observed in trials of ovarian and advanced prostate cancer patients treated with pertuzumab. In a study to evaluate safety and efficacy of pertuzumab with trastuzumab in patients with HER2-positive metastatic breast cancer, 11 patients were given 64 cycles of the mAb combination. Echocardiograms and magnetic imaging studies showed LVSD in six patients, three with grade 1, two with grade 2, and one with grade 3 reactions. Although the observed cardiotoxicity was asymptomatic and reversible (and trastuzumab-induced cardiac dysfunction is generally reversible), it must be said that the long-term cardiac effects of pertuzumab-trastuzumab treatments are yet to be determined. More recently, a database of 598 patients treated with pertuzumab was used to examine the incidence of asymptomatic LVSD and symptomatic heart failure. Of 331 patients treated with pertuzumab alone, 23 (6.9 %) developed LVSD and one (0.3 %) displayed symptomatic heart failure. The corresponding figures for 175 patients treated with pertuzumab in combination with a non-anthracycline cytotoxic agent were six (3.4 %) and two (1.1 %) and for 93 patients treated with trastuzumab, six (6.5 %) and one (1.1 %). The results revealed that pertuzumab given as a single agent or in combination with cytotoxic agents, including trastuzumab, showed a low incidence of cardiac dysfunction similar to trastuzumab. In response to the mAb, patients experienced relatively low incidences of both LVSD and symptomatic heart failure, and there was no significant increase in cardiac side effects when pertuzumab was given in combination with other anticancer agents. Importantly, no synergistic effect between the two mAbs was apparent.

Because of the recent approval of pertuzumab by the FDA and EMA, there have so far been relatively few postmarketing reports on the mAb's safety. At mid-2013, 234 reports to the FAERS were made up of adverse gastrointestinal symptoms (9 %), infection (6 %), and respiratory/cardiorespiratory/pulmonary vascular disorders (9.5 %). Of only 90 reports to the European pharmacovigilance database, most dealt with gastrointestinal signs; constitutional, respiratory, and hematologic disorders; and infections.

In the trial with HER2-positive metastatic breast cancer patients mentioned above, patients were randomized to receive pertuzumab in combination with trastuzumab and docetaxel or placebo in combination with trastuzumab and docetaxel. Eleven of 386 (2.8 %) patients in the pertuzumab-treated group and 23 of 372 (6.2 %) patients in the placebo-treated group developed antibodies reactive with pertuzumab, suggesting that the assay might have detected antibodies that cross-react with both mAbs. An additional uncertainty is the possibility that the presence of pertuzumab in patients' sera might interfere with the detection of anti-pertuzumab antibodies.

Trastuzumab

Trastuzumab (Herceptin®) (Tables 2.1 and 3.1), the first biologic agent developed and approved for the treatment of breast cancer, is a recombinant humanized IgG1κ mAb with high-affinity binding specificity for the extracellular domain of HER2. Pertuzumab (see above, section "Pertuzumab") is known to bind to the extracellular subdomain II of HER2, blocking ligand-induced HER2/HER3 dimerization (Fig. 3.5b). Trastuzumab binds to HER2 subdomain IV causing downregulation of PI3K/AKT signaling in tumor cells overexpressing HER2 in the absence of ligands, but it seemed accepted that this mAb did not block dimerization of HER2 with ligand-activated HER1 (EGFR) or HER3. In the early 2000s, it was shown that preventing the expression of HER3 or its interaction with HER2 produced an antiproliferative effect in HER2-positive breast cancer cells. Following up these observations, researchers at Genentech, Inc., San Francisco, used RNA silencing of HER receptor expression and a tumor-inducible HER3 knockdown model to show that EGFR has no role in HER2-positive breast cancer and, of greater significance, constitutively active HER2 is dependent on HER3 for mediating its effects. While it is clear that trastuzumab exerts its antitumor activity via a number of different mechanisms, it is also apparent that a detailed understanding of the mAb's mode of action is far from complete. A brief summary of the current understanding is as follows. Trastuzumab binds to the extracellular domain of HER2, downregulating the expression of HER2 on the cell surface and blocking cleavage of the domain preventing the formation of the constitutively active membrane-bound protein, p95-HER2. These events ultimately reduce PI3K and MAPK signaling. The cyclin-dependent kinase inhibitor p27kip1 has an important role in trastuzumab-induced cell cycle arrest and tumor inhibition by binding to, and inhibition of, cyclin E/Cdk2 complexes and inhibition of G1/S progression. It now seems that trastuzumab also selectively inhibits ligand-independent HER2–HER3 heterodimerization and disrupts the HER2-HER3-PI3K complex (Fig. 3.5) in both trastuzumab-sensitive and trastuzumab-insensitive cells. In the light of these findings, it was suggested that blocking the HER2/HER3 receptor complex with a combination treatment of trastuzumab and pertuzumab might be particularly efficacious in HER2-amplified breast cancer. Finally, in addition to the above actions, trastuzumab has anti-angiogenic effects lowering the threshold for tumor cell killing by cytotoxic drugs, and it displays significant ADCC activity by Fc interaction with FcγRIIIa receptors on immune effector cells such as NK cells and macrophages.

Despite the advance in treatment offered by trastuzumab for aggressive breast cancer, only about 30 % of patients with advanced disease benefit from monotherapy with the drug. Furthermore, a large proportion of patients (~70 %) who experience an initial beneficial response become resistant within a year. Resistance is especially marked in advanced gastric cancer. Possible mechanisms of resistance have been studied. One proposed mechanism, so-called epitope masking, involves overexpression of membrane mucins such as Muc4 which, to some extent, masks the epitope recognized by the mAb on HER2. Precipitation experiments demonstrated that HER2 and Muc4 interact, knockdown of Muc4 increases trastuzumab binding, and upregulation of Muc4 in HER2-positive breast cancer cells results in acquired resistance to trastuzumab. A second possible mechanism of trastuzumab resistance may be the failure of the antibody to suppress the progression of tumors expressing truncated HER2 isoforms. Trastuzumab blocks cleavage of the extracellular domain of HER2 and prevents the formation of constitutive p95-HER2, but oncogenic HER2 exists in multiple isoforms and in some cancers the extracellular domain-binding epitope may be lacking. Cross-signaling is another potential mechanism of trastuzumab resistance. Several molecules appear to activate HER2 in the presence of trastuzumab including insulin-like growth factor-1 receptor (IGF-1R), hepatocyte growth factor and its receptor Met, growth differentiation factor 15 (GDF15), and members of the ErbB family. The precise mechanisms underlying such cross-signaling generally remain unclear. Activation of, or increased downstream signaling, for example, by PI3K, focal adhesion kinase (FAK), and Src, may also counter the upstream inhibition exerted by trastuzumab. Activation of PI3K may occur as a result of mutations of reduced expression of the PTEN gene, deregulated signaling from upstream, or mutations of the PI3KCA (phosphatidylinositol-4,5-biphosphate 3-kinase, catalytic subunit alfa) gene. Little or no contribution from the normally present ADCC activity of trastuzumab is yet another possible mechanism of resistance to the mAb. This may occur because of impairment or loss of Fc and Fc gamma receptors and/or disruption of antibody—target cell or antibody—NK cell interaction (Chap. 2, section "Antibody-Dependent Cell-Mediated and Complement-Dependent Cytotoxicities").

With approved indications for breast cancer, metastatic gastric, and gastroesophageal junction adenocarcinomas overexpressing HER2, trastuzumab is the subject of an FDA black box warning for cardiomyopathy, infusion reactions, and pulmonary toxicity. For patients receiving trastuzumab as a single agent or in combination therapy, there is a four- to sixfold increase in the incidence of symptomatic myocardial dysfunction. Incidences are highest when an anthracycline drug is coadministered. Trastuzumab therapy can result in cases of left ventricular cardiac dysfunction, hypertension, arrhythmias, cardiomyopathy, cardiac failure, and cardiac death. Cardiac monitoring including determination of LVEF is recommended. Reported incidences of trastuzumab-related cardiotoxicity are 2–7 % for monotherapy, 2–13 % for trastuzumab combined with paclitaxel, and up to 27 % when given in combination with anthracyclines. In an early clinical trial, 136 of 844 patients (16 %) discontinued trastuzumab due to myocardial dysfunction or a decline in LVEF. In two other trials, 2.6 and 2.9 % discontinued the mAb due to cardiotoxicity. Infusion reactions may be relatively mild but serious and fatal reactions have been reported. Serious and fatal pulmonary toxicities following trastuzumab include dyspnea, interstitial pneumonitis, pulmonary infiltrates, pleural effusions, pulmonary insufficiency and hypoxia, non-cardiogenic pulmonary edema, ARDS, and pulmonary fibrosis. Other warnings and precautions refer to exacerbation of chemotherapy-induced neutropenia and febrile neutropenia and the possibility of fetal harm with trastuzumab increasing the risk of oligohydramnios during the second and third trimesters of pregnancy. Common and serious adverse events caused by trastuzumab are listed in Table 3.1. One important adverse event in need of closer analysis became apparent in early results from trials when a surprisingly high incidence of congestive heart failure was noticed in patients treated concurrently with the anthracycline doxorubicin. Early observations showed that approximately 5 % of patients treated with trastuzumab and chemotherapy developed evidence of systolic cardiac dysfunction with an incidence 4–5 times higher than controls. In a retrospective analysis of long-term cardiac tolerability of trastuzumab in metastatic breast cancer in 218 patients who received trastuzumab for at least 1 year, 28 % of patients experienced some type of cardiac event, 10.9 % had grade 3 cardiotoxicity, and there was one cardiac death. Forty-five percent of patients with breast cancer are older than 65 years, and bearing in mind that recent reports indicate that trastuzumab-induced cardiotoxicity is now more often seen than indicated in the early clinical trials, there is a concern that older patients are at higher risk. This concern was reinforced in 2012 by results of an Italian study of 490 patients treated with trastuzumab in the adjuvant setting; whereas the overall rate of congestive heart failure was 27 %, it was 38 % in patients over 68. In a larger study of trastuzumab-related cardiotoxicity in older patients, the incidence of congestive heart failure among those given trastuzumab was 29.4 % of 2203 patients (median age 71 years) compared with 18.9 % of those who did not receive the mAb. Patients given trastuzumab were also more likely to develop congestive heart failure, and in this group there was an increased risk of congestive heart failure in older patients, in those given trastuzumab weekly, and in patients with coronary artery disease and hypertension. Note that in the pivotal clinical trials, only 4 % of patients given trastuzumab experienced a cardiac event and this compared with 1.3 % in the control groups. The finding that the frequency of administration of trastuzumab, namely, once a week, was associated with a higher risk leaves some questions, in particular, is more frequent administration associated with more myocyte damage and is the finding unique to older patients? Of course, it must be remembered that trastuzumab cardiotoxicity usually appears to be reversible—89 % of patients showing an asymptomatic decrease in LVEF apparently recovered after stopping therapy with the mAb. As to the likely mechanism of trastuzumab-related cardiac dysfunction, it has been speculated that since HER2 may be involved in a number of different biological processes of cardiomyocytes, trastuzumab may have a direct effect on myocytes.

In relation to apparent anthracycline toxicity, some recent evidence indicates that patients treated with anthracycline-based chemotherapies exhibit poor performance of some cognitive skills compared to those who received no, or other, chemotherapies.

In the postmarketing setting for trastuzumab, the most frequently reported adverse events to the FAERS have been infections (5 %), gastrointestinal disorders (4.2 %), respiratory disorders (3.7 %), white blood cell abnormalities (3.5 %), and cutaneous reactions (2.9 %). Some attention has also been drawn to infusion reactions and oligohydramnios. Of 7116 reports to the European pharmacovigilance database, cardiac events (11 %), respiratory disorders (10 %), gastrointestinal disorders (6.8 %), cutaneous reactions (5.9 %), and infections (4.8 %) were most frequently reported. There were 95 reports of anaphylactic/anaphylactoid reactions, three cases of CRS, 39 cases of stomatitis, and 29 cases of pancreatitis. Adverse pulmonary events were made up largely of interstitial lung disease (181 cases) and lung infiltrations (49 cases). Of 673 hematologic events, 25 % were for neutropenia, 13 % for each of anemia and thrombocytopenia, and 11 % for febrile neutropenia.

Data on the immunogenicity of trastuzumab are hard to find although a number of assays for the detection of antibodies to trastuzumab in human sera are commercially available. The FDA reports the detection of human antihuman antibodies to the mAb in one of 903 women with metastatic breast cancer. The patient exhibited no adverse effects. Anti-trastuzumab antibodies were monitored in the so-called HannaH study, an investigation of a subcutaneous formulation of trastuzumab utilizing recombinant hyaluronidase to enhance tissue permeation and facilitate absorption of the antibody. The incidence of antibodies was found to be low and no clinical relevance was observed judging by efficacy and safety assessments that remained unaffected.

Ado-trastuzumab Emtansine

Ado-trastuzumab emtansine (Kadcyla®; trastuzumab emtansine, T-DM1) (Tables 2.1 and 3.1) is a humanized IgG1κ mAb-drug conjugate made by covalently coupling the microtubule inhibitory compound mertansine (DM1, *N*2′-deacetyl-*N*2′-(3-mercapto-1-oxypropyl)maytansine), a cytotoxic, thiol-containing maytansinoid (shown in red in Fig. 3.6), to trastuzumab via a bifunctional cross-linker, SMCC (succinimidyl *trans*-4-(maleimidylmethyl)cyclohexane-1-carboxylate) (shown in blue in Fig. 3.6). Each molecule of trastuzumab may be linked to as many as eight DM1 molecules although the average number is 3.5. In an examination of the drug (toxin) distribution profile of prepared DM1-trastuzumab conjugates by mass spectrometry and peptide mapping, up to 20 chemically modified lysine residues linked via ε-amino groups were found on

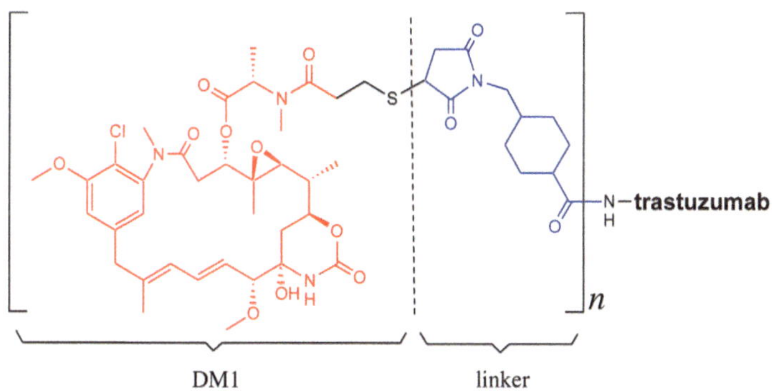

Trastuzumab emtansine

KADCYLA *(Where n ~ 3.5 DM1/Mab)*

Fig. 3.6 Structure of ado-trastuzumab emtansine formed by covalently coupling trastuzumab to the cytotoxic microtubule inhibitor, mertansine, or DM1 (shown in *red*), via the bifunctional cross-linker SMCC (succinimidyl *trans*-4-(maleimidylmethyl)cyclohexane-1-carboxylate) that contains a reactive succinimide ester and a reactive maleimide that links to thioethers (shown in *blue*). The succinimide reacts with a trastuzumab lysine; the maleimide group covalently links to the free sulfhydryl group of DM1. The mertansine structure of the resultant complex together with the linkage group is designated the emtansine component. Ado-trastuzumab emtansine contains an average of 3.5 DM1 molecules for every molecule of trastuzumab

both light (~7 residues) and heavy chains (~13 residues) of the conjugated mAb. The conjugated lysines were freely accessible and none were found in the antibody complementarity-determining regions (CDRs). Random conjugation of the mAb lysine residues led to heterogeneity reflected in the number of attached toxin molecules and the different sites of their linkage. Crystal structures of the Fab and Fc fragments (Fig. 3.7) containing the modified lysine residues revealed that most of the residues are structurally and solvent accessible in the surface loops of the mAb. Only two modified lysines, Lys323 and Lys 393 in the heavy chain, are located in β-sheets (Fig. 3.7). The methodology used in this study should prove applicable to the identification of modified amino acid residues in

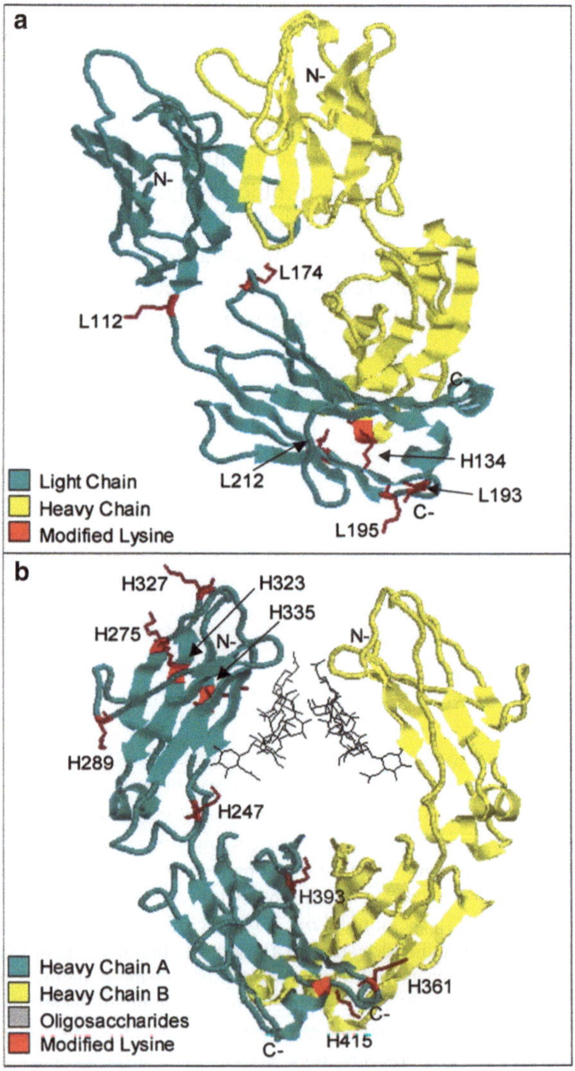

Fig. 3.7 Cartoon plots of the crystal structures of human IgG1 Fab (**a**) and Fc (**b**) fragments of the trastuzumab antibody-drug conjugate covalently linked to the cytotoxin mertansine (DM1) via lysine residues (shown in *red*). Modified lysines are shown on only one chain of the Fc fragment. N- and C-termini are labeled N and C, respectively. Images were created using Rasmol V2.6. Reproduced from Wang L et al. Structural characterization of the maytansinoid-monoclonal antibody immunoconjugate, huN901-DM1, by mass spectrometry. Protein Science 2005;14:2436–46. Reprinted with permission from John Wiley and Sons

other ADCs and different immunoconjugates. Linkage of DM1 to trastuzumab does not impair the binding of the mAb to HER2, it does not reduce its antitumor action, the linkage is nonreducible and non-cleavable, and there has been no evidence of toxic systemic exposure to free DM1 as a result of accumulation after repeated doses of the ADC. The covalent conjugate remains stable in both the circulation and tumor environment with the release of DM1 occurring only upon proteolytic degradation of the coupled mAb.

Ado-trastuzumab emtansine has a dual mechanism of action, combining trastuzumab's antitumor activity of inhibition of HER2-mediated signaling and ADCC (section "Trastuzumab") with the cytotoxic effect of DM1 on HER2-positive tumor cells. After binding of ado-trastuzumab emtansine to HER2, the HER2-T-DM1 complex enters the cell via receptor-mediated endocytosis where it is internalized by passage through endocytic vesicles to early endosomes and to mature lysosomes (Fig. 3.8). At the early endosome stage, some receptor and labeled mAb can be recycled back to the cell membrane. In the lysosome, the DM1 toxin released from T-DM1 as a result of proteolytic degradation of trastuzumab is liberated as a DM1-linker-lysine complex. DM1 inhibits the assembly of microtubules leading to cell cycle arrest and ultimately cell death.

Ado-trastuzumab emtansine is indicated as a single agent for the treatment of HER2-positive metastatic breast cancer patients who previously received trastuzumab and a taxane, separately or in combination. In its indications and usage, the FDA stipulates that patients should have received prior therapy for metastatic disease or developed disease recurrence during or within 6 months of completing adjuvant therapy.

An FDA black box warning for ado-trastuzumab emtansine mentions hepatotoxicity (including liver failure and death), cardiotoxicity (specifically reduction in LVEF), and the potential risk of fetal harm, plus a reminder that the ADC should not be substituted for, or with, trastuzumab. Warnings and precautions have also been issued for pulmonary toxicity, infusion-related and hypersensitivity reactions, thrombocytopenia, neurotoxicity, and extravasation. Signs and symptoms of dyspnea, cough, fatigue, and pulmonary infiltrates may indicate pulmonary toxicities such as interstitial lung disease, including pneumonitis, and cases of ARDS and death which were seen in clinical trials. The frequency of infusion-related reactions in clinical trials was 1.4 %; anaphylactic-like reactions may also occur. In two separate clinical trials, incidences of thrombocytopenia were as follows: any grade, 32 % and 31.2 %, and ≥grade 3, 11.7 and 14.5 %. In Asian patients the incidence of ≥grade 3 reactions in ado-trastuzumab emtansine-treated patients was 45.1 %. For peripheral neuropathy (predominately sensory), recorded incidences for any grade of reaction and ≥grade 3 reactions are ~20 % and 1.5–2 %, respectively. Reactions secondary to extravasation following infusion of the mAb may occur, usually within 24 h. No satisfactory treatment appears to be available, but reactions consisting of erythema, tenderness, skin irritation, pain, and swelling are usually mild. Commonly occurring adverse events following ado-trastuzumab emtansine are listed in Table 3.1.

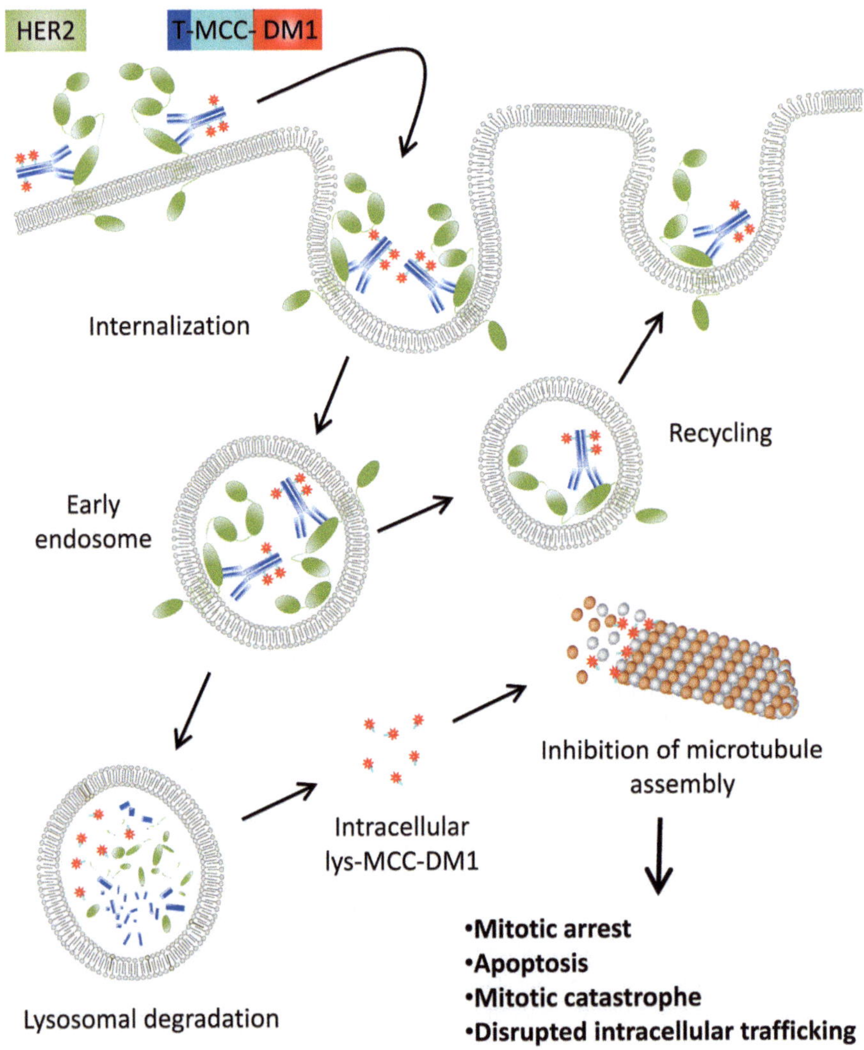

Fig. 3.8 Intracellular events involved in the antitumor action of ado-trastuzumab emtansine (T-MCC-DM1). After binding of the antibody-toxin conjugate to HER2, receptor-mediated endocytosis, some recycling of the complex back to the cell membrane, and transport via endosomes to lysosomes, the DM1-MCC linker-lysine complex released by proteolytic degradation acts to inhibit microtubule assembly. This results in cell cycle arrest, disruption of intracellular trafficking, and apoptosis. *MCC* 4-[*N*-maleimidomethyl] cyclohexane-1-carboxylate; *T* trastuzumab. Reproduced from Barok M et al. Trastuzumab emtansine: mechanisms of action and drug resistance. Breast Cancer Res 2014;16:209, an open-access article distributed under the terms of the Creative Commons Attribution License

Data on the immunogenicity of ado-trastuzumab emtansine are lacking and details of findings so far are vague. The FDA reports studies described as "tested at multiple time points for antitherapeutic antibody responses" to the ADC. Of 836 patients, 44 (5.3 %) tested positive "at one or more post-dose time points."

Denosumab

See also Chap. 4, section "Denosumab."

Denosumab (Xgeva®, Prolia®) (Tables 2.1 and 3.1), produced in transgenic mice, is a fully human IgG2κ mAb, MW ~ 147 kDa, with high specificity and affinity for receptor activator of nuclear factor kappa-B ligand, otherwise known as RANKL. Produced and expressed or released principally by cells of the osteoblast lineage and activated T cells, RANKL, a cytokine and member of the tumor necrosis factor (TNF) family (also called TNFSF11) (Chap. 5) responsible for bone resorption, is expressed in both soluble and membrane-bound forms. Expression is controlled by a number of cytokines and hormones, mainly IL-1, IL-6, TNF, $1,25(OH)_2$ vitamin D_3, parathyroid hormone, and parathyroid hormone-related peptide (Chap. 7, section "Parathyroid Hormone"), prolactin, prostaglandin E2, and corticosteroids. RANKL stimulates osteoclast formation, activation, adherence, survival, and ultimately resorption of the bone. Inhibition of RANKL results in an increase in bone density, volume, and strength. The actions of RANKL are effected via its cognate receptor RANK which is expressed on osteoclasts and their precursors. Binding of RANKL with its receptor occurs by interaction between extracellular domains of the ligand and extracellular cysteine-rich domains of the receptor. This activates several signaling pathways, in particular, protein kinase and NF-κB pathways, the latter upregulating c-fos in inducing gene transcription. The balance and coordination between the activities of osteoclasts and osteoblasts control bone formation, loss, and remodeling; osteoclasts first resorb the bone, osteoblasts follow, and regenerate it. A second member of the TNF family, osteoprotegerin, is also a player in the osteoclast-RANK-RANKL-induced balance of bone formation and loss. Osteoclast activity, at least in part, ultimately depends on the ratio of RANKL to osteoprotegerin since the latter binds RANKL preventing its binding and activation of RANK and leading to the inhibition of osteogenesis and resorption. Osteoprotegerin has been shown to markedly reduce the numbers of osteoclasts in bone lesions, for example, when neoplastic cells metastasize in the bone. In fact, superficially at least, osteoprotegerin and denosumab have similar mechanisms of action. Expression of osteoprotegerin is induced by the cytokines transforming growth factor-β (TGF-β) and platelet-derived growth factor (PDGF) and by estrogen, calcitonin, and calcium.

Marketed under two trade names, denosumab as Prolia® (Chap. 4, section "Denosumab") has approved cancer use indications for the treatment of men at high risk of fracture receiving androgen deprivation therapy for nonmetastatic prostate cancer and for the treatment of women at high risk of fracture receiving adjuvant aromatase inhibitor therapy for breast cancer. As Xgeva®, denosumab is approved for the prevention of skeletal-related events in patients with bone metastases from solid tumors and the treatment of giant cell tumor of the bone.

For Xgeva®, warnings and precautions consist of hypocalcemia (fatal cases have been recorded), osteonecrosis of the jaw, and embryofetal toxicity. The nadir for the level of calcium in serum after a dose of denosumab is reached on about day 10. More frequent monitoring of serum calcium concentrations at the baseline and during the first two weeks of treatment, especially in patients with an estimated glomerular filtration rate <30 mL/min, has been recommended. Osteonecrosis of the jaw may manifest as jaw pain, osteomyelitis, osteitis, bone erosion, tooth or periodontal infection, toothache, gingival ulceration, or gingival erosion. Persistent pain or slow healing of the mouth or jaw after dental surgery may also be seen. In clinical trials, 2.2 % of patients with osseous metastasis receiving denosumab after median exposure of 13 doses developed osteonecrosis of the jaw. This figure increased to 5.4 % in patients with prostate cancer, a non-approved indication. Taking results of studies in monkeys, fetal harm after administration of denosumab to pregnant women may manifest as fetal loss, stillbirths, postnatal mortality, the absence of lymph nodes, abnormal bone growth, and decreased neonatal growth. Commonly occurring adverse events are listed in Table 3.1. Cutaneous adverse events are not uncommon and include rashes, eczema, dermatitis, exanthema, photosensitivity and possibly rare eruptions, exfoliative reactions, and bullous conditions. For Prolia®, hypersensitivity, including anaphylaxis, is listed as a warning together with hypocalcemia, osteonecrosis of the jaw, serious infections, atypical femoral fractures and suppression of bone turnover, cutaneous reactions, and severe bone, joint, and muscle pain. The most common adverse events in prostate cancer and breast cancer patients with bone loss are arthralgia, back pain, pain in extremity, and musculoskeletal pain. RANKL is expressed on T and B cells and especially CD4+ T cells, and there has been speculation that, on the one hand, denosumab may be immunosuppressive leading to increased infections and, on the other hand, that RANKL may have an immune accessory role generating costimulatory signals for dendritic cells which activate T cells after the CD40L/CD40 primary signal. As yet, there is no compelling evidence for either suggestion.

Recent serious postmarketing surveillance reports for Xgeva® cover calcium and bone disorders, gastrointestinal symptoms, and musculoskeletal and dermatologic conditions. These events, plus respiratory disorders, are also the most commonly reported reactions. Reports on Prolia® relate to musculoskeletal disorders, dermatologic conditions, infections, and gastrointestinal disorders. The most frequently mentioned infections include pneumonia, cellulitis, urinary tract infections, and sepsis.

Immunogenicity of denosumab has not proved to be a major problem. The incidence of hypersensitivity reactions appears to be similar to control groups. Human antihuman antibodies were shown to have an incidence of <1 % in 13,000 tested patients. Of 2758 patients with osseous metastases treated with denosumab (as Xgeva®) for up to 3 years, seven (0.25 %) tested positive for antibodies to the mAb. None of the antibodies detected were neutralizing for denosumab. Similar results were obtained in assessments of the immunogenicity of Prolia®.

Ipilimumab

Ipilimumab (Yervoy®, MDX-010) (Tables 2.1 and 3.1) is a recombinant fully humanized IgG1κ mAb, MW ~148 kDa, that binds with high affinity to the extracellular domain of the protein receptor human cytotoxic T lymphocyte antigen 4 or CTLA-4 (also known as CD152), a member of the immunoglobulin superfamily. CTLA-4 has a critical role as an inhibitory regulator during the early stages of T-cell expansion. For T lymphocyte activation to occur, a naïve T cell needs to receive two precise signals from an antigen-bearing, antigen-presenting cell (APC) (dendritic cell, macrophage, or B cell). The first activation signal comes via the membrane-associated major histocompatibility complex (MHC) in its interaction with the T-cell receptor (TCR), while the second activation event is the provision of costimulatory signals effected by the APC membrane protein ligand CD80 (B7-1) working in tandem with membrane ligand CD86 (B7-2). The complementary receptor for these ligands is CD28, a protein constitutively expressed on naïve T cells that enhances IL-2 production and allows the cells to undergo clonal expansion. CTLA-4 is expressed on the lymphocyte surface but only becomes functional after the start of T-cell activation when it competitively interacts with the B7 ligands (Fig. 3.9) resulting in interference with IL-2 secretion and receptor expression and downregulation of the T-cell response. When bound to the B7 complex on APCs where it binds with greater affinity than CD28, CTLA-4 has been described as an immune "off" switch playing an important role in modulating overactivity of T cells and maintaining tolerance to self-antigens. However, another consequence of suppressing the immune response can be to allow cancer cells to be recognized as self and multiply in the absence of antitumor immune challenge. Recognition of the crucial role of CTLA-4 as an inhibitory regulator of the T lymphocyte response and studies of murine models of cancer led to the strategy of blockade of the receptor by specific antibody (Fig. 3.9) and this in turn resulted in the development of the mAb ipilimumab. Blocking antibodies such as ipilimumab, an example of immune checkpoint targeting, exert an antitumor action by at least two mechanisms: blockade-induced clonal expansion of activated CD3+ and CD4+ T lymphocytes (and also probably CD8+ cells) and depletion of regulatory T cells (Tregs) CD4+ CD25+ which are induced by the tumor and act to inhibit immune recognition of tumor antigens. In addition, an increase in Th17 CD4+ T cells producing cytokines IL-17 and IL-22 was found to be associated with fewer relapses in a trial of patients treated with ipilimumab. Ipilimumab does not exhibit CDC activity, but at high concentrations it exerts weak to moderate ADCC activity on activated T cells probably via IgG Fc FcγRI (CD64) rather than FcγRIII receptors.

With an approved indication for the treatment of unresectable or metastatic melanoma (Table 2.1), a number of adverse events induced by ipilimumab are related to the mAb's mode of action of T-cell-mediated immune aggression against cancer cells. In fact, the FDA has a black box warning for severe and fatal immune-mediated adverse reactions due to T-cell activation and proliferation. Although these reactions may involve any organ system, the most common ones include enterocolitis; hepatitis; neuropathy (e.g., Guillain-Barré syndrome, myasthenia gravis); endocrinopathy

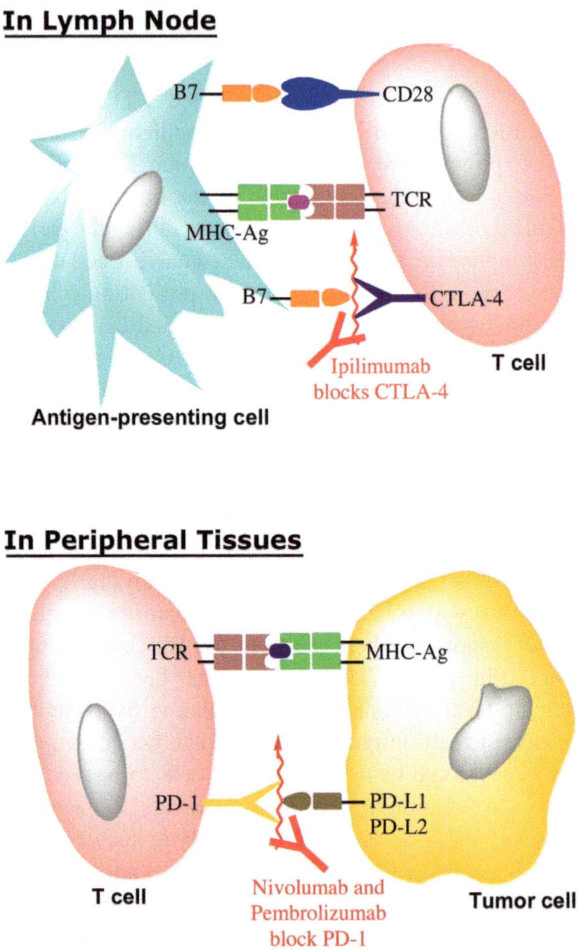

Fig. 3.9 Immune checkpoint targeting with anti-CTLA-4 and anti-PD-1 monoclonal antibodies. The cytotoxic T lymphocyte antigen 4 (CTLA-4) acts at the first checkpoint to modulate overactive T cells by competing for the B7 ligands. CTLA-4 only becomes functional after the start of T-cell activation when it competitively interacts with the B7 ligands resulting in downregulation of the T-cell response. Suppressing the immune response allowing cancer cells to be recognized as self and thus multiplying in the absence of antitumor immune challenge led to the strategy of blockade of the receptor by the anti-CTLA-4 immunostimulatory monoclonal antibody ipilimumab. The programmed cell death 1 (PD-1) receptor acts as another important immune checkpoint. Binding of PD-1 to its ligands PD-L1 and PD-L2 suppresses T-cell proliferation and activity allowing the tumor cells to avoid immune recognition and attack. Blocking at this second checkpoint by anti-PD-1 monoclonal antibodies nivolumab and pembrolizumab results in the reactivation of T cells including those with specificity for the tumor cells

(hypopituitarism, adrenal insufficiency, hypogonadism, hypothyroidism); severe cutaneous events such as Stevens-Johnson syndrome and toxic epidermal necrolysis; and immune-mediated ocular manifestations, uveitis, iritis, or episcleritis. A number of studies have led to and supported claims that ipilimumab elicits a high rate of immune-related adverse events. A recent review of records from 298 melanoma patients treated with ipilimumab revealed that 85 % experienced an immune-related adverse event of any grade, 19 % had to discontinue therapy due to the event, and 35 % of patients examined required systemic corticosteroids for their adverse event. In 31 of the 298 patients (10.4 %), symptoms did not resolve, necessitating the administration of anti-TNF therapy. One conclusion reached by the investigators was that ipilimumab can activate immune reactions against normal tissues leading to diarrhea, hepatitis, hypophysitis, and rash.

In late 2015, the FDA extended indications and usage of ipilimumab to adjuvant treatment of patients with cutaneous melanoma with pathological involvement of regional lymph nodes of more than 1 mm who have undergone complete resection, including lymphadenectomy.

The list of less common, but often potentially serious, adverse events provoked by ipilimumab is increasing as indicated by efforts to identify the spectrum of rare, often surprising, reactions. Patient files from 19 skin cancer centers revealed drug reaction (rash) with eosinophilia and systemic symptoms, otherwise known as DRESS, granulomatous inflammation of the central nervous system, and aseptic meningitis. Other rare, often life-threatening or even fatal, adverse events were myopathy, neuropathy, leukopenia, sarcoidosis, uveitis, and Guillain-Barré syndrome. Recent reports continue to highlight apparently rare reactions to ipilimumab, for example, the first case of organizing pneumonia; Guillain-Barré syndrome or acute polyradiculoneuritis; polyradiculoneuropathy with multifocal motor blocks; facial diplegia, tetraplegia, and areflexia; immune-mediated thrombocytopenia; autoimmune lymphocytic hypophysitis with anterior panhypopituitarism; hyponatremia; and Schwartz-Bartter syndrome. Usually less serious adverse events seen with a high frequency (≥5 %) following ipilimumab are diarrhea, fatigue, colitis, rash, and pruritus (Table 3.1).

At the end of 2012, gastrointestinal disorders, principally diarrhea and colitis, were the most frequently mentioned adverse events reported to the FAERS and European pharmacovigilance databases. Infections, mainly sepsis, pneumonia, and urinary tract infections, were near the top of the FAERS list, while in Europe, tumor progression, nervous and endocrine disorders, and skin reactions were prominent. There was one case of toxic epidermal necrolysis, 20 cases of intestinal perforation, 10 of hepatotoxicity, 11 of pancreatitis, two of anaphylaxis, and one case of CRS.

Immunogenicity does not appear to be a significant problem for ipilimumab therapy regardless of dosage levels. In a study of 1024 patients, 1.1 % had anti-ipilimumab serum antibodies, none neutralizing, and reactions suggestive of hypersensitivity were not seen.

Siltuximab

Siltuximab (Sylvant®, CNTO 328) (Tables 2.1 and 3.1) is a chimeric human-mouse IgG1κ mAb targeted to interleukin-6 (IL-6), forming high-affinity complexes with the cytokine's soluble bioactive forms. IL-6, produced by many different cells including lymphocytes, monocytes, fibroblasts, and endothelial and cancer cells, is a pleiotropic cytokine with a complex action producing both proinflammatory and anti-inflammatory effects. Promotion of inflammation by IL-6 results from its role in the activation and proliferation of T cells, stimulation of B cells, induction of acute phase proteins such as C-reactive protein (CRP) in the liver, and stimulation of hematopoietic precursor cell proliferation and differentiation. IL-6's anti-inflammatory role appears to be centered on inhibitory effects in turning off or modulating the synthesis of TNF, IL-1, and IL-10. The intricate interplay of pro- and anti-inflammatory effects of IL-6 suggests that the cytokine may be an important, if not sometimes crucial, participant in the parthenogenesis and/or response to some diseases. In fact, increased levels of IL-6 are known to be associated with a variety of diseases including rheumatoid arthritis, autoimmunities, Alzheimer's disease, neoplasia, atherosclerosis, Paget's disease, osteoporosis, and others. Cancers associated with increased production of IL-6 include renal cell carcinoma, prostate and bladder cancers, some neurologic cancers, and particularly multiple myeloma and the B-cell lymphoproliferative disorder, Castleman's disease. Castleman's disease may be localized or unicentric, or multicentric when it is characterized by generalized lymphadenopathy with systemic symptoms. Approximately half of the cases of multicentric Castleman's disease are caused by human herpesvirus-8 (HHV-8 or KSHV, Kaposi's sarcoma-associated herpesvirus), but the cause(s) of the other 50 % is unknown. Regardless, IL-6 has a central role in the pathophysiology of the multicentric form with its excess production leading to production of B lymphocytes and plasma cells, autoimmune reactions, and secretion of VEGF.

Interleukin-6 binds to a type I 80 kDa cytokine receptor (IL-6R) on target cells. The resultant complex associates with the protein gp130 (CD130), expressed on almost all cells, which dimerizes and initiates intracellular signaling. CD130 is the common signal transducer for several cytokines of the gp130 cytokine family including IL-11, leukemia inhibitory factor (LIF), and oncostatin M. Signaling proceeds through the JAK/STAT and Ras-Raf-MAPK signaling pathway after activation by phosphorylation of the phosphatase SHP-2 by JAK1. Cells that express gp130 only, do not respond to IL-6; IL-6 can only bind to cells expressing IL-6R. The receptor IL-6R, also found in soluble form (sIL-6R), binds IL-6 with an affinity comparable to the membrane-bound receptor. Importantly, cells expressing gp130, even in the absence of IL-6R, can respond to the complex of IL-6-sIL-6R. This process is called trans-signaling. Siltuximab blocks the binding of IL-6 to both the membrane-bound and soluble IL-6 receptors thereby preventing the formation of the signaling complex with gp130 on the cell surface.

Approved by both the FDA and EMA in 2014, siltuximab is indicated for the treatment of multicentric Castleman's disease in patients who are human immunodeficiency virus (HIV) negative and HHV-8 negative. Siltuximab has received orphan drug designation for multicentric Castleman's disease in both US and European Union. Warnings and precautions for the mAb issued by the FDA mention the risks inherent in administering siltuximab to patients with severe infections, including pneumonia and sepsis, and the possibility that siltuximab may mask signs of acute inflammation and suppress fever and acute phase reactants such as CRP. By extension, live vaccines should not be administered concurrently or within 4 weeks of commencing siltuximab therapy. Other warnings issued by the FDA are for infusion-related reactions and hypersensitivity and gastrointestinal perforation. An increased risk of malignancy and the development of hyperlipidemia and hepatic impairment are additional warnings issued by the EMA.

A summary of the safety profile of siltuximab reveals that anaphylaxis is the most serious adverse event associated with its use, while infections, including those of the upper respiratory tract, pruritus, and maculopapular rash are the most commonly occurring events (Table 3.1). In clinical trials, the incidence of infusion or hypersensitivity reactions was 4.8 % with 0.8 % of patients experiencing severe reactions. An evaluation of long-term exposure to siltuximab, namely, 3.4–7.2 years (median 5.1 years), showed that the most common adverse events (>20 %) were upper respiratory tract infections (63 %), diarrhea (32 %), and pain in extremities, arthralgia, and fatigue (21 % each). There were no deaths and no cumulative toxicities identified.

Sera from 411 patients on siltuximab mono- or combination therapy tested for anti-siltuximab antibodies proved positive in only one patient (0.24 %). Antibodies were present in low titer, they were non-neutralizing, and the patient's safety profile remained unaltered.

Monoclonal Antibodies Targeting Programmed Cell Death Protein 1 (PD-1): Pembrolizumab and Nivolumab

Pembrolizumab (Keytruda®) (Tables 2.1 and 3.1), approved by the FDA for advanced melanoma in September 2014 through a Breakthrough Therapy Designation, was also recently approved by the UK's Medicines and Healthcare Products Regulatory Agency (MHRA) via the UK's Early Access to Medicines Scheme (EAMS). FDA indications and usage are for patients with unresectable or metastatic melanoma and disease progression following ipilimumab. It is stated that the mAb should be administered with a BRAF inhibitor if the patient is positive for the BRAF V600 mutation. Pembrolizumab was approved by the EMA for marketing in Europe in July 2015. Breakthrough Therapy Designation for pembrolizumab in the USA for the treatment of non-small cell lung cancer was announced in October 2014, and the mAb is said to be showing encouraging trial results in the treatment of PD-L1-positive patients with advanced triple-negative breast cancer. In early June 2015, the FDA accepted for review the supplemental Biologics License Application for pembrolizumab for the treatment of advanced non-small cell lung cancer whose disease has progressed on or after platinum and EGFR therapy. The FDA granted priority review under its accelerated approval program with a target action date of early October 2015 and approval was finally granted on October 2. Nivolumab (Opdivo®), also targeted to PD-1 and approved in Japan in July 2014, was first granted breakthrough therapy, priority review, orphan product designation, and FDA approval in December 2014 before receiving accelerated approval for metastatic melanoma by the EMA in June 2015. Nivolumab appears to demonstrate improvement over other therapies including peginterferon alfa-2b, B-Raf enzyme inhibitors, and the mAbs ipilimumab and pembrolizumab. In early March 2015, nivolumab received its second FDA approval, this time for the treatment of patients with metastatic squamous non-small cell lung cancer with progression on or after platinum-based chemotherapy. In October 2015, approval was given for the treatment of non-squamous non-small cell lung cancer, and then in November, approval for nivolumab was extended to advanced metastatic renal cell carcinoma in patients who have received prior anti-angiogenic therapy.

Checkpoint Inhibitors and PD-1

In its vital balancing role of protecting against self-antigens while retaining an activated state to protect against foreign antigens, the immune system requires a number of checks and balances to maintain the necessary healthy equilibrium of self-tolerance, prevention of autoimmune reactions, the defense against invading organisms, and elimination of aberrant cells. The bases of these checks and balances are immune cells, particularly the T cell, together with an array of ligands and receptors that effect the many activation and inhibitory processes required for the stimulation and curtailment of immunostimuli. In relation to cancer, the terms "immune checkpoints" and "checkpoint inhibitors" refer to a number of molecules, usually proteins, on cell surfaces that exert a co-inhibitory effect on the immune response, for example, the cytotoxic T lymphocyte antigen 4 (CTLA-4) which acts at the first checkpoint (see section "Ipilimumab") to modulate overactive T cells by competing for the B7 ligands (Fig. 3.9). The programmed cell death protein 1 (PD-1, CD279) receptor which plays a critical role in cancer immunology acts as another important immune checkpoint. PD-1 is a transmembrane protein receptor of the Ig superfamily expressed on T cells during thymic development and on CD4+ and CD8+ T cells, B lymphocytes, NK cells, B cells, monocytes, and some dendritic cells during antigen signaling. PD-1 has two ligands, PD-L1 (CD274, B7-H1) and PD-L2 (CD273, B7-DC). Whereas the affinity of PD-L2 for PD-1 is three times higher than the affinity of PD-L1, it is expressed on fewer cells and cell types, for example, it is not present on resting cells but is inducibly expressed on dendritic cells, macrophages, and some B cells. PD-L1 is expressed on both hematopoietic and non-hematopoietic cells, including dendritic cells, macrophages, and B cells and cells of solid tumors. High expression of PD-L1 appears to be associated with increased aggressiveness of cancers and death. During PD-1 signaling in T cells, recruited phosphatases SHP-1 and SHP-2 cause dephosphorylation of associated signaling molecules downstream from the T-cell receptor (TCR) complex. Akt activation is inhibited via the PI3K pathway, and Akt signaling leads to suppression of T-cell proliferation, decreased protein synthesis, and survival and suppression of IL-2 production which further inhibits T-cell proliferation and survival.

Attempts to improve T cell function are proving promising in strategies aimed at complementing cancer immunotherapy. Reactivating CD8+ killer T cells , normally suppressed in many cancers, is yielding encouraging results. Enhanced proliferation of CD8+ but not CD4+ T cells was demonstrated by modulating cholesterol metabolism via inhibition of cholesterol esterification in T cells. In a mouse melanoma model, the esterification enzyme acetyl-CoA acetyltransferase 1 (ACAT1) was inhibited by the selective inhibitor avasimibe, leading to an increase in CD8+ membrane cholesterol levels and decreased melanoma growth and metastasis. When used with anti-PD-1 therapy, increased efficacy in controlling tumor progression was observed.

Targeting PD-1 in Cancer Therapy

Like CTLA-4, PD-1 negatively regulates T-cell activation. While remaining free of its complementary ligands, PD-1 does not interfere with the normal immune response, but, upon binding PD-L1 and PD-L2, signaling is induced, suppressing T-cell proliferation and activity. Malignant cells, including melanoma cells, often express PD-L1 which, upon binding, induces inhibitory signaling through the receptor preventing expansion of activated T cells and allowing the tumor cells to avoid immune recognition and attack. In other words, T-cell targeting of tumor cells can be subverted by cancer cells expressing PD-1 ligands and effectively using the PD-1 inhibitory pathway to blunt the immune antitumor response. Thus, it follows that selectively blocking the pathway should reverse the checkpoint inhibition and restore the T-cell-mediated response to the tumor. This strategy has, in fact, been adopted with the development and subsequent approval of mAbs designed to bind to PD-1 and prevent interaction of the receptor with its ligands (Fig. 3.9). Recent trials have shown that two of the mAbs, pembrolizumab and nivolumab, show encouraging clinical benefits with impressive high response rates and toxicity profiles superior to ipilimumab, all leading to the impression, so far, that PD-1 may be a better target for cancer therapy.

Pembrolizumab

Pembrolizumab (MK-3475, lambrolizumab) (Tables 2.1 and 3.1) is a humanized IgG4κ mAb, MW~149 kDa, targeted to PD-1 and indicated for the treatment of patients with unresectable or metastatic melanoma and disease progression following ipilimumab and, if BRAF V600 mutation positive, a BRAF inhibitor. Since the IgG4 subtype does not activate complement (Chap. 2, section "IgG Antibody Subclasses"), pembrolizumab is not cytotoxic when it binds to T cells. In addition to its approval for advanced non-small cell lung cancer whose disease has progressed on or after platinum-containing chemotherapy, FDA-approved usage for pembrolizumab extends to patients with EGFR or ALK (anaplastic lymphoma kinase) genomic tumor aberrations whose disease has progressed on FDA-approved therapy.

Pembrolizumab is generally well tolerated, but a spectrum of adverse events may be seen, and the drug's sizeable list of immune-related adverse events can be severe. Warnings and precautions issued by the FDA are for immune-mediated pneumonitis, colitis, hepatitis, hypophysitis, hyperthyroidism, hypothyroidism, renal failure, and nephritis, all seen with an incidence in the range ~0.5–3 %. Most of the data so far on these immune-mediated conditions were derived from an uncontrolled, open-label, multiple cohort trial involving 411 patients with unresectable or metastatic melanoma receiving the mAb. Adverse reactions that led to discontinuations were pneumonitis, renal failure, and pain. Serious effects occurred in 36 % of patients with the most frequent serious reactions being renal failure, dyspnea, pneumonia, and cellulitis. Some other immune-mediated adverse reactions, namely, uveitis, arthritis, myositis, pancreatitis, hemolytic anemia, and exfoliative dermatitis, were clinically significant but occurred in less than 1 % of patients. Based on its mechanism of action, a warning/precaution for pembrolizumab has also been issued for fetal harm. The most common adverse events reported are listed in Table 3.1.

In an early study of pembrolizumab's potential for immunogenicity, no treatment-emergent anti-pembrolizumab antibodies were detected in 97 patients examined.

Nivolumab

Nivolumab (BMS-936558, MDX-1106, ONO-4538) (Tables 2.1 and 3.1) is a fully human IgG4κ mAb, MW~149 kDa, targeted to PD-1 and indicated for the treatment of patients with unresectable or metastatic melanoma and disease progression following ipilimumab, and, if BRAF V600 mutation positive, a BRAF inhibitor should also be given. FDA approvals for the treatment of patients with metastatic squamous non-small cell lung cancer, metastatic non-squamous non-small cell lung cancer with progression on or after platinum-based chemotherapy, and advanced metastatic renal cell carcinoma have recently been announced. Like pembrolizumab, nivolumab is subject to warnings for immune-mediated adverse events of pneumonitis, colitis, hepatitis, hyperthyroidism, hypothyroidism, renal failure, and nephritis in melanoma patients. Other clinically significant, immune-mediated adverse events occurring in less than 1 % of treated melanoma patients were pancreatitis, uveitis, autoimmune neuropathy, demyelination, adrenal insufficiency, facial and abducens nerve paresis, hypophysitis, diabetic ketoacidosis, hypopituitarism, Guillain-Barré syndrome, and myasthenic syndrome. Awareness of the potential risk of embryofetal toxicity is a necessary additional precaution for pregnant women given nivolumab. Clinically significant nivolumab-induced adverse events were also assessed in 574 patients with solid tumors enrolled in clinical trials. Grade 3 and 4 adverse reactions occurred in 41 % of patients receiving nivolumab with abdominal pain, hyponatremia, and increases in aspartate transaminase and lipase reported in 2–5 % of patients. In addition to the adverse events listed here and in Table 3.1, ventricular arrhythmia, iridocyclitis, infusion-related reactions, neuropathies, and a number of different cutaneous disorders including exfoliative dermatitis, erythema multiforme, psoriasis, and vitiligo have been reported. Extensive safety data are not yet available for the use of nivolumab in patients with non-small cell lung cancer and renal cell carcinoma.

Of 281 nivolumab-treated patients, anti-nivolumab antibodies were detected in 24 patients (8.5 %) and neutralizing antibodies were found in two patients (0.7 %). There was no evidence that any of the antibodies influenced the effectiveness of the mAb or its toxicity profile.

Other mAbs directed against PD-1 are in the development pipeline. For example, in February 2016, the FDA granted Breakthrough Therapy Designation to the investigational human IgG1 mAb durvalumab for patients with inoperable/metastatic urothelial bladder cancer whose tumor has progressed on a platinum regimen.

Dinutuximab

Aberrant glycosylation is often characteristic of malignant cellular transformation. It is known that tumors with some carbohydrate antigens expressed at high levels show accelerated rates of metastasis and progression. The sialic glycosphingolipid (ganglioside), disialoganglioside (GD2), a short sialylated polysaccharide linked to ceramide through a β-glycosidic linkage (Fig. 3.10) and found highly expressed on neuroectoderm-derived cancers such as neuroblastoma, melanoma, brain tumors, osteosarcoma, and Ewing's sarcoma in children, is one such tumor-associated carbohydrate antigen. Significantly from the safety viewpoint, GD2 is not present on normal tissues and minimally expressed in the brain and peripheral nerves. The incidence of GD2-positive tumors in the USA is estimated to be in excess of 200,000 annually, and considering the high mortality rate of these cancers and the high priority given to GD2 as a potential target for cancer therapy, a number of strategies to exploit the therapeutic possibilities of this antigen have been investigated over the last three decades. One of the most promising approaches has been the development of mAbs targeted to GD2 for use in passive antibody therapy, but this strategy has faced the dual challenges of antibody affinity and tumor cell killing on the one hand and potential toxicities with pain inducement toward nerves and melanocytes on the other hand. Dinutuximab (ch14.18, Unituxin®) (Tables 2.1 and 3.1), a human-mouse chimeric IgG1κ mAb that binds GD2, received approval to treat neuroblastoma from the FDA in March 2015 and

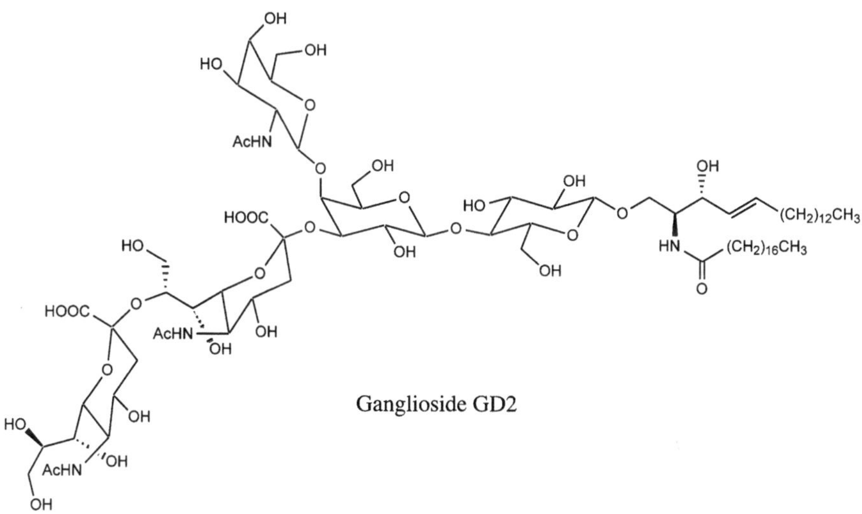

Ganglioside GD2

Fig. 3.10 Structure of ganglioside (G2), an antigen expressed on human tumors of neuroectodermal origin. The disialylated polysaccharide structure is linked to ceramide through a β-glycosidic linkage. The structure contains one molecule each of D-glucose, *N*-acetyl-D-galactosamine, and D-galactose with two attached *N*-acetylneuraminic acid residues. Only one of the various lipoforms is shown

from the EMA in May 2015. After being shown to prevent the outgrowth of experimental melanoma and neuroblastoma tumors, this mAb was found to have an acceptable safety profile in clinical trials on patients with neuroblastoma, melanoma, and osteosarcoma. Dinutuximab had been given an orphan product designation and received priority review and a pediatric disease priority review. A combination of the antibody with the cytokine granulocyte-macrophage colony-stimulating factor (GM-CSF) (Chap. 5, section "Colony-Stimulating Factors: Filgrastim, Sargramostim, and Tbo-filgrastim") was judged to enhance efficacy. Approved indications for dinutuximab now state that it should be used in combination with GM-CSF, IL-2, and 13-cis-retinoic acid for the treatment of pediatric patients with high-risk neuroblastoma who achieve at least a partial response to prior first-line multiagent, multimodality therapy.

FDA approval for dinutuximab comes with a boxed warning for potentially life-threatening infusion reactions requiring prior hydration and premedication and a warning for neuropathy, in particular severe neuropathic pain and nerve damage requiring prior, and subsequent, intravenous opioid treatment. In one study of more than 100 patients, severe (grade 3 or 4) infusion reactions occurred in just over one-quarter of the patients; urticaria was seen in 13 % and anaphylaxis resulted in 1 % of patients. One patient died from cardiac arrest within 24 h of receiving dinutuximab. Despite pretreatment with analgesics (including morphine), 85 % of 114 patients treated with dinutuximab experienced pain which was judged severe in 51 % of patients. Neuralgia and arthralgia occur but pain is most commonly reported as generalized or in the extremities, back, abdomen, and chest. Peripheral sensory neuropathy and severe peripheral motor neuropathy each occurred in 1 % of patients. The neuropathic effects of dinutuximab are generally more severe in adults than in children. Other warnings are for capillary leak syndrome (Chap. 1, section "Capillary Leak Syndrome"), found in one study in 23 % of patients (6 % severe reactions), hypotension (16 % grades 3 or 4), infection (bacteremia 13 % grades 3 or 4, sepsis 18 %), neurological eye disorders (blurred vision, photophobia, mydriasis, ptosis), bone marrow suppression (thrombocytopenia 39 %, anemia 34 %, neutropenia 34 %, febrile neutropenia 4 %), electrolyte abnormalities (hyponatremia, hypokalemia, hypocalcemia), atypical hemolytic uremic syndrome, and embryofetal toxicity. The most commonly seen adverse reactions to dinutuximab are listed in Table 3.1; the most common serious reactions are infections, infusion reactions, hypokalemia, hypotension, pain, fever, and capillary leak syndrome.

In two clinical studies in 284 and 103 patients, 18 % and 13 % of patients, respectively, tested positive for anti-dinutuximab antibodies. Neutralizing antibodies were found in 3.6 % of the patients.

Daratumumab

Daratumumab (HuMax-CD38, Darzalex®) (Tables 2.1 and 3.1), accepted for priority review by the FDA in September 2015 and approved by the agency in November 2015 for the treatment of multiple myeloma, is the first-in-class immunotherapy for this currently incurable disease in which almost all patients relapse or become resistant to therapy. Daratumumab, an IgG1κ human antibody, MW ~148 kDa, is the first mAb targeted to the 48 kDa glycoprotein CD38 (cyclic ADP ribose hydrolase), a surface antigen expressed by multiple myeloma cells and found on many immune cells including CD4+, CD8+, B lymphocytes, and natural killer (NK) cells. The mAb acts by inhibiting the growth of tumor cells expressing CD38, leading to apoptosis by Fc-mediated cross-linking and cell lysis induced via CDC, ADCC, and ADCP. Daratumumab is approved for the treatment of patients with multiple myeloma who have received at least three prior treatments including a proteasome inhibitor, an immunomodulatory agent, and who are double refractory to a proteasome inhibitor and an immunomodulatory agent.

Warnings and precautions for daratumumab are the occurrence of infusion reactions, interference with cross-matching and red blood cell antibody screening, and interference by the mAb when determining patient's response and disease progression in patients with IgG kappa myeloma (Table 3.1). As well as infusion reactions, adverse events recorded so far include infections, particularly pneumonia, thrombocytopenia, fatigue, and nausea. It is predicted that postmarketing usage will reveal cytopenias as one of the main adverse events (Table 3.1).

In an open-label clinical trial of patients with relapsed or refractory multiple myeloma treated with daratumumab, none of 111 patients proved positive for anti-daratumumab antibodies up to 8 weeks after completion of the treatment.

Elotuzumab

In September 2015, the humanized recombinant IgG1κ mAb elotuzumab (Empliciti®, HuLuc63) (~MW 148.1 kDa) targeting the cell surface glycoprotein receptor CS1, a member of the signaling lymphocytic activation molecule (SLAM) receptor family, was granted FDA priority review based on a phase III trial of the mAb together with lenalidomide and dexamethasone for the treatment of relapsed or refractory multiple myeloma. After its assisted pathway through orphan drug designation, breakthrough therapy, and priority review, formal approval for the mAb was granted by the FDA on November 30, 2015. Receptor CS1 (also known as CD2 subunit 1, SLAMF7, and CD319) is highly expressed on myeloma cells but not on other tissues, including hematopoietic stem cells. It has been predicted that treatment of elotuzumab with lenalidomide and dexamethasone may prove to be one of the new standards of care for patients with relapsed or refractory multiple myeloma with no additional toxic effects. In its approved indications for elotuzumab, the FDA specifies that the mAb should be given in combination with lenalidomide and dexamethasone to patients who have received one to three prior therapies. Elotuzumab targets SLAMF7 on myeloma cells and natural killer (NK) cells, facilitating the latter to kill myeloma cells through ADCC. The addition of lenalidomide to the mAb therapy results in enhanced NK cell-mediated killing.

Warnings and precautions issued by the FDA for elotuzumab are listed in Table 3.1. Infusion reactions have been reported in ~10 % of patients treated when the mAb is given in combination with lenalidomide and dexamethasone. As a result, premedication with dexamethasone, H1 and H2 antihistamines, and acetaminophen should be administered. Infections occurred commonly in clinical trials with opportunistic infections, including herpes zoster and fungal infections, seen at a significantly higher incidence in the elotuzumab-lenalidomide arm than the lenalidomide arm of the trial. Hematologic malignancies, solid tumors, and skin cancers reported in the elotuzumab-lenalidomide and lenalidomide arms of the trial were 1.6 % and 1.6 %, 3.5 % and 2.2 %, and 4.4 % and 2.8 %, respectively. Elevation in liver enzymes (ALT, AST), total bilirubin, and alkaline phosphatase consistent with hepatotoxicity indicates the need to periodically monitor liver enzymes during elotuzumab therapy. Being a humanized IgG kappa antibody, elotuzumab may interfere with assays used to monitor endogenous M-protein and determine the complete response of patients with IgG kappa myeloma protein. Other adverse reactions to elotuzumab determined so far are listed in Table 3.1.

Of 390 patients treated with elotuzumab, 72 (18.5 %) proved positive for anti-elotuzumab antibodies. In 63 of these patients, antibodies were detected within the first 2 months of treatment, but by 2–4 months, antibodies resolved in 49 of the 63 patients. In a multiple myeloma randomized trial, 19 of 299 patients (6.4 %) were found to have neutralizing antibodies.

Recent Approval: Atezolizumab

During the production stage of this book, the FDA granted mAb approval number 50 to the humanized IgG1 kappa antibody atezolizumab (Tecentriq™) for treatment of urothelial carcinoma. Targeted to PD-L1 and blocking interaction with PD-1 and B7.1 receptors, the mAb has a long list of warnings and precautions: immune-related pneumonitis, hepatitis, colitis, pancreatitis, endocrinopathies, and myasthenic syndrome as well as ocular inflammatory toxicity, infection, infusion reactions, and embryo-fetal toxicity.

Range of Side Effects of Monoclonal Antibodies Used for Cancer Therapy

Although mAbs used for cancer immunotherapy are generally better tolerated than small molecule chemotherapeutic drugs, their range of adverse effects is still wide, varying from headaches, mild gastrointestinal symptoms, and transient rashes to severe cytopenias; anaphylaxis; autoimmunity; pulmonary, cardiac, hepatic, kidney, neurological, and embryofetal toxicities; and rare life-threatening toxidermias. Some adverse events are clearly immune-mediated, but many others do not have, or appear to not have, an immune basis. Direct cytotoxic effects account for some reactions.

Types I–IV Hypersensitivities and Cytopenias

As discussed in Chap. 2, immunogenicity is always a safety concern for mAbs, even those that are fully humanized, since the possibility of generating anti-idiotype antibodies remains. The likelihood of immune-mediated adverse events following mAb administration therefore cannot be totally eliminated, and this covers the full range of hypersensitivity responses from type I IgE antibody-mediated immediate reactions such as anaphylaxis, urticaria, and angioedema to type II drug-induced thrombocytopenia, hemolytic anemia, and agranulocytosis; type III serum sickness and drug-induced vasculitis; a range of type IV cutaneous hypersensitivities mediated by Th1, Th2, and Th17 lymphocytes; and effector mechanisms involving cytotoxic lymphocytes, macrophages, eosinophils, and a number of other cell types. In fact, "hypersensitivity" is a much misused and often poorly understood term, and there is a pressing need for the acceptance of a common definition across the medical disciplines (see Chap. 1, section "Hypersensitivities"). It is not always clear that a reaction is a true hypersensitivity response or if any immune process, direct or indirect, is involved. In the first instance, the terminology used to describe hypersensitivities needs to be standardized to enable accurate interpretations of many adverse events.

Because of their immunogenic potential, mAbs generally carry warnings of immune reactions, especially anaphylaxis, but the observed incidences of such reactions are actually quite small. IgE-mediated reactions to chimeric proteins used for cancer therapy containing mouse and/or rat sequences (catumaxomab, blinatumomab, ibritumomab tiuxetan, brentuximab vedotin, cetuximab, rituximab, siltuximab, and dinutuximab) (Table 3.1) are considered to be of greater risk, and overall, and not unexpectedly, this has proved to be true. Anaphylaxis has been reported for, at least, cetuximab, rituximab, brentuximab, bevacizumab, trastuzumab, pertuzumab, ibritumomab, and dinutuximab, but the real incidences of reactions for each of the mAbs are hard to establish for a number of reasons including common misunderstanding and misuse of the term hypersensitivity, the clinician's ability/inability to distinguish true type I IgE-mediated anaphylaxis from some severe infusion and anaphylactoid reactions, the frequent failure to test for the presence of mAb-specific IgE antibodies, and differences in the frequency and extent of the use of different mAbs. For example, a study of 901 patients showed that 79 (9 %) experienced what was described as an immediate hypersensitivity reaction, while 76 % developed symptoms during their initial infusion leading the authors to conclude that immediate hypersensitivity to rituximab commonly occurs during or after the first infusion. Nonetheless, no evidence for the presence or absence of specific IgE antibodies either as skin test or serum immunoassay results was presented.

Cytopenias occur in some patients treated with mAbs during anticancer immunotherapy, but the underlying mechanisms frequently remain unexplored. Type II and III hypersensitivities induced by mAbs may be underdiagnosed. The FDA has issued a boxed warning for the risk of severe cytopenias with ibritumomab tiuxetan and alemtuzumab, while general warnings and precautions are set down for obinutuzumab (thrombocytopenia, neutropenia), ofatumumab (cytopenias), brentuximab vedotin (neutropenia), trastuzumab (neutropenia), and ado-trastuzumab emtansine (thrombocytopenia).

Listed among the other warnings/adverse events for the 24 anticancer mAbs are cytopenias for catumaxomab, brentuximab vedotin, and pertuzumab; lymphopenia for elotuzumab; lymphopenia and leukopenia for blinatumomab; neutropenia for rituximab; thrombocytopenia for daratumumab; thrombocytopenia and anemia for trastuzumab; and thrombocytopenia, lymphopenia, and neutropenia for dinutuximab (Table 3.1). Autoimmune forms of thrombocytopenia and hemolytic anemia are type II hypersensitivities, and reductions in the platelet, erythrocyte, and neutrophil counts, especially in the lymphoproliferative diseases, may sometimes have an immune basis. Thrombocytopenia, well known following the use of many small molecule chemotherapeutic drugs, is much more rarely seen during and after mAb treatments. Rituximab, implicated in thrombocytopenia as often as any of the approved mAbs, appears to show an incidence of ~1.7 %, and this figure is similar for mono- and combination therapies. However, an incidence as high as 10.4 % was reported in one study of 72 patients with non-Hodgkin lymphoma given a total of 317 rituximab infusions. From the study of a case of thrombocytopenia induced by rituximab and a review of the literature, it was not possible to implicate rituximab-dependent antibodies and IL-1 and IL-6 were not increased, but complement levels were elevated. This led the authors to conclude that mAb-induced transient thrombocytopenia might be mediated by complement activation and associated with CRS. Two cases of transient severe thrombocytopenia during rituximab therapy, one in a patient with mantle cell lymphoma and the other with hairy cell leukemia, reversed a few days after the antibody was withdrawn, but, again, the underlying mechanisms were not investigated. Other mAbs implicated in treatment-related thrombocytopenia include trastuzumab and alemtuzumab. When the latter was given for early multiple sclerosis, ~3 % of patients developed potentially fatal thrombocytopenia, and of 11 patients with peripheral T-cell lymphoproliferative disorders given alemtuzumab, five developed lymphopenia, neutropenia, and thrombocytopenia.

Rituximab has been implicated in both early and late neutropenia. Late onset neutropenia manifests at least 4 weeks after the cessation of therapy; it occurs with a comparatively high incidence (4–23 %) and appears to be caused by a different mechanism than the early form of the disorder. While the mechanism underlying late onset neutropenia is poorly understood, results suggest that direct cytotoxicity is unlikely and immune mechanisms, including autoantibodies, may be responsible for the rituximab-induced disease.

Incidences of 1.1 and 5.2 % for severe anemia have been reported for patients receiving rituximab monotherapy. Other serious cases involving rituximab include severe autoimmune hemolytic anemia in a patient with a lymphoproliferative disorder; a case of intravascular hemolysis, rhabdomyolysis, renal failure, and bone marrow necrosis; and multiple organ ischemia due to an anti-Pr cold agglutinin in a patient with mixed cryoglobulinemia after treatment with rituximab. At least two mAbs, rituximab and alemtuzumab, have been implicated in the induction of pure red cell aplasia and autoimmune hemolytic anemia.

Hypersensitivity vasculitis induced by drugs is a manifestation of a type III response, and a few mAbs including rituximab have been implicated in cases of cutaneous vasculitis. Serum sickness reactions to mAbs, another type III hypersensitivity, are probably underdiagnosed and reported. Chimeric antibodies in particular have the potential to induce the reactions, and, again, rituximab has been the mAb most implicated. It has been claimed that rituximab serum sickness-like reactions can occur in up to 20 % of treated patients, especially in those with autoimmune diseases (particularly autoimmune thrombocytopenia) and hypergammaglobulinemia.

Although autoimmune diseases induced by anticancer mAbs are rare, occasional cases occur with ipilimumab standing out as the most common cause. By targeting CTLA-4 and acting as an immunostimulatory agent, ipilimumab produces an antitumor response and augments T-cell activation that sometimes leads to immune-mediated colitis, hepatitis, nephritis, hypothyroidism, and hyperthyroidism. By blocking the PD-1 receptor, mAbs nivolumab and pembrolizumab may provoke a similar range of autoimmune reactions as well as autoimmune pneumonitis.

Type IV cutaneous reactions to drugs, including mAbs, generally become apparent 7–21 days after exposure, but subsequent reactions may appear only a day or two after reexposure. Specificity of the culprit antigen is established by patch and intradermal testing, usually read after a delay of at least 48 h, although great caution, or preferably avoidance, should be exercised in skin testing cases of cutaneous toxidermias such as Stevens-Johnson syndrome and toxic epidermal necrolysis. Besides these two rare and potentially life-threatening conditions, type IV cutaneous reactions include allergic contact dermatitis, psoriasis, maculopapular exanthema, fixed drug eruption, acute generalized exanthematous pustulosis, erythema multiforme, and drug reaction with eosinophilia and systemic symptoms (DRESS). Being mediated by lymphocytes, sensitivity to the provoking antigen(s) can be transferred by lymphocytes in type IV reactions. Such delayed cutaneous hypersensitivity reactions to anticancer mAbs are rare with most reported cases restricted mainly to ibritumomab tiuxetan, brentuximab vedotin, and rituximab. Lichenoid dermatitis, vesiculobullous dermatitis, and paraneoplastic pemphigus have also occurred in response to rituximab, and a case of lichenoid eruption was recently seen after obinutuzumab (section "Obinutuzumab"). A number of other mAb-induced cutaneous manifestations with features seemingly common to a type IV response may be true type IV hypersensitivities, but mechanisms remain to be established, for example, cases of dermatitis induced by catumaxomab, bevacizumab, denosumab, ipilimumab, and panitumumab.

Infusion Reactions and Cytokine Release Syndrome

The FDA has issued boxed warnings for the possibility of serious or even fatal infusion reactions to ibritumomab vedotin, rituximab, alemtuzumab, cetuximab, panitumumab, trastuzumab, and dinutuximab and a general warning for the risk of infusion reactions during or following treatment with obinutuzumab, ofatumumab, brentuximab vedotin, bevacizumab, ramucirumab, pertuzumab, ado-trastuzumab emtansine, and siltuximab (Table 3.1). Infusion reactions provoked by mAbs usually begin within hours of the initial infusion. Reactions are typically mild to moderate manifesting as "flu"-like symptoms of fever, chills, rigor, headache, nausea, asthenia, pruritus, and rash. In a small number of patients, severe, life-threatening symptoms common to type I IgE antibody-mediated anaphylaxis, in particular, hypotension, bronchospasm, cardiac arrest, and urticaria, may occur, usually during the first or second infusion (Chap. 1, section "Hypersensitivities"). The similarity of the signs and symptoms can make it difficult to distinguish an infusion reaction from a true allergic hypersensitivity although IgE-mediated reactions generally have a faster and more severe onset, usually within minutes. Severe reactions have been reported for all, or almost all, the mAbs although some show a much higher incidence with the chimeric rituximab and humanized trastuzumab antibodies being the leading offenders. The incidence of reactions for cetuximab, another human-mouse chimera, is ~15–20 % (grade 3–4, 3 %); for trastuzumab, first infusion ~40 % (grade 3–4, <1 %); and for rituximab, first infusion ~77 % (grade 3–4, 10 %). Approximately 80 % of fatal infusion reactions to rituximab occurred after the first infusion, and 30 and 14 % of patients still reacted after the fourth and eighth infusions, respectively. Even though trastuzumab is a humanized mAb, it induces a relatively high incidence of infusion reactions, but bevacizumab, another humanized antibody, shows a reaction incidence of only <3 % (grades 3–4, 0.2 %) which is similar to the fully humanized panitumumab (3 %, grades 3–4, ~1 %). Elotuzumab, recently approved for the treatment of multiple myeloma, provokes infusion reactions in a large proportion of patients and needs to be given with premedication (Table 3.1). The mechanisms of mAb-induced infusion reactions are not yet fully understood. Cytokines, especially TNF and interleukins such as IL-6, may be involved since the symptoms they produce resemble those seen in infusion (and type I allergic) reactions. An important finding was the observation that the severity of infusion reactions is related to the number of circulating lymphocytes. For example, a severe reaction is thought to require a lymphocyte count of $>50 \times 10^9$/L. CRS may be seen after the use of mAbs directed to malignant immune cells, for example, rituximab. It is thought that the systemic inflammatory response produced together with a high fever is a consequence of antibody binding to and activating the cells. The distinguishing features in the literature between CRS and severe infusion reactions are often not clear, and in many reported cases, the two designations may be interchangeable.

Pulmonary Adverse Events

Classified under the heading drug-induced lung diseases (DILDs), these pulmonary adverse events make up a heterogeneous group of diseases, most still of unknown, or poorly understood, mechanism of action. They have been grouped into four categories: interstitial pneumonitis and fibrosis, acute respiratory distress syndrome (ARDS), bronchiolitis obliterans organizing pneumonia (BOOP), and hypersensitivity pneumonitis. There are, however, a number of other classifications in the literature. Interestingly, hypersensitivity pneumonitis to some anticancer agents is

Table 3.3 Pulmonary adverse events caused by approved monoclonal antibodies used for cancer therapy

Monoclonal antibody	Pulmonary adverse events
Rituximab (MabThera®, Rituxan®)	ARDS, BOOP,[a] bronchiolitis obliterans,[b] bronchospasm, diffuse alveolar hemorrhage, hypersensitivity pneumonitis, interstitial lung disease,[b] interstitial pneumonitis[c]
Brentuximab vedotin (Adcetris®)	With bleomycin[d]: cough dyspnea, interstitial filtrations
	In Hodgkin lymphoma: pneumonitis, pneumothorax
Alemtuzumab (Campath®, MabCampath®)	Bronchospasm,[e] diffuse alveolar hemorrhage, pulmonary infection[f]
Cetuximab (Erbitux®)	Interstitial lung disease
Panitumumab (Vectibix®)	Interstitial lung disease,[g] lung infiltrates, pneumonitis, pulmonary fibrosis
Bevacizumab (Avastin®)	Bronchospasm/anaphylaxis, pneumonitis, pulmonary hemorrhage from the site of tumor
Trastuzumab (Herceptin®)	ARDS, BOOP, dyspnea, interstitial pneumonitis, pleural effusions, pulmonary infiltrates/fibrosis/edema/insufficiency and hypoxia
Ado-trastuzumab emtansine (Kadcyla®)	Interstitial lung disease—includes pneumonitis, ARDS, pulmonary infiltrates
Nivolumab (Opdivo®)	Severe pneumonitis or interstitial lung disease including fatal cases
Pembrolizumab (Keytruda®)	Immune-mediated pneumonitis

ARDS acute respiratory distress syndrome, *BOOP* bronchiolitis obliterans organizing pneumonia
[a]BOOP is the most common clinical diagnosis followed by interstitial pneumonitis, ARDS, and hypersensitivity pneumonitis
[b]Fatal cases have occurred
[c]Also called interstitial pneumonia or Hamman-Rich syndrome
[d]Concomitant use of brentuximab vedotin with bleomycin is contraindicated, e.g., with ABVD (Adriamycin, bleomycin, vinblastine, dacarbazine) combination therapy
[e]Serious fatal infusion reactions may include bronchospasm, ARDS, pulmonary infiltrates, and anaphylaxis
[f]For example, tuberculosis and aspergillosis
[g]Discontinue panitumumab in patients developing interstitial lung disease

increasingly looking like a combined type III and type IV hypersensitivity reaction in a Th1/Th17 response. Table 3.3 lists the mAbs used in cancer therapies that have been implicated in pulmonary adverse reactions. At least ten of the currently approved mAbs for cancer therapy have some recorded pulmonary toxicity in treated cancer patients. Once again, rituximab is the main offender inducing a wide range of adverse events with BOOP most often seen followed by interstitial pneumonitis, ARDS, and hypersensitivity pneumonitis. It has been suggested that early onset organizing pneumonia is a hypersensitivity reaction to the mAb, whereas the late onset condition is either related to mAb toxicity or to immune system restoration. ARDS symptoms appearing within a few hours of infusion may be a manifestation of CRS or TLS with no relationship to hypersensitivity although ARDS has also been linked to the release of proinflammatory cytokines. The incidence of rituximab-associated interstitial lung disease has been estimated to be 0.01–0.03 %. Its pathogenesis remains largely unknown although complement and TNF may be involved. Besides rituximab,

Table 3.4 Cardiac adverse events caused by approved monoclonal antibodies used for cancer therapy

Monoclonal antibody	Cardiac adverse events
Ibritumomab tiuxetan (Zevalin®)	Cardiac arrest related to infusions
Obinutuzumab (Gazyva®, Gazyvaro®)	Worsening of preexisting cardiac conditions leading to fatal cardiac events
Rituximab (MabThera®, Rituxan®)	Cardiac arrhythmias and angina,[a] fatal cardiac failure
Brentuximab vedotin (Adcetris®)	Supraventricular arrhythmia in systemic anaplastic large cell lymphoma
Alemtuzumab (Campath®, MabCampath®)	Cardiomyopathy, decreased LVEF,[b] cardiac arrhythmias associated with infusions[c]
Cetuximab (Erbitux®)	Cardiopulmonary arrest/sudden death[d]
Bevacizumab (Avastin®)	CHF: incidence of grade 3 reaction for LVD 1 %
Ramucirumab (Cyramza®)	Serious, sometimes fatal, myocardial infarction
Pertuzumab (Perjeta®)	Cardiomyopathy manifesting as CHF and decreased LVEF[b]
Trastuzumab[e] (Herceptin®)	Cardiomyopathy manifesting as CHF and decreased LVEF[b]
Ado-trastuzumab emtansine (Kadcyla®)	Decreased LVEF[b]

CHF congestive heart failure, *LVD* left ventricular dysfunction, *LVEF* left ventricular ejection fraction

[a]Can be life threatening. Discontinue infusions. Perform cardiac monitoring after each infusion for patients with arrhythmia/angina

[b]Incidence is highest when mAb is administered with cardiotoxic agents such as anthracyclines

[c]In ~14 % of previously untreated patients. Most reaction temporarily associated with infusions

[d]In patients treated with cetuximab and radiation therapy

[e]Patients receiving trastuzumab alone or in combination therapy show a four- to sixfold increase in the incidence of myocardial dysfunction

serious and/or fatal cases have occurred following administrations of alemtuzumab, trastuzumab, bevacizumab, panitumumab, and cetuximab.

Cardiac Adverse Events

Cardiac adverse events have occurred with at least 11 of the mAbs used for cancer therapy (Table 3.4). Cardiac arrhythmias and angina are reported for rituximab; obinutuzumab may lead to a worsening of preexisting cardiac conditions; brentuximab vedotin has been linked to supraventricular arrhythmia in some lymphoma patients; ramucirumab is implicated in serious and even fatal myocardial infarction; and bevacizumab is associated with congestive heart failure with an incidence of 1 % for LVD. Cardiopulmonary arrest and/or sudden death occurred in 4 (1.9 %) of 208 patients given cetuximab and radiation therapy, and cardiac arrest after ibritumomab tiuxetan and arrhythmia after alemtuzumab were each associated with infusions. Cardiomyopathy manifesting as congestive heart failure and decreased LVEF may occur following treatment with pertuzumab, trastuzumab, ado-trastuzumab emtansine, and alemtuzumab. Decreases in LVEF are well known for mAbs and other drugs that block HER2 activity, and this risk is increased in patients given anthracyclines or radiotherapy to the chest. In fact, patients administered trastuzumab show a four- to sixfold elevation in the incidence of myocardial infarction, and, again, this risk is highest when the mAb is given with an anthracycline. Interestingly, trastuzumab inhibits neuregulin 1, a growth factor for cardiac development and maintenance of heart structure and integrity. Necitumumab carries an FDA black box warning for cardiopulmonary arrest.

Mucocutaneous Reactions to Monoclonal Antibodies Targeted to Epidermal Growth Factor Receptor

These cutaneous reactions are not immune-mediated, that is, they are not genuine hypersensitivities. Skin reactions appear as a papulopustular eruption (sometimes less precisely called an acneiform rash), often in a large proportion of patients (50–100 %) and in a more severe form than seen with small molecule tyrosine kinase inhibitors. Eruptions tend to be confined to seborrheic regions (Fig. 3.3a–c), areas that maintain their integrity via EGFR expressed in the epidermis, sebaceous glands, and hair follicles. In the presence of inhibitors of EGFR, the epithelial barrier may be weakened allowing bacterial access and ultimately the development of the characteristic rash. Other adverse effects induced by mAbs targeted to EGFR include paronychia (Fig. 3.11), fissures, xerosis (Fig. 3.3h),

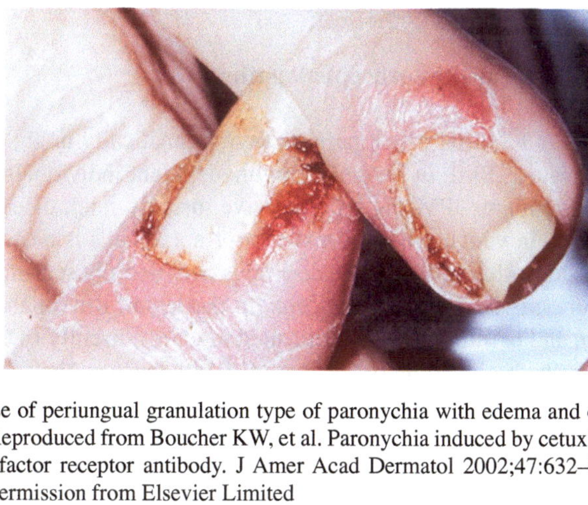

Fig. 3.11 A case of periungual granulation type of paronychia with edema and erythema caused by cetuximab. Reproduced from Boucher KW, et al. Paronychia induced by cetuximab, an antiepidermal growth factor receptor antibody. J Amer Acad Dermatol 2002;47:632–3. Modified and reprinted with permission from Elsevier Limited

palmar-plantar rash, hair changes, hyperkeratosis, mucositis (Fig. 3.3f, g), nail pyrogenic granuloma (Fig. 3.3i), and skin hyperpigmentation.

Other Rare Adverse Events Following Antitumor Monoclonal Antibody Therapy

See Chap. 1, sections "Tumor Lysis Syndrome" and "Progressive Multifocal Leukoencephalopathy," for a discussion of these two syndromes.

Tumor Lysis Syndrome

Depending on the anticancer agent used and the tumor load, within 48–72 h of starting the therapy, large numbers of malignant cells may be destroyed in a short time resulting in hyperkalemia, hypercalcemia, hyperphosphatemia, and hyperuricemia. Especially in patients with high tumor load, this can produce profound ionic imbalances in potassium, calcium phosphate, and uric acid and progress to acute renal failure, cardiac arrhythmias, seizures, and death. The condition is known as tumor lysis syndrome (TLS), and, unlike CRS, the response is easy to distinguish from type I immediate hypersensitivity reactions. TLS usually occurs in patients with leukemias and high-grade lymphomas and is rarely seen in association with solid tumors. The syndrome is well known to occur with the use of the CD20-targeted mAbs and brentuximab vedotin targeted to CD30, but the reaction elicited by rituximab appears to be somewhat atypical and remains to be further characterized. The FDA has issued a TLS boxed warning for rituximab and warning and precautions for obinutuzumab, brentuximab vedotin, and blinatumomab.

Progressive Multifocal Leukoencephalopathy

The polyomavirus JC virus which persists asymptomatically in about one-third of the population causes PML in severely immunodeficient individuals such as transplant and AIDS patients. PML is a progressive, usually fatal, disease resembling multiple sclerosis in which the myelin sheath of nerve cells is destroyed affecting nerve transmission. Although rare, the disease is occasionally seen upon the administration of some mAbs directed to B cells, in particular, rituximab, ofatumumab, obinutuzumab, and brentuximab vedotin, and there are currently FDA boxed warnings for potentially fatal PML in patients treated with these mAbs and a warning for ofatumumab (Table 3.1). In 2009, 57 cases of PML following rituximab therapy in HIV-negative patients were reported. A 2010 report of the WHO Collaborating Centre for International Drug Monitoring Adverse Event Data Bank revealed that rituximab was responsible for 114 of 182 cases of PML.

Summary

- Of the 50 monoclonal antibodies (mAbs) currently approved by the FDA and/or EMA, 24 are indicated for the treatment of hematologic, cutaneous, or solid tumor malignancies.
- **Catumaxomab** (**Removab**®), a mouse-rat hybrid mAb composed of a mouse kappa light chain and IgG2a heavy chain and a rat lambda light chain and IgG2b heavy chain, shows dual antigen recognition specificity for EpCAM and CD3, while the hybrid Fc fragment binds to FcγRI, FcγRIIa, and FcγRIIIa on macrophages and NK cells.
- Used to treat malignant ascites, catumaxomab may induce cytokine release syndrome (CRS), a reflection of T-cell activation, and the mAb's mode of action. Lymphopenia, reported in up to 14 % of patients, is reversible. Cutaneous reactions of rash, erythema, pruritus, and catheter-related reactions (erythema and infection) can be serious. Both human anti-mouse and human anti-rat antibodies are seen.
- **Blinatumomab** (**Blincyto**®), a bispecific T-cell-engaging (BiTE) fusion protein, is composed of two antibody single-chain variable fragments each from an H and L chain. One of the two binding specificities is directed to the B-cell antigen CD19, while the other targets CD3, part of the T-cell receptor. Reaction with both antigens is exploited to link malignant B cells of patients with acute lymphoblastic leukemia to cytotoxic T cells, activating them to destroy the tumor cells.
- Blinatumomab approved by the FDA is issued with a Risk Evaluation and Mitigation Strategy (REMS). Neurological toxicities range from confusion to tremors, convulsions, and speech disorders but appear to be reversible. CRS, sometimes life-threatening or even fatal, has been reported in up to 11 % of patients and may lead to hemophagocytic lymphohistiocytosis.
- CD20 is expressed on the surface of B cells, except for plasmablasts, at all stages of their development. It also occurs on B-cell lymphomas, B-cell chronic lymphocytic leukemia, hairy cell leukemia, and melanoma cancer stem cells. Due to its non-Hodgkin lymphoma (NHL) B-cell expression and the fact that it is not normally shed from cells and is internalized after binding to antibody, CD20 has been seen as an exploitable target for mAbs for the treatment of lymphomas. Four mAbs directed to CD20 are currently approved as antitumor agents—rituximab, ibritumomab, ofatumumab, and obinutuzumab.
- **Rituximab** (**MabThera**®, **Rituxan**®), a human-mouse chimeric IgG1κ antibody, was the first mAb approved to treat relapsed or refractory NHL and the first mAb approved specifically for cancer therapy. It is now a first-line therapy for several NHLs, including follicular lymphoma and diffuse large B-cell lymphoma. In 2002, the FDA authorized the use of rituximab as a component of ibritumomab therapy.
- A long list of side effects has been reported for rituximab ranging from the serious events detailed in an FDA black box warning, namely, infusion reactions, PML, TLS, and severe mucocutaneous reactions (all potentially fatal). Other adverse events are infections, hepatitis B reactivation, cardiac arrhythmias, lung

toxicities, and bowel problems plus a host of often less serious systemic and cutaneous events and toxicities.

- Infusion reactions to rituximab occur on the first infusion in up to 77 % of malignant patients and this decreases to approximately 10 % after the second infusion. Lymphocyte counts suggest that the appearance of CRS correlates with counts higher than 50×10^9/L.
- Postmarketing surveillance of rituximab administration has confirmed the importance of infections and respiratory and hematologic events in its safety profile. A 2010 report of the WHO Collaborating Centre for International Drug Monitoring Adverse Event Data Bank revealed that rituximab was responsible for 114 of 182 cases of the JC virus-induced PML.
- **Ibritumomab tiuxetan** (**Zevalin**®), a murine mAb covalently linked to the chelator tiuxetan by a stable thiourea bond, is radiolabeled with yttrium-90 for therapy or indium-111 for imaging. A boxed warning highlights severe, and potentially fatal, infusion reactions and severe cytopenias as the most serious adverse events experienced with ibritumomab therapy.
- Other serious events following ibritumomab tiuxetan are infections and potentially fatal myeloid malignancies or dysplasias. Hypersensitivity reactions, generally manifesting as bronchospasm or angioedema, are another potentially dangerous side effect. Of the non-hematologic events, gastrointestinal symptoms are commonly seen as well as rare toxidermias such as bullous dermatitis, erythema multiforme, Stevens-Johnson syndrome, and toxic epidermal necrolysis.
- The possibility of hepatitis B virus reactivation, cytopenias, intestinal obstruction, and PML comprise the warnings, precautions, and risks issued for **ofatumumab** (**Arzerra**®). Prolonged severe neutropenia and thrombocytopenia may occur, making necessary the regular monitoring of blood and platelet counts.
- Although exhibiting less CDC than rituximab, **obinutuzumab** (**Gazyva**®, **Gazyvaro**®) has demonstrated superior efficacy. Patients with CLL and coexisting illnesses unable to tolerate combined intravenous chemotherapy are likely to be more tolerant of obinutuzumab in its approved indication of the mAb's combination with chlorambucil.
- Grade 3–4 neutropenia, infusion reactions, anemia, thrombocytopenia, pyrexia, and musculoskeletal pain are the most common events induced by obinutuzumab. In addition to a boxed warning for hepatitis B virus reactivation and PML, warnings and precautions cover severe and life-threatening infusion reactions, TLS, the risk of infections, and cytopenias.
- **Brentuximab vedotin** (**Adcetris**®) is a chimeric mAb conjugated to the cytotoxic agent monomethyl auristatin E (MMAE) and targeted to CD30 which is overexpressed in Hodgkin lymphoma and some other lymphomas. After binding to cells expressing CD30, brentuximab's toxic payload MMAE is internalized. After reaching the lysosomes, MMAE is released, disrupting microtubules and ultimately inducing apoptosis.

- The list of warnings and precautions issued by the FDA for brentuximab vedotin covers 11 concerns related to neuropathies (predominately peripheral), infusion-related reactions, hematologic toxicities, serious infections, TLS, hepatotoxicity, PML, embryofetal toxicity, and serious dermatologic reactions. A small number of cases of anaphylaxis have been reported.
- Overall, the safety data for brentuximab vedotin shows that neutropenia, peripheral neuropathy, the risk of infections, and the occasional case of PML are perhaps the most important adverse events. Stevens-Johnson syndrome and toxic epidermal necrolysis constitute two other serious adverse events that, although rare, are potentially lethal.
- **Alemtuzumab (Campath®, MabCampath®)**, a humanized mAb with complementarity-determining regions from a rat mAb, is targeted to CD52 which is expressed on mature normal and malignant B and T lymphocytes, monocytes, macrophages, NK cells, a subpopulation of granulocytes, and dendritic cells.
- Approved for the treatment of B-cell CLL resistant to alkylating agents, alemtuzumab was widely used in cancer therapy until 2012 when it was withdrawn from US and European markets. Patients can still receive it through specific access programs and some off-label usage in cancer therapy remains.
- Cytopenia, resulting from alemtuzumab's mode of action in destroying white blood cells, is also the major adverse event, hence the FDA boxed warning. As well as a high incidence of neutropenia (75–85 %), febrile neutropenia and thrombocytopenia (serious in 57 % of cases) may occur. Infections related to alemtuzumab therapy have been reported with incidences of up to 80 % and serious events as high as 50 %.
- The destruction of T cells by alemtuzumab can lead to CRS, and potentially nephrotoxic TLS may occur as a result of the rapid and massive destruction of neoplastic cells. Infusion reactions to alemtuzumab occur most commonly during the first week of treatment. Serious reactions can be fatal.
- In a number of different tumors, EGFR and its ligands are associated with the growth of the cells, and elevated EGFR tyrosine kinase activity is found in many, if not most, solid tumors including breast, renal, head and neck, colon, and non-small cell lung cancer. **Anti-EGFR mAbs** such as cetuximab bind to the EGFR with higher affinity than the natural ligands, preventing subsequent activation of tyrosine kinase-mediated signal transduction pathways. Monoclonal antibodies targeting the EGFR were therefore seen as a new approach for treating a range of solid tumors.
- Patients with mutations in codons 12 and 13 of the *KRAS* gene almost never benefit from cetuximab and panitumumab treatments, but 10–30 % of patients with the wild-type gene respond to the mAbs. As a result, regulatory agencies restricted the use of cetuximab and panitumumab to colorectal cancer patients expressing the wild-type *KRAS* gene.
- Serious infusion reactions and the possibility of cardiopulmonary arrest make up an FDA black box warning for **cetuximab (Erbitux®)**. The severity and risk of many of the infusion reactions are underlined by the rapid onset of symptoms of

airway obstruction and other serious effects. Cardiopulmonary arrest and/or sudden death, the second subject of the boxed warning, occurred in 2–3 % of patients treated with cetuximab.

- Other warnings and precautions for cetuximab are for the possibility of interstitial lung disease; a variety of dermatologic events, for example, papulopustular rash (with an incidence of 76–88 %, up to 17 % serious); xerosis and fissuring; paronychial inflammation; hypertrichosis; infectious sequelae such as cellulitis and conjunctivitis; and hypomagnesemia and electrolyte abnormalities.
- **Panitumumab (Vectibix®)** binds the EGFR with high affinity. The antigenic structures recognized by cetuximab and panitumumab are not identical. This conclusion is supported by effective treatment with panitumumab of patients with disease progression under cetuximab and the development of resistance to cetuximab in a colorectal cancer patient who acquired a point mutation in the EGFR domain (Arg for Ser at position 468), while panitumumab binding and efficacy remained.
- At the first approval of panitumumab in 2006, the FDA issued a black box warning for dermatologic toxicity and infusion reactions. In the revised prescribing information issued in 2014, reference to infusion reactions in the boxed warning was removed, and dermatologic toxicities were stated to be severe in 15 % of patients.
- Mucocutaneous diseases provoked by panitumumab contribute to an extensive range of clinical manifestations including erythema, rash, pruritus, skin exfoliation, acneiform dermatitis, xerosis, paronychia, and skin fissures as well as life-threatening infectious complications such as necrotizing fasciitis and abscesses. Life-threatening bullous mucocutaneous diseases with erosions, blisters, and skin sloughing following panitumumab are known.
- Other warnings and precautions for panitumumab are for severe, including fatal, infusion reactions; severe hypomagnesemia, hypocalcemia, and hypokalemia; acute renal failure resulting from severe diarrhea and dehydration when panitumumab is used in combination with chemotherapy; interstitial lung disease and pulmonary fibrosis; ocular toxicities such as keratitis and ulcerative keratitis; dermatologic toxicities caused by sunlight exposure; and the possibility of increased toxicity and mortality when panitumumab is administered in combination with bevacizumab and chemotherapy.
- **Necitumumab (Portrazza®)** is a recombinant human mAb targeted to human EGFR. Indicated for first-line treatment of patients with metastatic squamous non-small cell lung cancer in combination with gemcitabine and cisplatin, necitumumab carries FDA black box warnings for cardiopulmonary arrest and/or sudden death and for hypomagnesemia.
- Five other warnings/precautions issued for necitumumab are the possible occurrence of venous and arterial thromboembolic events; some fatal, infusion-related reactions; embryofetal toxicity; increased toxicity and mortality in patients with non-squamous non-small cell lung cancer treated with the mAb plus pemetrexed and cisplatin; and dermatologic toxicities.
- **Bevacizumab (Avastin®)** binds to and inhibits the biological action of human vascular endothelial growth factor-A (VEGF-A). As well as its function of pro-

moting blood vessel formation in healthy subjects, VEGF-A-induced angiogenesis has a major role in the pathogenesis of a wide range of human diseases, for example, cancers, rheumatoid arthritis, and eye diseases.

- The relatively wide variety of tumors (about 30, including investigative and preliminary studies) treated by bevacizumab ensures a large number and variety of consequent adverse events. An extensive list of warnings and precautions is headed by a boxed warning for gastrointestinal perforation, complications of surgery and wound healing, and severe or fatal hemorrhage. Other serious adverse events seen are hypertension, arterial and venous thromboembolic events, proteinuria, PRES (<0.5 %), and anaphylaxis.
- **Ramucirumab** (**Cyramza**®) specifically binds VEGFR-2, the receptor-mediating angiogenesis, a major contributor, if not promoter, of tumor growth.
- Ramucirumab carries an FDA boxed warning for an increased risk of hemorrhage (including gastrointestinal hemorrhage that may be severe and sometimes fatal), gastrointestinal perforation, and impaired wound healing. Other warnings are for serious and sometimes fatal arterial thrombotic events including myocardial infarction, cardiac arrest, cerebrovascular accident, and cerebral ischemia; infusion-related reactions, usually during or following the first or second infusion; PRES; proteinuria including nephrotic syndrome; thyroid dysfunction; and the risk of deterioration in patients with cirrhosis.
- Overexpression of HER2 leads to constitutive activation of the growth factor signaling pathways with a consequent favorable environment for breast cancer cell growth. Overexpression, found in ~15–25 to 30 % of human breast cancers, tends to correlate with tumors that are more aggressive and show poorer prognosis. Three mAbs targeting HER2 are currently approved by the FDA and EMA, pertuzumab, trastuzumab, and the antibody-drug conjugate prepared by conjugating trastuzumab to the cytotoxin, mertansine.
- Although **pertuzumab** (**Perjeta**®) and **trastuzumab** (**Herceptin**®) target the same receptor, they recognize different binding sites. Trastuzumab inhibits ligand-independent signaling, complementing the mechanism of action of pertuzumab which inhibits ligand-dependent signaling between HER2 and HER3, a combination that potently activates cell survival and proliferation. The efficacy of pertuzumab alone is not impressive, but, together with trastuzumab, a more effective blockade of the HER2-driven signaling pathways results.
- There is an FDA boxed warning for **pertuzumab** for cardiomyopathy manifesting as congestive heart failure and decreased LVEF and for embryofetal toxicity that may result in fetal death and birth defects. Patients who have received prior treatment with anthracyclines or prior radiotherapy to the chest may be at higher risk of decreased LVEF. Infusion-related reactions and hypersensitivity responses including anaphylaxis are also the subject of a warning.
- Pertuzumab, given as a single agent or in combination with cytotoxic agents or trastuzumab, shows a low incidence of cardiac dysfunction similar to trastuzumab. Patients experience relatively low incidences of both LVSD and symptomatic heart failure. There was no significant increase in cardiac side effects

when pertuzumab was given in combination with other anticancer agents. No synergistic effect between pertuzumab and trastuzumab is apparent.

- **Trastuzumab** was the first biological agent developed and approved for the treatment of breast cancer, but only about 30 % of patients with advanced disease benefit from monotherapy with the drug. A large proportion of patients (~70 %) who experience an initial beneficial response become resistant within a year.
- Cross-signaling is a potential mechanism of trastuzumab resistance. Several molecules appear to activate HER2 in the presence of trastuzumab including insulin-like growth factor 1 receptor, hepatocyte growth factor and its receptor Met, growth differentiation factor 15, and members of the ErbB family. The precise mechanisms underlying such cross-signaling generally remain unclear.
- Trastuzumab is the subject of a boxed warning for cardiomyopathy, infusion reactions, and pulmonary toxicity. For patients receiving trastuzumab as a single agent or in combination therapy, there is a four- to sixfold increase in the incidence of symptomatic myocardial dysfunction. Incidences are highest when an anthracycline drug is coadministered. Trastuzumab therapy can result in cases of left ventricular cardiac dysfunction, hypertension, arrhythmias, cardiomyopathy, cardiac failure, and cardiac death. Trastuzumab may have a direct effect on myocytes.
- Other warnings and precautions for trastuzumab refer to exacerbation of chemotherapy-induced neutropenia and febrile neutropenia and the possibility of fetal harm with an increased risk of oligohydramnios during the second and third trimesters of pregnancy.
- **Ado-trastuzumab emtansine (Kadcyla®)** is an ADC made by covalently coupling the microtubule inhibitory compound mertansine (DM1), a cytotoxic, thiol-containing maytansinoid, to trastuzumab via a bifunctional cross-linker. DM1 inhibits the assembly of microtubules leading to cell cycle arrest and ultimately cancer cell death.
- An FDA black box warning for ado-trastuzumab emtansine mentions hepatotoxicity, cardiotoxicity (specifically reduction in LVEF), and the potential risk of fetal harm, plus a reminder that the ADC should not be substituted for, or by, trastuzumab. Warnings and precautions have been issued for pulmonary toxicity, infusion and hypersensitivity reactions, thrombocytopenia, neurotoxicity, and extravasation.
- **Denosumab (Xgeva®, Prolia®)** has high specificity and affinity for receptor activator of nuclear factor kappa-B ligand, otherwise known as RANKL. Responsible for bone resorption, RANKL stimulates osteoclast formation, activation, adherence, survival, and ultimately resorption of the bone. Inhibition of RANKL results in an increase in bone density, volume, and strength.
- Marketed under two trade names, denosumab as Prolia® has approved cancer indications for the treatment of men at high risk of fracture receiving androgen deprivation therapy for nonmetastatic prostate cancer and for the treatment of women at high risk of fracture receiving adjuvant aromatase inhibitor therapy for

breast cancer. As Xgeva®, denosumab is approved for the prevention of skeletal-related events in patients with bone metastases from solid tumors and the treatment of giant cell tumor of the bone.

- For Xgeva®, warnings and precautions consist of hypocalcemia, osteonecrosis of the jaw, and embryofetal toxicity. For Prolia®, hypersensitivity, including anaphylaxis, is listed as a warning together with hypocalcemia, osteonecrosis of the jaw, serious infections, femoral fractures and suppression of bone turnover, cutaneous reactions, and severe bone, joint, and muscle pain.
- Recent serious postmarketing surveillance reports for Xgeva® cover calcium and bone disorders, gastrointestinal symptoms, and musculoskeletal and dermatologic conditions; these events, plus respiratory disorders, are also the most commonly reported reactions. Reports on Prolia® relate to musculoskeletal disorders, dermatologic conditions, infections, and gastrointestinal disorders. The most frequently mentioned infections include pneumonia, cellulitis, urinary tract infections, and sepsis.
- **Ipilimumab (Yervoy®)** binds with high affinity to the extracellular domain of the protein receptor human cytotoxic T lymphocyte antigen 4 or CTLA-4. When bound to the B7 complex on APCs, CTLA-4 acts as an immune "off" switch modulating overactivity of T cells and allowing cancer cells to multiply in the absence of an antitumor challenge. This led to the development of ipilimumab, the first mAb example of immune checkpoint targeting.
- A number of adverse events induced by ipilimumab are related to the mAb's mode of action of T-cell-mediated immune aggression against cancer cells. These events are reflected in a boxed warning for immune-mediated reactions including enterocolitis, hepatitis, neuropathy, endocrinopathy, severe cutaneous events such as Stevens-Johnson syndrome and toxic epidermal necrolysis, and a number of immune-mediated ocular manifestations.
- **Siltuximab (Sylvant®)** targets interleukin-6 (IL-6), a proinflammatory and anti-inflammatory pleiotropic cytokine produced by many different cells including lymphocytes, monocytes, fibroblasts, and endothelial and cancer cells. Cancers associated with increased production of IL-6 include renal cell carcinoma, prostate and bladder cancers, some neurologic cancers, and particularly multiple myeloma and the B-cell lymphoproliferative disorder, Castleman's disease.
- Siltuximab is approved for the treatment of multicentric Castleman's disease in patients who are HIV and HHV-8 negative. Warnings and precautions mention the possibility of siltuximab masking signs of acute inflammation and suppressing fever and acute phase reactants such as CRP. By extension, live vaccines should not be administered concurrently or within four weeks of commencing siltuximab therapy. Other issued warnings are for infusion-related reactions and hypersensitivity and gastrointestinal perforation. An increased risk of malignancy and the development of hyperlipidemia and hepatic impairment are additional warnings issued by the EMA.

- The programmed cell death protein 1 (PD-1) receptor which plays a critical role in cancer immunology acts as another important immune checkpoint. PD-1 is expressed on CD4+ and CD8+ T cells, B lymphocytes, NK cells, B cells, and monocytes during antigen signaling. PD-1 has two ligands, PD-L1 and PD-L2. High expression of PD-L1 appears to be associated with increased aggressiveness of cancers and death.
- Like CTLA-4, PD-1 negatively regulates T-cell activation. PD-1 does not interfere with the normal immune response, but, upon binding PD-L1 and PD-L2, signaling is induced, suppressing T-cell proliferation and activity. Malignant cells often express PD-L1 which induces inhibitory signaling preventing expansion of activated T cells and allowing the tumor cells to avoid immune attack. Selectively blocking the pathway with anti-PD-1 antibodies reverses the checkpoint inhibition and restores the T-cell-mediated response to the tumor.
- **Pembrolizumab** (**Keytruda**®), targeted to PD-1, and indicated for metastatic melanoma and non-small cell lung cancer, can provoke a number of immune adverse reactions. Warnings have been issued for immune-mediated pneumonitis, colitis, hepatitis, hypophysitis, hyperthyroidism, hypothyroidism, renal failure, and nephritis.
- **Nivolumab** (**Opdivo**®), targeted to PD-1, is indicated for the treatment of patients with unresectable or metastatic melanoma and disease progression following ipilimumab, and, if BRAF V600 mutation positive, a BRAF inhibitor should also be given. Other indications are non-small cell lung cancer and metastatic renal cell carcinoma.
- Nivolumab is also subject to warnings for immune-mediated adverse events of pneumonitis, colitis, hepatitis, hyperthyroidism, hypothyroidism, renal failure, and nephritis in melanoma patients.
- The sialic ganglioside disialoganglioside (GD2), a short sialylated polysaccharide linked to ceramide through a β-glycosidic linkage, is found highly expressed on neuroectoderm-derived cancers such as neuroblastoma, melanoma, brain tumors, osteosarcoma, and Ewing's sarcoma in children. **Dinutuximab** (**Unituxin**®), a human-mouse chimeric mAb that binds GD2, has been shown to prevent the outgrowth of experimental melanoma and neuroblastoma tumors.
- Approval for dinutuximab comes with a boxed warning for potentially life-threatening infusion reactions and neuropathy. Severe neuropathic pain and nerve damage requiring prior, and subsequent, intravenous opioid treatment may occur.
- Other warnings issued for dinutuximab are for capillary leak syndrome, hypotension, infection, neurological eye disorders, bone marrow suppression, electrolyte abnormalities, atypical hemolytic uremic syndrome, and embryofetal toxicity.
- **Daratumumab** (**Darzalex**®) is the first-in-class immunotherapy for multiple myeloma targeted to CD38 (cyclic ADP ribose hydrolase), a surface antigen expressed by multiple myeloma cells and found on many immune cells including CD4+, CD8+, B lymphocytes, and natural killer (NK) cells. The mAb acts by inhibiting the growth of tumor cells expressing CD38, leading to apoptosis by Fc-mediated cross-linking and cell lysis induced via CDC, ADCC, and ADCP.

- Warnings and precautions for daratumumab are the occurrence of infusion reactions, interference with cross-matching and red blood cell antibody screening, and interference by the mAb when determining patient's response and disease progression in patients with IgG kappa myeloma. As well as infusion reactions, adverse events recorded so far include infections, fatigue, and nausea. It is predicted that postmarketing usage will reveal the occurrence of cytopenias.
- The humanized mAb **elotuzumab (Empliciti®)**, targeting the cell surface receptor CS1 (SLAMF7), a member of the signaling lymphocytic activation molecule receptor family, was recently approved for use with lenalidomide and dexamethasone for the treatment of relapsed or refractory multiple myeloma.
- Warnings and precautions associated with elotuzumab include infusion reactions (premedication is required), risk of infections, hepatotoxicity, and malignancies.
- In May 2016 FDA-approval was granted to the PD-L1-targeted mAb **atezolizumab** (Tecentriq™) for treatment of urothelial carcinoma.
- The mAbs approved for cancer therapy show the full range of hypersensitivity responses: type I IgE antibody-mediated immediate reactions such as anaphylaxis, urticaria, and angioedema; type II drug-induced thrombocytopenia, hemolytic anemia, and agranulocytosis; type III serum sickness and drug-induced vasculitis; and type IV cutaneous hypersensitivities mediated by Th1, Th2, and Th17 lymphocytes.
- Immunogenicity is always a safety concern for mAbs, even those that are fully humanized since the possibility of generating anti-idiotype antibodies remains. Warnings are often issued for anaphylaxis, but the observed incidences of such reactions are actually quite small. IgE-mediated reactions to chimeric proteins containing mouse and rat sequences are considered to be of greater risk. Anaphylaxis has been reported for, at least, cetuximab, rituximab, brentuximab, bevacizumab, trastuzumab, pertuzumab, ibritumomab, and dinutuximab.
- Cytopenias occur in some patients treated with mAbs during anticancer immunotherapy, but the underlying mechanisms frequently remain unexplored. Types II and III hypersensitivities induced by mAbs may be underdiagnosed.
- Incidences of 1.1 and 5.2 % for severe anemia have been reported for patients receiving rituximab monotherapy. At least two mAbs, rituximab and alemtuzumab, have been implicated in the induction of pure red cell aplasia and autoimmune hemolytic anemia.
- Hypersensitivity vasculitis induced by drugs is a manifestation of a type III response, and a few mAbs including rituximab have been implicated in its cause. Serum sickness reactions to mAbs are probably underdiagnosed and reported. Chimeric antibodies in particular have the potential to induce the reactions.
- Delayed, type IV cutaneous hypersensitivity reactions to anticancer mAbs are rare with most reported cases restricted mainly to ibritumomab tiuxetan, brentuximab vedotin, and rituximab. Mechanisms remain to be established for some type IV-like cutaneous reactions induced by catumaxomab, bevacizumab, denosumab, ipilimumab, and panitumumab.
- The mechanisms of mAb-induced infusion reactions are not yet fully understood. Cytokines, especially TNF and interleukins such as IL-6, may be involved

since the symptoms they produce resemble those seen in infusion and type I allergic reactions. An important finding was the observation that the severity of infusion reactions is related to the number of circulating lymphocytes.

- Cytokine release syndrome, also called cytokine storm, may be seen after the use of mAbs directed to malignant immune cells, for example, rituximab.
- At least ten of the currently approved mAbs for cancer therapy have some recorded pulmonary toxicity in treated cancer patients. Adverse pulmonary events provoked by mAbs include interstitial pneumonitis and fibrosis, ARDS, bronchiolitis obliterans organizing pneumonia (BOOP), and hypersensitivity pneumonitis. Hypersensitivity pneumonitis to some anticancer agents appears to be a combined type III and type IV hypersensitivity reaction in a Th1/Th17 response.
- Cardiac adverse events have occurred with at least 11 of the mAbs used for cancer therapy.
- Cutaneous reactions elicited by mAbs targeted to EGFR are generally not immune-mediated. Skin reactions appear as a papulopustular eruption, often in a large proportion of patients (50–100 %) and in a more severe form than seen with small molecule tyrosine kinase inhibitors. Eruptions tend to be confined to seborrheic regions such as the epidermis, sebaceous glands, and hair follicles. Other adverse effects induced by mAbs targeted to EGFR include paronychia, fissures, xerosis, palmar-plantar rash, hair changes, hyperkeratosis, mucositis, pyrogenic granuloma, and skin hyperpigmentation.
- Tumor lysis syndrome usually occurs in patients with leukemias and high-grade lymphomas. Large numbers of malignant cells may be destroyed in a short time by the mAb resulting in hyperkalemia, hypercalcemia, hyperphosphatemia, and hyperuricemia. This may produce profound ionic imbalances in potassium, calcium phosphate, and uric acid and progress to acute renal failure, cardiac arrhythmias, seizures, and death.
- Progressive multifocal leukoencephalopathy, caused by the polyomavirus JC virus in severely immunodeficient individuals, is occasionally seen upon the administration of some mAbs directed to B cells, in particular, rituximab, ofatumumab, obinutuzumab, and brentuximab vedotin.

Further Reading

Ahmed M, Hu J, Cheung N-KV. Structure based refinement of a humanized monoclonal antibody that targets tumor antigen disialoganglioside GD2. Front Immunol. 2014;5:372. doi:10.3389/fimmu.2014.00372.

Alinari L, Lapalombella R, Andritsos L, et al. Alemtuzumab (Campath-1H) in the treatment of chronic lymphocytic leukemia. Oncogene. 2007;26:3644–53.

Baldo BA. Adverse events to monoclonal antibodies used for cancer therapy. Focus on hypersensitivity responses. OncoImmunology. 2013;2(10):e26333. doi:10.4161/onci.26333

Baldo BA, Pagani M. Adverse events to nontargeted and targeted chemotherapeutic agents. Immunol Allergy Clin N Am. 2014;34:565–96.

Bargou R, Leo E, Zugmaier G, et al. Tumor regression in cancer patients by very low doses of a T cell-engaging antibody. Science. 2008;321:974–7.

Boross P, Leusen JHW. Mechanism of action of CD20 antibodies. Am J Cancer Res. 2012;2:676–90.

Bryan LJ, Gordon LI. Releasing the brake on the immune system: the PD-1 strategy for hematologic malignancies. Oncology. 2015;29:431–9.

Callahan MK, Wolchok JD. At the bedside: CTLA-4- and PD-1-blocking antibodies in cancer immunotherapy. J Leukoc Biol. 2013;94:41–53.

Castellano D, Sepulveda JM, Garcia-Escobar I, et al. The role of RANK-Ligand inhibition in cancer: the story of denosumab. Oncologist. 2011;16:136–45.

Chavez-MacGregor M, Zhang N, Buchholz TA, et al. Trastuzumab-related cardiotoxicity among older patients with breast cancer. J Clin Oncol. 2013;31:4222–8.

Cheung CH. Managing premedications and the risk for reactions to infusional monoclonal antibody therapy. Oncologist. 2008;13:725–32.

Clarke JM, Hurwitz HI. Targeted inhibition of VEGF Receptor-2: an update on ramucirumab. Expert Opin Biol Ther. 2013;13:1187–96.

De Roock W, Claes B, Bernasconi D, et al. Effects of KRAS, BRAF, NRAS, and PIK3CA mutations on the efficacy of cetuximab plus chemotherapy in chemotherapy-refractory metastatic colorectal cancer: a retrospective consortium analysis. Lancet Oncol. 2010;11:753–62.

Duffy MJ, O'Donovan N, Crown J. Use of molecular markers for predicting therapy response in cancer patients. Cancer Treat Rev. 2011;37:151–9.

Ellis LM. Mechanisms of action of bevacizumab as a component of therapy for metastatic colorectal cancer. Semin Oncol. 2006;33 Suppl 10:S1–7.

Goltsov A, Deeni Y, Khalil HS, et al. Systems analysis of drug-induced receptor tyrosine kinase reprogramming following targeted mono- and combination anti-cancer therapy. Cells. 2014;3:563–91.

Holmes K, Roberts OLI, Thomas AM, et al. Vascular endothelial growth factor receptor-2: structure, function, intracellular signalling and therapeutic inhibition. Cell Signal. 2007; 19:2003–12.

Holstead Jones D, Nakashima T, Sanchez OH, et al. Regulation of cancer cell migration and bone metastasis by RANKL. Nature. 2006;440:692–6.

Jelinek T, Hajek R. Monoclonal antibodies -- A new era in the treatment of multiple myeloma. Blood Rev. 2016;30:101–10.

Junttila TT, Akita RW, Parsons K, et al. Ligand-independent HER2/HER3/PI3K complex is disrupted by trastuzumab and is effectively inhibited by the PI3K inhibitor GDC-0941. Cancer Cell. 2009;15:429–40.

Kimby E. Tolerability and safety of rituximab (MabThera®). Cancer Treat Rev. 2005; 31:456–73.

Lioté H, Lioté F, Séroussi B, et al. Rituximab-induced lung disease: a systematic literature review. Eur Resp J. 2010;35:681–7.

Liu Y-C, Szmania S, van Rhee F. Profile of elotuzumab and its potential in the treatment of multiple myeloma. Blood Lymphat Cancer. 2014;4:15–27. doi:10.2147/BLCTT.S49780.

Lokhorst HM, Plesner T, Laubach JP, et al. Targeting CD38 with daratumumab monotherapy in multiple myeloma. N Engl J Med. 2015;373:1207–19. doi:10.1056/NEJMoa1506348.

Lonberg N. Fully human antibodies from transgenic mouse and phage display platforms. Curr Opin Immunol. 2008;20:450–9.

Marmor MD, Skaria KB, Yarden Y. Signal transduction and oncogenesis by ErbB/HER receptors. Int J Radiat Oncol Biol Phys. 2004;58:903–13.

McDermott DF, Atkins MB. PD-1 as a potential target in cancer therapy. Cancer Med. 2013;2:662–73.

McKinney MS, Beavan AW. Yttrium-90 ibritumomab tiuxetan in the treatment of non-Hodgkin lymphoma. Blood Lymphat Cancer Targets Ther. 2014;4:45–59.

Merelli B, Massi D, Cattaneo L, et al. Targeting the PD-1/PD-L1 axis in melanoma: biological rationale, clinical challenges and opportunities. Crit Rev Oncol Hematol. 2014;89:140–65.

Miao H-Q, Hu K, Jiminez X, et al. Potent neutralization of VEGF biological activities with a fully human antibody Fab fragment directed against VEGF receptor 2. Biochem Biophys Res Commun. 2006;345:438–45.

Ohaegbulam KC, Assal A, Lazar-Molnar E, et al. Human cancer immunotherapy with antibodies to the PD-1 and PD-L1 pathway. Trends Mol Med. 2015;21:24–33.

Rose-John S. IL-6 trans-signaling via the soluble IL-6 receptor: importance for the pro-inflammatory activities of IL-6. Int J Biol Sci. 2012;8:1237–47.

Peggs KS, Quezada SA, Chambers CA, et al. Blockade of CTLA-4 on both effector and regulatory T cell compartments contributes to the antitumor activity of anti-CTLA-4 antibodies. J Exp Med. 2009;206:1717–25.

Portera CC, Walshe JM, Rosing DR, et al. Cardiac toxicity and efficacy of trastuzumab combined with pertuzumab in patients with trastazumab-insensitive human epidermal growth factor receptor 2-positive metastatic breast cancer. Clin Cancer Res. 2008;14:2710–6.

Sartore-Bianchi A, Martini M, Molinari F, et al. PIK3CA mutations in colorectal cancer are associated with clinical resistance to EGFR-targeted monoclonal antibodies. Cancer Res. 2009;69:1851–7.

Senter PD, Sievers EL. The discovery and development of brentuximab vedotin for use in relapsed Hodgkin lymphoma and systemic anaplastic large cell lymphoma. Nat Biotechnol. 2012;30:631–7.

Tarhini A. Immune-mediated adverse events associated with ipilimumab CTLA-4 blockade therapy: the underlying mechanisms and clinical management. Scientifica. 2013;Article ID 857519:19 p. doi:10.1155/2013/857519

Teeling JL, Mackus WJM, Wiegman LJJM, et al. The biological activity of human CD20 monoclonal antibodies is linked to unique epitopes on CD20. J Immunol. 2006;177:362–71.

Thatcher N, Hirsch FR, Luft AV, et al. Necitumumab plus gemcitabine and cisplatin versus gemcitabine and cisplatin alone as first-line therapy in patients with stage IV squamous non-small-cell lung cancer (SQUIRE): an open-label, randomised, controlled phase 3 trial. Lancet Oncol. 2015;16:763–74.

van der Donk NW, Moreau P, Plesner T, et al. Clinical efficacy and management of monoclonal antibodies targeting CD38 and SLAMF7 in multiple myeloma. Blood. 2016;127:681–95.

Voigt M, Braig F, Göthel M, et al. Functional dissection of the epidermal growth factor receptor epitopes targeted by panitumumab and cetuximab. Neoplasia. 2012;14:1023–31.

Voskens CJ, Goldinger SM, Loquai C, et al. The price of tumor control: an analysis of rare side effects of anti-CTLA-4 therapy in metastatic melanoma from the ipilimumab network. PLoS One. 2013;8(1):e53745. doi:10.1371/journal.pone.0053745.

Zhang B. Ofatumumab. mAbs. 2009;1:326–31.

Chapter 4
Other Approved Therapeutic Monoclonal Antibodies

Monoclonal Antibodies Targeted to Human Tumor Necrosis Factor: Adalimumab, Certolizumab Pegol, Infliximab, and Golimumab

Tumor necrosis factor (formerly tumor necrosis factor alfa) is a proinflammatory cytokine (Chap. 5) that exists in membrane-bound and soluble forms both of which bind to the TNF complementary receptors TNFR1 and TNFR2, initiating the expression of several other proinflammatory cytokines including IL-1, IL-6, interferon-gamma (IFN-γ), and adhesion molecules such as intercellular adhesion molecule 1 (ICAM-1, CD54). TNF, expressed as a 26 kDa type II membrane-bound protein, self-associates to form a bioactive homotrimer. The soluble homotrimeric form is released via proteolytic cleavage by the metalloprotease TNF-alfa-converting enzyme (TACE, ADAM17). TNF is secreted by a number of different cells, particularly activated macrophages, T cells, and monocytes but also endothelial cells, fibroblasts, cardiac myocytes, neurons, and adipose tissue, following toll-like receptor or cytokine receptor activation and NF-κB (nuclear factor kappa-light-chain-enhancer of activated B cell)-mediated gene transcription.

Although the pathogenesis of many of the immune-mediated inflammatory diseases such as rheumatoid arthritis, Crohn's disease, ulcerative colitis, ankylosing spondylitis, psoriatic arthritis, plaque psoriasis, and inflammatory bowel disease remains incompletely understood, nonclinical studies, and several animal and disease models, have demonstrated that TNF has a pivotal place as a mediator of the inflammatory process in, at least, rheumatoid arthritis, Crohn's disease, and psoriasis. This understanding was the forerunner for the development of monoclonal antibodies targeted to human TNF, namely, adalimumab, certolizumab pegol, infliximab, and golimumab, each produced as antagonistic therapies for the chronic inflammatory diseases. The aim of this therapy is to specifically reduce the levels of TNF leading to clinical improvement without also immunosuppressing the patient.

B.A. Baldo, *Safety of Biologics Therapy*, DOI 10.1007/978-3-319-30472-4_4

Boxed Warnings and Precautions for Adalimumab, Certolizumab Pegol, Infliximab, and Golimumab

Serious infections and the risk of malignancy are the subjects of an FDA black box warning for all four approved anti-TNF mAbs (Table 4.1). Attention is drawn to opportunistic infections due to bacteria, fungi, viruses, and parasitic organisms with particular mention of aspergillosis, blastomycosis, candidiasis, coccidioidomycosis, histoplasmosis, legionellosis, listeriosis, pneumocystosis, and tuberculosis; diseases already reported to occur with TNF blockers. When infection occurs, patients treated with the TNF-targeted mAbs are more likely to present with disseminated rather than localized disease. Needless to say, patients with an active infection should not be treated with anti-TNF mAbs, and patients over 65 years of age, and those taking immunosuppressants, may be at greater risk of infection. Tuberculosis, either reactivated or as a new infection, is given special emphasis with recommendations to test for latent infection prior to initiating mAb therapy, to administer antituberculosis therapy in patients with a past history of the disease, and to closely monitor patients during treatment.

The possibility of malignancies, some fatal, is also referred to in the boxed warning for the anti-TNF mAbs. Reported among children, adolescents, and young adults, approximately half of the cases associated with anti-TNF agents have been lymphomas; most patients were receiving concomitant immunosuppressants and malignancies occurred after a median of 30 months. Note, however, that even in the absence of TNF-blocking therapy, patients with Crohn's disease, rheumatoid arthritis, and plaque psoriasis (particularly on immunosuppressant therapy) may be at a higher risk than the general population for the development of lymphoma. That also appears to be the case for leukemia. Fatal cases of the rare but aggressive hepatosplenic T-cell lymphoma have occurred in mostly young adult males being given TNF blockers such as adalimumab or infliximab for Crohn's disease or ulcerative colitis. All of the patients had received immunosuppressants. Melanoma and Merkel cell carcinoma have also been reported in patients treated with TNF blockers.

Adalimumab

Created using phage display technology, adalimumab (Humira®) (Tables 2.1 and 4.1), is a fully human recombinant IgG1κ mAb, MW ~ 148 kDa, that binds human soluble and transmembrane TNF with high affinity, blocking its interaction with the TNFR1 p55 and TNFR2 p75 cell surface TNF receptors. In addition to its action of TNF blockade, adalimumab demonstrates apoptotic activity, complement-dependent cytotoxicity (CDC), and antibody-dependent cell cytotoxicity (ADCC) (Chap. 2, section "Antibody-Dependent Cell-Mediated and Complement-Dependent Cytotoxicities"). Adalimumab also modulates the expression of some adhesion molecules (e.g.,

Table 4.1 Adverse events associated with approved[a] monoclonal antibodies used for non-cancer therapies as at June 2016

Monoclonal antibody[a] INN and trade names	Target[b]	Warnings, precautions, risks and safety concerns	Other adverse events,[c] serious and common
Adalimumab (Humira®)	TNF	*Boxed warning*: Serious infections[d,e]; malignancy *Other*: Anaphylaxis, serious allergic reactions; hepatitis B reactivation; demyelinating disease; cytopenias[f]; heart failure; lupus-like syndrome	*Systemic*: Infections[g]; isr; ILD; sarcoidosis; liver failure *Cutaneous*: SJS; EM; psoriasis; cutaneous vasculitis; alopecia
Certolizumab pegol (Cimzia®)	TNF	*Boxed warning*: Serious infections[d,e]; lymphoma and other malignancies *Other*: Heart failure; serious allergic reactions; hepatitis B reactivation; demyelinating disease; cytopenias; lupus-like syndrome	*Systemic*: URTI; cardiac disorders[h]; eye disorders[i]; isr; hepatitis and ↑ liver enzymes; nephrotic syndrome; renal failure; thrombophlebitis; vasculitis *Cutaneous*: Dermatitis; erythema nodosum; urticaria
Infliximab (Remicade®)	TNF	*Boxed warning*: Serious infections[d,e]; malignancy *Other*: Hepatitis B reactivation; hepatotoxicity; cytopenias; demyelinating disease; lupus-like syndrome	*Systemic*: Infections[g,j]; pancytopenia; anemia; cellulitis; serum sickness; thrombophlebitis; intestinal obstruction; ILD; anaphylaxis; IRs *Cutaneous*: Cutaneous vasculitis; SJS; EM; psoriasis;
Golimumab (Simponi®)	TNF	*Boxed warning*: Serious infections[d,e]; lymphoma and other malignancies *Other*: Invasive fungal infections; heart failure; hepatitis B reactivation; demyelinating disease; hypersensitivity	*Systemic*: URTI[j]; viral infections; bronchitis; ↑ liver enzymes; sarcoidosis; ILD; paresthesia *Cutaneous*: Skin exfoliation; rash
Abciximab (ReoPro®)	Glycoprotein IIb/IIIa	↑ Risk of bleeding; thrombocytopenia[k]	*Systemic*: Bleeding[l]; intracranial hemorrhage or stroke; GI; CV[m]; anemia; NS[n]; respiratory[o]; urinary disorders[p] *Cutaneous*: Pruritus; generalized exanthema[q]
Alemtuzumab (Lemtrada®)	CD52	*Boxed warning*: Autoimmunity; IRs; malignancies[r] *Other*: Other immune cytopenias; glomerular nephropathies; thyroid disorders; delay therapy in cases of infections; pneumonitis	*Systemic*: Headache; pyrexia; nausea; UTI; herpes virus infection; extremity and back pain; dizziness; flushing; cough; chills; vomiting; dyspnea *Cutaneous*: Rash; urticaria; pruritus; dermatitis

(continued)

Table 4.1 (continued)

Monoclonal antibody[a] INN and trade names	Target[b]	Warnings, precautions, risks and safety concerns	Other adverse events,[c] serious and common
Basiliximab (Simulect®)	IL-2 receptor α chain (CD25)	*Boxed warning*: General risk of immunosuppressive therapy *Other*: Immunogenicity; hypersensitivity	*Systemic*: GI; viral infection; peripheral edema; UTI; URTI; dyspnea; wound complications; hypertension; anemia; hypo- and hyperkalemia and hyperuricemia; headache; tremor *Cutaneous*: Rash; pruritus; hypertrichosis
Belimumab (Benlysta®)	BLyS	Mortality[s]; serious infection[t]; malignancy; hypersensitivity including anaphylaxis; IR; depression[u]; immunization[v]	*Systemic*: Nausea; diarrhea; pyrexia; pain in extremity; bronchitis; depression; migraine *Cutaneous*: Rash; pruritus
Canakinumab (Ilaris®)	IL-1β	Increased risk of serious infections[v]; immunization[w]; MAS[x]; hypersensitivity; immunosuppression	*Systemic*: CAPS—Nasopharyngitis; diarrhea; influenza; headache; nausea; dizziness/vertigo; SJIA—Nasopharyngitis; URTI; abdominal pain; isr
Denosumab (Prolia®)	RANKL	Hypersensitivity; hypocalcemia; serious infections; osteonecrosis of jaw; atypical femoral fractures; severe bone, joint, muscle pain; suppression of bone turnover; dermatologic reactions	*Systemic*: Post-menopausal osteoporosis—Back extremity, musculoskeletal pain; hypercholesterolemia; cystitis. Male osteoporosis—Back pain; arthralgia; nasopharyngitis *Cutaneous*: Rash; pruritus; dermatitis; eczema
Eculizumab (Soliris®)	C5	*Boxed warning*: Serious meningococcal infections[y]	*Systemic*: PNH—Headache; nasopharyngitis; back pain; nausea. AHUS—Hypertension; URTI; GI; abdominal pain; anemia; cough; pyrexia; peripheral edema *Cutaneous*: Rash; pruritus
Natalizumab (Tysabri®)	α4 integrin (binds to α4β1 and α4β7 integrins)	*Boxed warning*: PML[z] *Other*: Hypersensitivity; hepatotoxicity; immunosuppression/infections; IRIS	*Systemic*: MS—Headache; fatigue; arthralgia; urinary tract infection; URTI; gastroenteritis; vaginitis; diarrhea. CD—Headache; URTI; nausea *Cutaneous*: Rash; urticaria
Vedolizumab (Entyvio®)	α4β7 integrin	Hypersensitivity/infusion reactions; infections; PML; liver injury	*Systemic*: Headache; arthralgia; nausea; pyrexia; URTI; cough; bronchitis; influenza; back pain; pain in extremities; nasopharyngitis *Cutaneous*: Rash; pruritus

Omalizumab (Xolair®)	IgE	Anaphylaxis; malignancy; acute asthma; ↓ CSs gradually; eosinophilia[aa]; SS-like reaction[ab]; parasitic infection[ac]	*Systemic*: Allergic asthma—Arthralgia; pain; dizziness; fracture; earache. CIU—Nausea; pharyngitis; URTI; sinusitis; arthralgia; headache; cough; virus infections *Cutaneous*: Pruritus; dermatitis.
Palivisumab (Synagis®)	RSVF	Anaphylaxis; delay administration during moderate-severe infections; give with caution in cases of thrombocytopenia or coagulation disorders	*Systemic*: isr; pyrexia; apnea; cough; dizziness thrombocytopenia *Cutaneous*: Rash; itching; erythema
Ranibizumab (Lucentis®)	VEGF-A	Endophthalmitis and retinal detachment, increase in intraocular pressure and risk of arterial thromboembolic events after intravitreal injection	Conjunctival haemorrhage; eye pain; vitreous floaters; vitreous detachment; increased intraocular pressure; cataracts
Raxibacumab (ABthrax®)	*Bacillus anthracis* PA	IR	*Systemic*: Pain in extremity; somnolence; headache; URTI; nausea; cough; arthralgias. *Cutaneous*: Rash; pruritus; urticaria
Obiltoxaximab (Anthim®)	*Bacillus anthracis* PA	*Boxed warning*: Hypersensitivity including anaphylaxis	*Systemic*: Headache; URTI; cough; bruising; isr; nasal congestion; pain in extremity. *Cutaneous*: Pruritus; urticaria
Secukinumab (Cosentyx®)	IL-17A	Infections; tuberculosis activation; exacerbation of Crohn's disease; hypersensitivity; avoid live vaccines	*Systemic*: Nasopharyngitis; diarrhea; URTI; rhinitis *Cutaneous*: Urticaria
Ixekizumab (Taltz®)	IL-17A	Infections; tuberculosis activation; inflammatory bowel disease including exacerbations; hypersensitivity; avoid live vaccines	URTI; isr; nausea; tinea infections; nasopharyngitis
Tocilizumab (Actemra®; RoActemra®)	IL-6R	*Boxed warning*: Serious infections *Other*: GI perforation; avoid live vaccines; hypersensitivity; laboratory monitoring[ad]	*Systemic*: Nasophraryngitis; nausea; ↑ liver enzymes; infusion reactions; hypertension; thrombocytopenia; neutopenia; headache; dermatologic reactions
Ustekinumab (Stelara®)	IL-12 IL-23	Infections; tuberculosis; PRES; malignancies; anaphylaxis; avoid live vaccines	*Systemic*: Nasopharyngitis; headache; dental infections; URTI; isr; arthralgia; GI. *Cutaneous*: Pruritus
Alirocumab (Praluent®)	PCSK9	Allergic reactions (pruritus, urticaria, rash) including some serious (including hypersensitivity vasculitis)[ae]	Nasopharyngitis; isr; influenza; UTI; diarrhea; myalgia; bronchitis; muscle spasms, sinusitis; cough; contusion; musculoskeletal pain; liver enzyme abnormalities[ad,ae]

(continued)

Table 4.1 (continued)

Monoclonal antibody[a] INN and trade names	Target[b]	Warnings, precautions, risks and safety concerns	Other adverse events,[c] serious and common
Evolocumab (Repatha®)	PCSK9	Patients with renal and hepatic impairments have not yet yet been adequately studied. Cover of prefilled syringe and pen contain latex which may cause allergic reactions	Nasopharyngitis; isr; influenza; URTI; back pain; allergic reactions (rash, hives); nausea; arthralgia; hypertension[ae]
Idarucizumab (Praxbind®)	Dabigatran	Thomboembolic risk; re-elevation of coagulation parameters[af]; risk in patients with hereditary fructose intolerance due to sorbitol[ag]; hypersensitivity reactions	Hypokalemia; delerium; pyrexia; pneumonia, constipation; headache[ae]
Mepolizumab (Nucala®)	IL-5	Hypersensitivity[ah]; opportunistic infection of Herpes zoster; not to be used to treat acute asthma symptoms; reduce dose of corticosteroids gradually; parasitic (Helminth) infection[ai]	Headache; isr; back pain; fatigue[ae]
Reslizumab (Cinqair®)	IL-5	*Boxed warning*: Anaphylaxis. *Other*: Malignancy; parasitic infection[ai]; corticosteroid dose reduction	Oropharyngeal pain; elevated creatinine phosphokinase; myalgia

AHUS atypical hemolytic uremic syndrome, *BLyS* B lymphocyte stimulator, also known as B cell activating factor, BAFF, *C5* complement component 5, *CAPS* cryopyrin-associated periodic syndrome, *CD* Crohn's disease, *CIU* chronic idiopathic urticaria, *CSs* corticosteroids, *CV* cardiovascular, *EM* erythema multiforme, *GI* gastrointestinal, *ILD* interstitial lung disease, *IR* infusion reaction, *IRIS* immune reconstitution inflammatory syndrome, *isr* injection site reaction, *MAS* macrophage activation syndrome, *MS* multiple sclerosis, *NS* nervous system, *PA* protective antigen of *B. anthracis* toxin, *PCSK9* proprotein convertase subtilisin/kexin type 9, *PML* progressive multifocal leukoencephalopathy, *PNH* paroxysmal nocturnal hemoglobinuria, *PRES*, posterior reversible encephalopathy syndrome, *RANKL* receptor activator of nuclear factor kappa-B ligand (CD254), *REMS* Risk Evaluation Mitigation Strategy, *RSVF* human respiratory syncytial virus (F (fusion) protein coat antigen,), *SJIA* active systemic juvenile idiopathic arthritis, *SJS* Stevens-Johnson syndrome, *SS*, serum sickness, *URTI* upper respiratory tract infection, *UTI* urinary tract infection, *VEGF-A* vascular endothelial growth factor A

[a]Approved by FDA or EMA or both

[b]Specificity of antibody

[c]Adverse events in addition to those mentioned as occurring, or potentially likely to occur, and shown in column 3

[d]Including tuberculosis, tuberculosis reactivation, bacterial sepsis and invasive fungal infections

[e]Not to be given to patients with active infections and opportunistic pathogens, e.g., histoplasmosis. Concurrent administration with anakinra or abatacept was associated with a greater proportion of serious infections

[f]Also rare reports of pancytopenia including aplastic anemia
[g]Upper respiratory tract and sinusitis
[h]Including angina pectoris, arrhythmias, atrial fibrillation, hypertensive heart disease, myocardial infarction, myocardial ischemia, pericarditis, stroke
[i]Optic neuritis, retinal hemorrhage, uveitis
[j]Not to be given with live vaccines and therapeutic infectious agents
[k]Monitor platelet counts
[l]Particularly in the presence of anticoagulants such as heparin
[m]Including system-ventricular tachycardia, pseudoaneurysm, palpitation, arteriovenous fistula
[n]Including dizziness, anxiety, agitation, confusion
[o]Including pneumonia, rales, pleural effusion, bronchitis, bronchospasm
[p]Including urinary retention, dysuria, abnormal renal function
[q]T cell-mediated reaction
[r]Lemtrada® available only through the restricted program, Lemtrada REMS Program
[s]More deaths with belimumab than placebo in three clinical trials
[t]Including PML
[u]Depression and suicide reported in trials
[v]IL-1 blockade may interfere with immune response to infections
[w]Live vaccines should not be given concurrently
[x]Life-threatening disorder may develop with SJIA
[y]Available only through a REMS program. Use with caution with any other systemic infection
[z]Available only under a special restricted distribution program, TOUCH® Prescribing Program (MS TOUCH® or CD TOUCH®). Patients who are anti-JC virus antibody-positive have a higher risk of developing PML
[aa]Asthmatic patients may respond with severe eosinophilia with vasculitis/Churg-Strauss syndrome. May result from reduction of CS therapy
[ab]Signs and symptoms include arthritis/arthralgia, rash, fever and lymphadenopathy within 1–5 days of omalizumab administration
[ac]Monitor patients at high risk of geohelminth infections
[ad]Laboratory monitoring for neutropenia, thrombocytopenia, elevated liver enzymes, lipid abnormalities
[ae]The relatively limited range and number of warnings, precautions, and adverse events should be viewed with caution at this early stage after recent regulatory approvals
[af]Some patients may require an additional 5 g of Praxbind® (e.g., if bleeding reappears together with elevated coagulation parameters)
[ag]Recommended dose of Praxbind® contains 4 g of sorbitol as excipient
[ah]Reactions may show delayed onset
[ai]Eosinophils may be involved in the immunological response to some helminth infections

ICAM-1) and, via its transmembrane receptor blockade, suppresses the activity of some proinflammatory cytokines. It is thought that activated T cells and macrophages in the inflamed gut express high amounts of transmembrane TNF, and modulation or killing of these cells is at least partly responsible for the efficacy of anti-TNF antibody therapy. Adalimumab and infliximab have been shown to induce apoptosis of monocytes in vitro and to downregulate lipopolysaccharide-induced IL-10 and IL-12 production. Other mAb TNF blockers do not necessarily have exactly the same mechanisms of action (see certolizumab pegol below), and this may account for the sometimes different clinical efficacies shown by the antibodies in the treatment of different immune-mediated inflammatory diseases as well as the occasional variation in the adverse event profiles of the four approved mAbs.

Indications, Warnings, Precautions, and Adverse Events

The FDA list of approved indications for adalimumab covers a number of the important immune-mediated inflammatory diseases that have long proved difficult to manage: rheumatoid arthritis, juvenile idiopathic arthritis, ankylosing spondylitis, Crohn's disease, pediatric Crohn's disease, ulcerative colitis, and plaque psoriasis. In 2012, the EMA extended the indications for adalimumab to ulcerative colitis, moderate forms of Crohn's disease, and other inflammatory bowel diseases and to adult patients with axial spondyloarthritis with or without signs of ankylosing spondylitis.

As outlined in section "Boxed Warnings and Precautions for Adalimumab, Certolizumab Pegol, Infliximab, and Golimumab," serious infections, part of an FDA boxed warning for adalimumab, are the most frequent adverse events during or following the administration of the mAb with opportunistic infections and reactivation, or new episodes, of tuberculosis being of particular concern. Commonly seen infection-related events are pneumonia, urinary tract infections, gastroenteritis, appendicitis, cellulitis, and herpes zoster infections. Of 240 reports of histoplasmosis, ~20 % were fatal. Other warnings and precautions for adalimumab therapy, listed in Table 4.1, cover the possibility of reactivation of the hepatitis B virus; association of the mAb with adverse outcomes in patients with heart failure; rare cases of leukopenia, neutropenia, thrombocytopenia, and pancytopenia including aplastic anemia; hypersensitivities including type I and type IV reactions; rare neurological reactions manifesting as CNS demyelinating disorders (including multiple sclerosis and optic neuritis) and peripheral demyelinating disorders (such as Guillain-Barré syndrome); and autoimmunity which may rarely develop as a lupus-like syndrome. Given the risk of infection for patients receiving adalimumab, it is recommended that live vaccines should not be given concurrently with the mAb. In a similar vein, concurrent administration of abatacept (Chap. 6) has been associated with an increased risk of infections, and serious infections and neutropenia have been seen with concurrent use of anakinra (Chap. 5) and etanercept (Chap. 6).

In rheumatoid arthritis patients, injection site reactions are the most commonly encountered adverse event, occurring in ~20 % of adalimumab-treated patients. Reactions are usually mild and transient, manifesting as erythema, urticarial plaques, and pruritus. Anaphylaxis and angioneurotic edema have occurred following adalimumab administration. The involvement of IgE antibodies in at least some of the systemic reactions can be inferred by the finding of positive skin tests to the mAb.

Being a fully human antibody, adalimumab was early seen as a possible substitute for infliximab in patients intolerant to that mAb. This indeed proved to be the case in seven patients who had experienced hypersensitivities to infliximab. Over the years since its first approval in 2002, a wide range of apparent hypersensitivities induced by adalimumab have become apparent. As well as anaphylaxis, these include bronchospasm, asthma, autoimmune hepatitis, urticaria, psoriasis, exacerbation of palmoplantaris pustulosa psoriasis, and a range of other cutaneous reactions including rare cases of the severe toxidermias erythema multiforme, Stevens-Johnson syndrome, and toxic epidermal necrolysis.

Postmarketing reports in the FDA Adverse Event Reporting System (FAERS) database reveal an infection rate of 7 % from more than 77,000 reports with many reports of hepatitis B virus reactivation. Other reported, less commonly seen but interesting, events include the gastrointestinal disorders, diverticulitis, large bowel perforation, perforations associated with appendicitis, and pancreatitis, sarcoidosis, systemic vasculitis, and interstitial lung disease, including pulmonary fibrosis and pulmonary embolism. Of more than 21,000 reports in the EurdraVigilance Reporting System (EVRS) database, infections and malignancies account for ~46 % and ~17 % of the reports, respectively. Hepatosplenic T-cell lymphomas make up 0.05 % (11 cases) of the total reports.

An intriguing event seen in some patients treated with TNF inhibitors, including all four approved anti-TNF mAbs, is the paradoxical induction of psoriasis (Fig. 4.1) with many of the patients showing a palmoplantar distribution together with tiny pustules. Given the involvement of other mAb TNF blockers and etanercept (Chap. 6, section "Cutaneous Events") as well as adalimumab, these reactions seem to be a class effect. Increased production of interferon alfa (IFNα) in the affected patients has been demonstrated in the vasculature of the dermal lesions and in the lymphocytic infiltrate suggesting that TNF inhibitors may stimulate aberrant IFNα expression, an increased Th17 function, and a reduction in regulatory T cells. Other paradoxical skin eruptions associated with TNF inhibitors including lupus-like cutaneous events, dermatomyositis-like eruptions, and cutaneous vasculitis (Fig. 4.2), as well as cases of induced colitis, have been described and linked to an increased production of IFNα. It has been suggested, however, that the delay between the commencement of anti-TNF therapy and the subsequent adverse events indicates that other factors and mechanisms are involved in the appearance of the paradoxical lesions.

Table 4.1 summarizes the warnings issued for adalimumab together with the spectrum of serious and common systemic and cutaneous adverse events currently recorded for the mAb.

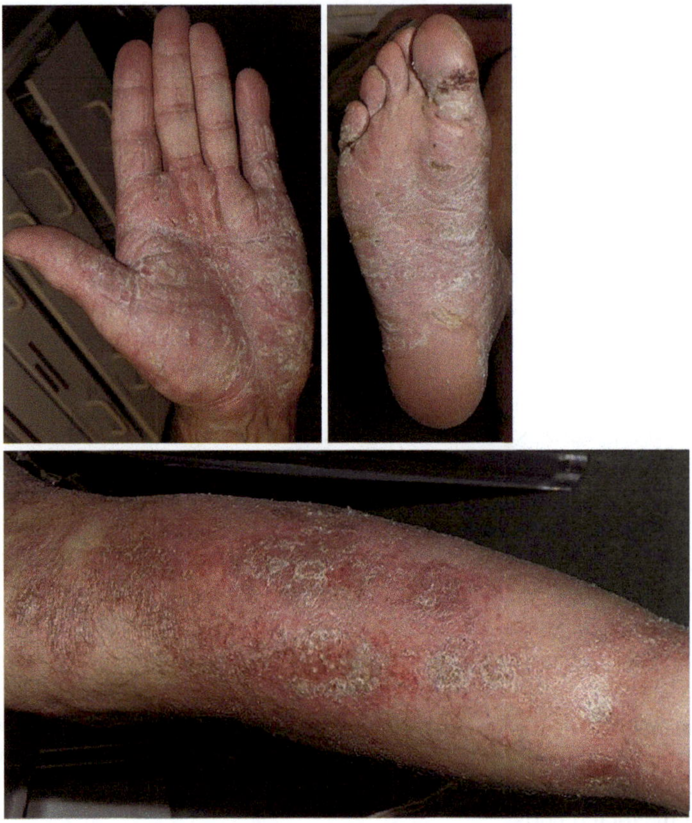

Fig. 4.1 Paradoxical induction of psoriasiform eruptions during treatment of rheumatoid arthritis with adalimumab. Many of the patient reactions show a palmoplantar distribution together with tiny pustules. Reproduced from Aubin F et al. The complexity of adverse side effects to biological agents. J Crohn's Colitis 2013;7:257–62. Distributed under a Creative Commons License; reproduced with permission from Elsevier Limited

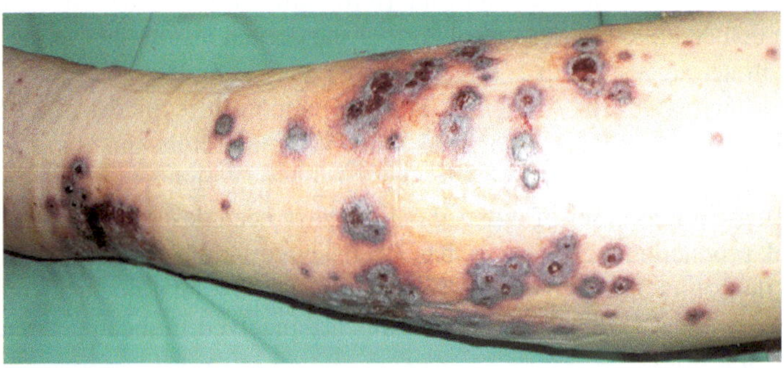

Fig. 4.2 Cutaneous vasculitis, a type III immune complex-mediated reaction, induced by adalimumab treatment. Reproduced from Aubin F et al. The complexity of adverse side effects to biological agents. J Crohn's Colitis. 2013;7:257–62. Distributed under a Creative Commons License; reproduced with permission from Elsevier Limited

Immunogenicity of Adalimumab

Although adalimumab is a fully human mAb, human antihuman antibody responses have been detected, but, until recently, good data has been hard to find. In a vague summary of adalimumab immunogenicity by the EMA, anti-adalimumab antibodies were found in 8 of 218 (3.7 %) patients with ulcerative colitis who had previously been exposed to the mAb and in 11 of 269 (4.1 %) patients who had never been exposed. In 269 patients with Crohn's disease who received adalimumab for up to 56 weeks, seven patients (2.6 %) proved antibody-positive. A recent UK-based prospective study conducted by the Biologics in Rheumatoid Arthritis Genetics and Genomics Study Syndicate (BRAGGSS) assessed the influence of immunogenicity on the efficacy of long-term treatment of rheumatoid arthritis with adalimumab. Patients who satisfied the criteria and had active disease at baseline were selected to begin adalimumab treatment. Serum samples for antibody determinations were collected at 3, 6, and 12 months following the start of therapy. Anti-adalimumab antibodies were detected in 31 of the 125 patients (24.8 %) at one or more time points over the 12-month treatment period. After 12 months, patients with antibodies showed less improvement in disease activity than patients without antibodies, and patients who did not develop antibodies were more likely to be on methotrexate than patients who did. The investigators concluded that anti-adalimumab antibodies are associated with reduced effectiveness of adalimumab in rheumatoid arthritis and that concomitant methotrexate may reduce antibody formation. The relationship between clinical response, adalimumab levels, and anti-adalimumab antibody titers was examined in a joint Dutch-Taiwanese study of 115 patients with ankylosing spondylitis treated with the mAb. At the end of a 24-week follow-up period, 31 (27 %) patients had detectable levels of anti-adalimumab antibodies; serum levels of the mAb were significantly higher in antibody-negative patients than in antibody-positive patients, and a significant association was seen between adalimumab and anti-adalimumab levels. High antibody titers and no detectable levels of adalimumab were found in 11 (9.6 %) patients, and these results were matched by elevations of C-reactive protein levels and the erythrocyte sedimentation rate. It was concluded that adalimumab levels are related to clinical response in ankylosing spondylitis patients and these are influenced by the presence of anti-adalimumab antibodies. The important question of whether adalimumab can be an effective treatment in patients with active disease who have lost response or are intolerant to infliximab was examined in a retrospective study of 30 patients with Crohn's disease. After 318 days (median range 83–632) of treatment with adalimumab, clinical response was 77 % (23 of 30), an escalation in dose was required in eight patients (27 %), and five patients (17 %) developed anti-adalimumab antibodies. Four of the five patients were found to be nonresponders. By comparison, 17 (57 %) of the patients had anti-infliximab antibodies, and serum levels of antibody were significantly increased in patients nonresponsive to adalimumab.

Although data for certolizumab pegol and golimumab are scarce, the general conclusion on the immunogenicity of anti-TNF biologic therapies seems to be that antidrug antibodies are associated with all four of the mAb TNF blockers, especially adalimumab and infliximab. It appears that low levels of antidrug antibodies might not affect the efficacy of the mAb therapy, but high antidrug levels can reduce efficacy by lowering the levels of free anti-TNF mAb.

Certolizumab Pegol

Certolizumab pegol (CDP870, Cimzia®) (Tables 2.1 and 4.1) is a recombinant humanized Fab' antibody fragment specific for human TNF conjugated to a polyethylene glycol macromolecule of MW ~ 40 kDa to increase solubility, stability, and half-life of the Fab'. This allows a subcutaneous injection to be given every 2–4 weeks and is said to reduce the molecule's immunogenicity. The certolizumab Fab' is made up of a light chain with 214 amino acids and a heavy chain of 229 amino acids and, being a pepsin-generated antibody fragment, lacks the Fc portion, giving a final MW of 91 kDa for the pegylated molecule.

Mechanism of Action

Certolizumab pegol binds both soluble- and membrane-bound TNF and does so with higher affinity than adalimumab and infliximab. Like these other two TNF blockers, certolizumab pegol inhibits the release of IL-1β in response to lipopolysaccharide, but unlike adalimumab and infliximab, it does not promote apoptosis, and because it lacks the antibody Fc piece, it does not activate the complement pathway and mediate cell- and antibody-dependent cytotoxicities. While certolizumab pegol, infliximab, and adalimumab are effective in Crohn's disease, the fact that the latter two mAbs trigger apoptosis suggests that not only are there differences in the pathogenesis of the immune-mediated inflammatory diseases such as rheumatoid arthritis, Crohn's disease, inflammatory bowel disease, and so on, but differences also exist in the functional properties of the TNF inhibitors. In fact, it now seems that the mechanisms of action of different TNF inhibitors are more complex and currently less well explained than originally thought. In a recent analysis of transcriptional responses of reverse signaling induced by certolizumab pegol and infliximab, growth differentiation factor-1 (GDF-1) was found to be upregulated in the inflamed tissue from patients with Crohn's disease but downregulated after treatment with infliximab in patients whose condition improved following therapy. GDF-1 is a proinflammatory factor, inducing IL-6 and signal transducer and activator of transcription 3 (STAT3). Both certolizumab pegol and infliximab downregulate GDF-1 through non-apoptotic pathways by release of transforming growth factor-β (TGFβ). These potentially important findings suggest that investigations of the precise role of GDF-1 in inflammatory processes may further increase our understanding of some immune-mediated inflammatory disorders and perhaps provide a promising new therapeutic target.

Indications, Warnings, Precautions, and Adverse Events

Certolizumab pegol was granted approval by the FDA in 2008 for the treatment of Crohn's disease. The following year, the FDA, EMA, and Health Canada approved the mAb for the treatment of rheumatoid arthritis, but while the EMA has extended approvals to axial spondyloarthritis, ankylosing spondylitis, axial spondyloarthritis without radiographic evidence of ankylosing spondylitis, and psoriatic arthritis, the agency has maintained its lack of approval for Crohn's disease.

The list of FDA warnings and precautions for certolizumab pegol, including an FDA boxed warning, is similar to those issued for adalimumab and comprises the risk of serious infections, malignancies, heart failure, hypersensitivity reactions, reactivation of hepatitis B infection, neurologic reactions, hematologic reactions, autoimmunity, the possible risk of secondary transmission of infection by live vaccines or attenuated organisms, the possibility of immunosuppression, and the possible higher risk of serious infection by the concomitant use of the IL-1 antagonist anakinra, abatacept, and rituximab (Table 4.1). As with the other TNF blockers, infections are the most common adverse event observed with certolizumab pegol therapy. Upper respiratory and urinary tract infections are the ones most often seen. In one 26-week study, serious adverse events and serious infections occurred with a frequency of 6 % and 3 %, respectively. The most frequently occurring serious infections in 4049 rheumatoid arthritis patients given certolizumab pegol were pneumonia and tuberculosis, the latter showing an incidence of 0.47 per 100 patient-years. Other serious infections seen include fungal esophagitis, pneumocystosis, nocardiosis, and herpes zoster infection. A 2011 Cochrane meta-analysis and overview demonstrated that anti-TNF agents showed no significant differences in rates of total adverse events, but certolizumab pegol was associated with a significantly higher rate of infections.

In an 80-week trial, serious adverse events were reported in 17–19 % of patients. Interestingly, and with the EMA's exclusion of Crohn's disease as an approved indication in mind, exacerbation of this disease was the most commonly reported adverse event. Apart from infections, some safety evaluations have been undertaken on malignancies, autoimmunity, injection site reactions, and the development of anti-certolizumab antibodies. In the rheumatoid arthritis trial involving 4049 patients, lymphoma and melanoma occurred with an incidence of 0.05 and 0.08 per 100 patient-years, respectively. Note, however, that rheumatoid arthritis patients are thought to show higher risks of malignancy and lymphoma in any case. With regard to autoimmune side effects, antinuclear and anti-double-stranded DNA antibodies developed in 16.7 % and 2.2 %, respectively, of certolizumab pegol-treated patients with rheumatoid arthritis, and a few cases of a lupus-like syndrome have been reported in patients given the mAb. The appearance of injection site reactions in 5.8 % of certolizumab pegol-treated patients in a placebo-controlled rheumatoid arthritis trial proved only marginally higher than the figure of 4.8 % seen in the placebo group. Symptoms recorded included erythema, pain, swelling, itching, hematoma, and bruising. The commonly seen adverse events are listed in Table 4.1.

In the postmarketing period and at April 2013, the FAERS database had recorded 14,800 reports with 7.2 % concerning infections, 9.4 % of reports were gastrointestinal related, 5 % were cutaneous disorders, and 4.6 % were injection site reactions. The most frequently reported infections were pneumonia, nasopharyngitis, and urinary and upper respiratory tract infections. There were 79 cases of tuberculosis. Rash, urticaria, and anaphylaxis/anaphylactic shock in that order were the most frequently seen hypersensitivities. Other notable reports were malignancies (39, including 25 cases of lymphoma), lung neoplasms (34), melanoma (20), and peripheral neuropathy (41 cases). At the end of 2012, the EVRS database contained 2797 reports (almost all serious), with gastrointestinal events (18.5 %), infections (14 %), cutaneous reactions (8 %), nervous disorders (6 %), and musculoskeletal problems being the most frequently reported events. The most common infections were pneumonia, urinary tract infections, sepsis, and nasopharyngitis; 15 cases of tuberculosis were recorded. Hypersensitivities included mainly rash, urticaria, and anaphylaxis/anaphylactoid reactions. There were 12 cases of peripheral neuropathy, six of optic neuritis, six lymphomas, and five cases of melanoma.

Immunogenicity of Certolizumab Pegol

As so often seems the case with published immunogenicity data, clear, convincing results and firm conclusions are lacking and/or vague. In a placebo-controlled trial of rheumatoid arthritis patients given certolizumab pegol, 9.6 % of patients developed antibodies to the mAb and approximately one-third of these showed neutralizing activity in vitro. According to the EMA, "antibody formation was associated with lowered drug plasma concentration and in some patients, reduced efficacy." Concomitant dosage with immunosuppressants such as methotrexate resulted in lower rates of antibody development. In two open-label studies extending over 5 years, transient formation of antibodies to certolizumab pegol developed in 8.4 % of exposed patients while 4.7 % of patients showed persistent antibody formation. The percentage of antibody-positive patients with a persistent reduction of mAb plasma concentration was 9.1 %, and this was said to be associated with reduced efficacy "in some patients." In phase III placebo-controlled trials of patients with axial spondyloarthritis or psoriatic arthritis, the percentages of patients with antibodies to certolizumab pegol on at least one occasion up to week 24 were 4.4 % and 11.7 %, respectively, and again, antibody formation was associated with a lowered mAb plasma concentration. In its prescribing information, the FDA states that 8 % of patients with active Crohn's disease in a randomized placebo-controlled study (662 patients) and a randomized treatment withdrawal study (428 patients) were antibody-positive to certolizumab pegol after continuous exposure to the mAb over 24 weeks. Approximately 6 % of the patients had antibodies that were neutralizing in vitro, but there was no apparent correlation of the development of antibodies with observed adverse events or efficacy. Patients also given immunosuppressants produced less antibody (3 % vs. 11 %). In a rheumatoid arthritis placebo-controlled study, 7 %

(105 of 1509) produced antibodies to certolizumab pegol on at least one occasion; in 39 of these patients (37 %), the antibodies had neutralizing activity and were associated with a reduction in plasma mAb levels and reduced efficacy. Again, antibody development was lower in patients receiving concomitant methotrexate immunosuppression.

Infliximab

Infliximab (Remicade®) (Tables 2.1 and 4.1), originally named cA2, is a chimeric human-mouse IgG1κ mAb, MW ~ 149 kDa, that binds human TNF with high affinity and specificity, inhibiting both transmembrane and soluble TNF by forming stable, biologically inactive, immune complexes. Infliximab binds to both monomer and trimer forms of soluble TNF with each infliximab molecule able to bind to two TNF molecules. In Crohn's disease, for example, infliximab reduces the levels of several disease markers such as IL-6, sTNFR1, and TNFR2; leads to reductions in intestinal mucosal lamina propria mononuclear cells expressing TNF, IL-19, and IFN-γ; and reduces the number of cells staining positive for CD4, CD8, and CD68. Similar findings of the suppression of inflammation markers such as IL-6, IL-1R, and TNF receptors were seen in rheumatoid arthritis patients treated with infliximab and in patients with psoriasis vulgaris on infliximab therapy; reductions in CD3-positive T cells and ICAM-1 expression were observed. Like adalimumab, but unlike certolizumab pegol, infliximab mediates CDC and ADCC and increases the proportion of cells undergoing apoptosis and the level of granulocyte degranulation. However, while infliximab-induced apoptosis has been suggested as an explanation for smaller numbers of synovial macrophages in patients with rheumatoid arthritis, some investigators have found no increase in apoptosis after treatment with infliximab. All three of the mAb TNF blockers inhibit lipopolysaccharide-induced IL-1β production.

Indications, Warnings, Precautions, and Adverse Events

Following the initial approvals by the FDA and EMA in 1998–1999 for active and fistulizing Crohn's disease, the list of approved indications for infliximab has steadily increased to now also include pediatric Crohn's disease, ulcerative colitis, rheumatoid arthritis (in combination with methotrexate), ankylosing spondylitis, psoriatic arthritis, and plaque psoriasis.

Serious infections and the risk of malignancy are the subjects of an FDA black box warning for infliximab along with the other warnings and precautions listed for the other three approved anti-TNF mAbs, adalimumab, certolizumab pegol, and golimumab (Table 4.1). The incidence of new and reactivated cases of tuberculosis has been estimated at 0.08 % in ~400,000 treated patients, while the overall risk of serious infections is twice as high in patients given infliximab. In a psoriasis longitudinal assessment and registry undertaken at dermatology centers in the USA, Canada, and Israel over the period 2007–2013 and designed to assess the risk of infection with biologic and systemic treatment of psoriasis, data from 11,466 patients (22,311 patient-years) were analyzed. The cumulative incidence rate of serious infections was 1.45 per 100 patients with the most commonly reported serious infections being pneumonia and cellulitis. Not only were infliximab and adalimumab associated with an increased rate of infection, they were associated with a higher risk of serious infections than nonbiologic and non-methotrexate therapies. By comparison, no increased risk was observed with the mAb ustekinumab (section "Ustekinumab") and the fusion protein etanercept (Chap. 6, sections "Etanercept" and "Etanercept").

The overall rate of malignancies appearing within one year in 5707 infliximab-treated patients was 0.08–0.1 %, an incidence three times higher than found in the general population. About 50 % of the cases were lymphomas and skin cancers. Of ~200 cases of hepatosplenic T-cell lymphoma, 28 (14 %) occurred in patients with Crohn's disease. Other warnings and precautions for infliximab therapy, listed in Table 4.1, cover the possibility of reactivation of the hepatitis B virus, hepatotoxicity (liver failure, jaundice, hepatitis, and cholestasis); association of the mAb with adverse outcomes in patients with heart failure; potentially fatal leukopenia, neutropenia, thrombocytopenia, and pancytopenia; hypersensitivity including a serum sickness-like reaction; rare neurological reactions manifesting as systemic vasculitis, CNS demyelinating disorders (including multiple sclerosis and optic neuritis) and peripheral demyelinating disorders (such as Guillain-Barré syndrome); and autoimmunity which may rarely develop as a lupus-like syndrome. Given the risk of infection for patients on infliximab, it is recommended that live vaccines and therapeutic infectious agents such as attenuated bacteria should not be given concurrently with the mAb. In a similar vein, concurrent administration of abatacept has been associated with an increased risk of infections, and serious infections and neutropenia have been seen with concurrent use of anakinra and etanercept.

The overall incidence of infusion reactions in 165 patients given a total of 479 infusions of infliximab was 6.1 % with 9.7 % of the patients affected. Severe reactions occurred in 1 % of the infusions. Acute reactions in 11 of 14 patients were determined to be nonimmediate hypersensitivities, while delayed reactions were seen in only 0.6 % of infusions. An examination of responses to 304 infliximab infusions in 52 adults and 34 children noted severe systemic reactions in 11 adults (21.2 %) and only one child (2.9 %). Reactions were characterized by hypotension and laryngospasm and required treatment with epinephrine, antihistamines, and/or corticosteroids. Delayed reactions seen as arthralgia, fever, and myalgia occurred in eight adults (9.3 %) but in no children. Of 84 patients with rheumatoid arthritis given infliximab, 16 (19 %) discontinued therapy because of an adverse reaction, and nine of these (11 %) experienced an immediate hypersensitivity reaction. In fact, a range of immune-mediated systemic and cutaneous adverse reactions have been reported following infliximab therapy. There are numerous reports of immediate reactions to infliximab including urticaria but particularly anaphylaxis, with over 650 cases of the latter at an incidence of ~0.9 % for the period 1999–2012. Nearly 70 % of the anaphylactic reactions occurred during the first month of therapy; females made up about 60 % of the reactors and children less than 10 years of age accounted for 7.5 % of reactors. Infliximab-reactive IgE antibodies have been identified in some cases of anaphylaxis, and successful desensitizations of patients experiencing an anaphylactic/anaphylactoid reaction to infliximab have been published. Type III hypersensitivities to infliximab are known to occur including some cases of vasculitis and a serum sickness-like reaction. Type IV hypersensitivities recorded include maculopapular rashes, psoriasis, erythema multiforme, and Stevens-Johnson syndrome. Intriguingly, there are reports of infliximab inducing rapid recovery of lesions in several cases of toxic epidermal necrolysis. Commonly seen and serious reactions are listed in Table 4.1.

Of over 57,600 postmarketing reports to the FAERS database, 6 % were for infections, 4 % for each of respiratory and cutaneous events, 3 % for injection site reactions, and 3 % each for neurological and gastrointestinal adverse events. Tuberculosis accounted for 1963 infections; skin cancers and gastrointestinal neoplasms were the most frequent malignancies; there were 321 cases of hepatosplenic T-cell lymphoma, 460 reports of anaphylaxis, 201 reports of a serum sickness-like reaction, 821 cases of systemic lupus erythematosus (SLE), 1460 cases of lupus-like syndrome, 96 cases of PML, and 37 reports of PRES. Of 47,300 reports to the EVRS database, infections constituted 13 % of the total, gastrointestinal disorders 9 %, respiratory problems 8 %, and adverse cutaneous events 7 %. Neoplasms accounted for 5 % of reports and there were 63 cases of hepatosplenic T-cell lymphoma, 4921 injection site reactions, 970 hypersensitivity responses, 35 cases of PML, and 22 of PRES.

Immunogenicity of Infliximab

Patients who develop antibodies to infliximab are 2–3 times more likely to experience infusion reactions. Without the concomitant administration of immunosuppressants, antibodies were found in 24 % of patients on infliximab therapy; with immunosuppression this percentage was halved. The corresponding figures without, and with, methotrexate for psoriatic arthritis were 26 and 4 % and for Crohn's disease, 13.3 and 3.3 %. The percentage for rheumatoid arthritis patients receiving infliximab and methotrexate was 8 %. The EMA states that some patients who developed high titers of anti-infliximab antibodies had evidence of reduced efficacy of the mAb. Psoriasis patients on a maintenance dose of infliximab developed antibodies in 28 % of cases. The FDA has reported that patients given infliximab in a three-dose induction regimen followed by maintenance dosing showed an incidence of antibody development of ~10 % assessed over a period of 1–2 years. Higher incidences were observed in patients with Crohn's disease and in 191 patients with psoriatic arthritis when anti-infliximab antibodies were found in 15 % of patients. Most positive responders had low-titer antibodies, but the presence of antibodies generally corresponded to higher rates of clearance of infliximab, reduced efficacy of the drug, and the manifestation of infusion reactions. As expected, immunosuppressant therapy with methotrexate, 6-mercaptopurine, or azathioprine resulted in the production of less antibody. In a Spanish study of the influence of immunogenicity on the efficacy of long-term treatment with infliximab in rheumatoid arthritis, 85 patients were treated with infliximab for a median of 4.42 years (interval 0.4–10.2 years). Anti-infliximab antibodies developed in 28 (32.9 %) patients and were found in all nonresponder patients classified on the European League Against Rheumatism (EULAR) scale of non-, moderate, and good responders. The highest antibody levels occurred in EULAR nonresponders throughout the study period. Nine patients (10.6 %), all with high antibody levels, developed infusion reactions. The presence of anti-infliximab antibodies often brought with it the need for an increased dose of infliximab. Methotrexate cotherapy did not produce fewer patients with antibodies, but it did reduce the levels of anti-infliximab antibodies. Despite the aforementioned findings and interpretations, some investigators doubt that anti-infliximab antibodies have any clinical importance for the mAb's efficacy and safety, believing that patient- and treatment-related factors involved in antibody development are complex and poorly understood. It has further been claimed that while antibodies may be weakly and variably associated with infusion reactions, they are generally not relevant to what has been called "clinical decision-making," since few of the antibodies have real clinical relevance. In the light of the perceived lack of evidence that anti-infliximab antibodies impact on drug efficacy and safety, it has been suggested that there is no need to measure or prevent them and that it would be preferable to measure circulating concentrations of the dosed drug.

Golimumab

Golimumab (CNTO148, Simponi®) (Tables 2.1 and 4.1) is a fully human, TNF-specific IgG1κ mAb existing in multiple glycoforms, MW 150–151 kDa, derived from transgenic mice immunized with human TNF and engineered to express human IgGs. Golimumab has a high capacity to neutralize TNF, binding soluble TNF with a 2.4-fold higher affinity than infliximab and a 7.1-fold higher affinity than adalimumab and binding transmembrane TNF with an affinity similar to infliximab but greater than adalimumab. Like other human IgG1 antibodies, golimumab binds to Fc receptors reflected in its properties of effecting CDC and ADCC, while its long terminal half-life of 2–3 weeks after intravenous and subcutaneous injection reflects FcRn binding at low pH (Chap. 2, section "Antibody-Dependent Cell-Mediated and Complement-Dependent Cytotoxicities"). Like infliximab, golimumab mediates lysis of cells overexpressing human TNF, but like both infliximab and adalimumab, it does not lyse lipopolysaccharide-stimulated human monocytes actively secreting TNF. Adhesion proteins such as E-selectin and ICAM-1, upregulated in the joint vasculature in rheumatoid arthritis, are inhibited in vivo by golimumab at IC_{50} vales significantly lower than the values for infliximab and adalimumab. Likewise, serum levels of proinflammatory cytokines and chemokines are reduced by golimumab in results similar to those reported for infliximab. In golimumab-treated mice, serum levels of granulocyte-macrophage colony-stimulating factor (GM-CSF), granulocyte colony-stimulating factor (G-CSF), keratinocyte chemoattractant, interferon-inducible protein-10 (IP-10), and IL-6, each show clear reductions at 23 days post treatment.

Golimumab was given FDA and EMA approval in 2009 for the treatment of moderate to severe rheumatoid arthritis in combination with methotrexate, for active psoriatic arthritis in adults in combination with methotrexate, and for ankylosing spondylitis in adults. Warnings, precautions, and serious and common adverse events for its use issued by the regulatory agencies are similar to the three other approved mAb TNF blockers. Emphasis is placed on a boxed warning for golimumab for serious infections and malignancy plus reference to the risk of hepatitis B reactivation, heart failure, demyelinating disease, hypersensitivity reactions, and the possible development of cytopenias (Table 4.1). Attention is drawn to opportunistic infections, particularly tuberculosis with the possibility of reactivated or new infections to invasive fungi and to hepatitis B virus reactivation. In controlled clinical trials, lymphoma was seen more often in patients receiving golimumab than in control patients, and during a phase III study, the incidence of malignancies other than lymphoma and non-melanoma skin cancer per 100 patient-years was 0.56 compared to 0 in the placebo group. Again, as with adalimumab, infliximab, and certolizumab pegol, golimumab should not be used with biologics such as anakinra and abatacept, agents that may increase the chance of serious infections, and the avoidance of the administration of live vaccines is recommended. Injection site reactions tend to be mild, and type I (anaphylaxis) and type III (serum sickness) hypersensitivities are extremely rare. A few cases of uveitis have recently been reported. Of 2125 reports to the FAERS database during the postmarketing period, most reports concern infections (18 %) with about six times as many reports as respiratory, neurological, and dermatologic events. Reports of adverse events such as sarcoidosis, interstitial lung disease, and skin exfoliation are therefore of interest being somewhat unexpected and unusual. Infections again predominate in the EVRS database with 41 cases of pneumonia and seven of tuberculosis. Four cases of anaphylaxis were reported. Overall, golimumab appears to show a better safety profile than the other registered mAb TNF inhibitors although further long-term monitoring is needed. A 3-year safety update of golimumab published in 2015 compared subcutaneously administered doses of 50 and 100 mg for up to year three in rheumatoid arthritis, psoriatic arthritis, and ankylosing spondylitis trials. Analyses showed that safety was consistent with other approved TNF inhibitors. The 100 mg dose produced a higher incidence of serious infections, demyelinating events, and lymphoma than 50 mg.

In phase III studies over 52 weeks with rheumatoid arthritis, psoriatic arthritis, and ankylosing spondylitis patients, anti-golimumab antibodies developed in 105 of 2115 patients (5 %). In vitro tests showed that most of the antibodies were neutralizing. In a comparison of treatments with golimumab-methotrexate and golimumab alone, antibodies were found in 3 % and 8 % of patients, respectively. In phase II and III ulcerative colitis studies over a 54-week period, antibodies to golimumab were detected in 26 of 946 patients (3 %). Neutralizing antibodies identified in vitro occurred in 68 % of antibody-positive patients. Fewer patients with antibodies (1 % compared to 3 %) were seen when the mAb was administered with azathioprine, 6-mercaptopurine, or methotrexate. A meta-analysis of original clinical trials to evaluate antibody levels after golimumab therapy concluded that levels were neither higher nor significant clinically, and there were no evident associations between golimumab-induced antibodies and lupus-like syndromes or infusion reactions.

Abciximab

Abciximab (monoclonal antibody 7E3, ReoPro®) (Tables 2.1 and 4.1) is the Fab fragment of a human-mouse chimeric IgG1κ mAb, MW ~ 47.6 kDa, targeted to the glycoprotein GPIIb/IIIa (CD41/CD61) receptor that aids platelet activation. The glycoprotein, also known as integrin αIIbβ3, is a receptor for fibrinogen and von Willebrand factor (Chap. 10, section "Platelet Activation and von Willebrand Factor"). In platelet activation, GPIIb and GpIIIa associate in a Ca^{2+}-dependent step. Activation by local agonists including ADP, epinephrine, collagen, serotonin, and thrombin induces a conformational change in the receptor that leads to fibrinogen binding, platelet aggregation, and thrombus formation. The GPIIb/IIIa receptor is the target of a number of different and important drugs including tirofiban (Aggrastat®) and eptifibatide (Integrilin®) (Chap. 10, section "Platelet Activation and von Willebrand Factor") and typically a number of 50,000–100,000 receptor molecules per platelet. Abciximab acts as a receptor antagonist preventing the binding of the natural ligands to the GPIIb/IIIa receptor by steric hindrance or a conformational effect rather than by direct interaction with the GPIIb/IIIa RGD (arginine-glycine-aspartic acid) binding site. Maximum inhibition of platelet aggregation is observed when ≥80 % of GPIIb/IIIa receptors are blocked by abciximab.

Integrin Recognition by Abciximab

The integrins, a family of receptors that share α- and β-subunits, are a class of cell adhesion molecules important in wound healing, inflammation, and hemostasis. Abciximab is not selective for platelet GPIIb/IIIa (CD41/CD61) but also binds with a similar affinity to the vitronectin receptor αvβ3 (CD51/CD61), a receptor present on platelets, leukocytes, vascular endothelial cells, and smooth muscle cells, that mediates procoagulant properties of platelets and cell proliferation. Abciximab blocks the αvβ3-mediated effects including cell adhesion. Abciximab also cross-reacts with macrophage-1 antigen (Mac-1, complement receptor 3 or CR3) an integrin designated as $\alpha_M\beta_2$ consisting of CD11b (integrin α_M) and CD18 (integrin β_2) found on monocytes and neutrophils. Cross-reaction occurs via the CD11b subunit of Mac-1, interrupting cell-cell and cell-extracellular matrix interactions and leading to the suggestion that this may influence the recruitment of circulating monocytes to sites of vessel injury reducing infiltration and inflammation in areas of myocardial ischemia. Together with the GPIIb/IIIa-targeted function of abciximab, the integrin recognition properties of the mAb Fab may contribute to the regulation of vascular repair and underlie the clinical benefits resulting from the use of abciximab after angioplasty.

Indications, Warnings, Precautions, and Adverse Events

Abciximab, the first fibrinogen receptor antagonist to be approved for clinical use, significantly reduces the incidence of repeat percutaneous coronary intervention and coronary artery bypass surgery in patients undergoing percutaneous transluminal coronary angioplasty. Abciximab is approved as an adjunct to percutaneous intervention for the prevention of cardiac ischemic complications in patients undergoing such intervention and in patients with unstable angina not responding to conventional therapy when percutaneous coronary intervention is planned within 24 h.

The main warning issued for abciximab is for bleeding, and because of this risk, it is contraindicated in a range of disorders where bleeding may be active or a threat, in thrombocytopenia, in cases of recent major surgery or trauma, in severe uncontrolled hypertension, in patients with a history of cerebrovascular accident or vasculitis, and in cases involving intracranial neoplasm, arteriovenous malformation, or aneurysm. Abciximab's potential to increase the risk of bleeding is particularly apparent in patients receiving anticoagulant therapy such as heparin, other anticoagulants, or thrombolytics. Thrombocytopenia, including severe cases, may occur indicating the need for platelet count monitoring before, during, and after treatment with abciximab. Thrombocytopenia may occur in up to 2 % of patients. A warning has also been issued for the possibility of allergic reactions including anaphylaxis which has sometimes proved fatal. Other serious and frequent adverse events are listed in Table 4.1. Postmarketing reports of adverse events have mainly centered on cardiovascular and hematologic disorders and death. Hypersensitivities, including anaphylaxis, accounted for up to 0.3 % of reports.

Immunogenicity of Abciximab

Readministration of abciximab may induce the production of human antichimeric antibodies potentially causing an allergic reaction, thrombocytopenia, or a reduced clinical benefit of the mAb Fab. In one study of over 1200 patients, human antichimeric antibodies were found in 6 % of patients prior to readministration of abciximab and in 27 % after readministration, but there were no reports of hypersensitivity. Of 36 patients receiving four or more exposures to abciximab, human antichimeric antibodies were found in 16 (44 %). Rates of thrombocytopenia tend to be higher in patients following readministration of the agent, and readministration may be associated with increased severity of thrombocytopenia. For example, in a study of 500 patients receiving abciximab for at least a second time during percutaneous coronary intervention, thrombocytopenia was found in 23 patients (4.6 %). Twelve (2.3 %) of these patients developed profound thrombocytopenia. Before the first readministration, human antichimeric antibodies were present in 22 of 454 patients (4.8 %); this percentage increased to 19 % (82 of 432) after a first readministration. Antibodies did not inhibit in vitro aggregation of platelets by abciximab or correlate with clinical events; no cases of major bleeding, hypersensitivity, or death were seen. Overall, the findings were consistent with the results of the randomized clinical trials of first treatments with abciximab except that there was a shift from mild to profound thrombocytopenia, and cases of delayed and recurrent thrombocytopenia were seen. In another study of the safety and efficacy of readministration of abciximab in coronary intervention, results obtained from 164 patients were retrospectively analyzed. Adverse events included death (2 %), intracranial hemorrhage (2 %), and myocardial infarction (3 %). Severe thrombocytopenia occurred in 4 % of patients after readministration of abciximab, but no hypersensitivities were observed, and patients receiving abciximab within two weeks of its first administration experienced a higher incidence of severe thrombocytopenia (12 % vs. 2 %). The authors concluded that abciximab remains clinically efficacious when readministered, but dose reduction may be necessary when it is readministered within days of the initial administration. A study in a small number (9) of patients who developed profound thrombocytopenia within hours of receiving abciximab for a second time developed IgG antibodies specific for murine sequences in abciximab, and it was suggested that these antibodies may cause life-threatening thrombocytopenia. Interestingly, antibodies found in healthy subjects that react with abciximab and don't cause thrombocytopenia are claimed to be specific for the papain cleavage site of any IgG Fab fragment.

Alemtuzumab

Alemtuzumab (Campath®, MabCampath®) (Tables 2.1 and 3.1), as therapy for B-cell chronic lymphocytic leukemia, is presented in Chap. 3, section "Alemtuzumab."

As Lemtrada® (Tables 2.1 and 4.1), alemtuzumab, a recombinant humanized IgG1κ mAb, MW ~ 150 kDa, directed against the 21–28 kDa glycoprotein CD52 expressed on normal and malignant T and B cells, monocytes, macrophages, and some granulocytes, NK, and dendritic cells, is marketed and approved for the treatment of adult patients with relapsing remitting multiple sclerosis. The detailed mechanism of alemtuzumab's therapeutic effect in multiple sclerosis is not yet understood although it is presumed to act via its target antigen CD52 on NK cells, macrophages, and monocytes, depleting and repopulating lymphocytes with the involvement of the mAb's CDC and ADCC actions. Reductions in the levels of circulating T and B cells followed by cell repopulation may delay disease progression and reduce relapses.

Autoimmunity, infusion reactions, and malignancies make up an FDA boxed warning for Lemtrada® which is available only through a restricted distribution arrangement called the Lemtrada® REMS Program. Serious autoimmune disorders, sometimes fatal, may manifest as thrombocytopenia and anti-glomerular basement membrane disease requiring periodic complete blood counts with differential serum creatinine levels and urine analysis over a 48-month period after the last dose. Incidences of 2 %, 0.3 %, and 34 % have been reported for thrombocytopenia, glomerular nephropathies, and thyroid disorders, respectively. Attention is also drawn to the possible occurrence in Lemtrada®-treated patients of other autoimmune cytopenias, namely, neutropenia and hemolytic anemia, as well as pancytopenia. It is thought that the increased risk of autoimmune conditions may be due to the broad range of autoantibody formation induced by alemtuzumab. Serious and life-threatening infusion reactions and anaphylaxis may occur making it necessary to administer the antibody in a suitable setting with appropriate personnel and equipment to manage reactions. Premedication with corticosteroids for the first 3 days of infusion together with antihistamines and/or antipyretics is recommended but reactions may occur despite pretreatment. Patients receiving Lemtrada® may be at increased risk of malignancies and lymphoproliferative disorders necessitating the performance of baseline and yearly skin examinations. Thyroid cancer and melanoma have each been reported with an incidence of 0.3 %. Other warnings and precautions draw attention to the risks of infections and pneumonitis, the latter seen in hypersensitivity form or as pneumonitis with fibrosis. In controlled clinical trials, infections occurred in 71 % of Lemtrada®-treated patients compared to 53 % in control patients treated with interferon beta-1a (Chap. 5, section "Interferon Beta"). Recorded serious infections include herpes viral infections, human papillomavirus, tuberculosis, fungal infections (especially candidiasis), and *Listeria* meningitis.

Table 4.1 lists some of the more frequently seen and occasionally serious adverse events provoked by alemtuzumab when it is used to treat multiple sclerosis. As yet, and distinct from information available for alemtuzumab when used to treat B-cell chronic lymphocytic leukemia at higher and more frequent dosage, there is little postmarketing information on Lemtrada® therapy for multiple sclerosis. During postmarketing use of alemtuzumab for leukemia treatment, additional autoimmune cytopenias including fatal anemias and cardiac disorders in patients previously treated with anthracyclines were reported (Chap. 3, section "Alemtuzumab"); this will be kept in mind as Lemtrada® therapy for multiple sclerosis is continued.

There is some data available on the immunogenicity of alemtuzumab used in the treatment of patients with multiple sclerosis. In the first treatment course, anti-alemtuzumab antibodies were found in 62 %, 67 %, and 29 % of Lemtrada®-treated patients at 1, 3, and 12 months, respectively, and in the second treatment course in 83 %, 83 %, and 75 % of patients at 13, 15, and 24 months. The corresponding percentages for neutralizing antibodies in the antibody-positive patients were 87, 46, and 5 % and 94, 88, and 42 %. Detected antibodies were associated with declines in alemtuzumab serum concentrations in the second treatment course but not the first, but the antibodies appeared to have no effect on clinical outcomes, lymphocyte counts, or adverse events. These findings are a challenge to interpret.

Basiliximab

Basiliximab (Simulect®) (Tables 2.1 and 4.1) is a recombinant human-mouse chimeric IgG1κ mAb, MW ~ 144 kDa, with specificity for the IL-2 receptor (IL-2R) alfa chain expressed on the surface of activated T lymphocytes.

The IL-2 Receptor and Mechanism of Action of Basiliximab

The biological effects of IL-2 are mediated mainly through the high-affinity IL-2R complex which is made up of the alfa chain IL-2Rα (CD25, p55, or Tac Ag), the beta chain IL-2Rβ (CD122, p75), and the common gamma chain γ_c (CD132, p65). Assembly of the IL-2R complex proceeds in a sequential manner initiated by the abundantly expressed IL-2Rα on activated T cells binding IL-2 and changing its conformation to a favorable IL-2Rβ binding state. IL-2Rβ and the IL-2/IL-2Rα complex are then brought into closer proximity by diffusion within the cell surface to form an IL-2/IL-2Rαβ complex. Recruitment of γ_c completes the biologically active final receptor complex that triggers the IL-2 signaling pathways Ras/Raf/MAPK, JAK/STAT, and PI3K/AKT that ultimately result in the growth, differentiation, and survival of cytotoxic T cells. Although IL-2Rα is not expressed on resting T cells, it is abundantly expressed on activated T cells and especially by T cells in some autoimmune diseases, T-cell leukemia, and organ allograft rejection. This is the rationale underpinning the development and application of basiliximab for the prevention of organ transplant rejection. Basiliximab blocks IL-2 signaling by binding IL-2Rα at residues 116–122 of the IL-2Rα D2 domain. Binding affinity of the mAb to the receptor is ~70 times higher than the natural ligand. Figure 4.3 is a diagrammatic representation of the events leading to the assembly of the IL-2R complex and shows the point of intervention of the mAb in the IL-2–IL-2R interaction sequence.

Basiliximab Indications, Warnings, Precautions, and Adverse Events

Basiliximab is approved for the prophylaxis of acute organ rejection in adult and pediatric patients (1–17 years) receiving allogeneic renal transplantation together with cyclosporin and corticosteroids. Effectiveness of basiliximab for acute rejection

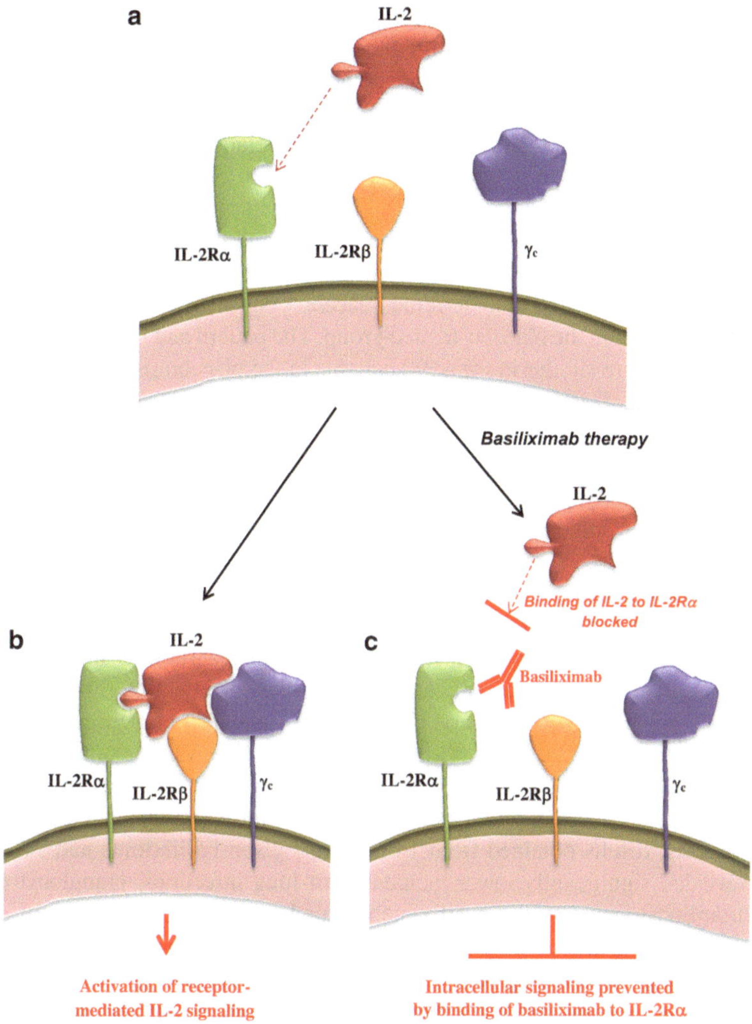

Fig. 4.3 Diagrammatic representation of the mechanism of action of basiliximab. (**a**) The IL-2 signaling pathway is initiated by the binding of IL-2 to the alfa chain IL-2Rα of its receptor. (**b**) This triggers the sequential assembly of the IL-2/IL-2Rα/IL-2Rβ/γ_c complex, the biologically active receptor form, which results after diffusion through the cell surface membrane of the receptor β-chain IL-2Rβ and the common γ_c chain. Signaling then proceeds via the activated receptor. (**c**) The binding of basiliximab to IL-2Rα blocks the binding of IL-2 to the receptor alfa chain, thus preventing IL-2 signaling

of other solid organ allografts has not been demonstrated, but it has been used successfully in heart transplantation to induce immunosuppression and for severe cases of rejection, as induction therapy in liver transplants, and for lung transplants. Overall, the safety profile has been found to be similar to the experience with basiliximab in its application to kidney transplant rejection.

An FDA boxed warning for basiliximab states that the mAb should be prescribed only by physicians experienced in immunosuppression therapy and management of organ transplant patients and that management should be undertaken in suitably equipped facilities with the necessary staff, laboratory facilities, and other supportive medical resources. Precautions are necessary in relation to infections because of increased susceptibility to infections and the possibility of fatal sepsis induced by the immunosuppressive regimens. Other warnings and precautions cover the potential risks of hypersensitivity, including anaphylaxis and immunogenicity.

Although adverse events were reported in 96 % of basiliximab-treated patients studied in four randomized, double-blind, placebo-controlled clinical trials, the same incidence was seen in the placebo-treated group. The pattern and incidence of serious adverse events were similar in the ~590 patients in the placebo (8–39 %) and treated (7–40 %) groups. Gastrointestinal disorders were the most frequently reported adverse event in both the treated patients (69 %) and placebo-treated patients (67 %), and similar incidences of serious infections (76 %) and malignancies (1 %) were also seen, but in one study, posttransplant lymphoproliferative disorder occurred in 18 % of the study group and only 5.4 % of the placebo patients. Table 4.1 lists the adverse events most often seen. Four cases of anaphylaxis occurred in pediatric patients receiving basiliximab compared to two cases in the placebo-treated patients.

An evaluation of the safety of basiliximab compared to antithymocyte globulin (ATG) was undertaken with 60 patients given ATG plus tacrolimus and steroids and 60 patients given the mAb. Significantly higher rates of infection (58 % vs. 33 %), including *Cytomegalovirus* infection (22 % vs. 5 %), and hematologic complications, namely, anemia (57 % vs. 20 %), thrombocytopenia (37 % vs. 2 %), and leukopenia (60 % vs. 23 %), were recorded in the ATG-treated group. In a Chinese retrospective study of the relative safety profiles of the two immunosuppressants in kidney transplant patients, results obtained from 146 patients given basiliximab and 116 given ATG revealed significantly lower incidences of lung infections, granulocytopenia, and thrombocytopenia in the basiliximab-treated group. There was no difference between the two groups in the incidences of heart dysfunction after transplantation.

During the postmarketing period so far, severe hypersensitivity reactions including anaphylaxis and urticaria have been reported as well as capillary leak and cytokine release syndromes (Chap. 1, sections "Capillary Leak Syndrome" and "Cytokine Release Syndrome"). By the end of 2012, infections, particularly viral infections, and renal disorders were the adverse events most often reported to the FAERS database, while in the EVRS database, infections and immune disorders were most prominent. There were 122 malignancies, 30 cases of posttransplant lymphoproliferative disorder, and 17 cases of anaphylaxis.

Immunogenicity of Basiliximab

Anti-idiotype antibodies detected in 4 of 339 (1.2 %) basiliximab-treated patients and human anti-mouse antibodies in 3.5 % of patients appeared to bring with them no adverse clinical effects such as accelerated clearance of the mAb. There are a number of reports of immediate type I hypersensitivities to basiliximab including the detection of anti-idiotype antibodies of the IgE class in a child who experienced an anaphylactic shock reaction after receiving a second course of basiliximab during renal transplantation. No IgE reactivity was detected to a control murine IgG mAb indicating exclusive recognition of basiliximab idiotypes. Anaphylaxis following first exposure to basiliximab has also been reported.

Belimumab

Belimumab (Benlysta®, previously known as LymphoStat-B) (Tables 2.1 and 4.1), approved for the treatment of systemic lupus erythematosus (SLE), is a recombinant fully human IgG1λ mAb, MW ~ 147 kDa, specific for soluble human B lymphocyte stimulator protein (BLyS), a B-cell survival factor.

BLys and Belimumab

BLys, a cytokine belonging to the TNF ligand family, is a 285-amino acid glycoprotein known under a range of different names including B-cell-activating factor (BAFF), tumor necrosis factor ligand superfamily member 13B (TNFSF13B), TNF- and APOL-related leukocyte expressed ligand (TALL-1), and CD257. Expressed as a homotrimeric transmembrane protein on monocytes/macrophages, activated T cells, dendritic cells, neutrophils, some B cells, and non-hematopoietic cells, BLyS can be cleaved into soluble homotrimers but also occurs in soluble form as a structure of 20 trimers or 60-mers. BLys is the natural ligand of three TNF receptors, BAFF-R (also known as BR3), TACI (transmembrane activator and calcium modulator and cyclophilin ligand interactor), and BCMA (B-cell maturation antigen). Each of these receptors is expressed mainly on mature B lymphocytes. BRAFF-R, expressed on all B cells except bone marrow plasma cells, is highly specific for BLys. TACI, expressed on memory B cells and bone marrow plasma cells, is a negative regulator of B-cell maturation. BCMA is the most important of the three receptors on long-lived plasma cells where it is important for cell survival. A proliferation-inducing ligand, APRIL, is another member of the TNF superfamily secreted by activated myeloid cells and with a stimulatory action on B cells. BlyS and APRIL share the TACI and BCMA receptors, while BLyS also interacts with BAFF-R. Both ligands enhance the development of autoimmune disease by sustaining B-cell activation, whereas their inhibition is thought to provide protection from autoimmunity by reducing survival of the reactive B cells (see also below, section "APRIL, Lupus, and Atacicept"). This understanding, the elevated levels of BLyS in patients with SLE, and the observation that transgenic mice overexpressing BLys develop conditions similar to SLE and to Sjögren syndrome, contributed to the targeting of the BLys/APRIL pathway in an attempt to modulate autoantibody production in SLE. In 2011, belimumab was approved by the FDA, EMA, and Health Canada for the treatment of SLE. It modulates the downstream signaling of BAFF-R expressed on mature B cells, memory B cells, and early plasma cells; it does not cause CDC and ADCC and does not kill B cells directly. Notwithstanding the mAb's beneficial effects in managing nonlife-threatening forms of SLE, its mechanism of action is still not fully understood.

Belimumab Warnings, Precautions, and Adverse Events

As with many agents affecting the immune response, infections, some serious, occur in patients given belimumab. In controlled clinical trials, the incidence of infections was 71 % (placebo 67 %) with upper respiratory tract and urinary tract infections, nasopharyngitis, sinusitis, bronchitis, and influenza predominating. The most frequently occurring serious infections were pneumonia, urinary tract infections, bronchitis, and cellulitis. Caution is recommended before using belimumab in patients with existing infections, interruption of therapy may be necessary for patients who develop an infection during treatment, and live vaccines should not be given for 30 days before, or with, belimumab therapy. Note that cases of JC virus-associated PML (Chap. 2, sections "Progressive Multifocal Leucoencephalopathy" and "Posterior Multifocal Leukoencephalopathy and Natalizumab"), some fatal, have been reported in SLE patients receiving belimumab. The mechanism of action of belimumab suggests the possibility of an increased risk of malignancies, but, as yet, there is no clear data that such a risk is real. In the controlled clinical trials, the incidence of malignancies was 0.4 % in both the belimumab and placebo groups. Hypersensitivity reactions, including anaphylaxis, associated with the use of belimumab are also included in the list of warnings and precautions issued by regulatory agencies. In the controlled clinical trials, reactions diagnosed as hypersensitivities, including anaphylaxis, were seen in 13 % and 0.6 %, respectively, of patients receiving belimumab and 11 % and 0.4 %, respectively, of patients receiving placebo. Infusion reactions, sometimes difficult to distinguish from true type I immediate hypersensitivities, showed similar incidences in the test and control groups—17 % (0.5 % serious reactions) in patients receiving the mAb verse and 15 % (0.4 % serious) in the controls. The finding of slightly higher incidences of psychiatric events in the belimumab-treated patients in the controlled trials has to be viewed against the need for further long-term investigations and the known occurrence of psychiatric disturbances in some SLE patients. Incidences labeled psychiatric events and depression-related events were 16 % and 6.3 % and 12 % and 4.7 %, respectively, in the belimumab-treated and placebo patients. Postmarketing surveillance data from the FAERS and EVRS databases reveal gastrointestinal reactions (nausea, vomiting, diarrhea), infections (especially pneumonia and urinary tract infections), and cutaneous reactions (pruritus, rash, and urticaria) as the most frequently reported events. By the end of 2012, 18 hypersensitivity reactions, eight cases of anaphylaxis, and one anaphylactoid reaction were reported in the EVRS database. Overall then, belimumab appears to have a more or less acceptable safety profile often with no significant differences in serious adverse events between placebo and treatment groups. Some of the more commonly seen and occasionally serious systemic and cutaneous adverse events provoked by belimumab not covered in the above discussion are listed in Table 4.1.

Immunogenicity of Belimumab

In keeping with its mechanism of action, belimumab induces a progressive decrease in total B-cell numbers up to a maximum reduction of about 60 % after one and a half years of therapy. As a consequence, circulating immunoglobulins also decrease with IgM and IgG levels falling by about 18 % and 6 %, respectively, after 1 year. In early clinical trials, anti-belimumab antibodies were detected in 1–13 % of patients. Subsequently, up to 4.8 % of patients were found to develop antibodies with the highest incidences matching the lower doses of the mAb, perhaps because of reduced assay sensitivity in the presence of high drug concentrations. Three patients receiving the lowest dose of belimumab (1 mg/kg) had neutralizing antibodies, and three patients with antibodies experienced mild infusion reactions with cutaneous manifestations.

APRIL, Lupus, and Atacicept

Belimumab binds to soluble, not membrane, BLyS, but the mAb does not bind to APRIL meaning that signaling through the receptors is not totally blocked. It was therefore suggested that neutralizing both BlyS and APRIL might be more efficacious than neutralizing BlyS alone. In fact, a correlation between APRIL serum levels and renal disease activity in patients with renal nephritis was claimed, and it was concluded that high levels of the cytokine might be a biomarker for lupus nephritis and APRIL antagonists might offer an effective treatment. However, whereas the role of BlyS in promoting B-cell-triggered autoimmune disease such as SLE is established, a similar role for APRIL remains controversial, and evidence is lacking that APRIL has a significant role in B-cell-mediated autoimmunity. The strategy of inhibiting APRIL as a therapeutic option for reinforcing inhibition of BLyS and inhibiting B-cell survival was investigated in a double-blind, placebo-controlled phase II/III study of patients with active lupus nephritis given the BlyS/APRIL-neutralizing agent atacicept. Atacicept is a fully humanized recombinant fusion protein (see Chap. 6, section "Atacicept") designed to block the activation of B cells by fusing the extracellular, ligand-binding portion of the TACI receptor to a modified Fc fragment of human IgG1. This resulted in a pronounced reduction in serum immunoglobulins and the occurrence of serious infections leading to premature termination of the trial. In another recent trial with atacicept, assessed for the prevention of flares in SLA patients, no differences between the fusion protein and placebo for flare rate, or time to first flare, were noted.

Canakinumab

Canakinumab (ACZ885, Ilaris®) (Tables 2.1 and 4.1) is a recombinant fully human IgG1κ mAb targeting human IL-1β with high affinity. In 2009, canakinumab was first approved by the FDA and EMA for the treatment of a group of rare hereditary autoinflammatory diseases of variable severity showing a spectrum of overlapping clinical features involving almost all organs and collectively known as cryopyrin-associated periodic syndromes or CAPS. The group of CAPS diseases has an estimated population frequency of 0.5–3 million. CAPS patients generally experience lifelong episodes of recurrent fever with systemic inflammation mediated by neutrophils.

CAPS and the Mechanism of Action of Canakinumab

IL-1β and IL-1α, the most studied members of the IL-1 family of 11 cytokines (Chap. 5), are both proinflammatory molecules and have a natural antagonist IL-1Ra. Anakinra (Chap. 5, section "Anakinra") is a recombinant form (IL-1RA) of IL-1Ra differing from the natural receptor by the addition of a single methionine at the amino terminal end. IL-1β, IL-1α, and IL-1Ra each bind to the IL-1 receptor, IL-1R which exists in two forms, types I and II. CAPS are the result of an autosomal dominant or new mutation of the cold-induced autoinflammatory syndrome-1 (CIAS1) (also known as nod-like receptor protein 3 [NLRP3]) gene on chromosome 1q44 that encodes cryopyrin and results in an overactivation of caspase-1, the enzyme that cleaves the precursors of the IL-1 family members, IL-1β, IL-18, and IL-33, into their active forms. This results in an increase in the production of IL-1β that shows a five times higher concentration in CAPS than in healthy subjects, and it is this cytokine that results in an overactive inflammasome driving inflammation. Support for this conclusion is indicated by results with the recombinant IL-1 receptor antagonist (IL-1RA) anakinra (Chap. 5, section "Anakinra") and the IL-1 trap fusion protein rilonacept (Chap. 6, section "Rilonacept") both of which, like IL-1β, act via IL-1RI. Therefore, by administering the human anti-IL-1β antibody canakinumab, the biological activity of IL-1β is neutralized by blocking its interaction with its receptor IL-1RI.

CAPS Diseases and Approved Indications for Canakinumab

CAPS includes three diseases each related to a defect in the same CIAS1/NLRP3 gene: familial cold autoinflammatory syndrome (FCAS), Muckle-Wells syndrome (MWS), and the most severe form, neonatal onset multisystem inflammatory disease (NOMID), also called chronic inflammatory (infantile) neurological cutaneous articular syndrome (CINCA). The major symptoms include fever, joint pain and swelling, urticarial-like rash without itching, red eyes, loss of hearing, and central nervous system damage. Onset of NOMID is almost always earlier than 3 months of age.

Whereas the list of FDA-approved indications for canakinumab covers CAPS in adults and children 4 years and older including FCAS and MWS and also active systemic juvenile idiopathic arthritis (SJIA), EMA's approved list of therapeutic indications is more extensive. Canakinumab is approved for CAPS in adults, adolescents, and children aged 2 years and older with body weight of 7.5 kg or above, including MWS, severe forms of FCAS/familial cold urticaria presenting with signs and symptoms beyond cold-induced urticarial skin rash, NOMID/CINCA, SJIA in patients aged 2 years and older, and gouty arthritis when nonsteroidal anti-inflammatory drugs, colchicine, and corticosteroids cannot be used.

Canakinumab's specificity, and high binding affinity for the proinflammatory IL-1β, makes it a likely choice as a potential treatment for a range of different inflammatory diseases, and with this in mind, the mAb is currently being investigated in a number of off-label studies and some clinical trials.

Warnings, Precautions, and Adverse Events for Canakinumab

The most common adverse event associated with the use of canakinumab is infection, usually nasopharyngitis, viral infections, and urinary tract and upper respiratory tract infections as well as occasional opportunistic infections of *Aspergillus*, atypical mycobacteria, *Cytomegalovirus*, and herpes zoster. Because of the risk of serious infections, canakinumab should not be administered with TNF inhibitors, and avoidance of live vaccines is recommended. There are also warnings for the possibility of hypersensitivity reactions which have been recorded and for immunosuppression, although the latter is precautionary due to potential immunosuppressive activity of IL-1 and TNF blockers rather than recorded experience. Neutropenia and leukopenia have been seen with canakinumab, and this should be kept in mind before treating patients with these cytopenias. This also highlights the need for white blood cell counts prior to, and after, initiation of therapy. A warning for the possibility of canakinumab-induced macrophage activation syndrome (MAS) (Chap. 1, section "Protein Therapeutics") is again not based on experience with the mAb but rather the knowledge that this life-threatening disorder is known to occasionally occur in SJIA (section "Approved Indications and Safety of Tocilizumab"). Besides infections, particularly nasopharyngitis and influenza, among the commonly reported adverse reactions to canakinumab in CAPS patients are headache, nausea, diarrhea, dizziness/vertigo, and injection site reactions (Table 4.1). Vertigo, reported with an incidence of 9–14 % and with at least two serious cases, appears to be confined to MWS patients. Subcutaneous injection site reactions have been observed in 9 % of patients. Adverse reactions associated with canakinumab in the treatment of SJIA are infections, abdominal pain, and injection site reactions. A wide range of organisms may cause serious infections (incidence ~4–5 %), such as pneumonia, varicella, gastroenteritis, measles, sepsis, otitis media, sinusitis, adenovirus, and pharyngitis.

To the end of 2012 in the postmarketing period for canakinumab, the most frequent reports to the EVRS database were infections (43 reports), respiratory disorders (30), skin reactions (28, including 5 cases of urticaria), and gastrointestinal disorders (24). There was one case of vertigo and one case of cytokine release syndrome. Of 462 reports to the FAERS database, 199 were for infections, 131 for cutaneous reactions with 16 for urticaria, and about 100 reports for both gastrointestinal and respiratory events. The most interesting reports of adverse events seen in a fewer number were 32 cases of myelosuppression, seven cases of vertigo, and six cases of disseminated intravascular coagulation.

Anti-canakinumab antibodies were detected in 1.5 and 3.1 % of CAPS and SJIA patients treated with canakinumab, but no neutralizing antibodies nor any correlation with clinical response or adverse events were found.

Denosumab

See also Chap. 3, section "Denosumab" and Table 3.1.

Denosumab (Xgeva®, Prolia®) (Tables 2.1 and 4.1), produced in transgenic mice, is a fully human IgG2κ mAb, MW ~ 147 kDa, with high specificity and affinity for receptor activator of nuclear factor kappa-B ligand, otherwise known as RANKL.

Marketed under two trade names, denosumab as Xgeva® is approved for the prevention of skeletal-related events in patients with bone metastases from solid tumors and the treatment of giant cell tumor of bone (Chap. 3, section "Denosumab"). As Prolia®, denosumab has regulatory approval for the treatment of men and menopausal women with osteoporosis at high risk of fracture, men at high risk of fracture receiving androgen deprivation therapy for nonmetastatic prostate cancer, and women at high risk of fracture receiving adjuvant aromatase inhibitor therapy for breast cancer. Safety issues concerning the indications and usage of denosumab in relation to osteoporosis and cancer treatment-related bone mass/fracture issues will be discussed here.

RANKL, a cytokine and member of the tumor necrosis factor (TNF) family (also called TNFSF11), is produced and expressed by cells of the osteoblast lineage and activated T cells. It stimulates osteoclast formation, activation, adherence, survival, and ultimately resorption of the bone. Inhibition of RANKL results in an increase in bone density, volume, and strength. For details of the interaction of RANKL with its receptor RANK and subsequent events including the role of osteoprotegerin, see Chap. 3, section "Denosumab."

As mentioned in Chap. 3, hypocalcemia may be exacerbated by denosumab therapy. Hypocalcemia, an added risk in patients with severe renal impairment, should therefore be treated before commencing treatment with Prolia®, and monitoring of calcium, magnesium, and phosphorus levels is recommended. Osteonecrosis of the jaw, generally associated with tooth extraction or local healing, is well known, hypersensitivity including anaphylaxis has been reported, and atypical femoral fractures commonly resulting from minimal or even no trauma may occur. Both osteonecrosis of the jaw and atypical fractures as well as delayed healing of fractures may be a consequence of significant suppression of bone remodeling, observed in some clinical trials of women with postmenopausal osteoporosis, treated with denosumab. Serious infections affecting the urinary tract, abdomen, ear, and skin, as well as endocarditis, have been observed in clinical trials. Trials also revealed cutaneous reactions in response to denosumab including rashes, dermatitis, and eczema at a significantly higher frequency than seen in the placebo group, and other reports of photosensitive responses, exanthema, skin eruptions, and bullous and exfoliative reactions, plus reports of relatively high incidences of dermatologic reactions in the postmarketing period (see below), suggest that long-term cutaneous adverse effects of denosumab may be a concern. Postmarketing surveillance data in the FAERS database reveal that the most common adverse events following Prolia® administration were dermatologic reactions, musculoskeletal disorders, and infections, with the latter reported more frequently for Prolia® than for Xgeva®. Serious events were musculoskeletal disorders, dermatologic reactions, infections, gastrointestinal disorders, and neurologic events. In the EVRS database at the end of 2012, gastrointestinal disorders (8.7 %), cutaneous reactions (7.3 %), and infections (7 %) were the highest reported adverse events. Pneumonia (61 reports), urinary tract infections (43), cellulitis (42), and sepsis were the most prominent infections, and there were 53 reports of hypersensitivity including 11 cases of anaphylaxis. Sixty-nine reports of rash, 38 of erythema, and 29 of urticaria again indicated that dermatologic events during/after denosumab treatment are proving to be more of a problem than originally appreciated. This tentative conclusion was reinforced by a recent analysis of FDA data appearing to demonstrate a consistent and high rate of adverse reactions related to Prolia® administration. Cases of severe musculoskeletal pain, severe symptomatic hypocalcemia, and a marked elevation in serum parathyroid hormone in patients with severe renal impairment or receiving dialysis after receiving Prolia® have been reported in the postmarketing period. A summary of the most common adverse events seen in Prolia®-treated women and men patients with osteoporosis is shown in Table 4.1.

Denosumab is a human antibody, so immunogenicity is expected to be low, and this has proved to be the case although there is always the potential for an anti-idiotype response. Fifty-five of 8113 patients (0.7 %) treated with Prolia® for up to 5 years developed antibodies to denosumab; none of the antibodies neutralized the biological action of the mAb.

Eculizumab

Eculizumab (h5G1.1 mAb, Soliris®) (Tables 2.1 and 4.1) is a humanized IgG2/4κ mAb, MW ~ 148 kDa, specific for complement component C5. Human kappa light chains are used, and the heavy chains are constructed from human IgG2 sequences

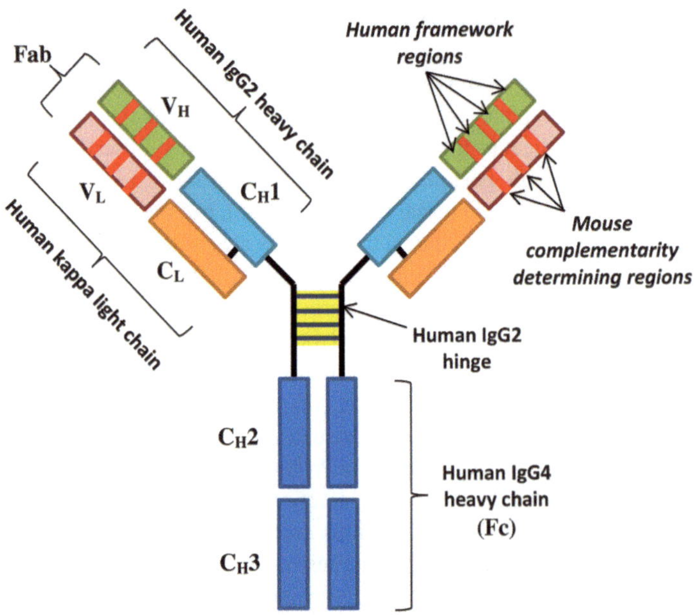

Fig. 4.4 Eculizumab (Soliris®) is a humanized IgG2/4κ mAb specific for complement component C5, engineered to reduce its immunogenicity by grafting mouse complementarity-determining regions into human antibody framework regions. A human IgG2 and IgG4 H chain hybrid constant region was constructed to reduce the potential of an antibody-induced inflammatory response by exploiting the properties of the human IgG2 isotype which does not bind Fc receptors and IgG4 which does not bind C1q and activate the complement cascade. By binding human C5 with high affinity, eculizumab blocks its cleavage to C5a and C5b and thereby prevents the C5-mediated downstream proinflammatory and cell lytic actions

in the constant region 1 (C_H1), the hinge, and the adjacent portion of the constant region 2 (C_H2) and from human IgG4 sequences in the remaining part of the C_H2 region and constant region 3 (C_H3). The variable heavy (V_H) and light (V_L) chains are constructed from human IgG2 heavy and light chain framework regions with grafted murine complementarity-determining regions from mouse antibody m5G1.1 to form the antibody combining sites (Fig. 4.4). In developing eculizumab, human IgG2 and IgG4 sequences were used to reduce the potential of an antibody-induced inflammatory response since the human IgG2 isotype does not bind Fc receptors and IgG4 does not bind C1q and activate the complement cascade.

Approved Indications

Orphan drug status for eculizumab was recognized by the FDA for the treatment of paroxysmal nocturnal hemoglobinuria (PNH) in 2003, for atypical hemolytic uremic syndrome (AHUS) in 2009, and for Shiga toxin-induced hemolytic uremic syndrome (HUS) in 2011. Orphan drug status was designated by the EMA in 2003 for PNH, in 2009 for AHUS, and in 2012 for infection-associated HUS. Eculizumab was approved by the FDA and EMA in 2007 for the treatment of PNH and then by Health Canada and the Australian Therapeutics Goods Administration in 2009. In 2011 the FDA and EMA extended the indications to AHUS. Specifically, eculizumab is indicated for the treatment of patients with PNH to reduce hemolysis and for patients with AHUS to inhibit complement-mediated thrombotic microangiopathy. Eculizumab is not indicated for the treatment of patients with Shiga-like toxin *E. coli*-related HUS (FDA) or infection-associated HUS (EMA).

Paroxysmal Nocturnal Hemoglobinuria, Atypical Hemolytic Uremic Syndrome, and Mechanism of Action of Eculizumab

PNH is a rare hemolytic disease originating from a somatic mutation of the *PIGA* gene that results in a deficiency of glycosylphosphatidylinositol, a glycolipid that attaches and anchors a number of different proteins to the cell membrane including CD59 (protectin, MAC-inhibitory protein [MAC-I]). CD59 inhibits the incorporation of C9 into the complement membrane attack complex (MAC, also called terminal complement complex [TCC]), so a deficiency of CD59 in PNH leads to unimpeded assembly of the MAC on the red cell surface resulting in chronic complement-mediated intravascular hemolysis. The classical, lectin, and alternative complement pathways of complement activation converge at the point of cleavage of C5 into C5a and C5b, initiating the terminal complement cascade. MAC, or C5b-9, the TCC, is formed from the recruitment of C6, C7, C8, and C9 by C5b. Hence, C5 is an attractive target and this was the strategy that led to the development of eculizumab (Fig. 4.5). Eculizumab is thought to bind to the contact interface between C5 and C5 convertase; the serine protease that converts C5 to C5a and C5b, preventing the close association of the enzyme and its substrate; and the eventual formation of the powerful anaphylatoxin C5a and the C5b-9 MAC. As well as preventing cell lysis in PNH, the prevention of C5 cleavage by the mAb blocks the formation of C5a, a potent proinflammatory molecule that mediates increased vascular permeability and leukocyte chemotaxis, alters smooth muscle tone, and induces the release of secondary inflammatory mediators including arachidonic acid metabolites and cytokines. AHUS, a rare, progressive, life-threatening disease affecting both children and adults, belongs to the category of diseases known as thrombotic microangiopathies (TMAs). The pathology of TMAs includes thickening of small blood vessel walls, endothelial swelling, clots in small blood vessels, and ultimately tissue ischemia with stroke, heart attack, kidney failure, and death. AHUS is associated with dysregulation of the alternative pathway of complement, and in approximately 60–70 % of patients, gene mutations involved in the regulation of the alternative pathway or autoantibodies to complement regulatory proteins

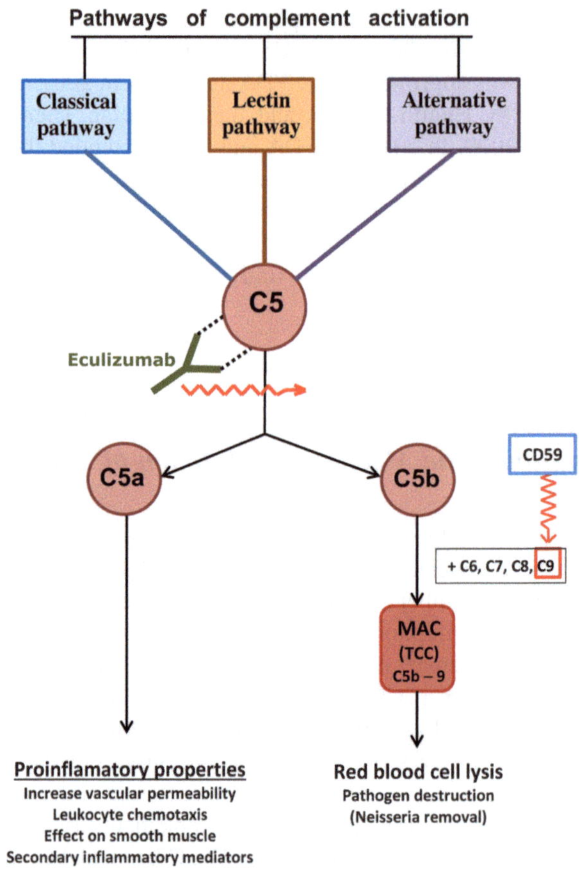

Fig. 4.5 The three complement activation pathways, the classical, lectin, and alternative complement pathways, converge at the point of cleavage of C5 into C5a and C5b, initiating the terminal complement cascade which generates molecules with proinflammatory and cell lytic actions. Assembly of the complement membrane attack complex (MAC, also called terminal complement complex [TCC]) results in chronic complement-mediated intravascular hemolysis. The formation of MAC, or C5b-9, formed from the recruitment of C6, C7, C8, and C9 by C5b, can be inhibited by CD59 which prevents the incorporation of C9. Targeting of C5 by eculizumab is therefore an effective way of preventing the inflammatory and cell lytic effects of terminal complement activation while preserving the protective and regulatory functions of proximal (pre-C5) complement

(e.g., anti-factor H) can be identified. Mutations in at least eight genes have been identified in AHUS, but current screening fails to detect mutations in 35–40 % of patients, and even when the genetics is well described, the cause of the disease in ~50 % of patients remains obscure. As for PNH, in the treatment of AHUS, eculizumab is thought to prevent the cleavage of C5 by C5 convertase, inhibiting the formation of C5a and C5b-9 and blocking the proinflammatory, prothrombotic, and lytic actions of complement. Eculizumab is now considered as a first-line therapy for adults with AHUS and for children with a first episode of the disease.

Patients with genetically determined C5 deficiencies are prone to recurrent infections, particularly Neisseria-induced meningitis. In clinical trials examining the responses of PNH patients to eculizumab, the Neisseria infection rate was 4.2 cases per 1000 patient-years. Importantly, however, antibody blockade at C5 does not impair the protective functions provided by earlier components necessary for complement-mediated opsonization and the clearance of immune complexes.

Warnings, Precautions, and Adverse Events

Perhaps at least partly due to its mechanism of action inhibiting C5 and its known association with Neisseria infections, a black box warning for serious meningococcal infections has been issued for eculizumab stating that life-threatening and fatal infections may occur or have occurred in treated patients. Consequences of this warning are the requirement that the drug be issued under a REMS Program; that, if possible (and subject to risk assessments), patients be immunized with meningococcal vaccine at least 2 weeks prior to the first dose of eculizumab; and that patients should be monitored for early signs of meningococcal infections. In clinical studies with PNH patients, three cases of Neisseria infections and two of meningococcal sepsis were seen, the latter over 474 patient-years of eculizumab therapy. With blockage of terminal complement activation, patients may have an overall increase in susceptibility to infections, especially encapsulated bacteria. Serious infections may occur due to *Aspergillus* sp., *Streptococcus pneumoniae*, and *Haemophilus influenzae*. Vaccinations for the latter two organisms should be considered. After discontinuing eculizumab, precautionary monitoring of patients for hemolysis for at least 8 weeks and AHUS patients for at least 12 weeks is recommended. Although no patients experienced infusion reactions requiring withdrawal from clinical trials, a warning of the possibility of such reactions, including true hypersensitivities, has been issued.

In patients with PNH, prominent adverse events (Table 4.1) seen include headache, nasopharyngitis, nausea, fatigue, sinusitis, URTI, constipation, myalgia, pain in extremity, *Herpes simplex* infections, and flu-like symptoms. In the AHUS pivotal study in which 37 of 40 patients received eculizumab, 54 % experienced serious reactions, mainly hypertension and infections, while the most common adverse events were gastrointestinal disorders, pyrexia, URTI, cough, nasal congestion, and tachycardia.

In the FAERS database of over 7500 reports, headaches and gastrointestinal, hematologic, and respiratory signs and symptoms with a frequency of 4–8 % were the most commonly reported adverse events. Of just over 3000 infections, 63 were meningococcal sepsis, 29 meningitis, nine Neisseria infections, eight streptococcal sepsis, and 16 cases of streptococcus pneumonia. There were 102 cases of flu-like syndrome, 78 cases of hypersensitivity (no cases of anaphylaxis), 11 cases of drug hypersensitivities, 49 cases of infusion reactions, seven cases of PML, and one case of systemic inflammatory response syndrome (SIRS) (Chap. 1, section "Systemic Inflammatory

Response Syndrome"). In the EVRS database at the end of 2012, infections (14.5 %), gastrointestinal events (13.6 %), and hematologic disorders (7.9 %) showed the highest frequencies of occurrence out of 6775 reported adverse events. Of 986 reports of infections, 90 were for pneumonia, 77 for sepsis (and 28 for septic shock), 56 for urinary tract infections, and 29 for viral infections. Twenty cases of sepsis were due to meningococcal infections, and Neisseria infected five patients. Twelve reports were for hypersensitivities, including four cases of anaphylaxis and two of serum sickness. There were 15 infusion reactions, two cases of SIRS, and one of PML.

Immunogenicity of Eculizumab

In patients with PNH given eculizumab, human antihuman antibodies to the mAb were detected in 1.5–3 % of 357 patients using two different assay methods. In AHUS patients, 3 % of 100 patients had antibodies. Of a total of eight antibody-positive patients from both groups, two from the PNH group and one from the AHUS group had neutralizing antibodies.

Monoclonal Antibody Integrin Inhibitors: Natalizumab and Vedolizumab

For the mAb integrin inhibitor abciximab, see section "Abciximab" above. Integrins are transmembrane glycoprotein heterodimers of non-covalently associated α- and β-subunits, typically comprising 1000 and 750 amino acids, respectively. There are 18 α- and eight β-subunits in vertebrates which, in different combinations, can generate 24 different integrin receptors. These demonstrate different binding properties and tissue distributions, mediate cell-cell and cell-extracellular matrix attachments, and transmit signals from outside to inside the cell by receptor tyrosine kinase signaling as well as signaling from inside to outside the cell.

Natalizumab

Natalizumab (Tysabri®) (Tables 2.1 and 4.1) is a recombinant humanized IgG4κ mAb raised against the human α4 integrin subunit found on T and B lymphocytes, NK cells, and most monocytes, macrophages, and granulocytes but not neutrophils. A forerunner to the development of the mAb was an earlier work showing that targeting α4 integrin prevented the development of demyelinating lesions in a mouse model of multiple sclerosis. Natalizumab is not absolutely specific in its targeting in that it interacts with both α4β1 (very late antigen 4, VLA-4; CD49d[α4], CD29[β1]) and α4β7 heterodimeric leukocyte adhesion molecules. Receptor α4β1 binds the cell adhesion molecule (CAM) vascular adhesion molecule-1 (VCAM-1, CD106), while VCAM-1 and the extracellular protein of the endothelium of venules, mucosal addressin-cell adhesion molecule-1 (MAdCAM-1, addressin), interact as ligands with α4β7, although VCAM-1 binds α4β7 with less affinity. Interaction between α4β1 and VCAM-1 occurs on endothelial cells, while the α4β7–MAdCAM-1 interaction takes place on leukocytes. VCAM-1, an immunoglobulin-like transmembrane molecule, mediates cell adhesion to vascular endothelium, including cerebrovascular endothelial cells, and, together with MAdCAM-1, it is upregulated on intestinal endothelium in Crohn's disease. The interaction between α4β1 and VCAM-1 initiates the emigration of lymphocytes from the blood through the venule walls and on into the tissues to inflammatory sites. Hence, in multiple sclerosis, by blocking adhesion of α4β1 to VCAM-1 expressed on inflamed cerebrovascular endothelial cells, natalizumab inhibits inflammation by preventing the movement of autoreactive leukocytes out of blood vessels and into target organs causing inflammation. Likewise, similar binding to the α4 subunit of leukocyte adhesion factors α4β1 and α4β7 by natalizumab, blocking adhesion of both VCAM-1 and MAdCAM-1 to intestinal endothelium, is believed to be the mechanism underlying the mAb's beneficial effect in Crohn's disease (Fig. 4.6).

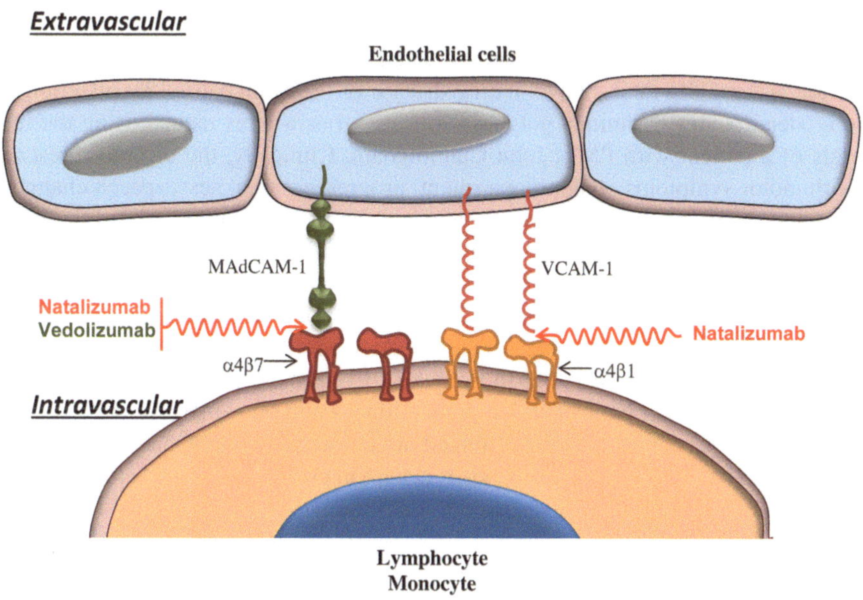

Fig. 4.6 VCAM-1 and MAdCAM-1 are upregulated on intestinal endothelium in Crohn's disease. Natalizumab prevents the adhesion of lymphocytes to endothelial cells by blocking the interaction between the integrin subunit α4β1 and VCAM-1 and between α4β7 and MAdCAM-1. This prevents inflammation by preventing the movement of autoreactive lymphocytes out of blood vessels and into target organs. Vedolizumab, used to treat ulcerative colitis and Crohn's disease, also acts by inhibiting the interaction between α4β7 and MAdCAM-1 thus preventing migration of memory T lymphocytes across the vascular epithelium and into the gut

Approved Indications for Natalizumab

Natalizumab, the first integrin inhibitor approved for the treatment of multiple sclerosis and Crohn's disease, was approved by the FDA for the treatment of relapsing multiple sclerosis in 2004 and for moderate to severe Crohn's disease in 2008. It was also approved for multiple sclerosis by Health Canada in 2006 and the Japanese regulatory authority in 2008. Although it was given EMA approval for relapsing remitting multiple sclerosis in 2006, that agency has not yet approved the mAb for the treatment of Crohn's disease.

As an integrin receptor antagonist, natalizumab is indicated as monotherapy to reduce clinical exacerbations and delay disability in the treatment of relapsing forms of multiple sclerosis and to treat Crohn's disease in adults in whom conventional therapies and TNF inhibitors are not suitable. Natalizumab should not be given with immunosuppressants or inhibitors of TNF. In relation to multiple sclerosis, in May 2015, the FDA stated: "Tysabri increases the risk of PML. When initiating and continuing treatment with Tysabri, physicians should consider whether the expected benefit of Tysabri is sufficient to offset the risk."

Posterior Multifocal Leukoencephalopathy and Natalizumab

PML (see also Chap. 1, section "Progressive Multifocal Leucoencephalopathy") is a rare and usually fatal demyelinating disease of the brain caused by infection of oligodendrocytes by human polyomavirus JC virus, a virus named using the initials of a patient with PML, John Cunningham. Clinically, the disorder presents with motor symptoms, vision impairment, progressive weakness, speech changes, and symptoms of cognitive impairment and generally progresses to blindness, dementia, coma, and death. JC virus, which seems to occur only in humans, is widespread with reported incidences of 44–77 % in the USA and UK and up to 92 % in Germany, Japan, and Brazil. Infection, possibly via the tonsils and/or gastrointestinal tract, seems to occur in childhood or adolescence; the virus is found in ~10 % of children 1–5 years old and ~65 % of 17-year-olds. It remains latent in the gastrointestinal tract but can reside in the kidneys and bone marrow and cross the blood-brain barrier. It has been linked to colorectal cancer but this remains unproven. While it is clear that JC virus is kept in check by the normally functioning immune system, immunodeficiency and immunosuppression aid the development of PML, and this is reflected in its relatively high incidence in acquired immune deficiency syndrome (AIDS) patients and patients on immunosuppressive medications including chemotherapy. The potential list of such patients includes those with transplants, autoimmune diseases, multiple sclerosis, psoriasis, and lymphomas. Prior to 1984, the 230 known cases were judged to be linked to transplantation, malignancies, and a range of hematologic and inflammatory disorders with only ~5 % due to HIV infection. In recent times, HIV involvement accounts for ~85 % of cases. Although immunosuppression or immunodeficiency is sometimes said to "reactivate" the virus to destroy oligodendrocytes, PML is probably not the consequence of a primary viral infection or viral reactivation in the brain but the result of infection by a mutated virus transported by B lymphocytes to the brain where it replicates. In a normally immune competent person, helper and cytolytic T cells contain the virus, but this does not occur in the immune compromised individual. As well as immunosuppressive therapies like corticosteroids, cytotoxic chemotherapeutics, and transplant drugs like tacrolimus, a range of modern biologic therapies may depress the immune response and allow mutated JC virus to avoid the normally active immune defenses and mediate unhindered the pathological processes leading to PML. A list of these biologics that carry some risk for the development of PML includes the mAbs, natalizumab, rituximab, obinutuzumab, ofatumumab, brentuximab vedotin, and efalizumab and the fusion protein belatacept (Chap. 6, sections "Belatacept" and "Belatacept Safety").

Early in 2005 after the reports of three cases of PML associated with the mAb (an incidence of ~one case per 1000 patients), natalizumab was suspended from the market by the FDA before being readmitted in late 2006 under restricted distribution with the drug available only through a REMS Program (see below). In 2015 in the light of accumulating knowledge of natalizumab, and the risk of PML, the European Commission has requested the EMA to review natalizumab and assess whether advice on managing risk factors for PML should be revised. The review will be carried out by the European Pharmacovigilance Risk Assessment Committee. Three key issues will be assessed: risk estimates, the diagnosis of PML before the development of clinical symptoms, and anti-JC virus antibodies.

Between the reintroduction of natalizumab in 2006 and early 2010, 42 cases of PML were reported in natalizumab-treated patients, and by February 2012, there were 212 confirmed cases among almost 100,000 patients, an incidence of 2.1 cases per 1000 patients. In the postmarketing period, 34 cases of PML were recorded in over 61,000 patients with Crohn's disease; 11 cases were reported in 2011 and eight in 2012. Following the FDA's 2005 suspension of natalizumab and its readmission to the market in 2006, the FDA issued a boxed warning on PML for the mAb. The warning, still current, states that natalizumab increases the risk of PML; draws attention to the risk factors such as the duration of therapy, prior use of immunosuppressants, and presence of anti-JV virus antibodies; and sets out the diagnostic needs to monitor patients for symptoms utilizing brain scan, gadolinium magnetic resonance imaging, and cerebrospinal fluid analysis for JC virus DNA. The boxed warning also covers the restricted availability program for natalizumab called the TOUCH® Prescribing Program which is divided into MS TOUCH® for patients with multiple sclerosis and CD TOUCH® for patients with Crohn's disease. Requirements for the program include regular evaluations of patients after the first infusion and after discontinuing natalizumab therapy; six monthly authorizations of treatment; submission of reauthorization and discontinuation questionnaires and reports of PML cases, infections, and deaths to the drug's manufacturer; and special certifications for pharmacies and infusion centers to dispense natalizumab.

Three factors that increase the risk of PML in natalizumab-treated patients are longer duration of treatment, for example, beyond 2 years; prior treatment with immunosuppressants such as azathioprine, methotrexate, and cyclophosphamide;

and the presence of JC virus antibodies. Since JC virus is necessary for the development of PML, patients with antibodies have a higher risk, but note that antibody-negative patients may still be at risk from a new infection or because of a false-negative antibody test result. Antibody-negative patients should therefore be retested periodically. Patients subjected to plasma exchange should not be tested for antibodies for at least 2 weeks because of antibody removal from the serum.

The reason why natalizumab increases the risk of PML and induces a relatively higher incidence of the syndrome than most other agents is the subject of speculation. Natalizumab exerts effects in numerous tissues including the CNS where it inhibits leukocyte trafficking. Natalizumab's efficacy in the CNS may be explained by its binding of α4β1 integrin leading to a decrease in immune surveillance by memory T cells in the CNS and predisposing some patients to PML.

FDA Warnings and Precautions for Natalizumab Including Immune Reconstitution Inflammatory Syndrome

As part of the FDA warnings and precautions for natalizumab, attention is drawn to the association between the drug and a pathological inflammatory response known as immune reconstitution inflammatory syndrome (IRIS) (Chap. 1, section "Immune Reconstitution Inflammatory Syndrome"). IRIS, also called immune recovery (or restoration or reconstitution) disease/syndrome, is a condition seen in some immunosuppressed and AIDS patients when immunity to infectious or noninfectious antigens is restored, leading to a paradoxical worsening of the patient's condition. Clinical manifestations of IRIS, described as "atypical exuberant inflammation" may be quite diverse and dependent upon the infectious or noninfectious agent involved, for example, reactions to mycobacterial infections, *Pneumocystis jirovecii* pneumonia, viral hepatitis, *Cytomegalovirus* retinitis, shingles, and PML. IRIS is seen in most PML patients taking natalizumab who subsequently discontinue the therapy with almost all occurring after plasma exchange undertaken to remove any remaining mAb from the circulation. Rare cases of IRIS have also been claimed to occur after withdrawal of natalizumab in patients who do not have PML and in the absence of plasma exchange. Patients' decline after stopping natalizumab therapy may be rapid and/or variable (days to weeks) and lead to serious neurologic complications and death. Mortality is estimated to be ~30 %. The pathogenesis of IRIS is far from settled with current theories and speculations including an antigenic stimulus provoking an accelerated restoration of specific immunity with a matching accelerated clinical response, the degree of immune restoration and the increase in CD4 memory T cells, and the hosts' genetic susceptibilities to the powerful immune response upon immune restoration. Involvement of increased levels of IL-6 has been suggested to be involved.

Other warnings and precautions issued by the FDA concern the risk of developing encephalitis and meningitis caused by herpes simplex and varicella zoster viruses, especially in multiple sclerosis patients given natalizumab. Liver injury, sometimes becoming manifest only a few days after receiving natalizumab, has been reported rarely in the postmarketing period; hypersensitivities (incidence <1 %), including anaphylaxis, have been reported; and infections, primarily a consequence of immunosuppression, are seen in both multiple sclerosis- and Crohn's disease-treated patients. Infections, seen more often in multiple sclerosis patients than control subjects, include pneumonia, urinary tract infection, gastroenteritis, and vaginal and herpes infections. Infections in Crohn's disease patients include URTI, urinary tract infections, influenza, and sinusitis. Apart from PML discussed above, the most common of the ~93,000 and ~7600 adverse events reported to the FAERS and EVRS databases, respectively, were neurological events and infections.

Immunogenicity of Natalizumab

In a trial of natalizumab for multiple sclerosis, patients were tested for anti-natalizumab antibodies every 3 months using an assay that was relatively insensitive. Approximately 9 % of patients developed antibodies at some time during treatment, and of those whose antibodies persisted, over 80 % showed the presence of antibodies by 3 months. The antibodies were found to be neutralizing in vitro. A comparison of anti-natalizumab antibody and natalizumab serum levels showed the mAb concentration in antibody-negative patients was 15 μg/mL and only 1.3 μg/mL in antibody-positive patients and persistence of serum antibodies was accompanied by a decrease in effectiveness of natalizumab. Infusion reactions to natalizumab, higher in multiple sclerosis than Crohn's disease patients, and other adverse reactions such as dyspnea, hypertension, tachycardia, and myalgia, were seen more often in antibody-positive patients.

Vedolizumab

Vedolizumab (Entyvio®) (Tables 2.1 and 4.1) is a humanized IgG1κ mAb that binds specifically to the α4β7 integrin. The mAb is a humanized version of the early mouse antibody Act-1, binding to a conformational determinant formed by the dimerization of the α4 and β7 subunits. It does not bind to, or inhibit the functions of, α4β1. Vedolizumab acts as a gut-selective immunosuppressive agent by virtue of its recognition of α4β7, a lymphocyte-homing receptor preferentially expressed on gut-homing T-helper lymphocytes. Binding of the mAb inhibits cell adhesion of the integrin to the cell adhesion molecule MAdCAM-1, mainly expressed on gut endothelial cells, but not to VCAM-1 (section "Natalizumab"). In ulcerative colitis and Crohn's disease, memory T cells expressing α4β7 preferentially migrate to the gastrointestinal tract where they provoke an inflammatory reaction. By inhibiting the interaction between α4β7 and MAdCAM-1, migration of the T lymphocytes across the vascular epithelium and into the gut is prevented (Fig. 4.6).

Approved by both the FDA and EMA in 2014 and by Health Canada in 2015 for moderate to severe ulcerative colitis and Crohn's disease in adults, vedolizumab is subject to warnings/precautions for infusion-related and hypersensitivity reactions, an increased risk of infections, liver injury, and PML. In clinical trials, one case of anaphylaxis was reported in 1434 patients and most hypersensitivity and infusion reactions were mild to moderate when they did occur. Premedication does not seem to be necessary prior to infusions of vedolizumab. Infections, some serious including sepsis, tuberculosis, *Listeria* meningitis, giardiasis, and cytomegaloviral colitis, were reported in the vedolizumab-treated patients at a frequency higher than the placebo group, indicating the need to withhold treatment from patients with existing infections. Live vaccines should be administered only if benefits outweigh the risks. Although no cases of PML appear to have been reported so far in patients treated with vedolizumab, the known association of the integrin inhibitor natalizumab and some immunosuppressives with the disease requires physicians to monitor patients for any new onset or worsening of neurological signs and to consider neurological referral if seen. Given the specificity of vedolizumab, PML is not expected to be a significant problem or to perhaps even occur. Some cases of elevations of transaminase and/or bilirubin in patients receiving vedolizumab have been reported.

In clinical trials with both ulcerative colitis and Crohn's disease patients, the incidences of adverse events were similar between the vedolizumab-treated and placebo groups. Events occurring in ≥3 % of the mAb-treated patients, and at incidence at least 1 % higher than seen in placebo patients, include (in order of highest incidence) nasopharyngitis, headache, arthralgia, nausea, pyrexia, URTI, fatigue, and cough. In the GEMINI II trial, an induction and maintenance phase study with Crohn's disease patients, incidences of serious adverse events were higher in the vedolizumab group than the placebo group (24.4 % vs. 15.3 %). Serious adverse events included some malignancies, infections, and several deaths.

In ulcerative colitis and Crohn's disease trials over a 52-week period of continuous treatment, 56 of 1434 (3.9 %) treated patients had anti-vedolizumab antibodies at some time; nine of the 56 were persistently positive and 33 developed neutralizing antibodies to the mAb. Six of the nine persistently antibody-positive patients had undetectable serum concentrations of vedolizumab and two had reduced levels. None of the nine achieved clinical remission at either week 6 or week 52 of the trial. The frequency of serum antibodies was 13 % 24 weeks after the final dosage of vedolizumab, that is, a length of time greater than five half-lives of the drug.

Omalizumab

Omalizumab (Xolair®) (Tables 2.1 and 4.1) is a recombinant humanized IgG1κ mAb, MW ~ 149 kDa, with specificity for human IgE antibodies.

Approved Indications

First approved in 2002 by the Australian Therapeutics Goods Administration (TGA) for the treatment of asthma resistant to inhaled steroids, the FDA and EMA followed suit in 2003 and 2005, respectively, approving the mAb for moderate to severe asthma in adults and adolescents whose symptoms remained uncontrolled by standard treatment, namely, inhaled corticosteroids. In 2005, the TGA extended the indications to asthma already being treated with inhaled steroids, and in 2009, the EMA approved add-on therapy to children aged 6–12 years. Currently, the approved indications in the USA and Europe for asthma in patients 12 years of age and older refer to persistent asthmatics with a positive skin test or in vitro reactivity to a perennial aeroallergen and symptoms inadequately controlled with inhaled corticosteroids and a long-acting beta2-agonist. In 2014, omalizumab was approved by the FDA, EMA, and a number of other national regulatory agencies for chronic idiopathic (spontaneous) urticaria in adults and adolescents with an inadequate clinical response to H1 antihistamines.

Mechanism of Action of Omalizumab

Omalizumab binds to a determinant in the Cε3 region of human IgE, the same region of the molecule involved in binding to the FcεRI receptor on the mast cell. It binds to free, circulating IgE antibodies and to IgE molecules bound to some membranes, in particular B cells expressing the antibody, but it does not bind to IgE already bound to mast cells, basophils, and dendritic cells. Since the interaction of IgE with the latter three cell types also involves the Cε3 region of the antibody, this region of cell-bound IgE becomes inaccessible to omalizumab. Cross-linking of cell-bound IgE by anti-IgE, an event normally producing mast cell activation and release of allergic and inflammatory mediators, therefore cannot eventuate, thus averting an allergic reaction including anaphylaxis. Importantly, omalizumab depletes free circulating IgE to almost negligible levels leading to a decline in the number of FcεRI receptors on mast cells, basophils, and dendritic cells, rendering these cells less sensitive to allergen stimulation. As a second consequence of depletion of free IgE, antigen trapping by IgE and subsequent presentation by dendritic cells are reduced, resulting in a marked reduction of activated allergen-specific Th2 cells. Omalizumab also interferes, apparently by steric hindrance, with the binding of IgE to the low-affinity IgE receptor FcεRII (CD23), expressed on airways smooth muscle cells, mature B lymphocytes, macrophages, monocytes, dendritic cells, and eosinophils. By virtue of its capacity to bind a range of different ligands, the FcεRII receptor controls multiple functions. Hence, interference with the binding of IgE to FcεRI and FcεRII receptors has a profound effect on immediate type I IgE antibody-mediated allergic responses.

Safety of Omalizumab

The FDA has issued a black box warning for anaphylaxis to omalizumab pointing out that reactions have been recorded after the first exposure to the drug and as late as following a dose given 1 year after treatment began. Most reactions occur within 2 h of administering the dose. In an assessment of omalizumab's safety and tolerability of immune effects, data from 7500 patients with asthma, rhinitis, and other conditions revealed an incidence of anaphylaxis of 0.14 %. A review of the period June 2003 to December 2005 carried out by a joint task force of the major US allergy societies identified 41 cases of anaphylaxis in 39,510 patients given omalizumab, a rate of 0.1 %. A further 83 episodes of anaphylaxis were recorded in 2006. These findings, together with the higher incidence of anaphylaxis to omalizumab in the postmarketing period, led the Omalizumab Joint Task Force to issue the following guidelines for the administration of the agent: (1) prior to administering omalizumab, patients should be assessed for vital signs, asthma control, and lung function; (2) informed consent should be obtained; (3) patients should be advised how to recognize anaphylaxis, how to use an epinephrine auto-injector, and how to ensure that the injector is always available during the administration of omalizumab; (4) omalizumab should only be administered in a facility that has both the staff and equipment to treat anaphylaxis; and (5) patients should be observed for 2 h after each of the first three injections and for 30 min after subsequent injections of the mAb. Other immune system disorders seen, or possible, in patients treated with omalizumab are a serum sickness-like type III hypersensitivity reaction, systemic eosinophilia and vasculitis consistent with Churg-Strauss syndrome, and helminth infection. Symptoms of arthritis, arthralgia, fever, rash, and lymphadenopathy 1–5 days after injection of omalizumab may indicate serum sickness although immune complexes and/or skin biopsy consistent with a type III hypersensitivity are not necessarily seen. Because IgE may be involved in the human immune response to parasitic infections, caution should be exercised with patients at high risk of helminth infections and in areas where such infections are endemic. A placebo-controlled trial showed a small increase in the helminth infection rate although there appeared to be no other clinical differences.

Malignant neoplasia has been recorded in 20 of 4127 (0.5 %) omalizumab-treated adult and adolescent asthmatic/allergic patients compared with five of 2236 (0.2 %) control subjects. The risk of malignancy over a longer period of exposure is not yet known. Other safety concerns warranting warnings and precautions are the reminders that omalizumab is not indicated for alleviating acute asthma exacerbations, acute bronchospasm, or status asthmaticus, and inhaled corticosteroids should not be discontinued abruptly at the start of omalizumab therapy. Table 4.1 lists the adverse reactions commonly seen when treating allergic asthma and chronic idiopathic urticaria with omalizumab. Few studies on the immunogenicity of the mAb seem to have been undertaken. Anti-omalizumab antibodies were detected in one of 1723 patients in the clinical studies on allergic asthma. No antibodies appear to have been detected and/or reported in trials/studies on chronic idiopathic urticaria.

Palivisumab

Palivizumab (Synagis®) (Tables 2.1 and 4.1) is a recombinant humanized IgG1κ mAb, MW ~ 148 kDa, directed to a determinant in the A antigenic site of the F protein of the *Pneumovirus* human respiratory syncytial virus (RSV), a highly contagious and major cause of lower respiratory tract infections such as pneumonia, bronchiolitis, and tracheobronchitis during infancy and childhood. First infections usually occur within the first 2 years of life, but adult patients with cancer, chronic obstructive pulmonary disease, asthma, and heart disease are also at risk. Palivizumab is targeted to the surface F protein, the attachment protein of the RSV protein coat. The F or fusion protein, a disulfide-linked heterodimer and one of the three transmembrane proteins, is homologous to both RSV subtypes and antibodies to neutralize the capacity of both virus subtypes to enter the cell and form syncytia.

First approved by the FDA in 1998 and by the EMA in 1999, palivizumab was the first mAb licensed for application to an infectious disease. The drug is now approved widely for therapy throughout the world in more than 50 different countries. Palivizumab is used prophylactically as a series of five injections each one month apart beginning prior to the RSV infection season. Candidates for injection are mainly infants at high risk, especially those that are premature or have cardiac or lung disease. In their statement of indications and usage, the FDA and EMA specifically mention children with bronchopulmonary dysplasia and premature infants and children with hemodynamically significant congenital heart disease.

Overall, the safety profile of palivizumab is distinguished by mild-moderate adverse events with most reactions in controlled studies not directly attributable to the mAb. A Cochrane analysis of a study of 186 infants up to 2 years of age receiving five monthly doses of palivizumab found an equal distribution of adverse events and drug-related adverse events in the patient and placebo groups although serious adverse events were slightly higher in the treated patients (20.7 % vs. 17 %). No other definitive conclusions were reached. Table 4.1 lists some of the more commonly reported adverse events to palivizumab. Incidences of these events are often marginally higher than seen in control subjects.

Palivisumab does not appear to be particularly immunogenic. In an early trial, incidences of anti-palivizumab antibodies after the fourth injection were 0.7 % in the palivizumab group and 1.1 % in the placebo group. Only one of 56 children receiving a second course of palivizumab developed antibodies and these were transient and of low titer. A trial involving 379 high-risk preterm children up to 2 years old found only one participant with antibodies to lyophilized palivizumab, an incidence of 0.3 %.

Ranibizumab

A vascular endothelial growth factor (VEGF) inhibitor, ranibizumab (Lucentis®) (Tables 2.1 and 4.1), designed for intraocular use, is essentially derived from the full mAb bevacizumab. It is a recombinant humanized IgG1κ Fab fragment composed of a 214-amino acid light chain disulfide linked at the C-terminus to the 231 residue *N*-terminus of the H chain (MW ~ 48 kDa) and targeted to VEGF-A. Ranibizumab binds with high affinity to all of the VEGF-A isoforms, including the biologically active cleaved form $VEGF_{110}$, thereby preventing the interaction of VEGF-A with its receptors VEGFR-1 and VEGFR-2 (see also Chap. 3, bevacizumab section "Bevacizumab" and Chap. 6, aflibercept sections "Aflibercept" and "Aflibercept Safety") and the subsequent receptor-mediated proliferation of endothelial cells, neovascularization, and vascular leakage. Each of these is thought to contribute to the progression of neovascular (wet) age-related macular degeneration (AMD) and the pathophysiology of diabetic macular edema and macular edema following retinal vein occlusion. Each of these three conditions is FDA- and EMA-approved indications for ranibizumab, and in July 2013, the EMA broadened the indications to include visual impairment due to choroidal neovascularization secondary to pathologic myopia.

Compared to the related whole mAb bevacizumab with its half-life of ~20 days, ranibizumab has a shorter half-life (~9 days) but provides better tissue penetration without losing efficacy while retaining a reasonably good safety profile. Intravitreal injection, including ranibizumab injection, may result in endophthalmitis and retinal detachment, and these conditions, plus increases in intraocular pressure and thromboembolic events in patients treated with the mAb fragment, are the subjects of warnings. In a pooled trial involving 759 patients, marginally higher incidences of arterial thromboembolic events and death rates of 7.2 % and 4.4 %, respectively, for the 0.5 mg dose and 5.6 % and 2.8 % for the 0.3 mg dose, were of some concern when compared to control incidences of 5.2 % and 1.2 %. These rates, however, were not statistically significant. The possibility of hypertension, a safety concern with systemic VEGF inhibition, should be kept in mind. The most common ocular events seen with ranibizumab include conjunctival hemorrhage, eye pain, vitreous floaters, increased ocular pressure, vitreous detachment, and cataracts. Although diabetics receiving ranibizumab may be considered at greater risk of adverse events due to their increased risk of cardiovascular events and infections, an events profile in diabetics similar to the use of ranibizumab for other indications has been observed. Postmarketing FAERS surveillance data on ranibizumab revealed ~10,500 reports with the main categories being vision abnormalities (11 %), ocular hemorrhage (9 %), what was called ocular structural changes (9 %), cardiovascular disorders (5 %), and infections (4 %). Deaths accounted for 5 % of reports. Of ~6300 reports to the EVRS database, 99 % were serious, eye disorders 39 %, and infections 8 %. Eye disorders included vision abnormalities (10 %), hemorrhage and pain (2.3 %), cataract (1.5 %), and retinal tear (0.6 %). Antibodies that react with ranibizumab have been detected in up to 5 % of patients prior to treatment with the mAb. After monthly injections with the drug for 6–24 months, antibodies were found in 1–9 % of patients. Some AMD patients with the highest levels of antibodies had iritis or vitritis but linkage was not established.

Raxibacumab

Raxibacumab is a human antibody for the treatment and prophylaxis of inhalational anthrax.

Anthrax and Background to the Development of Raxibacumab

The origins of the development of the mAb raxibacumab (Abthrax®) (Tables 2.1 and 4.1) lie in the US bioterrorist attacks of 2001 which resulted in five deaths from inhalational anthrax and the subsequent legislative response, the Project Bioshield Act of 2004, that funded the research and development of the mAb. Anthrax is a highly contagious and potentially fatal disease caused by *Bacillus anthracis*, an aerobic bacterium that forms highly resistant, long-surviving spores. Anthrax infections may manifest in four different forms: cutaneous, gastrointestinal, anthrax resulting from parenteral administration (e.g., from drug use), and inhalational anthrax, a rapidly developing and lethal toxin-induced disease. After inhalation, patients initially develop influenza-like symptoms over a short incubation period of 4 days, but this rapidly progresses to respiratory failure with a high death rate.

Because anthrax is such a lethal disease, human studies in developing a mAb are obviously not feasible. In the path leading to final approval, the FDA granted fast-track designation, priority review, and orphan drug status and applied the Animal Rule, a "regulation concerning the approval of new drugs when human efficacy studies are not ethical or feasible." The Animal Rule states that "for drugs developed to ameliorate or prevent serious or life-threatening conditions caused by exposure to lethal or permanently disabling toxic substances, when human challenge studies would not be ethical to perform and field trials to study effectiveness after accidental or intentional human exposure have not been feasible, FDA may grant marketing approval based on adequate and well-controlled animal efficacy studies when the results of those studies establish that the drug is reasonably likely to produce clinical benefit in humans."

Mechanism of Action of Raxibacumab

Raxibacumab is a fully human recombinant IgG1λ mAb, MW 146 kDa, that targets the protective antigen (PA) produced by rapidly dividing *B. anthracis*. The antibody does not have a direct antibacterial action but works by binding PA, thereby preventing its binding to its cell receptors. *B. anthracis* secretes two bipartite anthrax toxins produced by a plasmid, pXO2. These toxins consist of PA combined with two enzymes, edema factor and lethal factor, that produce, respectively, the AB-type exotoxins, edema toxin and lethal toxin. These toxins both contain PA as their receptor-binding B-moiety but have different catalytic A-moieties. PA mediates binding to its low-affinity receptor ANTXR1 (formerly tumor endothelial marker) and high-affinity receptor ANTXR1 (formerly human capillary morphogenesis protein 2). This results in pore formation and facilitates entry of edema factor and lethal factor into the cell. The A-moiety of edema toxin, edema factor, a calmodulin-dependent adenylate cyclase that catalyzes an increase in intracellular c-AMP levels, produces cell and tissue edema seen as pulmonary edema, lung congestion, and pleural effusions. The A-moiety of lethal toxin, lethal factor, a zinc metalloproteinase that inactivates mitogen-activated protein kinases, suppresses the immune response, allowing rampant bacterial division, hemorrhagic mediastinitis, septic shock, and eventual cell death. By blocking the binding of PA to its cell receptors, subsequent pore formation and cell entry of edema and lethal toxins are prevented. This strategy, the simplest and most effective way of neutralizing the toxicity of inhalational anthrax, thus provided the rationale for the development of raxibacumab.

Indications and Usage of Raxibacumab

Approved by the FDA in 2012, raxibacumab is licensed for the treatment of adult and pediatric patients with inhalational anthrax in combination with appropriate antibacterial drugs and for prophylaxis of inhalational anthrax when alternative therapies are not available or not appropriate. Raxibacumab is not yet approved by the EMA. Inhalational anthrax is estimated to affect less than 0.01 in 10,000 people in the European Union, well below the ceiling for orphan drug designation which is five per 10,000 people. However, in late 2014, orphan designation for raxibacumab was granted by the European Commission.

Although antibiotics effective against *B. anthracis* and an anthrax vaccine (anthrax vaccine adsorbed, BioThrax®; Chap. 11, Table 11.1) are available to treat inhalational anthrax infections, patients die due to the lethal bacterial toxin which reaches high levels at a very early stage, and the bacteria's short incubation period preceding toxin release is too rapid for the 4-week delay in the development of sufficient protective immunity. It is for these reasons that raxibacumab is given in combination with anthrax vaccine to augment immunity and with antimicrobial drugs to help overcome the early and rapid release of toxin which antibiotics cannot neutralize.

Safety of Raxibacumab

Raxibacumab has been evaluated for efficacy and safety in animals and a few human studies. The only operative safety warning is for infusion reactions which were seen to manifest as rash, urticaria, and pruritus in some patients in clinical trials. Premedication with diphenhydramine within 1 h of commencing infusion is recommended. Other reactions recorded include pain in extremity, somnolence, headache, URTI, nausea, cough, and arthralgias. Other reactions, seen at low incidence in test patients but more commonly than in subjects given a placebo, include fatigue, vertigo, palpitations, lymphadenopathy, peripheral edema, back pain, muscle spasms, insomnia, flushing, and hypertension. The development of antibodies to raxibacumab was looked at in a preliminary way in a total of 326 patients given the mAb in three different clinical trials. Anti-raxibacumab antibodies were not detected in any of the treated patients given either a single or repeat dose of raxibacumab.

Secukinumab

Secukinumab (Cosentyx®) (Tables 2.1 and 4.1) is a recombinant fully human IgG1κ mAb targeted to the proinflammatory cytokine IL-17A and used to treat plaque psoriasis.

Mechanism of Action of Secukinumab

Effector Th17 cells, induced from naïve T cells by the cytokines TGF-β and IL-6 and enhanced by IL-23, produce a family of molecules, IL-17A, B, C, D, E, and F. Although important for protection from invasive pathogens, IL-17A provokes tissue inflammation, damage, and autoimmune reactions, including psoriasis, by inducing the proinflammatory cytokines IL-6 and TNF and chemokines including CCL2, CXCL1, and CXCL2 that activate macrophages and granulocytes. Besides T cells, a variety of other cells including macrophages, dendritic cells, and natural killer cells produce IL-17. The functions of IL-17A and IL-17F are mediated through a heterodimeric receptor complex of IL-17RA and IL-17RC; a lack of either receptor prevents the inflammatory action of both cytokines. Downstream signaling following IL-17A interaction with IL-17R leads to activation of the adaptor protein, Act1, TRAF-6-dependent and TRAF-6-independent NF-κB activation and mitogen-activated protein kinase (MAPK) involvement resulting in production of proinflammatory cytokines and chemokines, and subsequent recruitment of myeloid cells to the inflamed sites. The accumulating evidence that IL-17 and the IL-17 pathway have an important role in the pathology of psoriasis and other immune-mediated inflammatory diseases led to efforts to specifically target IL-17A with a mAb. By binding IL-17A and blocking its interaction with its receptor expressed on many cell types including keratinocytes, secukinumab inhibits the release of proinflammatory cytokines, chemokines, and other mediators of tissue damage, including damage in the skin. Results are seen as a reduction in erythema, induration, and desquamation in the lesions of plaque psoriasis.

Approved Indications and Safety of Secukinumab

Having received its first approval in Japan in late 2014, secukinumab was approved by both the FDA and EMA in 2015 for the treatment of moderate to severe plaque psoriasis in adult patients who are candidates for systemic therapy or phototherapy.

Infections head the list of safety warnings for secukinumab, and this risk is supported by results obtained in phase III placebo-controlled trials with infections reported in 28.7 % of 1382 patients receiving secukinumab and 18.9 % of 694 subjects receiving placebo. Common infections seen included nasopharyngitis, URTI, and *Candida* mucocutaneous infections. Over a 52-week period, infections occurred in 47.5 % of the 3430 patients given secukinumab in any dose. Caution should therefore be exercised in deciding whether or not to prescribe the mAb to patients with a history of infections, it should not be given to those with an existing active infection, pretreatment screening for tuberculosis should be undertaken, and live vaccines should not be administered to patients receiving secukinumab therapy. Other warnings relate to the possibilities of secukinumab-induced exacerbation of Crohn's disease after three of the 3430 clinical trial patients experienced such an event and allergic reactions including anaphylaxis and urticaria. In addition to the infections mentioned above, adverse events occurring at an incidence higher than 1 %, and higher than the placebo result, were diarrhea, rhinitis, oral herpes, pharyngitis, urticaria, and rhinorrhea. Adverse events seen at a low incidence were mainly infections. The list includes sinusitis, conjunctivitis, tonsillitis, oral candidiasis, tinea pedis, otitis media, impetigo, inflammatory bowel disease, neutropenia, and elevated liver transaminases. Over 52 weeks of treatment, less than 1 % of patients treated with secukinumab developed anti-secukinumab antibodies. Approximately half of these antibodies were neutralizing but no loss of drug efficacy was noticed. Recent results with secukinumab showed low numbers of potential T cell epitopes and low T cell response rates in two separate in vitro assays. In summary, although there is a long list of potential adverse events induced by immunomodulating mAbs, for example, infections, vascular events, autoimmune disorders, malignancies, neutropenia, hypersensitivities, and infusion and injection site reactions, the frequencies of adverse events seen with secukinumab were only higher than the placebo results for infections, hypersensitivities, exacerbation of Crohn's disease, and neutropenia.

Tocilizumab

Tocilizumab (Actemra®, RoActemra®) (Tables 2.1 and 4.1), a recombinant humanized IgG1κ mAb, MW ~ 148 kDA, targeted to the IL-6 receptor IL-6R, provides a new and promising approach to the treatment of rheumatoid arthritis and related immunoinflammatory conditions.

IL-6 and Mechanism of Action of Tocilizumab

IL-6 is a pleiotropic cytokine produced by monocytes and macrophages after stimulation of toll-like receptors in the early phases of infectious and noninfectious inflammation. By stimulating various cells including B and T cells, hematopoietic progenitor cells, and hepatocytes, IL-6 plays an important role in host defense and induces the expression of acute-phase proteins, notably C-reactive protein (CRP), serum amyloid A, and fibrinogen. As mentioned above (section "Mechanism of Action of Secukinumab"), IL-6 together with TGF-β promotes differentiation of Th17 helper T cells involved in autoimmune reactivity, but IL-6 also inhibits differentiation of regulatory T cells (Tregs) induced by TGF-β. The resultant Th17:Treg balance can have clinical consequences in the form of development of various inflammatory and autoimmune diseases. IL-6 also promotes the generation of cytotoxic T cells from CD8+ T cells and RANKL in bone marrow stromal cells and induces VEGF in inflamed tissue, for example, in rheumatoid arthritis. The biological effects of IL-6 are mediated via its receptor IL-6R which exists in two forms, a transmembrane 80 kDa form (mIL-6R, CD126) and a soluble (sIL-6R) form derived from proteolytic cleavage of mIL-6R. The IL-6–IL-6R complex formed after IL-6 binds to the transmembrane receptor associates with glycoprotein (gp)130 to form a hexameric structure made up of two molecules each of IL-6, IL-6R, and gp130 (Fig. 4.7). Signaling through this pathway is referred to as classical signaling. IL-6 can affect a wide variety of cells by also complexing with sIL-6R and transducing signals to cells that do not express the transmembrane receptor but do express gp130. This is referred to as the trans-signaling mechanism. Under normal circumstances, negative regulation of IL-6 is effected by the removal of the source of stress, infection, etc. leading to the normalization of levels of IL-6 and CRP. However, in a range of chronic inflammatory and autoimmune diseases, IL-6 production persists, and animal models of rheumatoid arthritis, SLE, scleroderma, and some autoimmune diseases have demonstrated a pathological association between IL-6 and disease development. It is now known that IL-6 production is implicated in the pathogenesis of the human diseases, rheumatoid arthritis, Castleman's disease, and various other inflammatory, autoimmune, and malignant diseases. This prompted the development of the humanized mAb tocilizumab to block IL-6R, the binding of IL-6 to its receptor, and subsequent IL-6-mediated signal transduction (Fig. 4.8). It is not yet known what events occur following binding of the mAb to cells; speculations include simply receptor blockade, apoptosis, or eventual phagocytosis.

Approved Indications and Safety of Tocilizumab

Tocilizumab is currently the only mAb IL-6 receptor blocker/signal pathway inhibitor with regulatory approval and therefore also the only anti-IL-6R mAb approved for the treatment of immunoinflammatory diseases. FDA-approved indications and

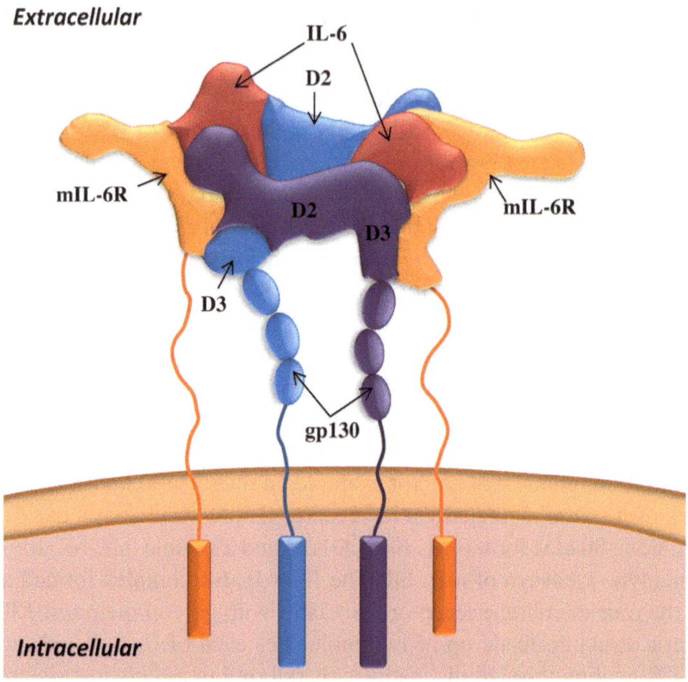

Fig. 4.7 Diagrammatic representation of the hexameric complex between IL-6, IL-6R, and the extracellular binding and activation molecule gp130 that associates with the Il-6–IL-6R binary complex. The extracellular region of IL-6R contains an *N*-terminal domain characteristic of the Ig superfamily and a cytokine-binding domain (CBD) of two fibronectin type III-like domains D2 and D3. The IL-6–IL-6R binary complex interacts with the cytokine-binding homology (CBH) region D2D3 of gp130. The resultant ternary complex forms a hexamer containing two IL-6, two IL-6R, and two gp130 molecules that assemble sequentially and cooperatively. IL-6 is shown in *red*, mIL-6R is in *orange*, and two molecules of gp130 are in *purple* and *blue* to distinguish the molecule at the front of the three-dimensional structure (in *purple*) from the partly obscured molecule (*blue*) at the rear. See also Fig. 4.8. Diagram has been drawn from the 3.65 Å-resolution structure derived from crystallographic data presented by Boulanger MJ, Chow D-c, Brevnova E, and Garcia KC. Hexameric structure and assembly of the interleukin-6/IL-6Rα-receptor/gp130 complex. Science. 2003;300:2101–4

usage of tocilizumab are for treatment of rheumatoid arthritis in patients who have had an inadequate response to one or more disease-modifying antirheumatic drugs and for polyarticular juvenile idiopathic arthritis (PJIA) and systemic juvenile idiopathic arthritis (SJIA), both in patients 2 years of age and older. The EMA has approved the mAb to be given for the above conditions in combination with methotrexate although it is stated that monotherapy may be used in cases of intolerance or where methotrexate therapy is inappropriate.

Tocilizumab is subject to an FDA black box warning of the risk of serious infections including tuberculosis and other bacterial, fungal, viral, and other opportunistic infections. Antituberculosis therapy prior to the commencement of antibody therapy is recommended for patients considered at risk of a tuberculosis

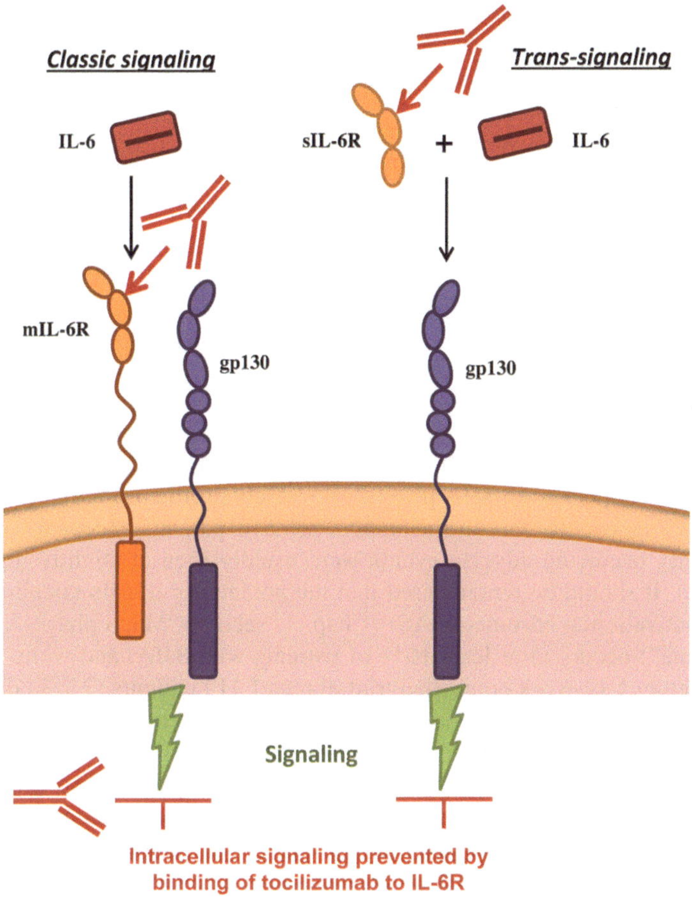

Fig. 4.8 Diagrammatic representation of classic and trans-signaling by IL-6 via the membrane-bound (mIL-6R) and soluble (sIL-6R) receptors and subsequent blocking of signaling by tocilizumab, a monoclonal antibody targeted to IL-6R. In classic signaling, IL-6 binds to mIL-6R on hepatocytes and some leukocytes allowing this binary complex to form a tertiary complex with the signal-transducing protein gp130. Neither IL-6 nor IL-6R alone has any gp130 binding capacity; binding can take place only between gp130 and the binary complex. In trans-signaling, the membrane-bound metalloprotease ADAM17 cleaves mIL-6R forming sIL-6R which binds IL-6 and then associates with membrane-bound gp130 to initiate signaling. The IL-6–sIL-6R complex associates with gp130 on cells that do not express mIL-6R. Note that the activated IL-6–IL-6R–gp130 ternary complex triggers downstream signaling as a hexameric structure comprising two molecules each of IL-6, IL-6R, and gp130 (see Fig. 4.7), but for clarity, the interactions between only one molecule of each are shown here. Tocilizumab inhibits both the classic and trans-signaling pathways by binding to mIL-6R and sIL-6R, thereby blocking formation of the Il-6–IL-6R complex

infection and administration of live vaccines should be avoided. Other warnings and precautions relate to gastrointestinal perforations (mainly as a complication of diverticulitis already seen in clinical trials), hypersensitivity reactions, and the possibilities of viral reactivations (cases of herpes zoster exacerbation have been observed), immunosuppression and the development of malignancies, and demyelinating disorders. In addition, treatment with tocilizumab has been associated with a higher incidence of neutropenia (3.4 % vs. 1 % in controls), reductions in platelet counts (1.7 % vs. <1 %), increases in lipid parameters, and transaminase elevations. Consequently, tocilizumab is not recommended for patients with active hepatic disease or hepatic impairment.

When tocilizumab was used to treat rheumatoid arthritis, the most common adverse events in response to monotherapy were infections. The rate of serious infections was 3.6 events per 100 patient-years of exposure compared to 1.5 events in the oral methotrexate group. The most commonly reported infections were pneumonia, cellulitis, gastroenteritis, herpes zoster, and sepsis. Infusion reactions occurred in 7–8 % of patients with headache, rash, urticaria, and pruritus the main adverse reactions. The most frequent event during the infusion period was hypertension when tocilizumab was being given intravenously. Hypersensitivity reactions requiring treatment discontinuation occurred in 0.2 % of patients. Fatal anaphylaxis to tocilizumab has been reported in the postmarketing period. In SJIA patients, the safety profile of tocilizumab proved similar to the profile seen in rheumatoid arthritis patients except that the most commonly occurring adverse events were usually seen at slightly higher frequencies. It should be remembered that the potentially life-threatening macrophage activation syndrome (MAS) (Chap. 1, section "Macrophage Activation Syndrome") occurs in at least 10 % of patients with SJIA, and while no cases were seen in a 12-week controlled trial, three of 112 patients (2.7 %) developed MAS during open-label treatment with tocilizumab and two more patients developed MAS during the long-term treatment extension. MAS is characterized by an explosive inflammatory reaction caused by excessive activation of macrophages and T lymphocytes producing fever, severe cytopenias, hepatosplenomegaly, lymphadenopathy, and disseminated coagulation. MAS resembles a spectrum of diseases collectively known as hemophagocytic lymphohistiocytosis in which there is uncontrolled proliferation of T cells and macrophages, decreased NK-cell and cytotoxic T-cell functions, and abnormal expression of perforin. In relation to infusion reactions, within 24 h of the completion of tocilizumab infusion, 16 % of SJIA patients and 5 % of control patients developed reactions of rash, diarrhea, epigastric discomfort, arthralgia, headache, and urticaria, the latter considered serious.

In PJIA patients, clinical trial results showed an infection rate of 163.7 per 100 patient-years with nasopharyngitis and URTI, the most common events. Rates of serious infections were higher in patients given higher doses of the mAb and who were less than 30 kg in weight. Infusion reactions, mainly headache, nausea, and hypotension, occurred in 11 patients (5.9 %) during the infusion of tocilizumab and in 38 patients (20.2 %) within 24 h of infusion. The main post-infusion symptoms were dizziness and hypotension.

Postmarketing surveillance reports to the FAERS database again showed that infections occurred with the highest incidence (11 %) followed by respiratory disorders (3 %), dermatologic reactions (2.8 %), and hepatic adverse events (2 %). Reports in the EVRS database for the period 2005–2012 showed infections at the head of the list of adverse events with an incidence of 21 % and with pneumonia, sepsis, cellulitis, diverticulitis, and herpes zoster infections being implicated most often. Other reports with a relatively high incidence dealt with neutropenia (4.8 %), thrombocytopenia (3.5 %), infusion reactions (3 %), and hypersensitivities (4–5 %). There were 36 cases of disseminated intravascular coagulation, 20 cases of tuberculosis, 11 cases of necrotizing fasciitis, six cases of encephalopathy, five cases of demyelinating disorders, and five cases of leukoencephalopathy.

In a 24-week controlled study of the efficacy and safety of intravenous tocilizumab for rheumatoid arthritis involving 2876 patients, 46 patients (1.6 %) developed anti-tocilizumab antibodies and these were found to be neutralizing in 30 patients (1 %). Five patients with antibodies developed a hypersensitivity reaction sufficiently concerning to force withdrawal from the trial. When tocilizumab was administered subcutaneously over a 24-week period, 13 of 1454 patients (0.9 %) developed anti-tocilizumab antibodies; neutralizing antibodies were present in 12 (0.8 %) of the patients. Of 112 JIAP patients treated intravenously with tocilizumab, two developed antibodies to the mAb: one withdrew from the trial with anaphylaxis, angioedema, and urticaria; the other developed MAS while on escape therapy and also withdrew.

Ustekinumab

Ustekinumab (Stelara®) (Tables 2.1 and 4.1) is a recombinant fully human, first-in-class IgG1κ mAb that binds to IL-12 and IL-23, two interleukins that modulate lymphocyte function. Ustekinumab was developed using human immunoglobulin transgenic mice immunized with human IL-12, fusion of antibody-producing B-cell splenocytes, and human T cells to select IL-12-specific antibodies.

IL-12 and IL-23 and Immune-Mediated Diseases

IL-12, a key cytokine in the differentiation of Th1 cells, has an important role in cell-mediated immunity, inducing killer cells and activating NK cells and T cells. IL-23, produced by keratinocytes, Langerhans' cells, dendritic cells, and macrophages, is, along with a number of other cytokines, critical for Th17 cell differentiation. Human Th1 cells produce IFN-γ and TNF and Th17 cells also produce several proinflammatory cytokines, including IL-17A, IL-17 F, TNF, IL-22, IL-26, and IFN-γ. IL-12 and IL-23 exist only as heterodimeric cytokines of two protein subunits named according to their molecular weight, one of which, p40, is shared by both. The other protein components of the IL-12 and IL-23 heterodimers are p35 and p19 subunits, respectively (Fig. 4.9). The p40 subunit is also part of the IL-6 receptor and p35 is homologous to IL-6 and G-CSF. The IL-12 receptor is a heterodimer of two receptor chains IL-12Rβ1 and IL-12Rβ2 expressed on the

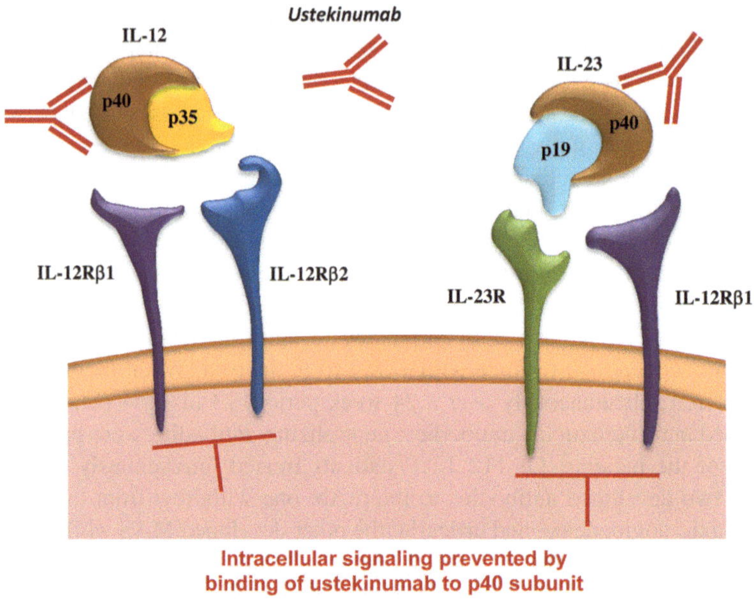

Fig. 4.9 Ustekinumab has demonstrated efficacy in the treatment of plaque psoriasis by inhibiting both the IL-12 and IL-23 pathways. IL-12 and IL-23 exist only as heterodimeric cytokines of two protein subunits (named according to their molecular weight) one of which, p40, is shared by both. The other protein components of the IL-12 and IL-23 heterodimers are p35 and p19 subunits, respectively. The IL-12 receptor is a heterodimer of two receptor chains IL-12Rβ1 and IL-12Rβ2 expressed on the surface of NK and T cells. Each receptor chain provides a separate function for the activity of IL-12; IL-12Rβ1 binds the p40 subunit while IL-12Rβ2 interacts with p35 as a prelude to intracellular signaling. IL-23 also recognizes a heterodimeric receptor of two receptor chains, IL-12Rβ1 and IL-23R. Again, IL-12Rβ1 is involved in binding the p40 subunit, while IL-23R is responsible for intracellular signaling. Ustekinumab binds IL-12 and IL-23 due to recognition of the common p40 subunit preventing the interaction of both cytokines with the IL-12Rβ1 receptor chain on the surface of NK and T cells. This results in the inhibition of downstream signaling, gene activation, cytokine production, and effector cell function

surface of NK and T cells. Each receptor chain provides a separate function for the activity of IL-12; IL-12Rβ1 binds the p40 subunit, while IL-12Rβ2 interacts with p35 (Fig. 4.9) as a prelude to intracellular signaling via activation of transcription proteins STAT4 and STAT6 that leads to cytokine production, including IFN-γ and NK-cell lytic activity. IL-23 also recognizes a heterodimeric receptor of two receptor chains, IL-12Rβ1 and IL-23R. Again, IL-12Rβ1 is involved in binding the p40 subunit, while IL-23R is responsible for intracellular signaling via phosphorylation of STAT3 leading to activation of lymphocytes and cytokine production, such as IL-17A.

Observations of some immunoinflammatory diseases in humans and studies in animal models have demonstrated a clear association between dysregulation of the Th1 and Th17 pathways and some systemic and cutaneous diseases, in particular, psoriasis, rheumatoid arthritis, Crohn's disease, and multiple sclerosis. For example, administration of IL-12 exacerbated a mouse model of psoriasis, and in humans, expressed levels of IL-12, IL-23, and IFN-γ are elevated in skin lesions from psoriasis patients, and the p40 and p35 subunits of IL-12 may be found in gastrointestinal tissue of patients with Crohn's disease.

Mechanism of Action of Ustekinumab

Although formed in response to IL-12 immunization, ustekinumab binds IL-12 and IL-23 equally well due to recognition of the common p40 subunit. Antibody binding to IL-12 and IL-23 prevents the interaction of both cytokines with the IL-12Rβ1 receptor chain, and downstream signaling, gene activation, cytokine production, and effector cell function are thus prevented (Fig. 4.9). Along with Th17 cells, IL-12 and IL-23 appear to have a critical role in psoriasis. IL-23 stimulates Th17 cells which regulate the production of inflammatory cytokines IL-17, IL-22, IL-21, IL-6, and TNF. IL-12 is involved in the development of Th1 cells which produce IFN-γ, TNF, and IL-2. Ustekinumab cannot bind to IL-12 and IL-23 already bound to receptor, and it appears to lack Fc-mediated ADCC and CDC effector functions. In vitro, the mAb neutralizes IL-12-mediated production of the cytokine IFN-γ and IL-23-mediated production of IL-17A, IL-17F, and IL-22.

Indications and Safety of Ustekinumab

Both the FDA and EMA license ustekinumab to treat moderate to severe plaque psoriasis and active psoriatic arthritis. For the former condition, the FDA stipulates adult patients who are candidates for phototherapy or systemic therapy, whereas the EMA requires adult patients who failed to respond to, or were unable to receive, other therapies such as cyclosporin and methotrexate. For psoriatic arthritis, the FDA simply requires that ustekinumab be used alone or in combination with methotrexate, while the EMA adds that the mAb, either alone or in combination with methotrexate, be used in patients with a previous inadequate response to other non-biologic disease-modifying antirheumatic drugs.

Interference with IL-12 signaling decreases the production of IFN-γ and therefore may increase the possibility of immunosuppression and the risk of infections and malignancies. This is reflected in regulatory agency warnings that ustekinumab may cause new infections, reactivate latent infections, and increase the risk of serious infections such as diverticulitis, cellulitis, pneumonia, appendicitis, cholecystitis, osteomyelitis, sepsis, gastroenteritis, and viral and urinary tract infections. Linked with such warnings are the needs to avoid live vaccines, to avoid treatment of patients with active tuberculosis, and to evaluate patients for tuberculosis prior to the commencement of therapy. In psoriasis clinical trials representing 8998 subject-years of exposure and with a median follow-up of 3.2 years, malignancies excluding non-melanoma skin cancers occurred in 1.7 % of ustekinumab-treated patients, while non-melanoma skin cancers were seen in 1.5 % of patients. With the concern in mind of the possibility of increased immunosuppression due to ustekinumab-induced reduction of IFN-γ, cytokine production by memory T cells, differentiation of naïve T cells, and T-cell diversity were evaluated in ustekinumab-treated psoriasis patients. Results showed clinical improvements without affecting cytokine production or T-cell maturation or diversity suggesting that the T-cell immune response is not diminished in patients treated with ustekinumab. The most common adverse reactions identified in clinical evaluations of patients treated with ustekinumab and listed by the EMA in its Summary of Product Characteristics are URTI, nasopharyngitis, dental infections, headache, diarrhea and nausea, arthralgia, fatigue, injection site reactions, and pruritus. Not surprisingly, infections dominated the postmarketing FAERS and EVRS reports of adverse events to ustekinumab issued at the end of 2012, but there were a number of interesting/unusual reports including eight cases of PRES, 64 cases of myocardial infarction and ten of cardiac failure, two cases each of anaphylaxis and serum sickness, and 38 cases of depression. Three percent of reports to the FAERS database were for psoriasis exacerbation and the EVRS received 183 reports of malignancies. As is the case with many of the mAbs used for therapy, extensive data on immunogenicity of ustekinumab is lacking. The FDA states that 6 % of patients with psoriasis/psoriatic arthritis treated with ustekinumab developed antibodies, generally of low titer. Antibodies in patients with psoriasis were generally neutralizing.

Proprotein Convertase Subtilisin/Kexin Type 9 (PCSK9) Inhibitors: Alirocumab and Evolocumab

Figures released by the Centers for Disease Control and Prevention reveal that about 610,000 people die of heart disease in the USA every year, equivalent to about one in every four deaths. Elevated plasma levels of low-density lipoprotein cholesterol (LDL-C) are known to be a risk factor for the development of ischemic cardiovascular disease such as myocardial infarction and stroke. Familial hypercholesterolemia (FH) is a common genetic cause of premature coronary heart disease. Those with the condition have lifelong elevated levels of LDL-C. Individuals heterozygous for FH (~1 in 500 whites) typically develop coronary heart disease before ages 55–60, while homozygotes (~1 in one million) develop the disease early in life and generally die before age 20 if left untreated. Central to the metabolism of LDL is its receptor LDLR which binds cholesterol derived from LDL and clears it from the circulation. The enzyme PCSK9, a major regulator of cholesterol homeostasis, acts as a negative regulator of LDLR, promoting internalization and lysosomal degradation of the receptor by binding to it and preventing its recycling back to the membrane where it can bind more LDL. PCSK9 has therefore become a target for lipid-lowering therapy since its net effect is to reduce the number of receptors in the liver that remove LDL-C from the blood. Blocking PCSK9 results in increased cell surface expression of LDLRs; increased uptake of LDL-C from the blood, especially by the liver; and an overall reduction of circulating LDL-C levels.

Two mAbs, alirocumab and evolocumab, that bind PCSK9 (Tables 2.1 and 4.1) have been developed and are now approved for therapy. The FDA approved alirocumab (Praluent®) in July 2015 and evolocumab (Repatha®) in August 2015. Also in July 2015, the EMA approved evolocumab and recommended the granting of a marketing authorization for alirocumab, an intermediary step for that drug's final approval for patient use which will follow the adoption of a decision on an EU-wide marketing authorization. Alirocumab is a fully human, recombinant IgG1κ mAb, MW ~ 146 kDa; evolocumab is a fully human recombinant IgG1λ mAb, MW ~ 144 kDa. Both mAbs, produced in CHO cells, bind PCSK9 near the catalytic site that interacts with the LDLR.

Alirocumab: Indications and Safety

Indications for, and usage of, alirocumab laid down by the FDA are as an adjunct to diet and maximally tolerated statin therapy for the treatment of adults with heterozygous FH or clinical atherosclerotic cardiovascular disease, who require additional lowering of LDL-C.

Allergic reactions constitute the main warning/precautions for alirocumab. Reactions can be mild, as with pruritus and rash, or severe, e.g., hypersensitivity vasculitis, nummular eczema, or severe hypersensitivities requiring hospitalization. The safety of alirocumab was evaluated in nine placebo-controlled trials involving 2476 patients; 37 % of the patients had heterozygous FH and 66 % had clinical atherosclerotic cardiovascular disease. Adverse events reported in at least 2 % of alirocumab-treated patients and occurring more frequently than reactions seen in placebo-treated patients are listed in Table 4.1. Adverse events seen most often were nasopharyngitis, injection site reactions, and influenza. The most common adverse reactions leading to discontinuation of treatment with alirocumab were allergic reactions (0.6 %) and elevated liver enzymes (0.3 %). At this early stage in the marketing of alirocumab, the yet-to-be-determined effect of alirocumab on cardiovascular morbidity and mortality is an important precautionary reminder.

Pooled placebo-controlled trial data showed that 4.8 % of patients treated with alirocumab had anti-alirocumab antibodies compared to 0.6 % of control patients. The treated patients with antibodies also experienced a higher incidence of injection site reactions (10.2 %) compared to patients who did not develop antibodies (5.9 %). Neutralizing antibodies were detected in 1.2 % of patients treated with alirocumab, and 0.3 % of patients with neutralizing antibodies experienced some loss of treatment efficacy.

Evolocumab: Indications and Safety

Evolocumab is indicated in adults with primary hyperlipidemia (heterozygous familial and nonfamilial) and mixed dyslipidemia as an adjunct to diet to reduce LDL-C and other lipids. Evolocumab may be given in combination with a statin or with a statin and other lipid-lowering therapies (e.g., ezetimibe), or alone or in combination with other lipid-lowering therapies in patients who are statin intolerant, or alone or in combination with other lipid-lowering therapies in patients for whom a statin is not considered clinically appropriate. Evolocumab is also indicated in adults and adolescents (12 years and older) with homozygous FH to reduce LDL-C and other lipids in combination with other lipid-lowering therapies, for example, statins and LDL-C apheresis.

The safety of evolocumab was assessed in 12-phase two and three studies and two open-label extension studies of primary hyperlipidemia and mixed dyslipidemia. The integrated safety analyses involved more than 6000 patients with primary hyperlipidemia and mixed dyslipidemia. Adverse events were generally nonserious and mild to moderate in severity. Adverse events occurring in 2 % or more of patients and more frequently than reactions seen in placebo-treated patients were determined to be significant adverse events. These events are set out in Table 4.1. Most reactions were mild to moderate and safety profiles in the homozygous FH and primary and mixed dyslipidemias were similar. The incidence of hypersensitivities was low and with reactions mostly mild to moderate cases of rash or urticaria. The incidence of anti-evolocumab antibodies was also low (0.1 %), and antibodies, when they did occur, were non-neutralizing and not associated with adverse events. As with alirocumab, the use of evolocumab may lead to very low cholesterol levels where long-term safety has not yet been established and the effect of the mAb on cardiovascular morbidity and mortality has not yet been determined.

Idarucizumab

Approved by the FDA in October 2015, submitted for approval in Australia and Canada, and already received a positive opinion from EMA's Committee for Medicinal Products for Human Use in September 2015, idarucizumab (Praxbind®) is a humanized mAb Fab fragment of the IgG1 isotype targeted to the thrombin inhibitor dabigatran (Tables 2.1 and 4.1). The molecule, MW ~ 48 kDa, is composed of a κ L chain of 219 amino acids and a H chain of 225 amino acids covalently linked between cysteine 225 of the H chain fragment and cysteine 219 of the L chain.

Mechanism of Action of Idarucizumab

Novel anticoagulants such as dabigatran etexilate (Pradaxa®, Prazaxa®, Pradax®) are effective alternatives to vitamin K antagonists, but the therapy carries the risk of bleeding. In cases where drug discontinuation is not successful in stopping

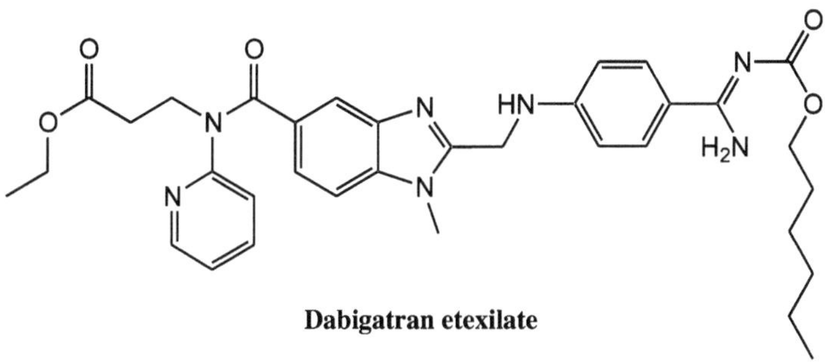

Fig. 4.10 Dabigatran etexilate, a synthetic benzamidine-based nonpeptide anticoagulant acting by direct inhibition of thrombin, is administered to prevent strokes, deep vein thrombosis, and pulmonary embolism. Dabigatran, a very polar compound lacking bioavailability after oral administration, is made suitable for oral absorption by conversion to the more hydrophobic prodrug dabigatran etexilate by masking the amidinium group as a carbamate ester and converting the carboxylate group to an ester. Dabigatran is the target of the newly approved monoclonal antibody idarucizumab, developed to overcome the absence of a specific means of reversing the drug's anticoagulant effect in cases of uncontrolled bleeding

bleeding, rapid reversal of anticoagulation by directly targeting the drug would be desirable. Idarucizumab, given intravenously, binds specifically with high affinity to the benzamidine-based thrombin inhibitor dabigatran (Fig. 4.10) and its acyl glucuronide metabolites, reversing its anticoagulant effect by displacing the drug from its complex with thrombin. Idarucizumab does not shorten clotting time or affect the coagulation cascade (Chap. 10).

Indications and Safety of Idarucizumab

Idarucizumab is indicated by the FDA when reversal of the anticoagulant effects of dabigatran is needed for emergency surgical and/or other urgent procedures and in cases of life-threatening or uncontrolled bleeding. The FDA has issued four warnings and precautions for the mAb (Table 4.1): by reversing dabigatran therapy, patients may be at thrombotic risk from their underlying disease; bleeding may require an additional dose of idarucizumab, hypersensitivity reactions, and the risk in patients with hereditary fructose intolerance of a serious reaction to the sorbitol excipient present in the mAb formulation. Adverse reactions to idarucizumab in healthy volunteers seem to be confined to headache, while in treated patients, reactions reported so far include hypokalemia, delirium, pyrexia, pneumonia, and constipation.

In an immunogenicity study, preexisting low-titer antibodies cross-reactive with idarucizumab were found in 36 of 283 patients (12.7 %) (224 treated with the mAb). After treatment, nine of the 224 patients (4 %) had anti-idarucizumab antibodies. Investigations to identify the antigenic determinants revealed that 35 of the 36 preexisting antibodies were located at the nonbinding C-terminus of the mAb. For the antibodies that appeared after treatment, five recognized the C-terminus of idarucizumab, two were specific for the variable region, one had mixed specificity, and one was indeterminate.

Mepolizumab

Approved by the FDA in November 2015, mepolizumab (Nucala®) (Tables 2.1 and 4.1) is a recombinant humanized IgG1κ mAb, MW ~ 149 kDa, targeted to IL-5.

Mechanism of Action of Mepolizumab

IL-5 is a cytokine affecting the growth, differentiation, activation, and survival of eosinophils, one of the cell types involved in inflammation including inflammation associated with asthma. Mepolizumab binds to IL-5, blocking its interaction with the alfa chain of its receptor on the eosinophil surface and thus inhibiting IL-5 signaling. Although this inhibits the production and survival of eosinophils, the precise events involved in mepolizumab's beneficial effect in some asthmatic patients are not yet identified or understood.

Indications and Safety of Mepolizumab

Mepolizumab is indicated for add-on maintenance treatment of patients 12 years and older with severe asthma of the eosinophilic type. Mepolizumab was found to reduce eosinophil counts in blood and sputum and reduce the exacerbation rate in patients with recurrent asthma and evidence of eosinophilic inflammation despite high doses of inhaled steroids. The FDA has set out limits for the use of this mAb in asthma patients (see below).

Warnings and precautions issued for mepolizumab concern the possibility of hypersensitivity reactions including urticaria, angioedema, bronchospasm, hypotension, and rash, the need to not discontinue corticosteroids abruptly upon initiation of therapy, and the reminder that mepolizumab should not be used to treat acute bronchospasm or status asthmaticus. Two other warnings relate to infections: herpes zoster has occurred in patients given mepolizumab indicating that varicella vaccination, if needed, should be given prior to the initiation of therapy, and patients with preexisting helminth infections should be treated before beginning therapy with the mAb. In a 52-week clinical trial, adverse reactions occurring with an incidence ≥3 % and more common than placebo included headache, injection site reactions, back pain, fatigue, influenza, urinary tract infection, upper abdominal pain, allergic rhinitis, asthenia, bronchitis, cystitis, dizziness, dyspnea, gastroenteritis, nasopharyngitis, and nausea.

Fifteen of 260 subjects (5.8 %) treated with mepolizumab developed antibodies to the mAb. Antibodies, neutralizing in only one subject, slightly increased the clearance of mepolizumab although there was no evidence of a correlation between antibody titers and eosinophil levels. As with many other mAbs used therapeutically, the clinical relevance of the presence of anti-mepolizumab antibodies remains unknown.

Recent Approvals: Obiltoxaximab, Ikexizumab, Reslizumab

After completion of the manuscript for this book, three more mAbs, **obiltoxaximab, ixekizumab, and reslizumab**, were approved by the FDA.

Obiltoxaximab (Anthim®), a chimeric IgG1 kappa mAb directed against the PA of *B. anthracis*, is indicated for treatment and prophylaxis of inhalational anthrax. Hypersensitivity, including anaphylaxis, constitutes a boxed warning for this mAb. Most frequently reported adverse reactions are headache, URTI, cough, injection site reactions including bruising, pruritus, and urticaria (Table 4.1).

Ixekizumab (Taltz®), a humanized recombinant IgG4 mAb, selectively binds IL-17A blocking its interaction with its receptor and thus inhibiting the release of proinflammatory cytokines and chemokines. Ixekizumab is indicated for treatment of plaque psoriasis. Issued warnings cover the possibility of infections including tuberculosis, hypersensitivity, and inflammatory bowel disease. Common adverse reactions are listed in Table 4.1.

Reslizumab (Cinqair®) is a humanized recombinant IgG4 kappa mAb produced in NSO cells and, like mepolizumab, is indicated for severe asthma. Reslizumab binds IL-5, a cytokine inducing growth, differentiation, activation, and survival of eosinophils. Antibody binding blocks interaction with the IL-5 receptor on the eosinophil surface, reducing cell survival and inflammation. The precise mechanisms of reslizumab's beneficial effects in asthma are still to be established. Reslizumab is subject to a boxed warning for anaphylaxis and warnings/precautions for malignancy, parasitic infection and the need to avoid abrupt decrease of corticosteroid therapy (Table 4.1).

Summary

- The pivotal place TNF has as a mediator of the inflammatory process in rheumatoid arthritis, Crohn's disease, and psoriasis underlies the development of the mAbs **adalimumab (Humira®)**, **certolizumab pegol (Cimzia®)**, **infliximab (Remicade®)**, and **golimumab (Simponi®)**. Each is targeted to TNF with the aim of bringing clinical improvement without immunosuppressing the patient.
- Serious infections and the risk of malignancy are the subjects of an FDA black box warning for all four approved anti-TNF mAbs. Fatal cases of hepatosplenic T-cell lymphoma have occurred mostly in young adult males being given TNF blockers. All of the patients had also received immunosuppressants.
- **Adalimumab** binds human soluble and transmembrane TNF with high affinity, blocking its interaction with the TNFR1 p55 and TNFR2 p75 cell surface TNF receptors. Adalimumab demonstrates apoptotic activity, complement-dependent cytotoxicity (CDC), and antibody-dependent cell cytotoxicity (ADCC).
- In rheumatoid arthritis patients, injection site reactions are the most commonly encountered adverse event, occurring in ~20 % of adalimumab-treated patients.

A wide range of apparent hypersensitivities induced by adalimumab have become apparent including anaphylaxis, urticaria, bronchospasm, asthma, psoriasis, exacerbation of palmoplantaris pustulosa psoriasis, autoimmune hepatitis, and a range of different cutaneous reactions including rare cases of the severe toxidermias erythema multiforme, Stevens-Johnson syndrome, and toxic epidermal necrolysis.

- An intriguing event seen in some patients treated with TNF inhibitors, including all four approved anti-TNF mAbs, is the paradoxical induction of psoriasis with many of the patients showing a palmoplantar distribution together with tiny pustules. Other paradoxical skin eruptions associated with TNF inhibitors include lupus-like cutaneous events, dermatomyositis-like eruptions, and cutaneous vasculitis.
- Although adalimumab is a fully human mAb, human antihuman antibody responses have been detected. Anti-adalimumab antibodies are associated with reduced effectiveness of adalimumab in rheumatoid arthritis. Concomitant methotrexate may reduce antibody formation. Antidrug antibodies are associated with all four of the mAb TNF blockers, especially adalimumab and infliximab.
- While certolizumab pegol, infliximab, and adalimumab are effective in Crohn's disease, the latter two mAbs trigger apoptosis suggesting not only differences in the pathogenesis of the immune-mediated inflammatory diseases but also differences in the functional properties of the TNF inhibitors.
- **Infliximab** reduces the levels of several disease markers, IL-6, sTNFR1, and TNFR2; leads to reductions in TNF, IL-19, and IFN-γ; and reduces the number of cells staining positive for CD4, CD8, and CD68. Like adalimumab, but unlike certolizumab pegol, infliximab mediates CDC and ADCC and increases the proportion of cells undergoing apoptosis and the level of granulocyte degranulation.
- There are numerous reports of immediate reactions to infliximab including urticaria but particularly anaphylaxis, with over 650 cases of the latter at an incidence of ~0.9 % for the period 1999–2012. Type III hypersensitivities to infliximab occur including some cases of vasculitis and a serum sickness-like reaction. Type IV hypersensitivities occur in the form of maculopapular rashes, psoriasis, erythema multiforme, and Stevens-Johnson syndrome. Intriguingly, there are reports of infliximab inducing rapid recovery of lesions in several cases of toxic epidermal necrolysis.
- Anti-infliximab antibodies have been found in ~10–15 % of patients. Most positive responders had low-titer antibodies, but the presence of antibodies generally corresponded to higher rates of clearance of infliximab, reduced efficacy of the drug, and the manifestation of infusion reactions.
- **Golimumab** has a high capacity to neutralize TNF, binding soluble TNF with a 2.4-fold higher affinity than infliximab and a 7.1-fold higher affinity than adalimumab and binding transmembrane TNF with an affinity similar to infliximab but greater than adalimumab. Golimumab binds to Fc receptors reflected in its properties of effecting CDC and ADCC.
- As with adalimumab, infliximab, and certolizumab pegol, golimumab should not be used with biologics such as anakinra and abatacept, agents that may increase the chance of serious infections.

- The Fab fragment **abciximab** (**ReoPro®**) targets the glycoprotein GPIIb/IIIa receptor (integrin αIIb β3), a receptor for fibrinogen and von Willebrand factor that aids platelet activation. Abciximab also binds the vitronectin receptor αvβ3, a receptor present on platelets, leukocytes, vascular endothelial cells, and smooth muscle cells, which mediates procoagulant properties of platelets and cell proliferation. Abciximab blocks the αvβ3-mediated effects including cell adhesion.
- The main warning issued for abciximab is for bleeding, and because of this risk, it is contraindicated in a range of disorders where bleeding may be active or a threat. A warning has also been issued for the possibility of allergic reactions, including anaphylaxis, which has sometimes proved fatal.
- Readministration of abciximab may induce the production of human antichimeric antibodies potentially causing an allergic reaction, thrombocytopenia, or a reduced clinical benefit of the mAb Fab.
- As **Lemtrada®**, **alemtuzumab**, directed against CD52 expressed on T and B cells and a number of other cells, is marketed and approved for the treatment of relapsing remitting multiple sclerosis. The mAb is presumed to act via its target antigen CD52 on NK cells, monocytes, and monocytes, depleting and repopulating lymphocytes with the involvement of its CDC and ADCC actions.
- Lemtrada® is available only through a restricted distribution arrangement called the Lemtrada® REMS Program. Serious autoimmune disorders, sometimes fatal, may manifest as thrombocytopenia and anti-glomerular basement membrane disease.
- **Basiliximab** (**Simulect®**) blocks IL-2 signaling by binding its receptor IL-2Rα. Although IL-2Rα is not expressed on resting T cells, it is abundantly expressed on activated T cells and especially by T cells in some autoimmune diseases, T-cell leukemia, and organ allograft rejection. This is the rationale underpinning the development and application of basiliximab for the prevention of organ transplant rejection.
- An FDA boxed warning for basiliximab states that the mAb should be prescribed only by physicians experienced in immunosuppression therapy in a suitably equipped facility. Other warnings and precautions cover the potential risks of hypersensitivity, including anaphylaxis, and immunogenicity.
- A comparison of basiliximab and antithymocyte globulin in kidney transplant patients revealed significantly lower incidences of lung infections, granulocytopenia, and thrombocytopenia in the basiliximab-treated group.
- During the postmarketing period for basiliximab, severe hypersensitivity reactions, including anaphylaxis and urticaria, have been reported as well as capillary leak and cytokine release syndromes. Infections, particularly viral infections, and renal disorders are the most often reported adverse events.
- **Belimumab** (**Benlysta®**) is specific for soluble human B lymphocyte stimulator (BLyS) protein, a B-cell survival factor. BLyS is the natural ligand of three TNF receptors, BAFF-R, TACI (transmembrane activator and calcium modulator and cyclophilin ligand interactor), and BCMA (B-cell maturation antigen). Each of these receptors is expressed mainly on mature B lymphocytes. Elevated levels of BLyS in patients with SLE have contributed to the targeting of the BLys/APRIL

(a proliferation-inducing ligand secreted by activated myeloid cells and with a stimulatory action on B cells) pathway in an attempt to modulate autoantibody production in SLE.

- Infections, some serious, occur in patients given belimumab. The most common serious events are pneumonia, urinary tract infections, bronchitis, and cellulitis. Some fatal cases of PML have been reported in SLE patients.
- The anti-IL-1β antibody **canakinumab (Ilaris®)** is approved for the treatment of a group of rare hereditary autoinflammatory diseases of variable severity showing a spectrum of overlapping clinical features involving almost all organs and collectively known as cryopyrin-associated periodic syndromes or CAPS. Patients generally experience lifelong episodes of recurrent fever with systemic inflammation mediated by neutrophils.
- CAPSs are the result of an autosomal dominant or new mutation of the cold-induced autoinflammatory syndrome-1 (CIAS1) gene that encodes cryopyrin and results in an overactivation of caspase-1, an enzyme that ultimately results in an increase in the production of IL-1β and inflammation. Canakinumab neutralizes the biological activity of IL-1β by blocking its interaction with its receptor IL-1RI.
- The most common adverse events associated with the use of canakinumab are infections, particularly nasopharyngitis, viral infections, and urinary tract and upper respiratory tract infections.
- As **Prolia®**, **denosumab**, which is specific for RANKL, has regulatory approval for the treatment of men and menopausal women with osteoporosis at high risk of fracture, men at high risk of fracture receiving androgen deprivation therapy for nonmetastatic prostate cancer, and for the treatment of women at high risk of fracture receiving adjuvant aromatase inhibitor therapy for breast cancer.
- Since hypocalcemia may be exacerbated by denosumab therapy and is an added risk in patients with severe renal impairment, it should be treated before commencing treatment with Prolia®. Adverse events associated with denosumab are osteonecrosis of the jaw, hypersensitivity including anaphylaxis, atypical femoral fractures, endocarditis, and serious infections affecting the urinary tract, abdomen, ear, and skin. Cutaneous reactions include rashes, dermatitis, eczema, photosensitive responses, exanthema, skin eruptions, and bullous and exfoliative reactions.
- **Eculizumab** (**Soliris®**) is a mAb specific for complement component C5 constructed from human IgG2 and human IgG4 sequences with grafted murine complementarity-determining regions to form the antibody combining sites. Human IgG2 and IgG4 sequences are used to reduce the potential of an antibody-induced inflammatory response since the human IgG2 isotype does not bind Fc receptors and IgG4 does not bind C1q and activate the complement cascade.
- Eculizumab is indicated for paroxysmal nocturnal hemoglobinuria (PNH), atypical hemolytic uremic syndrome (AHUS), and for Shiga toxin-induced hemolytic uremic syndrome (HUS).
- Eculizumab inhibits cleavage of C5 to C5a (a potent anaphylatoxin with proinflammatory activity) and C5b. This prevents the formation of the membrane attack complex, the terminal complement component, thus blocking the proin-

flammatory, prothrombotic, and lytic actions of complement and the clinical consequences seen in PNH and AHUS.

- A black box warning for serious meningococcal infections has been issued for eculizumab stating that life-threatening and fatal infections may occur or have occurred in treated patients. A consequence of this warning is the requirement that the drug be issued under a Risk Evaluation and Mitigation Strategy (REMS).
- With blockage of terminal complement activation, patients taking eculizumab may have an overall increase in susceptibility to infections, especially encapsulated bacteria. Serious infections may occur due to *Aspergillus* sp., *Streptococcus pneumoniae*, and *Haemophilus influenzae*.
- **Natalizumab** (**Tysabri**®), a recombinant IgG4 mAb, targets both α4β1 and α4β7 heterodimeric leukocyte adhesion molecules. Receptor α4β1 binds the cell adhesion molecule vascular adhesion molecule-1, VCAM-1. VCAM-1 mediates cell adhesion to vascular endothelium and is upregulated on intestinal endothelium in Crohn's disease. The interaction between α4β1 and VCAM-1 initiates the emigration of lymphocytes from the blood through the venule walls and into the tissues to inflammatory sites. In multiple sclerosis, by blocking adhesion of α4β1 to VCAM-1 expressed on inflamed cerebrovascular endothelial cells, natalizumab inhibits inflammation by preventing the movement of autoreactive leukocytes out of blood vessels and into target organs causing inflammation.
- Similar binding to the α4 subunit of leukocyte adhesion factors α4β1 and α4β7 by natalizumab blocking adhesion of both VCAM-1 and MAdCAM-1 to intestinal endothelium is believed to be the mechanism underlying the mAb's beneficial effect in Crohn's disease.
- Natalizumab should not be given with immunosuppressants or inhibitors of TNF. Under a boxed warning for natalizumab, the FDA warns of the risk of PML.
- Three factors that increase the risk of PML in natalizumab-treated patients are longer duration of treatment, for example, beyond 2 years; prior treatment with immunosuppressants such as azathioprine, methotrexate, and cyclophosphamide; and the presence of JC virus antibodies.
- The reason why natalizumab increases the risk of PML is the subject of speculation. Natalizumab inhibits leukocyte trafficking in the CNS presumably by binding the α4β1 integrin and the resultant decrease in immune surveillance by memory T cells in the CNS may predispose patients to PML.
- Natalizumab may be associated with immune reconstitution inflammatory syndrome (IRIS). IRIS, described as "atypical exuberant inflammation," is a condition seen in some immunosuppressed patients when immunity to infectious or noninfectious antigens is restored leading to a paradoxical worsening of the patient's condition.
- Other warnings and precautions for natalizumab concerning the risk of developing encephalitis and meningitis caused by herpes simplex and varicella zoster viruses, liver injury, hypersensitivities, and infections, are primarily a consequence of immunosuppression.
- In a trial of natalizumab for multiple sclerosis, persistence of serum antibodies was accompanied by a decrease in effectiveness of natalizumab.

- **Vedolizumab** (**Entyvio**®), a humanized IgG1 mAb, binds the $\alpha4\beta7$ integrin. It does not bind to, or inhibit the functions of, $\alpha4\beta1$. Vedolizumab acts as a gut-selective immunosuppressive agent by virtue of its recognition of $\alpha4\beta7$, a lymphocyte-homing receptor, preferentially expressed on gut-homing T-helper lymphocytes. Binding of the mAb inhibits cell adhesion of the integrin to the cell adhesion molecule MAdCAM-1, mainly expressed on gut endothelial cells. In ulcerative colitis and Crohn's disease, memory T cells expressing $\alpha4\beta7$ preferentially migrate to the gastrointestinal tract where they provoke an inflammatory reaction. By inhibiting the interaction between $\alpha4\beta7$ and MAdCAM-1, migration of the T lymphocytes across the vascular epithelium and into the gut is prevented.
- Approved for moderative to severe ulcerative colitis and Crohn's disease in adults, vedolizumab is subject to warnings/precautions for infusion-related and hypersensitivity reactions, an increased risk of infections, liver injury, and PML.
- **Omalizumab** (**Xolair**®) is approved for the treatment of persistent chronic asthma and idiopathic urticaria.
- Omalizumab binds to a determinant in the Cε3 region of human IgE, the same region of the molecule involved in binding to the FcεRI receptor on the mast cell. It binds to free, circulating IgE antibodies, but it does not bind to IgE already bound to mast cells, basophils, and dendritic cells. Since the interaction of IgE with these cells also involves the Cε3 region of the antibody, this region of cell-bound IgE becomes inaccessible to omalizumab. Cross-linking of cell-bound IgE by the anti-IgE mAb therefore cannot eventuate, thus averting an allergic reaction.
- A black box warning for anaphylaxis has been issued for omalizumab. The incidence of anaphylaxis to omalizumab has been estimated to be ~0.1–0.14 %. Other immune system disorders with omalizumab are a serum sickness-like type III hypersensitivity reaction, systemic eosinophilia, vasculitis consistent with Churg-Strauss syndrome, and helminth infection.
- **Palivizumab** (**Synagis**®) is directed to a determinant in the A antigenic site of the F protein of the *Pneumovirus* human respiratory syncytial virus (RSV), a highly contagious and major cause of lower respiratory tract infections such as pneumonia, bronchiolitis, and tracheobronchitis during infancy and childhood.
- Commonly seen adverse events following palivizumab therapy include pyrexia, apnea, injection site reactions, cough, dizziness, rash, erythema, and itching.
- **Ranibizumab** (**Lucentis**®), essentially derived from the full mAb bevacizumab, is targeted to VEGF-A and designed for intraocular use. Approved indications for the mAb are neovascular (wet) age-related macular degeneration, diabetic macular edema, and macular edema following retinal vein occlusion. Ranibizumab blocks the interaction of VEGF-A with its receptors VEGFR-1 and VEGFR-2 preventing the subsequent receptor-mediated proliferation of endothelial cells, neovascularization, and vascular leakage.
- Intravitreal injection of ranibizumab may result in endophthalmitis and retinal detachment. These conditions, plus increases in intraocular pressure and

thromboembolic events in patients treated with the mAb fragment, are the subjects of warnings.

- The most common ocular events seen with ranibizumab include conjunctival hemorrhage, eye pain, vitreous floaters, increased ocular pressure, vitreous detachment, and cataracts.
- **Raxibacumab (Abthrax®)** is a human antibody for the treatment and prophylaxis of inhalational anthrax. Raxibacumab targets the protective antigen PA produced by rapidly dividing *B. anthracis*. The antibody does not have any direct antibacterial action but works by binding PA, thereby preventing its binding to its cell receptors.
- Raxibacumab has been evaluated for efficacy and safety in animals and a few human studies. The only operative safety warning is for infusion reactions which manifested as rash, urticaria, and pruritus in some patients in clinical trials.
- **Secukinumab (Cosentyx®)** is a fully human IgG1 mAb targeted to the proinflammatory cytokine IL-17A and used to treat plaque psoriasis. By binding IL-17A and blocking its interaction with its receptor expressed on many cell types including keratinocytes, secukinumab inhibits the release of proinflammatory cytokines, chemokines, and other mediators of tissue damage, including damage in the skin.
- Infection is the main safety warning for secukinumab. Common infections seen include nasopharyngitis, URTI, and *Candida* mucocutaneous infections.
- Other warnings relate to the possibilities of secukinumab-induced tuberculosis activation, the exacerbation of Crohn's disease, and allergic reactions.
- **Tocilizumab (Actemra®, RoActemra®)**, targeted to the IL-6 receptor IL-6R, provides a new and promising approach to the treatment of rheumatoid arthritis and related immunoinflammatory conditions. IL-6 helps to promote differentiation of Th17 helper T cells involved in autoimmune reactivity, promotes the generation of cytotoxic T cells from CD8+ T cells and RANKL in bone marrow stromal cells, and induces VEGF in inflamed tissue, for example, in rheumatoid arthritis.
- IL-6 production is implicated in the pathogenesis of the human diseases, rheumatoid arthritis, Castleman's disease, and various other inflammatory, autoimmune, and malignant diseases. This prompted the development of the humanized mAb tocilizumab to block IL-6R, the binding of IL-6 to its receptor, and subsequent signal transduction.
- Approved indications for tocilizumab are the treatment of rheumatoid arthritis, polyarticular juvenile idiopathic arthritis, and systemic juvenile idiopathic arthritis.
- Tocilizumab is subject to an FDA boxed warning of the risk of serious infections including tuberculosis and other bacterial, fungal, and viral opportunistic infections. Antituberculosis therapy prior to the commencement of antibody therapy is recommended for patients considered at risk of a tuberculosis infection. Administration of live vaccines should be avoided.
- Other warnings and precautions relate to gastrointestinal perforations, hypersensitivity reactions, the possibilities of viral reactivations and immunosuppression, and the development of malignancies and demyelinating disorders. Tocilizumab has been associated with a higher incidence of neutropenia, reductions in platelet counts, increases in lipid parameters, and transaminase elevations.

- When tocilizumab is used to treat rheumatoid arthritis, the most common adverse event in response to monotherapy is infection, in particular, pneumonia, cellulitis, gastroenteritis, herpes zoster, and sepsis. Infusion reactions and hypersensitivity also occur.
- Macrophage activation syndrome (MAS) occurs in at least 10 % of patients with systemic juvenile idiopathic arthritis. No cases were seen in a 12-week controlled trial, but three of 112 patients (2.7 %) developed MAS during open-label treatment with tocilizumab and two more patients developed MAS during the long-term treatment extension.
- In polyarticular juvenile idiopathic arthritis patients, nasopharyngitis and URTI were the most common infections seen in clinical trials with tocilizumab.
- **Ustekinumab** (**Stelara**®), licensed to treat moderate to severe plaque psoriasis and active psoriatic arthritis, is a human mAb that binds IL-12 and IL-23, two interleukins that modulate lymphocyte function. IL-12 is a key cytokine in the differentiation of Th1 cells while IL-23 is critical for Th17 cell differentiation. Human Th1 cells produce IFN-γ and TNF and Th17 cells produce several proinflammatory cytokines, including IL-17A, TNF, and IFN-γ.
- Although formed in response to IL-12 immunization, ustekinumab binds IL-12 and IL-23 equally well due to recognition of the common p40 subunit. Antibody binding to IL-12 and IL-23 prevents the interaction of both cytokines with the IL-12Rβ1 receptor chain thus preventing downstream signaling, gene activation, cytokine production, and effector cell function.
- Interference with IL-12 signaling decreases the production of IFN-γ and therefore may increase the possibility of immunosuppression and the risk of infections and malignancies. This is reflected in regulatory agency warnings that ustekinumab may cause new infections, reactivate latent infections, and increase the risk of serious infections.
- PCSK9 (proprotein convertase subtilisin/kexin type 9) has become a target for lipid-lowering therapy since its net effect is to reduce the number of receptors in the liver that remove low-density lipoprotein cholesterol (LDL-C) from the blood. Blocking PCSK9 results in increased cell surface expression of LDLRs, increased uptake of LDL-C from the blood, and an overall reduction of circulating LDL-C levels.
- Two fully human mAbs, **alirocumab** (**Praluent**®) and **evolocumab** (**Repatha**®), that bind PCSK9 have been developed and are now approved for therapy.
- Alirocumab is indicated for the treatment of adults with heterozygous familial hypercholesterolemia or clinical atherosclerotic cardiovascular disease, who require additional lowering of LDL-C.
- Allergic reactions, the most common adverse reaction leading to discontinuation of treatment with alirocumab, constitute the main warning/precautions for the mAb. Reactions can be mild or severe. Adverse events seen most often in clinical trials were nasopharyngitis, injection site reactions, and influenza.

- Evolocumab is indicated in adults with primary hyperlipidemia (heterozygous familial and nonfamilial) and mixed dyslipidemia. It may be given in combination with a statin or with a statin and other lipid-lowering therapies. Evolocumab is also indicated in adults and adolescents with homozygous familial hypercholesterolemia to reduce LDL-C and other lipids. Adverse events to evolocumab are generally nonserious and mild to moderate in severity.
- **Idarucizumab (Praxbind®)** is a humanized mAb Fab fragment of the IgG1 isotype targeted to the thrombin inhibitor dabigatran.
- Idarucizumab binds specifically and with high affinity to the benzamidine-based thrombin inhibitor dabigatran, reversing its anticoagulant effect by displacing the drug from its complex with thrombin.
- Idarucizumab is indicated when reversal of the anticoagulant effects of dabigatran is needed for emergency surgical and/or other urgent procedures and in cases of life-threatening or uncontrolled bleeding. The FDA has issued four warnings: thrombotic risk, bleeding may require an additional dose of idarucizumab, hypersensitivity reactions, and the risk in patients with hereditary fructose intolerance of a serious reaction to the sorbitol excipient present in the mAb formulation.
- **Mepolizumab (Nucala®)** is indicated for add-on maintenance treatment of patients 12 years and older with severe asthma of the eosinophilic type. Mepolizumab binds to IL-5, blocking its interaction with the alfa chain of its receptor on the eosinophil surface and thus inhibiting IL-5 signaling. IL-5 is a cytokine affecting the growth, differentiation, activation, and survival of eosinophils, one of the cell types involved in inflammation including inflammation associated with asthma.
- Warnings and precautions issued for mepolizumab concern the possibility of hypersensitivity reactions, the need to not discontinue corticosteroids abruptly upon initiation of therapy, the reminder that mepolizumab should not be used to treat acute bronchospasm or status asthmaticus and warnings related to possible infections of Herpes zoster and helminth infections.
- Three mAbs, **obiltoxaximab, ixekizumab, and reslizumab**, were approved by the FDA in late March, 2016. **Obiltoxaximab** (Anthim®), directed against the PA of *B. anthracis*, is indicated for treatment and prophylaxis of inhalational anthrax. Hypersensitivity, including anaphylaxis, constitutes a boxed warning for this mAb. **Ixekizumab** (Taltz®), indicated for plaque psoriasis, binds IL-17A, inhibiting the release of proinflammatory cytokines and chemokines. Warnings cover the possibility of infections including tuberculosis, hypersensitivity, and inflammatory bowel disease. **Reslizumab** (Cinqair®), indicated for severe asthma, binds IL-5, blocking interaction with its receptor on the eosinophil surface and reducing cell survival and inflammation. Reslizumab is subject to a boxed warning for anaphylaxis.

Further Reading

Benson JM, Peritt D, Scallon BJ, et al. Discovery and mechanism of ustekinumab a human monoclonal antibody targeting interleukin-12 and interleukin-23 for treatment of immune-mediated disorders. mAbs. 2011;3:535–45.

Campbell ID, Humphries MJ. Integrin structure, activation, and interactions. Cold Spring Harb Perspect Biol. 2011;3:a004994.

Chang J-H, Garg NK, Lunde E, et al. Corneal neovascularization: an anti-VEGF therapy review. Surv Ophthalmol. 2012;57:415–29.

de Gannes GC, Ghoreishi M, Pope J, et al. Psoriasis and pustular dermatitis triggered by TNF-α inhibitors in patients with rheumatologic conditions. Arch Dermatol. 2007;143:223–31.

Dhaliwal B, Yuan D, Pang MOY, et al. Crystal structure of IgE bound to its B-cell receptor CD23 reveals a mechanism of reciprocal allosteric inhibition with high affinity receptor FcεRI. Proc Natl Acad Sci U S A. 2012;109:12686–91.

Du J, Yang H, Zhang D, et al. Structural basis for the blockage of IL-2 signaling by therapeutic antibody basiliximab. J Immunol. 2010;184:1361–8.

Eikelboom JW, Quinlan DJ, van Ryn J, et al. Idarucizumab: The antidote for reversal of dabigatran. Circulation. 2015;132:2412–22.

Elloso MM, Gomez-Angelats M, Fourie AM. Targeting the Th17 pathway in psoriasis. J Leukoc Biol. 2012;92:1187–97.

Ferenczy MW, Marshall LJ, Nelson CDS, et al. Molecular biology, epidemiology, and pathogenesis of progressive multifocal leukoencephalopathy, the JC virus-induced demyelinating disease of the human brain. Clin Microbiol Rev. 2012;25:471–506.

Ferrante M, Vermeire S, Rutgeerts PJ. Drug safety evaluation of certolizumab pegol. Expert Opin Drug Saf. 2014;13:255–66.

Hilvering B, Xue L, Pavord ID. Evidence for the efficacy and safety of anti-interleukin-5 treatment in the management of refractory eosinophilic asthma. Ther Adv Respir Dis 2015;9:135–45.

Keene DL, Legare C, Taylor E, et al. Monoclonal antibodies and progressive multifocal leukoencephalopathy. Can J Neurol Sci. 2011;38:565–71.

Kummerfeldt CE. Raxibacumab: potential role in the treatment of inhalational anthrax. Infect Drug Resist. 2014;7:101–9.

La Cava A. Targeting the Blys-APRIL signaling pathway in SLE. Clin Immunol. 2013;148:322–7.

Lachmann HJ, Lowe P, Felix SD, et al. In vivo regulation of interleukin 1β in patients with cryopyrin-associated periodic syndromes. J Exp Med. 2009;206:1029–36.

Lambert G, Sjouke B, Choque B, et al. The PCSK9 decade. J Lipid Res. 2012;53:2515–24.

Leonardi C, Matheson R, Zachariae C, et al. Anti-interleukin-17 monoclonal antibody ixekizumab in chronic plaque psoriasis. N Engl J Med 2012;366:1190–9.

Lewiecki EM. Safety and tolerability of denosumab for the treatment of menopausal osteoporosis. Drug Healthc Patient Saf. 2011;3:79–91.

McInnes IB, Mease PJ, Kirkham B, et al. Secukinumab, a human anti-interleukin-17A monoclonal antibody, in patients with psoriatic arthritis (FUTURE 2): a randomised, double-blind, placebo-controlled, phase 3 trial. Lancet. 2015;386:1137–46. doi:10.1016/S0140-6736(15)61134-5.

Nesbitt A, Fossati G, Bergin M, et al. Mechanism of action of certolizumab pegol (CDP870): in vitro comparison with other anti-tumor necrosis factor alpha agents. Inflamm Bowel Dis. 2007;13:1323–32.

Nickoloff BJ, Nestle FO. Recent insights into the immunopathogenesis of psoriasis provide new therapeutic opportunities. J Clin Invest. 2004;113:1664–75.

Nordestgaard BG, Chapman MJ, Humphries SE, et al. Familial hypercholesterolemia is underdiagnosed and undertreated in the general population: guidance for clinicians to prevent coronary heart disease. Consensus statement of the European Atherosclerosis Society. Eur Heart J. 2013;34:3478–90.

Onishi RM, Gaffen SL. Interleukin-17 and its target genes: mechanisms of interleukin-17 function in disease. Immunology. 2010;129:311–21.

Resch B. Respiratory syncytial virus infection in high-risk infants—an update on palivizumab prophylaxis. Open Microbiol J. 2014;8:71–7.

Robinson KA, Odelola OA, Saldanha IJ, et al. Palivizumab for prophylaxis against respiratory syncytial virus infection in children with cystic fibrosis. Cochrane Database Syst Rev. 2013;5:CD007743. doi:10.1002/14651858.CD007743.

Robinson JG, Farnier M, Krempf M, et al. Efficacy and safety of alirocumab in reducing lipids and cardiovascular events. N Engl J Med. 2015;372:1489–99.

Rother RP, Rollins SA, Mojcik CF, et al. Discovery and development of the complement inhibitor eculizumab for the treatment of paroxysmal nocturnal hemoglobinuria. Nat Biotechnol. 2007;25:1256–64.

Scholz JL, Crowley JE, Tomayko MM, et al. BLys inhibition eliminates primary B cells but leaves natural and acquired humoral immunity intact. Proc Natl Acad Sci U S A. 2008;105:15517–22.

Shealy D, Cai A, Staquet K, et al. Characterization of golimumab, a human monoclonal antibody specific for human tumor necrosis factor α. mAbs. 2010;2:428–39.

Simon DI, Xu H, Ortlepp S, et al. 7E3 monoclonal antibody directed against the platelet glycoprotein IIb/IIIa cross-reacts with the leukocyte integrin Mac-1 and blocks adhesion to fibrinogen and ICAM-1. Arterioscler Thromb Vasc Biol. 1997;17:528–35.

Soler D, Chapman T, Yang L-L, et al. The binding specificity and selective antagonism of vedolizumab, an anti-α4β7 integrin therapeutic antibody in development for inflammatory bowel disease. J Pharmacol Exp Ther. 2009;330:864–75.

Stewart MW. The expanding role of vascular endothelial growth factor inhibitors in ophthalmology. Mayo Clin Proc. 2012;87:77–88.

Stohl W, Hiepe F, Latinis KM, et al. Belimumab reduces autoantibodies, normalizes low complement levels, and reduces select B cell populations in patients with systemic lupus erythematosus. Arthritis Rheum. 2012;64:2328–37.

Tanaka T, Narazaki M, Kishimoto T. Therapeutic targeting of the interleukin-6 receptor. Annu Rev Pharmacol Toxicol. 2012;52:199–219.

van Schouwenburg PA, Rispens T, Wolbink GJ. Immunogenicity of anti-TNF biologic therapies for rheumatoid arthritis. Nat Rev Rheumatol. 2013;9:164–72.

Vichyanond P. Omalizumab in allergic diseases, a recent review. Asian Pac J Allergy Immunol. 2022;29:209–19.

Walsh GM. Mepolizumab-based therapy in asthma: an update. Curr Opin Allergy Clin Immunol. 2015;15:392–6.

Chapter 5
Cytokines

Introduction

General Characteristics

Cytokines, currently known to be more than 130 in number, are relatively small signaling proteins of MW < 30 kDa, usually glycosylated, and produced by a variety of different cells including those of the immune system, epithelia, endothelia, and stroma. Cytokines are key modulators of the immune and inflammatory responses functioning in an autocrine, paracrine, or endocrine manner stimulating or suppressing cellular activities in infection, innate and adaptive immunity, autoimmunity, inflammation, and malignancy. Key to an understanding of these regulatory proteins is the recognition of their pleiotropism and sometimes overlapping activities, functional redundancies, and side effects. Their secretion may be induced by an array of different stimuli associated with infection, inflammation, or tumorigenesis, first releasing waves of (for example) proinflammatory molecules followed by anti-inflammatory cytokines to restore homeostasis. Cytokines therefore induce a diverse range of biological responses including proliferation, differentiation, activation, inflammation, chemotaxis, and cell death and the nature of an immune or inflammatory stimulus determines whether an immune response is humoral- or cell-mediated, cytotoxic, immunosuppressive or allergic.

Classification of Cytokines

The many attempts to classify cytokines over the last three decades and the complexities in devising classifications based on structural and/or functional parameters are not hard to understand given the sheer number of imprecisely defined "factors" identified in the early years and the difficulties and work involved in trying to accumulate details on functions and diseases. In their discussion of the evolution of cytokine biology and

B.A. Baldo, *Safety of Biologics Therapy*, DOI 10.1007/978-3-319-30472-4_5

nomenclature, Steinke and Borish draw attention to three phases of development in the identification and classification of cytokines. The identification of cytokines by their biologic activities (e.g., T-cell growth) occurred in the first, or factor, stage. The production of recombinant cytokines and demonstration of their pleiotropism and redundancy led to much of our current understanding and this can be called the recombinant-cloning or second phase. Currently, we are experiencing the third, or genomic phase, where cytokines are being identified on the basis of homology with known, characterized cytokines. In the more recent progressive assemblages published by Tato and Cua detailing each cytokine's receptor(s), source, targets, major function, and disease association, the first 16 interleukins were grouped in order of their discovery. Many of these interleukins form homodimeric structures and have the γc and/or βc chains in their receptors. More recently discovered interleukins have proven more difficult to classify in relation to their function in health and disease due to the complexity of their heterodimeric ligands and receptors. For example, a homotrimeric motif for ligands and receptors and bidirectional signaling were found to be important features of the TNF family. Many original names are still in use and many of the originally described "factors" share receptors with other interleukins.

Here, we focus on the 23 FDA-approved cytokine products from the CDER-approved Biologic Products list. The cytokine classification presented is based on the Kyoto Encyclopedia of Genes and Genomes with input from Vacchelli et al. (http://dx.doi.org/10.4161/onci.20459). Nine main families are recognized (Table 5.1) with most of the cytokines of interest classified in the hematopoietic growth factor, interferon (IFN), platelet-derived growth factor (PDGF), and transforming growth factor β (TGFβ) families. In the hematopoietin family, approved cytokines manufactured by recombinant DNA technology are aldesleukin (rh-interleukin-2 [IL-2]), oprelvekin (rhIL-11), filgrastim and tbo-filgrastim (rh-granulocyte colony-stimulating factor [G-CSF]), sargramostim (rh-granulocyte macrophage [GM]-CSF), metreleptin (rh-leptin) and rh-erythropoietins, epoetin, and darbepoetin alfa. Anakinra, a recombinant receptor antagonist for IL-1, is a representative of the IL-1 cytokine family; recombinant interferons alfa-1, alfa-2, beta-1, and gamma-1 make up the interferon family; palifermin (rh-keratinocyte growth factor [KGF]) and becaplermin (rhPDGF-BB) are in the PDGF family; and rh-bone morphogenetic protein [BMP]-2 and rhBMP-7 represent the TGFβ family. Chemokines, placed here in group 9 (Table 5.1), behave as regulatory molecules for leukocytes and lymphoid tissue and have an important role in infectious, inflammatory, allergic and autoimmune responses as well as angiogenesis, hematopoiesis, and tumor growth. No members of the chemokine family are yet approved for therapy.

Adverse Effects of Individual Approved Recombinant Cytokine Analogs

A number of the characteristics and properties of cytokines provide an insight into the possibility of adverse effects when these "natural" agents are used therapeutically. These include, in particular, their pleiotropic nature; relatively short

Table 5.1 Family classification of cytokines[a] relevant to this review

Family	Members
Hematopoietin[b]	IL-2; IL-6[c]; IL-11[c,d]; IL-12[e]; G-CSF[c]; GM-CSF; leptin[c,f]; EPO[g]; TPO[g]; SCF
IL-1	IL-1α; IL-1β; IL-18[h]
IL-10[i]	IL-10[j]
IL-17[k]	IL-17; IL-17B; IL-17C; IL-17D; IL-17E; IL-17F
Interferon[l]	IFNα-1; IFN α-2; IFNβ-1; IFNγ-1
PDGF[m]	EGF; KGF; M-CSF; PDGFA-D; PGF; VEGFA-D
TGFβ	BMP-2[n]; BMP-7[o]; TGFβ1[p]; TGFβ2[p]; TGFβ3[p]
TNF	TNF; TNFSF4[q]; TNFSF5[r]; TNFSF6[s]; TNFSF10[t], TNFSF11[u]; TNFSF12[v]
Chemokines	CC subfamily; CXC subfamily; C subfamily; CX3C subfamily[w]

BMP bone morphogenetic protein, *EGF* epidermal growth factor, *EPO* erythropoietin, *G-CSF* granulocyte colony-stimulating factor, *GM-CSF* granulocyte macrophage colony-stimulating factor, *IFN* interferon, *IL* interleukin, *KGF* keratino-cyte growth factor, *M-CSF* macrophage colony-stimulating factor, *PDGF* platelet-derived growth factor, *PGF* placenta growth factor, *SCF* stem cell factor, *TGFβ* transforming growth factor β, *TNF* tumor necrosis factor, *TNFSF* tumor necrosis factor ligand superfamily member, *TPO* thrombopoietin, *VEGF* vascular endothelial growth factor

From Baldo BA. Side effects of cytokines approved for therapy. Drug Saf 2014;37:921–43. Adapted and reproduced with permission from Springer Science+Business Media

[a]Based on the Kyoto Encyclopedia of Genes and Genomes, www.genome.jp/kegg/ and Vacchelli et al. OncoImmunology 2012;1:493–506. http://dx.doi.org/10.4161/onci.20459

[b]Class I cytokines

[c]Member of IL-6 receptor subfamily that also includes IL-11, G-CSF, and leptin. IL-6 involved in cytokine storm reactions

[d]Also called AGIF, adipogenesis inhibitory factor. Promotes platelet recovery after chemotherapy-induced thrombocytopenia

[e]Promotes Th1 responses and stimulates production of IFNγ and TNF from T and NK cells

[f]Homologous in structure to a cytokine. Included here according to Vacchelli et al. (see above text) but often described as a hormone. Produced primarily in adipose tissue; regulates fat storage

[g]Member of single chain subfamily

[h]Proinflammatory but suppresses metastasis surveillance by NK cells

[i]Class II cytokines. Interferons sometimes classified in this family. Family also includes IL-19, -20, -22, -24, and -26

[j]Anti-inflammatory and immunosuppressive

[k]Proinflammatory cytokines; stimulate release of other cytokines, for example, IL-1β, IL-6, GM-CSF, TGFβ, and TNF

[l]Class II cytokines. Comprise three types: type I (IFNα, IFNβ, IFNω1, IFNκ1, and FNτ1), type II (IFNγ), type III (IL-28A, -28B, and -29)

[m]PDGFs, PGF, and VEGFs belong to subclass I of cysteine-knot growth factors. M-CSF is included in the 4-helix bundle growth factors

[n]BMP2 subfamily

[o]BMP5 subfamily

[p]Member of TGFβ subfamily

[q]Also called OX40L and CD252, the ligand for CD134. Expressed on the surface of activated B, T, dendritic and endothelial cells

[r]Also called CD40L and CD154. Costimulatory molecule with T-cell receptor in activation of antigen presenting cells

[s]Also called FASL or Fas ligand. Binding with its receptor induces apoptosis

(continued)

Table 5.1 (continued)

[t]Also called CD253 or TRAIL,TNF-related apoptosis-inducing ligand
[u]Also called RANKL, receptor activator of nuclear factor kappa-β ligand
[v]Also called TWEAK, TNF-related weak inducer of apoptosis
[w]Small peptides divided into four subfamilies on the basis of a cysteine motif

half-lives; the presence of other cytokines; their capacity to release other cytokines producing a cytokine "cocktail"; and the existence of multiple receptors on different cells that bind the same cytokine with different affinities. Overall, and as one might expect with biological systems involving genetically diverse patients; the diverse range of biological activities of cytokines; their action in causing the release of additional cytokines; the knock-on pharmacological effects of these secondarily released agents; and different disease statuses of patients, side effects of cytokines are not unusual, are to be expected, and patient-to-patient spectra of these effects will be variable.

For the common side effects of cytokines used as therapeutic agents, as well as for the less common but important hematologic, psychiatric, endocrine, neurologic, pulmonary, and dermatologic adverse effects, space constraints and the many hundreds of relevant studies do not always allow individual consideration of the many pertinent report. Instead, general summaries and one or more selected examples or studies that are particularly germane are provided.

Individual Approved Cytokines

The main physicochemical features, FDA-approved indications, modes of action and side effects, as well as warnings, are summarized for the cytokine preparations approved by the FDA CDER (Table 5.2). They will now be considered individually.

Interferon Alfa

Interferons are a class of broad-spectrum antiviral cytokines, seven of which occur not only in humans and which have overlapping, but also in some individual, activities. They can be divided into three classes, designated types I, II, and III. Of most interest for therapy are interferons alfa, beta, and gamma. The former two, classified as type I interferons, bind to the interferon alfa receptor (IFNAR) consisting of IFNAR1 and IFNAR2 chains; interferon gamma, a type II interferon, binds the interferon gamma receptor (IFNGR) consisting of IFNGR1 and IFNGR2.

It is said that virtually all patients treated with interferon alfa experience some adverse effect(s) at some time during therapy. In fact, the literature on side effects to interferons is voluminous and probably greater than all the other approved, non-mAb biologics literatures put together. Three interferon alfa preparations are

Table 5.2 Cytokines approved for human therapy[a]: properties, approved indications[a], mechanisms, and side effects

Generic and trade names	Properties	Approved indications[a]	Mechanism(s) of action relevant to indications	Warnings and side effects, serious and common
Peginterferon alfa-2a[b] (Pegasys®)	Covalent complex of recombinant interferon alfa-2a 127 amino acids MW ~ 20 kDa with PEG linked by an amide bond to lysine	Chronic hepatitis C[c,d]; chronic hepatitis B[c] (HBeAg[e] ± patients)	Not fully known. [f] IFNα binds to its receptor, activating JAK1 and Tyk2 which phosphorylate receptors which bind STAT1 and STAT2. These combine with IRF-9[g] leading to expression of multiple interferon stimulated genes. Type I IFNs have antiviral and proliferative effects and modulate immune responses but their relative potencies differ. IFNα binds IFN receptors less stably than IFNβ	*Boxed warnings*: Neuropsychiatric, autoimmune, ischemic, and infectious disorders[h] and ribavirin-associated effects[i]; other effects: fatigue/asthenia; pyrexia; headache; myalgia; cytopenias; autoimmunity; infection; colitis; pulmonary, CV, and cutaneous disorders
Interferon alfa-2b (Intron A®)	Recombinant protein MW ~19 kDa 165 amino acids with Arg 23; similar to leukocyte IFN	Chronic hepatitis B and C; MM; HCL; A-RKS; FL; condylomata acuminata		*Boxed warnings*: Neuropsychiatric, autoimmune, ischemic, and infectious disorders.[h] Other effects: Flu-like symptoms of fever, fatigue, chills, headache, and myalgia; neutropenia; less common/PM period[i]
Peginterferon alfa-2b (Pegintron®)	Recombinant protein linked to PEG	Chronic hepatitis C with or without ribavirin		*Boxed warnings*: Neuropsychiatric, autoimmune, ischemic, and infectious disorders[h] and ribavirin-associated effects[j]; Other effects: fatigue/asthenia; fever; nausea; rigor; myalgia; and less common/PM period[k]
(Sylatron®)	Recombinant protein linked to PEG	Adjuvant treatment of melanoma		*Boxed warnings*: Depression and other neuropsychiatric disorders. Other effects: as above plus ↑ ALT and AST and less common/PM period[l]

(continued)

Table 5.2 (continued)

Generic and trade names	Properties	Approved indications[a]	Mechanism(s) of action relevant to indications	Warnings and side effects, serious and common
Interferon beta-1a (Avonex®; Rebif®) Peginterferon beta-1a (Plegridy®)	Recombinant 166 amino acid glycoprotein MW 22.5 kDa; amino acid sequence identical to natural protein	Relapsing forms of multiple sclerosis	Not fully understood. IFNβ binds to receptor leading to complex events including ↑ expression of anti-inflammatory agents and ↓ proinflammatory cytokines; gene products and markers include 2', 5 oligoadenylate synthetase, neopterin; CD56 killer cells increase	Warnings and precautions: Depression; ↓ blood count; hepatic injury; anaphylaxis; AI disorders; seizures; monitor patients with CHF. Other effects: ISR; flu-like symptoms—chills, fever, myalgia; asthenia; depression; immunogenicity; anaphylaxis; and pruritus; rash
Interferon beta-1b (Betaseron®; Extavia®)	Recombinant 165 amino acid protein MW 18.5 kDa; gene contains ser for cys at position 17	Relapsing forms of multiple sclerosis		Flu-like symptoms; lymphopenia, leukopenia and neutropenia; ISR; myalgia; depression; hypertonia; abdominal pain; asthenia; rash; ↑ liver enzymes; immunogenicity; and anaphylaxis
Interferon gamma-1b (Actimmune®)	Recombinant 140 amino acid polypeptide; noncovalent dimer of two identical 16.465 kDa monomers of 6 α-helices	Chronic granulomatous disease; malignant osteopetrosis	Interacts with heterodimeric receptor IFNγR1 and IFNγR2 activating JAK-STAT pathways and altering transcription of up to 30 genes	Most common: flu-like symptoms—fever; headache, chills, fatigue; ISR; rash; and diarrhea. Other effects: neutropenia; thrombocytopenia; hepatotoxicity; CV, pulmonary, CNS and GI events; and pulmonary toxicity
Filgrastim (Neupogen®; Nivestim®; Zarzio®) Pegfilgrastim[m] (Neulasta®)	Recombinant hu-G-CSF; 175 amino acid MW 18.8 kDa nonglycosylated protein; differs from natural by an *N*-terminal methionine	Cancer patients receiving: chemotherapy for AML, myelosuppression or BMT; patients with chronic neutropenia or undergoing PBPCCT	Acts via G-CSF receptors on progenitor cells of neutrophil-granulocyte linage. Enhances phagocytosis, chemotaxis, cytotoxicity of mature neutrophils.[n] Signals via JAK/STAT, Ras/MAPK, PI3K/PKB	Warnings: Splenic rupture and sickle cell crisis. Other effects: nausea/vomiting; fever; bone pain; hypersensitivity; ARDS; ISR; alveolar hemorrhage; immunogenicity; osteoporosis; rash; cutaneous vasculitis; and Sweet's syndrome
Tbo-filgrastim (Granix®; Tevagrastim®)	Recombinant biosimilar nonglycosylated G-CSF expressed in *E. coli*. Formulated for short action	Severe neutropenia in patients with nonmyeloid malignancies receiving myelosuppressive anticancer drugs	As for filgrastim	Warnings: Splenic rupture; ARDS; allergic reactions; and sickle cell crisis. Other effects: bone pain; nausea/vomiting; fever; diarrhea; immunogenicity; cutaneous vasculitis; and Sweet's syndrome

Generic and trade names	Properties	Approved indications[a]	Mechanism(s) of action relevant to indications	Warnings and side effects, serious and common
Sargramostim (Leukine®)	Recombinant hu-GM-CSF, three molecule species MWs 19.5, 16.8 15.5 kDa 127 amino acids; leu23 differs from natural factor	Patients receiving: chemotherapy for AML, BMT; or undergoing PBPCCT; myeloid recovery in NHL, ALL, and BMT	Induces progenitor cells to proliferate → neutrophils, monocytes, and macrophages and enhances neutrophil function via specific receptors and signaling through JAK2, STAT, MAPK, and PI3K pathways and transcriptional changes	Warnings and precautions: Fluid retention; respiratory, CV, renal, and hepatic symptoms. Other effects: fever; headache; nausea/vomiting; myalgia; malaise; anorexia; bone pain; diarrhea; alopecia; stomatitis; rash; and Sweet's syndrome
Oprelvekin (Neumega®)	Recombinant IL-11, nonglycosylated 177 amino acids MW 19 kDa; lacks *N*-terminal proline of 178 amino acid natural IL-11	Prevention of thrombocytopenia and reduction of need for platelet transfusion after myelosuppressive chemotherapy	Stimulates megakaryocytopoiesis and thrombopoiesis → ↑platelet production. Binds to IL-11Rα; gp130 activates JAK which phosphorylates tyr on gp130	*Boxed warning*: Allergic reactions including anaphylaxis. Warnings: fluid retention[p]; dilutional anemia; CV events[q]; papilledema; and stroke. Other effects: nausea; vomiting; asthenia; abdominal and bone pain; myalgia; anorexia; chills; and alopecia
Becaplermin (Regranex®)[r]	Recombinant PDGF[s] MW ~ 25 kDa; homodimer of two identical peptide chains of 109 amino acids -S-S-joined at cys43 and 52	Treatment of diabetic neuropathic ulcers that extend into subcutaneous tissue[t]	Binds to and activates PDGF receptors by dimerization and autophosphorylation binding SH_2 sites and activating signal pathways	*Boxed warning*: Increased rate of mortality secondary to malignancy.[u] Other effects: erythematous skin rash; burning at application site; infection; URTI; skin ulceration; cellulitis; osteomyelitis; skin hypertrophy; and bullous eruption
Palifermin (Kepivance®)	Truncated recombinant human KGF[v] 140 amino acids, nonglycosylated, MW 16.3 kDa	Severe oral mucositis in patients with hematologic malignancies	Binds to fibroblast growth factor receptor activating Ras-MAPK signaling and transcriptional activation of cell growth and survival	*Warning*: Potential for stimulation of tumor growth. Other effects: fever; dysesthesia; tongue discoloration/thickening; arthralgias; ↑serum amylase; edema; rash; erythema; hand–foot syndrome; and pruritus

(continued)

Table 5.2 (continued)

Generic and trade names	Properties	Approved indications[a]	Mechanism(s) of action relevant to indications	Warnings and side effects, serious and common
Aldesleukin (Proleukin®)	Recombinant analog of human IL-2 MW 15.3 kDa; unlike IL-2, not glycosylated, ser for cys at position 125 and no *N*-terminal ala	Metastatic renal cell carcinoma; metastatic melanoma	Binds to IL-2 receptor → heterodimerization of IL-2Rβ and -2Rγ → activation JAK3; phosphorylation of tyr on IL-2Rβ → activated receptor, signaling molecules and T cell stimulation	*Boxed warning*: Restrict to patients with normal cardiac and pulmonary functions; administer in hospital with ICU facility and specialists; CLS[w]; impaired neutrophil function[x]; and withhold in cases of lethargy and somnolence. Other effects: chills; diarrhea; hypotension; oliguria; thrombocytopenia; erythema; and rash
Anakinra (Kineret®)	Recombinant receptor antagonist for IL-1 (IL-1RA), 153 amino acids MW 17.3 kDa; has met added to amino terminal	Rheumatoid arthritis; cryopyrin-associated periodic syndrome	Binds to IL-1RI receptor blocking activity of IL-1α and β and acting as a biological response modifier, for example, for cartilage degradation and bone resorption	ISR; worsening rheumatoid arthritis; upper respiratory and other infections; headache; nausea; diarrhea; flu-like symptoms; arthralgia; abdominal pain; hypersensitivity (including anaphylaxis and angioedema); and sinusitis
Epoetin alfa (Epogen®; Procrit®; Eprex®; Erypo®)	Recombinant human erythropoietin; glycoprotein, 165 amino acids (identical to natural product) MW 30.4 kDa	Treatment of anemia due to: chronic kidney disease; zidovudine in HIV patients; effects of chemotherapy; reduction of allogeneic red blood cells in surgery	Binds receptors on erythroid progenitor cells triggering conformational change, activation of JAK2 by transphosphorylation, Src signaling, STAT regulation of genes for cell division and differentiation	*Boxed warning*: ESAs increase the risk of death, myocardial infarction, stroke, venous thromboembolism, thrombosis of vascular access and tumor progression or recurrence. Other effects: pyrexia; arthralgias; nausea; hypersensitivity; headache; cough; ISR; hypertension; rash; pruritus; stomatitis; myalgia; and pure red cell aplasia[y]
Darbepoetin alfa (Aranesp®)	Recombinant human erythropoietin, 165 amino acids, MW ~ 37 kDa; two amino acids substituted to enhance glycosylation	Treatment of anemia due to: chronic kidney disease; effects of concomitant myelosuppressive chemotherapy	As for epoetin alfa	*Boxed warning*: As for epoetin alfa. Other effects: hypertension; dyspnea; peripheral edema; cough; abdominal pain; pure red cell aplasia; thrombovascular events; seizures; hypersensitivity (including anaphylaxis, angioedema, and bronchospasm); and rash/erythema

Generic and trade names	Properties	Approved indications[a]	Mechanism(s) of action relevant to indications	Warnings and side effects, serious and common
Bone morphogenetic protein 2 (InFUSE® Bone Graft/ LT-Cage®)[z]	Recombinant human BMP-2 (rhBMP-2 and dibotermin alfa); a disulfide-linked homodimer; glycosylated subunits 114 and 131 amino acids	Spinal fusion procedures in patients with degenerative disc disease	BMP binds to ser/thr kinase types I and II receptors forming activated complexes. SMAD proteins, part of type I receptors, relay BMP signal to target genes in the nucleus. This in turn induces transcription of osteogenic genes leading to cell proliferation and differentiation	Erythema; swelling over implant site; immunogenicity; ectopic/heterotopic ossification; myositis ossificans; wound-related complications; osteolysis; infections; radiculitis; compression of airways after spine fusion; urogenital events; retrograde ejaculation; and allergy
Bone morphogenetic protein 7 (OP-1 Putty[27], OP-1 Implant[aa]; Opgenra®[ab]; Osigraft®[ab])	Recombinant human BMP-7 (rhBMP-7; OP-1; and eptotermin alfa). 30 k Da homodimeric glycoprotein produced by CHO cells; two 139 amino acid peptides correspond to 293–431 of full length BMP-7[ac]	Opgenra: posterolateral lumbar spinal fusion with spondylolisthesis and failed autograft Osigraft: tibial nonunions of at least 9 months		
Metreleptin[ad] (Myalept®)	Recombinant analog of leptin, 147 amino acids, nonglycosylated, MW 16.14 kDa; 1 more met than leptin at NH_2 terminal; 1 -S-S- at cys97-cys147	Complications of leptin deficiency in patients with congenital and acquired generalized lipodystrophy	Binds and alters conformation of homodimer receptor[ae] activating JAK2 which phosphorylates other tyr residues within receptor JAK2 complex to mediate downstream signaling	*Boxed warning*: Antimetreleptin antibodies with neutralizing activity worsening metabolic control and/or infection; T-cell lymphoma. *Warnings*: hypoglycemia with concomitant insulin/insulin secretagogues; autoimmunity; hypersensitivity; and benzyl alcohol toxicity. Other effect: immunogenicity

(continued)

Table 5.2 (continued)

Generic and trade names	Properties	Approved indications[a]	Mechanism(s) of action relevant to indications	Warnings and side effects, serious and common
Ancestim[af] (Stemgen®)	Recombinant human soluble SCF[ag] with *N*-terminal met, 166 amino acids, 18.5 kDa monomer[ah]	Use with filgrastim for PBPC transplantation to increase number of collected PBPCs	Growth factor stimulating hemopoietic progenitor cells, mast cells and melanocytes. SCF binds to c-Kit receptor[ai] activating signaling pathways PI3K, ras/ERK, Src kinase, and JAK/STAT	Warnings: Need for premedication; history of allergy/asthma; caution with chemotherapy/radiotherapy; and growth factor potential. Other effects: ISR; tachycardia; respiratory symptoms; paresthesia; headache; dizziness; allergic reactions; nausea; rash; and distant skin reactions

AI autoimmune, *ALL* acute lymphoblastic leukemia, *ALT* alanine aminotransferase, *AML* acute myeloid leukemia, *ARDS* acute respiratory distress syndrome, *A-RKS* AIDS-related Kaposi's sarcoma, *AST* aspartate aminotransferase, *BMP* bone morphogenetic protein, *BMT* bone marrow transplantation, *CHF* congestive heart failure, *CHO* Chinese hamster ovary, *CLS* capillary leak syndrome, *CNS* central nervous system, *CV* cardiovascular, *EM* erythema multiforme, *EMA* European Medicines Agency, *ESA* erythropoiesis-stimulating agent, *FDA* US Food and Drug Administration, *FL* follicular lymphoma, *GI* gastrointestinal, *HCL* hairy cell leukemia, *hu-G-CSF* human granulocyte colony-stimulating factor, *hu-GM-CSF* human granulocyte macrophage colony-stimulating factor, *IFN* interferon, *ISR* injection site reactions, *JAK* Janus-activated kinase, *KGF* keratinocyte growth factor, *MAPK* mitogen-activated protein kinase, *MM* malignant melanoma, *NHL* non-Hodgkin lymphoma, *PI3* phosphoinositide 3-kinase, *PBPC* peripheral blood progenitor cell, *PBPCCT* peripheral blood progenitor cell collection therapy, *PDGF* platelet-derived growth factor, *PEG* bis-monomethoxy polyethylene glycol, *PI3K* phosphatidylinositol 3-kinase, *PKB* protein kinase B, *PM* postmarketing, *SJS* Stevens–Johnson syndrome, *SLE* systemic lupus erythematosus, *STAT1*, *STAT2* signal transducer and activator of transcription proteins 1 and 2, *SCF* stem cell factor, *TEN* toxic epidermal necrolysis, *Tyk2* tyrosine kinase 2

*Cytokines (and date) approved by:

FDA—Peginterferon alfa-2a (2002); Interferon alfa-2b (1986); Peginterferon alfa-2b, Pegintron® (2001), Sylatron® (2011); Interferon beta-1a, Avonex® (1996), Rebif® (2002); Interferon beta-1b, Betaseron® (1993), Extavia® (2009); Interferon gamma-1b (1999); Filgrastim, Neupogen® (1991), Nivestim® (2010); Pegfilgrastim (2002); Tbo-filgrastim (2012); Sargramostim (1991); Oprelvekin (1997); Becaplermin (1997); Palifermin (2004); Aldesleukin (1992); Anakinra (2001); Epoetin alfa (1989); Darbepoetin alfa (2001); Bone morphogenetic protein 2 (2002, 2004, and 2007 for different indications); Bone morphogenetic protein 7 (2001); Metreleptin (2014)

EMA—Peginterferon alfa-2a (2002); Interferon alfa-2b (2000); Peginterferon alfa-2b, Pegintron® (2001); Interferon beta-1a, Avonex® (1997), Rebif® (1998); Interferon beta-1b, Betaferon® (1995), Extavia® (2008); Filgrastim, Neupogen® (1991), Nivestim® (2010); Pegfilgrastim (2002); Tbo-filgrastim (2008); Becaplermin (1999); Palifermin (2005); Aldesleukin (2006); Anakinra (2002); Epoetin alfa (2007); Darbepoetin alfa (2001); Bone morphogenetic protein 2 (2002); Bone morphogenetic protein 7 (2001). Metreleptin designated an orphan drug in 2012

From Baldo BA. Side effects of cytokines approved for therapy. Drug Saf 2014;37:921-43. Adapted and reproduced with permission from Springer Science+Business Media

[a]Approved by FDA CDER or EMA or both *

[b]Peginterferon alfa-2a and ribavirin (Copegus®) are indicated for the treatment of adults not previously treated with interferon alfa and with chronic hepatitis C and liver disease. This drug combination is the only FDA-approved regimen for the treatment of chronic hepatitis C infected with both hepatitis C virus and HIV

[c]In adults with compensated liver disease

[d]Combination therapy with ribavirin recommended

[e]HBeAg, hepatitis B "e" antigen circulating in blood when the virus is replicating

[f]All type 1 interferons have antiviral, antiproliferative, and immunomodulatory activities

[g]IFN-regulatory factor 9

[h]Fatal or life threatening

[i]Reactions less commonly seen and/or seen during PM period include nephrotic syndrome; renal insufficiency and failure; pancreatitis, SJS, TEN, injection site necrosis, myositis, immune-mediated disorders including thrombocytopenia

[j]Ribavirin may cause birth defects; avoid pregnancy. It is a potential carcinogen

[k] Less commonly seen and during PM period: thrombocytopenia; cardiac disorders; renal insufficiency and failure; hearing and eye disorders; infections; immune disorders including anaphylaxis, angioedema, urticaria, SJS, TEN, SLE, and EM; nervous system disorders such as peripheral neuropathy and seizures

[l]Less commonly seen and during PM period: CV; endocrinopathies; hepatic failure; retinopathy; ear, eye, pulmonary, and immune (thrombocytopenic purpura, SLE, EM, SJS, and TEN) disorders; pancreatitis; colitis; and psoriasis

[m]Indicated to decrease the incidence of infections in patients with nonmyeloid malignancies receiving myelosuppressive anticancer drugs and associated with febrile neutropenia. Extends half-life to 42 h from ~3.5 h for filgrastim

[n]Mechanisms still poorly understood

[o]Other actions: regulation of intestinal epithelium growth; inhibition of adipogenesis and proinflammatory cytokines; and induction of acute phase protein synthesis (e.g., fibrinogen)

[p]Pulmonary and peripheral edema, dyspnea, and CLS

[q]Arrhythmias, pulmonary edema

[r]Used topically as a gel

[s]PDGF also known as PDGF-BB

[t]Promotes chemotactic recruitment and proliferation of cells for wound healing and formation of granulation tissue

[u]Gel should only be used when benefits are expected to outweigh the risks and used with caution in cancer patients with known malignancy

[v]rhKGF differs from endogenous protein by truncation of the *N*-terminal amino acid to increase stability

[w]CLS results in hypotension, reduced organ perfusion and possibly death and may be associated with cardiac arrhythmias, angina, myocardial infarction, respiratory insufficiency, edema, etc.

(continued)

Table 5.2 (continued)

[x]Reduced chemotaxis therefore treat pre-existing infection prior to aldesleukin therapy; patients with indwelling central lines particularly at risk

[y]Severe anemia with erythrocyte count 1 and 0.5 % mature erythroblasts in bone marrow

[z]InFUSE® Bone Graft consists of rhBMP-2 absorbed to a collagen sponge. The LT-Cage® titanium alloy device is a small, hollow, perforated machined cylinder with one end closed and the other open for addition of the InFUSE® Bone Graft component

[aa]Acquired by Olympus Biotech from Stryker Corp

[ab]Equivalent to OP-1 Putty and Implant preparations, respectively

[ac]Several different recombinant mature forms starting at positions 293, 300, 315, and 316 have been identified

[ad]Leptin, often called a hormone, shows some structural homology to cytokines

[ae]Leptin receptors are members of the IL-6 class I cytokine receptor family

[af]Approved in Australia, New Zealand, and Canada

[ag]Also known as Kit-ligand or steel factor

[ah]Produced in *E. coli*. Amino acid sequence is identical to human sequence except for *N*-terminal methionine. Normally exists as noncovalently linked dimer

[ai]CD117, a receptor tyrosine kinase expressed on hematopoietic and germ cells, mast cells, and melanocytes

in the CDER Biologic Products List. Peginterferon alfa-2a together with the guanosine analog and nucleoside inhibitor, ribavirin (1-β-D-ribofuranosyl-1*H*-1,2,4-triazole-3-carboxyamide Copegus®; Fig. 5.1) are indicated for the treatment of chronic hepatitis C in adults who have compensated liver disease and were not previously treated with interferon alfa. This drug combination is also the approved treatment of patients infected with hepatitis C and HIV and peginterferon alfa-2a alone is approved for the treatment of patients with chronic hepatitis B who have compensated liver disease, viral replication, and liver inflammation. Interferon alfa-2b is administered extensively for hepatitis B and C as well as several malignancies (Table 5.2). It upregulates the expression of MHC I proteins enhancing activation of CD8+ T cells and cytotoxic lymphocyte-mediated killing as well as inducing synthesis of several other antiviral agents including protein kinase R. Peginterferon alfa-2a and peginterferon alfa-2b are covalent conjugates of the recombinant interferon with a single branched bis-monomethoxy polyethylene glycol (PEG) chain, MW 40 kDa. Pegylation, which is FDA approved, nontoxic and contributes to water solubility, helps to protect the protein from immune recognition, that is, it reduces the immunogenicity and antigenicity and increases the molecule's size thus extending protein half-life and circulatory time and reducing renal clearance. For interferon alfa-2a, adverse events in patients treated with the pegylated form and ribavirin occur with a similar, or significantly less, frequency than those treated with standard interferon/ribavirin. For interferon alfa-2b, a number of adverse events occur more frequently with pegylated interferon/ribavirin. Premedication with acetaminophen is often recommended prior to the first dose of peginterferon alfa-2b; thereafter, premedication is undertaken as needed. In some reports on side effects, especially in the earlier literature, interferon alfa is often not distinguished as alfa-2a or alfa-2b although this can be important as demonstrated by some of the different effects induced by alfa-2a and alfa-2b interferons mentioned below.

Interferon alfa-induced neuropsychiatric disorders, particularly depression, cognitive dysfunction, and mania are well known and have been intensively studied. Other symptoms include altered sleep pattern, anorexia, and fatigue. Of the patients who develop severe depressive symptoms, most occur within the first 3 months of

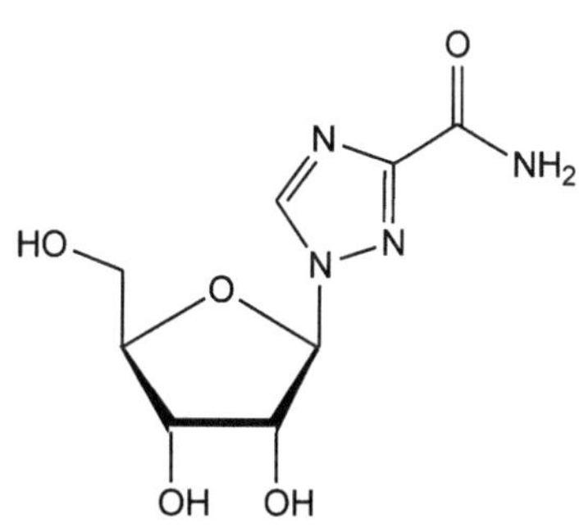

Ribavirin

Fig. 5.1 Structure of ribavirin (1-β-D-ribofuranosyl-1*H*-1,2,4-triazole-3-carboxyamide; Copegus®), the guanosine analog and nucleoside inhibitor which, in combination with interferon alfa, is the mainstay of treatment of hepatitis C infection

treatment and the incidence of depressive disorders has been estimated to be 23–41 %. Symptoms may be prolonged for 6 months or more after the cessation of therapy. There is some evidence that the serotonergic system is involved in the pathophysiologic mechanism although the central opioid, dopamine, and glutamate neurotransmitter systems may also be involved. A positive correlation between depression scores and serum concentrations of soluble ICAM (intracellular adhesion molecule)-1 in patients who received interferon alfa led to the suggestion that the cytokine may induce the adhesion molecule which then increases the permeability of the blood–brain-barrier, allowing the interferon to more easily enter the brain. A number of susceptibility factors have been suggested including a history of depression; high dose of interferon; long treatment duration; female sex; and possession of the apolipoprotein Eε4 allele, said to be associated with some neuropsychiatric disorders.

The appearance of autoantibodies and development or exacerbation of autoimmune diseases are known to occur in response to interferon alfa therapy. In one study, seven cases of autoimmune disease, including one of hypothyroidism, two each of immune-mediated hemolysis and systemic lupus erythematosus, one of Raynaud's disease, and one case of mixed connective tissue disease were identified in 76 patients after a median of 19 months of treatment. Reports of autoimmune reactions to interferon alfa or its combination with ribavirin are not rare and include cases of Hashimoto's thyroid-toxicosis followed by type 1 diabetes, autoimmune thyroiditis, and development and exacerbation of a lupus-like syndrome. See also "Endocrine effects" below.

In addition to their neuropsychiatric and immune effects, interferon alfas occasionally provoke an extensive range of adverse reactions including cardiovascular, respiratory, endocrine, hematologic, metabolic, urinary tract, and skin adverse events as well as adverse effects on the nervous and sensory systems. Cardiovascular complications such as pericarditis and cardiomyopathy with left ventricular dilatation in patients with malignancies improved after withdrawal of the interferon and thereafter treatment with lower doses proved possible. Pegylated interferon alfa-2b has been associated with acute myocardial infarction, pericarditis, pericardial effusion with tamponade, and sick sinus syndrome producing arrhythmias. An orthotopic heart transplant patient died after allograft failure with death attributed to interferon toxicity. Interstitial lung disease, reported for both interferon alfa-2a and 2b, is seen more frequently with the former agent and with high doses of the latter. Potentially fatal interstitial pneumonitis, secondary to interferon alfa-ribavirin therapy for hepatitis C infection, is said to have an incidence of 0.03–0.3 % although an incidence of ~1.1 % was found in 558 Japanese patients. Fatal interstitial pulmonary disease can occur with pegylated interferon alfa-2b as shown by a patient with interstitial pneumonitis who also developing adult respiratory distress syndrome. Cases of bronchiolitis obliterans organizing pneumonia (BOOP), some fatal, are also known.

Interferon alfa may have adverse effects on the nervous system in the form of seizures in patients with no history of epilepsy, involuntary facial movements and weakness, features resembling multiple sclerosis, restless legs syndrome, 17 reports of sensorimotor polyneuropathy, and Bell's palsy. Adverse effects on sensory systems, mainly not only the eyes but also the ears, occur particularly to interferon alfa-2b. Ocular complications include occlusive vasculitis, central retinal artery occlusion, and anterior ischemic optic retinopathy. Twenty seven of 42 patients taking interferon alfa-2b/ribavirin developed a retinopathy: cotton wool spots occured in 27 patients, retinal hemorrhage in six, subconjunctival hemorrhage in two, and optic nerve edema in one patient. Other ocular complications described in patients treated with interferon alfa-2b include permanent loss of sight due to combined retinal artery and central retinal vein obstruction, development of an epiretinal membrane, and the T-cell-mediated autoimmune syndrome, Vogt-Koyanagi-Harada disease.

Endocrine effects of interferon alfa are probably best illustrated by thyroid dysfunction which is not yet fully understood but may have an autoimmune mechanism. Thyroid dysfunction occurs with an incidence of 5–14 % in patients treated for chronic hepatitis C. Hypothyroidism occurs more often than hyperthyroidism and resolution occurs in about 60 % of cases. Interferon alfa-2b can cause both conditions. Although an autoimmune reaction is the most likely mechanism, some patients develop hypothyroidism without autoimmunity. A direct inhibitory effect of thyrocytes has been suggested as the possible mechanism.

Neutropenia induced by interferon alfa is fairly commonly seen while other reported hematologic side effects include acute and autoimmune thrombocytopenia, pernicious anemia, bone marrow hypoplasia which may be immune mediated, and pure red cell aplasia.

A number of acute renal complications in response to interferon alfa have been well documented and include renal thrombotic microangiopathy, acute nephrotic syndrome, hemolytic-uremic syndrome, renal insufficiency due to interstitial nephritis, tubular necrosis, and IgA nephropathy.

The list of cutaneous reactions to interferon alfa is extensive and includes injection site reactions (erythema, necrosis, and vasculitis), pruritus, xerosis, urticaria, hyperpigmentation, psoriasis, alopecia, lichen planus, pityriasis rosea, sarcoid nodules, eosinophilic fasciitis, livedo reticularis, vitiligo, and fixed drug eruption. Interferon alfa is well known for exacerbating pre-existing psoriasis but cases of new onset, and extensive, psoriasis have been reported for both interferon alfa-2a (Fig. 5.2) and interferon alfa-2b. In one example of extensive psoriasis induced by interferon alfa-2b, an adult patient being treated for chronic hepatitis C developed a mild form of psoriasis during the third month of therapy. The condition became worse by the fifth month at which time the patient was hepatitis C virus RNA-negative. Therapy was completed at 6 months and one month later the patient's psoriasis receded spontaneously and completely with no recurrence after 4 years.

The apparent association of vitiligo with interferon therapy for hepatitis C is interesting. Vitiligo is an idiopathic acquired skin disease characterized by loss of skin pigment due to destruction, probably by apoptosis and not necrosis, of melanocytes (Fig. 5.3). Vitiligo often manifests during the first 6 months of interferon therapy, but there is conflicting evidence on the question of a relationship between presence of the virus and the skin response. Although the exact pathogenesis of vitiligo remains unclear, an autoimmune process has been implicated, perhaps with

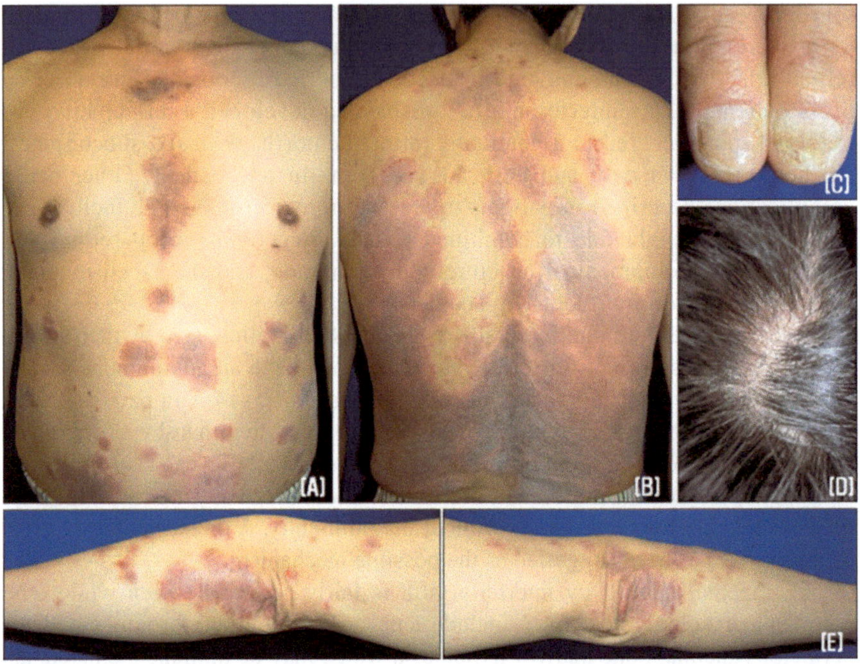

Fig. 5.2 Extensive psoriasis in a patient with no previous history of psoriasis, treated for chronic hepatitis C with interferon alfa-2a and ribavirin. Patient presented with clearly defined erythematous plaques with scales on the chest and abdomen (**a**), the back (**b**), scalp (**d**), and extremities (**e**). The finger nails (**c**) showed signs of pitting and onycholysis. Reproduced from Kim G-W, Jwa S-W, Song M, et al. Ann Dermatol. 2013;25:479–82. doi:10.5021/ad.2013.25.4.479, an open-access article distributed under the terms of the Creative Commons Attribution License)

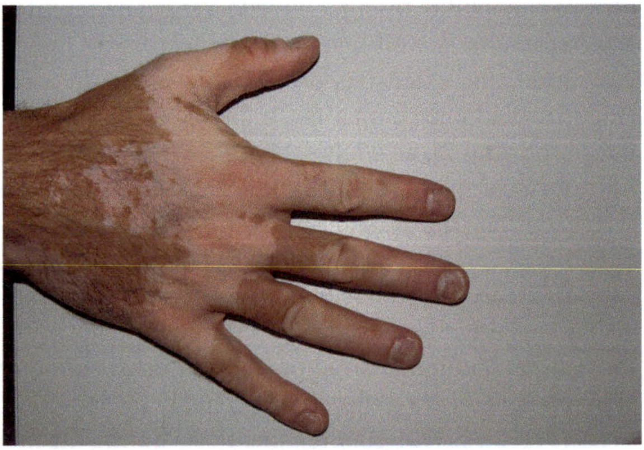

Fig. 5.3 An example of vitiligo in an adult male showing loss of skin pigment of the hand due to destruction of melanocytes. The exact pathogenesis of vitiligo remains unclear (Photograph kindly provided by Dr. R. Spiewak)

the involvement of cytokines such as interferons, IL-2 (section "Aldesleukin"), soluble IL-2 receptor (sIL-2R), IL-10, IL-13, and IL-17A. Recent research on the pathophysiology of the disorder indicates an involvement of cytotoxic T lymphocytes expressing interferon gamma that ultimately leads to melanocyte apoptosis (section "Interferon Gamma").

Interferon Beta

The transcriptional response to interferons beta-1a and beta-1b appear to be indistinguishable, but the biological and clinical responses may vary with the dosage schedules. A flu-like illness is the most commonly occurring adverse event following administration of the interferon beta proteins (Table 5.2) and injection site reactions are common. A comparison of interferon beta-1a, 30 μg, given intramuscularly (im) once per week with interferon beta-1b, 44 μg, subcutaneously (sc) every other day, showed that injection site reactions and antibodies were significantly more frequent in patients given the beta-1b preparation but after 2 years, clinical outcomes to this agent were superior. The questions of the production of neutralizing antibodies to interferon beta and whether they reduce the therapeutic effectiveness in treated patients, especially in the treatment of multiple sclerosis, are important ones. Such antibodies are found in about a quarter of patients treated with sc administered interferon beta-1b and the consensus is that they neutralize or reduce the cytokine's activity. Some believe that this has the potential to significantly reduce the effectiveness of the therapy and it has been suggested that the immunogenic potential of interferon beta should therefore be considered as well as its safety. Other immunologic effects observed are cases of a lupus-like syndrome to both beta interferons and cutaneous lymphocytic vasculitis to sc interferon beta-1b.

Unlike interferon alfa, results from studies do not support an association of interferon beta with depression, but the FDA mention depression, suicide, and psychotic disorders in their warnings and precautions for the cytokine. Interferon beta can induce thyroid disorders notably hyperthyroidism and a severe case of hypothyroidism to interferon beta-1a resembling Hashimoto's encephalopathy has been described. Skin reactions reported include urticaria to interferon beta-1a and an acneiform eruption to interferon beta-1b.

In August 2014 the FDA granted approval for Plegridy®, a pegylated preparation of interferon beta-1a produced as a glycosylated protein in Chinese hamster ovary cells and then covalently attached via the *N*-terminal residue to a linear 20 kDa methoxypolyethylene glycol molecule, giving the complex a total molecular mass of approximately 44 kDa. The amino acid sequence of the recombinant cytokine is identical to its human interferon beta counterpart. The apparent molecular mass of Plegridy® in solution is more than 300 kDa, that is, more than 13-fold increase compared to interferon beta-1a. This ensures a significantly reduced patient clearance of the pegylated preparation. In placebo-controlled clinical studies, the most common adverse reactions to Plegridy® were similar to the nonpegylated form of the cytokine with injection site reactions, an influenza like illness, asthenia, arthralgia, and pruritus seen most commonly. Issued warnings and precautions for the preparation are also similar to the nonpegylated form (Table 5.2). Whereas less than 1 % of patients given Plegridy® every 14 days for 1 year developed neutralizing antibodies, 7 % of treated patients developed antibodies to PEG.

Interferon Gamma

Interferon gamma, structurally distinct from other interferons, is produced predominately by NK (TCR not expressed) and NKT cells and by CD4 and CD8 cytotoxic T lymphocytes in antigen-specific immunity. The cytokine shows a different biological activity spectrum, in particular in its action of differentiating normal and B lymphocytes, and as an immunomodulator of macrophage activity. It also has an important role in dealing with intracellular pathogens, including viruses, and tumor control.

Early phase I studies of the biological activity of, and tolerance to, recombinant interferon gamma showed the common appearance of flu-like symptoms and granulocytopenia. In another early study, a 30 % fall in peripheral blood lymphocytes was seen after 10 days of interferon gamma therapy. The occurrence of fatal acute respiratory failure in four patients treated with interferon gamma-1b for advanced idiopathic pulmonary fibrosis prompted further investigation in the form of a double blind study of the effect of the cytokine in 330 patients with that condition. No significant differences were found in lung function, gas exchange, or quality of life, but the patients experienced more frequent upper respiratory infections and pneumonia. However, acute respiratory insufficiency has been reported in a single patient with idiopathic pulmonary fibrosis 4 months after receiving interferon gamma. Cardiovascular toxicity to interferon gamma, particularly at higher doses, and including hypotension, arrhythmias, coronary vasospasm, and ventricular tachycardia and renal toxicity, namely acute renal failure, nephrotic syndrome, and tubular necrosis, have been recorded.

There appears to be few reports of cutaneous reactions to interferon gamma, but severe erythroderma occurred in 5 of 10 bone marrow transplant patients given the drug. Recent studies have found increased levels of interferon gamma mRNA in skin of patients with vitiligo (Fig. 5.3) and inhibitors of the cytokine have proved to be beneficial treatments in some cases. In an investigation of 50 patients with vitiligo, the frequency of interferon gamma-producing cells in skin and peripheral blood was determined. Significant expansions of CD8+ cytotoxic T lymphocytes expressing interferon gamma were detected and, when examined in vitro, the cytokine directly induced melanocyte apoptosis leading the authors to conclude that the CD8+ cells have a pivotal role in the induction and maintenance of the skin disease.

Colony-Stimulating Factors: Filgrastim, Sargramostim, and Tbo-Filgrastim

CSFs, produced by most tissues and cell types, are glycoprotein cytokines with multiple actions on hematopoietic cells. Described by Metcalf as "the master regulators of granulocyte and macrophage populations," the CSFs are used to treat chemotherapy-induced neutropenia, mobilize stem cells for transplantation, and enhance the immune response to cancer. Despite dissimilarities in amino acid sequences, human granulocyte-colony-stimulating factor (G-CSF) and human granulocyte-macrophage colony-stimulating factor (GM-CSF) show three-dimensional structural similarities (Figs. 5.4 and 5.5) to each other and a number of other signaling proteins, for example, human growth hormone (Chap. 7, section "Human Growth Hormone"), interferon beta (section "Interferon Beta"), IL-2 (section "Aldesleukin"), and IL-4. This conservation of tertiary structure suggests similar binding of the different ligands to their respective receptors. Currently, approved members of the CSF family are filgrastim and pegfilgrastim, both G-CSFs, sargramostim, a GM-CSF, and tbo-filgrastim, a short acting biosimilar (Chap. 13) G-CSF (Table 5.2). The latter is used for severe neutropenia in patients with lung cancer receiving platinum drug chemotherapy. GM-CSF, used as an immunostimulant following bone marrow transplantation and chemotherapy, is also viewed as a potential immunoadjuvant for anticancer vaccines. The E. coli-derived GM-CSF molgramostim, seen as a potential immunostimulant, showed a higher incidence of adverse effects than sargramostim and was not granted FDA approval.

As well as the most common, and usually mild and transient reactions of headache, bone pain, myalgia, fever, flushing, and rash for filgrastim and sargramostim, other more severe, but rare, respiratory, cardiovascular, hematologic, and cutaneous

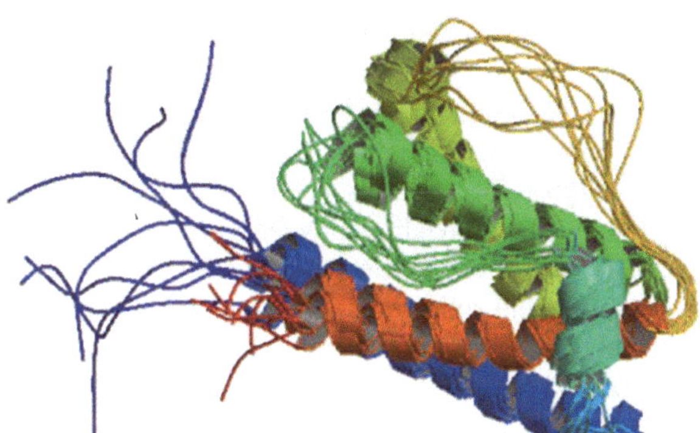

Fig. 5.4 Structure of recombinant human G-CSF (rhu-G-CSF) determined by NMR spectroscopy. Structure is predominately helical with 106 of the 175 amino acids forming a left-handed four-alfa-helix bundle. Helices are composed of helix A, residues 11–41 and helix B, residues 71–95, aligned parallel to each other (*up*, *up*), and helx C, residues 102–125 and helix D, residues 145–170, which are antiparallel (*down*, *down*). The structure is from Protein Data Bank RCSB PDB file 1GNC (Zink T, Ross A, Luers K, et al. Biochemistry. 1994;33:8453–63)

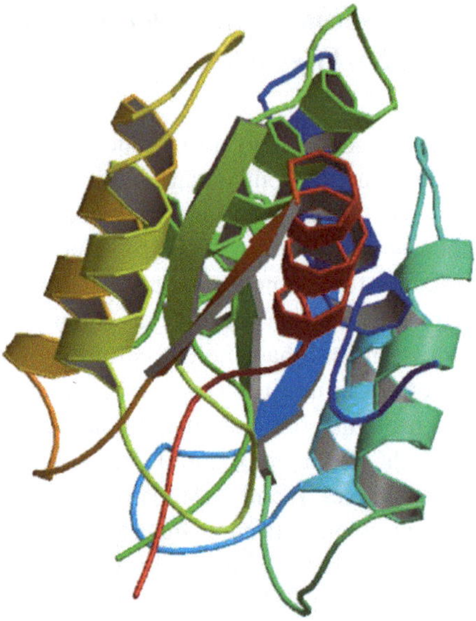

Fig. 5.5 Three-dimensional crystal structure of recombinant GM-CSF (rhu-GM-CSF) determined at 2.8 Å. Like G-CSF, the main structural feature is a four-alfa-helix bundle (helices A, B, C, and D) which makes up nearly half the structure. Helices are in a left-handed antiparallel arrangement with two overhand connections containing a two-stranded antiparallel beta sheet. The structure is from Protein Data Bank RCSB PDB file 1CSG (Walter MR, Cook WJ, Ealick SE, et al. J Mol Biol. 1992;224:1075–85)

reactions occur. Adult respiratory distress syndrome (ARDS) following G-CSF is more likely when a rapid rise in the white cells occurs in patients taking pulmonary toxic drugs, when there is concomitant infection, and in patients with HLA-B51 or HLA-B52 antigens. Other occasional respiratory side effects are pulmonary toxicities, particularly pulmonary edema which has proved fatal, and interstitial pneumonitis. There has been speculation that GM-CSF might contribute to the development of acute coronary syndrome. In fact, cardiovascular complications have been observed. These include fluid retention, pulmonary edema and weight gain, aortitis to molgramostim, and capillary leak syndrome (Chap. 1, section "Capillary Leak Syndrome") following G-CSF which can be severe and even fatal. Recorded hematologic side effects to CSFs consist mainly of a number of cases of thrombocytopenia, some with an immune mechanism, splenomegaly, and splenic rupture (note FDA issued warning, Table 5.2). There is a belief that G-CSF may be a risk for the progression of myelodysplastic syndrome (MDS), but this has not been unequivocally established. MDS has been reported after G-CSF treatment and the incidence of MDS or acute myeloid leukemia (AML) was found to be 11 % in patients treated with G-CSF, but only 5.8 % in patients receiving immunosuppression alone. In another more recent study, patients who received G-CSF showed a 2.5-fold increased risk. Interpretation of results relevant to the alleged risk of G-CSF is not straight forward however. Findings that there is no significant relationship between G-CSF therapy and MDS/AML onset are at odds with the belief that the risk of leukemia in severe congenital neutropenia patients increases with the G-CSF therapy. Two other potentially life-threatening responses to CSFs, both the subject of warnings, are anaphylactic/anaphylactoid reactions and severe adverse events such as acute chest syndrome, vaso-occlusive episodes, multiorgan failure, and death seen in patients with sickle cell disease.

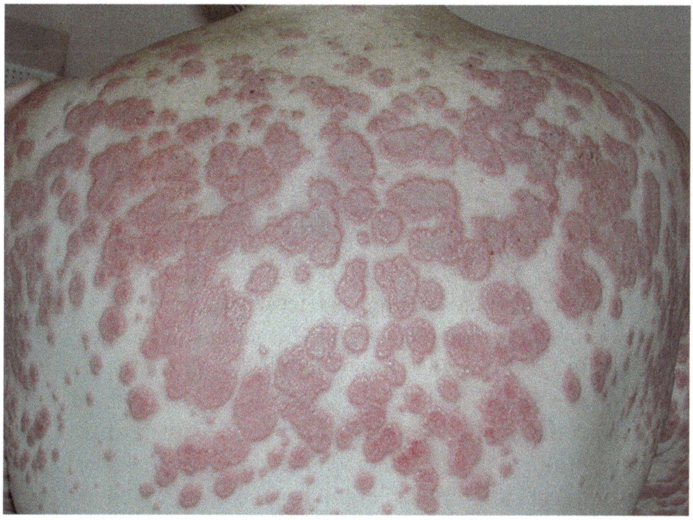

Fig. 5.6 Cutaneous eruption mimicking Sweet's syndrome or acute febrile neutrophilic dermatosis following the administration of filgrastim (recombinant granulocyte colony-stimulating factor [G-CSF]). Sweet's syndrome is characterized by an elevated neutrophil count and the presence of neutrophils in the upper dermis. Reproduced from Aubin F et al. The complexity of adverse side-effects to biological agents. J Crohn's Colitis. 2013;7:257–62. Distributed under a Creative Commons License; reproduced with permission from Elsevier Limited (CCC License number 3662450405678)

There is a long list of adverse skin reactions provoked by CSFs. The most commonly occurring cutaneous reaction is Sweet's syndrome seen after therapy with filgrastim (Fig. 5.6) and sargramostim. In fact, these two colony-stimulating factors are the most frequently implicated drugs in Sweet's syndrome. Other adverse cutaneous events to CSFs include psoriasis flare, pyogenic granulomas, pruritic erythematous maculopapular eruptions, palmoplantar pustulosis, erythema multiforme, and neutrophilic dermatoses.

Oprelvekin

Recombinant human IL-11, or oprelvekin (Table 5.2), is used to prevent chemotherapy-induced thrombocytopenia and reduce the need for platelet transfusions in patients with nonmyeloid malignancies. The most commonly occurring adverse events seen in placebo-controlled studies were edema, dyspnea, tachycardia, palpitations, atrial fibrillation/flutter, pleural effusions, conjunctival injection, and oral moniliasis. Fluid retention and an increase in plasma volume underlie many of the adverse events, for example, edema, dyspnea, pleural effusions, arrhythmia, dilutional anemia, and renal failure and indicate that oprelvekin should be used with caution in patients with congestive heart failure. No evidence of cumulative toxicity or bone marrow exhaustion has been observed after sequential cycles of the cytokine and no proliferative effect on tumors has been noted. Two other clinically important adverse reactions reported are papilledema and periosteal bone formation. An incidence of 3–4 % was found for antioprelvekin antibodies in treated patients.

Becaplermin

Becaplermin is a recombinant human platelet-derived growth factor (PDGF), a homodimer made up of two disulfide-bonded B chains and hence written as rhPDGF-BB. Naturally occurring PDGF has A and B chains in homodimeric or heterodimeric form. The PDGF-A chain binds to the α receptor, whereas the PDGF-B chain binds to both the α and β. rhPDGF-BB promotes the growth of granulation tissue and wound healing via interaction with receptors on fibroblasts (α and β) and endothelial cells (β receptors). Becaplermin has therefore found use in gel form as a topical application for patients with lower extremity diabetic neuropathic ulcers, lesions that are notoriously difficult to heal and a major cause of morbidity (Fig. 5.7).

Growth factors cause cell proliferation so the possibility of increased cancer rates is considered for drugs with a cell growth-promoting property. In a retrospective study by the FDA of a medical claims database, cancer rates and deaths were compared for 1622 becaplermin users and 2809 matched nonusers. The incidence rate ratios of becaplermin to matched controls for all cancers and for mortality from all cancers were 1.2 and 1.8, respectively, and the incidence rates for mortality among patients who received three or more tubes of becaplermin and controls were 3.9 and 0.9 per 1000 patient-years, respectively. The rate ratio for cancer mortality in the patient group receiving 3 or more tubes was 5.2 (95 % CI 1.6–17.60). Following an

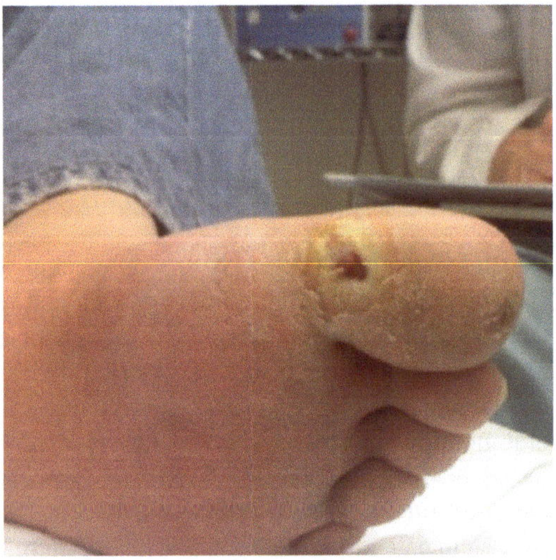

Fig. 5.7 Diabetic pressure-induced foot ulcer. Reproduced from Shapiro J, Koshimune D, Moellmer R. Diabetic foot ulcers—treatment and prevention. In: Masuo K, editor. Type 2 diabetes. InTech; 2013. Ch 12. http://www.intechopen.com/books/type-2-diabetes/diabetic-foot-ulcers-treatment-and--prevention Accessed October 13, 2015. An open-access article distributed under the terms of the Creative Commons Attribution License

earlier safety study in 2001, where more cancers were found in the becaplermin group than a nonuser group, the FDA in 2008 issued a boxed warning for Regranex® Gel stating that "malignancies distant from the site of application have occurred in becaplermin users… and an increased rate of death from systemic malignancies was seen in patients who have received 3 or more tubes." As a consequence, it was stated that "becaplermin should be used with caution in patients with known malignancy" and only used "when the benefits can be expected to outweigh the risks." In 2010, the EMA's Committee for Medicinal Products for Human Use recommended that becaplermin should not be used in patients with a pre-existing cancer but, at the same time, admitted that there was no evidence either way to establish, or rule out, a link between therapeutic use of the cytokine and cancer. In July 2012, the European Commission withdrew the marketing authorization for Regranex®.

In studies on the safety of becaplermin gel in the treatment of neuropathic diabetic foot ulcers, clinical findings showed little difference between the drug and placebo in relation to cardiovascular, respiratory, musculoskeletal, and nervous system disorders. No neutralizing antibodies were detected. Rash was seen in 2 % of patients treated with becaplermin and 1 % receiving placebo. Apart from the possibility of becaplermin-induced cancers and the drug's known side effects listed in Table 5.2, there is a dearth of subsequent studies on the safety of becaplermin, including case reports. This is probably because clinical experience with the agent did not live up to the initial high expectations, the concerns related to cancer, and the high cost of the agent.

rhPDGF-BB, together with beta-tricalcium phosphate, is a component of Augment® Bone Graft, a combination device/drug product developed for bone repair and regeneration. Intended for the treatment of foot and ankle fusions, a major claimed advantage of the product is the elimination both of the need to harvest autologous bone and the associated risks of ongoing pain and infection. Augment® Bone Graft is indicated for use as an alternative to autograft in hindfoot and ankle fusion procedures that require supplemental graft material, for example, in tibiocalcaneal, talonavicular, and calcaneocuboid fusions. In October 2014, the FDA approved the Premarket Approval Application for Augment® Bone Graft subject to a preapproval facilities inspection. Already approved in Canada, the rationale for the product's action is the inclusion of rhPDGF-BB for the promotion of growth and proliferation of osteoblasts and beta-tricalcium phosphate, a resorbable synthetic bone matrix, as the framework for new bone growth. Each component is packaged separately and mixed immediately before use. Already issued warnings and precautions for the product include its as yet unknown effect on fetal development; whether or not it is excreted in milk; its safety at sites other than the ankle and foot; the need for its use on well-vascularized bone; unknown safety of repeat applications; and the product's safety in patients less than 18 years old. In the primary clinical study, Augment® Bone Graft was compared to autologous bone graft as the "gold standard" in a representative foot and ankle fusions clinical model. Safety studies revealed 973 treatment-emergent adverse events with no significant differences between the two treatment groups. Overall, the Augment group showed fewer serious adverse events and fewer complications associated with surgery and infections. In a prospective, open-label, multicenter trial undertaken in Canada and designed to evaluate Augment® Bone Graft, 60 patients requiring hindfoot, midfoot, or ankle fusions, were followed for 36 weeks. A total of 22 adverse events, none serious, was recorded, most arising from the surgery; 15 were general and administration site disorders (swelling, feeling hot, tenderness, and impaired healing), 5 were due to injury and procedural complications, and 2 were musculoskeletal and connective tissue disorders (muscle spasms, and pain).

Palifermin

Palifermin, a recombinant human keratinocyte growth factor produced by mesenchymal cells and fibroblasts, stimulates differentiation, proliferation, and migration of epithelial cells via interaction with its complementary receptors on epithelial cells widely distributed in numerous tissues including skin, hair follicles, tongue, stomach, intestine, lung, liver, kidney, lens of the eye, and many other tissues and organs. The recombinant molecule is a nonglycosylated, 16.3 kDa, 140 amino acid protein belonging to the fibroblast growth factor family that has been genetically modified to increase stability by shortening the natural protein at the *N*-terminal end. Palifermin is an important agent in oncological supportive care, aiding the management of mucositis in cancer patients by protecting the mucosal epithelium and aiding its regeneration after chemotherapy- and radiation-induced injury.

Reported adverse events following palifermin administration in a phase III double-blind, placebo-controlled trial were rash, pruritus, erythema, paresthesia, edema, taste alteration, rhinitis, arthralgia, thickening of the tongue, and numbness. The keratinocyte growth stimulation properties of palifermin may underlie a number of cutaneous reactions seen following its administration. Cases of palmoplantar erythrodysesthesia (acral erythema and hand-foot syndrome), a papulopustular (acne-like) eruption on the head and trunk, hyperpigmented papillomatous plaques in the axillae and inguinal areas, and a case of lichenoid papules have been described. The latter reaction consisted of a cutaneous eruption of planar papules resembling lichen planus, together with erythema, mainly in an intertriginous distribution, and confluent white plaques on the oral mucosa. Being a growth factor, palifermin carries a warning of potential stimulation of tumor growth (Table 5.2).

Aldesleukin

Interleukin-2 (IL-2) is one of the best studied cytokines after its discovery as an activator of T lymphocytes nearly 40 years ago. Because it possesses a wide range of immune effects regulating T cells and immune activation and homeostasis, IL-2 was one of the first cytokines characterized at the molecular level. The recombinant form, called aldesleukin, differs from the natural cytokine by absence of glycan residues and at position 125 and the end terminal amino acid (Table 5.2). X-ray and NMR studies have shown that the IL-2 fold is similar to that seen in the myelopoietic stimulatory factors G-CSF (Fig. 5.4) and GM-CSF (Fig. 5.5), but it has three minor structural differences to the other four-helix bundles (Fig. 5.8).

Aldesleukin has been applied clinically in a number of ways, particularly for melanoma and renal cell carcinoma, and from its earliest applications showed a wide range of the sort of side effects often seen with cytokines including fever, chills, myalgia, nausea, vomiting, diarrhea, hypotension, oliguria, and edema (Table 5.2) plus a number of more severe cardiovascular, hematologic, endocrine, kidney, central nervous system, infectious, and cutaneous toxicities.

Cardiovascular adverse events are the main dose-limiting toxicities of aldesleukin with recorded cases of hypotension, tachycardia, peripheral edema, pleural effusions, myocarditis, myocardial infarction, heart block, arrhythmias, cardiac eosinophilic infiltration, and coronary ischemic changes. An important occasional and serious adverse event of IL-2 therapy is capillary (sometimes called vascular) leak syndrome (Chap. 1, section "Capillary Leak Syndrome") which causes hypovolemia and fluid accumulation in the extravascular spaces and may lead to oliguria, ischemia, and confusion. Aldesleukin therapy can induce increased vascular permeability, interstitial edema, and ultimately organ failure seen as an increase in body weight, fluid retention, peripheral edema, ascites, pleural and pericardial effusions, and ultimately pulmonary and cardiovascular failure. Pulmonary side

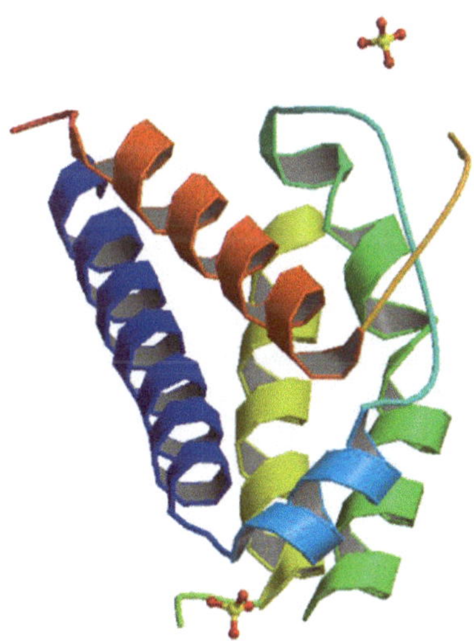

Fig. 5.8 Secondary structure of IL-2 consists of four alfa helices A, B, C, and D, with an up–up and down–down arrangement, a structure similar to G-CSF (see Fig. 5.4) and GM-CSF (Fig. 5.5). IL-2, however, differs in that it has an irregular one turn helix in the AB loop, a distortion in the B helix, and a two-stranded antiparallel beta structure formed from parts of the AB and CD loops. The structure is from Protein Data Bank RCSB PDB file 1M47 (Arkin MA, Randal M, DeLano WL, et al. Proc Natl Acad Sci USA. 2003;100:1603–8)

effects to aldesleukin are usually related to capillary leak syndrome and are more likely, and more severe (e.g., as pulmonary edema and respiratory distress), in patients with existing cardiac problems. Hematologic adverse effects, particularly anemia, leukopenia, and thrombocytopenia occur but are rarely severe or dose limiting. Thrombocytopenia is a common toxicity of high dose IL-2 therapy but rapidly reverses upon cessation of treatment. Eosinophilia may occur in the later stages of therapy accompanied by rash and pruritus. Figure 5.9 shows the hematologic changes, often rapid and dramatic, occurring in response to aldesleukin challenges over a 6 day period. Within hours of the first dose, platelets and lymphocytes fall rapidly to low levels and eosinophil numbers rise slowly, whereas neutrophils and hemoglobin are maintained at stable levels. At the cessation of treatment, lymphocytes and platelets quickly return to normal and, in fact, may exceed their baseline levels by 2- to 5-fold. Eosinophils persist and may do so for many weeks. Endocrine effects usually manifest as hypothyroidism which may affect up to one-third of patients, or as the far less common hyperthyroidism. Renal toxicity, especially oliguria, and gastrointestinal toxicities are also seen, the latter being particularly common in the form of nausea, vomiting, diarrhea, anorexia, gastritis, and mucositis. Gastrointestinal perforation has also been reported. IL-2-induced infectious toxicities may occur at venous catheter sites and in the urinary tract. Such infections, usually due to *Staphylococcus* species, are thought to arise from the known affect of IL-2 on neutrophil chemotaxis. Neurological effects, especially to high doses during IL-2 therapy, include anxiety, depression, altered sleep patterns, somnolence, emotional fragility, vivid dreams, and confusion. The list of aldesleukin-induced cutaneous reactions is extensive, ranging from mild erythema, pruritus, injection site reactions, and vitiligo, to urticaria, angioedema, reactivation

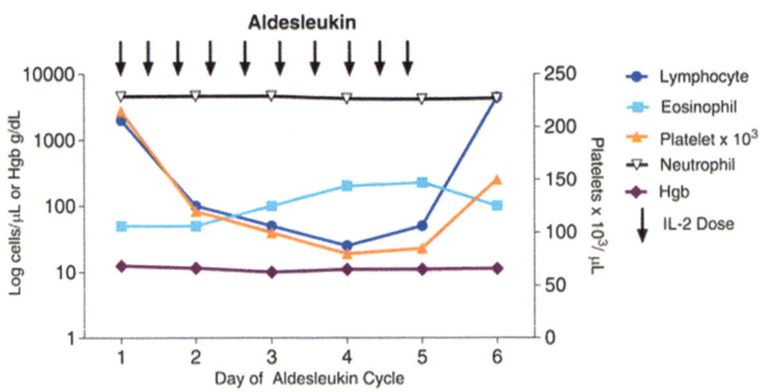

Fig. 5.9 Plot of hematologic responses to a series of IL-2 injections over a 5 day period. Platelet and lymphocyte numbers fell rapidly, the eosinophil count showed a steady but undramatic increase and the neutrophil count remained constant. After the completion of treatment, platelets and lymphocytes recovered quickly. The elevated eosinophil count may persist for weeks. Hemoglobin levels remained constant throughout (Reproduced from Dutcher JP, Schwartzentruber DJ, Kaufman HL, et al. J Immunother Cancer 2014;2:26. doi:10.1186/s40425-014-0026-0, an open-access article distributed under the terms of the Creative Commons Attribution License)

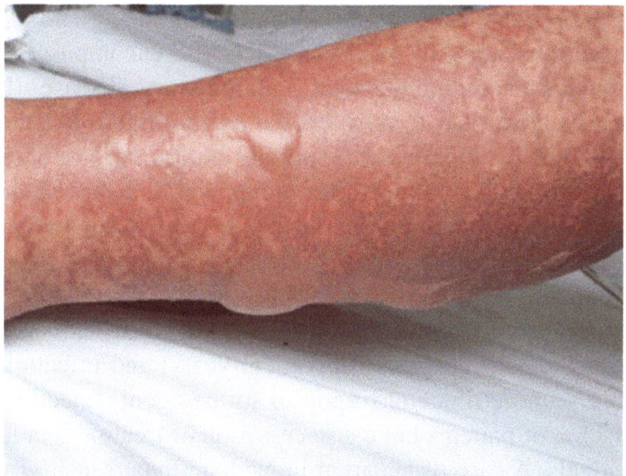

Fig. 5.10 A severe cutaneous reaction of erythema with blisters and bullae resembling acute scalding induced by IL-2 in an immunocompromised adult patient (Reproduced from Zhang J, Meng L-H, Zhang L-T, et al. J Clin Exp Dermatol Res. 2014;5:1. doi:10.4172/2155-9554.1000204, an open-access article distributed under the terms of the Creative Commons Attribution License)

of eczema, exacerbation of psoriasis, generalized erythema followed by desquamation, vasculitis, and severe manifestations like pemphigus, IgA bullous dermatosis, and toxic epidermal necrolysis. An acute blistered scalded skin-like reaction after IL-2 therapy in an immunocompromised patient (Fig. 5.10) suggests that the cytokine should be given with great caution, and on an individualized basis, to immunocompromised patients. A curious case involved the implication of high-dose IL-2 therapy in the occurrence of multifocal fixed drug eruptions after the administration of other drugs, namely, ondansetron, granisetron, acetaminophen (paracetamol), and indomethacin.

Anakinra

Interleukin-1 (IL-1) is a cytokine produced in response to inflammatory stimuli in a number of immunological reactions including rheumatoid arthritis. The receptor for IL-1 (IL-1R), in membrane or soluble form, is widely expressed on tissues and organs and exists as two types, type I which is responsible for the expression of the inflammatory effects of IL-1 and type II which may compete for IL-1 and act as a suppressor of the cytokine. Anakinra is a recombinant-specific receptor antagonist (IL-1RA) for IL-1 differing from the natural receptor by the addition of a single methionine at the amino terminal end (Table 5.2). Anakinra therefore acts as a biological response modifier in the treatment of diseases like rheumatoid arthritis and the spectrum of autoinflammatory syndromes collectively known as cryopyrin-associated periodic syndrome (CAPS) (Chap. 4, sections "CAPS

Diseases and Approved Indications for Canakinumab" and "Canakinumab: Warnings, Precautions, and Adverse Events"). CAPS encompasses Muckle–Wells syndrome, neonatal-onset multisystem inflammatory disease (NOMID), and familial cold-induced urticaria, now known as familial cold autoinflammatory syndrome (FCAS). All symptoms of Schnitzler syndrome, a rare, underdiagnosed systemic disease with many features in common with the above autoinflammatory syndromes including nonitching urticarial-like lesions, fever, and bone/joint pain, can be relieved within hours of the first injection of anakinra. Symptoms recur if treatment is stopped.

The side effects profile of anakinra is not large with two adverse events, injection site reactions (122 events per 100 patient years) and infection episodes, the most commonly seen detrimental responses to the agent. Injection site reactions occur in up to 73 % of patients but cause cessation of treatment in less than 5 % of affected individuals. A case of apparent immediate hypersensitivity was reported in a 25-year-old woman with familial Mediterranean fever who developed urticaria and angioedema on the face and diffuse erythema over the entire body after the 12th subcutaneous daily dose of anakinra (100 mg/day). The reactions responded well to antihistamines and intradermal tests with the drug proved positive, suggesting a type I hypersensitivity response. A desensitization protocol was employed after a premedication dose of 10 mg of cetirizine given 1 h before the first sc injection of anakinra. Six doses, starting at 1.5 mg, were administered at 1 h intervals. Subsequent doses were 3, 5.5, 12.5, 25, and 52.5 mg, totaling a cumulative dose of 100 mg of anakinra. After desensitization, the intradermal test was negative and the patient continued on daily anakinra without problems. A similar cutaneous reaction involving itching, erythema on the face and abdomen, shortness of breath, and abdominal pain was seen 3 h after administration of anakinra. The patient proved skin prick test-positive to the recombinant cytokine, again indicating a type I allergic reaction. Successful desensitization to anakinra has also been reported in a 34-year-old man who developed a delayed local injection site reaction to the protein. Systemic reactions to anakinra are rare, but an anaphylactic reaction occurred in a patient with rheumatoid arthritis and severe systemic symptoms including urticaria, angioedema, and pruritic tongue were reported in a 7-year-old girl with juvenile idiopathic arthritis.

Infections, particularly URTI, involving a wide variety of organisms have been reported, but it has been suggested that the risk of infection is associated with high doses of anakinra and in patients with comorbidities. Septicemia due to *S. aureus*, hemolytic streptococci, and *E. coli* occurred after anakinra was added to prednisolone for rheumatoid arthritis. Anakinra provoked reactivation of pulmonary tuberculosis, adenovirus, gastroenteritis, varicella pneumonitis, and visceral leishmaniasis and acute Epstein–Barr virus infection occurred in juvenile idiopathic arthritis patients treated with the cytokine. A patient with Still's disease given anakinra developed systemic inflammatory response syndrome (SIRS) (Chap. 1, section "Systemic Inflammatory Response Syndrome") together with ARDS and some other patients with this disease had the cytokine withdrawn because of infections or severe skin reactions. Other reported side effects of anakinra include progression of rheumatoid arthritis, exacerbation of Crohn's disease, anaphylaxis with a positive skin test to the cytokine, cellulitis at injection sites, and an interstitial granulomatous reaction which resolved after withdrawal of anakinra and recurred on challenge.

Epoetins

Erythropoietin (EPO) is a heavily glycosylated cytokine with three *N*-linked and one *O*-linked oligosaccharide chains that are important for the protein's biological activity and stability. Activity is also dependent on two disulfide bonds between cysteines 7 and 160 and 29 and 32. In both native human EPO and rhEPO (epoetin alfa), the originally secreted molecule is a 166 amino acid peptide before the carboxy-terminal arginine is removed to give the final active protein of 165 amino acids.

In an early study of rhEPO in anemic patients with end-stage renal disease, the main observed adverse effects and their incidences were myalgias 5 %, iron deficiency 43 %, elevated blood pressure 35 %, and seizures 5.4 %. Hypertension is a common side effect with approximately one-third of dialysis patients affected. Hypertension and increased viscosity due to rhEPO may lead to encephalopathy, convulsions, cerebral edema, and seizures. Thromboembolism is said to be a potential outcome from EPO therapy, but controlled studies have not always provided support for this claim. Nevertheless, controlled studies on cancer patients revealed a 1.55-fold higher risk of thromboembolic events with rhEPO therapy than controls. Cerebral ischemia with increased metabolic rate and blood viscosity is a potentially severe side effect of EPO therapy and it has been pointed out that this could limit or halt the use of EPO for neurovascular diseases. EPO receptors have been demonstrated in tumor tissue and the cytokine may assist with tumor angiogenesis, suggesting the possibility of EPO initiating tumor growth or aiding tumor progression. The FDA has issues a boxed warning for both epoetin alfa and darbepoetin alfa related to a possible increased risk of death, cardiac and thromboembolic events, and tumor progression or recurrence (Table 5.2).

Pure red cell aplasia (PRCA), caused by neutralizing antibodies to epoetin that cross-react with natural erythropoietin, produces a rapid decline in hemoglobin concentration, severe anemia, low reticulocyte count, and an almost total absence of red cell precursors. In cases of transfusion-dependent PRCA with neutralizing serum antibodies to EPO, patients should not be switched to another epoetin such as darbepoetin alfa. Development of wheals at former epoetin alfa subcutaneous injection sites on a patient with PRCA following intravenous injection with epoetin beta and darbepoetin alfa, provoked a systemic anaphylaxis/anaphylactoid response and anti-EPO antibodies cross-reactive with epoetin beta and darbepoetin alfa were detected in the serum. Other cases of anaphylaxis to epoetin alfa have occurred and a delayed hypersensitivity reaction in the form of acute exanthematous pustulosis following replacement of epoetin alfa with darbepoetin alfa was reported.

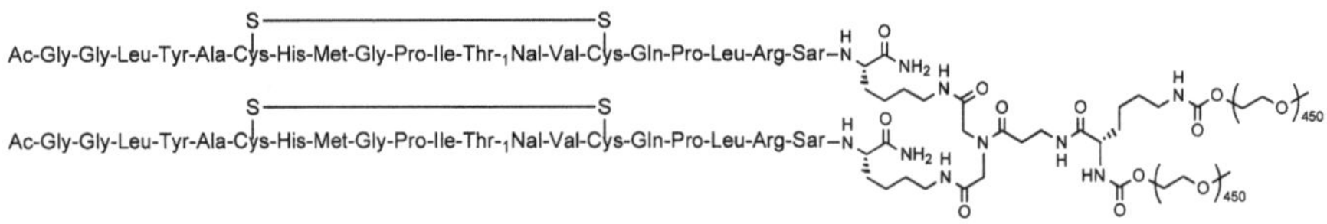

OMONTYS (peginesatide)

Fig. 5.11 The erythropoiesis-stimulating agent peginesatide (Omontys®) is a synthetic dimeric peptide of two identical 21 amino acid chains, pegylated via linkage of lysine residues to methoxy-polyethylene glycol. Note the presence of the natural amino acid, sarcosine (*N*-methylglycine). *Ac* acetyl

In a large randomized, double-blind controlled trial comparing two different administration schedules of darbepoetin alfa for the treatment of chemotherapy-induced anemia, serious adverse events that were treatment related occurred in 3 % of the 672 patients. Deep vein thrombosis was seen in 1.1–1.7 % of patients, pulmonary embolism in 0.8 % and hypertension in 0.3–1.1 %. No antibodies to darbepoetin alfa were found in the serum of any patient. An investigation of the effect of darbepoetin alfa on exercise tolerance in anemic patients with symptomatic chronic heart failure revealed three events with a >5 % difference in incidence between the treatment and placebo groups namely, neurological signs and symptoms, upper respiratory tract infections, and joint-related signs and symptoms.

Approved in 2012 and voluntarily recalled in February 2013, **peginesatide** (Omontys®) is an erythropoiesis-stimulating agent with no amino acid sequence homology to erythropoietin. Not a natural cytokine, peginesatide is a dimeric peptide of two identical chains of 21 amino acids, MW ~ 4.9 kDa, covalently attached via lysine residues to a structure formed from β-alanine and iminodiacetic acid (Fig. 5.11). The peptide so formed is then pegylated by linking to methoxypolyethylene glycol to give a large molecule of MW ~ 45 kDa. Approved indications and usage were as an erythropoiesis-stimulating agent for the treatment of anemia due to chronic kidney disease in adult patients on dialysis. Warnings and precautions were extensive with a black box warning of the risk of death, myocardial infarction, stroke, thromboembolism, thrombosis, and tumor progression or recurrence. Precautions included the risk of serious allergic reactions and the need for careful use of the drug in patients with existing cardiovascular disease and stroke. Uncontrolled hypertension was listed as a contraindication. Adverse reactions to peginesatide observed in controlled studies of dialysis patients were diarrhea, nausea, vomiting, dyspnea, cough, arteriovenous fistula, headache, muscle pains and spasms, arthralgia, pyrexia, hypotension, hypertension, hyperkalemia, and URTI. Seizures were seen in some patients. Antibodies to peginesatide were detected in 1.2 % of patients (29 of 2357); 0.9 % (21) were neutralizing antibodies and in approximately half of these patients hemoglobin levels declined. As a result of postmarketing reports of serious hypersensitivity reactions including potentially life-threatening anaphylaxis, the FDA announced a voluntary recall of Ormontys® Injection. Fatal reactions were reported in ~0.02 % of patients usually within 30 min of the first intravenous dose. The overall rate of hypersensitivity reactions was estimated to be approximately 0.2 % with about one third of these being serious reactions.

Bone Morphogenetic Proteins

BMPs are growth factors inducing the formation of bone and cartilage and important signaling proteins in some disease states such as adenocarcinoma and the progression of colon cancer. Of the 20 BMPs so far identified, six, numbers 2–7, belong to the TGFβ cytokine family (Table 5.1). BMP-2 and BMP-7 promote the differentiation of osteoblasts and, on the basis of this action, recombinant forms of both cytokines (rhBMP-2 and rhBMP-7) are approved by the FDA for specific uses in orthopedic, oral, and maxillofacial surgery and implant dentistry although up to 85 % of their usage is said to be off-label.

Safety studies on BMPs have a curious and troubling history of discrepancies due to the possible involvement of inadequate peer review and editorial oversight. More recent results with rhBMP-2 indicate a much higher incidence of side effects and complications than reported in the original peer-reviewed and industry-sponsored work. No adverse events following rhBMP-2 administration were reported in 13 of the original studies involving analyses of 780 patients due, it seems, to methodological bias against the control group. Identification of previously unpublished adverse effects, study inconsistencies, and a comparative review of FDA material, revealed an adverse event frequency associated with rhBMP-2 in spine fusion of 10–50 %. In a retrospective review of adverse events associated with the use of rhBMP-2 in spinal surgery, a search of the Manufacturer and User Facility Device Experience Database for the period July 2002 to August 2011 was undertaken. Only 4 of 834 reports described procedures using rhBMP-2 in accordance with the approved indication while 370 reports (44.4 %) stated that the patient required revision surgery or other invasive interventions to deal with the adverse event. The adverse events reported were swelling; fluid collection; osteolysis; pain/radiculopathy; heterotropic bone; pseudarthrosis; surgical site infections and other wound complications; thromboembolic events; respiratory distress; cancer; and some others. In an examination of the prevalence and complications of BMP use in spinal fusion procedures, the following complications and incidences were identified: vertebral osteolysis 44 %; graft migration 31 %; graft subsidence 27 %; formation of neutralizing antibodies to BMP-2 26 %; ectopic/heterotopic bone formation 7 %; and hematoma 3 %.

In accordance with the classification followed in the recent review of rhBMP-2-associated complications by Tannoury and Howard, the main adverse events to this cytokine are considered under the headings of those occurring during lumbar spine surgery and those seen in or after cervical spine surgery. In posterior and transforaminal interbody fusions in particular, postoperative radiculitis may occur after BMP-2 use in lumbar spine surgery, occurring, it seems, without neural compression and possibly because of the formation of ectopic bone. Postoperative radiculitis was estimated to occur in 11.4 % of patients who underwent a minimally invasive transforaminal interbody fusion procedure. Postoperative nerve injury and ectopic bone formation with the use of rhBMP-2 have been reported, the latter with an incidence of 20.8 % compared to 8.3 % in the absence of the cytokine. Other major adverse events seen following BMP-2 use in lumbar spine surgery include vertebral osteolysis, edema, and retrograde ejaculation. Although the formation of neutralizing antibodies to the bone growth protein is a theoretical concern, there is so far no clinical evidence that that has occurred. Reviews of complications after the use of rhBMP-2 in cervical spine surgery have revealed an incidence of 43 % for osteolysis and graft subsidence and 5.5–17 % for dysphagia and swelling (in particular, the neck) with respiratory difficulties. Other adverse events include hematoma with high doses of rhBMP-2; lucency and subsidence of fusion levels amongst allograft and demineralized bone matrix patients; and complications in posterior cervical fusion with BMP such as neurologic decline, wound complications, and asymptomatic heterotopic ossification.

Besides being a growth factor, BMP-2 receptors are expressed on some tumor cells and it is therefore not surprising that the cytokine has been investigated as a potential carcinogen in studies on breast cancer cells, malignant human gastric epithelial cells, oral cell carcinoma, and the risk of subsequent pancreatic cancer. Concern for the carcinogenic potential of rhBMP-2 was somewhat reinforced by a 2010 FDA Orthopedic and Rehabilitation Devices Advisory Panel report on the Amplify™ rhBMP-2 matrix of increased cancer rates among BMP-2-treated patients. At ≤24 months, cancer incidences were patients 5 % and controls 0.9 %; at 60 months, patients 5 %, controls 2.1 %. Other studies have reported tumor-enhancing, tumor-suppressing, or no dependence effects so, in this situation of uncertainty, it would be prudent to very carefully consider the question of the use of BMPs together with the risk-to-benefit ratio for cancer patients requiring spinal fusion.

Less widely used than rhBMP-2 which promotes better bone growth, rhBMP-7, also known as osteogenic protein 1 (OP-1), is a multifunctional growth factor thought to have other possible therapeutic applications besides bone and cartilage growth and development. These hoped-for potential applications include identification and treatment of cancer, and a beneficial role in Parkinson's disease, ankylosing spondylitis, diabetes, and asthma, as well as some diseases of the kidney, liver, intestine, brain, adipose tissue, and cardiovascular system. Apart from an FDA Public Health Notification of life-threatening complications associated with rhBMP (including rhBMP-7) in cervical spine fusion, and the reminder that rhBMPs are contraindicated in skeletally immature or pregnant patients and those with hypersensitivity to the protein, studies on, and reports of, adverse events to BMP-7 are not yet extensive. This is in contrast to the large and rapidly growing literature on rhBMP-2 induced adverse events. In an early clinical trial designed to evaluate rhBMP-7 in the treatment of tibial reunions, adverse events were reported to be mild or moderate and nonserious, for example, fever, nausea and vomiting, leg edema and discomfort, and hematoma at the operative site. Low levels of anti-BMP-7 antibodies were detected in 10 % of the treated patients but all titers were low with no related adverse events. With the possibility in mind that risks following the use of rhBMP-7 in anterior cervical fusion procedures might not be as high as seen with BMP-2, the safety of rhBMP-7 was examined in 123 patients undergoing anterior cervical discectomy and fusion using interbody cages. Assessed over the first 30 days, there were no deaths or reoperations but 2.4 % of patients experienced brachialgia and dysphagia. Although a slight increase on post-operative prevertebral swelling was seen on radiological evaluation, this was judged to be not clinically significant. The authors concluded that BMP-7 can be used safely in anterior cervical fusion surgery.

Pleomorphic sarcomas around heterotropic bone nodules were found in some animals during a study of rhBMP-7 in rats and five cancers, four nonosseous, and one recurrence of chondrosarcoma, occurred in 570 humans receiving OP-1. Note, however, published material on BMP-7 and carcinogenesis that is not related to manufacturers does not appear to be available.

Metreleptin

Leptin, a 167 amino acid protein of MW 16 kDa produced by a number of different cells in different organs but primarily adipocytes, helps to control energy homeostasis and body weight by adjusting hunger and energy expenditure to regulate fat stores. It also regulates some neuroendocrine functions and other physiological processes, many yet to be defined. In February 2014, the FDA approved metreleptin, a recombinant analog of leptin (Table 5.2) as replacement therapy to treat leptin deficiency in patients with congenital or acquired generalized lipodystrophy. Metreleptin is not to be used in patients with general obesity, for HIV-related lipodystrophy or in patients with metabolic diseases (e.g., diabetes mellitus) without concurrent generalized lipodystrophy. Neutralizing antibodies may develop to metreleptin and because of this and the possibility of the occurrence of T-cell lymphoma in patients with acquired generalized lipodystrophy, the protein is only available under the Myalept Risk Evaluation and Mitigation Strategy Program.

Kinetic studies on metreleptin in relation to age, sex, production and clearance, demonstrated that the recombinant cytokine's half-life was 3.4 ± 1.5 h, older subjects show decreased production and clearance rates, and females have higher baseline levels which increase with increasing adiposity. In fact, an increased body mass is associated with higher endogenous leptin levels, a higher rate of production, and a longer half-life.

Common side effects observed in early clinical trials were headache, weight loss, hypoglycemia, and abdominal pain. In a randomized, double-blind study designed to evaluate the weight-lowering effect in human obesity of an amylin/leptin drug combination using pramlintide/metreleptin, adverse events specifically due to metreleptin occurring with an incidence of ≥5 % were injection site reactions 66.7 %, nausea 25.9 %, nasopharyngitis 7.4 %, headache 7.4 %, hypersensitivity 7.4 %, and vomiting 7.4 %. Injection site reactions often include inflammation, erythema, and ecchymoses. Other potentially more serious reported adverse events to metreleptin include the worsening of renal disease, the production of antimetreleptin antibodies, and development of T-cell lymphomas.

Ancestim

Ancestim (Table 5.2) is a recombinant human stem cell factor (SCF). Produced in *E. coli*, it is nonglycosylated but, after expression, retains an *N*-terminal methionine and is therefore also referred to as r-met-hSCF. SCF is produced by fibroblasts and endothelial cells in soluble and transmembrane forms, both of which bind to the c-Kit receptor and are biologically active. Sometimes referred to as a pluripotent growth factor, the cytokine is important for hematopoiesis, spermatogenesis, and melanogenesis; nonlethal point mutations in its receptor can cause anemia, impaired fertility, and pigmentation. The c-Kit receptor, also referred to as the stem (or mast) cell growth factor receptor, proto-oncogene c-Kit, tyrosine-protein kinase kit, and CD117, is a receptor tyrosine kinase type III expressed on a number of different cells including mast cells, melanocytes, and germ cells but, importantly for ancestim, on a range of early to mature hematopoietic progenitor cells. Although ancestim shows only weak colony-stimulating activity in vitro, it acts in synergy with some other hematopoietic growth factors such as G-CSF, GM-CSF, erythropoietins, and IL-2 to stimulate multiple hematopoietic lineages in humans and some other animals. It also activates mast cells and stimulates melanocyte development and the production of pigment.

Given in combination with filgrastim, ancestim produces increases in circulating peripheral blood progenitor cells (PBPCs) including CD 34^{+} cells compared to filgrastim alone although when given as a single agent, the effect appears to be minimal to weak. Ancestim is therefore used with filgrastim to effect a sustained mobilization of PBPCs and achieve a reduction in the number of apheresis required to reach the PBPC number target. This has resulted in the Australian Therapeutic Goods Administration (TGA) approving the combination to increase the number of PBPCs for transplant patients at risk of poor PBPC mobilization.

The TGA has issued a number of important warnings and precautions for ancestim. In the first place, for reasons of efficacy and the accumulated findings on combination therapy but not monotherapy, the cytokine should not be used alone. Together with filgrastim, ancestim should only be administered to patients who are at risk of inadequate PBPC mobilization. Care should also be exercised in the simultaneous use of the cytokine combination in patients given chemo/radiotherapy; administration should be avoided 24 h before and after the cytotoxic therapy.

Allergy may be a problem. Because SCF increases mast cell proliferation, adhesion, and survival and promotes the release of histamine and tryptase, allergic-like symptoms sometimes occur in treated patients. Ancestim should therefore only be administered in a setting with the appropriate staff, facilities, and medications to respond to a possible life-threatening anaphylactic/anaphylactoid reaction. In addition, patients should be premedicated with H_1 and H_2 antihistamines and a bronchodilator. Patients with a history of anaphylaxis, asthma, recurrent urticaria and/or angioedema, and mast cell diseases such as systemic mastocytosis, urticaria pigmentosa, or diffuse cutaneous mastocytosis may be at particular risk.

The carcinogenic potential of ancestim has so far remained unstudied. Being a growth factor, it may stimulate the growth of a range of possible different tumors, particularly melanomas, mast cell or basophil leukemia, small cell lung cancers, and myeloid malignancies. Other precautions relate to the collection by apheresis of malignant cells, their stimulation and subsequent reinfusion into patients; and leukocytosis (the white cell count should be monitored frequently).

Bearing in mind that almost all reports of adverse events following administration of ancestim originate from treatments in which it is given in combination with myelopoietic-stimulating agents such as filgrastim, there is usually an element of doubt as to which of the two cytokines, or both, are responsible for the observed effects. Injection site reactions, occurring within 1–24 h, are the most commonly observed adverse events with up to 84 % of patients given ancestim showing mild-to-moderate reactions. Reported reactions consisted of erythema (59 %), pruritus (25 %), and urticaria (16 %) with occasional cases of hyperpigmentation and rash at the injection site. Erythema at a previous injection site has been seen in a few patients following an injection of ancestim at a different site. Rash, pruritus, and urticaria have occurred in 18 % of patients given ancestim/filgrastim and 5 % given ancestim alone. Other commonly seen reactions consisting of central/peripheral nervous system, gastrointestinal, cardiac, and respiratory events following ancestim are listed in Table 5.2. Respiratory problems, mainly cough, pharyngitis, and dyspnea, affected 25 % of recipients of combination therapy and 14 % of those on ancestim alone. Systemic allergic reactions, generally moderate to severe but not life threatening, occur more often at higher dosages; at <30 μg/kg/day of ancestim, 5 % of patients had such a reaction while 27 % of patients given 30–100 μg/kg/day had reactions. With regard to immunogenicity of ancestim, 23 of 258 (9 %) produced serum antibodies to the cytokine but no adverse clinical consequences, including a reduced therapeutic effect or serum sickness, have been recorded.

Concluding Remarks

Cytokines have already had a revolutionary impact on our understanding of cellular functions and extracellular messaging but although their biological effects seem to offer great potential for the treatment of a wide range of human conditions, their pleiotropism, potency, and complexity to produce cytokine "cocktails" with signaling cascades and accompanying side effects, demands caution in attempts to introduce individual members into the clinic. The range of biological events set in motion even by individual cytokines, warns of the possibility of unwanted side effects and the resultant caution is reflected by the relatively small number of cytokines currently approved by regulatory agencies and reviewed here. Good examples of the sort of doubts that exist and why clinical developments proceed so cautiously have been illustrated with interferons, aldesleukin, becaplermin, palifermin, and bone morphogenetic proteins. A glance at Table 5.2 shows that 16 of the 23 listed FDA-approved cytokine preparations (19 different cytokines with four also in pegylated form) carry warnings with 10 of these being black box warnings. Having been used in human therapy for many years, interferon alfa preparations are well known for a number of often widely different, potentially serious side effects specified in boxed warnings. The diverse nature of these side effects including neuropsychiatric, autoimmune, ischemic, and infection adverse events, together with their therapeutic benefits, provides a good illustration of the two-edged nature of cytokine pleiotropism. Becaplermin, a growth factor, causes cell proliferation so the possibility of malignancy with its continued use needs to be kept in mind, especially in patients with known cancers. Likewise, palifermin, another growth factor and a valuable treatment for mucositis in cancer patients, has with it the potential for stimulation of tumor growth, especially since its complementary receptors occur widely on many different cell types in the body. Aldesleukin, the recombinant IL-2, is a potent activator of T lymphocytes and stimulates immune responses to cancer, producing regression of tumors in metastatic renal cell carcinoma and melanoma. However, this activity can also lead to a range of adverse cardiac and pulmonary events. Perhaps, the best example of the safety uncertainties and benefits-to-risk ratios associated with these heterogeneous, pleiotropic cell regulators, is seen with the bone morphogenetic proteins BMP-2 and BMP-7. Already with a troubled history of underreported adverse events, in the postmarketing period, these growth factors are currently a focus of attention and speculation as potential carcinogens. BMP-2 receptors are expressed on some tumor cells and increased rates of cancer following its use have been reported but, in keeping with the complexity of cytokine-induced responses, and the difficulty of ascribing many adverse effects to causes, tumor-suppressing effects, or no dependence, have also been reported.

In any consideration of adverse event profiles of approved biologics, two other potentially important contributing factors need to be recognized. Any drug brought to market under the Orphan Drug Designation program where development was mediated because of the rarity of a condition, may not reveal its full spectrum of adverse effects until well into its postmarketing period since a relatively smaller number of administrations results from a smaller pool of patients. The dose of a particular cytokine may also be of critical importance in avoiding dangerous side effects by narrowing the spectrum of activity of the pleiotropic agent and tipping the balance to a specific biological activity.

Lest the attention drawn in this review to the known and potential toxicities of cytokines obscures their often substantial benefits and the improved outcomes they can produce, readers are reminded that the focus here on adverse effects does not negate the clear clinical improvements each of the approved cytokines can bring. Cytokines may indeed sometimes provoke a wide range and number of toxicities and adverse events but, overall, the second edge of their pleiotropism often offsets the side effects profiles and this is reflected in their lists of indications and approved regulatory status. In fact, in some cases, toxicities correspond with improved outcomes. For example, in an assessment of the significance of autoimmunity in melanoma patients treated with interferon alfa-2b, interferon-induced autoimmunity was found to be a prognostic marker for improved relapse-free, and overall, survival.

Together with monoclonal antibodies, chimeric fusion proteins, vaccines, a range of recombinant enzymes, hormones, clotting factors, various receptor proteins, a few purified approved toxins and some cell-based and nonspecific adjuvant therapies, the pool of over 130 cytokines seems to offer, via genetically engineered or modifications of the natural proteins, the potential of a major expansion of biologic therapies, some revolutionary, over the forthcoming decade or less. Meanwhile, the relatively few currently approved recombinant cytokines are already revealing their true natures in relation to their efficacy and side effects, influenced above all by their pleiotropism, redundancies and potencies. The large range of activities displayed by the family of cytokine proteins, together with their potential for the treatment of many different diseases and our steadily accumulating knowledge and experience with the small number currently used clinically, may indeed end up helping to achieve the prediction that the future of therapy belongs to the emerging biologics.

Summary

- Cytokines, currently known to be more than 130 in number, are relatively small signaling proteins of MW < 30 kDa. They are usually glycosylated and produced by a variety of different cells including those of the immune system, epithelia, endothelia, and stroma. Cytokines are key modulators of the immune and inflammatory responses functioning in an autocrine, paracrine, or endocrine manner in infection, innate and adaptive immunity, autoimmunity, inflammation, and malignancy. Key to an understanding of these regulatory proteins is the recognition of their pleiotropism and sometimes overlapping activities, functional redundancies, and side effects.
- In the current genomic phase, cytokines are identified on the basis of homology with known, characterized cytokines. Many original names are still in use and many of the originally described "factors" share receptors with other cytokines, for example, some interleukins.
- For the 23 FDA-approved cytokine products from the CDER-approved Biologic Products list, the cytokine classification presented is based on the Kyoto Encyclopedia of Genes and Genomes.
- Nine main families are recognized with most of the cytokines of interest classified in the hematopoietic growth factor, interferon (IFN), platelet-derived growth factor (PDGF), and transforming growth factor β (TGFβ) families. The approved cytokines are manufactured by recombinant DNA technology.
- Hematopoietin family: aldesleukin (rh-interleukin-2 [IL-2]), oprelvekin (rhIL-11), filgrastim and tbo-filgrastim (rh-granulocyte colony-stimulating factor [G-CSF]), sargramostim (rh-granulocyte macrophage [GM]-CSF), metreleptin (rh-leptin), rh-erythropoietins, epoetin and darbepoetin alfa, and stem cell factor (r-met-hSCF); IL-1 cytokine family: anakinra, a recombinant receptor antagonist for IL-1; interferon family: recombinant interferons alfa-1, alfa-2, beta-1, and

gamma-1; PDGF family: palifermin (rh-keratinocyte growth factor [KGF]) and becaplermin (rhPDGF-BB); and TGFβ family: rh-bone morphogenetic protein [BMP]-2 and rhBMP-7.

- Interferons are a class of broad-spectrum antiviral cytokines with overlapping, but also some individual, activities. Of most interest for therapy are interferons alfa, beta, and gamma. Virtually all patients treated with interferon alfa experience some adverse effect(s) at some time during therapy.
- Interferon alfa preparations occasionally provoke an extensive range of adverse reactions including cardiovascular, respiratory, endocrine, hematologic, metabolic, urinary tract, and skin adverse events as well as adverse effects on the nervous and sensory systems.
- Peginterferon alfa-2a and peginterferon alfa-2b are covalent conjugates of the recombinant interferon with a single branched bis-monomethoxy polyethylene glycol (PEG) chain. Pegylation helps to protect the protein from immune recognition and increases the molecule's size thus extending protein half-life and circulatory time and reducing renal clearance.
- Peginterferon alfa-2a together with ribavirin is indicated for the treatment of chronic hepatitis C in adults who have compensated liver disease and for patients infected with hepatitis C and HIV. Peginterferon alfa-2a alone is approved for patients with chronic hepatitis B.
- Interferon alfa-2b is administered extensively for hepatitis B and C as well as several malignancies.
- Interferon alfa-induced neuropsychiatric disorders, particularly depression, cognitive dysfunction, and mania are well known and have been intensively studied. Of the patients who develop severe depressive symptoms, most occur within the first 3 months of treatment. The incidence of depressive disorders is estimated to be 23–41 %. Symptoms may be prolonged for 6 months or more after the cessation of therapy.
- Autoantibodies and development or exacerbation of autoimmune diseases including hypothyroidism, immune-mediated hemolysis, systemic lupus erythematosus, Raynaud's disease, and mixed connective tissue disease are known to occur in response to interferon alfa therapy.
- Pegylated interferon alfa-2b has been associated with acute myocardial infarction, pericarditis, pericardial effusion with tamponade, and sick sinus syndrome producing arrhythmias.
- Interstitial lung disease, reported for both interferon alfa-2a and 2b, is seen more frequently with the former agent and with high doses of the latter. Cases of fatal interstitial pneumonitis, adult respiratory distress syndrome, and bronchiolitis obliterans organizing pneumonia (BOOP) following pegylated interferon alfa-2b are known.
- Interferon alfa may have adverse effects on the nervous system in the form of seizures in patients with no history of epilepsy, involuntary facial movements and weakness, features resembling multiple sclerosis, restless legs syndrome, sensorimotor polyneuropathy, and Bell's palsy.

- Adverse effects on sensory systems, mainly not only the eyes but also the ears, occur particularly to interferon alfa-2b. Ocular complications include occlusive vasculitis, central retinal artery occlusion, anterior ischemic optic retinopathy, retinal hemorrhage, subconjunctival hemorrhage, and optic nerve edema. Other ocular complications described in patients treated with interferon alfa-2b include permanent loss of sight due to combined retinal artery and central retinal vein obstruction; development of an epiretinal membrane; and the T-cell-mediated autoimmune syndrome, Vogt–Koyanagi–Harada disease.
- Endocrine effects of interferon alfa, best illustrated by thyroid dysfunction, may have an autoimmune mechanism. It occurs with an incidence of 5–14 % in patients treated for chronic hepatitis C. Hypothyroidism occurs more often than hyperthyroidism. Interferon alfa-2b can cause both conditions.
- Neutropenia induced by interferon alfa is fairly commonly seen. Other hematologic side effects include acute and autoimmune thrombocytopenia, pernicious anemia, bone marrow hypoplasia which may be immune mediated, and pure red cell aplasia.
- Renal complications to interferon alfa include renal thrombotic microangiopathy, acute nephrotic syndrome, hemolytic-uremic syndrome, renal insufficiency due to interstitial nephritis, tubular necrosis, and IgA nephropathy.
- The list of cutaneous reactions to interferon alfa is extensive and includes injection site reactions (erythema, necrosis, and vasculitis), pruritus, xerosis, urticaria, hyperpigmentation, psoriasis, alopecia, lichen planus, pityriasis rosea, sarcoid nodules, eosinophilic fasciitis, livedo reticularis, vitiligo, and fixed drug eruption. Interferon alfa is well known for exacerbating pre-existing psoriasis but cases of new onset psoriasis occur with both interferon alfa-2a and interferon alfa-2b.
- A flu-like illness is the most commonly occurring adverse event following administration of interferon beta and injection site reactions are also common.
- Neutralizing antibodies are found in about a quarter of patients treated with subcutaneously administered interferon beta-1b and the consensus is that they neutralize or reduce the cytokine's activity. Other immunologic effects observed to both beta interferons are some cases of a lupus-like syndrome and cutaneous lymphocytic vasculitis.
- Unlike interferon alfa, results from studies do not support an association of interferon beta with depression but the FDA mention depression, suicide, and psychotic disorders in their warnings and precautions for the cytokine.
- Interferon beta can induce thyroid disorders notably hyperthyroidism and a severe case of hypothyroidism resembling Hashimoto's encephalopathy has been reported.
- Skin reactions include urticaria to interferon beta-1a and an acneiform eruption to interferon beta-1b.
- In 2014, the FDA granted approval for Plegridy®, a pegylated preparation of interferon beta-1a. Common adverse reactions to Plegridy® are similar to the nonpegylated form of the cytokine, viz., injection site reactions, an influenza-like illness, asthenia, arthralgia, and pruritus.

- Cardiovascular toxicity to interferon gamma, particularly at higher doses, include hypotension, arrhythmias, coronary vasospasm and ventricular tachycardia, and renal toxicity, namely acute renal failure, nephrotic syndrome, and tubular necrosis.
- The occurrence of fatal acute respiratory failure in some patients treated with interferon gamma-1b for advanced idiopathic pulmonary fibrosis prompted further investigation of the condition. No clear evidence for the involvement of interferon gamma-1b was found.
- Described as "the master regulators of granulocyte and macrophage populations," the colony-stimulating factors (CSFs) are used to treat chemotherapy-induced neutropenia, mobilize stem cells for transplantation, and enhance the immune response to cancer. Currently, approved members of the CSF family are filgrastim and pegfilgrastim, both G-CSFs, sargramostim, a GM-CSF, and tbo-filgrastim, a short acting biosimilar G-CSF.
- CSF-induced adverse events are usually mild and transient including headache, bone pain, myalgia, fever, flushing, and rash. More severe, and rare, events are adult respiratory distress syndrome, pulmonary toxicities, particularly pulmonary edema and interstitial pneumonitis, fluid retention, aortitis, capillary leak syndrome, thrombocytopenia, splenomegaly, and spleen rupture. G-CSF may be a risk for the progression of myelodysplastic syndrome. Other potentially life-threatening responses to CSFs, the subject of warnings, are anaphylactic/anaphylactoid reactions and severe adverse events such as acute chest syndrome, vaso-occlusive episodes, multiorgan failure, and death seen in patients with sickle cell disease.
- There is a long list of adverse skin reactions provoked by CSFs. The most commonly occurring cutaneous reaction is Sweet's syndrome seen after therapy with sargramostim and filgrastim. Other adverse cutaneous events include psoriasis flare, pyogenic granulomas, pruritic erythematous maculopapular eruptions, palmoplantar pustulosis, erythema multiforme, and neutrophilic dermatoses.
- Recombinant human IL-11, or oprelvekin, is used to prevent chemotherapy-induced thrombocytopenia and reduce the need for platelet transfusions in patients with nonmyeloid malignancies. Fluid retention and an increase in plasma volume underlie many of the adverse events, for example, edema, dyspnea, pleural effusions, arrhythmia, dilutional anemia, and renal failure, and indicate that oprelvekin should be used with caution in patients with congestive heart failure.
- Becaplermin is a recombinant human platelet-derived growth factor (PDGF), a homodimer made up of two disulfide-bonded B chains and hence written as rhPDGF-BB.
- rhPDGF-BB promotes the growth of granulation tissue and wound healing via interaction with receptors on fibroblasts and endothelial cells. Becaplermin has therefore found use in gel form as a topical application for patients with difficult to heal diabetic neuropathic ulcers.
- The FDA has issued a boxed warning for becaplermin gel to the effect that becaplermin should be used with caution in patients with known malignancy and only used when the benefits can be expected to outweigh the risks.

- Palifermin, a recombinant human keratinocyte growth factor produced by mesenchymal cells and fibroblasts, stimulates differentiation, proliferation, and migration of epithelial cells via interaction with its complementary receptors on epithelial cells widely distributed in numerous tissues. Palifermin is an important agent in oncological supportive care, aiding the management of mucositis in cancer patients by protecting the mucosal epithelium and aiding its regeneration after chemotherapy- and radiation-induced injury.
- Adverse events following palifermin administration include rash, pruritus, erythema, paresthesia, edema, taste alteration, rhinitis, arthralgia, thickening of the tongue, and numbness. Numerous cutaneous reactions include acral erythema, a papulopustular eruption on the head and trunk, hyperpigmented papillomatous plaques in the axillae and inguinal areas, and a case of lichenoid papules. Being a growth factor, palifermin carries a warning of potential stimulation of tumor growth.
- Aldesleukin, a recombinant human IL-2, differs from the natural cytokine by absence of glycan residues and at position 125 and the end terminal amino acid. X-ray and NMR studies have shown that the IL-2 fold is similar to that seen in the myelopoietic-stimulatory factors G-CSF and GM-CSF.
- Cardiovascular adverse events are the main dose-limiting toxicities of aldesleukin with cases of hypotension, tachycardia, peripheral edema, pleural effusions, myocarditis, myocardial infarction, heart block, arrhythmias, cardiac eosinophilic infiltration, and coronary ischemic changes. Vascular leak syndrome causes hypovolemia, fluid accumulation in the extravascular spaces, oliguria, and pulmonary side effects. Hematologic adverse effects, particularly anemia, leukopenia, and thrombocytopenia occur but are rarely severe or dose limiting.
- Endocrine effects of aldesleukin usually manifest as hypothyroidism or, far less commonly, hyperthyroidism. Gastrointestinal perforation has been reported and IL-2-induced infectious toxicities, usually due to *Staphylococcus*, may occur at venous catheter sites. Neurological effects, especially to high doses during IL-2 therapy, include anxiety, depression, altered sleep patterns, somnolence, emotional fragility, vivid dreams, and confusion.
- The list of aldesleukin-induced cutaneous reactions is extensive, ranging from mild erythema, pruritus, injection site reactions and vitiligo, to urticaria, angioedema, reactivation of eczema, exacerbation of psoriasis, generalized erythema followed by desquamation, vasculitis, and severe manifestations like pemphigus, IgA bullous dermatosis, and toxic epidermal necrolysis.
- Anakinra is a recombinant specific receptor antagonist (IL-1RA) for IL-1 differing from the natural receptor by the addition of a single methionine at the amino terminal end. Anakinra acts as a biological response modifier in the treatment of diseases like rheumatoid arthritis and the spectrum of autoinflammatory syndromes collectively known as cryopyrin-associated periodic syndrome (CAPS).
- The most common adverse events to anakinra are injection site reactions which occur in up to 73 % of patients and infection episodes. Other reported side effects include progression of rheumatoid arthritis, exacerbation of Crohn's disease, anaphylaxis, systemic inflammatory response syndrome, ARDS, an interstitial granulomatous reaction, and some severe skin reactions.

- The main adverse effects of epoetin (rhEPO) in anemic patients with end-stage renal disease are myalgias, iron deficiency, elevated blood pressure, and seizures. Hypertension is a common side effect with approximately one-third of dialysis patients affected and hypertension and increased viscosity due to rhEPO may lead to encephalopathy, convulsions, cerebral edema, and seizures. Controlled studies on cancer patients revealed a higher risk of thromboembolic events with rhEPO therapy than controls.
- Cerebral ischemia with increased metabolic rate and blood viscosity is a potentially severe side effect of EPO. EPO receptors have been demonstrated in tumor tissue and the cytokine may assist with tumor angiogenesis, suggesting the possibility of EPO initiating tumor growth or aiding tumor progression.
- Both epoetin alfa and darbepoetin alfa carry a boxed warning related to a possible increased risk of death, cardiac, and thromboembolic events and tumor progression or recurrence.
- In cases of transfusion-dependent pure red cell aplasia (PRCA) with neutralizing serum antibodies to EPO, patients should not be switched to another epoetin such as darbepoetin alfa since a systemic anaphylaxis/anaphylactoid response mediated by anti-EPO antibodies cross-reactive antibodies may result.
- An investigation of the effect of darbepoetin alfa in anemic patients with symptomatic chronic heart failure revealed three main adverse events: neurological signs and symptoms, upper respiratory tract infections, and joint-related signs and symptoms.
- Bone morphogenetic proteins (BMPs) are growth factors inducing the formation of bone and cartilage and important signaling proteins in some disease states such as adenocarcinoma and the progression of colon cancer. BMP-2 and BMP-7 promote the differentiation of osteoblasts and, on the basis of this action, recombinant forms of both cytokines (rhBMP-2 and rhBMP-7) are approved by the FDA for specific uses in orthopedic, oral, and maxillofacial surgery and implant dentistry although up to 85 % of their usage is said to be off-label.
- Recent results with rhBMP-2 indicate a much higher incidence of side effects and complications than originally reported. Identification of previously unpublished adverse effects, study inconsistencies, and a comparative review of FDA material has revealed an adverse events frequency associated with rhBMP-2 in spine fusion of 10–50 %. Reported adverse events were swelling, fluid collection, osteolysis, pain/radiculopathy, heterotropic bone, pseudarthrosis, surgical site infections and other wound complications, thromboembolic events, respiratory distress, and cancer.
- In an examination of the prevalence and complications of BMP use in spinal fusion procedures, the following complications were identified: vertebral osteolysis, graft migration, graft subsidence, formation of neutralizing antibodies to BMP-2, ectopic/heterotopic bone formation, and hematoma.
- Concern for the carcinogenic potential of rhBMP-2 was reinforced by a 2010 FDA Orthopedic and Rehabilitation Devices Advisory Panel report on the Amplify™ rhBMP-2 matrix of increased cancer rates among BMP-2-treated patients. At ≤60 months cancer incidences were patients 5 % and controls 2.1 %. Other studies have reported tumor-enhancing, tumor-suppressing, or no dependence effects.

- rhBMP-7, also known as osteogenic protein 1 (OP-1), is a multifunctional growth factor. Apart from an FDA Public Health Notification of life-threatening complications associated with rhBMP (including rhBMP-7, in cervical spine fusion), and the reminder that rhBMPs are contraindicated in skeletally immature or pregnant patients and those with hypersensitivity to the protein, studies on, and reports of, adverse events to BMP-7 are not yet extensive.
- In a clinical trial designed to evaluate rhBMP-7 in the treatment of tibial reunions, adverse events were mild or moderate and nonserious, for example, fever, nausea, vomiting, leg edema and discomfort, and hematoma at the operative site. Low levels of anti-BMP-7 antibodies were detected in 10 % of the treated patients, but all titers were low with no related adverse events.
- Leptin, a 167 amino acid protein of MW 16 kDa produced primarily by adipocytes, helps to control energy homeostasis and body weight by adjusting hunger and energy expenditure to regulate fat stores. It also regulates some neuroendocrine functions and other physiological processes. Metreleptin, a recombinant analog of leptin is used as replacement therapy to treat leptin deficiency in patients with congenital or acquired generalized lipodystrophy. It is not to be used in patients with general obesity, for HIV-related lipodystrophy or in patients with metabolic diseases (e.g., diabetes mellitus) without concurrent generalized lipodystrophy.
- Neutralizing antibodies may develop to metreleptin. Because of this, the possibility of worsening of renal disease, and the occurrence of T-cell lymphoma in patients with acquired generalized lipodystrophy, the protein is only available under the Myalept Risk Evaluation and Mitigation Strategy Program.
- Ancestim is a recombinant human stem cell factor (SCF) important for hematopoiesis, spermatogenesis, and melanogenesis. It acts in synergy with some other hematopoietic growth factors such as G-CSF, GM-CSF, erythropoietins, and IL-2 to stimulate multiple hematopoietic lineages.
- Together with filgrastim, ancestim should only be administered to patients who are at risk of inadequate peripheral blood progenitor cell (PBPC) mobilization. Ancestim is used with filgrastim to effect a sustained mobilization of PBPCs and achieve a reduction in the number of apheresis required. Care should be exercised in the simultaneous use of the cytokine combination in patients given chemo/radiotherapy; administration should be avoided 24 h before and after the cytotoxic therapy.
- Patients with a history of anaphylaxis, asthma, recurrent urticaria, angioedema, and mast cell diseases such as systemic mastocytosis who are to be given ancestim may be at particular risk and should be premedicated with H1 and H2 antihistamines and a bronchodilator.
- The range of biological events set in motion even by individual cytokines, warns of the possibility of unwanted side effects and the resultant caution is reflected by the relatively small number of cytokines currently approved by regulatory agencies. Sixteen of the 23 listed FDA-approved cytokine preparations carry warnings with ten of these being black box warnings.
- The diverse nature of interferon side effects, including neuropsychiatric, autoimmune, ischemic, and infection adverse events, together with their therapeutic

benefits, provides a good illustration of the two-edged nature of cytokine pleiotropism. Despite this, cytokine adverse events profiles do not generally negate benefits and sometimes observed toxicities may even correspond with improved outcomes.

Further Reading

Baldo BA. Side effects of cytokines approved for therapy. Drug Saf. 2014;37:921–43.

Bessa PC, Casal M, Reis RL. Bone morphogenetic proteins in tissue engineering: the road from the laboratory to the clinic, part I (basic concepts). J Tissue Eng Regen Med. 2008;2:1–13.

Blijlevens N, Sonis S. Palifermin (recombinant keratinocyte growth factor-1): a pleiotropic growth factor with multiple biological activities in preventing chemotherapy- and radiotherapy-induced mucositis. Ann Oncol. 2007;18:817–26.

Broudy VC. Stem cell factor and hematopoiesis. Blood. 1997;90:1345–64.

Carragee EJ, Hurwitz EL, Weiner BK. A critical review of recombinant human bone morphogenetic protein-2 trials in spinal surgery: emerging safety concerns and lessons learned. Spine J. 2011;11:471–91.

Cottle TE, Fier CJ, Donadieu J, et al. Risk and benefit of treatment of severe chronic neutropenia with granulocyte colony-stimulating factor. Semin Hematol. 2002;39:134–40.

Dusheiko G. Side effects of alpha interferon in chronic hepatitis C. Hepatology. 1997;26 Suppl 1:112S–21S.

Dutcher JP, Schwartzentruber DJ, Kaufman HL, et al. High dose interleukin-2—expert consensus on best management practices—2014. J Immunother Cancer. 2014;2:26. doi:10.1186/s40425-014-0026-0.

Elliott SG, Foote M, Molineux G, editors. Erythropoietins, erythropoietic factors, and erythropoiesis: molecular, cellular, preclinical, and clinical biology. 2nd ed. Basel: Birkhauser Verlag: 2009.

Fattovich G, Giustina G, Favarato S, et al. A survey of adverse events in 11,241 patients with chronic viral hepatitis treated with alfa interferon. J Hepatol. 1996;24:38–47.

Feld JJ, Hoofnagle JH. Mechanism of action of interferon and ribavirin in treatment of hepatitis C. Nature. 2005;436:967–72.

Fried MW. Side effects of therapy of hepatitis C and their management. Hepatology. 2002;36 Suppl 1:S237–44.

Geertsen PF, Gore ME, Negrier S, et al. Safety and efficacy of subcutaneous and continuous intravenous infusion rIL-2 in patients with metastatic renal cell carcinoma. Br J Cancer. 2004;90:1156–62.

Gogas H, Ioannovich J, Dafni U, et al. Prognostic significance of autoimmunity during treatment of melanoma with interferon. N Engl J Med. 2006;354:709–18.

Heinrich PC, Behrmann I, Haan S, et al. Principles of interleukin (IL)-6-type cytokine signalling and its regulation. Biochem J. 2003;374:1–20.

Hercus TR, Broughton SE, Ekert PG, et al. The GM-CSF receptor family: mechanism of activation and implications for disease. Growth Factors. 2012;30:63–75.

Kelesidis T, Kelesidis I, Chou S, et al. Narrative review: the role of leptin in human physiology: Emerging clinical applications. Ann Intern Med. 2010;152:93–100.

Leader B, Baca QJ, Golan DE. Protein therapeutics: a summary and pharmacological classification. Nat Rev Drug Discov. 2008;7:21–39.

Margetic S, Gazzola C, Pegg GG, et al. Leptin: a review of its peripheral actions and interactions. Int J Obes. 2002;26:1407–33.

Metcalf D. The CSFs and cancer. Nat Rev Cancer. 2010;10:425–34.

Negro F. Adverse effects of drugs in the treatment of viral hepatitis. Best Pract Res Clin Gastroenterol. 2010;24:183–92.

Poreaux C, Bronowicki J-P, Debouverie M, et al. Managing generalized interferon-induced eruptions and the effectiveness of desensitization. Clin Exp Allergy. 2014;44:756–64.

Schoenborn JR, Wilson CB. Regulation of interferon-gamma during innate and adaptive immune responses. Adv Immunol. 2007;96:41–101.

Smith 2nd JW. Tolerability and side-effect profile of rhIL-11. Oncology. 2000;14 Suppl 8:41–7.

Steed DL. Clinical evaluation of recombinant human platelet-derived growth factor for the treatment of lower extremity diabetic ulcers. J Vasc Surg. 1995;21:71–81.

Swart JF, Barug D, Möhlmann M, et al. The efficacy and safety of interleukin-1-receptor antagonist anakinra in the treatment of systemic juvenile idiopathic arthritis. Expert Opin Biol Ther. 2010;10:1743–52.

Tannoury CA, An HS. Complications with the use of bone morphogenetic protein 2 (BMP-2) in spine surgery. Spine J. 2014;14:552–9.

Turner MD, Nedjai B, Hurst T, et al. Cytokines and chemokines: at the crossroads of cell signalling and inflammatory disease. Biochim Biophys Acta. 1843;2014:2563–82.

Vacchelli E, Galluzzi L, Eggermont A, et al. Immunostimulatory cytokines. Oncoimmunology. 2012;1:493–506. http://dx.doi.org/10.4161/onci.20459

Chapter 6
Fusion Proteins

There are currently 11 chimeric fusion proteins on the market with regulatory approval and many more are at different stages of clinical development with some already in phase III clinical trials. It has already become apparent that the concept of fusion proteins for human therapy is realistic, versatile, and successful with the preparations introduced so far proving effective and safe. Continued success of this form of biologics therapy is therefore widely anticipated. Etanercept (Enbril®), the most commercially successful fusion protein, generated worldwide sales of $7.3 billion in 2010 and $8.37 billion in 2012 making it the second best-selling drug after the monoclonal antibody adalimumab (Humira®), one of the biggest selling drugs of all time. Forecasts for 2013–2017 estimate revenue of $58 billion for adalimumab and $45 billion for etanercept.

Desired Properties and Composition of Chimeric Fusion Proteins

Chimeric fusion proteins used as biologic therapeutic agents consist of a carefully selected and relatively short-lived effector domain, generally a peptide, coupled to a "carrier," usually protein or peptide that also contributes to the functional properties of the resultant fusion protein. Fusion proteins are produced by genetic engineering, linking genes for the separate proteins involved to give a new polypeptide formed from the incorporated separate domains together with their functional properties. The linked effector peptide may have widely varying properties, for example, contributing to recognition, binding, and toxicity, while its fused partner may aid stability and targeting of the chimeric polypeptide. Effector peptides employed until now have been limited to ligand-binding portions of receptors of a few cytokines and growth factors, extracellular domains of some lymphocyte antigens, coagulation factors, a glucagon-like peptide, and a fragment of a protein toxin.

B.A. Baldo, *Safety of Biologics Therapy*, DOI 10.1007/978-3-319-30472-4_6

As with mAbs, three main requirements in the preparation of an effective chimeric fusion protein are to endow the macromolecule with: (1) stability, that is, produce a polypeptide with a suitably extended half-life; (2) effective targeting and subsequent specific binding; and (3) cytotoxicity or at least the capacity to inhibit the deleterious processes underlying the treated condition. Most peptides likely to be considered as effector peptides have a short half-life due to proteolytic degradation and are usually rapidly cleared via the kidneys within minutes. Conjugation to polyethylene glycol (PEG), or pegylation, can extend the half-life by increasing the hydrodynamic radius and decreasing filtration in the kidneys but safety concerns surround pegylation mainly because of lack of biodegradability. Other means of extending half-life have therefore been sought. Human serum albumin fusions has been employed as a fusion partner for some therapeutic chimeric fusion proteins (see below, Sect. "Albumin Fusion Proteins") and transferrin is another possible fusion partner being studied but the crystallizable Fc region of the human IgG antibody has been the most commonly employed fusion partner.

Fc Fusion Proteins

The Fc portion of IgG consists of the CH2 and CH3 domains of the immunoglobulin heavy (H) chain, the hinge region, and the two disulfide bridges connecting the H chains. In most chimeric fusion proteins, the C-terminus of the effector molecule, often a peptide, is fused to the N-terminus of the hinge region (Fig. 6.1). The effector molecule can be fused to one (Fig. 6.2c) or both of the Fc H-chains (Fig. 6.2a, b, e–h) creating a monomeric (Fig. 6.2c), dimeric (Fig. 6.2a), or heterodimeric (Fig. 6.2e) fusion protein or more than one ligand-binding domain may be employed to form a "trap" (Fig. 6.2b). In another configuration, effector peptides can be fused at the carboxyl terminus of the Fc fragment in the form of the so-called peptibody as seen with the thrombopoietin receptor agonist romiplostim or fusions may be undertaken at both the *N*- and C-termini (Fig. 6.2h).

Receptor-mediated recycling via interaction with the salvage neonatal FcRn receptor protects Fc-containing molecules from lysosomal degradation. At low pH (<6.5) in the endosome, FcRn salvages the Fc fragment by binding, recycling, and then releasing the protein in the blood at neutral pH thus extending the Fc half-life by avoiding breakdown in the lysosomes. Fc fusion proteins also interact with Fc receptors on immune cells and have become the most frequently employed and successful structures in the preparation of chimeric fusion protein drugs with 10 of the 12 proteins that have been registered being Fc fusion proteins. Half-lives of the 13 originally approved fusion protein preparations are listed in Table 6.1 where it can be seen that denileukin diftitox, the one fusion protein lacking an attached Fc piece or not bound to albumin, has easily the shortest half-life of only about 70–80 min compared to a number of days for the Fc fusion proteins, factor IX fusion protein rIX FP, and albiglutide, the albumin fusion protein, which has an extended half-life of up to 8 days. Note that the half-life of fusion proteins is typically significantly shorter than half-lives of the mAbs (Chap. 2, section "IgG Antibody Subclasses").

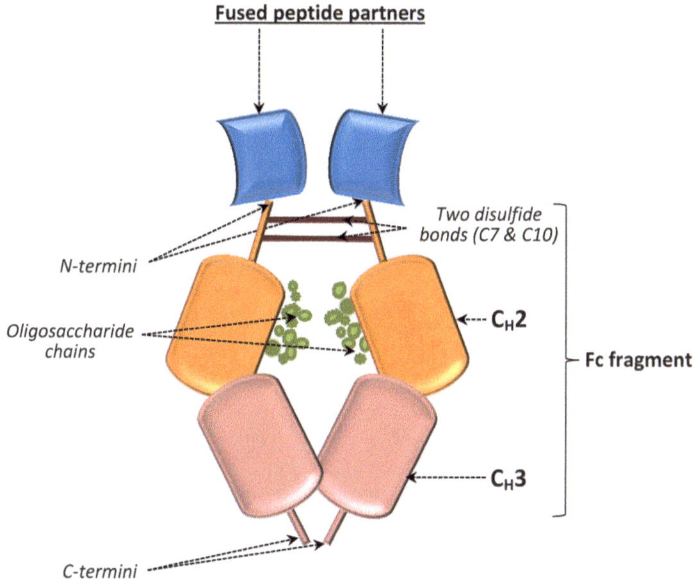

Fig. 6.1 Diagrammatic representation of the general structure of a chimeric human IgG Fc fusion protein linked to its effector peptide partner via the Fc *N*-terminus. The effector molecule(s) may be linked to one or both of the Fc chains creating a monomeric or dimeric fusion protein, respectively. For example, in etanercept the fused peptide partner is the extracellular ligand-binding protein of the 75 kDa human tumor necrosis factor receptor (TNFR); in abatacept the fusion protein is the lymphocyte antigen CTLA-4 fused to an antibody Fc portion mutated to lose its ADCC and CDC activities. Reproduced from Baldo BA. Chimeric fusion proteins used for therapy: Indications, mechanisms, and safety. Drug Saf 2015;38:455–79. Reproduced with permission from Springer Science+Business Media

Fc Fusion Proteins as Glycoproteins

A conserved *N*-linked glycosylation site occurs at asparagine 297 near the hinge region of the CH2 domain of the human Fc fragment of each of the four human IgG subclasses. The attached glycan structure is biantennary comprising a core heptasaccharide of *N*-acetyl-D-glucosamine and D-mannose with variable extensions formed from the addition of the monosaccharides L-fucose, *N*-acetyl-D-glucosamine, D-galactose, and terminal sialic acid (Chap. 2, section "Glycosylation of Monoclonal Antibodies"). By contrast, glycosylation sites on linked effector proteins such as receptor domains can be *N*- and/or *O*-linked and may contain more sialic acid residues, imparting more charge heterogeneity on Fc fusion proteins than, for example, monoclonal antibodies. About 70 % of purified recombinant preparations are glycoproteins and while cells such as Chinese hamster ovary (CHO) cells can be engineered to express some human glycosyltransferases, variations from human glycan patterns can be obtained from a range of other organisms including bacteria, fungi, plants, insects, and mammals. For example, although *N*-glycolylneuraminic acid cannot be synthesized in humans, CHO

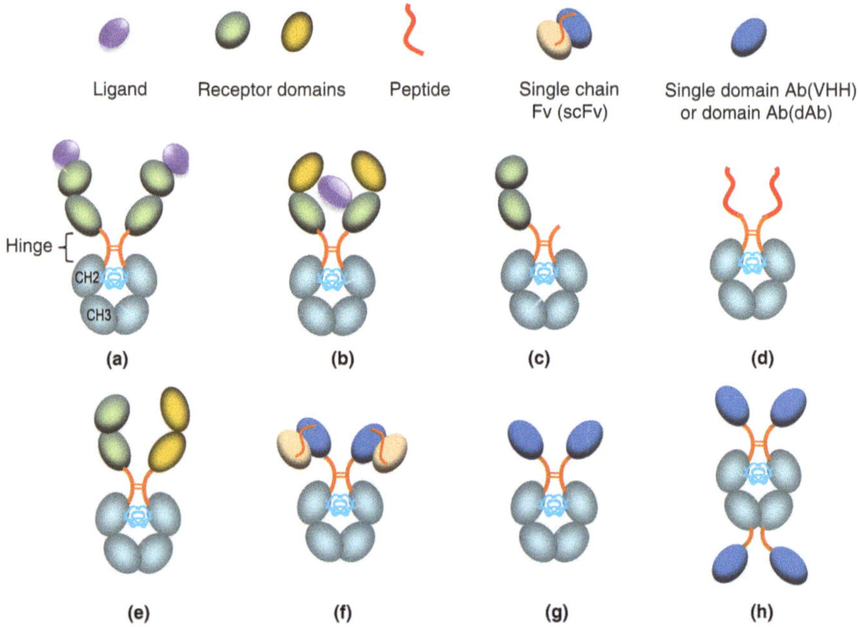

Fig. 6.2 Diagrammatic representations of different fusion protein constructs. (**a**) Commonly employed form in which the effector protein (e.g., an extracellular domain of a receptor) is fused to the Fc CH2 domain *N*-terminal creating a monomeric dimer. (**b**) Linked different effector proteins fused to the CH2 domains. This form is seen in Fc TRAP configurations, for example, rilonacept. (**c**) A monomeric Fc fusion with one protein fused to only one of the CH2 domains. (**d**) A mimetibody construct in which a bioactive peptide is linked to each CH2 domain of the Fc region. (**e**) Two different effector proteins fused to the Fc fragment to form a heterodimer. (**f**) Single-chain Fv (scFv) fused to each of the CH2 domains. (**g**) Protein fused to both CH2 Fc domains. (**h**) Effector protein fused to both the *N*- and *C*-termini. The construct in which fusion is with the *C*-terminus is seen in romiplostim. Reproduced from Huang C. Receptor-Fc fusion therapeutics, traps, and Mimetibody™ technology. Curr Opin Biotechnol. 2009;20:692–9. Reprinted with permission from Elsevier Ltd.

cells can add this sialic acid during glycoprotein expression. For post-translational modifications, attention to the cell expression system may allow recombinant proteins to be glycosylated in a predetermined and controlled manner. Structurally complex glycoproteins are produced mainly by mammalian expression systems since they require post-translational modifications of some complexity such as specific glycosylations, disulfide-bonded domains, γ-carboxylation, and sulfation. Glycosylation can have a major influence on the effectiveness of biologics therapy via its effects on actions, protein solubility, stability, serum half-life, immunogenicity, and selectivity of receptor binding. For example, the terminal sugars of glycans in the CH2 domain of human Fc fragments help to determine antibody-dependent cellular cytotoxicity (ADCC) and complement-dependent cytotoxicity (CDC) (Chap. 2, section "Antibody-Dependent Cell-Mediated and Complement-Dependent Cytotoxicities"). Shielding by glycan structures can protect peptide sequences from proteolysis and the hydrophilic

Table 6.1 Half-life of fusion proteins used for therapy

Fusion protein[a]	Trade name	MW (kDa)	Half-life[b] in days/h
Etanercept[c]	Enbrel®	150	3–6
Belatacept	Nulojix®	90	8–10
Abatacept	Orencia®	92	8–25
Rilonacept	Arcalyst®	251	9
Aflibercept	Zaltrap®; Eylea®	115	5–6
Romiplostim	Nplate®	60	1–34
Alefacept	Amevive®	92	11
Factor VIII Fc fusion protein[d]	Eloctate®; Elocta®	220	19.7 h[e]
Factor IX Fc fusion protein[f]	Alprolix®	98	~3.4
Dulaglutide[g]	Trulicity®	63	4–5
Albiglutide[h]	Tanzeum®; Eperzan®	73	6–8
Denileukin diftitox[i]	Ontak®[j]	58	1.2–1.3 h
Factor IX fusion protein rIX-FP[k]	Idelvion®[j]	125	3.6–3.9

From Baldo BA. Chimeric fusion proteins used for therapy: indications, mechanisms, and safety. Drug Saf 2015;38:455–79. Adapted and reproduced with permission from Springer Science + Business Media

[a]All Fc fusion proteins except albiglutide, denileukin diftitox, and factor IX fusion protein rIX-FP

[b]In adults

[c]The half-lives of the mAbs targeting TNF (infliximab, adalimumab, golimumab, and certolizumab pegol are ~2–3 times longer than the half-life of etanercept

[d]Also known as B-domain deleted recombinant factor VIII Fc fusion protein (BDDrFVIIIFc) or rFVIIIFc

[e]16.4 h in 12–17 year olds; 14.6 h in 6–11 year olds; 12 h in 2–5 year olds

[f]Also known as recombinant factor IX Fc fusion protein, rFIXFc

[g]Fc fusion partner is from IgG4 unlike the other Fc fusion proteins which utilize the Fc fragment from the IgG1 subclass

[h]Fusion partner is human serum albumin not IgG Fc. See also Chap. 7 and Fig. 7.8

[i]Not an Fc or albumin fusion protein and shows a significantly shorter half-life than other fusion proteins

[j]Discontinued; listed in the CDER Discontinued Therapeutic Biologic Products list January 30, 2014

[K]Approved March 2016. Fusion partner is human serum albumin not IgG Fc. See also Chap. 10 and Table 10.1

nature of the oligosaccharide chains aids solubility, helps to inhibit aggregation, and protects proteins against physical denaturation. The importance of post-translational modifications of *N*- and O-linked glycosylations can be illustrated by the unexpectedly rapid clearance of an Fc fusion protein due to incompletely formed glycan structures with terminal N-acetylglucosamine or galactose on the Fc fusion partner.

Due to their complex nature and the limited clinical experience with biologics such as chimeric fusion proteins and biosimilars before approval, a high level of characterization is demanded for their continued development. This characterization needs to cover protein and peptide mapping, and glycan analyses utilizing state-of-the-art analytical methods for the characterization of glycoforms. Liquid chromatography-mass spectrometry (MS) and capillary electrophoresis-MS, as well as classical electrophoretic and chromatographic methods, are playing an increasingly important role in this respect.

IgG Subclasses of Fc Fusion Proteins: Increasing and Decreasing Effector Function

Like almost all of the mAbs approved for therapy at one time or another, the Fc region utilized in nine Fc fusion proteins belong to the human IgG1 subclass. The IgG4 subclass is utilized in dulaglutide, the fusion protein used for glycemic control (sect. "Dulaglutide" below). The antibody subclass used is an important consideration, especially in cancer treatment, because the biological properties of the four different IgGs differ.

Human IgG1 and IgG3 bind to all FcγRs, IgG2 bind to FcγRIIA and FcγRIIIA and IgG4 binds to FcγRI, FcγRIIA, FcγRIIB, FcγRIIC, and FcγRIIIA. IgG1 (especially in non-fucosylated form) is the subclass of choice for ADCC (utilized in, for example, rituximab, alemtuzumab, and trastuzumab) whereas human IgG2 or IgG4 which do not aid cytotoxicity may be used when killing is not wanted. CDC, involving the binding of complement component C1q, activation of the complement cascade, and ultimately cell death, is another mechanism by which mAbs, such as Ibritumomab, kill tumor cells. IgG3 followed by IgG1 are the most effective activators of the complement cascade while IgG2 and IgG4 are relatively poor activators. Sometimes, fixation and stimulation of complement needs to be avoided or is unnecessary, for example, in reactions involving cytokines, and IgG2 and IgG4, which show lower affinities for Fc and complement receptors, have found use in mAbs such as tositumomab (murine IgG2 with ^{131}I for killing), denosumab (IgG2), panitumumab (IgG2), natalizumab (IgG4), and eculizumab (IgG2/4; targets complement protein C5). Until recently, the Fc domains of either of these two IgG subclasses have not been used in fusion proteins but investigations indicate that they may sometimes impart superior or optimal performance. For the recently approved fusion protein dulaglutide (approved by the FDA, September 2014), the developers selected to employ a modified IgG4 Fc fragment to minimize ADCC and CDC effects of the resultant fusion construct. IgG3's more effective activation of complement and FcγR-mediated functions appears to make it the subclass of choice for immunotherapy but the antibody shows a significantly decreased half-life (~1 week compared to 3 weeks for the other subclasses), and it is therefore generally not considered as a suitable Fc fusion partner. Even so, studies with a human IgG3 containing a His at position 435 instead of the usual Arg show that the half-life of the His435-IgG3 allotype is comparable to IgG1.

Origin, Nature, Mechanism of Action, and Usage of Fc Fusion Proteins

Etanercept

Etanercept (Enbrel®), a recombinant, engineered, fully human dimeric Fc fusion protein linked to the ligand-binding portion of the human TNF receptor (TNFR) (Fig. 6.1) was the first chimeric fusion protein to gain regulatory approval when in

1998 it was approved by the FDA for the treatment of rheumatoid and other forms of arthritis. Like the mAbs infliximab, certolizumab, adalimumab, and golimumab (Table 2.1; Chap. 4, section "Monoclonal Antibodies Targeted to Human Tumor Necrosis Factor: Adalimumab, Certolizumab Pegol, Infliximab, and Golimumab"), etanercept binds tumor necrosis factor (TNF) (Table 6.2) thereby inhibiting the interaction of this cytokine with cell surface TNF receptors and ultimately reducing the ensuing inflammatory response. As with infliximab, etanercept is sometimes administered to patients with rheumatoid arthritis when other treatments have failed and, as well as its indication for this disease, the protein is useful for the treatment of other autoimmune diseases including ankylosing spondylitis, plaque psoriasis and, according to some, Crohn's disease. An interesting successful therapeutic application of etanercept that may be due to its recognition of TNF ligand was seen in its treatment of a so far small number of patients with toxic epidermal necrolysis (TEN)-like acute lupus erythematosus, a syndrome, unlike TEN, where there is generally no identified provoking drug.

Belatacept

Interference with, or prevention of, T cell costimulation can be an effective strategy for immunomodulation, and this approach has been utilized in the development of two Fc fusion proteins, belatacept (Nulojix®) and abatacept (Table 6.2). For naïve T cells to be activated, two signals are required: signal 1 from an antigenic peptide—major histocompatibility complex expressed on antigen-presenting cells which interacts with the T cell receptor; and signal 2, the so-called costimulatory stimulus which is antigen nonspecific and activated when B7-1 (CD80) and B7-2 (CD86) on the surface of dendritic cells bind CD28 on T cells. CD80/86 can also bind a homolog of CD28, cytotoxic T lymphocyte-associated antigen 4 (CTLA-4), which in fact binds with higher affinity. Given the high affinity of CTLA-4 for CD80/86, it was utilized as the peptide partner to prepare abatacept (see below) by fusing the extracellular domain of the lymphocyte antigen to the *N*-terminal of the Fc fragment of human IgG1 (Fig. 6.1) to increase the half-life of the chimeric fusion protein. The rationale for this approach was to block stimulation of T cells via CD28 and although abatacept proved efficacious for some T cell-mediated autoimmune disorders such as rheumatoid arthritis, it proved less efficacious as an immunosuppressant in transplantation. This was because although abatacept showed high affinity for CD80/86, CTLA-4 is a much less potent inhibitor of CD86-dependent than CD80-dependent costimulation. This information led on to the realization that a modified CTLA-4 molecule with higher avidity for CD86 should be sought. Codon-based mutagenesis and surface plasmon resonance studies identified amino acid residues 24 and 104 as critical for binding. The most avid molecule, later named belatacept, differed from abatacept by two amino acid substitutions (L104E, A29Y). It is a homodimeric fusion protein, MW 92.3 kDa, with two polypeptide chains of 357 amino acids. Belatacept binds four times more avidly to CD86 than abatacept resulting in an overall 10-fold increase in biological activity compared to unmodified CTLA-4.

Table 6.2 Chimeric Fc fusion proteins approved for human therapy[a]: Properties, approved indications[a], mechanisms, and side effects (as at June 2016)

Generic and trade names	Properties	Approved indications[a]	Mechanism(s) of action relevant to indications	Side effects, serious and common
Etanercept (Enbrel®)	A dimeric fusion protein, MW 150 kDa, of the extracellular ligand-binding portion of 75 kDa human TNFR with Fc portion of human IgG1[b]	Rheumatoid arthritis; polyarticular juvenile idiopathic arthritis; psoriatic arthritis; plaque psoriasis; ankylosing spondylitis	Acts as a soluble decoy receptor for the inflammatory cytokine TNF blocking its interaction with its natural receptor	*Boxed warnings*: Serious infections; malignancies. Other effects: fever; injection site reactions; cutaneous vasculitis; hypersensitivity (including anaphylaxis, angioedema, urticaria); pruritus; demyelinating disease; cytopenia; lupus syndrome
Belatacept (Nulojix®)	Differs from abatacept by two amino acid substitutions in the ligand-binding region of CTLA-4: alanine 29 to tyrosine and leucine 104 to glutamic acid	Prophylaxis of organ rejection in adult patients receiving a kidney transplant[c]	Blocks CD28-mediated T cell activation and production of cytokines[d] by binding CD80/CD86 on antigen-presenting cells	*Boxed warnings*: ↑ risk of PTLD[e]; ↑ susceptibility to infections and malignancies. Other effects: anemia; diarrhea; peripheral edema; hypertension; urinary tract infections; cough; hypo- and hyperkalemia; graft dysfunction
Abatacept (Orencia®)	Homodimeric fusion protein of Fc fragment of human IgG1 and extracellular domain of CTLA-4. Fc mutated to lose its ADCC and CDC actions	Adult rheumatoid arthritis; juvenile idiopathic arthritis	As for belatacept but lower affinity for CD80/86, slower dissociation rate and less potent and prolonged action	Infections; malignancies; immunogenicity; hypersensitivity; reactions in patients with COPD[f]; injection site reactions; upper respiratory tract infection; headache; nausea
Rilonacept (Arcalyst®)	A dimeric fusion protein of the extracellular ligand-binding domains of IL-1RI and IL-1RAcP linked in line to human IgG1 Fc	Cryopyrin-associated periodic syndromes (CAPS)[g]	Acts as IL-1 trap blocking IL-1β signaling thereby preventing its binding to its cell receptors and reducing inflammation	Injection site reactions; urti; immunogenicity

Generic and trade names	Properties	Approved indications[a]	Mechanism(s) of action relevant to indications	Side effects, serious and common
Aflibercept (Zaltrap®[h]; Eylea®)	Fusion protein of Fc portion of IgG1 and ligand-binding domain 2 of VEGFR1 and domain 3 of VEGFR2	Zaltrap: Metastatic colorectal cancer in combination with FOLFIRI. Eylea: Wet macular degeneration	Acts as a soluble decoy VEGFR1 and VEGFR2 trap binding multiple isoforms of VEGF-A and placental growth factor thereby inhibiting angiogenesis	*Zaltrap Boxed warnings*: Hemorrhage; compromised wound healing; GI perforation. Other effects: cytopenias; proteinuria; hypertension; ↑ serum creatinine; acral erythema; stomatitis. *Eylea*: eye pain; cataract; conjunctival hemorrhage; vitreous detachment; ↑ intraocular pressure
Romiplostim (Nplate®)	Dimeric fusion peptibody[i] MW ~60 kDa; four copies of thrombopoietin-mimetic peptide fused to C-terminus of aglycosylated human IgG1 Fc	Thrombocytopenia in patients with chronic ITP[j]	A thrombopoietin receptor agonist. Stimulates JAK2 and STAT5 pathways[k] → megakaryocytes and ultimately platelets	Arthralgia; dizziness; insomnia; abdominal and shoulder pain; myalgia; pain in extremity; dyspepsia; paresthesia; headache
Alefacept (Amevive®)[l]	A dimeric fusion protein MW 91.4 kDa; consists of the first extracellular domain of LFA-3[m] fused to human IgG1 Fc	Moderate-severe chronic plaque psoriasis in candidates for systemic therapy or phototherapy	Binds CD2 on T cells blocking interaction of CD2 with LFA on APCs thereby inhibiting T cell activation	*Warnings and precautions*: lymphopenia; malignancies; infections; hypersensitivity; hepatic injury; immunosuppression. Other effects: headache; chills; pharyngitis; urti; dizziness; cough; nausea; infections; pruritus; inject site reactions

(continued)

Table 6.2 (continued)

Generic and trade names	Properties	Approved indications[a]	Mechanism(s) of action relevant to indications	Side effects, serious and common
Factor VIII Fc fusion protein[n] (Eloctate®; Elocta®)	Large Fc fusion monomer, 1890 amino acids MW ~ 330 kDa made up of BDD-rFVIII linked to IgG1Fc. Produced in HEK cells	Control and prevention of bleeding episodes in adults and children with hemophilia A; for routine prophylaxis and surgical prophylaxis (perioperative management) to prevent or reduce the frequency of bleeding episodes	Replaces factor VIII in patients with hemophilia A; acts as cofactor for factor IXa in conversion of prothrombin to thrombin which then converts fibrinogen to fibrin and clot formation	*Warnings and precautions*: Hypersensitivity reactions[p]; neutralizing antibodies; monitor factor VIII activity[q] and inhibitors[r]. Other effects: arthralgia; malaise; rash[s]
Factor IX Fc fusion protein (Alprolix®)	Recombinant factor IX expressed in HEK cells covalently linked to Fc domain of human IgG1; fusion protein has 867 amino acids, MW ~ 98 kDa	Indicated in adults and children with hemophilia B for control and prevention of bleeding; perioperative management; routine prophylaxis to prevent or reduce frequency of bleeding episodes	Replaces factor IX in patients with hemophilia B; activated by factor XIa, forms complex with Ca^{2+}, factor VIII and phospholipids to activate factor X ultimately resulting in conversion of prothrombin to thrombin and fibrin clot	*Warnings and precautions*: Hypersensitivity including anaphylaxis; development of neutralizing antibodies; thromboembolic complications. Other effects: headache; oral paresthesia
Dulaglutide (Trulicity®)	IgG4[t] Fc[u] fusion dimer of GLP-1 analog[v] linked via small peptide MW ~ 63 kDa; expressed in HEK cells	As an adjunct to diet and exercise to improve glycemic control in adults with type 2 diabetes mellitus	GLP-1 receptor agonist increases intracellular cAMP via interaction with receptor in beta cells leading to glucose-dependent insulin release. Also decreases glucagon release and gastric emptying	*Warnings and precautions*: Risk of thyroid C-cell tumors; pancreatitis; hypoglycemia[w]; macrovascular outcomes; hypersensitivity; severe GI disease; renal impairment. Other effects: nausea; abdominal pain; diarrhea; vomiting; decreased appetite

APC antigen-presenting cells, *BDDrFVIII* B domain deleted recombinant factor VIII, *COPD* chronic obstructive pulmonary disease, *CTLA-4* cytotoxic T lymphocyte-associated antigen 4, *FOLFIRI* folinic acid (leucovorin), 5-fluorouracil, irinotecan, *GI* gastrointestinal, *GLP-1* glucagon-like protein-1, *HEK* human embryonic kidney, *IL-1RAcP* interleukin 1 receptor accessory protein, *IL-1RI* interleukin 1 receptor type I, ITP immune thrombocytopenia, *JAK2* Janus kinase 2, *LFA-3* lymphocyte function-associated antigen 3, *PTLD* post-transplant lymphoproliferative disorders, *STAT5* signal transducer and activator of transcription 5, *TNF* tumor necrosis factor; *TNFR* tumor necrosis factor receptor, *VEGFR1 and VEGFR2* vascular endothelial growth factor receptors 1 and 2, *urti* upper respiratory tract infection

From Baldo BA. Chimeric fusion proteins used for therapy: Indications, mechanisms, and safety. Drug Saf 2015;38:455–79. Adapted and reproduced with permission from Springer Science+Business Media

[a]Approved by FDA or EMA or both. Two albumin fusion proteins, albiglutide (Tanzeum®, Eperzan®; see Table 6.1, Chap. 6, section "Albiglutide", Chap. 7, "Glucagon-Like Peptide-1", Fig. 7.8) and Factor IX albumin fusion protein rIX-FP (Idelvion®; see Table 6.1, Chap. 6, section "Factor IX Fusion Proteins", Table 10.1, Chap. 10, sections "Factor IX Albumin Fusion Protein" and "Factor IX Fusion Proteins") are also approved

[b]The Fc portion of IgG contains the CH2 and CH3 domains and the hinge region but not the CH1 domain

[c]To be used in combination with basiliximab induction, mycophenolate mofetil and corticosteroids and only in patients who are EBV seropositive

[d]Blocks production of IL-2, IL-4, IFNγ, TNFα

[e]Especially of the CNS. Note: recipients without EBV immunity are more at risk; therefore use in EBV-positive patients only

[f]Effects include exacerbation, cough, rhonchi, dyspnea

[g]Including familial cold autoinflammatory syndrome (FCAS) and Muckle-Wells syndrome (MWS) in adults and children. Note that a supplemental biologics license application for Arcalyst® for the treatment of gout flares was recommended for rejection by the FDA's Arthritis Advisory Committee in May 2012. The FDA has requested additional clinical data and details of proposed dosage

[h]Ziv-aflibercept in the USA

[i]IUPAC name: L-methionyl[human immunoglobulin heavy chain constant gamma-1(227 C-terminal residues)-peptide (Fc fragment)] fusion protein with 41 amino acids peptide (7-7′; 10-10′) -bisdisulfide dimer. Lack of glycosylation negates F_c functionality

[j]Not for thrombocytopenia due to myelodysplastic syndrome or causes other than ITP

[k]Janus kinase 2 is a non-receptor tyrosine kinase. STAT5 consists of two related proteins STAT5A and STAT5B

[l]In FDA Center for Drug Evaluation and Research's (CDER) Discontinued Therapeutic Biologic Products list. Discontinued date, September 28, 2012

[m]Also called CD58

[n]B-domain deleted recombinant factor VIII Fc fusion protein (BDD-rFVIIIFc). Also known as efraloctocog alfa

[o]The B-domain portion of the factor VIII portion of the fusion protein is covalently attached to the Fc fragment by a 14 amino acid linkage

[p]Including anaphylaxis, pruritus, rash, urticaria, facial swelling, dizziness, hypertension, nausea, chest discomfort, cough, dyspnea, wheezing, flushing

[q]By the one-stage clotting assay

[r]Use Bethesda units (BU) to titer antibody inhibitors

[s]Surprisingly few adverse events. Results from a trial with only 164 patients; more experience needed with more patients

[t]To minimize antibody-dependent cell-mediated cytotoxicity and complement-dependent cytotoxicity

[u]Modified to prevent half antibody formation and to reduce potential for interaction with high affinity Fc receptors that may result in immunologic cytotoxicity

[v]Shows 90 % homology to GLP-1(7-37); linked to *N*-terminal of Fc

[w] With concomitant use of insulin secretagogues or insulin

Belatacept acts as an immunosuppressant reducing reliance on calcineurin inhibitors and corticosteroids and appears to show better allograft function and improved cardiovascular and metabolic risk profiles than cyclosporin. This has led to its approval by the FDA for prophylaxis of organ rejection in adult kidney transplant patients.

Abatacept

For a description, the development, and mechanism of action of abatacept, see above under belatacept. Abatacept (Orencia®) (Table 6.2), a homodimeric Fc fusion protein containing the extracellular domain of CTLA-4, is approved as a first-line treatment of adult rheumatoid arthritis and juvenile idiopathic arthritis but preliminary results indicate that the drug may also have a role in the treatment of other autoimmune diseases such as psoriatic arthritis and psoriasis.

Rilonacept

Rilonacept (Arcalyst®) is a dimeric Fc fusion protein in which the IL-1R accessory protein (IL-1RAcP) ligand-binding region is fused via its C-terminus to the N-terminus of the interleukin receptor IL-1RI extracellular domain, and these linked peptides are then fused via IL-1RI to the N-terminus of each of the Fc chains of human IgG1 (Fig. 6.3). Rilonacept, also known as IL-1 trap (target-related affinity profiling), captures IL-1β preventing activation of IL-1 receptors and thus reducing the inflammation and other effects due to overproduction of IL-1 (Table 6.2). Rilonacept was granted orphan drug status and approved for the treatment of cryopyrin-associated periodic syndromes (CAPS) (Chap. 4, sections "CAPS and the Mechanism of Action of Canakinumab," and "CAPS Diseases and Approved Indications for Canakinumab"), a group of rare inflammatory diseases with an incidence of about 1 in one million in the USA. CAPS is generally caused by mutations in the NLRP-3 (nucleotide-binding domain, leucine-rich family [NRL], pyrin domain containing 3) gene. Cryopyrin regulates caspase-1 which controls the activation of IL-1β involved in activation of the immune and inflammatory responses. Mutations in NLRP-3 lead to excess release of IL-1β and the resultant inflammatory symptoms seen in CAPS.

Aflibercept

Aflibercept (Zaltrap®; Eylea®) (Table 6.2), or VEGF trap, is a human recombinant protein made by fusing domain 2 from vascular endothelial growth factor receptor-1 (VEGFR-1) to domain 3 of VEGFR-2 and attaching this combination to the hinge

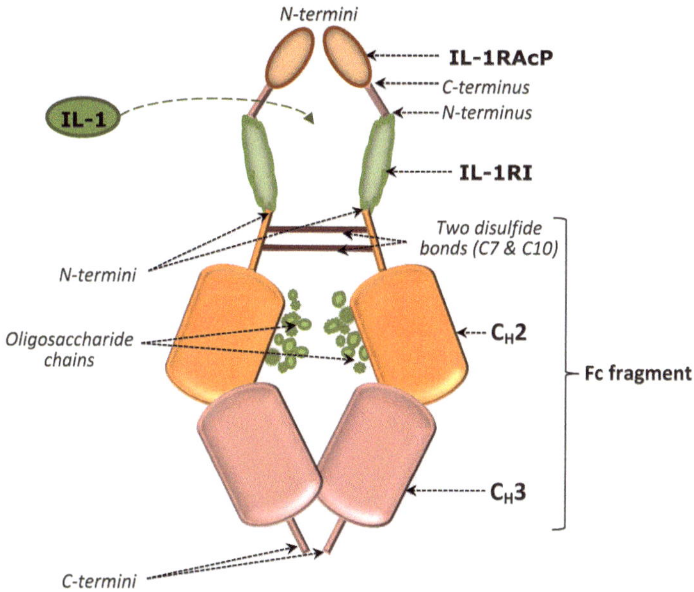

Fig. 6.3 Diagrammatic representation of the structure of the dimeric human IgG1 Fc fusion protein rilonacept or "IL-1trap" which captures IL-1β preventing activation of IL-1 receptors. Reproduced from Baldo BA. Chimeric fusion proteins used for therapy: Indications, mechanisms, and safety. Drug Saf 2015;38:455–79. Reproduced with permission from Springer Science+Business Media

region of the Fc domain of human IgG1. VEGF trap acts as a circulating antagonist-preventing receptor binding by VEGF and placental growth factor (PIGF). Aflibercept, as ziv-aflibercept or Zaltrap®, is used in combination with 5-fluorouracil, leucovorin, and irinotecan (FOLFIRI) for the treatment of oxaliplatin-resistant metastatic colorectal cancer and, as Eylea®, as an ophthalmic intravitreal injection for the treatment of neovascular (wet) age-related macular degeneration and for macular edema following central retinal vein occlusion.

Romiplostim

Romiplostim (Nplate®) (Table 6.2), a ~60 kDa so-called peptibody, is formed by the fusion of four identical copies of a thrombopoietin mimetic peptide to the C termini of aglycosylated human IgG1 Fc chains produced in *E. coli*. Each H chain of the Fc-protein is attached at residue 228 by a pentaglycine bridge to a molecule of the thrombopoietin mimetic peptide linked to another molecule of the peptide by an octaglycine bridge (Fig. 6.4). Note that unlike the Fc fusion proteins reviewed above, the biologically active peptide of romiplostim is linked to the C-terminal not the N-terminal end of the Fc fragment. The peptide employed was identified by

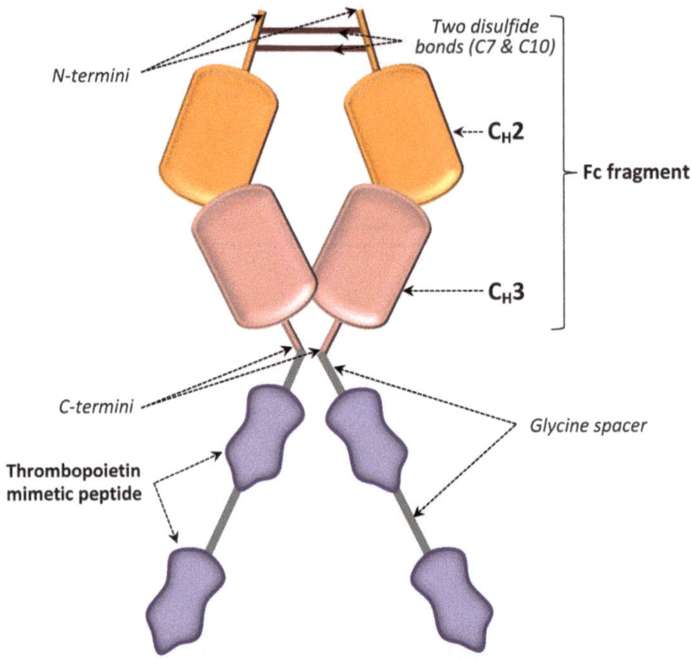

Fig. 6.4 Diagrammatic representation of the structure of the dimeric human IgG1 Fc fusion protein romiplostim which acts as a thrombopoietin receptor agonist. Note that unlike the other 6 approved chimeric Fc fusion proteins, the effector peptide partner (two linked thrombopoietin mimetic peptide molecules attached to each Fc H-chain) of romiplostim is linked to the *C*-terminal, not the *N*-terminal, end of the Fc fragment. Note also that the oligosaccharide chains normally attached to the IgG1 C_H2 domains are absent in romiplostim. Reproduced from Baldo BA. Chimeric fusion proteins used for therapy: Indications, mechanisms, and safety. Drug Saf 2015;38:455–79. Reproduced with permission from Springer Science + Business Media

screening phage libraries and optimized for its capacity to displace thrombopoietin (TPO) from its receptor hTPOR and to stimulate the proliferation of cells specifically engineered to express hTPOR.

Alefacept

Interaction of CD2 with human cell surface LFA-3 (lymphocyte function-associated antigen 3; CD58) produces signaling events involved in the regulation of T cell responses (Table 6.2). In 1993, G. T. Miller et al. localized the CD2-binding site to the first domain of LFA-3, used as the effector peptide to prepare a fusion protein with IgG1, and showed that this *N*-linked fusion protein could inhibit cell proliferation and mixed lymphocyte reactions. Alefacept (Amevive®), which can be viewed as an anti-CD2 (Table 6.2), selectively targets effector memory T cells for both CD4+ and CD8+ but not naïve T cells and central memory T cells in psoriasis vulgaris. The increased expression of CD2 on effector T cells, their decrease induced by alefacept, and their concentration in psoriatic skin lesions, suggest a role for the effector T cells in disease pathogenesis. However, although circulating memory T cells decrease in all alefacept-treated psoriasis patients, a good correlation does not exist between clinical response and circulating memory T cell reductions. The mechanism of T cell reduction also remains to be established.

Factor VIII Fc Fusion Protein

Factor VIII (Chap. 10, section "Factor VIII") synthesized as a large single-chain protein of 2332 amino acids (MW ~ 300 kDa) with several distinct domains A1-A2-B-A3-C1-C2 can lose most of the B domain without any diminution of factor VIII activity. This finding led to the construction of the so-called B-domain deleted recombinant factor VIII protein, designated BDDrFVIII, which is covalently linked to the human IgG1 Fc to prepare the factor VIII fusion protein (rFVIIIFc) (Eloctate®; Elocta®) (Table 6.2; see also Chap. 10, section "Factor VIII Fc Fusion Protein" and Fig. 10.9) with a somewhat extended half-life of ~19.7 h (Table 6.1). The fusion protein homodimer was not well secreted but the active Fc fusion monomer showed good pharmacodynamic and pharmacokinetic properties.

See Chap. 10, section "Factor VIII Fc Fusion Protein" for a description of factor VIII Fc fusion protein and section "Factor VIII Fc Fusion Protein" and Tables 6.2 and 10.1 for safety data on this recently approved fusion protein.

Factor IX Fc Fusion Protein

Factor IX, licensed since 1997 for the treatment of hemophilia B, is a ~55 kDa, vitamin K-dependent serine protease with a relatively short half-life of 14–34 h. Recombinant factor IX Fc fusion protein (rFIXFc) (Alprolix®) (Table 6.2; Chap. 10, section "Factor IX Fc Fusion Protein") is a monomeric construct with a single factor IX ligand covalently linked to the Fc fragment of human IgG1 and an extended half-life of approximately 3.5 days (Table 6.1). Again, as with rFVIIIFc, the monomeric form proved pharmacodynamically and pharmacokinetically superior to a dimeric factor IX- Fc fusion construct. Recombinant factor IX has an amino acid sequence identical to one allelic form of plasma-derived factor IX and structural and functional properties similar to the natural coagulation factor.

See Chap. 10, section "Factor IX Fc Fusion Protein" for a description of factor IX Fc fusion protein and Tables 6.2 and 10.1 for safety data on this recently approved fusion protein.

Dulaglutide

Dulaglutide (Trulicity®), used to treat type 2 diabetes, is a modified, long-acting receptor agonist analog of human GLP-1 covalently attached to each of the two chains of a modified human IgG4 Fc fragment by a 16 amino acid peptide linker (Chap. 7, sections "GLP-1 Receptor Agonists" and "Safety of GLP-1 Receptor Agonists"). To prevent inactivation by dipeptidyl peptidase 4 (DPP-4) (Chap. 7, section "Glucagon-Like Peptide 1 and The Incretin Effect") and reduce immunogenicity, amino acid modifications were made at positions 8, 22, and 36 of GLP-1 (see Fig. 7.11) while a modified IgG4 Fc was chosen as the fusion partner to minimize ADCC and CDC of the fusion construct, prevent half-antibody formation, and reduce interaction with high affinity Fc receptors. With as half-life of approximately 4.7 days, dulaglutide can be administered once a week.

Atacicept

Atacicept (see Chap. 4, section "APRIL, Lupus, and Atacicept"), a fully humanized recombinant fusion protein is composed of the extracellular, ligand-binding portion of the TACI receptor fused to a modified Fc fragment of human IgG1. TACI (transmembrane activator and calcium modulator and cyclophilin ligand interactor), expressed on memory B calls and bone marrow plasma cells, is a negative regulator of B cell maturation and one of the three TNF receptors for the B cell survival factor soluble B lymphocyte stimulator protein (BLyS). TACI also binds APRIL (a proliferation-inducing ligand), another member of the TNF superfamily secreted by activated myeloid cells and with a stimulatory action on B cells. Developed as a potential therapeutic agent to bind APRIL as well BLyS and, as yet, not approved by the regulatory agencies, atacicept initially proved a disappointment in trials designed to assess its efficacy in treating lupus.

Safety of Approved Fc Fusion Proteins

As outlined above, 10 of the 12 chimeric fusion proteins (etanercept, belatacept, abatacept, rilonacept, aflibercept, romiplostim, alefacept, dulaglutide, the coagulation protein factor VIII and one of the factor IX fusion proteins), registered and approved for therapy at one time or another, are Fc fusion proteins (Table 6.1). The safety of the IgG4 Fc construct dulaglutide (Chap. 7, section "GLP-1 Receptor Agonists") is summarized in Table 6.2 and further considered in Chap. 7, sections "Safety of GLP-1 Receptor Agonists" and "Immunogenicity of GLP-1 Receptor Agonists". Issued warnings and precautions for factor VIII and factor IX Fc fusion protein are summarized in Tables 6.2 and 10.1 and dealt with in chapter 10, sections "Factor VIII Fc Fusion Protein" and "Factor IX Fusion Proteins," respectively.

Safety considerations for the albumin fusion protein albiglutide (section "Albiglutide") are summarized in Chap. 7, section "Safety of GLP-1 Receptor Agonists." Denileukin diftitox, a genetically engineered recombinant fusion construct consisting of a toxin fused to IL-2 and therefore quite different to the other fusion proteins, is considered separately below (section "Denileukin Diftitox").

The remaining seven approved Fc fusion proteins are discussed here with their properties, approved indications, mechanisms of action, and adverse effects summarized in Table 6.2.

Etanercept

Early studies in particular reported that the most common adverse events associated with etanercept administration are relatively mild, for example, fever, headache, injection site reactions, mild allergic reactions, and pruritus but, as so often occurs after extended usage over time, a wide variety of less often seen and sometimes rare and serious reactions are now known. In a retrospective safety and effectiveness assessment of etanercept given to 118 patients, 51 patients (43.2 %) experienced adverse events with one quarter being infections, mainly those of the upper respiratory tract. A long-term examination of responses of 182 pediatric patients (4–17 years) with plaque psoriasis, revealed at least one adverse event in 80 % of the participants with upper respiratory tract infections showing the highest incidence (25 %), followed by nasopharyngitis, streptococcal pharyngitis (each 13 %), and sinusitis (11 %). No opportunistic infections were recorded. Continuous treatment with etanercept over 2 years of 110 patients with psoriatic arthritis recruited from 22 Canadian clinical practices, resulted in 20 serious adverse events in 14 patients. These events included abdominal abscess, appendicitis, malignant lung neoplasm, pneumonia, streptococcal infection, angina, cardiac arrest, and cerebral hemorrhage.

The list of etanercept adverse events issued in order of frequency by the FDA comprises infection, dermatologic, neurologic, musculoskeletal, pulmonary, cardiac, and vascular effects but before discussing each of these categories, it is important to point out a list of exclusion criteria formulated after careful consideration of the etanercept adverse event profile. One such list presented on behalf of the British Society for Rheumatology Standards, Guidelines and Audit Working Group (SGAWG), refer to six different exclusion criteria: (1) Women who are pregnant or breast feeding; (2) the presence of active infection; (3) septic arthritis of a joint in the last year; (4) sepsis of a prosthetic joint; (5) grade 3 or 4 congestive cardiac failure; (6) a clear history of demyelinating disease.

Etanercept and Infection Risk

As referred to above, some retrospective assessments and trial results clearly show that a high incidence of infection often accompanies etanercept therapy. In fact, the infection risk in placebo-controlled trials has been estimated at ~35 %

and serious infection is one of the two black box FDA warnings for the drug (Table 6.2). Specifically, the warnings refer to: (1) Increased risk of serious infections or death due to tuberculosis, bacterial sepsis, and other opportunistic pathogens; (2) the need to discontinue the drug if infection or sepsis develops; (3) the need to perform tests for latent TB; (4) the need to monitor all patients for active TB during treatment. Note that the TNF-targeted mAb infliximab carries with it a 3–4 higher risk than etanercept. With reference to granulomatous infections associated with etanercept or infliximab where tuberculosis, histoplasmosis, candidiasis, and listeriosis make up more than 80 % of the reported cases, analysis of the reports received by the FDA's Adverse Event Reporting system revealed that of 639 episodes of granulomatous infections, 556 were associated with infliximab and 83 with etanercept. In contrast to infliximab, etanercept may generally be associated with less severe infections but severe cases of viral pneumonia, pneumococcal sepsis, and osteoarticular tuberculosis have occurred and fatalities reported. The range of organisms implicated in infections associated with etanercept therapy is wide and includes bacteria (tuberculosis, streptococcus, listeria, Actinobacillus), viruses (varicella), fungi (aspergillus), protozoa (toxoplasma), and cestodes (echinococcus).

Cutaneous Events

The most common skin manifestation seen with etanercept is an injection site reaction (Fig. 6.5) which may be seen in ~20–50 % of patients. These are usually mild to moderate and involve erythema, edema, pain, and pruritus. Reactions occur within the first 2 months of treatment and may reoccur 1–2 days after the final injection. Interestingly from an immunological viewpoint, some patients develop recall reactions while continuing etanercept and skin biopsy and immunohistological examinations revealed an inflammatory infiltrate made up mainly of lymphoid cells with some eosinophils and without evidence of leukocytoclastic vasculitis. The lymphoid cells were predominantly activated mature cytotoxic HLA-DR/CD3+/CD4-/CD8+ T lymphocytes. Biopsy of a recall reaction showed epidermal keratinocytes with strong expression of HLA-DR. The results were seen as consistent with a T cell-mediated delayed hypersensitivity reaction which waned over time due to tolerance. Cases of cutaneous vasculitis either as sole manifestation of etanercept therapy or concomitant with other conditions such as severe glomerulonephritis or accelerated nodulosis have been reported and necrotizing vasculitis with eosinophils shown by biopsy and described as an autoimmune skin rash has been described. Other cases of nodules developing during etanercept therapy have also been published. There are rare reports of discoid lupus erythematosus, clinically subacute cutaneous lupus erythematosus, and anti-neutrophil cytoplasmic antibody (ANCA)-associated vasculitis induced by anti-TNF therapy including etanercept.

In a comprehensive survey of 126 study reports, Lecluse and coworkers in 2011 identified 72 separate adverse skin events provoked by etanercept from

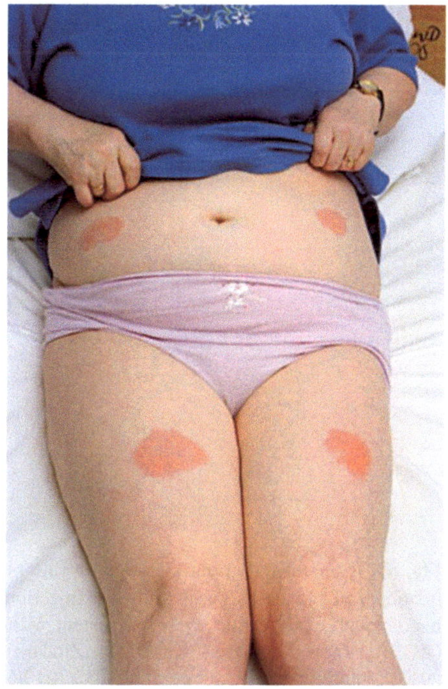

Fig. 6.5 Injection site reactions at multiple sites in a rheumatoid arthritis patient treated with etanercept. Injection site reactions are the most common skin manifestation seen with etanercept and occur in ~20–50 % of patients given the drug. Reproduced with permission from Rajakulendran S, Deighton C. Delayed multiple injection site reaction in a rheumatoid arthritis patient treated with etanercept. Rheumatology. 2004;43:1588–9

approximately 65 specific diagnoses in case studies involving 153 patients. Various rashes of unknown pathology and urticaria were the most commonly occurring reactions. Overall, etanercept was found to be associated with a wide variety of adverse dermatological events, most mild but some serious and a few life threatening. Recorded cases include new onset and exacerbations of psoriasis and atopic dermatitis, induction of psoriasis, lichenoid reaction, erythema multiforme, angiokeratomata, pemphigus vulgaris, and palmoplantar pustulosis. There are two interesting, and rare, reports of acute generalized exanthematous pustulosis (AGEP) induced by etanercept. Within two days of the initiation of etanercept treatment (50 mg s.c. twice weekly) for severe plaque psoriasis, an adult male developed a widespread maculopapular rash. By day four of etanercept treatment plus antihistamines and corticosteroids, the rash had progressed to a generalized edema with tiny pustules (Fig. 6.6). Histological examination revealed neutrophils, no infection and mild spongiosis consistent with AGEP. The rash improved upon withdrawal of etanercept. Following interruption and reintroduction of etanercept therapy, a 51-year-old female patient developed multiple erythematous and edematous facial lesions with small non-follicular, sterile pustules accompanied by fever above 38 °C (Fig. 6.7a). After a few days, diffuse erythema and swelling with postpustular desquamation spread to the limbs (Fig. 6.7b). Lesion morphology, the course of the disease and histological findings led to a clear diagnosis of AGEP.

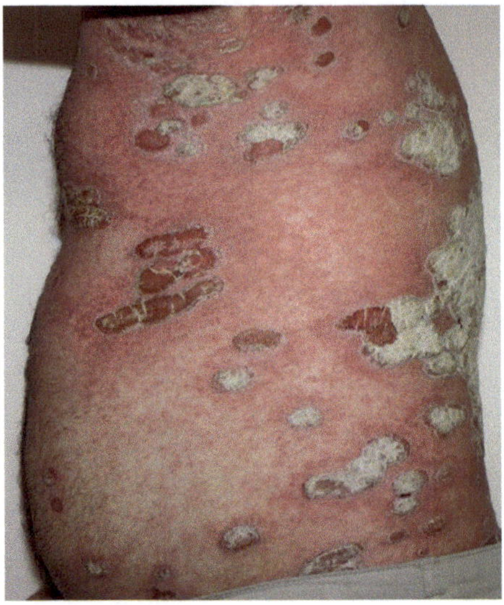

Fig. 6.6 Generalized erythroderma with tiny pustules in a male adult 4 days after initiation of treatment of plaque psoriasis with two injections of etanercept. Clinical features and results of histological examination were consistent with acute generalized exanthematous pustulosis (AGEP). Reproduced from Kavala M, Zindanci I, Türkoglu Z et al. Case Rep Dermatol Med. 2013; article ID 601412. doi.org/10.1155/2013/601412, an Open Access article distributed under the terms of the Creative Commons Attribution License

Neurologic Events

Included in exclusion criteria for etanercept, for example by the British Society for Rheumatology SGAWG (see section "Etanercept" above), is the reminder of the risk of demyelinating disorders. In an early double-blind, placebo-controlled phase II study conducted on 168 patients, most with relapsing, remitting multiple sclerosis (MS), lenercept, now discontinued, a recombinant fusion protein similar to etanercept, combining two extracellular domains of the human p55 kDa TNF receptor with one IgG1 heavy chain, was investigated for a possible effect on the induction of new lesions detected by MRI scans and clinical assessments. Results showed an increase in the frequency of attacks and the severity of attacks appeared to worsen. Demyelination occurring during anti-TNF therapy for inflammatory arthritides and reported to the FDA includes 17 cases associated with etanercept. Symptoms of one case included paresthesia, optic neuritis, and confusion. Transverse myelitis accompanied by antinuclear and anti-cardiolipin antibodies occurred abruptly in a 45-year-old woman nine days after the commencement of etanercept, and there are reports of a number of demyelinating disorders during etanercept therapy including the so-called demyelinating syndrome; relapsing remitting MS; demyelination of the spinal cord and cerebral cortex; progressive multifocal leukoencephalopathy (PML); posterior reversible encephalopathy syndrome (PRES); demyelinating disease that resolved after withdrawal of the drug; and central nervous system demyelination producing ophthalmic manifestations.

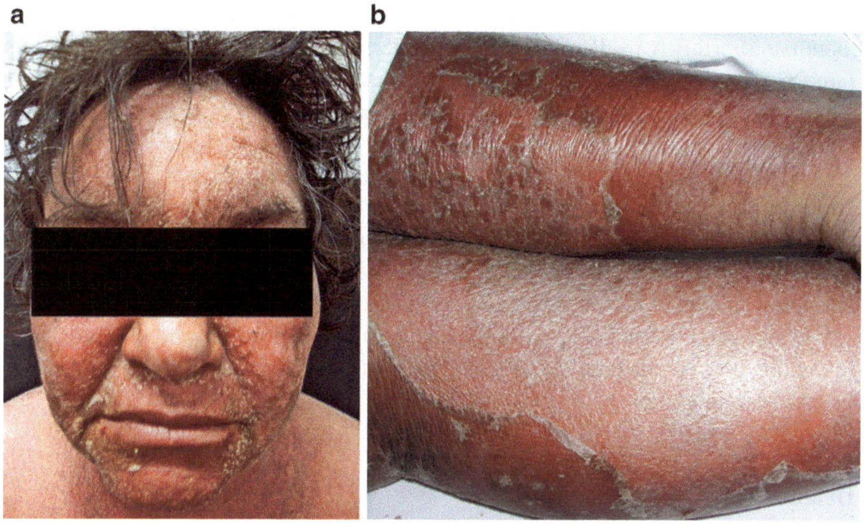

Fig. 6.7 (**a**) Edematous erythema with perioral non-infected pustules on the face of a female patient with refractory erythrodermic psoriasis and psoriatic arthritis following reintroduction of etanercept therapy. (**b**) Progression after a few days of diffuse erythema and swelling of the limbs showing characteristic postpustular desquamation. Figures (**a**) and (**b**) both reproduced from Vasconcelos LMF, Teixeira FM, Francelini EV et al. J Pharmacovigilance. 2014;2:120. doi:10.4172/2329-6887.1000120, an Open Access article distributed under the terms of the Creative Commons Attribution License

Tumorigenicity of Etanercept

The possibility of the development of lymphoma and other malignancies in children and adolescents treated with TNF blockers is the subject of a second FDA black box warning (Table 6.2). A number of studies suggest an increased lymphoma risk for patients with rheumatoid arthritis, particularly those with severe disease, and the risk appears to be slightly increased in those patients treated with TNF antagonists. If this is true, the cause remains poorly understood. Analysis of data on 8614 patients, 29 with lymphoma, from the US National Data Bank for Rheumatic Diseases, revealed a standardized incidence ratio (SIR) for lymphoma of 3.8 for etanercept. A second cohort study of data registered in the South Swedish Arthritis Treatment Group compared 757 patients treated with etanercept with 800 patients on conventional antirheumatic therapy. In the former group, 16 tumors, including 5 lymphomas, occurred in 1603 person-years at risk (SIR 1.1, relative risk for lymphoma 11.5); in the comparison group, 69 tumors (2 lymphomas) occurred in 3948 person-years (SIR 1.4, relative risk 1.3). Total tumor relative risk excluding lymphoma was 0.79 and 1.39, respectively. These results led to the overall conclusion that TNF blockers do not increase overall tumor risk in patients with rheumatoid arthritis but may be associated with an increased risk of lymphomas. As this debate continues, a number of methodological questions about the design, execution, and analysis of the different studies has been raised. These questions include whether or not the risk following anti-TNF treatment is explained by the risk of the disease, rheumatoid arthritis itself, rather than the therapy; do all of the patients carry the same risk, e.g., those with more severe disease?; the influence of previous treatments; and is the risk related to cumulative dose and length of therapy? As well as lymphoma, there are a few reports of other malignancies associated with etanercept therapy. In the so-called Wegener's Granulomatosis Etanercept Trial Research Group study involving 180 patients with the active disease, six solid malignancies occurred. Although all patients treated with etanercept who developed tumors were also treated with cyclophosphamide, it was concluded that the combination of TNF inhibition and cyclophosphamide may increase the risk of cancer over and above the risk from cyclophosphamide alone. There are a few reports of patients developing squamous cell carcinomas after treatment with etanercept including two separate cases of carcinoma of the penis.

Hematologic Events

The incidence of etanercept-induced hematologic disorders is not clear. A post-marketing survey of 820 patients found an incidence of 3.4 per 1000 patient-years for patients treated with etanercept but this figure is certain to be influenced by the treatment of many of the subjects with other antirheumatic drugs, particularly methotrexate. After a reported ten cases which ended in fatal sepsis, regulatory agencies issued post-marketing warnings of the possibility of pancytopenia or aplastic anemia. Subsequence experience, however, has shown hematological cases occur only rarely, for example, a case of reversible aplastic anemia after 16 weeks of etanercept therapy and myelopoiesis diagnosed as exacerbation of macrophage activation syndrome in a young woman with adult-onset Still's disease. After multiple transfusions, intravenous immunoglobulin, and granulocyte-macrophage colony-stimulating factor, she was successfully treated with methylprednisolone and cyclosporin. Of 267 patients with rheumatoid arthritis, ankylosing spondylitis or psoriatic arthritis receiving etanercept therapy, 49 (18.4 %) developed at least one episode of neutropenia, although only about 1 % of patients developed severe infections secondary to neutropenia. Other reports of cytopenias following the fusion protein include cases of neutropenia, leukopenia, thrombocytopenia, and leukopenia alone. In 2010, the British Society of Rheumatology updated their consensus guidelines recommending regular complete blood cell counts for patients undergoing TNF inhibitor therapy.

Respiratory Events

Possible nodulosis after etanercept is well known (Sect. 5.3.1.2), but pulmonary nodulosis is an unusual manifestation of rheumatoid disease. Pulmonary nodules have been diagnosed after etanercept treatment but it is not always necessary to discontinue the drug. The condition demonstrates the need for careful monitoring of the drug treatment and for testing to achieve a differential diagnosis of tuberculosis. Pulmonary granulomas after etanercept may also be difficult to distinguish from other lung complications including tuberculosis. Infection as a cause was not ruled out in two cases examined by lung biopsy—non-caseating granulomas containing birefringent particulates in one and caseating necrosis in the other. Two patients, both previously given methotrexate and with preexisting lung disease, developed acute respiratory symptoms within 3–6 weeks of beginning etanercept therapy. Rapid deterioration into accelerated interstitial lung disease ensued and one patient died despite aggressive treatment. The authors concluded that caution is needed with rheumatoid patients taking methotrexate and with preexisting lung disease when etanercept is added. Three other cases of exacerbation of preexisting interstitial lung disease after administration of etanercept and a case of organizing pneumonia in a patient with rheumatoid arthritis treated with etanercept have been described.

Immunologic Events

Any discussion of immunological events associated with etanercept therapy, and in fact any therapy with protein biologics, needs to recognize that many events considered under a range of different headings, for example, cutaneous, hematologic, pulmonary, renal, endocrine, and probably others, are either already known to have an immune basis or component or such an association is suspected. Thus, a number of dermatologic events provoked by the fusion protein and diagnosed as psoriasis, atopic dermatitis, erythema multiforme, and acute generalized exanthematous pustulosis are almost certainly type IV hypersensitivity responses. Likewise, the occasional recall skin responses and identification of CD8+ cytotoxic T cells seen in injection site reactions indicate a delayed hypersensitivity mechanism while some early responses in such reactions may be true immediate, type I, hypersensitivities. At least some cases of etanercept-induced cytopenias and vasculitis may be type II and type III hypersensitivities, respectively, and some pulmonary events caused by etanercept may ultimately be shown to be type III or combined type III/type IV reactions. There are increasing indications that etanercept can provoke autoimmune reactions such as hyperthyroidism and the development of anti-synthetase syndrome, Crohn's disease, and Henoch–Schönlein purpura (Fig. 6.8). In one case of the latter, increased concentrations of IgA rheumatoid factor resulted in IgA immune complexes; in another case, Henoch–Schönlein purpura occurred with acute renal failure. In fact, biologics-induced autoimmune renal disorders are being increasingly recognized, especially in relation to etanercept which has been the biological agent most frequently identified with the condition. A total of 20 vasculitic adverse events after etanercept, particularly hypersensitivity vasculitis and necrotizing vasculitis, were recorded by the FDA Adverse Events Reporting System. Most cases developed within 3 months but evidence to establish underlying mechanisms of the reactions was generally not sought, let alone published. This subject, together with the question of whether some adverse reactions to etanercept and other fusion

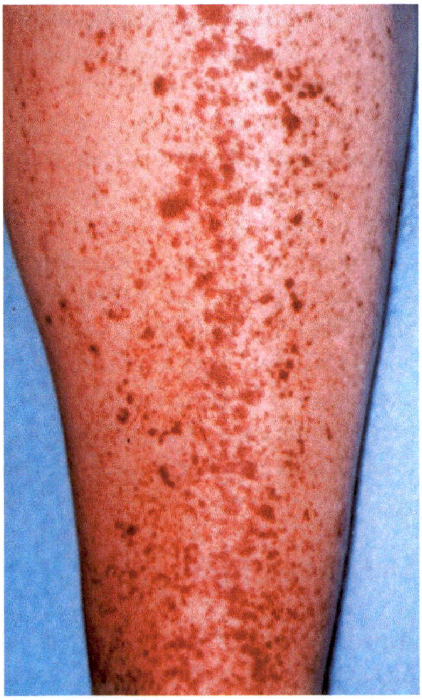

Fig. 6.8 Henoch–Schönlein purpura, usually confined to the legs and buttocks, is a systemic vasculitis affecting the arterioles, capillaries, and venules. Reproduced from Feldmann R, Rieger W, Sator P, et al. BMC Dermatology. 2002;2:1. doi:10.1186/1471-5945-2-1. http://www.biomedcentral.com/1471-5945/2/1 An Open Access article permitting copying and redistribution

proteins are true type I, II, III, or IV hypersensitivity responses, is taken up below in section "Diagnosis of Hypersensitivities to Fusion Proteins, Premedication and Desensitization."

The occurrence of anti-etanercept antibodies is well known but there appears to be no evidence so far that these antibodies reduce the clinical efficacy of the drug. The antibodies occur with frequencies of 3–5.6 %, 0 %, 8 %, and 18 % in rheumatoid arthritis, ankylosing spondylitis, children with juvenile idiopathic arthritis and psoriasis, respectively. It has been known for some time that patients given etanercept commonly develop antinuclear and/or anti-double-stranded DNA antibodies and cases of a lupus-like syndrome, cutaneous lupus erythematosus (Fig. 6.9), and acute discoid lupus have been reported. In at least one patient, the antinuclear and anti-DNA antibodies appeared to be associated with treatment failure. There appears to be at least a half dozen possible/probable reports of immediate hypersensitivity/anaphylaxis to etanercept including two cases of anaphylaxis in children with juvenile idiopathic arthritis and two episodes of angioedema. In two other reported cases, one patient experienced urticaria and swelling of the tongue and periorbital regions within hours of administration of etanercept; the second patient, who had Still's disease, experienced facial swelling, periorbital edema, diffuse pruritic rash, and difficulty swallowing, again within hours of the injection. In a study of hypersensitivity reactions to the anti-TNF agents infliximab, adalimumab, and etanercept, nine patients had reactions to etanercept, five with urticaria/angioedema, and four local reactions. Positive intradermal tests to the drug were seen in two of the five patients with urticaria/angioedema and three of the patients with local reactions. Two patients

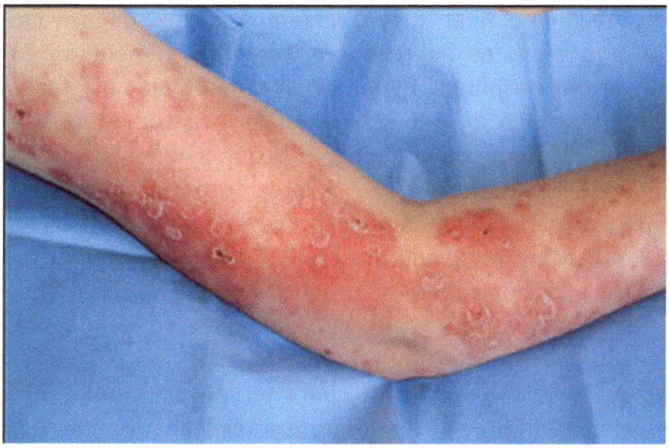

Fig. 6.9 Erythematous macular eruption on the arms of a patient diagnosed with cutaneous lupus erythematosus induced by etanercept. Reproduced with permission from Abourazzak FE, Guggenbuhl P, Perdriger A, et al. La Rev Med Interne. 2008; 29:744–7

were found to react to etanercept and both mAbs; one proved skin test-positive only to infliximab, the second positive to all three agents. Although an injection site reaction is the most common adverse event of etanercept therapy, the mechanism(s) of the reaction(s) has not been well studied and suggestions that at least some of the reactions are type I hypersensitivities is not always supported by definitive evidence. Positive skin tests to etanercept, both prick and intradermal, were obtained in two patients treated with the drug but so far there are no reports of specific IgE antibodies. A patient who developed a severe generalized pruriginous exanthema 2 h after receiving etanercept proved patch and prick test-negative to the agent but reacted positively upon intradermal testing. As noted, some injection site reactions, sometimes at multiple sites, may be delayed hypersensitivities mediated by CD8+ lymphocytes but biopsies from two patients with rheumatoid arthritis who developed recall reactions during etanercept therapy revealed a predominance of CD4+ T cells in the inflammatory infiltrate. In another variant finding, biopsy specimens from a pruritic, erythematous, and edematous reaction on the thigh of a woman treated with etanercept demonstrated papillary edema and a polymorphous infiltrate with a predominance of eosinophils and scattered flame figures. This was interpreted as a case of eosinophilic cellulitis proceeding via a Th2-mediated response.

Belatacept Safety

Approximately 20 % of patients undergoing therapy with belatacept develop adverse effects, the most common of which are listed in Table 6.2. In the two major phase III trials termed BENEFIT and BENEFIT-EXT, acute infusion reactions, defined as a reaction within the first hour of an infusion, occurred in a total of 24 of the 804 patients (3 %) receiving the more intense and less intense belatacept dosage regimens. All reactions were mild to moderate except for one which was a serious prolonged hypotensive event. Data on infections from the phase III trials, revealed mild to severe urinary tract infections in 263 of 949 patients (27.7 %) receiving the drug, upper respiratory infections in 8.5 %, cytomegalovirus in 10.1 % (compare 11.7 % in cyclosporin patients), and pneumonia in 2.5 % of patients. Immunosuppression with belatacept-based and corticosteroid-avoiding regimens in de novo kidney transplant recipients was studied in a one-year controlled, open-label study in which recipients of renal allografts were randomized to receive belatacept-mycophenolate mofetil, belatacept-sirolimus, or tacrolimus-mycophenolate mofetil. Infection of any sort was seen in 79 %, 77 %, and 67 %, and a serious infection in 21 %, 15 %, and 17 % of patients in the three different groups, respectively. These findings indicate that with respect to infection rate, belatacept compares favorably to cyclosporin. However, in addition to an FDA black box warning stating that "only physicians experienced in immunosuppressive therapy and management of kidney transplant patients should prescribe Nulojix," the warning also states that "increased susceptibility to infection…may result from immunosuppression" and the drug's use in liver transplant patients is not recommended due to an increased risk of graft loss and death.

Fatal PML (Chaps. 1 and 4, sections "Progressive Multifocal Leukoencephalopathy" and "Posterior Multifocal Leukoencephalopathy and Natalizumab") occurred in one kidney and one liver transplant patient receiving the more intense belatacept regimen. This lead to this condition being included in the FDA's "Warnings and precautions" for the fusion protein (Table 6.2). In the two phase III belatacept trials, 11 out of 445 patients (2.5 %) in one trial developed malignancy and 8 out of 359 (2.2 %) developed malignancy in the other. Comparative incidences for cyclosporin were 0.5 and 3.3 %. The possibility of the development of malignancies is included in the current FDA black box warning for belatacept (Table 6.2). Of additional concern was the diagnosis of five cases of post-transplant lymphoproliferative disorder (PTLD) (two involving the CNS) in the BENEFIT trial and five cases (all five involving the CNS) in the BENEFIT-EXT trial. Taking into account the major trials and the follow-up period, 13 cases of PTLD occurred, while for cyclosporin only two cases (0.4 %) were recorded. Interestingly, Epstein–Barr virus (EBV or human herpesvirus 4 [HHV-4]) seronegativity is a risk for developing PTLD from belatacept therapy for kidney transplantation. EBV-seronegative transplant recipients developed PTLD with an incidence of 7.3 % compared to 0.6 % for EBV-seropositive patients. The importance of a transplant recipient's EBV-serum antibody status appears to be borne out by the apparent absence of cases of PTLD in a trial excluding EBV-seronegative transplant recipients. In June 2011, the FDA issued a document entitled Risk Evaluation and Mitigation Strategy (REMS) for Nulojix® in which the stated goals were to inform healthcare providers of the increased risk of post-transplant PTLD, predominately in the CNS, associated with Nulojix®; the increased risk of PML associated with the drug; and to inform patients of serious risks associated with

Nulojix®. To ensure the effectiveness of the REMS, the provision of a medication guide with each Nulojix® infusion and a communication plan for healthcare workers were required. PTLD and the relevance of EBV-negativity/positivity are also included in the current FDA black box warning.

Development of antibodies to belatacept was assessed in 372 treated patients, many for up to 2 years. Of 29 who tested positive, 13 had antibodies to the modified CTLA-4 fusion protein. Anti-belatacept antibodies were not implicated in altered clearance of the drug.

Abatacept

Adverse Events Identified in Clinical Trials

The most frequently reported adverse reactions to abatacept recorded in a number of trials include nasopharyngitis, headache, nausea, diarrhea, upper respiratory tract infections, and arthralgia. The safety of abatacept was assessed in multicenter, randomized, double-blind, placebo-controlled studies of patients with methotrexate- or anti-TNF-resistant rheumatoid arthritis. Acute infusion reactions occurred more frequently in the abatacept-treated groups than the placebo groups although there was no apparent relationship between serious reactions and the number of infusions. Two patients discontinued participation because of a hypersensitivity rash and one discontinued because of hypotension. Infections, particularly pneumonia, nasopharyngitis, sinusitis, upper respiratory infections, and bronchitis, were seen more often in the abatacept groups. The incidences of benign and malignant neoplasms and hematological disorders were similar in the drug and placebo groups and no major autoimmune disorders such as lupus or multiple sclerosis occurred. Immunogenicity studies found only small numbers of patients developed antibodies to abatacept—six (1.4 %) in one trial and, in another, 3 of 234 (1.3 %) with one against the Fc portion and two against CTLA-4. An integrated safety analysis carried out on the abatacept clinical trials assessed results on 1955 patients treated with the fusion protein during the double-blind periods and 2688 during the cumulative double-blind and open-label periods, yielding 4764 patient-years of exposure in total. Acute infusion reactions were mostly mild to moderate, the overall frequencies of adverse events, serious adverse events and malignancies were similar in the treated and control patients and abatacept was associated with low immunogenicity with no associated safety or efficacy issues. In a similar safety assessment that included open-label, long-term extension of exposure (3–5 years) data, incidence rates defined as events/100 patient-years were determined for serious events (8.76), infections (44.8), serious infections (1.72), malignancies (1.19), and autoimmune events (1.31). Twenty-seven patients (2 %) experienced injection site reactions, all except one of which were mild.

Other Reported Adverse Events to Abatacept

Adverse skin reactions to abatacept appear to be rare but there are a number of cases of paradoxical psoriasiform eruptions to the agent. A meta-analysis of trials in which rheumatoid arthritis patients received abatacept as monotherapy (1332 patients), or together with other antirheumatic drugs (1945 patients), revealed 4 (0.3 %), and 13 (0.67 %) cases of psoriasis, respectively. A case of erythema elevatum diutinium has been reported in a juvenile idiopathic arthritis patient treated with abatacept.

The recent interesting observations that abatacept produced partial or complete remissions of proteinuria in five patients with focal segmental glomerulosclerosis (FSGS), that four of the patients had recurrent FSGS after transplantation, and patients had proteinuria with B7-1 immunostaining of podocytes in kidney biopsy specimens, suggests that B7-1 may be a useful marker for the treatment of some glomerulopathies.

Rilonacept

Perhaps reflecting its orphan drug status and therefore consequent lighter usage, the list of adverse events provoked by rilonacept (Table 6.2) is relatively short compared to most therapeutic biologics. Clinical trials, as well as results of a 72-week open-label extension study in patients with CAPS, (including familial cold autoinflammatory syndrome and Muckle-Wells syndrome), revealed few concerning side effects. Adverse events were generally mild to moderate, the most common being injection site reactions and upper respiratory tract infections. Other reactions observed included sinusitis, cough, nausea, diarrhea, hypoesthesia, and urinary tract infections. A phase III, randomized, placebo-controlled trial of rilonacept for gout flare prevention also concluded that these two events were the most frequently occurring reactions. No clear increase was seen in rilonacept-associated infections and no tuberculosis or opportunistic infections were reported. Gout exacerbation and neutropenia were responsible for two discontinuations from the trial. Headache and dizziness were the most common adverse events in a randomized, controlled trial of rilonacept in the treatment of acute gouty arthritis. Warnings and precautions for rilonacept issued by the FDA state that IL-1 blockade may interfere with the immune response to infections and live vaccines should not be given concurrently with the drug. The impact of treatment with rilonacept on the development of malignancy is not known. Immunogenicity examinations showed that 19 of 55 (35 %) subjects who received rilonacept for at least 6 weeks developed antibodies to the fusion protein but no correlations between the presence of antibodies and clinical and safety effectiveness were found.

Aflibercept Safety

Early clinical trial results with aflibercept in patients with advanced solid tumors showed that the most common adverse events were fatigue, nausea/vomiting, and toxicities associated with treatment were dysphonia, hypertension, and proteinuria. No patients developed antibodies to the fusion protein. In a phase III multicenter, randomized, controlled trial designed to compare aflibercept/ FOLFIRI with placebo, adverse events led to discontinuation of 26.6% and 12.1 % of patients in the aflibercept and placebo groups, respectively. Grade 3 and 4 adverse events in the patients receiving aflibercept with a more than 5 % incidence compared to placebo were, in order of highest to lowest incidence, neutropenia, hypertension, diarrhea, asthenia, stomatitis, infections, and proteinuria. Febrile neutropenia showed an incidence of 3.9 %. A 13-fold increase was seen in the hypertension rate and a 6.5-fold increase in the proteinuria rate. As set out in Table 6.2, warnings, precautions, and known side effects of ziv-aflibercept are quite extensive. A boxed warning covers potentially fatal hemorrhage including GI hemorrhage, GI perforation, and compromised wound healing. Other warnings and precautions draw attention to fistula formation, hypertension, arterial thrombotic events, proteinuria, neutropenia and its associated complications, diarrhea and dehydration, and PRES. A recent phase II multicenter lung cancer trial of ziv-aflibercept with cisplatin and pemetrexed in 42 patients was closed prematurely because of three confirmed and two suspected cases of PRES. Although PRES has been noted in other ziv-aflibercept clinical assessments, the rate observed in this study was higher than previously seen. The incidence of antibodies to ziv-aflibercept after its intravenous administration was found to be 3.1 %. Neutralizing antibodies were found in 17 of 48 patients but their impact, if any, on the efficacy and safety of the drug was not assessed.

After the monoclonal antibody ranibizumab, aflibercept was the second agent to be approved by the FDA for macular edema secondary to central retinal vein occlusion. Ranibizumab (Chap. 4 section "Ranibizumab") was developed specifically for intraocular use and approved for age-related macular degeneration, central retinal vein occlusion, and macular edema. Bevacizumab (Chap. 3, section "Bevacizumab"), another monoclonal antibody targeted to VEGF and approved for colorectal cancer, has been increasingly used off-label for these disorders. Aflibercept appears to offer pharmacological advantages over the two monoclonal antibodies due, in particular, to its higher binding affinity and longer action. Apart from the necessary warnings and precautions associated with intravitreal injections, the FDA prescriber's information for the aflibercept ophthalmic preparation Eylea® draws attention to a potential risk of arterial thrombotic events following intravitreal use of VEGF inhibitors. The most common adverse reactions occurring with a frequency of ≥5 % are listed in Table 6.2. Whichever of the three preparations is administered, intravitreal injection remains a prominent risk factor that may lead to endophthalmitis, retinal detachment, and an increase in intraocular pressure. Much of the data on adverse events were obtained in clinical trials designed to evaluate the efficacy and safety of aflibercept injection in the treatment of wet age-related macular degeneration and macular edema secondary to central retinal vein occlusion. In the studies on both of these conditions, immunoreactivity to the drug was 1–3 %. Again, no differences in efficacy or safety were seen in patients with or without antibodies.

Romiplostim

As a thrombopoietin receptor agonist, romiplostim is indicated for the treatment of thrombocytopenia in patients with chronic immune thrombocytopenia (ITP) who have had an insufficient response to corticosteroids, immunoglobulins, or splenectomy. It is not used for any cause of thrombocytopenia other than ITP. In the main, adverse events in clinical trials were rated mild to moderate (Table 6.2) but there are regulatory agency warnings for thrombotic/thromboembolic complications, bone marrow reticulin formation, and bone marrow fibrosis following the occurrence of cases in clinical trials. A warning has also been issued of the risk of progression of myelodysplastic syndromes to acute myelogenous leukemia. Adverse events seen in a trial with children were mild to moderate with reports of headache, cough, vomiting, and epistaxis. Thromboembolic events were not seen. No correlation between antibody activity (preexisting antibodies or antibodies developed during treatment), and clinical effectiveness or safety has been observed. Romiplostim appears to be relatively free of adverse cutaneous effects with only one report of a systemic rash to a high dose in a patient in a phase two clinical trial and a recent case of a grade 3 maculopapular rash 24 h after administration of the drug for chronic thrombocytopenia accompanying the rare disorder autoimmune lymphoproliferative syndrome. The rash resolved in about a week after withdrawal of romiplostim and did not reappear when therapy was resumed in the following week or subsequent weeks.

Alefacept

The safety and tolerability of alefacept has been detailed in a number of clinical trials where the most common side effects, including injection site reactions, were mild to moderate (Table 6.2) and there appeared to be no correlation between decreased CD4+ T cell counts and the incidence of infections. An analysis of 13 trials examining the long-term safety of alefacept in 1869 psoriasis patients who had received up to nine courses of therapy showed a similar spectrum of adverse events and their incidences, for example, headache (0–14.2 %), nasopharyngitis (7.7–25 %), influenza (0–8.1 %), upper respiratory tract infection (0–12.5 %), and pruritus (0–7.5 %). Less than 1 % of patients developed an infection, no opportunistic infections were seen, and infections that did occur appeared to be unrelated to CD4+ T lymphocyte counts. Incidences of serious adverse events (0–4.8 %), serious infections (0–0.9 %), malignancies (0–4.8 %), and discontinuations due to adverse events (0–4.8 %) showed no apparent increase with repeated exposure. Four of 1869 patients (0.2 %) experienced angioedema while urticaria occurred in six patients. Lymphopenia with reductions in CD4+ and CD8+ lymphocytes, malignancies, serious infections, and hypersensitivity reactions are each the subject of warnings and precautions by the FDA (Table 6.2) and each are listed as the most serious adverse reactions of alefacept. Thirteen malignancies were detected in 11 alefacept-treated patients in a placebo-controlled study. The incidence of malignancy was 1.3 % and 0.5 % in the placebo group. Hepatic injury caused by the drug is also listed in post-marketing reports and the potential for excessive immunosuppression is another warning issued by the FDA. In this respect, it has been suggested that alefacept should be used with caution in patients with known mycosis fungoides or an unclassified atypical lymphocytic skin infiltrate. Approximately 3 % of 1357 patients developed low titer antibodies to the fusion protein but the long-term effect, if any, of these antibodies is not known. After a decision by the manufacturer to cease production of Amevive®, FDA approval for alefacept was discontinued in September 2012. The decision does not seem to have been associated with any safety or risk concerns. The drug was never approved for the European market.

Albumin Fusion Proteins

Human serum albumin, the protein present in the greatest concentration in plasma, has a relatively long half-life of more than 14 days and this, together with its safety, lack of enzymatic activity and normally low immunogenic potential, makes the protein a prime candidate for consideration in the development of some biopharmaceuticals, including fusion proteins. Effector peptides can be attached at either the *N*-or *C*-terminus or both, and overall, albumin fusion technology using recombinant proteins offers a versatile and relatively safe drug development platform for a wide array of potentially useful, but short acting, biological agents. Like Fc fusion proteins, serum albumin also interacts with the FcRn; FcRn recycling occurs by binding at acid pH and release takes place at neutral pH.

Perhaps the most commonly occurring drawback seen with the use of recombinant human serum albumin fusion technology is interference by the albumin carrier with the specific activity of effector molecule(s) and/or its target. This may occur despite the use of linkers and efforts to orient the relevant domains. In the development of the recently approved factor IX-albumin fusion protein (rIX-FP, Idelvion®; section "Factor IX Fusion Proteins" above), a linker sequence derived from the natural cleavage site of factor IX responsible for proteolytic activation was exploited (Chap. 10, section "Factor IX Albumin Fusion Protein") while for a factor VIIa-fusion protein (in development), a flexible glycine-serine linker was utilized. Despite the low inherent immunogenicity of recombinant human albumin (rHA), this risk and the possibility of interference with activity always need to be considered in the development of any fusion protein employing the protein as fusion partner.

Of the currently 11 licensed fusion proteins, only two, albiglutide and factor IX fusion protein rIX-FP, are albumin fusion constructs. At least two other chimeric albumin preparations are, however, in late stage study as therapeutic agents and more than a dozen others are in an early stage of development.

Albiglutide

The GLP-1 receptor agonist albiglutide (Tanzeum®; Eperzan®) is a recombinant dipeptidyl peptidase-4-resistant dimer fused to human serum albumin and used to treat type 2 diabetes. Prepared in *Saccharomyces cerevisiae*, albiglutide is composed of two 30 amino acid GLP-1(7-36) monomers coupled in tandem and fused to rHA. The amino acid Gly is substituted for Ala at position 8 to impart resistance to DPP-4 attack (see Chap. 7, sections "GLP-1 Receptor Agonists," "Safety of GLP-1 Receptor Agonists," "Immunogenicity of GLP-1 Receptor Agonists" and Fig. 7.8). With a half-life of approximately seven days (Table 6.1), albiglutide can be administered once a week.

Factor IX Albumin Fusion Protein

The recombinant factor IX-recombinant human serum albumin construct, designated rIX-FP (Idelvion®), was registered by the FDA in March 2016. It has an extended half-life of ~ 87–93 hours (Table 6.1). A discussion and its adverse effects are set out in Chap. 10, sections "Factor IX Albumin Fusion Protein" and "Factor IX Fusion Proteins".

Albumin Fusion Proteins in Late Stage Development

Interferon Alfa-2b Albumin Fusion Protein

The cytokine interferons (IFN) alfa-2a and 2-b (Chap. 5, section "Interferon Alfa"), administered for viral infections hepatitis B and C and some malignancies, have short half-lives requiring dosages to be given every 1–3 days. Pegylated IFN alfa-2a and alfa-2b preparations are used to prolong half-life but, with nearly half of all patients showing a disappointing response, the half-life of IFN alfa-2b was extended by forming albinterferon alfa-2b (Albuferon®), a recombinant fusion protein of the cytokine with the *C*-terminal of rHA. Produced in *S. cerevisiae* and with an MW of ~85.7 kDa, the fusion protein retained its antiviral activity for 8 days and demonstrated a half-life of 93 h, more than twice the mean half-life of pegylated IFN alfa-2b and 18 times that of IFN alfa-2b. Adverse events provoked by the fusion protein were similar to those seen with IFN alfa but high doses, for example, 1200 μg, were associated with a high incidence of serious pulmonary adverse reactions resulting in treatment discontinuations.

Filgrastim Albumin Fusion Protein

Filgrastim, or recombinant human granulocyte colony-stimulating factor (rhG-CSF) (Chap. 5, section "Colony-Stimulating Factors: Filgrastim, Sargramostim and Tbo-Filgrastim") is approved for the treatment of myelosuppression and patients with chronic neutropenia. The short half-life of filgrastim necessitating its daily administration, prompted the development of Albugranin™, formed by fusing rhG-CSF to the *C*-terminal of rHA. The resultant increase in stability of filgrastim in the fusion complex form enables it to be given as a weekly injection. Albugranin™ is currently in phase III clinical trials for chemotherapy-induced neutropenia in breast cancer patients.

Albumin Fusion Proteins in the Early Stage of Development

A wide variety of albumin fusion proteins are in development although many of them are at an early stage in animal studies prior to human trials. The following effector proteins fused to albumin are some of the more interesting fusion constructions under investigation.

Interleukin-2 (IL-2) as aldesleukin (rhIL-2) (Chap. 5, section "Aldesleukin") has been linked to rHA to produce an 81.8 kDa chimeric protein (Albuleukin®) in a *S. cerevisiae* fusion system. Studies in mice showed an extended half-life of 6–8 h compared to 19–57 min for rhIL-2.

Insulin (Chap. 7, section "Insulin") linked by a dodecapeptide to the *N*-terminal of rHA (Albulin) was expressed in yeast or CHO cells. Elimination half-life in mice was 7 h compared to 10 min with rhInsulin.

Human growth hormone (Chap. 7, section "Human Growth Hormone") with a very short half-life of less than 20 min, has been attached to the *C*-terminal of rHA to form the fusion protein Albutropin® with the fusion gene being expressed in *S. cerevisiae*. Half-life of the fusion construct in monkeys is six times longer than human growth hormone, clearance was eight times slower and a single subcutaneous dose maintains IGF-1 levels (Chap. 7, section "Insulin-Like Growth Factor 1") for 7 days, the equivalent of seven daily injections of the natural hormone.

Recombinant human erythropoietin (rhEPO) (Chap. 5, section "Epoetins") has a half-life of 4–13 h after intravenous injection meaning that 2–3 injections of the protein are required per week. To avoid a possible cross-interaction between rhEPO and the carrier in an albumin fusion protein, the effector was attached to rHA via a $(GGSGG)_4$ linker. The fusion gene was expressed in CHO cells. The observed increased half-life of the fusion protein in rodents was similar to darbopoietin. The linker was later modified to avoid cleavage.

The blood coagulation factors VIIa (Chap. 10) produced in human embryonic kidney cells, is being investigated as an albumin fusion protein to introduce important post-translational modifications including *N*- and *O*-glycosylation, γ-carboxylation, and β-hydroxylation. The recombinant factor VIIa fusion protein (rFVIIaF) contains a 31 amino acid flexible Gly/Ser cleavable linker connecting the *C*-terminal of recombinant factor VIIa to rHA and like the recently approved factor IX-albumin fusion construct, increased effectiveness results from improvements in its glycosylation. Compared to the uncomplexed forms, both coagulation fusion proteins demonstrate improved pharmacokinetic parameters in mice, rats, and rabbits.

Thymosin-alfa1, a thymic peptide with immunomodulatory activity is administered for some virus infections, malignancies, and immunodeficiencies. With a short half-life of only 2 h, the yeast *Pichia pastoris* was used for protein expression in the production of an albumin fusion protein with the recombinant thymic peptide fused at the *C*-terminal of rHA. Half-life of the fusion protein proved to be 32–36 h.

B-type natriuretic peptide (BNP), used to treat decompensated congestive heart failure, requires continuous intravenous infusion because of its short half-life. A BNP-albumin fusion protein (AlbuBNP; Cardeva™) generated in human embryonic kidney cells has an elimination half-life of 12–19 min in rats compared to 3 min for the natural peptide.

Other promising albumin fusion proteins currently being studied include chimeric constructs with effector proteins/peptides such as the GLP-1 receptor agonist exenatide; butyrylcholinesterase (used to prevent nerve toxicity to organophosphates); cocaine hydrolase; the platelet aggregation inhibitor barbourin; infestin-4 (used for Chargas' disease); thioredoxin with its anti-inflammatory effects; and the thrombin inhibitor, hirudin.

Denileukin Diftitox

This now discontinued genetically engineered recombinant fusion protein introduced in 1992 as DAB389IL2 was the first fusion toxin to be approved. Made up of the full length IL-2 molecule and the catalytic domain of diphtheria toxin, that is a single polypeptide chain of 388 amino acids obtained by deleting the 147 amino acid receptor-binding domain from the 535 amino acid full length diphtheria toxin, the resultant fusion protein, named denileukin diftitox (Ontak®), retains the ADP-ribosyltransferase and membrane translocating domains of native diphtheria toxin. Once bound to the IL-2 receptor (IL-2R), the fusion toxin needs to undergo endocytosis to effect cell killing by inhibiting protein synthesis. This is achieved by binding only to cells that have intermediate or high affinity IL-2R receptors. Human IL-2Rs may have low, intermediate, or high affinity. High affinity IL-2R ($K_d \sim 10^{-11}$) results from the complex of three different proteins, the α chain of MW 55 kDa (CD25, p55, TAC), β chain of MW 75 kDa (CD122, p75), and the γ chain, MW 64 kDa (CD132, p64). By itself, CD25 acts as the low affinity receptor; CD122 and CD132 together function as an IL-2R receptor of intermediate affinity ($K_d \sim 10^{-9}$).

Perhaps the most extensive examination and analysis of safety data for denileukin diftitox is contained in the results of the Pivotal phase III trial of the fusion protein for the treatment of cutaneous T cell lymphoma. In this study, adverse effects were first seen, and occurred in most patients, during the first treatment course. In the first 24 h, acute hypersensitivity-like reactions involving dyspnea, hypotension, chest tightness, and back/chest pain occurred and approximately one-third of patients experienced cutaneous infusion-related events including flushing and pruritus. The possibility of serious and even fatal infusion reactions was included in an FDA black box warning for denileukin diftitox. Temporary interruption of treatment, a decrease in the infusion rate and/or the administration of antihistamines and corticosteroids, were used to alleviate or prevent the acute symptoms. The most frequently seen adverse events were flu-like symptoms (55 of 65, 85 %) and gastrointestinal symptoms (65 of 71, 92 %) consisting of chills, fever, headache, nausea/vomiting, diarrhea, asthenia, myalgia, arthralgia, and anorexia. Vascular leak syndrome, also included in the FDA black box warning, usually seen in the first 14 days of treatment, and defined as at least two of edema, hypoalbuminemia and/or hypotension, was reported by 25 % of patients. Infections occurred in 56 % of patients but were considered typical of advanced stage cutaneous T cell lymphoma patients and unrelated to treatment. Leukopenia, neutropenia, and thrombocytopenia were reported in from one to three of the 71 patients and although lymphopenia occurred in 70 % of patients, 24 % had lymphopenia at baseline. Rashes not related to infusions manifested in 25 % of patients but overall 35 % had cutaneous reactions classified as maculopapular, petechial, vesicular-bullous, urticarial, and/or eczematous. Cutaneous reactions classified as delayed hypersensitivities and including a case of exfoliative dermatitis were reported in 3 of 35 patients with psoriasis participating in a denileukin diftitox dose escalation study. There was at least one report of toxic epidermal necrolysis (TEN) to denileukin diftitox; the case proved fatal. Results of two studies of the immunogenicity of denileukin diftitox have been summarized by the FDA. In the first, 66 % of 95 treated patients tested positive for antibodies at baseline, probably due to previous exposure to diphtheria organism or vaccine. By treatment courses 1, 2, and 3, the percentage had risen to 94 %, 99 %, and 100 %, respectively, while pharmacokinetic parameters decreased significantly and clearance increased two- to eight-fold. In the second study, 39 % of 131 patients had antibodies at baseline, and this increased to 66 % after one course of treatment and 97 % after three courses. Assessment of neutralizing antibodies showed that inhibited functional activity increased from 45 % at baseline to 97 % after three courses. In the Pivotal phase III trial, the authors concluded that development of antibodies to the fusion toxin did not appear to impair the response to treatment and no clinical correlation was observed between levels of antibodies to IL-2 and any adverse event. In fact, higher levels of antibody to denileukin diftitox were associated with lower incidences of rash and hypoalbuminemia and higher transaminase levels. Loss of visual acuity, with or without retinal pigment mottling, was a third possible adverse event listed in the FDA black box warning. Although some patients recovered, visual impairment persisted in most. Ontak® was included in the CDER discontinued biologic products list in January 2014.

The Immunogenicity of Therapeutic Fusion Proteins: Attempts to Help Recognize Patients at Risk

The possibility of an immune response to a therapeutic protein must always be kept in mind since such a response has the potential to adversely affect the safety as well as the efficacy of the treatment by inhibiting or blocking the protein's therapeutic action. There are a number of immunologically based or influenced adverse events that can eventuate after administration of a fusion protein such as anaphylaxis manifesting as cardiovascular collapse, bronchospasm, angioedema, urticaria, and erythema; infusion reactions; cytokine release syndrome; autoimmune reactions; cytotoxic type II and immune complex type III hypersensitivities; and delayed cell-mediated type IV hypersensitivities, often manifesting as cutaneous reactions. In February 2013, the FDA issued a draft guidance entitled "Immunogenicity assessment for therapeutic protein products" which, it was stated, would, in final form, represent the FDA's thinking on the topic. The guidelines basically represent a risk-based approach to evaluating and mitigating immune responses to therapeutic proteins that might reduce, change, or eliminate the intended therapeutic action and/or induce adverse events otherwise not seen in the absence of an antibody or cellular targeted response. The immunogenicity of an administered protein such as the fusion proteins discussed here can be affected by both patient-specific and product-specific factors. Patient-related factors include: (1) the immunological status and competence of the patient which is relevant to, for example, the patient's age and whether the patient is immune suppressed or immune activated (as in infection); (2) prior sensitization and/or a history of allergy; (3) the route, dose, frequency, and length of administration; (4) the genetic status of the patient, e.g., some HLA haplotypes may predispose patients to an adverse event; (5) the patient's status of immune tolerance to endogenous proteins, e.g., the presence of autoantibodies to cytokines and growth factors and the possibility that a recombinant therapeutic protein might induce, or break, tolerance to an endogenous protein. Some important product-specific factors include: (1) The product's origin—non-human protein always has the potential to be immunogenic; (2) primary molecular structure and post-translational modifications—the primary sequence is especially important for fusion proteins where a potential exists for new antigens to form from the linking of foreign and endogenous proteins; (3) quaternary structure: protein aggregates—Fc fusion proteins are well known to aggregate and misfold and disulfides may not pair as desired; (4) pegylation and glycosylation—clearance may be accelerated, due, for example, to incomplete glycan chains with terminal D-galactose or *N*-acetyl-D-glucosamine (see Chap. 2, section "Glycosylation of Monoclonal Antibodies"); (5) immunomodulatory properties of the protein, e.g., IL-2 is both immunogenic but also up-regulates immune responses to endogenous proteins and may induce clinical autoimmunity.

Since interactions between the Fc domain and its receptors have immunological consequences, attention has been drawn to Fc fusion as a platform technology for modulating immunogenicity. It has been claimed that while administration of a suitably engineered Fc fusion partner may improve both the disease outcome and the safety profile of the fusion partner, interactions between the Fc domain and its receptors raises concerns about the long-term use of Fc fusion proteins.

The higher incidence of reactions to the anti-TNF mAb infliximab in rheumatoid arthritis patients compared to patients with ankylosing spondyloarthritis and vasculitis demonstrates that the disease itself can also be important in the development of an immune response to a biological agent. From this brief consideration of the complexities of immunogenicity of protein biologics, it is not hard to see that, at the moment, the risk of an immune response to a fusion protein cannot be estimated or eliminated. The risk can, however, be managed.

Diagnosis of Hypersensitivities to Fusion Proteins, Premedication, and Desensitization

Apart from a still small but growing body of investigation on a few of the mAbs approved for human therapy, little data are available on appropriate diagnostic and desensitization procedures for the investigation of hypersensitivity responses to the steadily increasing number of approved cytokines, enzymes, toxins, and chimeric fusion proteins. What information there is on the fusion proteins tends to be limited almost exclusively to etanercept. For the diagnosis of immediate and delayed hypersensitivities to fusion proteins, no tests, validated or otherwise, for the detection of specific IgE antibodies are generally available leaving skin testing, as yet unstandardized, as the prime diagnostic procedure for detecting IgE-mediated reactions as well as T cell-mediated adverse events. For prick testing, concentrations of 5–25 mg/mL of etanercept have been used; for intradermal testing, 0.0025–0.25 mg/mL and 0.1–5 mg/mL. Patch testing has been undertaken with etanercept at concentrations of 1 and 5 % in petrolatum. It is clear that many reactions induced by individual fusion proteins remain inadequately investigated and given their chimeric nature, the list of hypersensitivity reactions fusion proteins provoke is likely to expand. Diagnostic recommendations have been made for immune thrombocytopenia and although immunoassays to detect platelet-reactive antibodies are available in a few laboratories, the tests are not standardized, do not detect the drug-dependent antibodies, are technically difficult, and may produce false positives. The situation is similar for other type II hypersensitivities such as immune neutropenia where the monoclonal antibody immobilization of granulocyte antigens assay (MAIGA) seems to be the test of choice. Diagnostic shortcomings, due in some cases to lack of laboratory markers and absence of routine tests for type III hypersensitivity immune complex reactions such as serum sickness and vasculitis, extend to some fusion protein-induced liver and lung injuries and delayed cutaneous reactions. The pathomechanisms underlying some delayed skin reactions to fusion proteins have been little studied. Apart from what appears to be a few cell-mediated, type IV true hypersensitivities, some fusion protein-induced skin responses may represent direct targeting events that are not genuine hypersensitivities and are similar to, for example, agents that bind EGFR causing non-immune-mediated adverse cutaneous events or biologics that provoke exacerbation of psoriasis.

Premedication along with infusion rate adjustment may be used to reduce the incidence of adverse reactions to biological agents although opinion on the effectiveness of the former is divided. For mAbs such as rituximab, acetaminophen, and antihistamines are often given. A commonly used premedication protocol employs a corticosteroid, usually dexamethasone or prednisolone, and H1 and H2 antihistamines such as diphenhydramine and cimetidine/ranitidine/famotidine are often administered 12 and 6 h before infusion. Corticosteroids are sometimes given for several days before infusion. There are a few reports of successful desensitization of acute injection site reactions to etanercept. After the finding of a positive skin test in a patient who developed pruritus, redness and swelling following the 22nd injection of etanercept and a strong positive challenge to the fusion protein that persisted for a month, a 4-day desensitization regimen was undertaken. One hour prior to the commencement of the desensitization procedure, the patient who was on daily cetirizine throughout, was given oral aspirin 325 mg, montelukast 10 mg, diphenhydramine 25 mg, and famotidine 40 mg. On days 1 and 2, a starting dose of 0.25 mg was given, and this was built up slowly at regular intervals until a total dose of 12.5 mg was administered. On day 3, a similar schedule starting with 0.5 mg was increased at intervals up to a maximum of 24.75 mg of etanercept. On day 4, a total of 25 mg of the protein was administered in a single dose. The patient was maintained on twice weekly etanercept injections with diphenhydramine 25 mg 1 h before each injection and daily cetirizine. A closely similar protocol was used successfully in another patient with a severe injection site reaction to etanercept. Two further cases of successful desensitization in etanercept-sensitive patients have been reported. A patient with a progressively severe injection site reaction who developed drug-induced lupus was given etanercept by incremental s.c. injections in the range 0.025–12.5 mg every 30 min on day 1 and started on 7.5 mg twice weekly on day 3. A second patient also with a severe injection site reaction to etanercept received four incremental s.c. injections in the range 0.25–5 mg every 30 min on day 1, 7.5 mg, 10 mg, and 15 mg each 30 min apart on day 3 and 20 mg on day 5. From day 8 the patient was maintained symptom free on 12.5 mg twice weekly. In what was claimed as a novel method of "desensitization," immunosuppressant therapy with methylprednisolone together with methotrexate (7.5 mg) given together as a single dose was used successfully to treat an etanercept-induced urticarial eruption. One week later when the patient's condition had markedly improved, a second dose of the two-drug combination was administered. After a further week, the urticaria had completely abated and thereafter etanercept was continued at 50 mg a week.

Following the successful desensitization of two patients who experienced an injection site reaction to etanercept, the same investigators standardized their methodology in successfully desensitizing a further seven patients, six of whom had an injection site reaction to etanercept and one with an immediate systemic reaction. Each of the injection site reactions consisted of local pruritus, erythema, and swelling or edema. The immediate reaction included swollen face and lips, urticaria, wheezing, dyspnea, nausea, vomiting, and hypotension. At least some injection site reactions are thought to be T cell mediated, delayed type IV hypersensitivities sometimes involving, or developing into, an IgE-mediated immediate reaction, making continued therapy with any provoking drug problematic. All seven patients were skin test-positive to etanercept; the six with injection site reactions responded to intradermal test concentrations of 0.025–0.5 mg/mL. The desensitization protocol, described as rapid, was carried out over three days with the administration of six subcutaneous injections (0.5, 1, 2, 4, 8, 9 mg etanercept) at 30 min intervals for a total of 24.5 mg on days 1 and 2. On day 3, seven doses (0.5, 1, 2, 4, 8, 16, 18.5 mg) for a total of 50 mg were given at 30 min intervals. A maintenance weekly injection of etanercept with cetirizine premedication enabled all seven patients to continue with etanercept therapy. Minor local erythema resolved within 1–2 h.

Concluding Remarks

Realization that the interaction of human immunoglobulin G with the salvage neonatal FcRn receptor can be exploited to prepare and deliver therapeutically useful proteins in chimeric form, has seen Fc fusion proteins become increasingly recognized as a valuable and safe form of biologics therapy. Advances in knowledge and application of molecular cloning, antibody engineering techniques, protein design, development of cell lines, and bioprocessing have all contributed to the general strategy of fusing selected effector domains of short-lived but otherwise therapeutically promising macromolecules with a suitable fusion partner, often the Fc fragment of human IgG but sometimes human serum albumin or transferrin. In fact, because many peptides and proteins have short half-lives due to rapid renal clearance, the prime stimulus for the development of fusion proteins is usually plasma therapeutic half-life extension. Other benefits of IgG Fc fusion include the capacity of the attached Fc domain to recognize Fc receptors on immune cells (potentially important for vaccines and cancer therapy); the conferring of immunomodulatory properties such as the induction of tolerance to immunogenic effector proteins; improved solubility and stability of proteins; and facilitation of the manufacturing process by protein A affinity purification. Looking at the efficacy and safety of the fusion proteins already approved; the promising clinical pipeline of candidates; the impending appearance of biosimilars; the range of potential effector proteins that can be used as fusion partners; the promise of Fc fusion proteins as vaccines and as therapies for a much wider range of disorders; and the commercial success of the first generation of the chimeric macromolecules, subsequent generations of therapeutic fusion proteins are anticipated with considerable optimism.

Summary

- There are currently (2016) 11 chimeric fusion proteins (nine Fc fusion proteins and two albumin constructs) on the market with regulatory approval and many more are at different stages of clinical development. The Fc fusion protein alefacept and the fusion toxin denileukin diftitox are now no longer marketed.
- Chimeric fusion proteins used as biologic therapeutic agents consist of a relatively short-lived effector domain, generally a peptide, coupled to a "carrier," usually protein or peptide, that also contributes to the functional properties of the resultant fusion protein. Fusion proteins are produced by genetic engineering, linking genes for the separate proteins involved to give a new polypeptide formed from the incorporated separate domains together with their functional properties.
- Three main requirements in the preparation of an effective chimeric fusion protein are to endow the macromolecule with: (1) stability, that is, produce a polypeptide with a suitably extended half-life; (2) effective targeting and subsequent specific binding; and (3) cytotoxicity or at least the capacity to inhibit the deleterious processes underlying the treated condition. Most peptides likely to be

considered as effector peptides have a short half-life due to proteolytic degradation and are usually rapidly cleared via the kidneys within minutes.

- The crystallizable Fc region of the human IgG antibody consisting of the CH2 and CH3 domains of the immunoglobulin heavy (H) chain has been the most commonly employed fusion partner but recombinant human serum albumin and transferrin have also been used.
- The effector molecule can be fused, generally to the *N*-terminal of one or both of the Fc H-chains creating a monomeric, dimeric, or heterodimeric fusion protein, or more than one ligand-binding domain may be employed to form a "trap." In another configuration, effector peptides can be fused at the carboxyl terminus of the Fc fragment in the form of the so-called peptibody.
- Receptor-mediated recycling via interaction with the salvage neonatal FcRn receptor protects Fc-containing molecules from lysosomal degradation.
- Fc fusion constructs have become the most frequently employed and successful fusion proteins making up 10 of the 13 current, or once licensed preparations.
- Denileukin-diftitox, the one fusion protein lacking an attached Fc piece, or not bound to albumin, has easily the shortest half-life of only about 70–80 min compared to an average of at least 3–5 days for the Fc fusion proteins and up to 8 days for the albumin fusion protein albiglutide.
- A conserved *N*-linked glycosylation site occurs at asparagine 297 of the CH2 domain of the human Fc fragment of each of the four human IgG subclasses. Glycosylation sites on linked effector proteins such as receptor domains can be *N*- and/or *O*-linked and may contain more sialic acid residues.
- Structurally complex glycoproteins are produced mainly by mammalian expression systems since they require post-translational modifications such as specific glycosylations, disulfide-bonded domains, γ-carboxylation, and sulfation. Glycosylation can have a major influence on the effectiveness of biologics therapy via its effects on actions, protein solubility, stability, serum half-life, immunogenicity, and selectivity of receptor binding. For example, the terminal sugars of glycans in the CH2 domain of human Fc fragments help to determine antibody-dependent cellular cytotoxicity (ADCC) and complement-dependent cytotoxicity (CDC).
- The Fc region utilized in nine Fc fusion proteins belong to the human IgG1 subclass. For the recently approved fusion protein dulaglutide, a glucagon-like peptide-1 (GLP-1) receptor agonist used to treat type 2 diabetes, a modified IgG4 Fc fragment was used to minimize ADCC and CDC effects of the resultant fusion construct.
- Two of the 11 currently approved fusion proteins, albiglutide and factor IX fusion protein rIX-FP are albumin fusion constructs. Albiglutide is a recombinant dipeptidyl peptidase-4-resistant GLP-1 dimer fused to human albumin and used to treat type 2 diabetes. The coagulation factor fusion protein rIX-FP is used to treat hemophilia B.
- Etanercept, a recombinant, engineered, fully human dimeric Fc fusion protein linked to the ligand-binding portion of the human TNF receptor (TNFR) was the first chimeric fusion protein to gain regulatory approval when in 1998 it was

licensed by the FDA for the treatment of rheumatoid and other forms of arthritis. Like the mAbs that also target TNF, etanercept is sometimes used for the treatment of other autoimmune diseases including ankylosing spondylitis, plaque psoriasis, and Crohn's disease.

- Early studies reported that the most common side effects of etanercept are relatively mild, for example, fever, headache, injection site reactions, mild allergic reactions, and pruritus but a wide variety of less often seen and sometimes rare and serious reactions are now known.
- The list of etanercept adverse events issued by the FDA in order of frequency of occurrence, comprises infection, dermatologic, neurologic, musculoskeletal, pulmonary, cardiac, and vascular effects.
- Serious infections is one of two black box FDA warnings for the drug (the other is malignancies). Organisms implicated include bacteria (tuberculosis, streptococcus, listeria, actinobacillus), viruses, fungi (aspergillus), protozoa (toxoplasma), and cestodes (echinococcus).
- The most common skin manifestation seen with etanercept is an injection site reaction. Overall, etanercept has been associated with a wide variety of adverse dermatological events including onset and exacerbations of psoriasis and atopic dermatitis, lichenoid reaction, erythema multiforme, and palmoplantar pustulosis.
- There are reports of a number of demyelinating disorders during etanercept therapy including the so-called demyelinating syndrome; relapsing remitting MS; demyelination of the spinal cord and cerebral cortex; PML; PRES; demyelinating disease that resolved after withdrawal of the drug; and central nervous system demyelination producing ophthalmic manifestations.
- The possibility of the development of lymphoma and other malignancies in children and adolescents treated with TNF blockers is the subject of a second FDA black box warning. A number of studies suggest an increased lymphoma risk for patients with rheumatoid arthritis and the risk appears to be slightly increased in those patients treated with TNF antagonists.
- Pulmonary nodules, pulmonary granulomas, exacerbation of preexisting interstitial lung disease and organizing pneumonia have been reported after administration of etanercept.
- A number of dermatologic events induced by etanercept including psoriasis, atopic dermatitis, erythema multiforme, and acute generalized exanthematous pustulosis are almost certainly type IV hypersensitivity responses while some injection site reactions may be true immediate, type I, hypersensitivities. At least some cases of etanercept-induced cytopenias and vasculitis may be type II and type III hypersensitivities, respectively, and some pulmonary events caused by etanercept may ultimately be shown to be type III or combined type III/type IV reactions.
- There are increasing indications that etanercept can provoke autoimmune reactions such as hyperthyroidism and the development of anti-synthetase syndrome, Crohn's disease, and Henoch–Schönlein purpura.

- The occurrence of anti-etanercept antibodies is well known but there appears to be no evidence so far that these antibodies reduce the clinical efficacy of the drug. Patients given etanercept commonly develop antinuclear and/or anti-double-stranded DNA antibodies and cases of a lupus-like syndrome, cutaneous lupus erythematosus and acute discoid lupus have been reported. There have been a number of possible/probable reports of immediate hypersensitivity/anaphylaxis to etanercept.
- Given the high affinity of CTLA-4 for CD80/86, the former was utilized as the peptide partner to prepare abatacept by fusing the extracellular domain of the lymphocyte antigen to the *N*-terminal of the Fc fragment of human IgG1.
- Abatacept is approved as a first-line treatment of adult rheumatoid arthritis and juvenile idiopathic arthritis.
- Although abatacept shows high affinity for CD80/86, CTLA-4 is a much less potent inhibitor of CD86-dependent than CD80-dependent costimulation. This information led to the realization that a modified CTLA-4 molecule with higher avidity for CD86 should be sought. Belatacept, which differs from abatacept by two amino acid substitutions, binds four times more avidly to CD86 resulting in an overall 10-fold increase in biological activity compared to CTLA-4.
- Belatacept acts as an immunosuppressant and appears to show better allograft function and improved cardiovascular and metabolic risk profiles than cyclosporin. This led to its approval by the FDA for prophylaxis of organ rejection in adult kidney transplant patients.
- Belatacept is subject to an FDA black box warning stating that "increased susceptibility to infection…may result from immunosuppression." The drug's use in liver transplant patients is not recommended due to an increased risk of graft loss and death.
- Fatal PML is included in the FDA's "Warnings and precautions" for belatacept. Epstein–Barr virus seronegativity is a risk for developing post-transplant lymphoproliferative disorder (PTLD) from belatacept therapy for kidney transplantation. PTLD and the relevance of EBV-negativity/positivity are also included in the drug's current FDA black box warning.
- Rilonacept, also known as IL-1 trap (target-related affinity profiling), captures IL-1β preventing activation of IL-1 receptors and thus reducing the inflammation and other effects due to overproduction of IL-1. It is a dimeric Fc fusion protein in which the IL-1R accessory protein (IL-1RAcP) ligand-binding region is fused via its C-terminus to the N-terminus of the interleukin receptor IL-1RI extracellular domain and these linked peptides are then fused via IL-1RI to the N-terminus of each of the Fc chains of human IgG1.
- Rilonacept was granted orphan drug status and approved for the treatment of cryopyrin-associated periodic syndromes (CAPS). Adverse events following its administration are generally mild to moderate. Warnings and precautions for rilonacept issued by the FDA state that IL-1 blockade may interfere with the immune response to infections and that live vaccines should not be given concurrently with the drug.

- Aflibercept, or VEGF trap, is a human recombinant protein made by fusing domain 2 from vascular endothelial growth factor receptor-1 (VEGFR-1) to domain 3 of VEGFR-2 and attaching this combination to the hinge region of the Fc domain of human IgG1. VEGF trap acts as a circulating antagonist preventing receptor binding by VEGF and placental growth factor.
- Aflibercept, as ziv-aflibercept or Zaltrap®, is used in combination with 5-fluorouracil, leucovorin, and irinotecan (FOLFIRI) for the treatment of oxaliplatin-resistant metastatic colorectal cancer and, as Eylea®, as an ophthalmic intravitreal injection for the treatment of neovascular (wet) age-related macular degeneration and for macular edema following central retinal vein occlusion.
- Warnings, precautions, and known side effects of ziv-aflibercept are quite extensive. A black box warning covers potentially fatal hemorrhage including GI hemorrhage, GI perforation, and compromised wound healing. Other warnings and precautions draw attention to fistula formation, hypertension, arterial thrombotic events, proteinuria, neutropenia and its associated complications, diarrhea and dehydration, and PRES. For the ophthalmic preparation Eylea®, attention is drawn to a potential risk of arterial thrombotic events following intravitreal use of VEGF inhibitors.
- Romiplostim, the so-called peptibody, is formed by the fusion of four identical copies of a thrombopoietin mimetic peptide to the C termini of aglycosylated human IgG1 Fc chains. Each H chain of the Fc-protein is attached at residue 228 by a pentaglycine bridge to a molecule of the thrombopoietin mimetic peptide linked to another molecule of the peptide by a glycine bridge.
- As a thrombopoietin receptor agonist, romiplostim is indicated for the treatment of thrombocytopenia in patients with chronic immune thrombocytopenia (ITP) who have had an insufficient response to corticosteroids, immunoglobulins, or splenectomy. It is not used for any cause of thrombocytopenia other than ITP.
- For romiplostim, there are regulatory agency warnings for thrombotic/thromboembolic complications, bone marrow reticulin formation, and bone marrow fibrosis. A warning has also been issued of the risk of progression of myelodysplastic syndromes to acute myelogenous leukemia.
- The first domain of LFA-3 was used as the effector peptide to prepare alefacept, a fusion protein with the Fc of IgG1. Alefacept, which can be viewed as an anti-CD2, selectively targets effector memory T cells for both CD4+ and CD8+ but not naïve T cells and central memory T cells in psoriasis vulgaris.
- Lymphopenia with reductions in CD4+ and CD8+ lymphocytes, malignancies, serious infections, and hypersensitivity reactions are each the subject of warnings and precautions by the FDA and each are listed as the most serious adverse reactions of alefacept. Hepatic injury caused by the drug is also listed in post-marketing reports and the potential for excessive immunosuppression is another warning issued by the FDA. Alefacept was discontinued in September 2012. The drug was never approved for the European market.

- Recombinant factors VIII (rFVIIIFc) and IX (rFIXFc) Fc fusion proteins, used to treat hemophilia A and B, respectively, are monomeric constructs with single ligands covalently linked to the Fc fragment of human IgG1. Both have the desired extended half-life necessary to improve treatment.
- Denileukin diftitox, used for the treatment of cutaneous T cell lymphoma, is made up of the IL-2 molecule and the catalytic domain of diphtheria toxin, a single polypeptide chain of 388 amino acids obtained by deleting the 147 amino acid receptor-binding domain from the 535 amino acid full length diphtheria toxin.
- The possibility of serious and even fatal infusion reactions and vascular leak syndrome was included in an FDA black box warning for denileukin diftitox. Loss of visual acuity was a third possible adverse event listed in the FDA black box warning. Denileukin diftitox was discontinued in January 2014.
- The immunogenicity of an administered protein such as the fusion proteins can be affected by both patient-specific and product-specific factors. Patient-related factors include: (1) The immunological status and competence of the patient; (2) prior sensitization and/or a history of allergy; (3) the route, dose, frequency, and length of administration; (4) the genetic status of the patient; (5) the patient's status of immune tolerance to endogenous proteins. Some important product-specific factors include: (1) The product's origin—non-human protein always has the potential to be immunogenic; (2) primary molecular structure and post-translational modifications; (3) quaternary structure; and (4) pegylation and glycosylation.
- Apart from etanercept, little data are available on appropriate diagnostic and desensitization procedures for the investigation of hypersensitivity responses to the fusion proteins.
- For the diagnosis of immediate and delayed hypersensitivities to fusion proteins, no tests, validated or otherwise, for the detection of specific IgE antibodies are generally available leaving skin testing, yet to be validated, as the prime diagnostic procedure for detecting IgE-mediated reactions as well as T cell-mediated adverse events. For prick testing, concentrations of 5–25 mg/mL of etanercept have been used; for intradermal testing, 0.0025–0.25 mg/mL and 0.1–5 mg/mL concentrations are recommended. Patch testing has been undertaken with etanercept at concentrations of 1 and 5 % in petrolatum.
- Although immunoassays to detect platelet-reactive antibodies in cases of suspected fusion protein-induced immune thrombocytopenia are available in a few laboratories, the tests are not standardized, do not detect the drug-dependent antibodies, are technically difficult and may produce false positives. The situation is similar for other type II hypersensitivities such as immune neutropenia where the monoclonal antibody immobilization of granulocyte antigens assay (MAIGA) seems to be the test of choice
- Diagnostic shortcomings, due in some cases to lack of laboratory markers and absence of routine tests for type III hypersensitivity immune complex reactions such as serum sickness and vasculitis, extend to some fusion protein-induced liver and lung injuries and delayed cutaneous reactions.

- Some fusion protein-induced skin responses may represent direct targeting events that are not genuine hypersensitivities and are similar to, for example, agents that bind EGFR causing non-immune-mediated adverse cutaneous events or biologics that provoke exacerbation of psoriasis.
- Some successful premedication and desensitization protocols have been devised and published for injection site and immediate reactions to etanercept.

Further Reading

Baldo BA. Chimeric fusion proteins used for therapy: indications, mechanisms and safety. Drug Saf. 2015;38:455–79.

Baker K, Qiao SW, Kuo T, et al. Immune and non-immune functions of the (not so) neonatal Fc receptor, FcRn. Semin Immunopathol. 2009;31:223–36.

Beck A, Diemer H, Ayoub D, et al. Analytical characterization of biosimilar antibodies and Fc-fusion proteins. Trends Anal Chem. 2013;48:81–95.

Chamow SM, Ryll T, Lowman HB, Farson D, editors. Therapeutic Fc-fusion proteins. Weinheim: Wiley-Blackwell; 2014.

Czajkowsky DM, Hu J, Shao Z, et al. Fc-fusion proteins: new developments and future prospects. EMBO Mol Med. 2012;4:1015–28.

Heier JS, Brown DM, Chong V, et al. Intravitreal aflibercept (VEGF trap-eye) in wet age-related macular degeneration. Ophthalmology. 2012;119:2537–48.

Isenberg D, Gordon C, Licu D, et al. Efficacy and safety of atacicept for prevention of flares in patients with moderate-to-severe systemic lupus erythematosus (SLE): 52-week data (APRIL-SLE randomised trial). Ann Rheum Dis. 2015;74:2006–15. doi:10.1136/annrheumdis-2013-205067.

Kuter DJ, Rummel M, Boccia R, et al. Romiplostim or standard of care in patients with immune thrombocytopenia. N Engl J Med. 2010;363:1889–99.

Leader B, Baca QJ, Golan DE. Protein therapeutics: a summary and pharmacological classification. Nat Rev Drug Discov. 2008;7:21–39.

Lecluse LLA, Dowlatshahi EA, Limpens CEJM, et al. Etanercept. An overview of dermatologic adverse events. Arch Dermatol. 2011;147:79–94.

Ledingham J, Deighton C, on behalf of the British Society for Rheumatology Standards, Guidelines and Audit Working Group (SGAWG). Update on the British Society for Rheumatology guidelines for prescribing TNFα blockers in adults with rheumatoid arthritis (update of previous guidelines of April 2001). Rheumatol. 2004;44:157–63.

Rath T, Baker K, Dumont JA, et al. Fc-fusion proteins and FcRn: structural insights for longer-lasting and more effective therapeutics. Crit Rev Biotechnol. 2015;35:235–54. doi:10.3109/07388551.2013.834293.

Schmidt SR, editor. Fusion protein technologies for biopharmaceuticals. Applications and challenges. Hoboken: Wiley; 2013.

Wang T-F, Lockhart AC. Aflibercept in the treatment of metastatic colorectal cancer. Clin Med Insights Oncol. 2012;6:19–30.

Wekerle T, Grinyó JM. Belatacept: from rational design to clinical application. Transpl Int. 2012;25:139–50.

Zeltser R, Valle L, Tanck C, et al. Clinical, histological, and immunophenotypic characteristics of injection site reactions associated with etanercept: a recombinant tumor necrosis factor alpha receptor: Fc fusion protein. Arch Dermatol. 2001;137:893–9.

Further Reading

Chapter 7
Peptide Hormones

Hormones, synthesized and released in response to often quite specific biochemical signals, are signaling molecules or messengers, communicating between organs and tissues via specific receptors to regulate a wide range of physiological and behavioral activities in the human body. Hormones are numerous; show great diversity in structure and physiological activities; often exhibit complex interactions with other hormones as well as cells; and have been closely studied and utilized as therapeutic agents. Key studies over many decades of the roles of hormones in health and disease were Banting and Best's reversal of symptoms of diabetes mellitus with pancreas extract in 1922, important mechanistic insights by Loewi in 1921 and Sutherland in 1962, and the determination of the amino acid sequence of insulin by Sanger in 1953. Because of their number, regulatory role in so many biological processes, and the inevitable opportunities for defects to occur in myriad interactions at the molecular and cellular levels, many human hormone-related disorders have been identified. It inevitably follows, therefore, that with the variety of hormone therapies applied and their many induced physiologic and pharmacologic effects, side effects are extensive in range and sometimes severe and/or unique in nature.

Although published reports on so-called hormone "hypersensitivity" are not hard to find, they often deal with adverse reactions to the steroid hormones, estrogens and, particularly, progesterone. Intolerance to sex hormones is said to be not uncommon, manifesting with symptoms such as dermatitis, premenstrual syndrome, dysmenorrhea, asthma, rhinitis, headache, arthralgia, acne, pruritus, and a range of skin lesions. Acute and delayed responses resembling type I and type IV hypersensitivities have been recorded to both exogenous and endogenous progesterone and the finding of IgG antibodies to the steroid has been interpreted as evidence of an autoimmune basis for the induced syndrome. In fact, there are more than two dozen reports of what has been termed autoimmune progesterone dermatitis, a reaction thought to occur in response to endogenous progesterone, particularly during the luteal phase of the menstrual cycle. Progesterone-induced urticaria has been described in addition to a range of other skin lesions including eczema, angioedema, erythema multiforme,

B.A. Baldo, *Safety of Biologics Therapy*, DOI 10.1007/978-3-319-30472-4_7

fixed drug eruption, folliculitis, papulovesicular eruptions, and vulvovaginal pruritus. Anaphylaxis to progesterone has been reported and the cyclical cases of anaphylaxis seen in catamenial anaphylaxis occurring during menstruation may be due to this steroid hormone and/or released prostaglandins. For the peptide and glycoprotein hormones, there is a relatively small number of convincing studies indicating the occurrence of hypersensitivity responses to the agents. Peptide and glycoprotein hormones with regulatory approval for therapeutic administration to humans comprise a number of important recombinant and natural hormone preparations including insulin, glucagon, glucagon-like peptide-1, human growth hormone, insulin growth factor-1, somatostatin, vasopressin, oxytocin, adrenocorticotropic hormone (ACTH), gonadotropin-releasing hormone (GnRH), parathyroid hormone, thyrotropin, and members of the gonadotropin family of hormones. These will be examined in this chapter and in Chap. 8 with a focus on their structures, mechanisms of action, approved indications, and safety issues.

Peptide hormones may be small peptides such as oxytocin and vasopressin, each with only nine amino acids, or somewhat larger such as insulin with 51 and growth hormone with 191 amino acids. Peptide hormones are often prepared by cleavage of larger inactive precursors or prohormones before being released for action. Being secretory products, they are synthesized on the rough endoplasmic reticulum, packaged and processed in the Golgi apparatus, and stored in secretory vesicles in the cytoplasm. Their release is regulated by signals initiating the fusion of vesicles with plasma membrane prior to the release of hormone into the blood and extracellular fluid for delivery to the target tissue(s). In addition to the most well-known peptide hormone insulin, the first protein to be sequenced and the first to be synthesized by recombinant DNA technology, other peptide hormones approved for use as important therapies will be discussed including glucagon, glucagon-like protein-1, growth hormone, somatostatin, vasopressin, oxytocin, ACTH, GnRH, and parathyroid hormone.

Insulin

Insulin is a peptide hormone essential for the regulation of the metabolism of carbohydrates and fats. It is produced in a constant proportion necessary for the transport of glucose and the removal of potentially toxic excess levels of the sugar from the blood. Although some body tissues, for example, erythrocytes, nerve tissue, liver, kidney tubules, and cells of the intestine, do not require the intervention of insulin for the transfer of glucose, this is not the case for skeletal and cardiac muscle and adipose tissue. The liver stores glucose via insulin-facilitated phosphorylation of glucose to glucose-6-phosphate which is converted to glycogen for storage or for subsequent reconversion to glucose if needed. Insulin also promotes lipid synthesis through the uptake by adipose cells of blood lipids which are converted to triglycerides and inhibits both lipolysis and the release of free fatty acids from adipose cells. Some of the many other actions and effects of insulin include a decrease in the production of glucose from non-sugar substrates; an increased amino acid uptake and stimulation of protein synthesis; decreased proteolysis; decreased autophagy; an increase in the uptake of potassium by cells and decreased renal excretion of sodium; relaxation of arterial and microarterial wall muscle causing an increase in blood flow; and an increase in the secretion by stomach parietal cells of hydrochloric acid.

Diabetes Mellitus

Diabetes mellitus, often simply termed diabetes, is a group of metabolic diseases characterized by a high blood sugar level persisting over an extended period of time. Regardless of the specific cause, diabetes is due to an insufficient supply of insulin from the pancreas or a lack of response of body cells to the insulin produced. Diabetes mellitus can be classified into four categories: (1) type 1 diabetes (T1DM), formerly called insulin-dependent diabetes mellitus (IDDM) or juvenile diabetes; (2) type 2 diabetes, formerly non-insulin-dependent diabetes mellitus (NIDDM) or maturity-onset or adult-onset diabetes; (3) gestational diabetes; and (4) other diverse types such as those due to disease-induced pancreatic damage (e.g., chronic pancreatitis, cystic fibrosis), Cushing syndrome, genetic defects (e.g., of beta cells or in insulin processing), drugs (e.g., corticosteroids, statins, thiazides, streptozotocin [Zanosar®], β-adrenergic agonists), hormones antagonistic to insulin (growth hormone), infections such as coxsackievirus or cytomegalovirus, and cancers such as glucagonomas.

The most common and obvious symptoms of untreated diabetes mellitus include increased thirst, more frequent urination, increased hunger, dry mouth, and weight loss. Other detrimental effects include blurred and changed vision, headaches, fatigue, prolonged wound healing, skin rashes, and itching. More severe events and potential emergencies that may occur in patients with type 1 diabetes are nonketotic hyperosmolar coma and diabetic ketoacidosis characterized by abdominal pain, nausea, vomiting, heavy breathing, acetone-smelling breath, drowsiness, xeroderma, and possible loss of consciousness. Long-term complications are a definite risk for all forms of diabetes. Blood vessel damage in particular is a major contributor to cardiovascular problems particularly coronary artery disease, stroke, and peripheral vascular disease. Microvascular complications are also seen, manifesting as diabetic retinopathy, nephropathy, and neuropathy. The former can result in vision loss and potential blindness due to damage to blood vessels in the retina; diabetic nephropathy can lead to kidney scarring, chronic kidney disease and dialysis or kidney transplant; and diabetic neuropathy, the most common complication, can cause altered pain and other sensations, damaged skin, foot ulcers, muscle wasting, and weakness.

Type 1 diabetes mellitus, seen in ~10 % of diabetes cases, is caused by insulin deficiency due to loss of the insulin-producing beta cells in the islets of Langerhans in the pancreas. Loss of functioning beta cells may be immune-mediated or idiopathic in nature with the former mechanism being the most common cause. Most cases appear healthy at the onset of the disease. Type 1 diabetes may occasionally be difficult to manage with unpredictable hypoglycemia or hyperglycemia and ketosis. Multiple genes are involved in the occurrence of type 1 diabetes, some HLA genotypes are known to occur with high frequency and associations with some environmental factors such as viral infection have been suggested. The appearance of autoantibodies, often termed latent autoimmune diabetes of adults (LADA), has been shown to predict the appearance of type 1 diabetes but not everyone with the antibodies progresses to the disease. The relevant antibodies are islet cell autoantibodies (ICA), insulin antibodies, glutamic acid decarboxylase autoantibodies (GADA), insulinoma-associated autoantibodies (IA-2), and zinc transporter autoantibodies (ZnT8).

Type 2 diabetes, the most common type of diabetes, is characterized by resistance to insulin and sometimes by reduced secretion of insulin. Although all of the specific defects of the disease are yet to be identified and understood, unlike type 1 disease, type 2 diabetes is clearly related to lifestyle but there also appears to be a clear genetic component. Lifestyle factors identified include obesity, poor diet, physical inactivity, and stress. In relation to diet, saturated fats and trans fatty acids, excess consumption of drinks with a high sugar content, and a high intake of white rice, all seem to be risk factors. Intake of monounsaturated and polyunsaturated fats appear to decrease the risk. The long-term complications of type 2 diabetes are many and often severe with the list comprising, amongst others, cardiovascular disease (including stroke, ischemic heart disease and decreased blood flow to limbs leading to amputation), kidney failure, diabetic retinopathy, cognitive dysfunction, frequent infections, sexual dysfunction, and acanthosis nigricans. Impaired secretion of insulin, that is insufficient production by beta cells, and insulin resistance are characteristic of type 2 diabetes but the relative contributions of these two defects differs among individual patients. Other mechanisms may be involved, however, such as a high level of glucagon in the blood, increased breakdown of lipids in fat cells, abnormalities in the incretin gastrointestinal hormones (section "Glucagon-Like Peptide 1 and the Incretin Effect"), and retention of water and salt by the kidneys.

Production of Insulin

Insulin is produced by the beta cells of the islets of Langerhans in the pancreas not as the relatively small, compact, active protein essential for the regulation of blood glucose levels and the metabolism of fats but as preproinsulin, an inactive single polypeptide containing a 24 amino acid amino terminal signal sequence needed to direct the nascent peptide to the endoplasmic reticulum and through the membrane for posttranslational processing. Proteolytic removal of the signal peptide in the lumen of the rough endoplasmic reticulum forms proinsulin which is the prelude to the formation of three disulfide bonds and the molecule folding and being locked into its correct conformation. Soon after, proinsulin is transported to the trans-Golgi network where specific peptidases cleave the molecule to produce the final

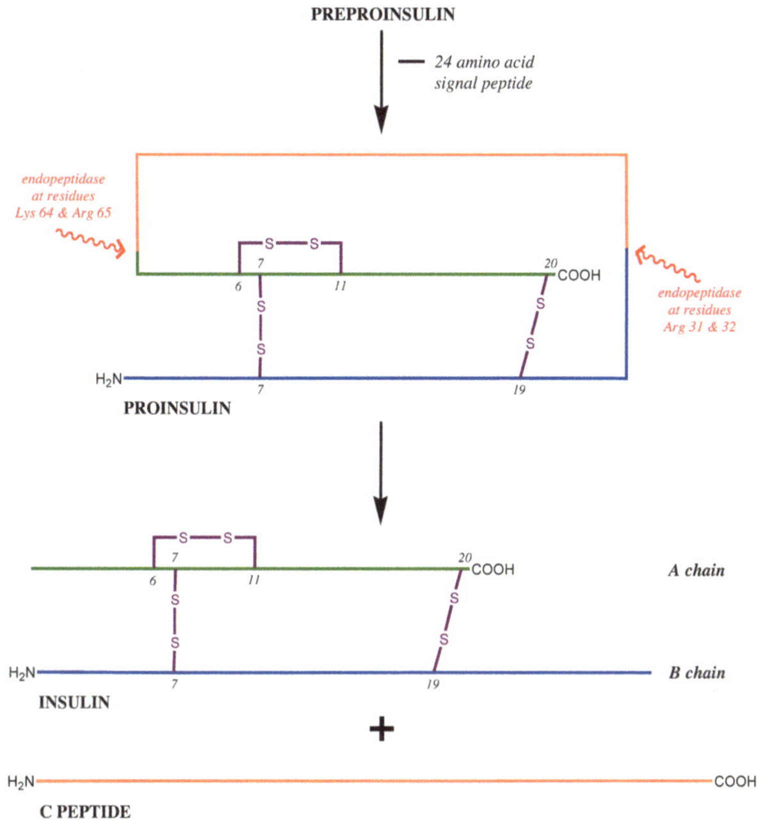

Fig. 7.1 Steps in the production of insulin from its biologically inactive precursor preproinsulin, the primary translational product of the insulin gene (INS). Proteolytic removal of the terminal 24 amino acid signal sequence from preproinsulin produces proinsulin which becomes locked into its correct conformation by the formation of three disulfide bonds. Proinsulin in turn is cleaved by endopeptidases at basic amino acids Lys64, Arg65 and Arg31 and 32 leaving two disulfide-linked chains A and B and the so-called C peptide. The final active hormone, mature insulin of 51 amino acids, results from removal of the now terminal basic amino acids by an exoprotease carboxypeptidase to produce an A chain of 21 amino acids and a B chain of 30 amino acids

product, mature and active insulin, a molecule of two disulfide bond-linked chains designated A and B. This is achieved through the action of cellular endopeptidases, prohormone convertases PC1 and PC2 which cleave off the so-called C peptide from proinsulin at sites after the basic amino acids Lys64 and Arg 65 and Arg31 and 32 (Fig. 7.1). These basic, now terminal, amino acids are subsequently removed by an exoprotease carboxypeptidase E giving a disulfide bond-linked structure of two polypeptide chains. Finally, the now biologically active hormone is packaged and stored in mature secretory granules in the cytoplasm awaiting triggering signals for its release.

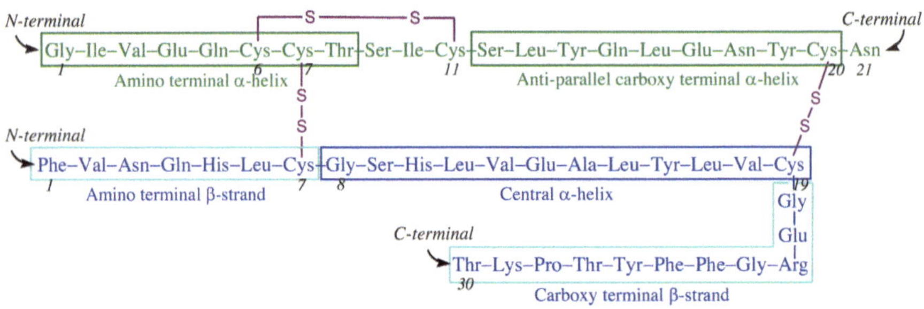

Fig. 7.2 Amino acid sequences of the A (*green*) and B (*blue*) chains of insulin linked together by two disulfide bridges at residues A7–B7 and A20–B19. A third disulfide bridge links cysteines 6 and 11 of the A chain. The A chain has two α-helices, an amino terminal helix (A1–A8) and an antiparallel carboxyterminal helix (A12–A20). The B chain forms an α-helix and a β-strand with the central helix (B8–B19) between amino and carboxy terminal strands, an arrangement called the T conformation

Structure of Insulin

Insulin, a protein of 51 amino acids, MW 5808 Da, is a dimer of an A chain of 21 amino acids and a B chain of 30 amino acids linked together by two disulfide bonds. A third disulfide bridge is internal, linking cysteines 6 and 11 of the A chain. The two intrachain disulfide bridges covalently lock the A and B chains at residues A7–B7 and A20–B19 (Figs. 7.1 and 7.2). The structure of insulin is remarkably conserved across the animal kingdom not only in terms of amino acid sequences with, for example, bovine insulin differing in three residues and porcine insulin in only one, but the three disulfide bonds are invariant across mammalian species.

Considering the secondary structure, the insulin A chain has two α-helices, an amino terminal helix (A1–A8) and an antiparallel carboxyterminal helix (A12–A20). The B chain forms an α-helix and a β-strand with the central helix (B8–B19) between amino and carboxy terminal strands (Fig. 7.2). This arrangement is called the T conformation. At micromolar concentrations, the two-chain disulfide-linked insulin structure associates with another insulin molecule forming a dimer and, in the presence of zinc ions, the dimer can associate with two more insulin dimers to form a hexamer (Fig. 7.3). The hexamer has two zinc ions at its center coordinately bound to a histidine residue from each monomer and all six monomers are in the T conformation. As well as the T conformation, there is an R conformation for the monomer in which the B chain helix extends from B1–B19. As a result of a high chloride concentration, both T and R conformations are found together in a four zinc ion hexamer where three of the monomers are in the T form and three are in the R form. Only the insulin monomer can bind to the insulin receptor (section "Insulin Binding to its Receptor and Ensuing Signaling"), so the dimer and the hexamer must each dissociate to the biologically active monomeric form. The solution structure of a phenol-stabilized 36 kDa insulin hexamer is in the R6 form. These facts are relevant to the formulation of insulin for therapeutic purposes since chloride ions are added to achieve isotonicity and phenol is used as a bacteriostatic.

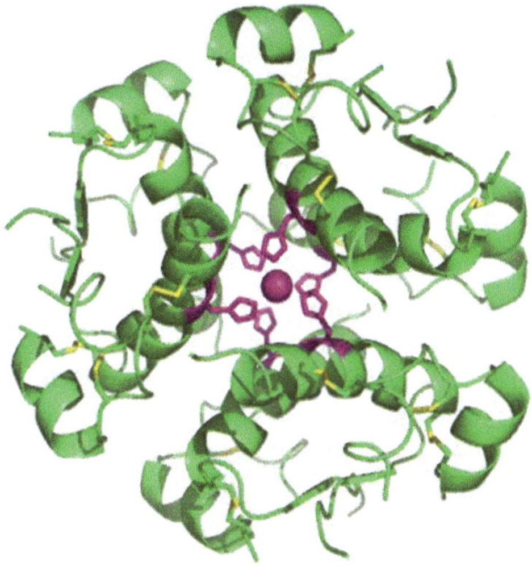

Fig. 7.3 Ribbon diagram of high resolution NMR structure of insulin monomer molecules in hexamer assemblage. Insulin hexamers have two centrally located zinc ions coordinately bound to a histidine residue from each monomer. The view of the threefold symmetry shown has at its center a spherical zinc atom and histidine residues (*purple shapes*). The insulin hexamer is inactive and is used in the body as a storage form for the active monomer. Created by Isaac Yonemoto (Wikimedia Commons, file InsulinHexamer.jpg) with Pymol (http://www.pymol.org/), Inkscape (https://inkscape.org/en/), and Gimp (http://www.gimp.org/) from NMR structure PDB 1ai0 in the RCSB Protein Data Bank (ref. Chang X, Jorgensen AM, Bardrum P, Led JJ. Biochemistry 1997;36:9409–22). Originally uploaded by Takometer at en.wikipedia

Release of Insulin

Insulin may be released rapidly in response to increased glucose concentrations or slowly, independent of sugar levels, by stimuli including, for example, incretins of intestinal origin, namely glucagon-like peptide-1 (GLP-1) and glucose-dependent insulinotropic peptide (GIP), the amino acids arginine and leucine, sulfonylurea and cholecystokinin via phospholipase C. Release triggered by an increase in glucose levels begins with transport of the sugar into beta cells by the glucose transporters GLUT2, phosphorylation of glucose by glucokinase glycolysis, formation of pyruvate, acetyl CoA and production of ATP via the Kreb's cycle leading to an increase in the ATP:ADP ratio and closure of the ATP-sensitive potassium channel. Closure of the channel prevents the efflux of K^+ and an increase in positive charge in the cell leading to depolarization of the cell. This results in the activation of voltage-gated channels that transport Ca^{2+} into the cell triggering the secretion of insulin by exocytosis and its rapid diffusion into blood vessels mediated by the extensive vascular network surrounding the pancreatic islet cells. Release of insulin is proportional to the amount of glucose in the blood, release rising due to high levels of the sugar and falling when the glucose level drops to normal physiological concentrations. If blood glucose falls to dangerously low levels, little or no insulin is released and hyperglycemic hormones, glucagon in particular, are released from alfa cells of the islets of Langerhans. Glucagon (section "Glucagon") induces glucose release from the stores of glycogen in the liver, thus preventing life-threatening hypoglycemia.

Insulin Binding to Its Receptor and Ensuing Signaling

Early studies identified the following insulin "classical binding surface" of amino acid residues on the A and B chains thought to be involved in receptor binding: Glu(A1), Gln(A5), Tyr(A19), Asn(A21), Val(B12), Tyr (B16), Gly(B23), Phe(B24), Phe(B25), and Tyr(B26). Subsequent investigations also implicated residues A21 and B23–B26 and then A13 and B17 which are involved in hexamer-forming surfaces. In addition, a "primary binding surface" made up of residues Ser(A12), Leu(A13), Glu(A17), His(B10), Glu(B13), and Leu(B17) was found to disrupt binding if mutated (Fig. 7.2). The insulin receptor, activated by insulin, IGF-1 and IGF-2, is a ~320 kDa disulfide-linked transmembrane structure (Fig. 7.4) belonging to the tyrosine kinase group of receptors. The dimeric insulin receptor binds, with high affinity, only one molecule of insulin beginning with the interaction of the insulin primary binding surface with the leucine-rich repeat L2 carboxy terminal region of the receptor and followed by interaction between the classical binding surface with the leucine rich L1 receptor binding site. This bivalent cross-linking triggers the ensuing signaling. Binding of insulin to the receptor induces structural changes within the receptor, autophosphorylation of various tyrosine residues and recruitment and phosphorylation of adapter proteins such as insulin receptor substrate 1 (IRS-1). This adapter protein has a key role in signaling to PI3K/Akt and Erk MAP intracellular pathways. Phosphorylated IRS-1 binds to and activates phosphoinositol-3-kinase (PI3K), which in turn catalyzes the formation of phosphatidylinositol-4,5-biphosphate (PIP2) to phosphatidylinositol-3,4,5-triphosphate (PIP3). PIP3 activates Akt or protein kinase B (PKB) which phosphorylates and inactivates glycogen synthase kinase (GSK) leaving glycogen synthase to produce more glycogen. Fusion of vesicles facilitated by PKB leads to an increase in the high affinity GLUT4 transporter molecules in the plasma membrane where it mediates the transport of glucose into the cell. Termination of insulin signaling is effected by degradation of the insulin-bound receptors, dephosphorylation of tyrosine residues, the action of serine/threonine kinases, and a decrease in the number of receptors.

Different Available Insulin Preparations

Insulin is destroyed in the gastrointestinal tract, so it must be administered parenterally either by subcutaneous or intramuscular injection after which it is absorbed into the blood. Insulin appears to be absorbed more rapidly following intramuscular

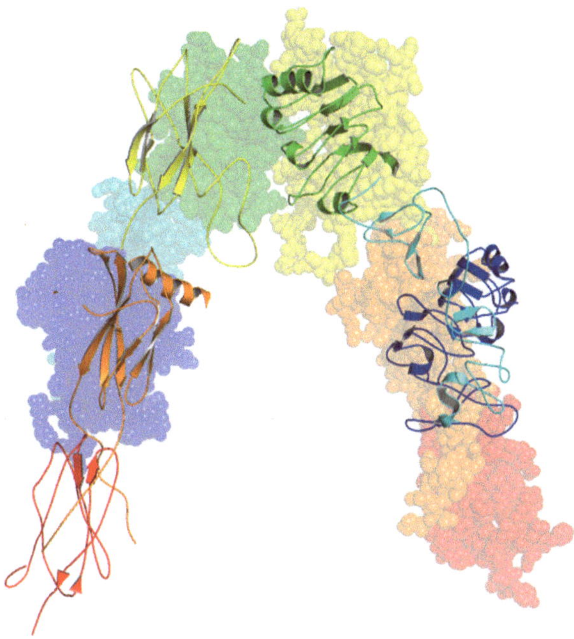

Fig. 7.4 Crystal structure of the dimeric insulin receptor ectodomain represented as a ribbon diagram (foreground) and as colored spheres (background). The leucine-rich repeat L2 carboxy terminal region of the receptor is shown in *green* and the leucine-rich repeat L1 receptor binding site in *blue*. The CR (cysteine-rich) ectodomain is in *cyan* and the fibronectin type III domains, FnIII-1 *yellow*, FnIII-2 *orange*, and FnIII-3 *red*. Image, derived from RCSB Protein Data Bank structure PDB 3LOH (Smith BJ, Huang K, Kong G, et al. Proc Natl Acad Sci USA 2010;107:6771–6) and generated by Pymol, is by Fletcher01 (Wikipedia Commons, file IR Ectodomain mod3LOH.png). Upon dimerization of the receptor and proteolytic cleavage into the α and β chains, an additional 12 amino acids, designated αCT, remain located at the C-terminus of the α chain. This tandem structural element (αCT-L1) defines the intact primary insulin-binding surface of the *apo*-receptor (Smith et al. 2010)

injection than when it is given subcutaneously, hence the route of administration is relevant to any times given for onset, peak levels, and duration of action. Times given are usually for subcutaneous administration. Note that times can vary between preparations and patients and can even vary in any given patient for a range of reasons, so it is not too surprising that onset, peak, and duration times in the literature sometimes vary significantly. Insulin preparations are generally classified as rapid-, short-, intermediate- and long-acting and premixed preparations containing different proportions of the first three of these are often used. Table 7.1 summarizes the main categories of the different insulins in common usage together with approximate time ranges for each preparation's onset, peak level reached and duration of action. Insulin therapy for diabetic patients is a complex and specialized clinical subject requiring careful patient management by experienced clinicians who tailor appropriate therapies for individual patients by assessing a range of factors before

Table 7.1 Summary of properties of different insulin preparations[a]

Type of insulin[b]	Generic name or description	Trade name	Onset[c]	Peak[d]	Duration[e]
Rapid-acting	Insulin aspart[f]	NovoLog®	12–18 min	up to 3 h[g]	3–5 h
	Insulin lispro[h]	Humalog®			
Short-acting	Insulin regular	Humulin® R Novolin® R	30–60 min	2–4 h	5–8 h
Intermediate-acting	Insulin NPH[i]	Humulin® N Novolin® N	1–3 h	6–10 h	12–16 h
Long-acting	Insulin detemir[j]	Levemir®	1–4 h	~9 h[l]	18–26 h
	Insulin glargine[k]	Lantus®			
Combination insulins	Premixed insulins with different durations of action	Humulin® 70/30[m] ; Novolin® 70/30[n];	30–60 min	2–10 hp	16–20 h
		Humalog® Mix 75/25[o]; NovoLog® Mix 70/30[q]	10–15 min	1–4 h[p]	

Refer to insulin structure Fig. 7.2
[a]Preparations approved by FDA or EMA or both
[b]With respect to duration of action and/or mixture of different insulin analogs/preparations
[c]Time taken to begin to lower blood glucose
[d]Time taken for insulin preparation to achieve maximum activity in lowering blood glucose level
[e]Time before completion of blood glucose lowering activity
[f]Recombinant human insulin with aspartic acid substituted for proline at position B28
[g]Some estimate 30–90 min
[h]Recombinant human insulin. Proline at B28 replaced by lysine and lysine at B29 replaced by proline
[i]NPH, neutral protamine Hagedorn or "isophane" (NPH) porcine insulin made by adding neutral protamine to regular insulin
[j]Recombinant insulin analog with threonine deleted at B30 and myristic acid bound to lysine at B29. Binds to serum albumin via the fatty acid and then slowly dissociates from this complex
[k]Microcrystalline for slow release. Recombinant insulin. Asparagine replaced at A21 by glycine and 2 arginines added at C-terminus of B chain
[l]Sometimes said to be "peakless"
[m]Recombinant human insulin suspension for s.c. injection; 70 % human insulin, 30 % isophane suspension
[n]70 % NPH human insulin isophane suspension and 30 % regular recombinant human insulin
[o]75 % insulin lispro protamine suspension and 25 % insulin lispro
[p]Varies
[q]70 % insulin aspartate protamine suspension and 30 % insulin aspartate

selecting individual insulin preparations, dosages, dosage schedules, and so on. In view of this, and because treatment selection, regimens, and dosages are beyond the scope of this book, no attempt has been made to go beyond a summary of the different approved insulin preparations available together with their pharmacokinetic and safety profiles.

The question of the best approach for insulin therapy has always been there but recent developments have enabled the focus to be narrowed to how can the optimal use of insulin be achieved? More specifically, should a patient use an insulin pump in a basal-prandial regimen to try and mimic the physiological delivery of insulin or follow a multidose injection schedule? Aside from the costs of pump technology, ongoing insulin supplies, glucose testing and education, all of which may be considerable or involve considerable effort, use of modern pumps, pen devices and continuous glucose monitoring has created the means of achieving a situation close to physiological insulin secretion, especially with the fast-acting insulin analogs such as lispro and glulisine that provide good glycemic control compared to the older regular insulin preparations and neutral protamine hagedorn. The fast-acting insulins are administered before each meal to manage the glucose surge. Additional benefits of pump therapy compared to multiple daily injections can include better absorption, a decrease in the amount of insulin used, and fewer unwanted glycemic fluctuations. Currently, about 20 % of diabetic patients in the USA are using pumps and the FDA has approved more than 50 pump devices. For both basal and prandial insulin needs, most patients on pump-delivered therapy use fast insulin analogs, aspart, lispro, or glulisin, and these, together with buffered regular insulin, are currently approved for use in pumps. Glulisine is approved for children more than 6 years old. The basal infusion caters for the glucose between meal intake while bolus doses are called upon to respond to the meal-induced relatively sudden glucose load. For next-generation pumps, insulin delivery is likely to be effected without patient input.

Not considered in Table 7.1 is the ultra-long-acting basal insulin analog, **insulin degludec** (Tresiba®) which differs from human insulin by a single deleted amino acid and the addition of hexadecanoic acid linked to lysine at position B29 (Figs. 7.1 and 7.2) by a γ-*L*-glutamyl spacer. Administered subcutaneously once a day, insulin degludec has an onset of action of up to ~90 min, a half-life of ~25 h, a so-called "peakless" activity level due to its slow release, and a duration of action of ~40 h. With once-daily dosing, a steady state is said to be reached after 2–3 days with a glucose-lowering effect steady after a 24 h period. Slow release is due to deposition of the hydrophobically modified insulin into subcutaneous tissues. Due to an increase in what were in fact nonsignificant cardiovascular events in degludec-treated patients, the preparation was not approved in 2013 by the FDA who requested additional cardiac safety data from a dedicated cardiovascular trial. As with all insulin preparations, hypoglycemia is a potential adverse effect. Rates of nocturnal hypoglycemia with insulin degludec have been seen in the range 3.7–5.1 events per patient year.

Recently (2014), the FDA granted approval for a rapid-acting, single use per cartridge, insulin powder known as **Technosphere insulin** (Afrezza®) administered as an inhalation in an individualized dose at the beginning of meals. Indicated for rapid glycemic control in adult diabetics, important limitations of the product are that it must be used together with a long-acting insulin in type 1 diabetes, it is not recommended for the treatment of diabetic ketoacidosis, and it should not be used in smokers. Adverse reactions to the preparation are the same as other insulins plus cough, throat pain or irritation and important warnings and precautions are for acute bronchospasm, a decline in pulmonary function, and the avoidance of usage in lung cancer. In fact, Afrezza® is the subject of a black box warning stating that: (1) acute

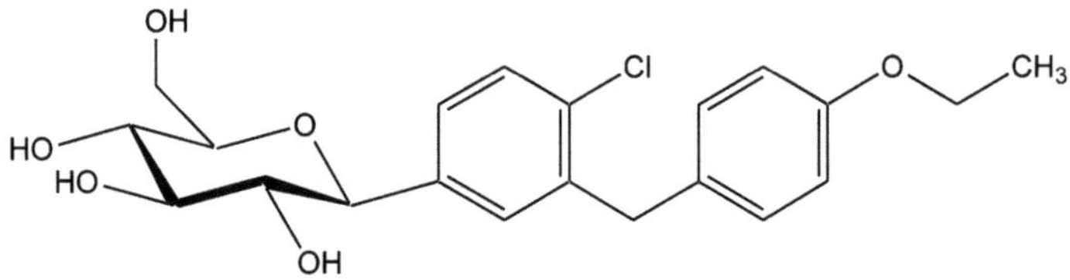

Dapagliflozin

Fig. 7.5 The 6-arylglucoside dapagliflozin, a first generation sodium-glucose co-transporter 2 (SGLT2) inhibitor, is the first SGLT2 inhibitor approved for human therapy. Often used in combination with metformin or a sulfonylurea for type 2 diabetes, dapagliflozin blocks glucose reabsorption in the kidneys. Two other approved SGLT2 inhibitors are canagliflozin and empagliflozin, the latter being closely related in structure to dapagliflozin

bronchospasm has been observed in patients with asthma and chronic obstructive pulmonary disease (COPD); (2) it is contraindicated in patients with chronic lung disease such as asthma or COPD; and (3) before initiating therapy, all patients should be subjected to a detailed review of medical history, a physical examination, and spirometry (FEV_1) to identify potential lung disease.

Sodium-Glucose Co-transporter 2 Inhibitors

In the continuing search for optimal glycemic control individualized to patient needs and preferences, a number of new agents are being developed and examined. Perhaps the most interesting and potentially useful of the new agents is a new class of antidiabetic therapeutics, the sodium glucose co-transporter 2 (SGLT2) inhibitors. These drugs enhance glucose excretion via the urine by reducing tubular glucose reabsorption, thus producing urinary excretion of the sugar. **Dapagliflozin** (Fig. 7.5), used to treat type 2 diabetes and known by the trade names of Farxiga® in the USA and Forxiga® in Europe, is the first SGLT2 inhibitor to be approved for human therapy. The FDA approved dapagliflozin in January 2014 for glycemic control, along with diet and exercise, in adult type 2 diabetics. The drug is now marketed in a number of European countries, the UK and Australia. In October 2014, FDA approval was given for the combination of dapagliflozin with metformin hydrochloride extended release (Xigduo XR®). Dapagliflozin is metabolized mainly to its 3-*O*-glucuronide derivative and eliminated by the kidneys indicating that it should not be used in patients with renal impairment. In general, clinical trials of up to 2 years appear to indicate that the drug is well tolerated with polyuria, glycosuria, weight loss, tiredness, nocturia, thirst, dehydration, hypotension, and genital and urinary tract infections, especially thrush, being the main reported adverse effects. However, in a recent report from Japan, a "considerable number" of serious adverse events were reported in patients receiving SGLT2 inhibitors in the first 3 months of their use in that country. Adverse events included urogenital infections, hypoglycemia, dehydration, and serious skin and subcutaneous reactions such as generalized rashes and skin eruptions. Concern was also expressed of potential SGLT2 inhibitor-induced serious adverse events in nonobese type 2 patients with reduced insulin secretion, an outcome often seen in East Asia. It is feared that such serious events may include severe hypoglycemia due to depletion of hepatic glycogen stores, ketosis/ketoacidosis, and accelerated diabetes-associated sarcopenia. The authors point out the needs for caution in the appropriate use of SGLT2 inhibitors and careful ongoing observation of the treated patients.

Two other approved SGLT2 inhibitors are canagliflozin (Invokana®; Silisent®) and empagliflozin (Jardiance®) which is closely related in structure to dapagliflozin. In 2013, canagliflozin became the first SGLT2 inhibitor approved in the USA; empagliflozin was approved in Europe and by the FDA in 2014. After early concerns regarding the cardiovascular safety of canagliflozin and ongoing investigations of this issue, in 2015 the FDA issued a warning that some SGLT2 inhibitors, including canagliflozin, may induce ketoacidosis. This warning was followed up by the FDA in 2015 with release of a Drug Safety Communication altering the label for canagliflozin to include the risk of decreased bone density and bone fracture. A reduction in cardiovascular mortality appears to be associated with empagliflozin.

Warnings, Precautions, and Adverse Events Associated with Insulin Use

Warnings and Precautions

Warnings and precautions issued by regulatory agencies for those administering insulin therapy are concerned with: (1) The need for glucose monitoring for all patients with diabetes since hypoglycemia is the most common adverse event of insulin therapy. (2) The possibility of insulin causing hypokalemia which can lead to respiratory paralysis, ventricular arrhythmia, and death. (3) The reduced requirement of insulin in some patients with renal or hepatic impairment. (4) Severe life-threatening hypersensitivities including anaphylaxis. (5) The possible occurrence of fluid retention and heart failure when insulin is used with peroxisome proliferator-activated receptor (PPAR)-gamma agonists such as thiazolidinediones. (6) Anti-insulin antibodies, some of them cross-reactive with different insulins, may be produced following insulin injection but their clinical significance is not always clear since their presence does not automatically result in impairment of insulin action. Also, the antibodies do not always increase on long-term exposure.

Hypoglycemia, Hyperglycemia, and Diabetic Coma

Hypoglycemic reactions may occur especially in the labile form of the disease or, for example, when patients fail to eat, receive too high a dose of insulin, or undergo a bout of unusually high exercise. Early symptoms may include sweating, a feeling of weakness, trembling, hunger, and tachycardia. These symptoms are more likely when a rapid fall of glucose occurs and epinephrine is released to compensate. With a slow fall of glucose, symptoms include headache, mental confusion, blurred vision, difficult to understand speech, diplopia and perhaps coma and convulsions. All symptoms may be present in some patients. In some cases, the onset of a hypoglycemic reaction may be characterized by hunger or nausea and bradycardia, mild hypotension and gastrointestinal discomfort may occur. If the period of hypoglycemia is prolonged, a rare event, irreversible damage to the brain, may occur manifesting as ataxia, aphasia, choreiform movements, epilepsy, parkinsonism, mental retardation, and incontinence. Severe hypoglycemia leading to unconsciousness, sometimes referred to as insulin shock, is one of three severe reactions in patients with diabetes mellitus sometimes termed diabetic coma, the other two being diabetic ketoacidosis and hyperosmolar hyperglycemic state (also called hyperosmolar nonketotic coma). Diabetic ketoacidosis is a medical emergency and occurs mainly in patients with type 1 diabetes but it can occur in type 2 patients. It results from a shortage of insulin which leads to an elevation of glucagon, release of glucose from the liver and osmotic diuresis with glucose, water, Na^+, and K^+ excreted in the urine causing dehydration. The absence of insulin also leads to lipolysis, that is the release of free fatty acids from adipose tissue, which in turn are converted in the liver to ketone bodies, namely acetoacetate and β-hydroxybutyrate. Although ketone bodies can be an energy source in the absence of glucose, in diabetic ketoacidosis they produce metabolic acidosis which overwhelms the bodies' bicarbonate buffering system. Treatment consists of isotonic fluids, intravenous electrolytes, and insulin. Hyperosmolar hyperglycemic state occurs predominately in diabetes type 2 patients causing polyuria, osmotic diuresis, high blood sugar levels, severe dehydration, a high risk of further complications, coma, and death. Unlike diabetic ketoacidosis, ketosis is absent because the presence of some insulin inhibits lipases involved in lipolysis. Treatment consists of intravenous fluids, electrolyte replacement, and insulin once potassium levels are high enough.

Hypersensitivity Reactions to Insulin

Adverse reactions to insulin resembling hypersensitivity responses were not uncommon in the past when porcine and bovine insulins were widely used but the introduction of recombinant human insulin decreased the incidence of such reactions. Local or systemic allergic reactions can occur in patients receiving insulin after a sensitization period but reactions on first exposure are also seen. Early studies showed that IgE antibodies developed in the sera of small numbers of patients treated with purified porcine insulin, mixed bovine-porcine insulin and even recombinant human insulin but the clinical relevance of the antibodies remained speculative since none of the patients developed clinical manifestations. A step forward was the finding of large local reactions in association with high levels of anti-human insulin IgE reactive with all three insulins in a patient with gestational diabetes who showed similar reactions to porcine and bovine insulins. Further investigations demonstrated positive skin tests and cross-reactive IgE antibodies to human, porcine, and bovine insulins in the same patients, supporting the conclusion of insulin allergy and the presence of similar or identical allergenic determinants on all three preparations. These results also demonstrated that the hope of eliminating allergic reactions by administering only human insulin would not be fulfilled. Numerous cases of IgE antibody-mediated immediate hypersensitivity to insulin since the early studies have revealed a range of associated symptoms from mild cutaneous reactions such as erythema, itching, swelling at the injection site, and pruritus of soles and palms, to generalized flushing, urticaria, and angioedema. Severe systemic cases with dyspnea, hypotension, and anaphylaxis, including rare deaths, have also been recorded.

As well as type I immediate allergic reactions to insulin, biphasic reactions and types III and IV hypersensitivities to the peptide have been recorded. In the biphasic reactions, wheal and flare responses were followed four to six hours later by an indurated lesion lasting up to 24 h. Lesions showed histopathological features similar to the late phase reactions seen with, for example, pollens and some drugs. Demonstration that the reactions were transferable by serum, that is by Prausnitz-Küstner (P-K) testing, supported the conclusion that the reactions were true late phase responses. Evidence of immune complex type III hypersensitivities was provided by the observation of a reaction developing four to six hours after bovine/porcine insulin injection, peaking at 12 h, not being transferable by P-K testing, and showing histological features of a true Arthus type reaction. In another example of a type III hypersensitivity response to insulin, a case of leukocytoclastic vasculitis was seen following injection of both semi-synthetic and recombinant human insulins. Symptoms of intense itching and redness but without a wheal and flare immediately after subcutaneous injection, were independent of the injection site. Skin biopsies from five hour and five-day-old lesions showed perivascular and interstitial infiltration with neutrophilic and eosinophilic granulocytes, granulocytic infiltration, fibrin deposition and erythrocytes in the vascular walls, indicating leukocytoclastic vasculitis.

In distinguishing the three types of hypersensitivity, a careful history is, as usual, important to give an indication of an allergic response and to determine whether the response is a type I, III, or IV hypersensitivity. Skin testing and specific IgE tests are valuable diagnostic aids but positive prick tests have been found in nonallergic diabetic patients who reacted to insulin. In relation to this, it is important to remember that a positive IgE antibody titer is, without supporting diagnostic data, an indication of sensitization only. Insulin-specific IgE and IgG antibodies without any apparent clinical relevance were found in up to 28 % of type 1 diabetic patients treated exclusively with recombinant human insulin. Insulin skin test kits are provided by Novo Nordisk, Bagsvaerd, Denmark, and Sanofi Aventis Pharma Deutschland, Bad Soden Am Taunus, Germany, and specific antibodies against

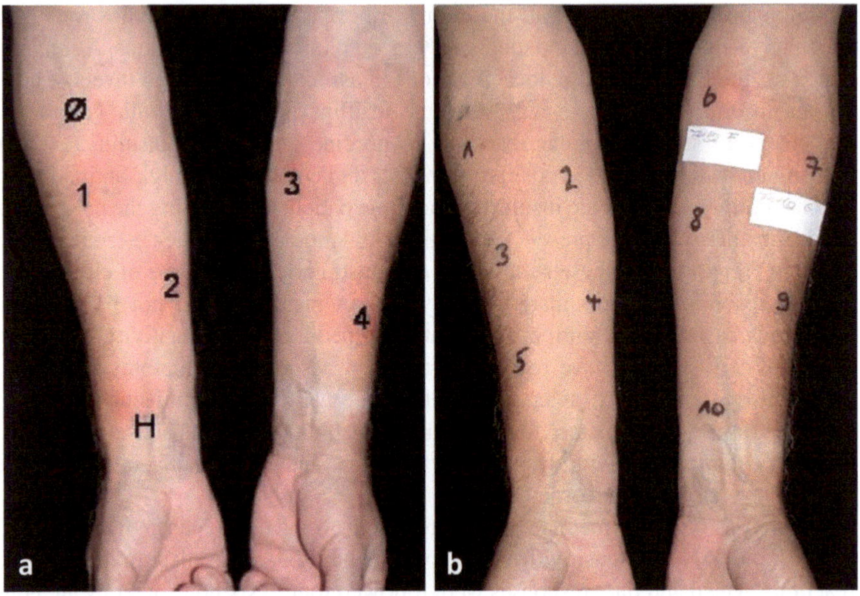

Fig. 7.6 (**a**) Intradermal tests showing positive reactions to: (*1*) Levemir® (insulin detemir), (*2*) Huminsulin Basal® (human insulin isophane), (*3*) Humalog® (insulin lispro), (*4*) Lantus® (insulin glargine). *H*, histamine positive control, *Ø*, vehicle negative control. See Table 7.1 for descriptions of the different insulins. (**b**) Results of intradermal tests to protamine test solutions (sites 6 and 7) and additives from the Sanofi Aventis Insuman® test kit, including phenol, cresol, glycerol, sodium acid phosphate, and zinc chloride (sites 1–5 and 8–10). Reactions to protamine were positive, all other sites were negative. All results were read 20 min after intradermal injection. Reproduced from Pföhler C et al. Successful desensitization with human insulin in a patient with an insulin allergy and hypersensitivity to protamine: a case report. J Med Case Reports 2008;2:283. doi 10.1186/1752-1947-2-283, an Open Access article distributed under the terms of the Creative Commons Attribution License

human insulin can be detected with the CentAK® anti-IA2 radioligand assay. In one interesting case, positive skin tests implicated human insulin and protamine (used to retard the absorption of insulin as neutral protamine Hagedorn [NPH]) in a patient who developed large (>15 cm diameter), persisting (days) pruritic plaques at the injection sites after receiving different insulin preparations. Intradermal tests showed positive reactions to different standard recombinant insulin and insulin isophane preparations, and to protamine solution (Fig. 7.6a, b) and although IgG antibodies to insulin were detected, IgE antibody tests were negative for both insulin and protamine. Patch tests were negative and skin biopsy revealed an Arthus-type type III reaction. The patient was successfully desensitized by subcutaneously administering insulin in a rush protocol (see below, section "Desensitization to Insulin").

In cases of type IV reactions to insulin, delayed hypersensitivities developing 8–12 h after insulin injection and peaking at 24 h were not preceded by a wheal and flare reaction and were not transferable by P-K testing.

Desensitization to Insulin

When symptomatic therapy is not sufficient or substituting a different insulin does not work or cannot be done, immunotherapy must be considered. There are now a number of reports of effective and well-tolerated subcutaneous injection or infusion desensitization treatments and one case of successful intravenous therapy when continuous subcutaneous injections of insulin and oral antiallergic agents did not prevent life-threatening symptoms. The intravenous route for desensitization was attempted when no reaction occurred following the injection of 0.05 units of regular insulin. The amount of insulin was then gradually increased until the required dose was reached using a central venous catheter, a subcutaneously embedded reservoir, and a portable infusion pump. Allergic symptoms disappeared as soon as intravenous injections began and, within a year, levels of anti-insulin IgE and IgG returned to normal without upsetting glucose control. After reversion to subcutaneous insulin injections resulted in exacerbation of allergic symptoms, continuous intravenous injections of insulin were resumed and this was supplemented with intravenous injections before meals. Antibody levels again declined reaching normal values 10 months after the start of intravenous treatment. Thus, the allergic response to the same insulin preparation differed depending on the route of administration.

In a successful attempt to overcome an allergy to human insulin, insulin lispo was administered as a continuous subcutaneous infusion via an insulin pump. The following schedule was adhered to: 0.7 IU/h for 2–8 h; 0.3 IU/h for 8–13 h; 0.6 IU/h for 13–18 h; 0.8 IU/h for 18–21 h; and 0.6 IU/h for 21–22 h, plus an additional bolus of 6 IU before breakfast, 5 IU before lunch, and 6 IU before dinner. Following this procedure, the allergic reaction did not reoccur and metabolic control was obtained with the patient tolerating insulin therapy. Although the patient remained clinically asymptomatic, the skin prick test to insulin remained positive 3 months later. In a successful treatment of patients presenting with severe insulin-induced allergic symptoms, Heinzerling and coworkers administered ascending single doses of insulin starting with a dose of 0.00001 units followed by progressive tenfold increases up to 1 unit and then 2, 4, 8, 12, 16, and 20 units. If local reactions occurred, the causitive dose was repeated until tolerated; for systemic reactions, the dose was halved. Blood sugar was monitored throughout and the procedures were carried out in an in-patient setting with emergency measures at hand if needed. A simple and successful procedure for desensitization of insulin allergy was undertaken with the long-acting, microcrystalline insulin analog, glargine in a type 2 diabetic patient who reacted to human isophane insulin with persisting 8–12 cm flares at the injection sites together with generalized pruritic erythema. Skin testing demonstrated positive reactions to regular, NPH, and lispro insulins but negative responses to insulin aspart and glargine. Administration of insulin aspart together with dextrochlorpheniramine proved disappointing with the rapid appearance of erythema, wheal, and pruritus at the injection site. After a 4-day interval, administration of 1 unit of glargine induced only a mild, non-pruritic reaction which decreased despite gradual increases in the daily dose. Reintroduction of insulin aspart proved possible on day five and the patient was able to continue to receive the therapy without concurrent administration of an antihistamine.

The mechanism of tolerance induction as a result of successful desensitization procedures remains unclear. In cases of suspected hypersensitivity it is usually only successful in IgE-mediated type I reactions but in the case discussed above (section "Hypersensitivity Reactions to Insulin," Fig. 7.6), desensitization to insulin was achieved after a series of subcutaneous injections of increasing doses of the hormone. On day 1, injections of 0.004, 0.01, 0.02, 0.04, 0.1, 0.2, 0.5, and 1 IU of human insulin were given at 30 min intervals. The antihistamine fexofenadine, 180 mg was given twice daily. On day 2, doses of 1, 2, 3, and 5 IU were given every 30 min and on day 3, two doses of Lantus® 6 IU were administered. All desensitizing doses were tolerated and the patient subsequently tolerated normal insulin therapy. Fexofenadine was reduced to 180 mg daily and withdrawn 6 months after desensitization.

Other Adverse Events

As well as hypoglycemia and true hypersensitivity reactions, rashes, pruritus, weight gain due to anabolic effects of insulin, edema due to sodium retention, and injection site irritant reactions that normally clear up in a few days, are amongst the most commonly seen insulin-induced adverse events. Lipoatrophy, that is, thinning of adipose tissue or a depression in the skin, and lipohypertrophy or enlargement and thickening of adipose tissue, are also common adverse effects. Rapid improvement in the control of glucose may result in transitory, reversible ophthalmologic refraction disorder, worsening of diabetic retinopathy, and peripheral neuropathy. Other adverse events occurring with a frequency of 5% or more in patients with type 1 diabetes include headache, influenza-like symptoms, dyspepsia, back pain, diarrhea, pharyngitis, rhinitis, skeletal pain, and upper respiratory tract infection. For type 2 diabetic patients, the equivalent list is upper respiratory tract infection, headache, diarrhea, neuropathy, pharyngitis, abdominal pain, and rhinitis.

Glucagon

Glucagon, a peptide hormone produced and stored in pancreatic alfa cells of the islets of Langerhans, is involved in glucose homeostasis. The hormone's secretion is coupled to levels of circulating glucose, its release being stimulated in hypoglycemia and inhibited in hyperglycemia. With a half-life of only a few minutes, glucogen acts in the liver where it stimulates the breakdown of glycogen to glucose. Together with insulin, it forms a feedback system controlling blood glucose levels. Glucagon belongs to a family of structurally related polypeptides that regulate G protein-coupled receptors (GPCRs) from the secretin receptor family. Within this superfamily of peptides, glucagon and the glucagon-like peptides, GLP-1 and GLP-2 (see section "Glucagon-Like Peptide 1"), are each encoded by a single proglucagon gene. Glucagon is generated from proglucagon in a posttranslational process that is tissue-specific. This is effected in the pancreatic α-cells by proprotein convertase 2 while in the intestine and brain, prohormone convertases liberate GLP-1 and GLP-2.

Structure and Mechanism of Action of Glucagon

Recombinant glucagon is a 29 amino acid, single chain polypeptide (Fig. 7.7), MW 3483 Da, identical to the human hormone. Brand names in use are Glucagon® and GlucaGen®. Glucagon acts via its receptor, a GPCR in the plasma membrane of pancreatic islet cells, liver, kidney, and brain causing a conformational change in the receptor, a protein with α, β, and γ subunits. Interaction of the G protein with the receptor causes the release of the β and γ subunits while the α subunit specifically activates adenylate cyclase which catalyzes the formation of cAMP. Protein kinase A is then activated followed in turn by the activation of phosphorylase kinase and the

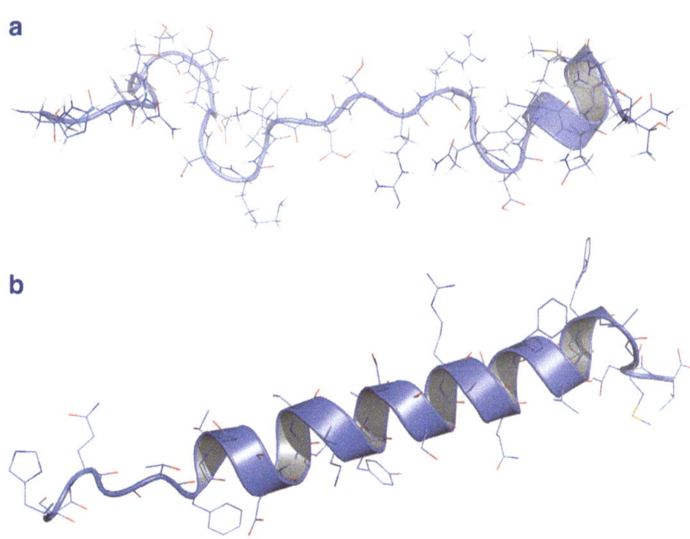

Fig. 7.7 Structure of glucagon determined by (**a**) nuclear magnetic resonance (NMR) and (**b**) X-ray crystallography. Images generated and supplied by Dr J. D. Cronk from structures from RCSB Protein Data Bank (PDB) files (http://www.rcsb.org/pdb/explore/explore.do?structureId=1KX6; http://www.rcsb.org/pdb/explore/explore.do?structureId=1GCN) and molecular graphics program PyMOL (http://www.pymol.org/; http://www.pymol.org/citing). Note that the NMR-generated structure is only one of 20 consistent with the generated data indicating flexibility of the glucagon structure (Cronk JD. 2014; Biochemistry dictionary. Glucagon. http://guweb2.gonzaga.edu/faculty/cronk/biochem/G-index.cfm?definition=glucagon)

phosphorylation of glycogen phosphorylase, subsequently converted to phosphorylase A, the enzyme that releases glucose-1-phosphate from glycogen. These signal pathways activated by glucagon lead to an increase in insulin secretion.

Indications and Adverse Effects of Glucagon

Given its role in the formation of glucose from glycogen, it is not surprising that glucagon is approved for the treatment of hypoglycemia. Another important, and useful, activity of glucagon is its capacity to relax smooth muscle when given parenterally, a property made use of when the hormone is employed as an inhibitor of gastrointestinal motility in some radiologic diagnostic examinations of the stomach, duodenum, small bowel, and colon.

Exogenous glucagon also stimulates the release of catecholamines, a property underlying an FDA warning of a possible sudden increase in blood pressure if given to patients with pheochromocytoma, a neuroendocrine tumor of the adrenal gland medulla cells that secretes an abnormally large amount of norepinephrine and a lesser amount of epinephrine. The FDA also warns that glucagon should be administered with caution to patients with insulinoma causing the release of insulin and subsequent hypoglycemia. Other warnings and precautions relate to: (1) the need for a sufficient amount of glycogen in the liver for glucagon to reverse hypoglycemia—glucagon should be used with caution in patients with conditions resulting in low levels of releasable glucose in the liver; (2) the need for caution when using glucagon as a diagnostic aid in diabetic patients; and (3) glucagon increases oxygen demand and should therefore be used with caution in patients with cardiac disease. Serious adverse reactions are rare but nausea and vomiting are occasionally seen and when reactions do occur they may be accompanied by hypoglycemia. A temporary increase in blood pressure may result after administration; other adverse events include rash, hypotension, and tachycardia.

Glucagon Hypersensitivity

Glucagon is commonly added to barium sulfate suspensions to diminish intestinal mobility in double-contrast radiologic procedures. Initial examinations in the late 1970s and 1980s evaluating the safety of glucagon did not report allergic or anaphylactic reactions but over the years occasional cases of reactions that appeared to be true hypersensitivities have been described. In one fairly typical case, a palpable purpuric rash appeared on the legs, and erythematous papules and plaques on the arms and trunk, of a patient given a barium enema together with intravenous glucagon. Skin biopsy revealed perivascular infiltrates of lymphocytes and eosinophils consistent with a drug eruption. Other more severe cases involved anaphylaxis to intravenous glucagon and cardiac arrest following a barium enema with glucagon. The latter reaction began with itching and a tingling sensation that rapidly progressed to vomiting and the patient becoming diaphoretic. Cardiopulmonary arrest and death soon followed. Hypersensitivity reactions to glucagon cover a wide spectrum of symptoms including skin rashes, urticaria, periorbital edema, erythema multiforme, breathing difficulties, and anaphylaxis. Of 11 hypersensitivity reports to a manufacturer of glucagon, 4 involved respiratory distress and/or hypotension following glucagon administration, five had previous exposure, and two had experienced a previous reaction to the hormone. Seven of the patients had received glucagon for treatment of hypoglycemia and four experienced the reaction during a radiologic procedure.

Glucagon-Like Peptide 1

The hyperglycemic hormone glucagon was discovered in pancreatic extracts in 1923 before the realization that hyperglycemic substances were present in extracts of gastrointestinal mucosa. This led Sutherland and DeDuve to correctly predict in 1948 that gastric extracts might also contain glucagon. Twenty years later, material that showed immunological cross-reactivity with glucagon was demonstrated in intestinal secretions in response to orally administered glucose and this "gut glucagon-like" substance(s) proved physicochemically different to glucagon. Two polypeptides, named oxyntomodulin and glicentin, both containing the 29 amino acid glucagon sequence, were subsequently shown to be present. Meanwhile, biosynthesis studies on pancreatic islet cells identified a proglucagon polypeptide secreted along with glucagon that did not contain the glucagon sequence but two glucagon-*like* sequences, now designated glucagon-like peptides 1 and 2 (GLP-1 and GLP-2). In the intestinal mucosa, however, the two glucagon-like peptides were shown to be formed and secreted separately. A fourth glucagon-related peptide hormone, the 29 amino acid gastric inhibitory polypeptide (GIP) shares a number of biological activities with the three proglucagon-derived peptides. Their amino acid compositions and sequences are compared in Table 7.2. All four peptides act through specific GPCRs of the glucagon receptor superfamily.

Table 7.2 GLP/GIP amino acid sequences of the proglucagon-derived peptide hormones glucagon, GLP-1 and GLP-2, and the glucagon-related peptide GIP

Proglucagon-derived peptide	Amino acid sequence
Glucagon	HSQGTFTSDYSKYLDSRRAQDFVQWLMNT
GLP-1	HAEGTFTSDVSSYLEGQAAKEFIAWLVKGRG
GLP-2	HADGSFSDEMNTILDNLAARDFINWLIQTKITD
GIP	YAEGTFISDYSIAMDKIRQQDFVNWLLAQ

Glucagon-Like Peptide 2

The 33 amino acid peptide GLP-2, which is attached at the carboxy terminal end of the GLP-1 sequence in the proglucagon fragment, is cosecreted with GLP-1, oxyntomodulin, and glicentin by gastrointestinal L cells. The peptide has been shown to exhibit intestinotrophic properties by stimulating nutrient absorption; inducing increases in mucosal thickening of the intestine; reducing secretion of gastric acid; inhibiting gastric emptying; reducing permeability of the small bowel; and stimulating intestinal hexose transport. The peptide also displays reparative and antiapoptotic actions like the reduction of mucosal damage and mortality in rodent models of bowel injury and increases nutrient absorption in both rats and humans with bowel disorders. Suggested speculative targets for GLP-2 and/or GLP-2 analogs include short bowel syndrome, Crohn's disease, and osteoporosis.

Glucagon-Like Peptide 1 and the Incretin Effect

The "incretin effect" describes the observation that glucose delivered orally produces a two- to threefold larger insulin response than the secretory response seen following intravenous administration. Note, however, that the effect is proportional to the challenge, being small with small amounts, and large with large amounts of glucose. GIP was the first incretin identified but studies showing that neutralization of this duodenal hormone did not completely prevent the incretin effect, ultimately led to the identification of the 31 amino acid peptide GLP-1 as the second major incretin. GLP-1 is often written GLP-1(7–37) with numbering relevant to the peptide precursor as the sequence 7–37, derived from cleavage of the proglucagon molecule, is the biologically active form. The relationship between numbering and the amino acid sequence of GLP-1 is shown in Fig. 7.8 alongside structures of some GLP-1 receptor agonists (see section "GLP-1 Receptor Agonists"). GLP-1 secretion is clearly meal-related and from hormone and glucose infusion studies in humans it is now known that following the ingestion of a meal, it, and GIP, enhance insulin secretion to an extent that explains both the insulin response and the incretin effect (Fig. 7.9). Although the effects of the two hormones on insulin secretion are additive, only GLP-1 inhibits glucagon secretion. There is now little doubt that the incretin effect has a prime role in the post-prandial secretion of insulin and glucose tolerance.

The action of GLP-1 in stimulating insulin secretion only when glucose levels are high while suppressing glucagon secretion (Fig. 7.9) means that the hormone carries with it a low risk of hypoglycemia. GLP-1 can be used to restore the insulin secretory response in type 2 diabetics whose responsiveness to the peptide is impaired. This property, as well as its actions of increasing insulin sensitivity of both alfa and beta cells of the pancreas and inhibiting acid secretion, stomach emptying, food intake, and apoptosis of beta cells, all suggest that the peptide should be

GLP-1:

N-terminal

His-Ala-Glu-Gly-Thr-Phe-Thr-Ser-Asp-Val-Ser-Ser-Tyr-Leu-Glu-Gly-Gln-Ala-Ala-Lys-Glu-Phe-
7(1)

Ile-Ala-Trp-Leu-Val-Lys-Gly-Arg-Gly
34(28) 37(31) *C-terminal*

Liraglutide:

CH_3-$(CH_2)_{14}$-C(=O)-Glu-OH

His-Ala-Glu-Gly-Thr-Phe-Thr-Ser-Asp-Val-Ser-Ser-Tyr-Leu-Glu-Gly-Gln-Ala-Ala-Lys-Glu-Phe-
7 26

Ile-Ala-Trp-Leu-Val-Arg-Gly-Arg-Gly
34

[-Lys^{34}]

Exenatide:

His-Gly-Glu-Gly-Thr-Phe-Thr-Ser-Asp-Leu-Ser-Lys-Glu-Met-Glu-Glu-Glu-Ala-Val-Arg-Leu-Phe-
1 10

Ile-Glu-Trp-Leu-Lys-Asn-Gly-Gly-Pro-Ser-Ser-Gly-Ala-Pro-Pro-Pro-Ser
38 39

Lixisenatide:

His-Gly-Glu-Gly-Thr-Phe-Thr-Ser-Asp-Leu-Ser-Lys-Glu-Met-Glu-Glu-Glu-Ala-Val-Arg-Leu-Phe-
1

Ile-Glu-Trp-Leu-Lys-Asn-Gly-Gly-Pro-Ser-Ser-Gly-Ala-Pro-Pro-Ser-Lys-Lys-Lys-Lys-Lys-Lys-NH_2
38 44

[-Pro^{38}]

Albiglutide:

(His-Gly-Glu-Gly-Thr-Phe-Thr-Ser-Asp-Val-Ser-Ser-Tyr-Leu-Glu-Gly-Gln-Ala-Ala-Lys-Glu-Phe-
7 8

Ile-Ala-Trp-Leu-Val-Lys-Gly-Arg)$_2$-Human serum albumin

Fig. 7.8 Amino acid sequences of GLP-1 and GLP-1 receptor agonists exenatide, liraglutide, and lixisenatide. Numbering of the GLP-1 sequence is based on the biologically active form (GLP-1(7–37)) of the hormone and is relevant to the parent proglucagon molecule. Amino acid residues/group changes to the GLP-1 sequence are shown in *red*. Changes in the lixisenatide structure compared to the exenatide structure are highlighted in *blue*. Liraglutide differs from GLP-1 at residue 26 (20) by the addition of an *N*-hexadecanoyl-*L*-glutamyl 16-C side chain on the Lys and at residue 34 (28) where an Arg is substituted for the Lys of GLP-1. Lixisenatide differs from exenatide by the omission of Pro at position 38 and the addition of six Lys residues to complete the sequence at residues 39–44. Albiglutide is a recombinant fusion protein composed of two 30 amino acid GLP-1(7–36) monomers with a Gly substituted for Ala at position 8, coupled in tandem and fused to human serum albumin. The substituted Gly confers resistance to DPP-4 enzymic attack

a valuable treatment for patients with type 2 diabetes mellitus. However, therapeutic applications of GLP-1 are not practical since the peptide is rapidly degraded (its half-life in plasma is only ~1–2 min) by the enzyme dipeptidyl peptidase-4 (DPP-4; also known as adenosine deaminase complexing protein 2 or CD26) (Fig. 7.10) which cleaves the two N-terminal amino acids leaving inactive GLP-1(9–36 amide)

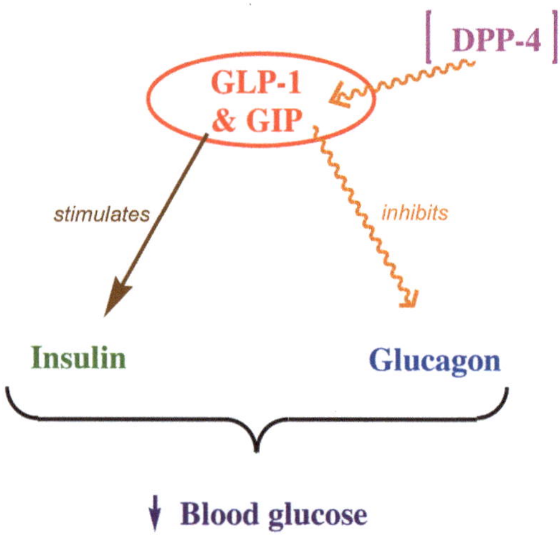

Fig. 7.9 Relationships between the actions of the incretin hormones GLP-1 and GIP and the dipeptidase DPP-4. GLP-1 stimulates insulin secretion only when glucose levels are high while suppressing glucagon secretion. DPP-4 cleaves the two *N*-terminal amino acids from GLP-1(7–37) inactivating the hormone. Thus, inhibitors of this enzyme maintain levels of GLP-1 that stimulate insulin secretion and inhibits glucagon leading to a decrease in blood glucose

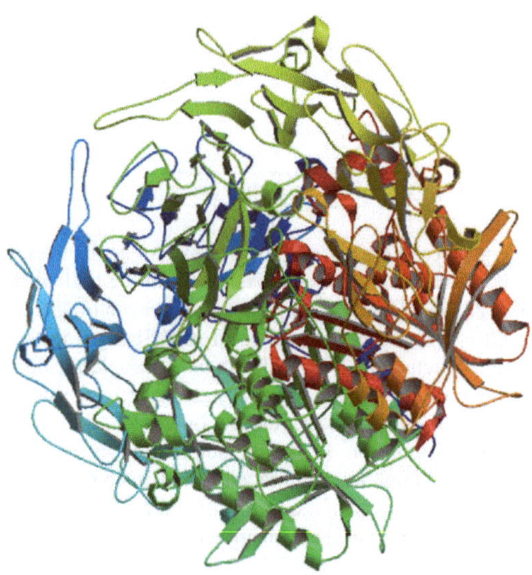

Fig. 7.10 Crystal structure of human dipeptidyl peptidase IV (DPP-4), the main GLP-1-degrading enzyme. The X-ray structure of the expressed and purified ectodomain of the enzyme was determined at 2.1Å resolution. DPP-4 is made up of two domains, a catalytic domain with an alfa/beta hydrolase fold and a beta domain with an eightfold repeat of a beta sheet motif. The structure is from Protein Data Bank RCSB PDB file 1NU6 (Thoma R, Loeffler B, Stihle M, et al. Structure 2003;11;947–59)

or GLP-1(9–37). DPP-4 selectively binds substrates containing proline at substrate cleavage position P1 and cleaves peptides (such as GLP-1 and GIP) which have proline or alanine at the second position. Removal of the subterminal alanine thus destroys the biological activity of GLP-1 and GIP.

GLP-1 receptors have been found in the heart and some findings already suggest that the hormone has some beneficial cardiac effects, for example, it increases myocardial performance after cardiac injury and improves left ventricular and endothelial functions in patients with high cardiac risk. It is type 2 diabetes, however, that is considered to be the main potential therapeutic application of GLP-1. In contrast to GIP, the insulinotropic effect of GLP-1 is retained in patients with type 2 diabetes but its potency in promoting glucose-induced insulin secretion is reduced by ~80 %. In other words, defects in the secretion/actions of both incretin hormones account for inadequate insulin secretion seen in the diabetic patients. The molecular events underlying the reduced effect of GLP-1 and the all but absent action of GIP in type 2 diabetes are yet to be worked out. Clearly, an understanding of the molecular mechanisms and clinical science of the actions of GLP-1 is a requirement if GLP-1 is to be fully exploited as an important component of type 2 diabetes therapy. Already, some GLP-1 receptor activators (agonists) or incretin mimetics are showing some clinical promise.

GLP-1 Receptor Agonists

The insulin secretagogue exendin-4 is a 39 amino acid agonist of the GLP-1 receptor (GLP-1R) isolated from the saliva of the Gila monster (*Heloderma suspectum*), a venomous lizard found in North America. Synthetic exendin-4, known as **exenatide** and marketed as an injection solution for subcutaneous administration under the trade name Byetta®, shows 53 % sequence identity with GLP-1 and was approved for the treatment of diabetes in 2005. The amino acid sequence of exenatide with residues differing from the corresponding positions in GLP-1(7–37) is shown in red in Fig. 7.8. Not being degraded by DPP-4 due to a glycine instead of an alanine at position 2, exenatide is reasonably stable with a half-life of ~30 min although peak plasma levels are reached after ~2 h and a significant concentration of drug remains in the plasma for up to 5 h after subcutaneous injection of the maximum dose. Exenatide therefore requires twice daily dosing, usually before breakfast and dinner. The drug lowers both fasting and post-prandial glucose levels. The promising clinical results with exenatide prompted a search for other GLP-1R agonists with sufficient stability for once-daily or even once-weekly administration. Formulation of **exenatide** as an extended release formulation (Bydureon®) has produced such a preparation suitable for weekly injections. Prolonged release is achieved by incorporating exenatide in an extended release microsphere formulation containing a lactide-glycolide copolymer. After 6–7 weeks of Bydureon® administration, mean exenatide concentrations of ~300 pg/mL were maintained over the weekly dosage period. The mean apparent clearance of exenatide is 9.1 L/h.

Thus far, endeavors seeking efficacious GLP-1R agonists with extended half-lives have resulted in FDA and/or EMA approvals for 3 more promising GLP-1 agonists: liraglutide, lixisenatide, and albiglutide.

Prepared in *Saccharomyces cerevisiae*, **Liraglutide** (Victoza®) is slightly modified from GLP-1 (i.e., GLP-1(7–37)) by substituting an arginine for lysine at position 34 and attaching palmitic acid (a C-16 fatty acid) via a glutamic acid spacer to the lysine at position 26 (Fig. 7.8). These modifications produce a peptide that self-associates thus delaying absorption. The molecule is stable to DPP-4, and binds reversibly to plasma proteins, particularly albumin, all contributing to a significantly longer plasma half-life of 13 h following subcutaneous injection. Liraglutide reaches peak plasma levels 9–12 h after injection, produces low rates of hypoglycemia and weight loss in most patients, and demonstrates therapeutic plasma concentrations for up to 24 h, making the peptide suitable for once-daily administration.

Lixisenatide (Lyxumia®), approved by the EMA in 2013, is modified from the exenatide structure by omitting proline at position 38 in exenatide and adding six lysine residues after the terminal serine to produce a peptide of 44 amino acids (Fig. 7.8). The terminal lysine at the C-terminal (position 44) is amidated. Following subcutaneous administration to patients with type 2 diabetes, lixisenatide is rapidly absorbed with the time taken to reach the maximum concentration (Tmax) in diabetic patients being ~3.5 h. Mean terminal half-life is ~3 h and mean apparent clearance rate ~35 L/h. The drug shows a moderate level of binding (55 %) to human serum proteins.

Albiglutide (Tanzeum®; Eperzan®), prepared in *Saccharomyces cerevisiae*, is a recombinant fusion protein (see Chap. 6, section "Albiglutide") made up of two copies of a modified 30 amino acid GLP-1 sequence fused in tandem to the *N*-terminus of human albumin to produce a large protein of ~73 kDa (Fig. 7.8). The two GLP-1(7–36) monomers have a glycine substitution for the naturally occurring alanine at position 8 to confer resistance to enzymic attack by DPP-4. This enzyme resistance together with the fused albumin component extends the protein's elimination half-life to ~5 days and together with a mean apparent clearance of 67 mL/h, albiglutide is suitable for once a week administration.

Dulaglutide (Trulicity®). Although GLP-1 receptor agonists provide glucose-dependent insulinotropic effects reducing glucagonemia, slowing gastric emptying, decreasing appetite, and promoting weight loss, their short half-lives due to inactivation by DPP-4 and excretion rates limits their therapeutic value. Like albiglutide, dulaglutide is a new long-acting, FDA-approved, human GLP-1 receptor agonist in fusion protein form (Chap. 6, section "Dulaglutide"). The dulaglutide molecule consists of a modified human GLP-1 analog sequence covalently attached to each of the two chains of a modified human IgG4 Fc fragment via a flexible 16 amino acid peptide with three Gly-Gly-Gly-Gly-Ser repeat sequences and an Ala residue (Fig. 7.11). The GLP-1 analog is approximately 90 % homologous to native human GLP-1 with amino acid substitutions at positions 8, 22, and 36 designed to protect the peptide from DPP-4 inactivation and reduce its immunogenicity. To reduce the potential for antibody-dependent cell-mediated cytotoxicity (ADCC) and complement-dependent cytotoxicity (CDC), IgG1 Fc was replaced with the Fc of

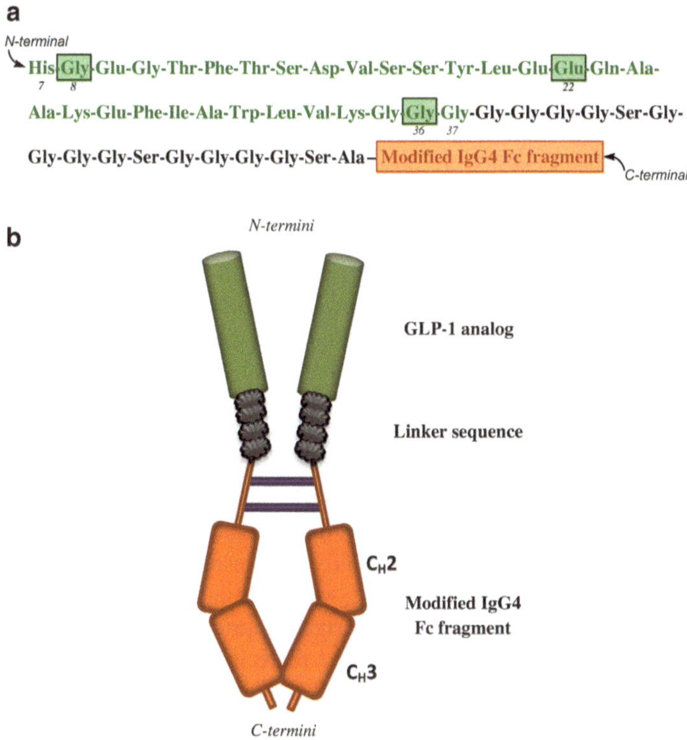

Fig. 7.11 Diagrammatic representation of the structure of the dimeric chimeric human IgG4 Fc fusion protein dulaglutide. The effector peptide partner, a modified human GLP-1 analog (amino acid sequence shown in *green*), is approximately 90 % homologous to native human GLP-1 with amino acid substitutions at positions 8, 22, and 36 (marked in *green boxes*). The GLP-1 analog is attached to a modified human IgG4 Fc fragment by a flexible 16 amino acid peptide (shown in *black*)

the IgG4 isotype (Chap. 2, section "IgG Antibody Subclasses") modified to prevent half-antibody formation and to reduce interaction with high affinity Fc receptors. The plasma half-life of dulaglutide of approximately 4.7 days makes it suitable for once a week administration.

Safety of GLP-1 Receptor Agonists

Both **exenatide** approved preparations, the solution for subcutaneous injection (Byetta®) and the extended-release suspension (Bydureon®), carry a number of warnings and precautions, some serious, issued by the FDA. Therapeutic use of both preparations is subject to warnings concerning pancreatitis, hypoglycemia, hypersensitivity, and a precautionary reminder that, as yet, there have been no clinical studies proving that the drugs reduce microvascular risk. There have been post-marketing reports of fatal and nonfatal hemorrhage or necrotizing pancreatitis leading to the warning that exenatide should not be used if pancreatitis is confirmed. Hypoglycemia is an increased risk when exenatide is used with a sulfonylurea; the drug should be avoided for patients with gastrointestinal disease, for example, gastroparesis; severe hypersensitivity reactions including anaphylaxis and angioedema have been reported; and the drug should not be used in patients with renal impairment or end-stage renal disease and only with caution in renal transplantation patients or those with moderate renal impairment. In addition, Bydureon®, but not Byetta®, is subject to an FDA black box warning drawing attention to the fact that the extended-release preparation causes thyroid C-cell tumors in rats although it is presently unknown whether this is true for C-cell tumors, and medullary thyroid carcinoma (MTC), in humans. The box warning also states that Bydureon® is contraindicated in patients with MTC, or a family history of MTC, or in patients with multiple endocrine neoplasia syndrome type 2 (MEN2). Adverse reactions to the exenatide preparations are generally not serious with the list comprising nausea, vomiting, hypoglycemia, diarrhea, jitters, dizziness, headache, asthenia, dyspepsia, and constipation. Nausea can be particularly troublesome (although it may decrease over time) and it, together with vomiting, were the main causes for withdrawal from treatment. Injection site reactions, in particular, pruritus and nodules, may occur with the extended-release preparation and there is at least one report of an IgE antibody-mediated immediate allergic reaction to the drug. This was recently reported in a 34-year-old woman with type 2 diabetes who took exenatide for 6 months before stopping when she became pregnant. Four months after delivery, she developed itching, generalized urticaria, and difficulty swallowing within 10 min of the first dose. Later, after an identical response following a rechallenge, the patient was skin prick tested with exenatide 0.25 mg/mL. Clear positive reactions were seen to the undiluted and 1–10 solutions but not to a 1–100 dilution. What appeared to be local allergic reactions to exenatide were seen in 5 of 58 patients after receiving the drug in a previous clinical trial. Two patients developed localized itching, one a localized rash, one a generalized rash, and one an eye allergy.

Liraglutide carries the same box warning of the risk of thyroid C-cell tumors and mention of MTC and MEN2 as exenatide as well as the warnings and precautions for pancreatitis and hypoglycemia. Its list of the most commonly seen induced adverse events is also similar to the exenatide safety profile with the addition of upper respiratory tract infection, influenza, sinusitis, and nasopharyngitis. Special warnings and precautions issued by the EMA for **Lixisenatide** (not yet approved by the FDA) cover acute pancreatitis, severe gastrointestinal disease, renal impairment, hypoglycemia, and dehydration. The most frequently reported adverse reactions during clinical studies were mostly mild and transient with nausea, vomiting, and diarrhea again featuring most prominently. Other commonly seen adverse events were upper respiratory tract infections including influenza and other viral infections, dizziness, somnolence, dyspepsia, back pain, and injection site pruritus. Anaphylactic reactions, including urticaria, are known but uncommon. Allergic reactions occurred in 0.4 % and anaphylaxis in 0.2 % of lixisenatide-treated patients.

The FDA black box warning issued for other GLP-1 receptor agonists also applies to **albiglutide** (Chap. 6, section "Albiglutide") as well as the warnings and precautions for pancreatitis, hypoglycemia, renal impairment, and hypersensitivity. The most common adverse events provoked by albiglutide are essentially the same as those seen with lixisenatide.

The Fc fusion protein **dulaglutide** (Chap. 6, section "Dulaglutide," Table 6.2) is also subject to an FDA black box warning drawing attention to the fact that the extended-release preparation causes thyroid C-cell tumors in rats and the warning which states that the drug is contraindicated in patients with MTC, or a family history of MTC, or in patients with MEN2. Other warnings relate to pancreatitis; a possible increased risk of hypoglycemia with concomitant use of insulin secretagogues (e.g., sulfonylurea); hypersensitivity reactions; renal impairment; and the possibilities of severe gastrointestinal disease and macrovascular outcomes. The most common adverse events reported to dulaglutide fusion protein in ≥5 % of treated patients are nausea, diarrhea, vomiting, abdominal pain, and decreased appetite. Sinus tachycardia was reported more frequently in patients treated with dulaglutide 1.5 mg (5.6 %), but not 0.75 mg, than placebo (3 %).

Immunogenicity of GLP-1 Receptor Agonists

In a 24 week monotherapy trial with **exenatide** subcutaneous solution (Byetta®), 28 % of patients developed low titer antibodies to the peptide. Of 210 patients tested, no cross-reactivity with GLP-1 and/or glucagon was detected. Measurement of antibodies to the extended-release injection (Bydureon®) in the sera of 918 patients at 4–14 week intervals revealed that 49 % had low titer antibodies to exenatide at any time during the trial and 45 % had antibodies at the end of the 30-month trial. The level of glycemic control in this low titer group was generally comparable to that observed in the patients treated with Bydureon® but without antibodies (43 %). Of the 12 % of patients who had high antibody titers, half also had an attenuated glycemic response at the conclusion of the trial. Over the 30-month trial, the mean antibody titer peaked at week 6 and declined by 56 % by week 30. Over a 26-week immunogenicity study of **liraglutide**, ~50–70 % of patients developed antibodies to the drug. At 52 weeks, 6.9 % of patients had antibodies that cross-reacted with GLP-1 and although antibodies were not tested for GLP-1-neutralizing effect, an in vitro assay detected neutralizing antibodies to liraglutide in 2.3 % of the subjects. The presence of antibodies was not associated with reduced efficacy of liraglutide. Examinations over 24 and 76 weeks found 69.8 % of patients with antibodies to **lixisenatide** but 6 months after the 76th week, the number of patients with antibodies had fallen by 30 %. No antibody cross-reactivity with GLP-1 or glucagon was detected and, apart from an increase in injection site reactions (4.7 % vs. 2.5 % in antibody-positive and antibody-negative patients, respectively), the presence of antibodies did not alter patient safety profiles. Trials with **albiglutide** over at least 26 weeks showed 5.5 % (116) of 2098 patients had antibodies to the peptide and

79 % of these (92 patients) were also antibody-positive for GLP-1. An in vitro bioassay failed to find albiglutide-neutralizing antibodies in any of the patients. Results from clinical trials showed a total of 64 patients (1.6 %) developed antibodies to **dulaglutide** when given the fusion protein. Thirty-four patients (0.9 % overall) had antibodies that neutralized dulaglutide, while 36 patients (0.9 % overall) developed antibodies to native GLP-1.

Dipeptidyl Peptidase-4 (DPP-4) Inhibitors (Gliptins)

Inhibitors of DPP-4, or gliptins, form a class of orally active hypoglycemics with relatively modest glucose-lowering activity that can be used to treat diabetes mellitus type 2 (usually in combination with other drugs). Since DPP-4 shows substrate selectivity for proline at the P1-position, chemists, in efforts to inactivate the enzyme, have prepared potential inhibitors containing structures mimicking proline, in particular, compounds with a pyrrolidine or pyrrolidine-like group. As outlined in section "Glucagon-Like Peptide 1 and the Incretin Effect" above, the dipeptidase DPP-4 cleaves the two *N*-terminal amino acids from GLP-1(7–37) inactivating the hormone. Thus, inhibitors of this enzyme maintain levels of GLP-1 that stimulate insulin secretion and inhibit glucagon leading to a decrease in blood glucose (Fig. 7.9). Results of recent studies suggest additional tissue-specific mechanisms may be involved in the glucose-lowering effect of gliptins, particularly in the gut and islet cells. Improvements in islet cell functioning resulting from the prevention of inactivation of other bioactive peptides such as stromal-derived factor 1-α and pituitary adenylate cyclase-activating polypeptide are another suggested mechanism. Although gliptins are not strictly "biologic" in the terms defined and adhered to in this volume (Chap. 1), their complementary use with other biologic-based treatments for diabetes justifies their examination here. Accordingly, the safety of some of the most frequently/widely used of these approved agents will be briefly examined.

Safety of Approved DPP-4 Inhibitors

Indicated by the FDA "as an adjunct to diet and exercise to improve glycemic control in adults with type 2 diabetes mellitus," **sitagliptin** (Januvia®) (Fig. 7.12), like all of the approved DPP-4 inhibitors, carries an FDA warning/precaution for acute pancreatitis even though a causal link is not yet proven. Other precautionary reminders concern acute renal failure, an increased risk of hypoglycemia, and serious allergic reactions such as anaphylaxis, angioedema, and cutaneous toxidermias including Stevens-Johnson syndrome. In trials, sitagliptin displayed few side effects, showing an overall incidence of adverse reactions similar to placebo. The most common reactions seen are upper respiratory tract infections, nasopharyngitis, and headache. Approved by the EMA in 2007 but not yet by the FDA, **Vildagliptin** (Galvus®,

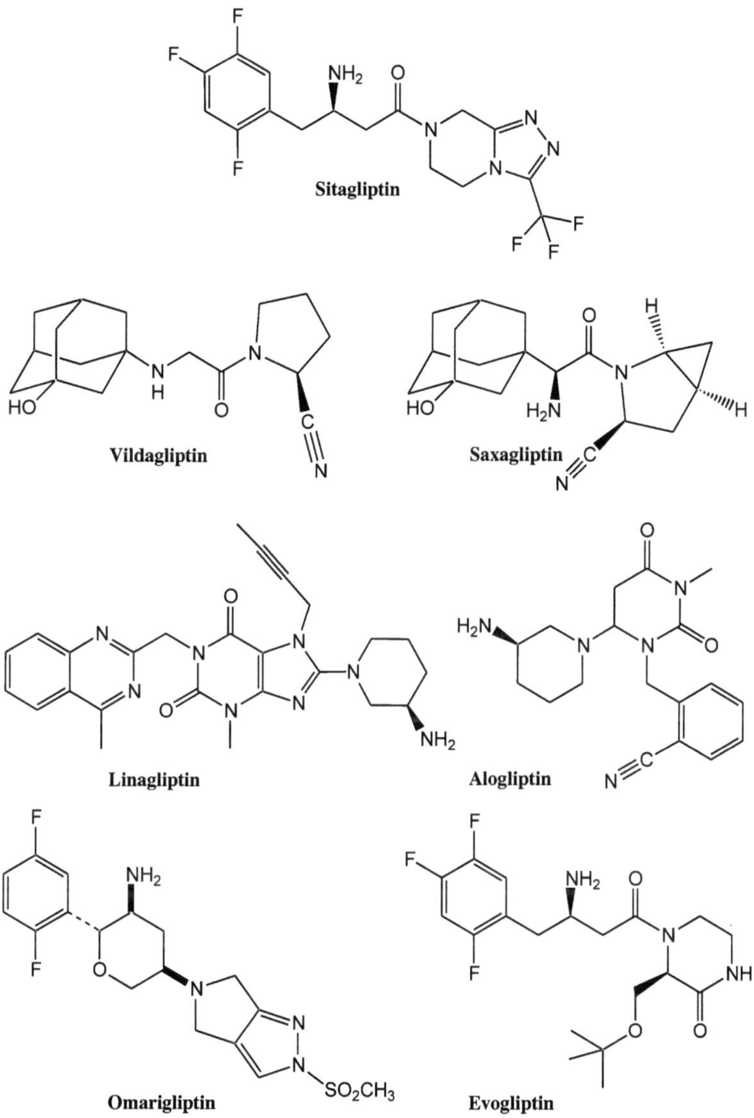

Fig. 7.12 Structures of dipeptidyl peptidase IV (DPP-4) inhibitors, also termed gliptins, used as incretin-based oral therapy for type 2 diabetes mellitus. Sitagliptin, the first-in-class gliptin, with a trifluorophenyl, a triazole and a pyrazine group, is structurally quite different to the other four FDA/EMA-approved compounds. Vildagliptin and saxagliptin are structurally similar, each containing an adamantyl and a carbonitrile group. Linagliptin has a dihydropurine dione and alogliptin a dihydropyrimidine dione substituent. Omarigliptin, an aminotetrahydropyran-based rigid analog of sitagliptin, approved in Japan, and evogliptin, a β-amino amide derivative with a piperazine-2-one moiety, approved in South Korea, like sitagliptin, each contain a fluorophenyl group

Zomelis®, Jalra®) (Fig. 7.12) inhibits the inactivation of GLP-1 and GIP. Special warnings and precautions issued by the EMA for the drug's use concern patients with renal impairment; hepatic impairment and the need to monitor liver enzymes at 3 month intervals during the first year and periodically thereafter; post-marketing reports of bullous and exfoliative skin lesions; and in common with other DPP-4 inhibitors, hypoglycemia and acute pancreatitis. Most adverse effects seen with vildagliptin are mild and transient, rarely, if at all, requiring discontinuation of treatment, and with no association with dose, treatment duration, and age. Adverse events commonly seen in patients in double-blind studies receiving vildagliptin in combination with metformin were hypoglycemia, nausea, tremor, headache, and dizziness. In a double-blind, monotherapy study involving 1855 patients, dizziness was the most common adverse response with hypoglycemia, headache, peripheral edema, arthralgia, and constipation occurring less often. Structurally related to vildagliptin, **saxagliptin** (Onglyza®) (Fig. 7.12) is, like the other DPP-4 inhibitors, a relatively safe drug. Again, attached FDA warnings concern acute pancreatitis and serious hypersensitivity reactions. In placebo-controlled monotherapy trials, as well as with glyburide and the add-on to metformin, thiazolidinedione, the overall incidence of adverse events in patients treated with saxagliptin was similar to placebo. The most common adverse events were lymphopenia, rash, and increases in blood creatinine and blood creatinine phosphokinase. In each case, however, incidences were very low (~0.1–0.5 %). Other adverse reactions in the treatment group with an incidence marginally above the placebo group were upper respiratory tract and urinary infections, headache, and nasopharyngitis. The number of warnings and precautions and adverse events listed for **linagliptin** (Tradjenta®, Trajenta®) (Fig. 7.12) is relatively few. As for saxagliptin, incidences of adverse reactions were found to be no more than, or marginally above, the placebo results. Low incidences of nasopharyngitis, cough, arthralgia, back pain, pancreatitis, and headache were found in monotherapy and drug combination clinical trials. A few cases of hypersensitivity and myalgia have been reported in other clinical studies. Linagliptin, sitagliptin, and vildagliptin have each been implicated in cases of bullous pemphigoid. **Alogliptin** (Nesina®, Vipidia®) (Fig. 7.12) is the latest gliptin to be registered for therapeutic use by the FDA, receiving regulatory approval in 2013. Acute pancreatitis, hypoglycemia, and hypersensitivity are listed as warnings and precautions along with the reminder that there have been some post-marketing reports of hepatic failure following administration of alogliptin where causality could not be excluded. Again, relative to placebo, adverse reactions are few with the most common being nasopharyngitis, upper respiratory tract infection, and headache.

First, and so far limited, regulatory approvals have recently been afforded to two more DPP-4 inhibitors. Structurally similar to sitagliptin, **omarigliptin** (Marizev®), unlike most gliptins which usually require once daily administration, can be given once a week. The drug received its first regulatory approval in September 2015 from the Japanese Pharmaceuticals and Medicinal Devices Agency for use in type 2 diabetes in adults after positive findings in phase III trials in Japan. Phase III clinical development is underway in several other countries. Omarigliptin, distinguished structurally by the presence of difluorophenyl, methylsulfonyl, dihydropyrrolo, and tetrahydropyran groups (Fig. 7.12), is a highly selective, potent inhibitor of DPP-4, being ~11-fold more potent than sitagliptin in in vitro DPP-4 inhibition studies in human plasma. DPP-4 inhibition of ~80 % 7 days after a final once-weekly dosage and a twofold increase in post-prandial 4 h GLP-1 levels compared to placebo have been reported. Safety and tolerability studies so far have shown a similar incidence of adverse events between the omarigliptin and placebo patient groups. Incidences of symptomatic hypoglycemia in treated patients were low and no severe hypoglycemia episodes or acute or chronic pancreatitis were reported in a recent multicenter, double-blind, 12-week study of 685 type 2 diabetes patients. In an extension of this study, infections, primarily due to nasopharyngitis, increased incidences of hyperglycemia and hypoglycemia, and a variety of respiratory, thoracic, and mediastinal adverse events occurred. A supratherapeutic dose did not produce any prolongation of corrected QT interval. There were no reports of serious hypersensitivity reactions or pancreatic cancer. **Evogliptin** (Suganon®), a piperazine derivative containing a trifluorophenyl group and also showing some structural similarities to sitagliptin (Fig. 7.12), is a highly selective and potent inhibitor of DPP-4. It received its first regulatory approval in South Korea in October 2015 for blood glucose control in type 2 diabetic patients. Safety and tolerability studies in healthy subjects revealed drug-associated adverse events of fatigue, insomnia, epigastric discomfort, erosion of oral mucosa, headache, and high urinary frequency. All the events were mild in severity and no serious adverse events were seen. In a phase II randomized, double-blind, placebo-controlled clinical trial in Korea with 157 type 2 diabetic patients, there were no statistically significant differences in the incidence rates of adverse events between the evogliptin treatment and placebo groups. In particular, no pancreatitis-related adverse events were reported. Published results of safety studies on evogliptin are so far few and sketchy and, as with omarigliptin, a longer post-marketing experience is needed before some safety issues are likely to emerge fully.

Cardiovascular and Pancreatic Safety of DPP-4 Inhibitors

It is well known that type 2 diabetes mellitus doubles the risk of major cardiovascular complications and that the majority of patients with diabetes die of cardiovascular disease. There is therefore a need to identify effective antihyperglycemic agents that can reduce cardiovascular complications and in 2008 the FDA and EMA revised their approval processes to require a demonstration of cardiac safety for new antihyperglycemic therapies. Findings in placebo-controlled trials and pooled analyses suggested that some DPP-4 inhibitors might show promise both for improving glycemic control and reducing the risk of major cardiovascular events. In relation to incretin-based drugs used in type 2 diabetes therapy, the DPP-4 inhibitor saxagliptin was recently selected for evaluation of efficacy and safety with respect to cardiovascular outcomes in diabetic patients at risk of cardiovascular events. The so-called Saxagliptin Assessment of Vascular Outcomes Recorded in Patients with Diabetes Mellitus (SAVOR)—Thrombolysis in Myocardial Infarction (TIMI) 53 study, a multicenter, randomized, double-blind, placebo-controlled, phase IV trial with 613 patients given saxagliptin and 609 given a placebo, reached the conclusion that saxagliptin did not increase or decrease the rate of ischemic events and that "although saxagliptin improves glycemic control, other approaches are necessary to reduce cardiovascular risk in patients with diabetes." Given this quite extensive study and its clear conclusion, as well as results from the 2013 Examination of Cardiovascular Outcomes with alogliptin versus Standard of Care (EXAMINE) trial and some other relevant but still far from conclusive findings, the FDA in its prescribing information continues to caution that there have so far been no clinical studies with any antidiabetic drug presenting conclusive evidence of a reduction of macrovascular risk.

With the increasing prevalence of diabetes worldwide, post-marketing reports of pancreatitis and pancreatic cancer in patients taking anti-diabetic medicines, for example, incretin-based drugs, raised concerns. During 2013, the FDA and EMA evaluated reports of pancreatitis and pancreatic cancer in patients on incretin-based drugs and toxicology studies on ~18,000 laboratory animals. Microscopic examinations carried out as part of the toxicology assessments showed no pancreatitis or obvious toxic effects and no drug-induced pancreatic tumors were seen. An analysis of data from more than 14,000 type 2 diabetes patients treated with sitagliptin revealed no evidence of an increased risk of pancreatitis or pancreatic cancer and reported rates of acute pancreatitis in the SAVOR and EXAMINE trials were low with similar incidences seen in the drug and placebo groups. In a short 2014 "Perspective" article in the *New England Journal of Medicine*, the FDA and EMA stated that: "Both agencies agree that assertions concerning a causal association between incretin-based drugs and pancreatitis or pancreatic cancer, as expressed recently in the scientific literature and in the media, are inconsistent with the current data. The FDA and EMA have not reached a final conclusion at this time regarding such a causal relationship." Therefore, until more data are available, both agencies are continuing to consider pancreatitis as a risk associated with incretin-based drugs.

Pramlintide

Pramlintide is a synthetic analog of human amylin, a neuroendocrine hormone synthesized by beta cells of the pancreas and cosecreted with insulin in the ratio of approximately 100:1 in response to food intake, thus contributing to glucose control in the postprandial period. Amylin, or islet amyloid polypeptide (IAPP), a 37 residue peptide hormone, is located with insulin in secretory granules. In healthy individuals the two hormones show similar patterns of release before and after food intake (Fig. 7.13) and in types 1 and 2 diabetes both are absent, or their secretion is reduced, in response to food. Amylin modulates glucose release by slowing gastric emptying, suppressing glucagon secretion (thus ultimately suppressing glucose output by the liver), and regulates appetite and therefore food intake. Pramlintide has a short half life of only about 48 min but like the naturally occurring hormone it slows gastric emptying, reduces postprandial rises in plasma glucagon, and decreases the intake of calories by reducing appetite.

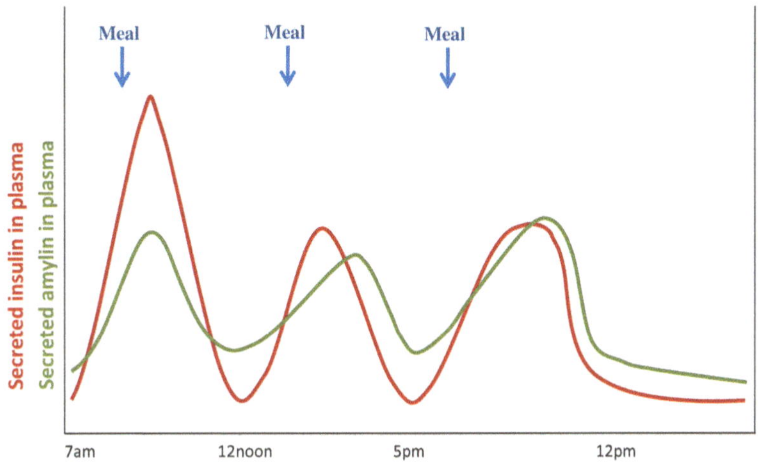

Fig. 7.13 Relative profiles of plasma levels of amylin and insulin before and after food intake. Both hormones show similar pre- and postprandial patterns of release in healthy individuals

S———S (disulfide bridge between Cys-2 and Cys-7)

Lys-Cys-Asn-Thr-Ala-Thr-Cys-Ala-Thr-Gln-Arg-Leu-Ala-Asn-Phe-Leu-Val-His-Ser-Ser-Asn-Asn-Phe-Gly-Pro-Ile-Leu-Pro-Pro-Thr-Asn-Val-Gly-Ser-Asn-Thr-Tyr-NH_2 *acetate*

25 28 29

SYMLIN (pramlintide acetate)

Fig. 7.14 Amino acid sequence of pramlintide, a synthetic analog of the 37 amino acid human hormone amylin. Amino acids alanine at position 25 and serine at positions 28 and 29 in the human sequence are each replaced by a proline residue (shown in *green*). Pramlintide, as the acetate salt, is marketed in injection form under the trade name Symlin®

Human amylin was identified in 1987 as the major component of islet amyloid deposits associated with diabetes. It forms amyloid fibrils in vitro and these fibers are cytotoxic to cell cultures. Amyloid formation can induce apoptosis of islet beta cells, an effect that may be involved in the development of type 2 diabetes. These properties of amylin were considered relevant in designing pramlintide. To avoid amylin formation, studies showed that some amino acids in rat amylin could be substituted without losing clinical activity. Amino acids alanine at position 25 and serine at positions 28 and 29 in the human sequence were therefore each replaced by a proline residue (Fig. 7.14). Pramlintide, as the acetate salt, is marketed as an anti-diabetic medication in injection form under the trade name Symlin®. It should be used with prandial insulin with the insulin doses reduced by half to prevent hypoglycemia. Pramlintide alone does not cause hypoglycemia but its addition to insulin increases the risk of hypoglycemia and this is reflected in an FDA black box warning stating that pramlintide use with insulin has been associated with an increased risk of severe hypoglycemia, particularly in patients with type 1 diabetes. The warning further states that the risk of hypoglycemia may be reduced by appropriate patient selection, careful patient instruction, and insulin dose reduction. Pramlintide should not be used in patients who may remain unaware of their hypoglycemic state or in patients with recurrent severe hypoglycemia, gastroparesis or hypersensitivity to pramlintide.

In addition to the warnings of the risk of hypoglycemia, other warnings and precautions for pramlintide cover the need for careful patient selection, the administration of certain other medications, and the possibility of allergic reactions. Proper patient selection is considered vital for safe and effective therapy with pramlintide. Patients who should not be considered for therapy include those showing poor compliance with insulin usage and blood glucose monitoring; patients with confirmed gastroparesis or who require drugs that stimulate gastrointestinal motility; patients with glycated hemoglobin (HbA1c) >9 %; and pediatric patients. Because pramlintide slows gastric emptying, other oral medications, especially those requiring rapid onset of action, should be administered at least 1 h prior to injection of pramlintide. Likewise pramlintide is not recommended for patients taking medications that alter gastrointestinal motility. Pramlintide and insulin should be given as separate injections not mixed before administration to avoid altering both the pharmacokinetics of each agent and subsequent glucose control. Safety data from placebo-controlled trials with patients with type 1 diabetes showed the most common adverse reactions (incidence ≥5 % compared to placebo) to be nausea, anorexia, vomiting, arthralgia, fatigue, and an allergic response. The corresponding list for type 2 diabetic patients included nausea, headache, anorexia, vomiting, abdominal pain, fatigue, dizziness, cough, and pharyngitis. The adverse events most often reported so far during the post-marketing period are injection site reactions and pancreatitis.

Human Growth Hormone

Human growth hormone, also called human pituitary growth hormone, somatropin (usually applied to recombinant forms) and somatotropin, is a peptide hormone synthesized by the somatotropic cells of the anterior pituitary gland. It has a wide range of physiological activities including protein synthesis, cell proliferation, and involvement in protein, carbohydrate, and lipid metabolism and immune regulation. These properties underlie the hormone's effects in promoting growth in childhood by stimulating metabolism of muscle, bone, and cartilage. Growth hormone may also be regarded as a stress hormone, stimulating production of insulin-like growth factor 1 (IGF-1) and raising the concentrations of glucose and free fatty acids. The effectiveness of human growth hormone treatment on growth was first demonstrated in 1958 after its isolation from pituitary glands of cadavers in 1956. Appreciation of the benefits of the hormone for growth hormone-deficient children soon followed and over the next two decades from the early 1960s, approximately 30,000–35,000 children worldwide were treated with the human pituitary-derived preparation. In 1985, supply of growth hormone was suspended and its therapeutic use abruptly ceased following reports of four cases of Creutzfeldt–Jacob disease (CJD) in young adults who had been treated with the hormone. Iatrogenic CJD continues to be seen occasionally due to the long incubation period of this prion infection (see section "Adverse Events to Somatropin" for recent developments).

Structure and Mechanism of Action of Growth Hormone

The growth hormone protein is encoded by the *GHN gene* (also called *GH1*) which, together with four related genes, forms a gene cluster at the growth hormone locus on chromosome 17. The five genes share a high degree of sequence identity with diversity arising from alternative splicing that generates additional isoforms. Gene mutations or deletions give rise to growth hormone deficiency and short stature. Determination of the structure of human growth hormone in 1972 and progress in gene cloning in 1979 were necessary steps in the evolution of a guaranteed supply of hormone and safe therapy was finally achieved in 1981 with the introduction of the first recombinant human growth hormone preparation. Because the natural hormone is a non-glycosylated protein, prokaryotic expression systems have been preferred for its production. The first recombinant preparations (rhGH; somatropin), prepared in *E. coli*, proved to have identical biological effects to the endogenous hormone and were the same chemically except for an additional methionine residue at the *N*-terminus. Some more recent recombinant somatotropin products have been prepared in yeast *Saccharomyces cerevisiae*. Human growth hormone is produced as a preprotein of 217 amino acids with a 26 amino acid *N*-terminal signal peptide essential for secretion. This is cleaved during secretion to give the mature 191 amino acid hormone. Human growth hormone is a single polypeptide chain with two disulfide bonds between Cys53-Cys165 and Cys182-Cys189 and molecular mass of 22.125 kDa. Secondary structure is a single chain peptide with a 45 % helical structure of 8 alfa-helices and X-ray studies show a four-helix core (Fig. 7.15). The receptor binding sites are residues 54 (Phe) and 74 (Gln).

Trade names of registered recombinant (somatropin) products currently, or formerly, used include: Accretropin®, BioTropin®, Genotropin®, Humatrope®, Norditropin®, Nutropin®, Omnitrope®, Protropin®, Saizen®, Serostim®, Tev-Tropin®, and Zorbtive®.

The actions of growth hormone are mediated via the growth hormone receptor, directly by tyrosine kinase activation and indirectly by IGF-1. Based on crystal structures of growth hormone bound to the extracellular domain of its receptor, that is the growth hormone—growth hormone binding protein complex (GH-GHbp), it was deduced that a single molecule of the hormone, via two of its distinct binding sites on opposite faces of the molecule, bound a single binding site on each of two receptor molecules (Fig. 7.16). In the original model, induced dimerization of the receptor activates receptor-associated JAK2 tyrosine kinase and tyrosyl phosphorylation of both JAK2 and growth hormone receptor. This was said to result in activation of a number of signaling events involving MAP kinases, protein kinase C, phosphatidylinositol 3′-phosphate kinase, diacylglycerol, intracellular Ca^{2+} and STAT

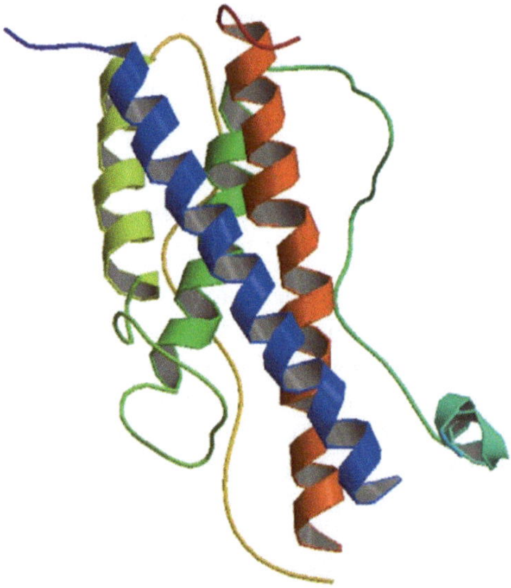

Fig. 7.15 Ribbon diagram of human growth hormone at 2.5Å resolution showing a crystal structure of a four-helix protein core with helices 2 and 3 connected by short loops and long cross-over connections linking helices 1 and 2 and helices 3 and 4. The structure shown is from Protein Data Bank RCSB PDB file 1HGU (Chantalat L, Jones ND, Korber F, et al. Protein Pept Lett 1995;2:333–40)

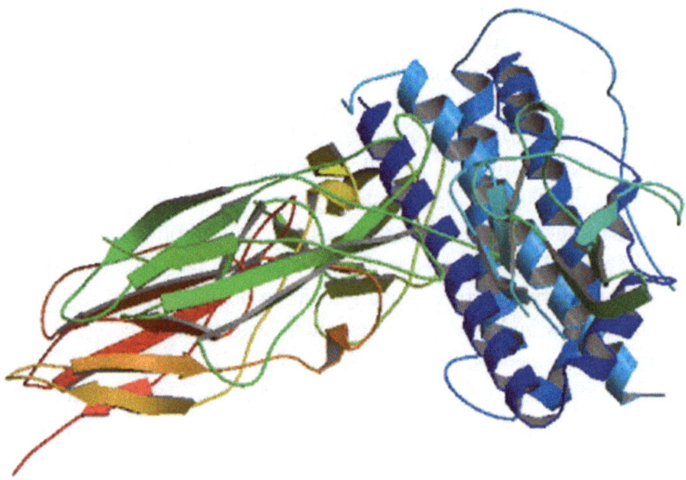

Fig. 7.16 Crystal structure of human growth hormone with its soluble binding receptor protein (GH-GHbp) in a 1:2 complex at 2.5Å resolution. The structure is from Protein Data Bank RCSB PDB file 1HWG (Sundstrom M, Lundqvist T, Rodin J, et al. J Biol Chem 1996;271:32197–203)

transcription factors that ultimately produce the activities related to metabolism and growth. More recent evidence indicates that before binding, the growth hormone receptor is *already* associated with inactive kinases, including JAK2 and Src kinases and, upon binding of the hormone, the receptor subunits are realigned by rotation and reorientation leading to activation of the associated kinases and ultimately phosphorylated cascades and transcriptional activity. Excessive activation of the growth hormone receptor by its hormone ligand can result in acromegaly and gigantism while autocrine activation of the receptor can lead to cell transformation and cancer.

Indications and Usage of Somatropin

With the availability of a ready and seemingly abundant commercial supply of recombinant product, usage of somatropin increased and with it the list of approved indications also grew, not only for growth hormone-deficient children but also for children without growth hormone deficiency and for adults. Approved indications for pediatric patients are treatment of short stature or growth failure associated with growth hormone deficiency; idiopathic short stature; SHOX deficiency (see below) and failure to catch up height after being small for gestational age birth; Turner syndrome; and Prader–Willi syndrome. An approved indication for adults is treatment of patients with either childhood- or adult-onset growth hormone deficiency. Deficiency of growth hormone may be congenital or acquired but the origin of the condition in most children is idiopathic.

Children suffering from chronic kidney disease may experience severe growth failure and in such cases growth hormone treatment has improved height, especially in young children who begin treatment earlier. Administration of growth hormone for chronic kidney disease with growth disorder was approved by the FDA in 1993.

Turner syndrome, a disorder associated with partial or complete absence of an X-chromosome, affects about one in 2000 female births. Affected females show short stature, typical somatic features, ovarian failure, and may experience other manifestations such as bone, cardiovascular, thyroid, and gastrointestinal involvement. Although growth hormone deficiency is not a factor in Turner syndrome, somatropin is administered to increase growth and height.

Growth hormone given as catch-up therapy for so-called "small for gestational age" children (defined as birth weight or length at least 2 SDs below the mean gestational age), and to children with idiopathic short stature, has shown benefit with patients reaching normal adult height. In 2006, the FDA approved the use of somatropin for patients with short stature homeobox (SHOX)-containing gene deficiency. The SHOX gene is located in the pseudoautosomal region of both the X and Y chromosomes. SHOX haploinsufficiency, that is presence of only one copy of the gene, or a mutated gene, is believed to be a cause of short stature in Turner syndrome, Léri–Weill dyschondrosteosis, and Langer mesomelic dysplasia. Growth hormone treatment improves the growth of patients with these various forms of SHOX insufficiency.

A recent approved indication of growth hormone is Noonan syndrome, an autosomal dominant heterogeneous congenital disorder affecting both males and females with an incidence of 1 in 1000–2500. Some of the main features of the syndrome include congenital heart defects, short stature, impaired blood clotting, pectus excavatum, characteristic dysmorphic facial and neck features, and learning difficulties. About 50 % of patients have a mutation in the *PTPN11* gene which encodes tyrosine phosphatase SHP-2, an enzyme involved in signal transduction pathways. Mutations in other genes including *SOS1*, *KRAS*, *RAF1*, *BRAF*, *NRAS*, and *HRAS* have also been associated with Noonan syndrome. Studies have shown that short-term growth hormone therapy (up to 4 years) can increase growth and height in patients.

Prader–Willi syndrome, due to the deletion, or lack of expression of genes of the paternal chromosome 15q11–q13, is a disorder of parent of origin imprinting in which the maternal copy of genes is silenced while the paternal copy is non-functional. The syndrome is characterized by hypotonia, hypogonadism, short stature, scoliosis, sleep disorders, delayed speech, and behavioral abnormalities. Chronic hunger often leads to obesity and diabetes mellitus, hypertension and cardiovascular disease may occur with age. Treatment of children with Prader–Willi syndrome with growth hormone promotes linear growth and a leaner body mass but an improved cardiovascular risk and increase in bone density are also thought to result.

The ready availability of human growth hormone in recombinant form has gone hand-in-hand with an expansion of the hormone's list of approved indications and this expansion appears to be still in progress with approvals for wasting associated with HIV infection and short bowel syndrome. In addition, there are reports of the application of somatropin therapy to osteoporosis, inflammatory bowel disease, juvenile rheumatoid arthritis, cystic fibrosis, and some patients requiring glucocorticoids and it seems likely that some functions of growth hormone may still be unknown. On the other hand, some present and proposed applications should be viewed with caution. Although it is not an approved indication and it is banned by the International Olympic Committee and many other official sporting bodies, growth hormone continues to be used as an anabolic agent in sport. Also, somatropin is sometimes prescribed for growth hormone-deficient older patients to increase vitality even though the efficacy and safety of this therapy are not yet established.

Adverse Events to Somatropin

Cases of true type I immediate hypersensitivity to human growth hormone are extremely rare with only two episodes of generalized urticaria to somatropin (Humatrope® and Protropin®) reported in the literature. Both patients were successfully desensitized with Humatrope®. Common adverse reactions to somatropin preparations listed by the FDA are injection site reactions and lipoatrophy (that is, localized loss of fat tissue), fluid retention, peripheral edema, arthralgia, myalgia, carpal tunnel syndrome/paresthesia (paraesthesias), headache, and hyperglycemia. These should not be considered without attention to accompanying issued warnings and precautions. For example, growth hormone stimulates the breakdown of glucose and lipids thus antagonizing the effect of insulin in glucose and lipid metabolism. It is therefore advisable to periodically monitor glucose levels and doses of antihyperglycemic drugs in patients as diabetes mellitus and impaired glucose tolerance may be unmasked. Growth hormone deficiency increases insulin sensitivity and is seen in neonates as hypoglycemia at birth. Other warnings and precautions relate to the possibility or development of unmasking latent hypothyroidism; fluid retention; intracranial hypertension; progression of scoliosis; hypopituitarism; and pancreatitis. In Prader–Willi syndrome, signs of upper airway obstruction and sleep apnea should be evaluated before initiation of growth hormone therapy. Three other rare adverse effects of somatropin therapy are self-limiting gynecomastia; slipped capital femoral epiphysis (a displacement of the proximal femoral epiphysis on the femoral neck); and benign intracranial hypertension. The first two of these effects result from growth hormone replacement therapy while intracranial hypertension is due to the antidiuretic effect of somatropin.

Growth hormone and IGF-1 show mitogenic and antiapoptotic effects, so the possibility of tumorigenesis needs to be considered from the viewpoints of the recurrence or stimulation of a former or preexisting tumor, the induction of a second tumor, and the induction of a tumor in a previously tumor-free patient. In cases of leukemia and brain and neck tumors, radiotherapy, and particularly intracranial surgery, may affect the pituitary gland, inducing growth hormone release. Although there is concern that somatropin may induce tumorigenesis, the current belief is that the hormone therapy does not increase the risk of malignancy. Whether or not meningiomas are increased is not yet clear but, even so, the FDA recommends the monitoring of patients with preexisting tumors and warns of an increased risk of a second neoplasm in childhood cancer survivors treated with somatropin, particularly meningiomas. In relation to long-term cancer-induced mortality, an extended epidemiological study undertaken in France (the Santé Adulte Gh Enfant or SAGhE study) was designed to assess the long-term mortality of patients treated with recombinant growth hormone during childhood. The investigation revealed a 30 % increased risk of death in the hormone-treated group (93 deaths) versus expected deaths (70) in the general population. Bone tumors and cardiovascular diseases were the main contributors to the increase in mortality.

In a surprising and unforeseen development, benzyl alcohol, a component of the diluent used for some preparations of somatropin, was unexpectedly implicated in a safety issue affecting pre- and full-term newborn babies. In 1982, the Center for Disease Control announced that 16 neonatal deaths caused by benzyl alcohol had been reported to the FDA. Deaths occurred in babies weighing up to 2.5 kg after being exposed to the alcohol via central intravascular catheters that were flushed periodically each day with a saline solution containing benzyl alcohol 9 mg/mL as bacteriostatic. Onset of illness followed in days to weeks with a typical clinical picture of metabolic acidosis; respiratory distress leading to gasping respiration; central nervous dysfunction with convulsions and intracranial hemorrhage; and hypotension leading to cardiovascular collapse and death. The FDA warns that for somatropin, and other medications containing benzyl alcohol, the combined daily metabolic load of the agent from all sources should be assessed and adjusted if necessary.

Worldwide, more than 200 subjects developed CJD as a result of treatment, usually in childhood, with human growth hormone prepared from human cadaver pituitary glands. Recent autopsies of the brains of eight people who died of the disease many years after contaminated growth hormone treatment revealed some surprising and potentially troubling findings. In addition to the damaging signs of CJD disease, six of the brains showed signs of amyloid-β pathology associated with Alzheimer's disease. This raises the possibility that more, perhaps many more, of the approximately

30,000 individuals injected with human growth hormone in the period 1958–1985 might be at risk of Alzheimer's disease, especially since Alzheimer's is a very common disease and amyloid-β would probably have been a more frequent contaminant of growth hormone than the rare CJD prion. A further concern is the known difficulty of deactivating prions because of their resistance to normal sterilization conditions and their property of sticking to metal surfaces. If Alzheimer's disease is transmitted in a prion-like way, the possibilities of it being passed on during neurosurgery or even by blood transfusion will have to be considered. In fact, amyloid-β "seeds," besides being long-lived in the brain, may be even more resistant to degradation than prions.

Immunogenicity of Growth Hormone

Some early recombinant growth hormone preparations prepared in *E. coli* induced antibodies to host cell proteins and, in such cases, higher incidences of antibodies to the hormone were sometimes also observed. Modifications to the manufacturing process largely eliminated the problem. The incidence of antibodies to Genotropin® in children after 12 months of treatment was 4 of 373 or 1.1 % suggesting low immunogenicity for this somatropin preparation. A comparison of antibodies developed to Genotropin® and Omnitrope® revealed incidences of 2.27 % for Genotropin® and 0 % for Omnitrope®. This compared with an incidence of 1.75 % for Genotropin® taken from the literature. It was concluded that there is no evidence for a clinically relevant difference in immunogenicity between Omnitrope® and Genotropin®. A comparison of the immunogenicity in children of a sustained release somatropin preparation and Genotropin® revealed a marked difference in the incidences of antibody formation. Seven of 87 (8 %) patients developed anti-growth hormone antibodies to Genotropin®, significantly less than the incidence of 41 % (37 of 91) seen with the sustained release preparation. Antibodies were mainly IgG1 and IgG4, most antibodies appeared at about 3 months of treatment, and the mean titer was maximal after 6 months, slowly declining thereafter. Importantly, the antibodies did not affect patient growth rates in any way and antibodies in adults developed much less frequently than in children.

Pegvisomant: A Growth Hormone Receptor Antagonist

There are two receptor binding sites on the growth hormone molecule with the binding affinity at site 1 greater than the affinity at site 2. The finding that a single mutation at site 2 where Gly-120 is replaced by an Arg (G120R), prevented binding of the site to the receptor was the first observation that led to the development, and ultimately the clinical application, of the growth hormone receptor antagonist pegvisomant (Somavert®). Eight mutations at binding site 1 were then introduced to increase the affinity by four- to fivefold. The resultant mutant growth hormone B2036 binds to a receptor dimer but does not induce the conformational changes required for signaling. The mutation in the second site also introduces a lysine residue which can be utilized for pegylation. A comparison of the crystal structures of the GH-GHbp complex with the B2306—GHbp complex showed that they crystallized under similar conditions and were remarkably similar with only minimal differences in conformation of hormone or receptor apparent.

Like the natural human hormone which has a short half-life (3.4 h and ~20 min after s.c. and i.v. administration, respectively), the half-life of B2036 is also too short to be used successfully as a growth hormone receptor antagonist. This was overcome by pegylation of the molecule which, although it decreased the polypeptide's antagonistic activity, the rate of clearance of the pegylated preparation was greatly reduced. This reduced clearance rate (eliminated from serum with a mean half-life of ~6 days) compensates somewhat for the decreased antagonistic effect, producing the efficacious recombinant drug pegvisomant that can be used in the clinic. Pegvisomant is composed of 191 amino acid residues, MW ~22 kDa, with 4–6 covalently attached polyethylene glycol (PEG) polymers (MW 5 kDa per PEG polymer) per molecule.

Pegvisomant is indicated for the treatment of acromegaly in patients who did not benefit sufficiently from surgery and/or radiation therapy/other therapies (such as somatostatin analogs). Issued precautions relate to the possible expansion of tumors that secrete growth hormone; an increase in glucose tolerance; a state of functional growth hormone deficiency; an elevation of liver enzymes; and lipohypertrophy. In pre-marketing clinical studies, substantial weight gain, marked transaminases, and lipohypertrophy were noted. Most reported treatment-induced adverse events were mild to moderate with infection, injection site reaction, flu syndrome, diarrhea, nausea, peripheral edema, and sinusitis most common. Approximately 17 % of patients developed low titer, non-neutralizing anti-growth hormone antibodies that did not appear to impair drug efficacy. In the post-marketing period, asymptomatic, transient elevations in transaminases 15 times upper limit of normal and with no apparent clinical consequences, were seen in 2 % of patients. Systemic hypersensitivity reactions including anaphylaxis, angioedema, laryngospasm, urticaria, rash, erythema, and pruritus have been recorded.

Insulin-Like Growth Factor 1

Insulin-like growth factor 1 (IGF-1) is a growth hormone with a growth stimulating effect independent of growth hormone. IGF-1 mediates the growth-promoting effect of growth hormone, in particular, its anabolic and mitogenic activities. Originally designated "sulfation factor" upon its discovery in 1957 because of its action of stimulating the incorporation of sulfate into cartilage, IGF-1 was found to have what was described as non-suppressible insulin-like activity and was later termed "somatomedin" because of its action of mediating the effects of growth hormone. Most IGF-1 is made in, and secreted by, the liver in response to stimulation by growth hormone and it can therefore be viewed as an endocrine hormone. However, it can also act locally as a paracrine hormone when secreted by other tissues, for example, cartilaginous cells. IGF-1 is produced throughout life, its production is highest during puberty, and undernutrition/fasting can rapidly reduce levels.

Structure and Mechanism of Action

IGF-1 belongs to a family of peptides related to insulin, a classification reflected in its structural similarity to the pancreatic hormone. IGF-1 and insulin share a homologous sequence, show some overlapping biological activity, and IGF-1 shows low affinity binding to the insulin receptor. Although their three-dimensional structures are similar, IGF-1 and insulin fold differently—insulin folds into a single stable tertiary structure while IGF-1 folds into two thermodynamically stable disulfide isomers. IGF-1 is a single polypeptide chain of 70 amino acids, MW 7649 Da, with three intramolecular disulfide bridges. From the *N*-terminus, the molecule can be divided into four distinct domains, B, C, A, and D. The IGF-1 B-domain of 29 amino acid residues is homologous to the B chain of insulin; the C-domain of 12 residues and the C-peptide of proinsulin are analogous but show no homology; and the 21 amino acid residue A-domain/chain of the two hormones are homologous. The 8 residue D-domain of IGF-1 has no counterpart in insulin. In plasma, almost 100 % of IGF-1 is protein-bound to IGF binding proteins (IGFBPs) (Fig. 7.17) with ~80 % bound to IGFBP-3, one of the six IGFBPs. This 140 kDa ternary complex circulates as one molecule of IGF-1, one molecule of IGFBP-3, and one molecule of an acid-labile protein, MW 88 kDa. The IGFBPs are thought to modulate the availability of free IGF-1.

The effects of IGF-1 are mediated via the hormone's complementary receptor IGF-1R and modulated by interactions with IGFBPs. The IGF-1R gene resembles the gene of the structurally related insulin receptor. The IGF-1R consists of two extracellular α subunits and two transmembrane β subunits that contain signaling tyrosine kinase domains. As with the insulin receptor, autophosphorylation results from ligand binding leading on to a cascade of tyrosine phosphorylation of multiple substrates that function as signaling adapter proteins. Two main signaling pathways, the P13K-AKT/PKB and Ras-MAPK pathways are activated. The former pathway stimulates protein synthesis and inhibits apoptosis; the latter increases cellular proliferation. IGF-1R also signals through the Janus kinase/signal transducer and activator of transcription pathway (JAK/STAT).

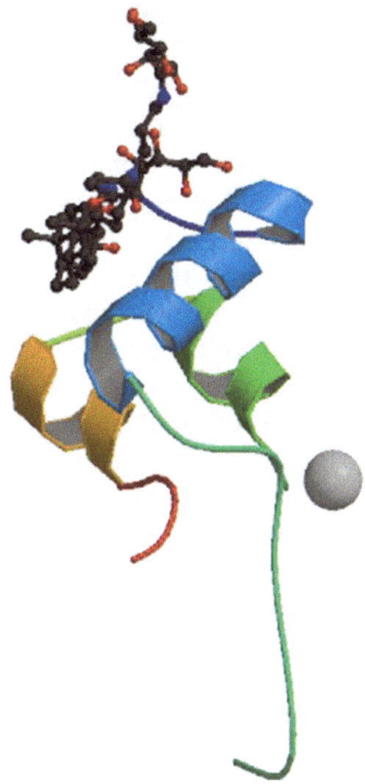

Fig. 7.17 Crystal structure of IGF-1 at 1.8 Å resolution. At physiological pH, IGF-1 is monomeric with three helical sequences linked by a 12 residue C-region and topology resembling insulin. The C-region residues in a type II beta turn are involved in receptor binding. The structure shown is from Protein Data Bank RCSB PDB file 1IMX (Vajdos FF, Ultsch M, Schaffer ML, et al. Biochemistry 2001;40:11022–9)

IGF-1 and Growth Stimulation

In the late 1960s, Laron and collaborators described cases of dwarfism with high serum levels of growth hormone but with characteristics indistinguishable from isolated growth hormone deficiency. The cause of this condition, named Laron syndrome, was shown to be insensitive to growth hormone. The syndrome results in failure to generate IGF-1 with a consequent reduction in synthesis of a number of substances including IGFBP-3. Studies with knockout mice confirmed the explanation that even in individuals with increased growth hormone levels, Laron syndrome with its pronounced growth retardation is due to IGF-1 deficiency. Subsequently, trials, first in adults and then in children, demonstrated that exogenous IGF-1 is an important growth hormone mediating the anabolic and linear growth effects of growth hormone and it also has its own stimulating effect independent of the pituitary hormone.

Mecasermin

Mecasermin (Increlex®) is a recombinant human IGF-1 (rhIGF-1) produced in *E. coli*. It is a 70 amino acid single chain polypeptide, MW 7649 Da, with three intramolecular disulfide bridges and an amino acid sequence identical to the sequence of endogenous IGF-1.

Clinical Significance

Diseases classified as a severe primary IGF-1 deficiency (primary IGFD) include different forms of growth hormone insensitivity due to mutations in the IGF-1 gene, in the growth hormone receptor, or in the receptor signaling pathway. Patients with such disorders are not deficient in growth hormone and are therefore unable to respond to exogenous growth hormone treatment. In late 2005, the FDA approved subcutaneous rhIGF-1 (Increlex®) for severe primary IGFD, specifically for the treatment of growth failure in children with this deficiency. A second approved indication is for the treatment of patients with a growth hormone gene deletion who have developed neutralizing antibodies to the hormone. Mecasermin was approved by the EMA in 2007. The FDA stated that mecasermin is not intended for use in patients with secondary IGF-1 deficiency such as growth hormone deficiency, malnutrition, or hypothyroidism, the latter two needing to be corrected before initiating treatment. Nevertheless, in the absence of evidence of efficacy and apparent disregard of established growth facts, it appears that substantial numbers of children are being treated with mecasermin in off-label usage. This has led to the Drug and Therapeutics Committee of the Lawson Wilkins Pediatric Endocrine Society to state that rhIGF-1 should only be used according to the FDA recommendations and that any other usage for proposed growth promotion should be investigational only. The complex of rhIGF-1 and rhIGFBP-3 appears to have a longer serum half-life and this, together with the belief/hope that adverse events may therefore be fewer or less severe, led to the introduction of the complex under the name mecasermin rinfabate. Some data indicated that the preparation was less effective than rhIGF-1 alone but, in any case, mecasermin rinfabate is no longer available in the USA and Europe due to legal factors.

Safety of Mecasermin

In addition to the reminder that intravenous usage of mecasermin is contraindicated, the FDA has issued a number of warnings and precautions for the drug. Because mecasermin has an insulin-like hypoglycemic effect, attention should be paid to adhering to the optimum time for its administration, namely, shortly before or after a meal, its use in young children, and avoidance of risk activities within 2–3 h of dosage. Other potential problems to be aware of are the rare occurrence of allergic reactions; intracranial hypertension with papilledema, visual changes, headache, and nausea; lymphoid tissue hypertrophy with complications of snoring and sleep apnea; slipped capital femoral epiphysis; progression of preexisting scoliosis in patients who experience rapid growth; and "gasping syndrome" in neonates induced by other medications containing benzyl alcohol (see section "Adverse Events to Somatropin"). Clinical trials experience with mecasermin identified hypoglycemia as the most commonly occurring adverse event with 42 % of participants experiencing it in mostly mild to moderate severity at least once. Hypoglycemia in severe form occurred in 5 (7 %) of patients with 4 experiencing seizures/loss of consciousness. Of

11 (15 %) patients who developed tonsillar hypertrophy, 7 (10 %) required tonsillectomy or tonsillectomy/adenoidectomy. Other adverse events seen with an incidence of 5 % or more were lipohypertrophy and bruising; arthralgia and extremity pain; cardiac murmur; otitis media, hypoacusis, ear pain, and fluid in the middle ear; headache, dizziness, and convulsions; vomiting; intracranial hypertension; and thymus hypertrophy. In the post-marketing period there have been reports of systemic and local allergic reactions along with injection site reactions of erythema, pain, induration, rash, and swelling; musculoskeletal and connective tissue disorders (e.g., osteonecrosis and avascular necrosis); and alopecia. At least one case of anaphylaxis to mecasermin is known and allergic-like reactions occurred in two other patients in the absence of IgE antibodies and positive skin tests. In one of the latter patients, a 3-year-old girl with primary IGF deficiency, local urticarial lesions progressing to generalized urticaria developed after 10 days of subcutaneous mecasermin treatment. Following challenge doses of mecasermin, urticarial lesions developed but no respiratory symptoms or blood pressure changes occurred. Because mecasermin was the only available treatment for the patient, desensitization was undertaken after premedication with dexchlorpheniramine (0.5 mg), montelukast (4 mg), and deflazacort (15 mg). Mecasermin was given subcutaneously in eight doses, each at 20 min intervals, starting with a dose of 0.0033 mg and then in doubling doses until the final dose of 0.4 mg and a cumulative dose of 0.8 mg was reached after 140 min. Following the procedure, the patient was able to continue receiving daily injections of mecasermin without adverse effects.

There seems to be limited data available on the immunogenicity of rhIGF-1 apart from a brief statement by the FDA of the detection of anti-IGF-1 antibodies in 14 of 23 children with primary IGFD treated with the hormone for 2 years. No affect of these antibodies on growth was observed.

Somatostatin

Somatostatin (SS) (also known as growth hormone-inhibiting hormone [GHIH] and somatropin release-inhibiting factor (or hormone) [SRIF]) was first detected and subsequently isolated by its capacity to inhibit the secretion of growth hormone from the pituitary. Not only does SS inhibit growth hormone, it also inhibits the release of a number of other hormones, some activities in the gastrointestinal tract, some functions in the central and peripheral nervous systems such as pain transmission and, importantly, the growth of a range of different gastroenteropancreatic neuroendocrine tumors (GEP-NETs). Other hormones inhibited include gastrin and prolactin and the secretion of insulin and glucagon by the pancreas, ACTH, and the glycoprotein hormones including thyroid stimulating hormone (TSH). Physiologically active SS is secreted in two forms of 14 (SS14) and 28 amino acids (SS28) (Fig. 7.18), after enzymatic posttranslational cleavage of prosomatostatin, a polypeptide of 92 amino acids. Prosomatostatin is derived from a larger precursor, preprosomatostatin of 116 amino acids. Preprosomatostatin contains SS14 and SS28 at its C-terminal end and a 24 amino acid signal peptide that is removed to give prosomatostatin. Both active forms of the hormone occur in tissues although the proportions present vary between tissues. On a molar basis, SS28 shows greater activity than SS14 in suppressing growth hormone, insulin, glucagon, prolactin, luteinizing hormone, follicle-stimulating hormone, and TSH, leading to the suggestion that SS14 is a part of the main hormone, SS28.

SS14 and SS28 each contain the receptor binding region and consequently both interact with the SS receptor (SSR). Note that the endogenous cortistatin neuropeptides also bind to the SS receptor. The receptor exists as five different G-protein-coupled subtypes (SSTRs 1–5) that are preferentially expressed in some well-differentiated neoplasia. Native SS (SS14 and SS28) binds to all five receptor subtypes but shows different affinities for each.

SS decreases splanchnic blood flow and portal venous pressure and although it has been used clinically, for example, as Stilamin®, to treat acute esophageal variceal bleeding in patients with cirrhotic liver disease, its rapid breakdown in the liver resulting in a short half-life in the circulation of only about 1–3 min means that its pharmacological effects are transient and it therefore needs to be given as a bolus followed by continuous infusion. SS is being, or has been, used under the trade names Ikestatina®, Somastat®, Somastin®, Somatin®, Somatrex®, Somatosan®, and others, in a number of countries including Canada, China, Malaysia, Singapore, Switzerland, Taiwan, and Thailand. A number of synthetic cyclic analogs, all containing fewer than 14 amino acids, have been prepared.

Synthetic Analogs of Somatostatin

Being resistant to peptidases, the octapeptide analogs octreotide (Sandostatin®) (Fig. 7.18), generally used as the acetate, the long-acting lanreotide (Somatuline®) (Fig. 7.18) and vapreotide (Sanvar®, Octastatin®), and the hexapeptide seglitide (MK678), have substantially longer half-lives, for example, 1.5–2 h, than the native hormone. Depot formulations of octreotide, Sandostatin LAR®, and lanreotide, Somatuline LA® (injected intramuscularly every 10–14 days) and Somatuline Autogel® and Somatuline Depot® (injected subcutaneously monthly), have largely eliminated the need for daily injections. Native SS and its synthetic analogs demonstrate different affinities for the five SSTR subtypes. Whereas the natural hormone recognizes all five subtypes, the first generation SS analogs octreotide (Fig. 7.18) and lanreotide interact with SSTRs 2 and 5, both drugs predominantly binding the SSTR2 receptor subtype with high affinity. The orphan drug pasireotide (Signifor®) (Fig. 7.18), a cyclic hexapeptide SS analog approved for the treatment of Cushing's disease by the EMA in 2009 and the FDA in 2012, binds with SSTRs 1, 2, 3, and 5 but not SSTR4. Different receptor recognition and binding profiles appear to reflect different pharmacological and clinical activities. For example, octreotide is significantly more potent in inhibiting growth hormone and glucagon secretions than SS and pasireotide with its high affinity for SSTR5, is a more potent inhibitor of insulin secretion than glucagon secretion.

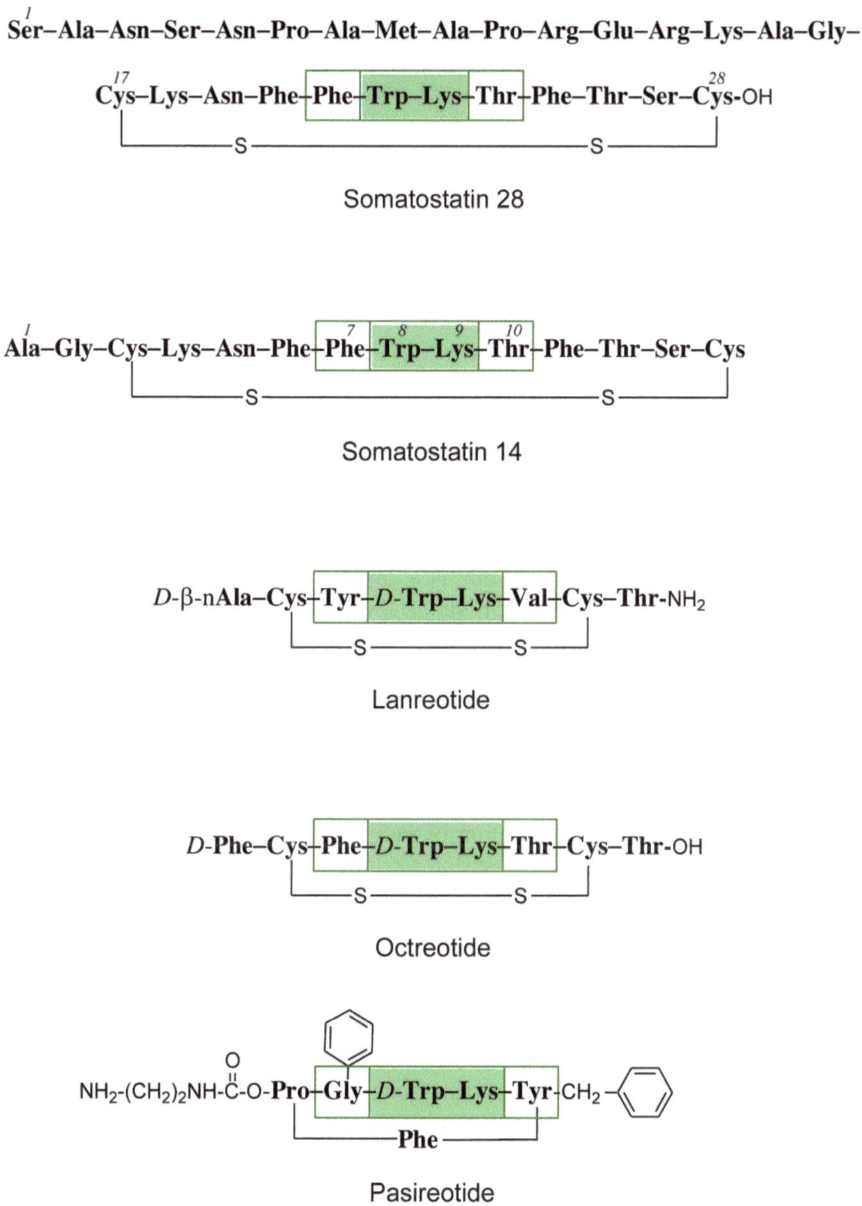

Fig. 7.18 Amino acid sequences of somatostatins 28 (SS28) and 14 (SS14) and synthetic analogs lanreotide, octreotide and the second generation drug, pasireotide. While the β turn sequence Phe-Trp-Lys-Thr (residues 21–24 in SS28 and 7–10 in SS14; outlined in *green*) is required for biologic activity, Trp (*L*- or *D*-form) and Lys (both shaded in *green*) are essential for activity and Phe and Thr can tolerate minor changes. These features of the necessity of Trp-Lys and the tolerance seen with the Phe and Thr residues are shown in the shaded and outlined areas of the analog structures, respectively. D-β-nAla, 2-naphthyl-*D*-alanyl

Approved Indications for Somatostatin Analogs

Octreotide acetate as Sandostatin® injection, first approved by the FDA in 1998, is indicated for the reduction of blood growth hormone and IGF-1 (somatomedin C) levels in patients with acromegaly who experience, in the FDA's words, an "inadequate response to or cannot be treated with surgical resection, pituitary irradiation, and bromocriptine mesylate at maximally tolerated doses." SS inhibits the secretion and the growth of different GEP-NETS, that is, tumors in the pituitary, thyroid, and pancreas endocrine organs or from neuroendocrine cells in lung or gut. GEP-NETs include carcinoid syndrome, Zollinger-Ellison syndrome, WDHA (watery diarrhea-hypokalemia-achlorhydria) syndrome, hypoglycemic syndrome, and glucagonoma syndrome. Stable, longer-acting SS analogs have therefore found application for both the diagnosis and therapy of these tumors. Specifically, approved indications and usage for the SS analog octreotide are carcinoid tumors where the peptide suppresses or inhibits the severe diarrhea and flushing and for the treatment of profuse watery diarrhea associated with vasoactive intestinal peptide (VIP) tumors (VIPomas; also called Verner Morrison syndrome). VIP, which has a half-life of only ~2 min, is a 28 amino acid neuropeptide hormone belonging to the glucagon-secretin superfamily. It causes vasodilation, lowers arterial blood pressure, contracts the heart, increases glycogenolysis, and relaxes stomach, gall bladder, and trachea smooth muscle. Lanreotide as Somatuline Depot® was approved by the FDA in December 2014 for the treatment of acromegaly in 2007 and for administration to patients with well, or moderately differentiated, locally advanced or metastatic GEP-NETs.

Octreotide, as part of the Octreoscan™ Kit, is radiolabeled with indium-111 (half-life 2.8 days) in the preparation of indium-111-pentetreotide and used in a procedure known as somatostatin receptor scintigraphy (SRS) to localize primary and metastatic GEP-NETs bearing SS receptors. The radioactive drug is injected into a vein to travel and attach to tumor cells. The positions of tumor cells are localized using a combination of octreotide scan, computed tomography scan, and positron emission tomography.

The second generation SS analog pasireotide is approved for the treatment of adult patients with Cushing's disease for whom surgery is not an option or when surgery has not been successful. Cushing's disease is caused by a tumor (adenoma) or hyperplasia of the pituitary gland that produces an excess of ACTH from the anterior pituitary or by excess production by the hypothalamus of corticotropin-releasing hormone (CRH). ACTH and CRH in turn stimulate the production and release of the stress hormone cortisol by the adrenal glands resulting in a wide range of symptoms including rapid weight gain (central obesity), so-called moonface and buffalo hump, hirsutism, immune suppression, sleep disturbances, hypertension, muscle and bone weakness, and mood disorders.

Warnings and Precautions for Somatostatin Analogs

Most clinical investigators and authors agree that SS analogs are reasonably safe drugs but for both octreotide and lanreotide, warnings and precautions have been issued by regulatory agencies for cholelithiasis and gallbladder sludge, hyperglycemia and hypoglycemia, thyroid function abnormalities, and cardiovascular abnormalities. Both drugs may reduce both gallbladder motility and bile secretion leading to gallbladder sludge or gallstones. Hyperglycemia and hypoglycemia can result as a consequence of an octreotide/lanreotide-induced altered balance between the counter-regulatory hormones insulin, glucagon, and growth hormone, and suppressed secretion of TSH may result in hypothyroidism. Bradycardia, arrhythmias, and conduction abnormalities have occurred during octreotide therapy as well as QT prolongation and other electrocardiogram changes but the connection of these changes to octreotide is not always established since many of the patients have cardiac disease. Sinus bradycardia, bradycardia, and hypertension are the most common cardiac adverse events reported for lanreotide. Both octreotide and lanreotide may decrease the bioavailability of cyclosporin and octreotide may alter absorption of fats and vitamin B12 and promote fluid loss with an accompanying rise in serum zinc levels in some patients.

Indium-111 pentetreotide should not be administered in parenteral nutrition admixtures to avoid the formation of complex glycosyl octreotide conjugates. To avoid a reduction in the sensitivity of indium-111 scintigraphy testing, octreotide acetate therapy should be temporarily suspended in patients before the administration of indium-111 pentetreotide and because octreotide acetate can produce hypoglycemia in patients with insulinomas, precautions should be taken to prevent this.

For patients with Cushing's disease, rapid and marked suppression of ACTH by pasireotide may lead to the depression of circulating levels of cortisol and transient hypocortisolism and hypoadrenalism with accompanying weakness, fatigue, nausea, vomiting, hypotension, hyperkalemia, and hypoglycemia. As for SS, inhibition of pituitary hormones other than ACTH may occur especially in patients who have undergone transsphenoidal surgery and pituitary irradiation. Patients treated with pasireotide may show alterations in blood glucose levels leading to hyperglycemia and, less often, hypoglycemia. Cholelithiasis has frequently been reported after long-term pasireotide therapy making 6–12 month ultrasonic examinations of the gallbladder a necessary precaution. Mild and transient increases in aminotransferases are common in patients on pasireotide. Elevated levels of alanine aminotransferase (ALT), aspartate aminotransferase (AST), and bilirubin may be seen and if they occur in patients with jaundice or other signs of liver dysfunction, the drug should be discontinued. Bradycardia and QT prolongation may occur following pasireotide making it important to monitor patients with cardiac disease and/or at risk of bradycardia.

Adverse Events to Somatostatin Analogs

SS analogs are generally well tolerated but in addition to the adverse events mentioned in the preceding section, clinical trials and post-marketing experiences have contributed to the recognition of a range of mostly non-severe adverse responses to the somatostatin analogs when used to treat acromegaly, GEP-NETs, and Cushing's disease.

Octreotide

Phase III double-blind, cross-over studies of the safety of octreotide in the treatment of acromegaly by the slow release form of the peptide, Sandostatin LAR®, revealed the following adverse events from the most to the least frequently occurring, in ≥10 % of patients: diarrhea, abdominal pain, flatulence, influenza-like symptoms, constipation, headache, anemia, injection site pain, cholelithiasis, hypertension, dizziness, fatigue. From a post-marketing randomized phase IV study, nausea, alopecia, and epistaxis can be added to this list. Overall, the most common adverse events are gastrointestinal. Mild to moderate diarrhea, abdominal pain, and nausea generally develop during the first month of therapy, dyspepsia, steatorrhea, feces discoloration, and tenesmus have been reported in 4–6 % of patients, and rare instances of effects resembling acute intestinal obstruction have occurred. The common gastrointestinal adverse events occur with significantly higher frequency in patients given octreotide subcutaneously three times a day compared to depot therapy every 28 days. Necrotizing enterocolitis, including one death, was reported in four neonates treated with octreotide. Three additional case reports and two further deaths were reported to the FDA in 2007. Pain at the injection site, reported in up to ~25 % of patients and usually mild to moderate and short-lived, is also a common side effect.

The incidence of gallbladder abnormalities following octreotide administration appears to be unrelated to dose, age, or sex but higher incidences parallel treatment durations. Biliary abnormalities of gallstones, sludge, microlithiasis, and dilatation were seen in 52 % of patients with acromegaly who received an octreotide depot preparation for 12 months or more. Cholelithiasis occurred in 22 % of the patients. Other adverse events and their incidences recorded in response to octreotide injection in acromegaly patients include hypoglycemia ~2 %, hyperglycemia ~15 %, hypothyroidism 12 %, and goiter 8 %; 4 % required thyroid replacement therapy. Incidences of hypothyroidism and goiter were each reduced to 2 % when a depot preparation of the drug was used. Reports of adverse cardiac events to octreotide therapy in acromegalics detail sinus bradycardia in 25 % of patients, conduction abnormalities in 10 %, and arrhythmias in 9 % but such findings are often difficult to interpret because many of the patients have underlying cardiac disease. The adverse events connected to gallbladder and cardiac abnormalities, glucose metabolism, and hypothyroidism seen in acromegaly patients also occur, and with similar frequencies, in patients treated for carcinoid tumors and VIPomas. Some other significant adverse events following octreotide depot therapy in each of the different patient groups have yet to be unequivocally linked to the drug. These events include malignant hyperpyrexia, cerebral vascular disorder, rectal bleeding, ascites, pulmonary embolism, pneumonia, and pleural effusion.

A large number of other diverse adverse events reported to the FDA in the post-marketing period include anaphylaxis, urticaria, facial edema, renal failure and insufficiency, atrial fibrillation, hepatitis, elevated liver enzymes, intestinal obstruction, pancreatitis, diabetes mellitus, thrombosis, gastrointestinal hemorrhage, thrombocytopenia, aphasia, peptic/gastric ulcer, pituitary apoplexy, diabetes insipidis, gallbladder polyp, epileptic seizures, decreased libido, and petechiae. Cutaneous effects of octreotide are usually restricted to the injection site or are related to transient hair loss but there is at least one report of a serious skin event. On the second day after the intramuscular injection of slow release octreotide, a 40-year-old woman with acromegaly developed round erythematous pruriginous plaques on the trunk and limbs accompanied by arthralgia and edema of the hands and joints. Erythema multiforme was diagnosed and despite corticosteroid treatment, the skin lesions persisted for 40 days. In another apparent hypersensitivity reaction of the skin, a 60-year-old man with acromegaly developed a papular rash on the abdomen, arms, and back together with pruritus and swollen lips shortly after starting subcutaneous octreotide. Rechallenges confirmed the association of the reactions with the drug. Treatment consisted of the development of a successful desensitization protocol for octreotide involving the successive administration of approximately doubling doses of the drug starting at 0.001 μg and given in 14 steps at 20 min intervals. Dosages for the subsequent 13 steps were, respectively, 0.002, 0.005, 0.01, 0.02, 0.05, 0.1, 0.2, 0.5, 1, 2, 5, 25, and 50 μg. The final dose of 50 μg was repeated after 8 h. Successful desensitization to octreotide was also achieved with a 12-year-old boy who reacted with flushing, erythema of the arms, periorbital and perioral swelling, abdominal pain, breathing difficulty, and cough. A subsequent dose of octreotide produced the same response. After intradermal testing with the drug produced positive reactions, a desensitization protocol without prior premedication therapy was commenced. Beginning with a dose of 1 μg of octreotide, increasing doses of the drug were given by intravenous infusion every 30 min until a cumulative dose of 76 μg was reached. No reactions occurred during the desensitization procedure and treatment was continued for 5 days without any reaction. At least three cases of lipoatrophy following subcutaneous octreotide have been recorded. The mechanism is not known. Beau's lines or onychomadesis, a reaction affecting the nails, was seen in a 72-year-old woman receiving monthly subcutaneous doses of octreotide. Pituitary apoplexy (hemorrhage or infarction), a rare and life-threatening syndrome, has been reported in a 51-year-old woman with high blood pressure after a single dose of long-acting octreotide. Some of the predisposing factors are thought to be pituitary stimulation tests, diabetes mellitus, anticoagulant or antiplatelet aggregation therapy, head trauma, and high blood pressure. Epileptic seizures occurred in a 6-year-old girl with acute lymphocytic leukemia and *L*-asparaginase associated pancreatitis and in another apparent effect on the nervous system, a high dose of octreotide seems to have been responsible for observed Parkinson-like symptoms.

Occasional hematologic reactions to octreotide have been noted, for example, a 55-year-old woman given an intravenous bolus followed by an infusion of octreotide for a splenic artery pseudoaneurysm expanding a pseudocyst, developed thrombosis of the pseudoaneurysm with segmental splenic infarction. Thrombocytopenia has also been attributed to octreotide in the same 53-year-old male patient on two separate occasions when the drug was administered for variceal bleeding. A complete atrioventricular block developed after an octreotide infusion for variceal bleeding. The block resolved 6 days after withdrawal of the drug.

A number of studies of antibody formation to subcutaneously administered octreotide have indicated that there is a low incidence of antibodies to the drug. In a comparison of antibody responses to octreotide delivered by three different routes, intranasal, subcutaneous, and intramuscular, incidences of antibody formation in patients treated intranasally for 9–12 months were 77 % (previously untreated) and 81 % (previously treated); incidences for subcutaneously treated patients were 27 % (mean exposure 3 years), 57 % (exposure >5 years), and 72 % (exposure 8 years); after intramuscular treatment with sustained release octreotide for a mean of 2.5 years, no patients were found to be antibody-positive. Octreotide antibodies did not cross-react with native SS and ~25 % of antibody-positive sera did not cross-react with lanreotide.

Adverse effects, generally transient, observed after the administration of indium-111 pentetreotide and recorded by the manufacturer, include dizziness, fever, flush, headache, hypotension, changes in liver enzymes, joint pain, nausea, sweating, and weakness. Since indium-111 pentetreotide is mainly eliminated via the kidneys, caution should be exercised when using Octreosan™ on patients with impaired renal function.

Lanreotide

Clinical trials experience with over 400 acromegalic patients showed the most common adverse reactions to lanreotide were gastrointestinal disorders (diarrhea, abdominal pain, nausea, constipation, flatulence, vomiting, loose stools), cholelithiasis, and injection site reactions. Other adverse events seen with an incidence of >3–5 % were arthralgia, headache, sinus bradycardia, hypertension, anemia, and alopecia. Pancreatitis occurred in <1 % of patients. As for octreotide, adverse events involving the gallbladder, glucose metabolism, and the heart were noted. Cholelithiasis and gallbladder sludge were reported in 20 % of patients and dysglycemia (hypoglycemia, hyperglycemia, diabetes) in 7 % of patients. Gastrointestinal, renal, and urinary disorders were seen more often in patients with hepatic impairment. Subcutaneous gluteal nodules detected by routine CT scanning and local inflammation were found in 29 of 43 patients with metastatic midgut carcinoid disease treated with SS analogs: 16 of 22 on lanreotide autogel, 5 of 12 on long-acting octreotide, and 8 of 9 treated with both drugs. There appeared to be no relationship between the presence of nodules and cumulative dose suggesting either injection technique or a granulomatous reaction to the drug as the possible cause. Lanreotide has been shown to cause foreign body granulomas and granuloma formation has been reported in a patient on octreotide LAR. Necrosis of fat at injection sites has been attributed to subcutaneous lanreotide depot. In one case, lesions with central necrosis surrounded by inflammatory tissue became apparent on both buttocks of an 80-year-old acromegalic woman after 4 months. The authors speculated that the patient was susceptible to, and developed, a foreign body-like inflammatory reaction aggravated by an anti-angiogenic effect of the drug.

The percentage of patients with antibodies to lanreotide following treatment with a depot preparation of the drug is low, being <1–4 %. Importantly, the antibodies do not appear to diminish efficacy or affect the drug's safety.

Pasireotide

Safety assessments on nearly 500 acromegalic patients given pasireotide, mostly via the intramuscular route in phase I, II, and III studies, showed the most common (frequency ≥10 %) adverse reactions were (in decreasing order), diarrhea, cholelithiasis, hyperglycemia, and diabetes mellitus. Most Common Toxicity Criteria grade 3 and 4 reactions were related to hyperglycemia and, compared to other SS analogs, hyperglycemia occurred more often when pasireotide was given intramuscularly. Adverse events commonly seen (frequency ≥1 % to < 10 %) were anemia; adrenal insufficiency; type 2 diabetes mellitus; impaired glucose tolerance; headache; dizziness; sinus bradycardia; QT prolongation; nausea; abdominal distention and pain; injection site reactions; alopecia and increases in glycosylated hemoglobin, ALT, blood glucose, and blood creatinine phosphokinase. Clinical trials experience from phase III studies when pasireotide was gived to patients with Cushing's disease was the observance of adverse events in >98 % of patients with the most frequent reactions (≥20 %) being diarrhea, nausea, hyperglycemia, cholelithiasis, headache, abdominal pain, fatigue, and diabetes mellitus. Adverse events commonly seen (frequency >5 %) were: injection site reactions, nasopharyngitis, alopecia, asthenia, increases in glycosylated hemoglobin, ALT and gamma-glutamyl transferase, peripheral edema, upper abdominal pain, decreased appetite, hypercholesterolemia, hypertension, dizziness, hypoglycemia, type 2 diabetes mellitus, anxiety, influenza, insomnia, and myalgia.

Although SS can cause hyperglycemia, it can also reduce glucose levels in some cases, for example, cases induced by neurotensin. SS analogs can cause hyperglycemia in acromegalic patients with normal glucose tolerance but not in those with impaired glucose tolerance or diabetes mellitus. In IDDM, SS analogs can increase insulin secretion, suppress glucagon secretion, and reduce hyperglycemia. Compared to other SS analogs, pasireotide is responsible for more frequent and severe problems of glucose homeostasis. In particular, despite a fall in serum cortisol in cases of Cushing's disease, and a reduction in growth hormone in acromegalics, pasireotide has been implicated in cases of paradoxical hyperglycemia during the management of these two disorders. Studies on the mechanism of pasireotide-induced hyperglycemia in healthy volunteers suggest that the increase in blood glucose levels is a result of decreased secretion of insulin and incretins (GLP-1 and GIP) (section "Glucagon-Like Peptide 1 and the Incretin Effect") and a fall in the mean value of glucagon while insulin sensitivity is unaffected. Suppression of insulin secretion is greater than that of glucagon which appears to be supplemented by postprandial secretion. Unlike the first generation SS analogs, octreotide and lanreotide which show a marked binding affinity for the SSTR2 receptor subtype with little adverse effect on the insulin/glucagon balance, pasireotide demonstrates high affinity for SSTR5 and more potent inhibition of insulin secretion than glucagon secretion. Thus, it has been suggested that the SSTR5:SSTR2 ratio is the major determining factor in pasireotide-induced hyperglycemia.

Vasopressin

Also known as arginine vasopressin (AVP), antidiuretic hormone or argipressin, vasopressin is, along with oxytocin (section "Oxytocin"), a neurohypophysial hormone, that is a hormone secreted by the neurohypophysis (the posterior pituitary gland). Vasopressin is synthesized by the magnocellular neurons of the hypothalamus and differs structurally from the closely related oxytocin by two amino acids (~80 % homology) although the hormones have very different physiologic activities. Virtually all vertebrate species make vasopressin, or a vasopressin-like peptide, and this is also true for oxytocin indicating that the hormones are evolutionarily highly conserved and necessary for survival.

Physiologic Actions of Vasopressin

Vasopressin has two main actions, both important regulatory functions in the mammalian body: regulation of water retention and the constriction of blood vessels. The first of these actions is effected by the hormone increasing water permeability and hence reabsorption of water in the collecting ducts of the kidney nephron and distal convoluted tubule. Some experimental data also suggests that vasopressin and its V_1 receptor plays a crucial role in the regulation of brain water and ion homeostasis. Unlike the antidiuretic effect of vasopressin which has been exploited in its application as therapy for diabetes insipidis, a condition involving hypernatremia, vasopressin's cardiovascular effects, although well known, are only now being used in a variety of disease states. By constricting blood vessels, vasopressin increases peripheral vascular resistance and, in turn, arterial blood pressure. Shock can increase the hormone's levels up to 200-fold causing profound vasoconstriction and helping to maintain end-organ perfusion. Vasopressin is also released into the brain by different neurons where it has been implicated in a variety of neurologic effects involved in the regulation of social cognition and behavior including aggression, blood pressure, temperature regulation, stress, pair-bonding, and other social interactions in some species. In fact, it has been suggested that both vasopressin and oxytocin are emerging as targets for novel treatment approaches for social function disorders such as autism, borderline personality and social anxiety disorders, and schizophrenia. Thus, vasopressin has a key role in the regulation of water, glucose, and salts in the blood and is essential for osmotic and cardiovascular homeostasis.

Vasopressin Gene and Hormone Structure

The vasopressin gene (AVP) and that of oxytocin are separate but both lie on chromosome 20p, the former at location 20p13, only 12 kilobases from the oxytocin gene. This close proximity and the structural similarity of the two peptides suggests recent gene duplication. At least 60 mutations in the AVP gene have been implicated in neurohypophyseal diabetes insipidis. Vasopressin in the parvocellular neurons of the hypothalamus, and CRH, are the main regulators of adrenocorticotropin hormone (ACTH) secretion. In studies designed to examine AVP gene regulation by glucocorticoids, cAMP increased AVP gene transcription and mRNA stability whereas glucocorticoids inhibited gene transcription and decreased mRNA stability suggesting that cAMP and glucocorticoids interact to regulate AVP gene expression.

Vasopressin (that is arginine vasopressin) is a nonapeptide with a disulfide bridge between cysteines at positions 1 and 6 and with the following amino acid sequence:

1   6   9
Cys-Tyr-Phe-Gln-Asn-Cys-Pro-Arg-Gly-NH_2

Lysine vasopressin replaces the arginine at position 8 with a lysine residue. The structure of oxytocin is similar with changes at position 3 (isoleucine for phenylalanine) and at position 8 (leucine for arginine or lysine).

Vasopressin Receptors

The physiologic actions of vasopressin are mediated by interaction with tissue-specific GPCRs. As with a number of other hormones (e.g., glucagon), signaling occurs through specific G proteins including $G_{q/11}$, G_s, and G_i and a receptor can activate more than one signaling pathway via more than one specific G protein. The three vasopressin receptor subtypes, V_1, V_2, and V_3, differing in their location, G protein coupling/signaling pathways, and function are summarized in Table 7.3. The oxytocin receptor (OTR), considered a non-selective vasopressin receptor, is recognized with equal affinity by both hormones. This is unlike the situation with the V receptors which show clear selectivity for vasopressin. OTRs occur on vascular endothelium, the heart, and reproductive tissues (section Oxytocin Receptors). They are coupled to $G_{q/11}$ proteins which stimulates the phospholipase C, inositol

Table 7.3 Vasopressin receptors

Receptor	Gene	Location	Selective G protein coupling	Signaling pathways	Functions
V_1[a]	AVPR1A[b]	Vascular smooth muscle of systemic, coronary, renal, splanchnic circulations; platelets; brain; testis; liver; superior cervical ganglion[c]	$G_{q/11}$	Ca^{2+} influx; phospholipases, IP_3, DAG, PKC	Vasoconstriction; platelet aggregation; glycogenolysis; uterine contraction
V_2	AVPR2[d]	Kidney; ?endothelium cells[e]	G_s	Inhibition adenylyl cyclase, ↑intracellular cAMP	Plasma volume and osmolality control by mobilization of aquaporin channels and increasing water reabsorption
V_3[f,g]	AVPR1B[h]	Anterior pituitary gland	Different G proteins: $G_{q/11}$, G_s and G_i[i]	Release of ACTH: activation of PKC; other cellular responses e.g., ↑DNA and ↑cAMP mediated via several pathways[i]	Release of ACTH. See note[i]

ACTH adrenocorticotropic hormone, *DAG* diacyl-glycerol, *IP_3* inositol triphosphate, *PKC* protein kinase C

[a]Previously known as $V_{1a}R$

[b]Located at chromosome region 12q14–15

[c]Role of vasopressin in many of the tissues not yet identified

[d]Located at chromosome region 10q28. Most cases of diabetes insipidis due to mutations in the V_2R gene

[e]Vasopressin and its analog 1-deamino-8-*D*-arginine vasopressin raise plasma levels of von Willebrand factor apparently by a direct effect on endothelial cells. However, unlike non-selective [^{3}H]arginine vasopressin, no binding of *selective* V_2 R radioligands with endothelial or liver cells was seen

[f]Previously known as $V_{1b}R$

[g]Shows 45 %, 39 %, and 45 % homology with the V_1, V_2, and the oxytocin receptors, respectively

[h]Located at chromosome region 1q32

[i]Depending on the level of receptor expression and vasopressin concentration

1,4,5-triphosphate (IP_3), 1,2-diacylglycerol (DAG), protein kinase C (PKC) cascades followed by a number of different cellular events in response to an increase in cellular Ca^{2+}, particularly endothelial nitric oxide synthase-catalyzed production of the potent vasodilator nitric oxide. Vasopressin's selective action on different vascular beds may be due to the different locations of the OTR and it has been suggested that V_1 and V_2 receptors on vascular epithelium might also lead to nitric oxide production and vasodilation. Note that some studies have suggested that additional vasopressin receptor and OTR subtypes exist.

Cell surface receptors for ADP, ATP, and UDP, referred to as P_2 purinoreceptors, function via G proteins. These receptors appear to have a role in cardiac contractility and, interestingly, vasopressin-induced cardiac effects have been shown to occur via activation of P_2 purinoreceptors expressed on cardiac endothelium. Coronary vasoconstriction and negative inotropy could be blocked by both vasopressin and P_2 purinoreceptor antagonists.

Safety of Vasopressin Therapy

Vasopressin intravenous injection (Vasostrict®) was approved in 2014 by the FDA to increase blood pressure in adults with vasodilatory shock who remain hypotensive despite fluids and catecholamines. With the warning that its use in patients with impaired cardiac response may worsen cardiac output, the FDA published the following list of adverse reactions drawn from the literature: bleeding and lymphatic disorders—hemorrhagic shock, intractable bleeding, decreased platelet count; cardiac disorders—right heart failure, atrial fibrillation, bradycardia, myocardial ischemia; gastrointestinal disorders—mesenteric ischemia; hepatobiliary disorders—increased bilirubin levels; renal/urinary disorders—acute renal insufficiency; vascular disorders—distal limb ischemia; metabolic disorders—hyponatremia; skin conditions—ischemic lesions.

There are a number of potential drug interactions when using Vasostrict®. In particular, it should be used with cautious with: catecholamines—may produce an additive effect on several hemodynamic parameters including mean arterial blood pressure; indomethacin—may prolong the effect on cardiac and systemic vascular resistance; ganglionic blocking agents—increased effect on mean arterial blood pressure; furosemide—increases the effect of vasopressin on osmolar clearance and urine flow; drugs suspected of causing syndrome of inappropriate antidiuretic hormone secretion (SIADH), for example, selective serotonin reuptake inhibitors, haloperidol, enalapril, methyldopa, cyclophosphamide—may increase pressor effect as well as antidiuretic effect; drugs suspected of causing diabetes insipidis, for example, demeclocycline, lithium, clozapine—may decrease pressor and antidiuretic effects.

Desmopressin

Desmopressin, with a disulfide bridge between positions 1 and 6, is 1-desamino-8-*D*-arginine vasopressin (or 1-(3-mercaptopropionic acid)-8-*D*-arginine vasopressin monoacetate trihydrate), a synthetic analog of vasopressin differing from vasopressin by deamination of the cysteine at position 1 to yield 3-mercaptopropionic acid (Mpr), and by replacement of *L*- with *D*-arginine at position 8. The amino acid sequence of desmopressin is therefore:

1 6 9
SCH_2CH_2CO-Tyr-Phe-Gln-Asn-Cys-Pro-*D*-Arg-Gly-$NH_2.CH_3COOH.3H_2O$

Desmopressin acetate is absorbed from the nasal mucosa and exhibits a biphasic elimination profile with half-lives of 7.8 and 75.5 min. This compares to initial and terminal phase half-lives for lysine vasopressin of 2.5 and 14.5 min. Desmopressin acetate is approved by the FDA for use as a nasal spray (DDAVP®, Minirin®) at a concentration of 0.1 mg/mL for antidiabetic replacement therapy in the management of central cranial diabetes insipidis and for management of temporary polyuria and polidipsia following head trauma or surgery in the pituitary region. Note that it is ineffective for treating nephrogenic diabetes insipidus. When it cannot be used as a nasal spray, the drug can be administered by injection (DDAVP® injection). Desmopressin has also been prescribed for bedwetting.

Forty years ago, infusion of desmopressin was found to increase levels of coagulation factor VIII, von Willebrand factor and plasminogen activator in the plasma of humans. Formulated at a concentration of 1.5 mg/mL, desmopressin acetate, again as a nasal spray (Stimate®), is indicated for hemophilia A with factor VIII coagulant activity levels >5 % and for patients with mild to moderate von Willebrand's disease (Type I) with factor VIII levels >5 %.

Warnings and precautions for DDAVP® relate to its action as a potent antidiuretic with the potential to cause water intoxication and/or hyponatremia and a small increase in blood pressure, an undesirable effect in patients with coronary artery insufficiency or hypertensive cardiovascular disease. Adverse reactions tend to be infrequent and usually provoked by high doses when they do occur. Most reports are for headache and nausea while nasal congestion, rhinitis, flushing, chills, abdominal pain/cramps, asthenia, upper respiratory tract infections, cough, sore throat, and epistaxis have been reported. There has been at least one fatal case of anaphylaxis to intravenous DDAVP®. The post-marketing period for DDAVP® has seen some rare reports of hyponatremic convulsions following concurrent use of imipramine and oxybutynin. This is a reminder that the concomitant administration of a number of different drugs may increase the risk of hyponatremia. As well as tricyclic antidepressants like imipramine, the list of drugs to be co-prescribed with caution include selective serotonin reuptake inhibitors, nonsteroidal anti-inflammatory drugs, opioid analgesics, chlorpromazine, lamotrigine, and carbamazepine. The catecholamine oxybutynin may have an effect on blood pressure. There are occasional reports of apparent decreased effectiveness of desmopressin, usually after more than 6 months usage. Neutralizing antibodies have been suggested but not found and local inactivation of the peptide has been put forward as the explanation.

The list of warnings and precautions issued by the FDA for Stimate® relate to the possibilities of slight blood pressure and heart rate changes in patients with cardiovascular disorders; hyponatremia in patients with fluid and electrolyte imbalances (e.g., cystic fibrosis, heart failure, renal disorders); thrombotic events in patients prone to thrombus formation; and severe allergic reactions. The recorded adverse reactions to Stimate® are similar to those listed immediately above with the addition of some adverse events seen in clinical trials with the nasal spay. These include somnolence, dizziness, edema, vomiting, itchy or light-sensitive eyes, insomnia, chest pain, dyspepsia, palpitations, tachycardia, agitation, and balanitis.

Terlipressin

Terlipressin, used as the acetate (Glypressin®, Teripress®), is *N*-[*N*-(*N*-glycylglycyl) glycyl]-8-*L*-lysinevasopressin, a dodecapeptide with three glycyl residues attached to the *N*-terminal of lysine vasopressin, a disulfide bridge between the cysteines at positions 4 and 9 and the following sequence of amino acids:

1 4 9 12
Gly-Gly-Gly-Cys-Tyr-Phe-Gln-Asn-Cys-Pro-Lys-Gly-NH_2

Terlipressin is a prodrug, the three glycines at the *N*-terminal being removed enzymically in vivo to produce the biologically active lysine vasopressin. Given i.v. to avoid local tissue necrosis, its indications for use are esophageal varices, norepinephrine-resistant septic shock, and hepatorenal syndrome. The drug is approved for use in a number of countries including India, Australia, the United Arab Emirates and much of Europe but is not currently available in the USA.

Like lysine vasopressin, terlipressin acts via the V_1 receptors on vascular smooth muscles of the splanchnic and portal circulations causing pronounced vasoconstriction. Acting via the V1 receptor, terlipressin also increases mean arterial pressure and a reduction in heart rate. It appears to show minimal recognition of the V_2 receptor demonstrating only 3 % of the antidiuretic effect of native arginine vasopressin and minimal effects on the fibrinolytic system in patients with cirrhosis. Given its cardiovascular effects, terlipressin should not be administered to patients with unstable angina or to those who experienced a recent myocardial infarction and it should be used with caution, and under strict monitoring, in cases of cardiac arrhythmias; uncontrolled hypertension; coronary artery disease or previous myocardial infarction; cerebral or peripheral vascular diseases; and patients with a history of QT interval prolongation. There should also be awareness of the possibility of cutaneous ischemia and necrosis unrelated to the injection site; constriction of smooth muscle in asthmatics and those with COPD; and electrolyte disturbances in patients with renal insufficiency. Terlipressin is contraindicated in pregnancy since it causes uterine contractions, increased uterine pressure, and may decrease uterine blood flow. Cardiac and vascular disorders feature as the main adverse effects provoked by the administration of terlipressin with bradycardia, peripheral vasoconstriction, peripheral ischemia, hypertension, and facial pallor the most commonly occurring events. Other common (incidence 1–10 %) events are headache, transient abdominal cramps, and transient diarrhea. Other less common (0.1–1 %) events include chest pain, tachycardia, atrial fibrillation, myocardial infarction, ventricular extrasystoles, fluid overload with pulmonary edema, hot flushes, intestinal ischemia, peripheral cyanosis, respiratory distress and failure, nausea/vomiting, injection site necrosis, and hyponatremia. Although terlipressin appears to show no consistent effect on serum sodium levels in healthy volunteers (and a very weak V_2 receptor-mediated antidiuretic effect), hyponatremia is considered to be a potential risk when the peptide is administered to patients with portal hypertension and actively bleeding esophageal varices. For example, in a retrospective cohort study, 67 % of 58 patients developed an acute decline in serum sodium over a 5 day treatment period. Recovery within ~4 days occurred upon discontinuation of therapy.

Vaptans: Vasopressin Receptor Antagonists

Hyponatremia is the most frequently occurring electrolyte imbalance seen in hospitalized patients with a reported prevalence as high as 28 %. The condition may be life-threatening and may lead to attention deficits and the risk of falls and fractures as well as being a prognostic factor in patients with heart failure and cirrhosis. Nonpeptide vasopressin receptor antagonists, or vaptans, promote excretion of electrolyte-free water by blocking binding of arginine vasopressin to its renal receptor and are important in the treatment of dilutional (euvolemic and hypernatremic) hyponatremia. Two vaptans have been approved for clinical use by the FDA and a number of others have been developed and/or are in clinical trials e.g., lixivaptan. **Conivaptan** (Vaprisol®) (Fig. 7.19) is a combined V_1/V_2 receptor antagonist approved for use in raising serum sodium levels in hospitalized patients with euvolemic and hypervolemic hyponatremia. Special care is advised if it is used to treat hypervolemic hyponatremia associated with heart failure and if a too-rapid correction of serum sodium needs to be avoided. The most commonly seen adverse events following its administration are infusion site reactions, including erythema, pain, and phlebitis; some serious reactions may occur with an incidence of up to 51 %. Other adverse reactions are pyrexia, hypokalemia, orthostatic hypotension, headache, anemia, constipation, nausea and vomiting, and peripheral edema. The second FDA-approved vaptan, **tolvaptan** (Samsca®) (Fig. 7.19), a selective V_2 receptor antagonist, is approved for the treatment of euvolemic and hypervolemic hyponatremia including patients with heart failure, cirrhosis, and SIADH. The drug carries an FDA black box warning advising that tolvaptan treatment should be initiated and re-initiated in hospital where serum sodium can be monitored and a too-rapid correction of hyponatremia (especially in patients with malnutrition,

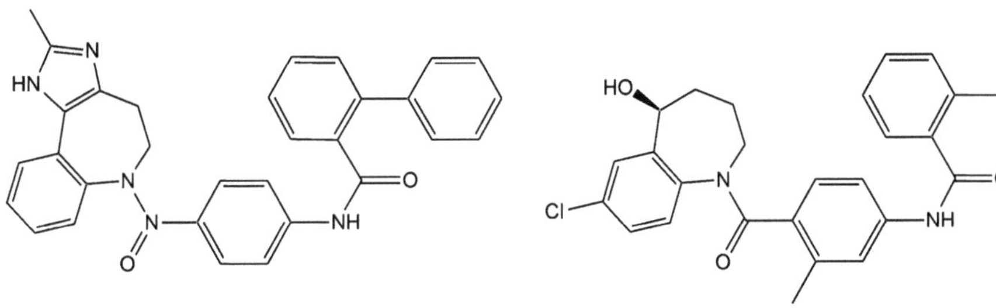

Fig. 7.19 The benzazepine analogs, conivaptan and tolvaptan, are non-peptide vasopressin receptor antagonists, or vaptans, approved for treatment of hyponatremia associated with congestive heart failure, cirrhosis, and syndrome of inappropriate antidiuretic hormone secretion (SIADH)

alcoholism, and advanced liver disease) must be avoided. In April 2013, the FDA announced that tolvaptan should not be used for more than 30 days and not in patients with underlying liver disease. Some clinicians believe, however, that the drug can be used for extended periods in SIADH where its benefits outweigh the risk, provided that liver function tests are carefully monitored. The most commonly reported adverse reactions are thirst, dry mouth, asthenia, constipation, pollakiuria or polyuria, and hyperglycemia.

Oxytocin

Oxytocin (Greek, "quick birth"), well known for its uterotonic and milk-ejecting activities, is an abundant neurohypophysial peptide hormone produced in the magnocellular neurons of the hypothalamic paraventricular and supraoptic nuclei. It is secreted into the circulation from the posterior pituitary gland in response to a number of stimuli including parturition, suckling, and stress but it is also synthesized in, and released from, a number of peripheral tissues such as corpus luteum, uterus, placenta, amnion, testis, and heart. The hormone has a wide range of actions ranging from modulation of neuroendocrine reflexes to involvement in social behaviors, bonding between individuals, reproduction, and care of offspring.

Indications

Oxytocin injection (Pitocin®), given intravenously, may be administered for antepartum and postpartum use. For the former purposes, the hormone is indicated for the initiation or improvement of uterine contractions to achieve vaginal delivery. In the postpartum situation, oxytocin is given to produce uterine contractions during the third stage of labor and to control postpartum bleeding or hemorrhage.

Oxytocin Structure

Oxytocin for injection is synthesized (as Pitocin®) to avoid possible contamination with vasopressin and other peptides that may occur in pituitary extracts. Like all neurohypophysial hormones, oxytocin is a nonapeptide with a disulfide bridge linking the cysteine residues at positions 1 and 6, thus giving the peptide a C-terminal three amino acid tail linked to a six amino acid cyclic structure. The sequence of amino acids is:

1 3 6 8
Cys-Tyr-Ile-Gln-Asn-Cys-Pro-Leu-Gly-NH_2

Neurohypophysial nonapeptide hormones are classified into the vasopressin family which all have a basic amino acid (lysine or arginine) at position 8 (section "Vasopressin Gene and Hormone Structure") or the oxytocin family with a neutral amino acid (leu) at this position. These amino acid substitutions at position 8 are necessary for appropriate recognition and activation of vasopressin receptors while activation of oxytocin receptors requires an isoleucine at position 3.

Oxytocin Receptors

The oxytocin receptor (OTR) belongs to the rhodopsin-like superfamily of GPCRs along with the three vasopressin receptor subtypes V_1, V_2, and V_3 (section "Vasopressin Receptors"). Although the vasopressin V_2 receptor shows a 40 % sequence identity with the OTR, the latter binds oxytocin and arginine vasopressin with similar high affinity whereas the V_2 receptor binds arginine vasopressin with ~400-fold higher affinity than oxytocin. This fact has been utilized in the formation of chimeric OTR/V_2 receptor constructs to identify ligand-binding receptor domains.

In terms of structural aspects of ligand-receptor recognition, the OTR is one of the best defined peptide hormone receptors. The OTR is functionally dependent on divalent cations such as Mg^{2+} or Mn^{2+} and cholesterol, existing in either a low or high affinity state for both agonists and antagonists. Cholesterol also aids the stability of the OTR, protecting it against proteolytic and thermal degradation. The binding domains of the OTR comprise the amino terminus E1 domain and E2 and E3 loops of the extracellular regions of the receptor. Position 34 within the E1 domain, the site of a conserved arginyl residue in all members of the neurohypophysial peptide hormone receptor family, is essential for high affinity binding of oxytocin and presumably for receptor binding of other neurohypophysial peptide hormone agonists. OTRs couple $G_{\alpha q}$ and $G_{\alpha i}$, stimulating phospholipase C and leading to an increase in intracellular Ca^{2+} concentration and activation of MAPK and PKC.

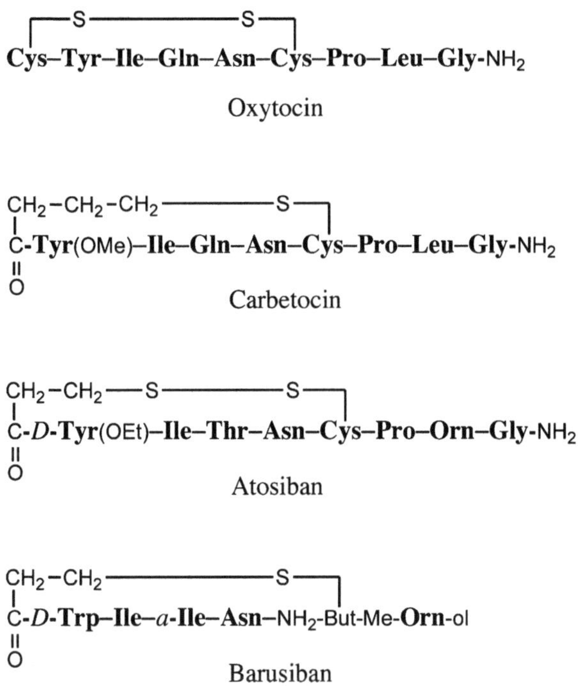

Fig. 7.20 Amino acid sequences of carbetocin, a synthetic peptide analog and agonist of oxytocin and the oxytocin receptor peptide antagonists, atosiban and barusiban. Carbetocin binds the OTR with a lower affinity than oxytocin but has a much longer half-life. The main structural differences between the two peptides are the replacements of the disulfide bridge of oxytocin with a thioether bridge and the tyrosine hydroxyl with a methoxy group. Atosiban is a nonapeptide desamino-oxytocin analog with an *O*-methyl-*D*-tyrosine and an ornithine amino acid. Barusiban is the most specific OTR antagonist showing even higher affinity and potency for the OTR than vasopressin. It is also the most structurally different of the four antagonists with an *O*-ethyl-*D*-tyrosine, an alloisoleucine residue at position 4, an *L*-2-aminobutanoyl group, and a terminal *N*-methyl-*L*-ornithinol group

Carbetocin, 1-deamino-1-monocarbo-(2-*O*-methyltyrosine)-oxytocin, is a long-acting synthetic oxytocin analog and agonist, used in obstetrics to control post-partum hemorrhage and bleeding. Structural differences between the two peptides are replacements of the free amino group with a hydrogen atom at carbon 1, the disulfide bridge of oxytocin with a thioether bridge, and the tyrosine hydroxyl with a methoxy group (Fig. 7.20). Carbetocin binds the OTR with a tenfold lower affinity than oxytocin and although it has a much longer half-life of 85–100 min compared to 3.5 min, its biological effect is only about half that of the natural hormone. Carbetocin (and other oxytocin agonists) appears to bind non-selectively at the extracellular *N*-terminus and loops E2 and E3 of the OTR. The interaction of oxytocin *antagonists* are used for the treatment of preterm labor or to halt premature labor by reducing myometrial contractions. The interaction of these agents with the OTR has, to some extent, also been investigated. The antagonist **atosiban**, a nonapeptide desamino-oxytocin analog with an *O*-methyl-*D*-tyrosine and an ornithine amino acid (Fig. 7.20), suppresses uterine contractions and shows moderate affinity for the OTR but lacks specificity. This peptide can act as a selective ligand, behaving as an antagonist in $G_{\alpha q}$ receptor coupling or an agonist in coupling $G_{\alpha i}$ proteins. **Barusiban** (Fig. 7.20), the most specific OTR antagonist, is also the most structurally different of the four antagonists with an *O*-ethyl-*D*-tyrosine, an alloisoleucine residue at position 4, an *L*-2-aminobutanoyl group, and a terminal *N*-methyl-*L*-ornithinol group. Barusiban shows higher affinity and potency for the OTR than vasopressin. Results from studies with chimeric receptor constructs indicate that transmembrane receptor domains 1 and 2 make direct contact with barusiban by hydrophobic interaction. In simple terms then, oxytocin agonists recognize and contact the amino terminal and extracellular loops of the receptor while oxytocin antagonists contact sites at the bottom of the pocket between transmembrane domains 1 and 2.

Safety of Oxytocin

For induction or stimulation of labor with oxytocin, intravenous infusion, drip method, should be used preferably with an infusion pump. For control of postpartum uterine bleeding, again intravenous infusion is used although intramuscular administration can be given after delivery of the placenta. Warnings and precautions for oxytocin include the possibility of fetal death due to a number of different causes and to maternal death due to rupture of the uterus, hypertensive episodes, and subarachnoid hemorrhage. Water intoxication should also be considered since the hormone has an antidiuretic effect, increasing the absorption of water from the glomerular filtrate. This possibility appears to be more likely when oxytocin is given via continuous infusion and the patient is receiving fluids by mouth. Other precautions to be kept in mind are the chances of increased blood loss and afibrinogenemia upon administration of oxytocin and the possibility of uterine problems including rupture following hypersensitivity responses or excessive dosage. Table 7.4 shows the adverse events induced by oxytocin in the mother and fetus or neonate and listed by the FDA.

Safety of Carbetocin

The oxytocin agonist carbetocin (Pabal®, Duratocin®, Lonactene®) is not registered and available in the USA but is approved for use in the UK, Canada, and Australia. Carbetocin primarily activates oxytocin receptors of the periphery and is used to stop bleeding after birth, particularly following cesarean section. Due to its structural similarity with vasopressin, carbetocin may show some low affinity binding

Table 7.4 Oxytocin adverse events following oxytocin administration[a]

In the mother	In the fetus or neonate
Anaphylaxis	*Due to induced uterine motility*
Postpartum hemorrhage	Bradycardia
Cardiac arrhythmia	Premature ventricular contractions and other arrhythmias
Fatal afibrinogenemia	Permanent CNS or brain damage
Nausea	Fetal death
Vomiting	Neonatal seizures
Premature ventricular contractions	*Due to use of oxytocin in the mother*
Pelvic hematoma	Low Apgar scores[c] at 5 min
Subarachnoid hemorrhage	Neonatal jaundice
Hypertensive episodes	Neonatal retinal hemorrhage
Rupture of the uterus[b]	

[a]Data from FDA: http://www.accessdata.fda.gov/drugsatfda_docs/label/2014/018261s031lbl.pdf
[b]May result from hypersensitivity or excessive dosage
[c]Apgar score (**A**ppearance; **P**ulse; **G**rimace; **A**ctivity; **R**espiration): A method to assess health of the newborn immediately after birth. A score of 7 and above, normal; 4–6, fairly low; 3 and below, critically low

with V_1 receptors in the uterus and, importantly, V_2 receptors in the kidney, meaning that the possibility of hyponatremia cannot be excluded, particularly in patients on intravenous fluids. The drug should therefore be used with caution in patients with conditions such as cardiovascular disease, asthma, and migraine. In clinical trials, intravenous carbetocin was associated with a high incidence (10–40 %) of adverse events, many the same as oxytocin, but since a high incidence of events also occurred in patients treated with placebo, it is likely that many of the events were associated with epidural anesthesia or cesarean section. Adverse effects that were very commonly seen (10 % or more) were headache, tremor, nausea, vomiting, abdominal pain, hypotension, flushing, a feeling of warmth, and pruritus. Other commonly occurring (1–10 %) effects were dizziness, metallic taste, chest pain, dyspnea, anemia, back pain, and chills.

Safety of Atosiban

Atosiban (Tractocile®, Antocin®, Atosiban SUN®), an inhibitor of oxytocin and vasopressin, is approved in the European Union where it is indicated for the delay of imminent pre-term birth in pregnant women. In clinical trials, 48 % of women treated with atosiban experienced adverse reactions that were generally mild. Nausea, the most common adverse effect, occurred in more than 10 % of patients while other adverse events occurring with a fairly high incidence (1–10 %) were headache, dizziness, vomiting, hypotension, hot flushes, hyperglycemia, and injection site reactions. Reactions that were uncommon to rare included insomnia, pyrexia, allergic reaction, uterine hemorrhage, uterine atony, pruritus, and rash. Post-marketing reports have identified respiratory events like dyspnea and pulmonary edema in patients given atosiban concomitantly with tocolytic (anti-contraction) drugs (e.g., calcium antagonists, beta-mimetics) or in women with multiple pregnancy.

Adrenocorticotropic Hormone

Adrenocorticotropic hormone (ACTH; also known as corticotropin) stimulates the secretion of cortisol from the adrenal glands. It is a polypeptide hormone produced and secreted by the anterior pituitary gland (adenohypophysis) in response to CRH released by the hypothalamus. ACTH and CRH are often produced in response to stress. Cortisol, the main secreted corticosteroid from the adrenals, also has a central role in the body's response to stress as well as in glucose metabolism. A deficiency in the production of cortisol leads to a chronic elevation of ACTH, a condition called primary adrenal insufficiency or Addison's disease. The signs and symptoms of prolonged exposure to high levels of cortisol are collectively called Cushing's syndrome.

Structure, Function, and Indications of ACTH

ACTH is generated from pre-proopiomelanocortin (pre-POMC) a 285 amino acid polypeptide precursor. Removal of a 44 amino acid signal peptide during translation produces proopiomelanocortin, POMC, a 241 amino acid polypeptide which undergoes a number of posttranslational changes to produce a range of fragments with different physiological activities. One of these fragments is ACTH, made up of 39 amino acids (MW 4540 Da) representing amino acids 138–178 of the 241 amino acid POMC molecule. ACTH stimulates secretion of corticosteroids after binding to its complementary surface receptor on cells of the adrenal cortex. The ACTH receptor, a seven membrane-spanning GPCR, undergoes a conformational change after binding its ligand. This initiates the stimulation of adenylyl cyclase, a consequent increase in intracellular cAMP and activation of protein kinase A (PKA).

Originally extracted from pituitary glands of mammals, usually pigs, ACTH as an injection was first approved for therapy by the FDA in 1957 but its marketing status is currently designated "discontinued." Since only the first 24 amino acids (from the *N*-terminal end) are required for biologic activity (and these are the same in humans, pigs, cows, and sheep), a synthetic open chain peptide named **cosyntropin** (trade names Synacthen and Cortrosyn) and composed of these 24 amino acids has been synthesized. The sequence of amino acids is:

1 10 20
Ser-Tyr-Ser-Met-Glu-His-Phe-Arg-Trp-Gly-Lys-Pro-Val-Gly-Lys-Lys-Arg-Arg-Pro-Val-Lys-Val-Tyr-Pro

Amino acids 1–20 is the minimum sequence for biological activity; any shortening from 20 residues leads to loss of activity, for example, even a decrease from 20 to 19 amino acids results in a 70 % loss of potency.

Originally approved by the FDA in 2008 but now also with a current marketing status of "discontinued," cosyntropin solution, intravenous, is still used in some countries for the ACTH stimulation test to diagnose or exclude primary and secondary adrenal insufficiency, Addison's disease and related conditions. Plasma cortisol levels are measured before and soon after injection with a subnormal response indicating adrenocortical insufficiency.

Adverse Events to Cosyntropin

Since cosyntropin is used as a diagnostic agent and not administered repeatedly and/or for long periods, adverse reactions to the agent are not generally anticipated. Adverse events recorded by the FDA with the comment that they have been neither confirmed nor refuted are bradycardia, tachycardia, hypertension, peripheral edema, and rash. Other adverse responses to cosyntropin, recorded as well as regarded as potential side effects, include nausea, dizziness, itching, swelling and/or erythema at the injection site, flushing, sweating, fainting, and headache.

With regard to immunogenicity of ACTH, the C-terminal portion (residues 22–39) of the polypeptide is said to be the most antigenic part of the structure both in humans and other mammals. Synthetic polypeptides containing the first 19 amino acids, or fewer, appear to have no detectable immunogenic activity while those containing the first 24 amino acids show little immunogenic activity but full biologic action. Cosyntropin would therefore be expected to exhibit less immunological activity than ACTH and be less of a risk for hypersensitivity reactions. This conclusion is supported by experience with patients who tolerate cosyntropin despite a history of hypersensitivity to ACTH and also by negative intradermal responses to cosyntropin in most patients who are skin test-positive to ACTH. Despite this, it must be assumed that the occurrence of hypersensitivity reactions, including anaphylaxis, remain possible in rare individuals and clinicians should keep this in mind.

Gonadotropin-Releasing Hormone

Gonadotropin-releasing hormone (GnRH; new nomenclature recommends GNRH), also known as luteinizing hormone-releasing hormone (LHRH), is a peptide neurohormone produced by the arcuate nuclei of the hypothalmus. GnRH is of central importance in the initiation of the reproductive cascade stimulating the synthesis and secretion of two gonadotropins, follicle-stimulating hormone (FSH) and luteinizing hormone (LH) (Chap. 8, sections "Follicle-stimulating Hormone," "Luteinizing Hormone"), by the anterior lobe of the pituitary gland. Porcine and ovine GnRH, identical in chemical structure to the natural human hormone and used to evaluate hypothalamic-pituitary function, were first approved for registration by the FDA in 1982 as gonadorelin hydrochloride (Factrel®) and gonadorelin diacetate tetrahydrate (Cystoreline®). These preparations are also used in veterinary medicine, for example, in cattle for cystic ovarian disease.

Structure of GnRH

Highly conserved across the vertebrates, GnRH is a linear decapeptide (pyroGlu-His-Trp-Ser-Tyr-Gly-Leu-Arg-Pro-Gly-NH_2) synthesized in humans as a prohormone, the 92 amino acid prepro-GnRH (Fig. 7.21) which is proteolytically cleaved to generate the free hormone along with a 23 amino acid signal sequence from the N-terminal end and a 56 amino acid GnRH-associated protein (GAP) from the C-terminal portion. In the preproprotein, GnRH is separated from GAP by a three amino acid cleavage site, Gly-Lys-Arg. GAP has been shown to be co-released with GnRH in some species and reported to inhibit prolactin release in rats and decrease intracellular Ca^{2+} levels in human adenoma cells secreting prolactin but conflicting experimental findings indicate that its physiological role has yet to be convincingly demonstrated. After enzymatic cleavage, GnRH is further modified within the secretory granule, although the exact steps and mechanisms are not completely clear. The *N*-terminal glutamine of liberated GnRH may be converted to a pyroglutamate, presumably by a glutaminyl cyclase, or the *N*-terminal of an intermediate may be processed to the pyroglutamate. Catalyzed by an amidating enzyme, the C-terminal glycine may donate its amide group or the terminal basic amino acids may be removed followed by amidation. Note that in addition to the above "classic" isoform of GnRH, often referred to as GnRH-I, a second isoform GnRH-II, which differs by three amino acids at positions 5, 7, and 8 (His^5,Trp^7,Tyr^8-GnRH-I) has been identified and cloned from human and monkey brain. Although GnRH-II differs in its distribution between tissues, it is widely distributed and conserved across the vertebrates implying a possible important physiological function.

GnRH Neurons and Secretion

Produced by neural cells derived from the olfactory placode, GnRH neurons share a common origin with olfactory neurons, a fact reflected by the inability of some patients with delayed puberty to detect odors. GnRH is secreted by the hypophyseal portal system which carries the hormone to the pituitary gland with its access to the

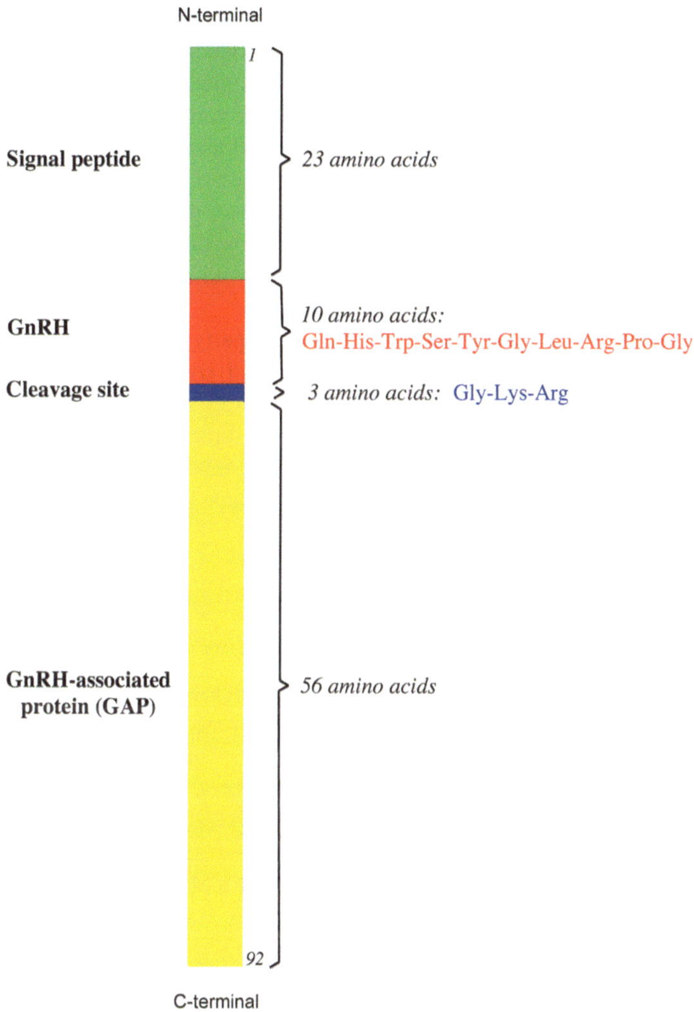

Fig. 7.21 Human GnRH is synthesized as the 92 amino acid prepro-GnRH before being proteolytically cleaved to generate the free hormone along with a 23 amino acid signal sequence from the N-terminal end and a 56 amino acid GnRH-associated protein (GAP) from the C-terminal portion. In the preproprotein, GnRH is separated from GAP by a three amino acid cleavage site, Gly-Lys-Arg. After cleavage, GnRH is further modified by conversion of the *N*-terminal glutamine to a pyroglutamate

gonadotrope cells and leads to the synthesis and secretion of the gonadotropins, FSH and LH. GnRH is absolutely required for reproduction and with its short half-life of only ~2–4 min, its absence results in the complete cessation of the secretion of LH and diminished secretion of FSH. To sustain gonadotropin secretion, secretion of GnRH from the neurons in pulses is required and interference with the generation of pulses, for example, by continuous infusion of GnRH, interferes with the secretion of LH and FSH; the former responds to faster pulse frequencies, the latter, to slower pulse frequencies. Secretion of the hormone differs in males and females. In males, GnRH is secreted at a constant frequency whereas in females, pulse frequency varies with the stage of the menstrual cycle.

Pulse generation is thought to be an intrinsic property of the neurons that contain the neuropeptides, kisspeptin, neurokinin B, and dynorphin (KND neurons). Kisspeptin is a 145 amino acid polypeptide which acts at the level of the hypothalamus stimulating GnRH secretion. It appears to act directly on gonadotrope cells of the pituitary to stimulate the release of LH. Some mutations that inactivate kisspeptin lead to hypogonadism and infertility in humans.

GnRH Receptor

The GnRH receptor (GnRHR) is a member of the rhodopsin-like GPCR superfamily with a hydrophobic domain spanning the membrane seven times, a hydrophilic extracellular domain, and an intracellular domain. The GnRHR signals through $G_{aq/11}$ proteins, activation of phospholipase C, release of IP_3 and DAG, activation of PKC, and mobilization of Ca^{2+}. PKC stimulated MAPK (mitogen-activated protein kinase) and ERK1/2 (extracellular signal-regulated kinase) pathways and Ca^{2+} effectors such as calmodulin ultimately results in stimulation of gonadotropins and their secretion.

GnRH Agonists and Antagonists

GnRH Agonists

The short half-life of GnRH, largely due to the breaking of the Gly-Leu bond between positions 6 and 7, has been utilized by substituting amino acids at this position to synthetically modify the natural decapeptide hormone. Application of this synthetic strategy to position 6 and to positions 9 and 10 with alkylation and deletion modifications while retaining the pyroGlu1 to Tyr5 sequence, has produced peptide analogs with increases in potency of up to 200 times and with extended half-lives seen with analogs such as **nafarelin** (Synarel®) and **triptorelin** (Trelstar®) (position 6), **leuprolide** (leuprorelin, Lupron®) and **goserelin** (Zoladex®). In the syntheses of the commonly used agonists leuprolide, nafarelin, triptorelin, histrelin, goserelin, and **buserelin**, *D*-amino acids or derivatives of *D*-amino acids are substituted for Gly in each case (Table 7.5). Figure 7.22 shows the structural formula of one of the agonists, the often-used decapeptide goserelin composed of pyroGlu-His-Trp-Ser-Tyr-*D*-Ser(tertBu)-Leu-Arg-Pro-AzGly-NH_2. Unusual features of the structure are the tertiary butyl-*D*-serine and an *L*-prolyl semicarbazide. GnRH agonists are used in controlled ovarian hyperstimulation in in vitro fertilization and in female disorders such as endometriosis to induce a hypoestrogenic state; to treat hormone-sensitive prostate and breast cancers; in the treatment of delayed puberty and precocious puberty; to suppress hormone levels in transsexuals, especially women; in cases of congenital adrenal hyperplasia; and in veterinary indications in dogs and livestock. GnRH agonists act by binding to GnRHRs producing an initial transient but intense surge in LH, FSH, testosterone, and dihydrotestosterone release from stored secretory granules ("flare effect") prior to saturation and down-regulation of the GnRHRs with a consequent decrease in hormone levels. In prostate cancer, the eventual inhibition of LH production in turn suppresses testosterone and dihydrotestosterone, necessary for the continued growth of cancer cells. It can take 3–4 weeks to suppress testosterone levels and during this time the testosterone surge may exacerbate symptoms in prostate cancer. Although anti-androgens such as flutamide and bicalutamide are used to limit the unwanted effects of the hormonal surge, they have their own adverse effects including hepato- and pulmonary toxicity, thrombocytopenia, and hot flushes and do not prevent the LH-induced testosterone and dihydrotestosterone surge.

GnRH Antagonists

GnRH antagonists such as **cetrorelix**, **abarelix**, **degarelix**, and **ganirelix** competitively and reversibly bind to GnRH receptors in the pituitary gland blocking the release of LH and FSH. The commonly used antagonists have major *D*- for *L*-amino acid derivative substitutions in the amino terminal domain which is involved in receptor activation and at position 6, important for receptor binding, where *D*-amino acid (or derivatives) are substituted (Table 7.5). The structural formula of cetrorelix is shown as an example of a GnRH antagonist (Fig. 7.22). Five of the ten amino acids of this synthetic decapeptide are in the *D*-isomeric form. The peptide sequence is acetyl-*D*-3-(2-naphthyl)-Ala-*D*-4-chloroPhe-*D*-3-(3-pyridyl)-Ala-Ser-Tyr-*D*-Citrulline-Leu-Arg-Pro-*D*-Ala-NH_2. GnRH antagonists produce quicker chemical castration than GnRH agonists and do so in the absence of testosterone surges and microsurges. As well as the approved indication of GnRH antagonists for advanced hormone-sensitive prostate cancer, the agents may be used prior in in vitro fertilization for the prevention of LH surges and endogenous ovulation, to treat women with hormone-sensitive breast cancer, endometriosis, and uterine fibroids, and in men for benign prostatic hyperplasia.

Table 7.5 Comparison of structures of gonadotropin-releasing hormone (GnRH) and its important agonists and antagonists

Amino acid position	1	2	3	4	5	6	7	8	9	10
GnRH-I[a]	pyroGlu[b]	His	Trp	Ser	Tyr	Gly	Leu	Arg	Pro	Gly-NH_2
GnRH-II[c]	pyroGlu	His	Trp	Ser	His	Gly	Trp	Tyr	Pro	Gly-NH_2
Agonists										
Leuprolide (Lupron®)	pyroGlu	His	Trp	Ser	Tyr	*D*-Leu	Leu	Arg	Proethyl-amide	
Nafarelin (Synarel®)	pyroGlu	His	Trp	Ser	Tyr	naphthyl *D*-Ala[d]	Leu	Arg	Pro	Gly- NH_2
Triptorelin (Trelstar®)	pyroGlu	His	Trp	Ser	Tyr	*D*-Trp	Leu	Arg	Pro	Gly- NH_2
Histrelin (Supprelin®)	pyroGlu	His	Trp	Ser	Tyr	*N*-benzyl -*D*-His	Leu	Arg	Proethyl-amide	
Goserelin (Zoladex®)	pyroGlu	His	Trp	Ser	Tyr	*t*-Bu-*D*-Ser	Leu	Arg	Pro	AzGly- NH_2
Buserelin (Suprefact®)	pyroGlu	His	Trp	Ser	Tyr	*t*-Bu-*D*-Ser	Leu	Arg	Proethyl-amide	
Antagonists										
Cetrorelix (Cetrotide®)	*N*-Ac-naph-*D*-Ala[e]	Cl-*D*-Phe[f]	Py-*D*-Ala[g]	Ser	Tyr	*D*-Cit[h]	Leu	Arg	Pro	*D*-Ala-amide
Abarelix (Plenaxis®)	*N*-Ac-naph-*D*-Ala	Cl-*D* -Phe	Py-*D*-Ala	Ser	*N*-Me -Tyr	*D*-Asp	Leu	*N*-iso-propyl -Lys[i]	Pro	*D*-Ala-amide
Degarelix (Firmagon®)	*N*-Ac-naph-*D*-Ala	Cl-*D* -Phe	Py-*D*-Ala	Ser	hopca -Phe[j]	aca-*D*-Phe[k]	Leu	MeEt -Lys[l]	Pro	*D*-Ala-amide
Ganirelix (Antagon®)	*N*-Ac-naph-*D*-Ala	Cl-*D* -Phe	Py-*D*-Ala	Ser	Tyr	*N,N*-Et_2 -*D*-Har[m]	Leu	*N,N*-Et_2 -*L*-Har[n]	Pro	*D*-Ala-amide

Amino acids are designated by the conventional three-letter abbreviations and are the *L*-form unless otherwise indicated. Amino acid variations at different positions of the human GnRH (isoform I) sequence are highlighted in red. All of the agonists and antagonists are used as the acetate salt. Triptorelin is also used as the pamoate

[a]Pyroglutamate

[b]Decapeptide original isoform isolated from human hypothalamus

[c]Second isoform more recently found in the brain of mammalian species including human. Widely distributed over the vertebrates

[d]3-(2-Naphthyl)-*D*-Ala

[e]Acetyl-*D*-3-(2′-naphthyl)-Ala

[f]*D*-4-Chlorophenyl-Ala

[g]*D*-3-(3′-Pyridyl)-Ala

[h]*D*-Citrulline

[i]*N*(ε)-Isopropyl-Lys

[j]4-[[[(4S)-Hexahydro-2,6-dioxo-4-pyrimidinyl]carbonyl]amino]-Phe

[k]4-[(Aminocarbonyl)amino]-*D*-Phe

[l]*N*6-(1-Methylethyl)-Lys

[m]N^9,N^{10}-Diethyl-*D*-homoarginyl

[n]N^9,N^{10}-Diethyl-*L*-homoarginyl

Safety of GnRH Analogs

Table 7.6 summarizes the approved indications, warnings, and precautions issued by the regulatory agencies and the most common and serious adverse effects of the commonly used GnRH agonists and antagonists. The lists of mild to moderate events for each analog are quite extensive but serious reactions are far less common. In general,

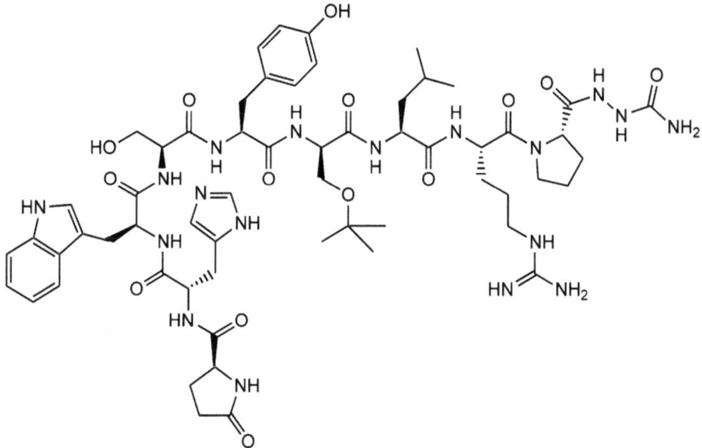

Goserelin

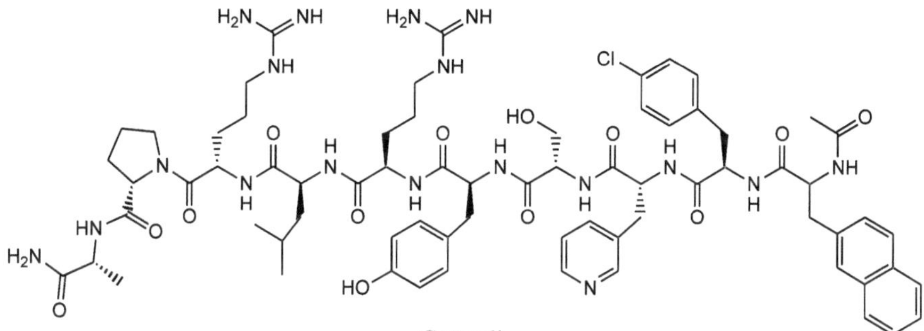

Cetrorelix

Fig. 7.22 Molecular structures of the synthetic decapeptides goserelin, a GnRH agonist, and the GnRH antagonist, cetrorelix

the agonists elicit adverse signs and symptoms of hypoestrogenism manifesting as hot flashes, headache, and osteoporosis. Depression and emotional lability are also common. Common adverse events following GnRH antagonists are hot flashes, headache, weight gain, abdominal pain, and ovarian hyperstimulation. For both groups, the risk of severe hypersensitivity with anaphylaxis is rare but real, especially with abarelix, the only GnRH analog with an FDA-issued box warning (Table 7.6).

Parathyroid Hormone

Parathyroid hormone is a key regulator of calcium levels in the blood with an important role in mineral homeostasis and bone maintenance. It is secreted by the principal cells of the parathyroid glands, cells that regulate intracellular and serum

Table 7.6 Indications and safety information for GnRH agonists and antagonists [a]

Generic and trade name	Indications[a]	Warnings and precautions	Adverse effects
Agonists			
Leuprolide[b] (Lupron®)	Injection[b]: Advanced prostatic cancer Depot[b]: Endometriosis; uterine leiomyomata (fibroids)[c]	Pregnancy; anaphylactoid symptoms. For norethindrone acetate co-treatment: vision disorders[d], pulmonary embolism, thrombophlebitis, risk of CV disease	Headache; hot flashes/sweats; asthenia; body pain; vaginitis; nausea/vomiting; depression, weight change; dizziness; insomnia/sleep disorders; libido changes; anaphylaxis; rash; urticaria; hypotension; peripheral neuropathy[e]
Nafarelin (Synarel®)	Endometriosis; central precocious puberty	Central precocious puberty: importance of diagnosis before treatment[f]; ovarian cysts in women with endometriosis	Acne; transient breast enlargement; vaginal bleeding; emotional lability; transient ↑pubic hair; seborrhea; hot flashes; pituitary apoplexy[g]
Triptorelin (Trelstar®)	Advanced prostatic cancer[h]	Transient ↑testosterone; bone pain; spinal compression; neuropathy; hypersensitivity; hematuria; urethral obstruction	Hot flashes; body pain; fatigue; hypertension; headache; diarrhea/vomiting; skeletal pain; insomnia; impotence; emotional lability; anemia; pruritus; urinary retention/infection
Histrelin (Supprelin® ; Vantas®)	Central precocious puberty; advanced prostatic cancer	Transient ↑estradiol/ testosterone; urethral obstruction; spinal compression; implant insertion/removal procedure; hyperglycemia and diabetes; CV disease	Hot flashes; fatigue; implant site reactions; testicular atrophy; erectile dysfunction; renal impairment; gynecomastia; ↓libido; insomnia; constipation; headache; depression, arthralgia
Goserelin (Zoladex®)	With flutamide for confined prostate cancer; advanced prostatic cancer	Tumor flare[i] ; hypersensitivity; hyperglycemia	Hot flashes; sexual dysfunction; urinary tract symptoms; general, pelvic, and bone pain; gynecomastia; tumor flare; asthenia; osteoporosis; hypersensitivity; hypercalcemia; rash; itch; acne; seborrhea
Buserelin (Suprefact®)	Endometriosis not requiring surgery; pituitary desensitization for ovulation induction; advanced prostatic cancer[j]	Monitor BP, glucose, depression, bone density in risk patients; ovarian hyperstimulation syndrome[k]; tumor flare at start of treatment	Hypersensitivity (anaphylaxis, urticaria); bone pain; tumor flare; tumor compression; sleep disorders; headache; thrombosis; depression; dizziness; GI disorders; osteoporosis; ↓libido; gynecomastia; skeletal pain; acne; alopecia

(continued)

Table 7.6 (continued)

Generic and trade name	Indications[a]	Warnings and precautions	Adverse effects
Antagonists			
Cetrorelix (Cetrotide®)	Inhibition of premature LH surges[l]	Before treatment exclude pregnancy; exclude women with severe allergies	Ovarian hyperstimulation syndrome; injection site reactions; hypersensitivity; headache; nausea
Abarelix (Plenaxis®)	Advanced symptomatic prostate cancer[m]; low flow priapism[n]	Boxed Warnings: Allergic reactions[o]; prescriber limits[p]; prostate cancer[q]; ↓effectiveness to suppress testosterone[r]. Other: QT interval prolongation[s]	Systemic allergic reactions; hot flashes; sleep disorders; breast enlargement/discomfort; pain; peripheral edema; headache; constipation; urti; dizziness; urinary retention/infection; fatigue
Degarelix (Firmagon®)	Advanced prostatic cancer	Hypersensitivity: anaphylaxis, angioedema, urticaria. QT interval prolongation; before treatment exclude pregnancy	Injection site reactions; hot flashes; back pain; weight gain;↑liver transaminases, GGT; chills; hypertension; arthralgia; urinary tract infection; erectile dysfunction; gynecomastia; testicular atrophy; hyperhidrosis;↓bone density; insomnia
Ganirelix (Antagon®)	Inhibition of premature LH surges[l]	Before treatment exclude pregnancy; exclude women with severe allergies	Abdominal pain (gynecological); fetal death; headache; ovarian hyperstimulation syndrome; vaginal bleeding; injection site reactions; nausea; abdominal pain (GI)

BP blood pressure, *CV* cardiovascular, *GGT* gammaglutamyltransferase, *GI* gastrointestinal, *LH* luteinizing hormone, *urti* upper respiratory tract infection

[a]Approved by FDA or EMA or both

[b]Available as depot preparation and acetate for subcutaneous injection

[c]Concomitantly with iron therapy for preoperative hematologic improvement of patients with anemia caused by uterine leiomyomata (fibroids)

[d]Proptosis, diplopia, migraine; withdraw drug if papilledema or retinal vascular lesions occur

[e]Rare events: thromboembolism, myocardial infarction, stroke, transient ischemic attack, pituitary apoplexy; convulsions; tenosynovitis-like symptoms, liver injury

[f]Especially in first 6–8 weeks to assure suppression of pituitary-gonadal function is rapid

[g]Rare. Most patients had pituitary adenoma. Presents as sudden headache, vomiting, visual changes, altered mental status, and sometimes CV collapse

[h]When orchiectomy or estrogen administration are either not indicated or acceptable to the patient

[i]May include urethral obstruction and spinal cord compression

[j]Not after bilateral orchiectomy

[k]May be higher risk combined with gonadotrophins. Withhold hCG if necessary

[l]In women undergoing controlled ovarian stimulation

[m]In whom GnRH agonist therapy is not appropriate and who refuse surgical castration and have one or more of (1) risk of neurological metastasis, (2) urethral or bladder outlet obstruction due to metastatic disease, (3) severe bone pain from skeletal metastases treated with opioid analgesics

[n]Orphan designation granted by EMA September 2010 for low flow priapism

(continued)

Table 7.6 (continued)

[o]Reactions may occur after initial dose
[p]Only physicians enrolled in the Plenaxis® Plus (Plenaxus user safety program) may prescribe the drug
[q]Indicated for palliative treatment of advanced prostate cancer (see Indications)
[r]Effectiveness of Plenaxis® to suppress testosterone decreases with continued dosage
[s]Especially in patients taking class IA (e.g., quinidine, procainamide) and class III (e.g., amiodarone, sotalol) antiarrhythmic medications

N-terminal

Ser–Val–**Ser–Glu–Ile–Gln–Leu–Met–His–Asn–Leu–Gly–Lys–His–Leu–Asn–Ser–**
1 *5* *10* *15*

Met–Glu–Arg–Val–Glu–Trp–Leu–Arg–Lys–Lys–Leu–Gln–Asp–Val–His–Asn–Phe
20 *25* *30* *C-terminal*

FORTEO [rhPTH(1-34)]

Fig. 7.23 Recombinant 34 amino acid peptide rhPTH(1–34), or teriparatide (Forteo®), the biologically active region of human parathyroid hormone, has an identical sequence to the 34 *N*-terminal amino acids of the mature 84 amino acid hormone. Sequence homology between the two peptides is restricted to 8 of the first 13 amino acids at the amino terminus with amino acids 1 and 2 (in *green*) being critical for activation of adenylate cyclase

changes in calcium concentrations. Parathyroid hormone acts primarily in kidney and bone tissues. The hormone promotes tubule resorption of calcium (but not phosphate) and synthesis of 1,25-dihydroxyvitamin D_3 in kidney tissues while increasing osteoclastic bone resorption and stimulating osteoblasts in bone tissues. Parathyroid hormone is thus essential for calcium homeostasis and increase in bone mass. This has led to the intermittent use of the hormone in the clinic as a treatment of osteoporosis and provided a stimulus for further research into its role in bone and calcium homeostasis.

Structure and Action of Human Parathyroid Hormone

Human parathyroid hormone is synthesized as a 115 amino acid precursor, preproparathyroid hormone, that is cleaved firstly to form the 90 amino acid proparathyroid hormone, and finally the mature 84 amino acid hormone of molecular mass 9425 Da. The 34 N-terminal amino acids (PTH(1–34); molecular mass 4117.8 Da) comprise the biologically active region of the mature hormone (Fig. 7.23). Known as **teriparatide**, this fragment has been manufactured by recombinant DNA technology, approved for therapy by the FDA, and marketed as Forteo®. Both the full length 84 amino acid hormone (FDA approved and marketed as Natpara®) and recombinant fragment rhPTH(1–34) are used in the clinic to treat osteoporosis. In addition to these biologically active full length and 34 amino acid *N*-terminal forms, a third biologically active peptide, parathyroid hormone-related peptide (PTHrP), was identified as the cause of a malignancy-associated hypercalcemia syndrome. This peptide, which can occur as isoforms of 139, 141, or 173 amino acids, has been shown to have an important role in bone formation during embryogenesis. Mature parathyroid hormone, PTH(1–34), and synthetic PTHrP(1–34) show the same receptor binding properties and affinity (see below) and, analogous with PTH(1–34), the bioactivity of PTHrP appears to reside in its *N*-terminal domain. Sequence homology between the three peptides (PTH(1–34), rhPTH(1–34), and PTHrP(1–34)) is restricted to 8 of the first 13 amino acids at the amino terminus with amino acids 1 and 2 being critical for activation of adenylate cyclase.

Parathyroid hormone has been described as a "double-edged sword" for bone metabolism with the pattern of exposure determining the balance between bone formation and bone loss. The anabolic or bone increase effects result from brief or intermittent exposures whereas catabolic effects are seen in pathological conditions of high levels or excess hormone. This occurs, for example, in chronic renal disease and hyperparathyroidism where the glands continuously secrete too much hormone. It is now accepted that parathyroid hormone shows pleiotropic effects; intermittent administration can be beneficial to patients with severe osteoporosis but long-term use of the hormone can be detrimental and is not yet fully understood.

Parathyroid Hormone Receptors and Signaling

The effects of parathyroid hormone are mediated primarily via binding to, and activation of, its complementary receptor PTH1R which is expressed in key target organs such as kidney and bone and cells such as osteoblasts. The hormone also binds to a second receptor PTH2R, expressed primarily in the central nervous system, pancreas, placenta, and testes. Both receptors belong to the class B GPCRs. PTH1R also recognizes, and is activated by, PTH(1–34) and PTHrP. Residues 1–14 of the *N*-terminal fragment of parathyroid hormone bind with low affinity to the transmembrane domain of the receptor activating it while the C-terminal fragment (residues 15–34) binds to the receptor's *N*-terminal extracellular domain. This two-domain recognition function confers both high affinity and specificity to the receptor. Binding of parathyroid hormone to PTH1R in the bone-forming osteoblasts activates two signal transduction pathways, the protein kinase A (PKA) and PKC pathways. In the former case, G_{as} stimulates the production of cAMP and activates PKA which phosphorylates transcription factors such as cAMP-response element-binding protein (CREB) and RUNX2 (runt-related transcription factor 2; also called core-binding factor subunit alfa-1[CBF-alfa-1]). RUNX2 is a key transcription factor associated with osteoblast differentiation. The bone anabolic effects of parathyroid hormone are thought to be mediated via the cAMP-PKA pathway. In the PKC pathway, G_{aq} activates phospholipase C leading to the formation of DAG, the activation of PKC, formation of IP_3, and an increase in intracellular Ca^{2+}.

Table 7.7 Safety of recombinant parathyroid hormone preparations[a]

Generic and trade names[a]	Description	Approved indications[a]	Warnings and precautions	Adverse effects
Parathyroid hormone (Natpara®; Preotact®; Preos®)	Recombinant parathyroid hormone produced in *E. coli*; 84 amino acids, MW 9424 Da	An adjunct to calcium and vitamin D therapy to control hypocalcemia in patients with hypoparathyroidism[b]	*Boxed warnings*: Potential risk of osteosarcoma[c], Other warnings: Severe hypercalcemia[d]; severe hypocalcemia[e]; digoxin toxicity[f]; available only through restricted Natpara REMS Program[g]	Clinical trials (in >10 % patients): Paresthesia; hypocalcemia; hypercalcemia; headache; nausea; hypoesthesia; hypercalciuria; vomiting; diarrhea; arthralgia
Teriparatide (Forteo®; Forsteo®)	Recombinant parathyroid hormone analog (1–34), rhPTH(1–34) produced in *E. coli*	Osteoporosis in menopausal women; primary or hypogonadal osteoporosis in men; and osteoporosis associated with sustained systemic glucocorticoid therapy in men and women at high risk of fracture	*Boxed warnings*: Potential risk of osteosarcoma[c], Other warnings: Not to be used for total of >2 years or for patients with bone metastases/ malignancies or hypercalcemia; may exacerbate urolithiasis; initial doses may produce orthostatic hypotension; may increase serum and urinary Ca^{2+} and serum uric acid	Clinical trials (in >2 % patients): Arthralgia; pain; nausea[h] Post-marketing: Osteosarcoma (rarely); hypercalcemia; allergic reactions; hyperuricemia; dyspnea, chest pain; leg/ back muscle spasms

[a]Approved by the FDA or EMA or both
[b]Limitations of use: because of potential risk of osteosarcoma, only patients not well controlled on calcium and vitamin D should be treated; hormone has not been studied in patients with hypoparathyroidism caused by calcium-sensing receptor mutations or in patients with acute post-surgical hypoparathyroidism
[c]Dose- and treatment length-dependent increases in osteosarcoma seen in rats. Should be prescribed only to patients not well controlled on calcium and vitamin D and/or when benefits outweigh the risks. Avoid use in patients with elevated baseline risk of osteosarcoma (e.g., Paget's disease of bone)
[d]Monitor serum calcium when starting or adjusting dose
[e]May occur when interrupting or discontinuing treatment; monitor levels
[f]Hypercalcemia increases risk of digoxin toxicity; monitor levels
[g]Preparation only available through Natpara REMS (risk evaluation and mitigation strategy) Program. Only certified providers can prescribe and pharmacies dispense
[h]Other adverse events with a higher incidence in the Forteo group than the control group were gastritis, dyspnea, pneumonia, insomnia, anxiety, herpes zoster

Safety of Parathyroid Hormone

Table 7.7 shows the warnings and precautions issued by the regulatory agencies for the approved recombinant full length and 34 amino acid *N*-terminal fragment of parathyroid hormone together with the main adverse effects recorded so far for these preparations. Note that in the clinical trials conducted so far, the overall number of patients participating was small and the majority were female. Differences in the frequency of side effects between the sexes have not been seen. This is also true in a comparison of patients above and below 65 years of age, however, only a limited number of patients over 65 have so far been examined. Although an FDA black box warning has been issued for the possibility of the development of bone cancer following parathyroid hormone, the warning is based on the finding of osteosarcoma in some rats; it is not currently known if these cancers are more likely to occur in patients taking the hormone. Consequently, to make known and closely follow this potential risk, the approved preparations are only made available through a Risk Evaluation and Mitigation Strategy (REMS) Program. Possible recorded serious side effects are hypercalcemia and hypocalcemia. Common adverse effects include paresthesia, headache, nausea, diarrhea, vomiting, and hypoesthesia (Table 7.7). Because parathyroid hormone may increase serum calcium levels and the inotropic effects of some drugs are affected by calcium levels, concomitant use of digitalis compounds such as digoxin may result in digoxin toxicity. Patients taking both drugs should therefore be monitored for serum calcium and digoxin levels with dosage adjustments if necessary.

Antibodies may be induced by parathyroid hormone therapy. The incidence of such antibodies in a small placebo-controlled study of patients with hypoparathyroidism was 8.6 % (3 of 35) following subcutaneous administration for 24 weeks. Treatment of patients with hypoparathyroidism with full length recombinant hormone for up to 2.6 years revealed low titer antibodies in 16.1 % (14/87) of immunized patients, only one of which had neutralizing antibodies. The antibodies did not seem to affect efficacy or safety of the hormone therapy. In clinical trials with teriparatide, antibodies to the parathyroid hormone analog were induced in 2.8 % of women (15/541). Antibodies generally appeared after 12 months, diminished in the absence of treatment, no adverse effects on serum calcium or bone density levels were noted, and there was no evidence of hypersensitivity reactions in the patients.

Intermittent administration of parathyroid hormone is an established osteoanabolic therapy, increasing bone formation and promoting bone vasculature. However, the long-term effects of the hormone therapy have not yet been determined and remain a concern. Parathyroid hormone receptors are found in the cardiovascular system including cardiomyocytes and blood vessel endothelial and smooth muscle cells, suggesting a function for the hormone beyond its established role in bone and mineral metabolism. A number of observations suggest that parathyroid hormone may also be implicated in the pathophysiology of cardiovascular disease. Firstly, hyperparathyroidism is associated with a range of cardiovascular disorders including hypertension, coronary microvascular dysfunction, and aortic valve calcification as well as a high risk of cardiovascular mortality. Hyperparathyroidism is also seen in cases of chronic kidney disease, causing bone disorders and, again, it is associated with an increased risk of cardiovascular mortality. In fact, even in the absence of hyperparathyroidism, parathyroid hormone is being increasingly recognized as a risk factor with predictive value for cardiovascular disease and mortality. In a recent study in Sweden of two community-based cohorts, parathyroid hormone was shown to be associated with subclinical and clinical atherosclerosis and the data supported a role for the hormone in the development of the disease. The findings are consistent with previous published conclusions on the association of parathyroid hormone with the severity of coronary artery disease and implication of elevated levels of the hormone as a cardiovascular risk factor in moderate chronic kidney disease.

It seems increasingly likely that as well as its role as a novel mediator of bone-renal interactions and its importance in bone disease, parathyroid hormone might also mediate vascular interactions in some cardiovascular disorders. This may lead to new therapeutic insights and preventive measures for disorders such as atherosclerosis, vascular and valvular calcification, cardiac hypertrophy, and hypertension but any new indication(s) for parathyroid hormone will no doubt reveal its own associated dose- and treatment length-dependent spectrum of adverse events.

Summary

- Hormones, synthesized and released in response to often quite specific biochemical signals, are signaling molecules or messengers, communicating between organs and tissues via specific receptors to regulate a wide range of physiological and behavioral activities in the human body. With the wide variety of hormone therapies employed, side effects are collectively numerous in number, wide in range and sometimes unique but important mechanistic insights have often been obtained in seeking to understand some adverse events.
- Peptide hormones may be small peptides such as oxytocin and vasopressin, each of which have only 9 amino acids, or somewhat larger, such as insulin with 51 and growth hormone with 191 amino acids. Peptide hormones are often prepared by cleavage of larger inactive precursors or prohormones before being released for action. In addition to insulin, the first protein to be sequenced and the first to be synthesized by recombinant DNA technology, other peptide hormones approved for use as important therapies include glucagon, glucagon-like protein-1, growth hormone, somatostatin, vasopressin, oxytocin, ACTH, GnRH, and parathyroid hormone.
- Insulin is a peptide hormone essential for the regulation of the metabolism of carbohydrates and fats. Type 1 diabetes mellitus, seen in ~10 % of diabetes cases, is caused by insulin deficiency due to loss of the insulin-producing beta cells in the islets of Langerhans in the pancreas. Type 2 diabetes, the most common type of diabetes, is characterized by resistance to insulin and sometimes by reduced secretion of insulin. Unlike type 1 disease, type 2 diabetes is clearly related to lifestyle including obesity, poor diet, physical inactivity, and stress. There also appears to be a clear genetic component.

- Insulin is produced in the pancreas as preproinsulin with a 24 amino acid signal sequence subsequently removed to form proinsulin. Proinsulin in turn is cleaved to produce mature, active insulin. Insulin is a protein of 51 amino acids (MW 5808 Da), a dimer of an A chain of 21 amino acids and a B chain of 30 amino acids linked together by two disulfide bonds. A third disulfide bridge is internal, linking two cysteines in the A chain.
- At micromolar concentrations, the two-chain disulfide-linked insulin structure associates with another insulin molecule forming a dimer and, in the presence of zinc ions, the dimer can associate with two more insulin dimers to form a hexamer. The hexamer has two zinc ions at its center coordinately bound to a histidine residue from each monomer and all six monomers are in a T conformation.
- The insulin receptor is a ~320 kDa disulfide-linked transmembrane structure belonging to the tyrosine kinase group of receptors. The adapter protein, insulin receptor substrate 1 (IRS-1), has a key role in signaling to PI3K/Akt and Erk MAP intracellular pathways.
- Insulin preparations are generally classified as rapid-, short-, intermediate- and long-acting and premixed preparations containing different proportions of the first three of these are often used.
- Use of modern pumps, pen devices, and continuous glucose monitoring has created the means of achieving a situation close to physiological insulin secretion, especially with the fast-acting insulin analogs such as lispro and glulisine that provide good glycemic control compared to the older regular insulin preparations and neutral protamine hagedorn. The fast-acting insulins are administered before each meal to manage the glucose surge. Recently, approval was granted for a rapid-acting insulin powder, Technosphere insulin (Afrezza®), administered as an inhalation in an individualized dose at the beginning of meals.
- Sodium glucose co-transporter 2 (SGLT2) inhibitors, for example, dapagliflozin used to treat type 2 diabetes, are a new class of apparently well-tolerated antidiabetic agents that enhance glucose excretion via the urine by reducing tubular glucose reabsorption and producing urinary excretion of the sugar.
- The worst of the insulin-induced adverse events are hypoglycemia, hyperglycemia and diabetic coma, and hypersensitivity reactions (types I, III, and IV).
- Glucagon, a peptide hormone produced and stored in pancreatic alfa cells of the islets of Langerhans, is involved in glucose homeostasis. Glucagon secretion is coupled to levels of circulating glucose, its release being stimulated in hypoglycemia and inhibited in hyperglycemia. With a half-life of only a few minutes, it acts in the liver where it stimulates the breakdown of glycogen to glucose. Together with insulin, glucagon forms a feedback system controlling blood glucose levels.
- Recombinant glucagon is a 29 amino acid, single chain polypeptide identical to the human hormone. Glucagon acts via its receptor in the plasma membrane of pancreatic islet cells, liver, kidney and brain. The signal pathways activated by glucagon lead to an increase in insulin secretion.
- Glucagon is approved for the treatment of hypoglycemia. Another important, and useful, activity of glucagon is its capacity to relax smooth muscle when

given parenterally, a property made use of when the hormone is employed as an inhibitor of gastrointestinal motility in some radiologic diagnostic examinations of the gastrointestinal tract.

- Exogenous glucagon stimulates the release of catecholamines that may lead to a sudden increase in blood pressure if given to patients with pheochromocytoma, a tumor of the adrenal gland medulla cells.
- Glucagon, added to barium sulfate suspensions in double-contrast radiologic procedures, occasionally provokes reactions that appear to be true hypersensitivities, namely, skin rashes, urticaria, periorbital edema, erythema multiforme, breathing difficulties, and anaphylaxis.
- The "incretin effect" describes the observation that glucose delivered orally, produces a two- to threefold larger insulin response than the secretory response seen following intravenous administration. Gastric inhibitory polypeptide (GIP) and the 31 amino acid peptide glucagon-like peptide-1 (GLP-1) are the two major incretins. GLP-1 secretion is meal-related and following the ingestion of a meal, it and GIP enhance insulin secretion to an extent that explains both the insulin response and the incretin effect. Although the effects of the two hormones on insulin secretion are additive, only GLP-1 inhibits glucagon secretion. The incretin effect has a prime role in the post-prandial secretion of insulin and glucose tolerance.
- GLP-1's insulin-related properties suggest that the peptide should be a valuable treatment for patients with type 2 diabetes mellitus. However, therapeutic applications of GLP-1 are not practical since the peptide is rapidly degraded (its half-life in plasma is only ~1–2 min) by the enzyme dipeptidyl peptidase-4 (DPP-4).
- Endeavors seeking efficacious GLP-1 receptor agonists with extended half-lives have resulted in regulatory approvals for GLP-1 agonists exenatide, liraglutide, lixisenatide, and albiglutide. All show some sequence identity with GLP-1 and are resistant to DPP-4.
- The GLP-1 receptor agonists carry an FDA black box warning drawing attention to the finding of thyroid C-cell tumors in rats, and the agonists are contraindicated in patients with medullary thyroid cancer (MTC), or a family history of MTC, or in patients with multiple endocrine neoplasia syndrome type 2 (MEN2). Other warnings and precautions are for pancreatitis, hypoglycemia, renal impairment, and hypersensitivity. Adverse events are generally mild and not severe.
- Inhibitors of dipeptidyl peptidase-4 (DPP-4), or gliptins, form a class of orally active hypoglycemics with relatively modest glucose-lowering activity that can be used to treat diabetes mellitus type 2. Warnings and precautions for these drugs include the possibility of acute pancreatitis, hypoglycemia, and hypersensitivity reactions.
- Pramlintide is a synthetic analog of human amylin, a neuroendocrine hormone synthesized by beta cells of the pancreas and cosecreted with insulin in the ratio of approximately 100:1 in response to food intake thus contributing to glucose control in the postprandial period.
- Pramlintide with a short half-life of ~48 min slows gastric emptying, reduces postprandial rises in plasma glucagon, and decreases the intake of calories by reducing appetite.

- Pramlintide alone does not cause hypoglycemia but its addition to insulin increases the risk of hypoglycemia and this is reflected in an FDA black box warning stating that its use with insulin has been associated with an increased risk of severe hypoglycemia, particularly in patients with type 1 diabetes.
- Human growth hormone is a single chain of 191 amino acid residues with two disulfide bonds between Cys53–Cys165 and Cys182–Cys189 and molecular mass of 22.125 kDa. The recombinant hormone (somatropin) has the same structure as the natural protein.
- Approved indications for growth hormone: Children—short stature or growth failure associated with growth hormone deficiency; idiopathic short stature; SHOX deficiency and failure to catch up height after being small for gestational age birth; Turner syndrome; and Prader–Willi syndrome. Adults—patients with either childhood- or adult-onset growth hormone deficiency.
- Common adverse reactions to somatropin preparations listed by the FDA are injection site reactions and lipoatrophy, fluid retention, peripheral edema, arthralgia, myalgia, headache, carpal tunnel syndrome/paresthesia, and hyperglycemia. Growth hormone stimulates the breakdown of glucose and lipids, antagonizing the effect of insulin in glucose and lipid metabolism.
- In relation to long-term cancer-induced mortality, an extended epidemiological study of patients treated with recombinant growth hormone during childhood revealed a 30 % increased risk of death in the hormone-treated group versus expected deaths in the general population.
- The growth hormone receptor antagonist pegvisomant, a pegylated, mutated, recombinant form of human growth hormone is indicated for the treatment of acromegaly in patients who did not benefit sufficiently from surgery and/or radiation therapy or other therapies such as somatostatin analogs. Issued precautions relate to the possible expansion of tumors that secrete growth hormone; an increase in glucose tolerance; a state of functional growth hormone deficiency; an elevation of liver enzymes; and lipohypertrophy.
- Insulin-like growth factor 1 (IGF-1), secreted by the liver in response to stimulation by growth hormone, is a growth hormone with a growth stimulating effect independent of growth hormone. IGF-1 mediates the growth-promoting effect of growth hormone, in particular, its anabolic and mitogenic activities. IGF-1 and insulin share a homologous sequence and show some overlapping biological activity.
- IGF-1 is a single polypeptide chain of 70 amino acids with three intramolecular disulfide bridges. In plasma, almost 100 % of IGF-1 is protein-bound to IGF binding proteins (IGFBPs) with ~80 % bound to one of the six IGFBPs, IGFBP-3. It circulates as a 140 kDa ternary complex of one molecule of IGF-1, one molecule of IGFBP-3, and one molecule of an acid-labile protein, MW 88 kDa. The effects of IGF-1 are mediated via the hormone's complementary receptor IGF-1R and modulated by interactions with IGFBPs.
- The cause of Laron syndrome, with its pronounced growth retardation, was shown to be insensitive to growth hormone. Laron syndrome results in failure to generate IGF-1. Recombinant IGF-1(rhIGF-1; mecasermin) has been approved

for severe primary IGF-1 deficiency, specifically for the treatment of growth failure in children with this deficiency. A second approved indication is for the treatment of patients with a growth hormone gene deletion who have developed neutralizing antibodies to the hormone.

- Hypoglycemia is the most commonly occurring adverse effect induced by mecasermin. Other adverse events include the rare occurrence of allergic reactions; intracranial hypertension with papilledema, visual changes, headache, and nausea; lymphoid tissue hypertrophy; and slipped capital femoral epiphysis.
- Somatostatin, also known as growth hormone inhibiting hormone (GHIH), inhibits the secretion of growth hormone from the pituitary as well as a number of other hormones including ACTH and the secretion of insulin and glucagon by the pancreas. Somatostatin also inhibits the release of a range of different gastroenteropancreatic neuroendocrine tumors (GEP-NETs).
- Physiologically somatostatin is secreted in two active forms of 14 and 28 amino acids. Both forms occur in tissues and each interact with the somatostatin receptor.
- Although somatostatin has been used clinically to treat acute esophageal variceal bleeding in patients with cirrhotic liver disease, it has a short half-life in the circulation of only about 1–3 min. Synthetic analogs, all containing fewer than 14 amino acids, with longer half-lives have therefore been developed and are used to treat acromegaly, carcinoid and vasoactive intestinal tumors and Cushing's disease.
- Native somatostatin and its synthetic analogs demonstrate different affinities for the five somatostatin receptor subtypes. Whereas the natural hormone recognizes all five subtypes, the analogs octreotide and lanreotide interact with receptor subtypes 2 and 5 and pasireotide, approved for the treatment of Cushing's disease, binds subtypes 1, 2, 3, and 5 but not 4.
- Somatostatin analogs are reasonably safe drugs but for both octreotide and lanreotide, warnings and precautions have been issued for cholethiasis and gallbladder sludge; hyperglycemia and hypoglycemia; thyroid function abnormalities; and cardiovascular abnormalities. Marked suppression of ACTH by pasireotide may lead to the depression of circulating levels of cortisol and transient hypocortisolism and hypoadrenalism with accompanying weakness, fatigue, nausea, vomiting, hypotension, hyperkalemia, and hypoglycemia.
- Vasopressin is a nonapeptide with a disulfide bridge between cysteines at positions 1 and 6. Lysine vasopressin replaces the arginine at position 8 with a lysine residue. The physiologic actions of vasopressin are mediated by three vasopressin receptor subtypes, V_1, V_2, and V_3.
- Vasopressin, also known as arginine vasopressin, has two main actions: regulation of water retention and the constriction of blood vessels. Water retention is effected by reabsorption of water in the collecting ducts of the kidney nephron and distal convoluted tubule. By constricting blood vessels, vasopressin increases peripheral vascular resistance and, in turn, arterial blood pressure.

- Vasopressin intravenous injection was approved in 2014 by the FDA to increase blood pressure in adults with vasodilatory shock who remain hypotensive despite fluids and catecholamines.
- Desmopressin, a synthetic analog of vasopressin differing from vasopressin by deamination of the cysteine at position 1 and by replacement of *L*- with *D*-arginine at position 8, is approved for use as a nasal spray for antidiabetic replacement therapy in the management of central cranial diabetes insipidis and temporary polyuria and polidipsia following head trauma or surgery in the pituitary region. Desmopressin acetate nasal spray is also indicated for hemophilia A and for patients with mild to moderate von Willebrand's disease (Type I). Warnings and precautions relate to the possibilities of slight blood pressure and heart rate changes in patients with cardiovascular disorders; hyponatremia in patients with fluid and electrolyte imbalances; and severe allergic reactions.
- Terlipressin, a dodecapeptide, is a prodrug with three glycyl residues attached to the *N*-terminal of lysine vasopressin. Enzymic removal of the glycyls in vivo produces the biologically active lysine vasopressin. Its indications for use are esophageal varices, norepinephrine-resistant septic shock, and hepatorenal syndrome. Acting via the V_1 receptor, terlipressin increases mean arterial pressure and a reduction in heart rate and should not be given (or given under strict monitoring) to patients with unstable angina and a number of other cardiovascular conditions. It is contraindicated in pregnancy since it causes uterine contractions, increased uterine pressure, and may decrease uterine blood flow. Cardiac and vascular disorders, e.g., bradycardia, peripheral vasoconstriction, peripheral ischemia, and hypertension, are the most common and serious adverse effects.
- Vasopressin receptor antagonists, or vaptans, are important in the treatment of dilutional hyponatremia, promoting excretion of electrolyte-free water by blocking binding of arginine vasopressin to its renal receptor.
- Two vaptans have been approved for clinical use. Conivaptan, a combined V_1/V_2 receptor antagonist, is approved for use in raising serum sodium levels in hospitalized patients with euvolemic and hypervolemic hyponatremia. An infusion site reaction is a commonly associated adverse event. Tolvaptan, a selective V_2 receptor antagonist, is approved for the treatment of euvolemic and hypervolemic hyponatremia including patients with heart failure, cirrhosis, and syndrome of inappropriate antidiuretic hormone secretion (SIADH). The drug carries an FDA black box warning advising that treatment should be initiated in hospital.
- Oxytocin injection may be administered for antepartum and postpartum use. For the former purposes, the hormone is indicated for the initiation or improvement of uterine contractions to achieve vaginal delivery. In the postpartum situation, oxytocin is given to produce uterine contractions during the third stage of labor and to control postpartum bleeding or hemorrhage.
- Like all neurohypophysial hormones, oxytocin is a nonapeptide with a disulfide bridge linking the cysteine residues at positions 1 and 6. Whereas nonapeptide hormones of the vasopressin family all have a basic amino acid (lysine or arginine) at position 8, oxytocin has a neutral amino acid, leucine. For recognition and activation of receptors, oxytocin requires an isoleucine at position 3.

- Although the vasopressin V_2 receptor shows a 40 % sequence identity with the oxytocin receptor, the latter binds oxytocin and arginine vasopressin with similar high affinity whereas the V_2 receptor binds arginine vasopressin with ~400-fold higher affinity than oxytocin.
- Warnings and precautions for oxytocin include the possibility of fetal deaths and maternal deaths due to rupture of the uterus, hypertensive episodes, and subarachnoid hemorrhage. Water intoxication should also be considered since the hormone has an antidiuretic effect, increasing the absorption of water from the glomerular filtrate.
- The oxytocin agonist carbetocin primarily activates oxytocin receptors of the periphery and is used to stop bleeding after birth particularly following cesarean section. Carbetocin may show some low affinity binding with V_2 receptors in the kidney meaning the possibility of hyponatremia cannot be excluded, particularly in patients on intravenous fluids
- Atosiban, an inhibitor of oxytocin and vasopressin, is indicated for the delay of imminent pre-term birth in pregnant women. Adverse reactions to atosiban are generally mild.
- Adrenocorticotropic hormone (ACTH), secreted by the anterior pituitary gland in response to corticotropin-releasing hormone (CRH), stimulates the secretion of cortisol from the adrenal glands.
- ACTH, a 39 amino acid fragment of the 241 amino acid polypeptide proopiomelanocortin (POMC), represents amino acids 138–178 of the POMC molecule.
- Since only the first 24 amino acids are required for biologic activity, a synthetic open chain peptide named cosyntropin composed of these 24 amino acids has been synthesized. Amino acids 1–20 is the minimum sequence for biological activity; any shortening from 20 residues leads to loss of activity. Cosyntropin given intravenously is used for the ACTH stimulation test to diagnose or exclude adrenal insufficiency.
- Gonadotropin-releasing hormone (GnRH), used to evaluate hypothalamic-pituitary function, is of central importance in the initiation of the reproductive cascade stimulating the synthesis and secretion of two gonadotropins, luteinizing hormone (LH) and follicle-stimulating hormone (FSH) by the anterior lobe of the pituitary gland.
- To sustain gonadotropin secretion, secretion of GnRH from the neurons in pulses is required. Interference with the generation of pulses interferes with the secretion of LH and FSH. Pulse generation is thought to be an intrinsic property of the neurons that contain the neuropeptides, kisspeptin, neurokinin B, and dynorphin (KND neurons).
- To overcome the short half-life of GnRH, peptide analogs with increases in potency of up to 200 times and with extended half-lives have been produced. In the syntheses of the commonly used agonists leuprolide, nafarelin, triptorelin, histrelin, goserelin, and buserelin, *D*-amino acids or derivatives of *D*-amino acids were substituted for glycine. GnRH agonists are used in controlled ovarian hyperstimulation in in vitro fertilization and in endometriosis; to treat

hormone-sensitive prostate and breast cancers; in the treatment of delayed puberty and precocious puberty; to suppress hormone levels in transsexuals; and in cases of congenital adrenal hyperplasia.
- GnRH antagonists such as cetrorelix and abarelix produce quicker chemical castration than GnRH agonists and do so in the absence of testosterone surges and microsurges. The antagonists may be used prior in in vitro fertilization for the prevention of LH surges and endogenous ovulation; to treat women with hormone-sensitive breast cancer, endometriosis, and uterine fibroids; and in men for benign prostatic hyperplasia.
- GnRH agonists elicit adverse signs and symptoms of hypoestrogenism manifesting as hot flashes, headache, and osteoporosis. Common adverse events following GnRH antagonists are hot flashes, headache, weight gain, abdominal pain, and ovarian hyperstimulation. The antagonist abarelix carries a black box warning for hypersensitivity.
- Parathyroid hormone is a key regulator of calcium levels in the blood with an important role in mineral homeostasis and bone maintenance.
- Of parathyroid hormone's 84 amino acids, the 34 *N*-terminal amino acids (PTH(1–34)) comprise the biologically active region of the mature hormone. Known as teriparatide, this fragment has been manufactured by recombinant DNA technology, approved for therapy by the FDA and marketed as Forteo®. Both the full length 84 amino acid hormone (FDA approved and marketed as Natpara®) and recombinant fragment rhPTH(1–34) are used to treat osteoporosis.
- An FDA black box warning has been issued for the possibility of the development of bone cancer following parathyroid hormone and approved preparations are only made available through a REMS Program. Possible recorded serious side effects are hypercalcemia and hypocalcemia. A range of common adverse effects include paresthesia, headache, nausea, diarrhea, vomiting, and hypoesthesia.
- The long-term effects of parathyroid hormone therapy have not yet been determined and remain a concern. A number of observations suggest that the hormone may also be implicated in the pathophysiology of cardiovascular disease. Parathyroid hormone is being increasingly recognized as a risk factor with predictive value for cardiovascular disease and mortality.

Further Reading

Insulin

Hilgenfeld R, Seipke G, Berchtold H, et al. The evolution of insulin glargine and its continuing contribution to diabetes care. Drugs. 2014;74:911–27.

Kahn CR, Weir GC, King GL, et al., editors. Joslin's diabetes mellitus. 14th ed. Boston: Lippincott Williams and Wilkins; 2005.

Kumar S, O'Rahilly S, editors. Insulin resistance. Insulin action and its disturbances in disease. Chichester: Wiley; 2005.
Menting JG, Whittaker J, Margetts MB, et al. How insulin engages its primary binding site on the insulin receptor. Nature. 2013;493:241–5.

Glucagon

Lefèbvre PJ. Glucagon and its family revisited. Diabetes Care. 1995;18:715–30.
Mayo KE, Miller LJ, Bataille D, et al. International union of pharmacology. XXXV. The glucagon receptor family. Pharmacol Rev. 2003;55:167–94.
Song WJ, Mondal P, Wolfe A, et al. Glucagon regulates hepatic kisspeptin to impair insulin secretion. Cell Metab. 2014;19:667–81.

Glucagon-like Peptide-1 (GLP-1)

Brubaker PL, Drucker DJ. Structure-function of the glucagon receptor family of G-protein-coupled receptors: The glucagon, GIP, GLP-1, and GLP-2 receptors. Recept Channels. 2002;8:179–88.
Chalmer T, Almdal TP, Vilsbøll T, et al. Adverse drug reactions associated with the use of liraglutide in patients with type 2 diabetes – focus on pancreatitis and pancreas cancer. Expert Opin Drug Saf. 2015;14:171–80.
Ferdinand KC, White WB, Calhoun DA, et al. Effects of the once-weekly glucagon-like peptide-1 receptor agonist dulaglutide on ambulatory blood pressure and heart rate in patients with type 2 diabetes mellitus. Hypertension. 2014;64:731–7.
Holst JJ. The physiology of glucagon-like peptide 1. Physiol Rev. 2007;87:1409–39.
Kim W, Egan JM. The role of incretins in glucose homeostasis and diabetes treatment. Physiol Rev. 2008;60:470–512.
Scheen AJ. A review of gliptins for 2014. Expert Opin Pharmacother. 2015;16:43–62.
Tibble CA, Cavaiola TS, Henry RR. Longer acting GLP-1 receptor agonists and the potential for improved cardiovascular outcomes: a review of current literature. Expert Rev Endocrinol Metab. 2013;8:247–59.

Growth Hormone

Brooks AJ, Waters MJ. The growth hormone receptor: mechanism of activation and clinical implications. Nature Rev Endocrinol. 2010;6:515–25.
Carter-Su C, Schwartz J, Smit LS. Molecular mechanism of growth hormone action. Annu Rev Physiol. 1996;58:187–207.
Jaunmuktane Z, Mead S, Ellis M, et al. Evidence for human transmission of amyloid-β pathology and cerebral amyloid angiopathy. Nature. 2015;525:247–50.
Ross J, Czernichow P, Biller BMK, et al. Growth hormone: health considerations beyond height gain. Pediatrics. 2010;125:e906–18.

Insulin-like Growth Factor-1 (IGF-1)

Cianfarani S, Clemmons DR, Savage MO, editors. IGF-1 and IGF binding proteins. Basic research and clinical management. Endocrin Dev, vol 9, Basel: Karger; 2005
Keating GM. Mecasermin. BioDrugs. 2008;22:177–88.
Rosenbloom AL. Insulin-like growth factor-1 (rhIGF-1) therapy of short stature. J Pediatr Endocrinol Metab. 2008;21:301–15.

Somatostatin

Baldelli R, Barnabei A, Rizza L, et al. Somatostatin analogs therapy in gastroenteropancreatic neuroendocrine tumors: current aspects and new perspectives. Front Endocrinol (Lausanne). 2014;5:7. doi:10.3389/fendo.2014.00007.eCollection2014.
Møller LN, Stidsen CE, Hartmann B, et al. Somatostatin receptors. Biochim Biophys Acta. 2003;1616:1–84.
Reubi JC, Schonbrunn A. Illuminating somatostatin analog action at neuroendocrine tumor receptors. Trends in Pharmacol Sci. 2013;34:676–88.

Vasopressin

Holmes CL, Patel BM, Russell JA, et al. Physiology of vasopressin relevant to management of septic shock. Chest. 2001;120:989–1002.
Kaufmann JE, Oksche A, Wollheim CB, et al. Vasopressin-induced von Willebrand factor secretion from endothelial cells involves V2 receptors and cAMP. J Clin Invest. 2000;106:107–16.
Nielsen S, Chou CL, Marples D, et al. Vasopressin increases water permeability of kidney collecting duct by inducing translocation of aquaporin-CD water channels to plasma. Proc Natl Acad Sci U S A. 1995;92:1013–7.

Oxytocin

Gimpl G, Fahrenholz F. The oxytocin receptor system: structure, function, and regulation. Pharmacol Rev. 2001;81:629–83.
MacDonald K, MacDonald TM. The peptide that binds: a systematic review of oxytocin and its prosocial effects in humans. Harv Rev Psychiatry. 2010;18:1–21.

Adrenocorticotropic Hormone (ACTH)

Hamilton DD, Cotton BA. Cosyntropin as a diagnostic agent in the screening of patients for adrenocortical insufficiency. Clin Pharmacol. 2010;2:77–82.

Gonadotropin-Releasing Hormone (GnRH)

Armstrong SP, Caunt CJ, Fowkes RC, et al. Pulsatile and sustained gonadotropin-releasing hormone (GnRH) receptor signaling. J Biol Chem. 2009;284:35746–57.
Flanagan CA, Millar RP, Illing N. Advances in understanding gonadotropin-releasing hormone receptor structure and ligand interactions. Rev Reprod. 1997;2:113–20.
Moreau J-P, Delavault P, Blumberg J. Luteinizing hormone-releasing hormone agonists in the treatment of prostate cancer: a review of their discovery, development, and place in therapy. Clin Ther. 2006;28:1485–508.
Schally AV. Luteinizing hormone-releasing hormone analogs: their impact on the control of tumorigenesis. Peptides. 1999;20:1247–62.

Parathyroid Hormone

Crane JL, Cao X. Bone marrow mesenchymal stem cells and TGF-β signaling in bone remodeling. J Clin Invest. 2014;124:466–72.
Hagström E, Michaëlsson K, Melhus H, et al. Plasma-parathyroid hormone is associated with subclinical and clinical atherosclerotic disease in 2 community-based cohorts. Arterioscler Thromb Vasc Biol. 2014;34:1567–73.
Jilka RL. Molecular and cellular mechanisms of the anabolic effect of intermittent PTH. Bone. 2007;40:1434–46.

Chapter 8
Glycoprotein Hormones

The glycoprotein hormone family is made up of three gonadotropins and a fourth non-gonadotropin member although a possible fifth member, named thyrostimulin, has been described. The gonadotropins are follicle-stimulating hormone (FSH), regarded as the prototypic member, luteinizing hormone (LH), and human chorionic gonadotropin (hCG); thyroid-stimulating hormone (TSH) is the fourth, and non-gonadotropin, glycoprotein member of the family. All belong to the cystine-knot growth factor superfamily and each is composed of a hormone-specific β subunit and a common α subunit with the same amino acid sequence. FSH, LH, and TSH are secreted from the anterior pituitary gland whereas hCG is secreted by the human placenta during early pregnancy. Gonadotropin-releasing hormone (GnRH) (Chap. 7, section "Gonadotropin-Releasing Hormone") controls the secretion of FSH and LF and thyroid-releasing hormone controls TSH. The four glycoprotein hormones act via their complementary receptors which belong to the guanine nucleotide-binding protein (G protein) membrane-coupled receptor family A, each with large *N*-terminal ectodomains. Like their hormone ligands, the receptors are also closely related. In fact, LH and hCG share the same receptor and the sequence homologies and functional similarities shared by the four hormones indicate their common evolutionary origin.

Follicle-Stimulating Hormone

Human FSH, produced by the anterior lobe of the pituitary gland exists in a variety of isoforms due to post-translational glycosylation. The naturally derived hormone prepared from the urine of post-menopausal women and available for more than 30 years (menopausal gonadotropin) is increasingly being replaced by recombinant preparations which are seen as purer products devoid of contaminants such as LH, unwanted proteins, and other allergenic materials. Recombinant FSH preparations such as follitropin alfa (Gonal-f®) is produced in genetically modified Chinese

B.A. Baldo, *Safety of Biologics Therapy*, DOI 10.1007/978-3-319-30472-4_8

hamster ovary (CHO) cells cultured in bioreactors and isolated using an FSH-specific antibody to give a highly purified product with a consistent isoform profile. The recombinant preparations contain no LH and their physicochemical, biological, and immunological properties have been shown to be comparable to pituitary- and urine-derived FSH, often termed urofollitropin. Studies have shown that urofollitropin and recombinant FSH are equally effective and well tolerated for the induction of ovulation and human menopausal gonadotropin is equally effective as urofollitropin for pregnancy outcomes.

For both females and males, FSH is necessary for the growth and maturation of gametes. In women, FSH stimulates growth and recruitment of ovarian follicles in the ovary and therefore ultimately the production of eggs. Although in most cases pure FSH, for example as follitropin alfa, can induce follicular growth on its own without the addition of LH, optimal follicular development seems to require LH, especially in LH-deficient females and most especially in women with hypogonadotropic hypogonadism. Excess LH can arrest follicular growth. When mature, the follicle begins to secrete estradiol which affects the hypothalamus and pituitary glands leading to rapid release of GnRH and LH. An increase in the level of estradiol inhibits GnRH production in the hypothalamus and leads to a decrease in FSH production. In menopause, FSH levels increase, down regulation of FSH receptors follows, and any remaining follicles no longer have FSH receptors. In males, FSH stimulates spermatocytes to secondary spermatocytes and interacts with Sertoli cells to initiate spermatogenesis.

Structure and Mechanism of Action

As for LH and TSH, FSH is secreted from the anterior pituitary gland as two non-covalently linked dissimilar glycoprotein subunits held together by ionic and hydrophobic interactions. The subunits are the common α, and hormone-specific β subunits composed of 92 (MW ~10.2 kDa) and 111 amino acids (MW ~10.5 kDa), respectively. The α subunit is glycosylated at Asn 51 and Asn 78, and has 3 methionines and 10 cysteine residues with 5 internal disulfide bonds; the β subunit has oligosaccharides at Asn 7 and Asn 24, 1 methionine, 12 cysteines, and 6 internal disulfide bonds. Follitropin alfa was the first recombinant form of FSH. It has an amino acid sequence identical to the natural hormone isolated from urine and the later recombinant follitropin beta. Both recombinant preparations have oligosaccharide chains that are similar, but not identical, to natural FSH. In addition to the two currently approved and marketed recombinant FSH preparations, Gonal-f® (Gonal-F® in some markets; follitropin alfa) and Puregon® (follitropin beta), a third recombinant FSH preparation, Ovaleap®, was approved by the EMA in 2013. Ovaleap® is a follitropin alfa biosimilar (Chap. 13), its reference medicine being Gonal-f®. In a comparison of the use of follitropins alfa and beta in controlled ovarian hyperstimulation for in vitro fertilization, the two recombinant preparations produced comparable numbers of retrieved oocytes although follitropin beta was

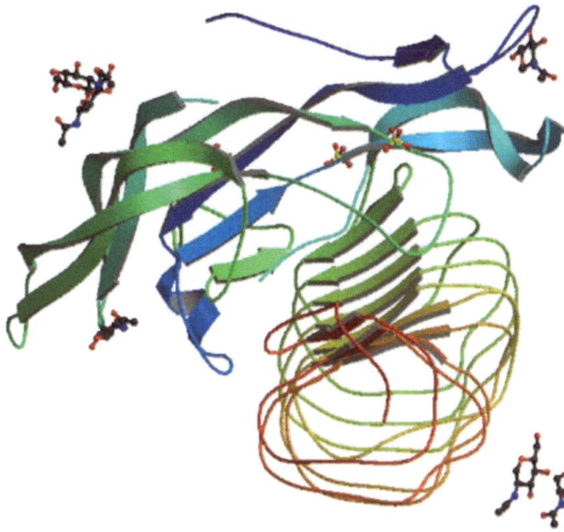

Fig. 8.1 Crystalline structure of a partially deglycosylated complex of human follicle-stimulating hormone (FSH) at 2.9Å resolution bound to the extracellular hormone-binding domain of its elongated, curved complementary receptor. The structure is from Protein Data Bank RCSB PDB file 1XWD (Fan QR, Hendrickson WA. Crystal structure of human follicle stimulating hormone complexed with its receptor. Nature. 2005;433:269–77)

associated with a lower clinical pregnancy rate and significantly higher estradiol levels despite the administration of a significantly lower total gonadotropin dose.

In inducing maturation of ovarian follicles and mature eggs, and the production of estrogen, FSH in target granulosa cells first binds to the large *N*-terminal ectodomain of its G protein-coupled receptor (Fig. 8.1), interacting with both the receptor's high affinity hormone-binding subdomain in the region of the *N*-terminus and a second subdomain at the C-terminal end involved in signal specificity. In what is believed to be a two-step monomeric receptor activation mechanism, binding to the high affinity site induces a conformational change in the FSH molecule producing a binding pocket specific for a sulfated tyrosine sTyr335 in the receptor and subsequently the sTyr is drawn into the pocket. Signaling is effected via many different signaling cascades including the extracellular regulated kinases (ERKs), p38 mitogen-activated protein kinases (MAPKs), and phosphatidylinositol-3-kinase (P13K). Protein kinase A (PKA) appears to initiate all these signaling events as well as phosphorylation of transcription factors like cAMP and promotion of chromatin remodeling by phosphorylating histone H3. PKA also enhances the activities of ERK, p38 MAPK, and P13K pathways. In all of this complexity, it is apparent that cross-talk is required between the different signaling cascades.

Indications and Usage of Follicle-Stimulating Hormone

In women, FSH is used for ovulation induction and in assisted reproductive technologies. As stated by the FDA, the hormone is indicated for the induction of ovulation and pregnancy in the anovulatory patient in whom the cause of infertility is functional and not due to primary ovarian failure. FSH is also indicated for the development of multiple follicles in the ovulatory patient participating in an assisted reproductive technology program. In men, FSH is used for the induction of spermatogenesis in patients with primary and secondary hypogonadotropic hypogonadism in whom infertility is not due to primary testicular failure.

The FSH receptor is selectively expressed on the surface of blood vessels of a wide range of tumors and the hormone is thought to promote vascularization, raising the possibility of the use of antagonists to FSH and its receptor as anti-tumor angiogenesis therapy.

Warnings, Precautions, and Adverse Events

Ovarian enlargement and/or hyperstimulation following FSH is the subject of a warning by regulatory agencies. Mild to moderate enlargement sometimes accompanied by abdominal symptoms occurs in about 20 % of patients given gonadotropins and human hCG but this condition usually subsides within 2–3 weeks. At this stage, hCG should not be administered to lessen the chance of the development of ovarian hyperstimulation syndrome (OHSS) and multiple ovulations. This warning is particularly important for patients with anovulation and hypothalamic hypogonadism. Clinical symptoms of mild OHSS include gastrointestinal problems, painful breasts, and enlarged ovaries. Severe OHSS may progress rapidly over only a few days with early warning signs of severe pulmonary pain, nausea, vomiting, and weight gain. Symptoms include abdominal pain and distension, gastrointestinal symptoms, dyspnea, weight gain, oliguria, hemoconcentration, decreased urinary output, ascites, pleural effusion, and ovarian enlargement due to ovarian cysts that are prone to rupture. Hepatic dysfunction may also be associated with OHSS. During induction of ovulation in clinical trials, OHSS occurred in 9 of 228 (3.9 %) women treated with Gonal-f®. One patient (0.4 %) was classified as severe. The incidence of OHSS in women given follitropin during in vitro fertilization treatment is said to be between 0 and 4.6 % with results from pooled studies showing an overall incidence of 2.6 %. Ovarian torsion after treatment with follitropin may be associated with OHSS and with pregnancy, ovarian cysts, and polycystic ovaries. In other complications of reproduction, ectopic pregnancies may occur, spontaneous abortion rates of 10–25 % have been reported and, in about 20 % of pregnancies, multiple ovulations lead to multiple births after treatment with gonadotropins and hCG. Thromboembolism with complications of thrombophlebitis, pulmonary embolism, pulmonary infarction, stroke, and arterial occlusion leading to limb amputation may occur in association with, or separate from, OHSS and serious pulmonary disorders such as acute respiratory distress syndrome, atelectasis, and exacerbation of asthma have also been reported.

In clinical trials of ovulation induction and assisted reproductive technologies, adverse events in women occurring in more than 10 % of patients were headache, ovarian cyst, nausea, and upper respiratory infection. Other adverse events experienced by more than 5 % of patients were intermenstrual bleeding, ovarian hyperstimulation, abdominal pain, flatulence, diarrhea, back pain, and sinusitis. Safety studies in men have generally been undertaken with far fewer patients. In 56 patients who received follitropin alfa for induction of spermatogenesis and fertility, 123 adverse events, 7 serious, occurred in 34 patients. The most frequently seen events were acne, breast pain, fatigue, gynecomastia, injection site pain, and varicocele.

From 1093 injections of follitropin alfa, cutaneous reactions made up of erythema, irritation, pain, and pruritus resulted from 20 (1.8 %) of the injections. No antibodies with adverse effects for any of the follitropin preparations appear to have been reported in treated patients.

Luteinizing Hormone

LH, produced by the gonadotroph cells of the anterior pituitary gland, is essential for reproduction in both females and males. In the menstrual cycle, LH and FSH increase in a midcycle surge producing a follicular and luteal phase. The increase in LH lasts 24–48 h, triggering ovulation and egg release and the conversion of the residual follicle into the corpus luteum which, in turn, produces progesterone in readiness for implantation of the fertilized embryo in the endometrium. Luteal function is maintained for the first 2 weeks and, if pregnancy results, luteal function is further maintained by hCG, switched on by the newly established pregnancy. In males, LH (sometimes called interstitial cell-stimulating hormone, ICSH) acts on the interstitial Leydig cells of the testes, producing testosterone.

Until the beginning of the twenty-first century, the only commercially available source of LH was obtained from the urine of postmenopausal women. The preparation contained LH and FSH but the suboptimal amount of LH often present required supplementing with hCG to achieve an LH:FSH action ratio of 1:1. The consequence of this was that different preparations of the so-called human menopausal gonadotropin contained different amounts of hCG. Lutropin alfa (Luveris®), first registered in the European Union in 2001, is so far the only recombinant human LH. The highly purified preparation is composed of a consistent isoform and its physicochemical, immunological, and biological activities are essentially identical to those of human pituitary LH.

Structure and Mechanism of Action

Lutropin alfa is a recombinant heterodimeric glycoprotein produced in genetically modified Chinese hamster ovary (CHO) cells and made up of two non-covalently linked subunits, the α subunit of 92 amino acids which is common to the other glycoprotein hormones FSH, hCG, and TSH, and the hormone-specific β subunit of 121 amino acids. Oligosaccharide chains are attached via *N*-linkages with *N*-glycosylation sites at Asn52 and Asn78 on the α subunit and Asn30 on the β subunit.

LH and hCG share the same receptor called luteinizing hormone/choriogonadotropin receptor (LHCGR), lutropin/choriogonadotropin receptor (LCGR) or simply luteinizing hormone receptor (LHR), a transmembrane receptor belonging to the G protein-coupled receptor family and located predominantly on the ovarian theca and granulosa cells and testis Leydig cells. The gene for the LHCGR is similar to the FSH receptor and TSH receptor genes. The receptor has 674 amino acids with a molecular mass of about 85–95 kDa. Binding of LH (or hCG) to the receptor ectodomains causes a conformational change in the receptor, activation and detachment of the G protein and activation of the cAMP system. Adenylate cyclase activates cAMP-dependent PKA which has two regulatory and two catalytic units. The catalytic units, released by the binding of cAMP to the regulatory units, initiate the phosphorylation of proteins and ultimately the physiological actions of steroid hormone production and follicle maturation processes. Although it now seems to be generally agreed that the LHR-mediated effects proceed by the G protein-adenylate cyclase-cAMP-PKA pathway, other pathways may be involved in other LHR-dependent events such as the proliferation and/or differentiation of target cells.

Indications and Usage of Lutropin Alfa

Lutropin alfa together with follitropin alfa is indicated for stimulation of follicle development in infertile hypogonadotropic hypogonadol women with profound LH deficiency. Note that the safety and efficacy of the administration of lutropin alfa with any other preparation or recombinant form of FSH or FSH derived from human urine is unknown. Both dose-finding studies and a double-blind, randomized trial demonstrated a significant increase in the rate of optimal follicular development in women with hypogonadotropic hypogonadism and profound LH deficiency in response to subcutaneous administration of lutropin alfa. Lutropin alfa may also benefit some normogonadotropic women but a study in women older than 35 years did not show any benefit following administration of the hormone.

Warnings, Precautions, and Adverse Events Associated with Lutropin Alfa

Regulatory authorities and the manufacturer of lutropin alfa warn that the safe and effective use of the recombinant hormone requires vigilant monitoring of the ovarian response on a regular basis. Careful observation and evaluations should be undertaken for ovarian enlargement; ovarian torsion (rotation of the ovary occluding the ovarian artery or vein); multiple pregnancy, pregnancy loss and ectopic pregnancy; congenital malformations; thromboembolic events; and reproductive system neoplasms, both benign and malignant.

The broad frequency of adverse reactions, common or rare, to lutropin alfa seen in the different organ and system categories can be summed up as follows: common—nervous system, gastrointestinal disorders, reproductive system, and breast disorders, skin; rare—immune system, vascular system. The safety of lutropin alfa (Luveris®) was assessed in clinical studies of 152 infertile women with hypogonadotropic hypogonadism who received lutropin alfa and follitropin alfa (Gonal-f®) in 283 treatment cycles. Adverse events occurring in at least 2 % of patients were, in order of the highest to the lowest frequency: headache, abdominal pain, nausea, ovarian hyperstimulation, breast pain, ovarian cyst, flatulence, injection site reaction, fatigue, dysmenorrhea, ovarian disorder, diarrhea. Medical events reported subsequent to pregnancies resulting from the administration of gonadotropins for the induction of ovulation include spontaneous abortion, ectopic pregnancy, premature labor, postpartum fever, and congenital abnormalities.

Human Chorionic Gonadotropin

The name chorionic gonadotropin is derived from the latin chordata or afterbirth, and gonadotropin because the molecule is gonad tropic. Produced by the villous syncytiotrophoblast of the placenta following implantation, hCG stimulates late follicular maturation and initiates rupture of the preovulatory ovarian follicle. Acting via the LHCGR (the same receptor recognized by LH) of the granulosa and theca cells of the ovary, hCG maintains the viability of the corpus luteum during the early stage of pregnancy allowing it to secrete progesterone and estrogen during the first trimester. Major differences between hCG and LH are the different sites of production (LH is made by cells of the anterior pituitary), the pIs of 3.5 and 8, respectively, and a circulating half-life of ~37 h for hCG and only 25–30 min for LH. In terms of its biological activity, hCG is sometimes called a "super LH." Although it is not always acknowledged, hCG has numerous biological functions including the promotion of angiogenesis in the uterine vasculature; promotion of the differentiation process in the formation of syncytiotrophoblast cells; stimulation of growth and differentiation of the umbilical cord; matching of uterine growth to fetal growth; prevention of an immune response to placental cells; signaling involved with implantation, and a number of other actions. In fact, hCG can be viewed as a term covering four independently functioning molecules each with a different cell origin. In addition to hCG produced by syncytiotrophoblasts, other forms of the hormone are a hyperglycosylated hCG produced by cytotrophoblast cells; pituitary hCG, a sulfated variant of hCG made by gonadotrope cells of the anterior pituitary; and free β subunit of hCG produced during primary non-trophoblastic malignancies. Hyperglycosylated hCG is an autocrine that promotes cell invasion and growth in the implantation stage; pituitary hCG acts like LH promoting follicular maturation, ovulation and progesterone production during the menstrual cycle; and the free β subunit promotes growth and malignancy of advanced cancers.

Because pregnant women are the most abundant source of the hormone, this has been exploited in preparing hCG for therapeutic use. Until the preparation of a recombinant form, choriogonadotropin alfa (rhCG, Ovidrel®), hCG was extracted from the urine of pregnant women (uhCG, Pregnyl®, Novarel®, Profasi®, Follutein®). In an early safety and efficacy study, an International Recombinant Human Chorionic Gonadotropin Study Group compared subcutaneously administered rhCG and uhCG in 198 WHO group II anovulatory women for ovulation induction after follicular stimulation with rhFSH. The two preparations showed equivalent efficacy in ovulation induction but the recombinant product was better tolerated. Note, however, that in analyzing and reporting such multicenter fertility trials, criticisms of possible biases in the interpretation of results have been leveled. In particular, it has been pointed out that variations between centers should be used as an estimate of error instead of the generally used "within center" variation. More recently, searches and assessments of the efficacy and safety of subcutaneous rhCG and high dose rLH compared with intramuscular uhCG for inducing final oocyte maturation triggering in IVF and intracytoplasmic sperm injection cycles were published by the Cochrane Database of Systematic Reviews. The authors found that there was no evidence of difference between rhCG or rhLH and uhCG in achieving follicular maturation in IVF; pregnancy rates and incidences of ovarian hyperstimulation syndrome were equivalent in the groups. The conclusion was that "uhCG is still the best choice for final oocyte maturation triggering in IVF and intracytoplasmic sperm injection treatment cycles."

Structure and Mechanism of Action

HCG is a heterodimeric glycoprotein of two non-covalently linked subunits, the α subunit of 92 amino acids common to FSH and LT, and the hormone-specific β subunit of 145 amino acids. Oligosaccharides are *N*-linked to Asn52 and Asn78 (Fig. 8.2) on the α subunit and *N*- or *O*-linked to Asn13, Asn30, Ser121, Ser127, Ser132, and Ser138 on the β subunit. Natural hCG has monantennary and biantennary *N*-linked oligosaccharides (composed of 8 and 11 sugar residues, respectively) and some *O*-linked trisaccharides, giving an MW of ~36 kDa. Twenty-five to thirty percent of the MW is derived from the oligosaccharide side chains. Hyperglycosylated hCG has mainly 15 sugar fucosylated triantennary *N*-linked oligosaccharides and hexasaccharide *O*-linked oligosaccharides, giving an MW of 40–41 kDa (35–41 % from sugar side chains) depending on the extent of glycosylation. Recombinant hCG (Ovidrel®), produced in CHO cells, is comparable to the urinary-derived human natural hormone differing in the oligosaccharide branching and extent of sialylation.

The LHCGR, a G protein-coupled receptor of 675 amino acids, responds to hCG, LH, and hyperglycosylated hCG but not hCG subunits. The α and β subunits of hCG are thought to be important for receptor binding and receptor specificity, respectively. Receptor binding by hCG activates membrane-bound adenylate cyclase which converts ATP to cAMP, forerunner events to phosphorylation and protein kinase activation

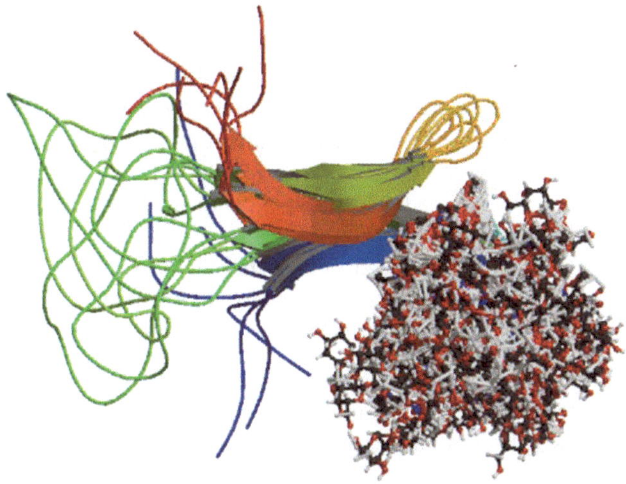

Fig. 8.2 Solution structure studied by NMR spectroscopy of the α-subunit of human chorionic gonadotropin with a di-antennary glycan at Asn78. NMR structures showed that the protein sterically hinders the glycan at Asn78 while N-acetyl-*D*-glucosamine bound to Asn78 shields the protein from the hydrophilic environment. The structure is from Protein Data Bank RCSB PDB file 1HD4 (Erbel PJA, Karimi-Nejad Y, Van Kuik JA, et al. Solution structure of the α-subunit of human chorionic gonadotropin (modeled with diantennary glycan at Asn78). Biochemistry. 2000;39:6012–21)

of the MAPK pathways and a Janus kinase signaling pathway. All of the endocrine functions involve DNA transcription or the generation of mRNA and the production of progesterone in the corpus luteum is aided by the involvement of a cholesterol side chain cleavage enzyme. Inositol phospholipid protein kinase C may also be involved in the hCG-mediated receptor events and some results indicate hCG and LH stimulate a phospholipase C which in turn stimulates protein kinase C and activation of the LHCGR.

Indications and Usage of Chorionic Gonadotropin

In the words of the FDA, choriogonadotropin alfa (Ovidrel®) is indicated for the induction of final follicular maturation and early luteinization in infertile women who have undergone pituitary desensitization and who have been appropriately pretreated with follicle-stimulating hormones as part of an assisted reproductive technology program such as in vitro fertilization and embryo transfer. The product is also indicated for the induction of ovulation and pregnancy in anovulatory infertile patients in whom the cause of infertility is functional and not due to primary ovarian failure. At the last labeling revision in April 2011, the FDA approved indications and usage for hCG derived from human pregnancy urine and for chorionic gonadotropin injection USP (Novarel®) were:

1. Prepubertal cryptorchidism not due to anatomic obstruction. In general, hCG is thought to induce testicular descent in situations when descent would have occurred at puberty. HCG thus may help to predict whether or not orchiopexy will be needed in the future. Although, in some cases, descent following hCG administration is permanent, in most cases the response is temporary. Therapy is usually instituted between the ages of 4 and 9.
2. Selected cases of hypogonadotropic hypogonadism (hypogonadism secondary to a pituitary deficiency) in males.
3. Induction of ovulation and pregnancy in the anovulatory, infertile woman in whom the cause of anovulation is secondary and not due to primary ovarian failure, and who has been appropriately pretreated with human menotropins.

Note that the FDA states that hCG has not been demonstrated to be effective adjunctive therapy for the treatment of obesity and there is no substantial evidence that it increases weight loss, affects fat distribution or reduces hunger associated with dieting.

The presence in urine or blood of hCG indicates an implanted embryo, so detection and measurement of hCG can be used as a pregnancy test. Test reagents employed, generally monoclonal antibodies, usually detect the hormone-specific β subunit to avoid a false positive conclusion by detecting LH or FSH.

Interestingly, LHCGR has been shown to occur in the brain of women including the hippocampus, hypothalamus, and brain stem, a finding causing some to speculate that this may be a possible explanation of morning sickness in pregnancy.

Warnings, Precautions, and Adverse Events

Warnings for the main serious adverse events seen during the use of uhCG are for the possibility of the induction of OHSS; enlargement of preexisting ovarian cysts and rupture of cysts with resultant hemoperitoneum; multiple births; arterial thromboembolism; the need to be watchful for gasping syndrome in premature babies due to the presence of benzyl alcohol as bacteriostatic in injection vehicles; and the possibility of anaphylaxis. The most important precautions to be kept in mind are the possibility of induction of precocious puberty in patients treated for cryptorchidism and androgen-induced fluid retention, particularly in patients with cardiac and renal disease, epilepsy, migraine, or asthma. The most commonly seen adverse reactions following uhCG are headache, irritability, restlessness, depression, fatigue, edema, precocious puberty, gynecomastia, and pain at the site of injection. Hypersensitivity responses, both systemic (anaphylaxis, angioedema, dyspnea, shortness of breath) and cutaneous (rash, erythema, urticaria), have been reported.

Warnings and precautions for rhCG center on the possibilities of ovarian enlargement and OHSS in women with and without pulmonary and vascular complications and the risk of vascular and pulmonary complications arising from thromboembolism. Commonly occurring adverse reactions seen in clinical studies of assisted reproductive technologies are injection site reactions (pain, erythema, bruising) and gastrointestinal disorders (abdominal pain nausea, vomiting). Other adverse events seen occasionally/rarely include ectopic pregnancy, breast pain, intermenstrual bleeding, vaginal discomfort and hemorrhage, hot flashes, genital moniliasis, urinary tract infection, headache, back pain, fever, malaise, and paresthesias. In trials of rhCG for ovulation induction, the most frequently seen adverse events were injection site, reproductive (ovarian cyst, ovarian hyperstimulation), and gastrointestinal disorders. Less common events were breast pain, abdominal enlargement, pharyngitis, upper respiratory tract infection, hyperglycemia, and pruritus.

Reports of controlled clinical studies subsequent to pregnancies resulting from hCG therapy named the following resultant adverse events: spontaneous abortion, ectopic pregnancy, premature labor, postpartum fever, and congenital abnormalities. Post-marketing experience revealed cases of allergic reactions including anaphylaxis and mild, transient skin rashes as well as thromboembolic events.

Thyroid-Stimulating Hormone

Thyrotropin-releasing hormone (TRH) stimulates the pituitary gland to produce TSH which, together with the three gonadotropins FSH, LH, and hCG discussed above, make up the glycoprotein hormone family. TSH is synthesized and secreted by the thyrotrope cells of the anterior pituitary. Binding of TSH to its receptor TSHR on thyroid epithelial cells or cancer tissues stimulates the uptake of iodine and synthesis and secretion of thyroglobulin and the tyrosine-based hormonal regulators of metabolism, thyroxine (T4) (3,5,3′,5′-tetraiodothyronine) and triiodothyronine (T3), both produced by the follicular cells of the thyroid gland. In quantitative terms, thyroxine (T4) is the major thyroid hormone in the blood being released in the ratio of ~20:1 compared to triiodothyronine (T3) but the latter is more active and more potent by a factor of 3–5 times. T4 has a longer half-life than T3 and can be viewed as a prohormone since it is converted into T3 by deiodinases but because a large proportion of T4 binds to plasma proteins, only a small fraction (0.02–0.03 %) is available for conversion to T3. Conversion of T4 to T3 by 5′-deiodination occurs via type 1 deiodinase, mainly in the thyroid gland, liver, and kidney and by type 2 deiodinase in the pituitary gland, skeletal muscle, placenta adipose tissue, and glial cells. The thyroid hormones act on almost all cells of the body, regulating protein, carbohydrate, and fat metabolism and are essential for cell differentiation and development. Production and release of TSH is controlled by a negative feedback mechanism. Low concentrations of T4 and T3 lead to an increase in the synthesis and secretion of the hormone whereas a high concentration of T4 and T3 is a prelude to decreased hormone release. Iodine is essential for the production of T4 and T3 with a deficiency leading to a fall in the hormone's production, enlargement of the thyroid gland, and ultimately the disease goiter.

Structure and Mechanism of Action

TSH is a heterodimeric glycoprotein made up of an α subunit of 92 amino acid residues containing two *N*-linked glycosylation sites (Asn52, Asn78) non-covalently linked to a β subunit of 118 residues containing one *N*-linked glycosylation site (Asn23). The α subunit is almost identical to the α subunits of hCG, LH, and FSH. The α subunit is involved in receptor binding and thought to be responsible for stimulation of adenylate cyclase involved the generation of cAMP. The β subunit is specific for TSH, and therefore determines its receptor specificity. As with the other glycoprotein hormones, the dimeric structure of TSH is essential for its biological action—the free subunits are inactive. Recombinant TSH (rhTSH, rhTSH alfa, thyrotropin alfa; trade name Thyrogen®) produced in CHO cells has an amino acid sequence identical to that of the natural human pituitary hormone. Both the recombinant and naturally occurring hormones are composed of a mixture of glycosylated forms. Pituitary TSH is secreted as a mixture of sialylated and sulfated forms with the sugars asparagine-linked in bi- and triantennary oligosaccharides whereas thyrotropin alfa is a sialylated glycoprotein only. Compared to sialylated TSH, asialo-TSH with terminal D-mannose, *D*-galactose, *L*-fucose, N-acetyl-D-glucosamine, and *N*-acetyl-*D*-galactosamine sulfate has a shorter half-life due to its removal by hepatocyte asialoglycoprotein receptors. On the other hand, the bioactivity of sialylated TSH seems to be less. Thyrotropin alfa has a longer half-life than pituitary TSH but a lower affinity and bioactivity at the TSHR.

TSH acts via the protein G-coupled TSHR, activating adenylate cyclase and phosphatidylinositol pathways that in turn activate a tyrosine kinase cascade. Stimulation of the pathways in the thyroid follicular cells results in iodine uptake and increased production and release of thyroglobulin, used by the thyroid gland to produce T4 and T3. Structural homologies between TSH and hCG and their receptors allows for cross-reaction of the TSHR with hCG and the production of thyroid hormones, an activity that may occur in the first trimester of pregnancy.

Antibodies may mimic TSH by stimulating the TSHR. Such antibodies, often referred to as thyroid-stimulating antibodies, occur in Graves' disease, otherwise known as primary hyperthyroidism or autoimmune thyroiditis type 3A. Graves' disease, classified as a type II hypersensitivity response (Chap. 1, section "Hypersensitivities"), is an autoimmune disease affecting ~2 % of women and characterized by an enlarged, overactive thyroid, an excess of circulating T4 and T3 and causing possible disorders of the heart, circulation, nervous system, skin, and eyes. Exophthalmos (proptosis) (Fig. 8.3), for example, is a characteristic disorder of the eyes seen in patients with the disease. Hypothyroidism is the name given to the condition resulting from a deficiency of T4 or T3 or both. Hashimoto's disease or primary (or congenital) hypothyroidism is perhaps the best known example.

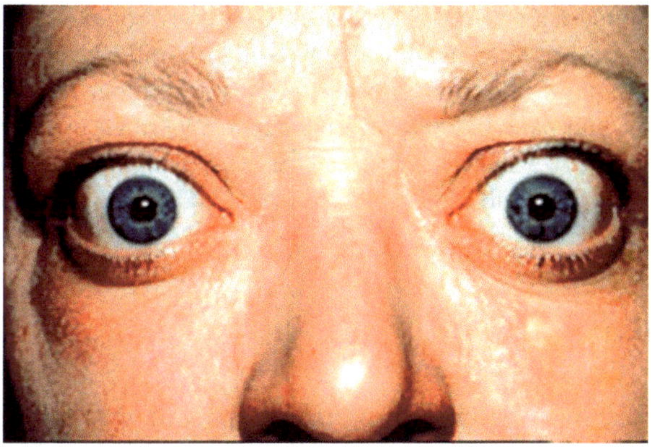

Fig. 8.3 Bilateral exophthalmos or proptosis and lid retraction as often seen in Graves'disease. Reproduced from "The eyes have it," created by Jonathan Trobe, University of Michigan Kellogg Eye Center, © 2009 The Regents of the University of Michigan. An open access site distributed under the terms of the Creative Commons Attribution 3.0 License

Indications and Usage of Thyrotropin Alfa

The recombinant TSH preparation thyrotropin alfa as Thyrogen® is registered by the FDA and EMA and approved for diagnostic and ablative use. Well-differentiated thyroid cancer which accounts for up to 90% of all thyroid cancers is initially treated by thyroidectomy and then followed up with thyroid hormone suppression therapy to prevent stimulation of remnant thyroid tissue and thyroid cancer. Thyrotropin alfa is used as a diagnostic tool in these patients, with or without radioactive imaging, to test for serum thyroglobulin. Thyrotropin alfa is also indicated for adjunctive treatment for radioactive ablation of any remaining thyroid tissue in patients who have undergone thyroidectomy and who show no evidence of metastatic thyroid cancer.

Warnings, Precautions, and Adverse Events

Thyrotropin alfa can cause a transient elevation of the serum concentration of thyroid hormones in patients with thyroid cancer metastases or some remaining thyroid tissue. Consequences may range from hyperthyroidism to death. Symptoms of acute hemiplegia, hemiparesis, loss of vision, laryngeal edema, pain at a distant site, and respiratory distress may result from a sudden and rapid enlargement of residual thyroid tissue or distant metastases following treatment with thyrotropin alfa. Stroke and stroke-like symptoms within three days of thyrotropin alfa administration have been the subject of some post-marketing reports, usually involving patients with known central nervous system metastases. Most cases appear to be young women taking oral contraceptives or with other risk factors for stroke.

The most common adverse reactions seen after thyrotropin alfa use for diagnostic or ablation purposes in clinical trials involving thyroid cancer patients who had undergone near—total thyroidectomy are nausea, vomiting, headache, dizziness, fatigue diarrhea, and paraesthesia. Adverse events identified in the post-marketing period include influenza-like symptoms and hypersensitivity reactions, chiefly urticaria, flushing, pruritus, and respiratory difficulties. Cases of stroke are rare.

Thyrostimulin

A new heterodimeric glycoprotein hormone named thyrostimulin composed of two new human glycoprotein hormone subunits was found to show high affinity for, and activate, the TSHR but not the LHR and FSHR. Identified and initially studied in the laboratory of AJW Hsueh, Stanford University School of Medicine, the heterodimer was shown to stimulate cAMP production and thymidine incorporation by cultured thyroid cells and increase serum thyroxine levels in vivo. The authors stated that the expression of thyrostimulin in the anterior pituitary suggests a paracrine mechanism and its discovery could facilitate the understanding of the physiological roles of extra-thyroid TSH receptor systems and the structural–functional basis of receptor signaling by related glycoprotein hormones. Lipopolysaccharide and inflammatory cytokines upregulate the expression of thyrostimulin and expression of its subunits has been observed in pituitary and adrenal glands, the central nervous system, gastrointestinal organs, testes, skin, and retina. A role for thyrostimulin in the Brokken–Wiersinga–Prummel loop, that is the ultra-short feedback control relating the concentration of thyrotropin to its secretion, has been suggested.

Summary

- The glycoprotein hormone family is made up of three gonadotropins and a fourth non-gonadotropin member. The gonadotropins are follicle-stimulating hormone (FSH), regarded as the prototypic member, luteinizing hormone (LH), and human chorionic gonadotropin (hCG); thyroid-stimulating hormone (TSH) is the fourth, and non-gonadotropin, glycoprotein member of the family.
- FSH is secreted from the anterior pituitary gland as two non-covalently linked dissimilar glycoprotein subunits held together by ionic and hydrophobic interactions. The subunits are the common α, and hormone-specific β, subunits composed of 92 (MW ~10.2 kDa) and 111 amino acids (MW ~10.5 kDa), respectively. Follitropin alfa was the first recombinant form of FSH. It has an amino acid sequence identical to the natural hormone isolated from urine and the later recombinant follitropin beta.

- FSH is used for ovulation induction and is also indicated for the development of multiple follicles in the ovulatory patient participating in an assisted reproductive technology program. In men, FSH is used for the induction of spermatogenesis in patients with primary and secondary hypogonadotropic hypogonadism in whom infertility is not due to primary testicular failure.
- Ovarian enlargement and/or hyperstimulation following FSH is the subject of a warning by regulatory agencies. The incidence of ovarian hyperstimulation syndrome (OHSS) in women given follitropin during in vitro fertilization treatment is between 0 and 4.6 %. Ovarian torsion after treatment with follitropin may be associated with OHSS and with pregnancy, ovarian cysts, and polycystic ovaries. Thromboembolism with complications of thrombophlebitis, pulmonary embolism, pulmonary infarction, stroke, and arterial occlusion leading to limb amputation may occur in association with, or separate from, OHSS.
- Lutropin alfa is a recombinant heterodimeric glycoprotein made up of two non-covalently linked subunits, the α subunit of 92 amino acids which is common to the other glycoprotein hormones FSH, hCG and TSH and the hormone-specific β subunit of 121 amino acids. LH and hCG share the same receptor.
- Lutropin alfa together with follitropin alfa is indicated for stimulation of follicle development in infertile hypogonadotropic hypogonadol women with profound LH deficiency.
- Regulatory authorities and the manufacturer of lutropin alfa warn that use of the recombinant hormone requires vigilant monitoring of the ovarian response on a regular basis. Careful observation and evaluations should be undertaken for ovarian enlargement; ovarian torsion; multiple pregnancy, pregnancy loss and ectopic pregnancy; congenital malformations; thromboembolic events; and reproductive system neoplasms, both benign and malignant.
- HCG has numerous biological functions including the promotion of angiogenesis in the uterine vasculature; promotion of the differentiation process in the formation of syncytiotrophoblast cells; stimulation of growth and differentiation of the umbilical cord; matching of uterine growth to fetal growth; prevention of an immune response to placental cells; and signaling involved with implantation.
- HCG is a heterodimeric glycoprotein of two non-covalently linked subunits, the α subunit of 92 amino acids common to FSH and LT, and the hormone-specific β subunit of 145 amino acids. Recombinant hCG is comparable to the urinary-derived human natural hormone differing in the oligosaccharide branching and extent of sialylation.
- Recombinant hCG is indicated for the induction of final follicular maturation and early luteinization in infertile women and for the induction of ovulation and pregnancy in anovulatory infertile patients in whom the cause of infertility is functional and not due to primary ovarian failure.
- FDA approved indications and usage for hCG derived from human pregnancy urine are prepubertal cryptorchidism, selected cases of hypogonadotropic

hypogonadism in males and induction of ovulation and pregnancy in the anovulatory, infertile woman in whom the cause of anovulation is secondary and not due to primary ovarian failure.

- Warnings for the main serious adverse events seen during the use of urinary-derived hCG are for the possibility of OHSS; enlargement of preexisting ovarian cysts and rupture of cysts with resultant hemoperitoneum; multiple births; arterial thromboembolism; the need to be watchful for gasping syndrome in premature babies due to the presence of benzyl alcohol as bacteriostatic in injection vehicles; and the possibility of anaphylaxis.
- Binding of TSH to its receptor on thyroid epithelial cells or cancer tissues stimulates the uptake of iodine and synthesis and secretion of thyroglobulin (Tg) and the tyrosine-based hormonal regulators of metabolism, thyroxine (T4) (3,5,3′,5′-tetraiodothyronine), and triiodothyronine (T3), both produced by the follicular cells of the thyroid gland.
- TSH is a heterodimeric glycoprotein made up of an α subunit of 92 amino acid residues containing a β subunit of 118 residues. The α subunit is almost identical to the α subunits of hCG, LH, and FSH and is involved in receptor binding; the β subunit is specific for TSH, and determines its receptor specificity.
- Antibodies may mimic TSH by stimulating the TSH receptor. Such antibodies occur in Graves' disease, otherwise known as primary hyperthyroidism or autoimmune thyroiditis type 3A. Graves' disease, classified as a type II hypersensitivity response, is an autoimmune disease affecting ~2% of women and characterized by an enlarged, overactive thyroid, an excess of circulating T4 and T3. Graves' disease causes possible disorders of the heart, circulation, nervous system, skin, and eyes.
- The recombinant TSH preparation thyrotropin alfa is registered by the FDA and EMA and approved for diagnostic and ablative use. After thyroidectomy, thyroid cancer patients may be followed up with thyroid hormone suppression therapy to prevent stimulation of remnant thyroid tissue and thyroid cancer. Thyrotropin alfa is used as a diagnostic tool in these patients, with or without radioactive imaging, to test for serum thyroglobulin. Thyrotropin alfa is also indicated for adjunctive treatment for radioactive ablation of any remaining thyroid tissue in patients who have undergone thyroidectomy and who show no evidence of metastatic thyroid cancer.
- Thyrotropin alfa can cause a transient elevation of the serum concentration of thyroid hormones in patients with thyroid cancer metastases or some remaining thyroid tissue. Consequences may range from hyperthyroidism to death. Symptoms of acute hemiplegia, hemiparesis, loss of vision, laryngeal edema, pain at a distant site, and respiratory distress may result from a sudden and rapid enlargement of residual thyroid tissue or distant metastases following treatment with thyrotropin alfa. Stroke and stroke-like symptoms within 3 days of thyrotropin alfa administration have been the subject of some post-marketing reports.

Further Reading

Follicle-Stimulating Hormone

Hunzicker-Dunn M, Maizels ET. FSH signaling pathways in immature granulosa cells that regulate target gene expression: Branching out from protein kinase A. Cell Signal 2006;18: 1351–9.
Jiang X, Dias JA, He X, et al. Structural biology of glycoprotein hormones and their receptors: Insights into signaling. Molec Cell Endocrinol 2014;382:424–51.
Jiang X, Liu H, Chen X, et al. Structure of follicle-stimulating hormone in complex with the entire ectodomain of its receptor. Proc Natl Acad Sci USA 2012;109:12491–6.
Practice Committee of the American Society for Reproductive Medicine, Birmingham, Alabama. Gonadotropin preparations: past, present, and future perspectives. Fertil Steril 2008;90: S13–20.

Luteinizing Hormone

Ascoli M, Fanelli, Segaloff DL. The lutropin/choriogonadotropin receptor, a 2002 perspective. Endocrin Rev 2002;23:141–74.
Dafau ML. The luteinizing hormone receptor. Annu Rev Physiol 1998;60:461–96.
Krause BT, Ohlinger R, Haase A. Lutropin alfa, recombinant human luteinizing hormone, for the stimulation of follicular development in profoundly LH-deficient hypogonadotropic hypogonadal women: a review. Biol Targets Ther 2009;3:337–47.

Human Chorionic Gonadotropin (hCG)

Cole LA. New discoveries on the biology and detection of human chorionic gonadotropin. Reprod Biol Endocrinol 2009;7:8 doi:10.1186/1477-7827-7-8.
Cole LA. Biological functions of hCG and hCG-related molecules. Reprod Biol Endocrinol 2010;8:102. doi: 10.1186/1477-7827-8-102.

Thyroid Stimulating Hormone

Dietrich JW, Landgrafe G, Fotiadou EH. TSH and thyrotropic agonists: key actors in thyroid homeostasis. J Thyroid Res 2012:351864, 29 p. http://dx.doi.org/10.1155/2012/351864.
Gramza A, Schuff KG. Recombinant human thyroid stimulating hormone in 2008: focus on thyroid cancer management. Onco Targets Ther 2008;1:87–101.
Szkudlinski MW, Fremont V, Ronin C, et al. Thyroid-stimulating hormone and thyroid-stimulating hormone receptor structure-function relationships. Physiol Rev 2001;82:473–502.

Chapter 9
Enzymes Approved for Therapy

Introduction

A large number and variety of enzyme defects have been identified in humans, many leading to diseases produced by enzyme deficiency or altered amounts of metabolites. Although it might be thought that the existing knowledge of enzymology together with evidence of an enzyme-based defect would lead to rapid progress in the development of successful targeted therapies, much progress in our understanding of cell and molecular biology was needed before significant advances leading to a range of registered efficacious therapeutic agents could occur. The so-called storage diseases provided both the best early examples of a clear association of enzyme deficiency and disease as well as candidates for a successful therapeutic approach to treatment. For treatment to be successful, a number of requirements were seen to be essential. Firstly, it was clear that not all storage diseases were suitable targets for therapy. Some diseases produce irreversible damage, for example, ganglioside GM2 storage diseases Tay–Sachs disease and Sandhoff's disease, where central nervous system damage that occurs early in life is untreatable. Other factors that need to be considered include the likelihood of producing sufficient specific enzyme; its stability; the possibility of delivering the enzyme to the cell target where it needs to act; the enzyme's likely antigenicity and the consequent antibody response in the treated patients; and the frequency of the disease—a rare disease with too few sufferers might not justify research and development efforts and costs, even under an orphan drug development program.

B.A. Baldo, *Safety of Biologics Therapy*, DOI 10.1007/978-3-319-30472-4_9

Toward Successful Enzyme Replacement Therapy: Gaucher Disease

Early evidence obtained from experiments using cultured skin fibroblasts from patients with a mucopolysaccharide storage disease or mucopolysaccharidosis (MPS) together with cells of a different genotype, indicated that post-translational modification of lysosomal enzymes could correct previously defective glycosaminoglycan catabolism. Subsequent investigations showed that the corrective modification of these lysosomal storage diseases resulted from the recognition by complementary receptors of *D*-mannose-6-phosphate on the modified lysosomal enzymes. During the same period as these studies, lysosomal enzymes injected into rats were shown to avoid rapid clearance and be specifically taken up by liver reticuloendothelial cells via recognition of mannose residues. Importantly, the complementary mannose receptors were demonstrated on the surface of macrophages and their recognition spectrum was shown to extend beyond this sugar to *N*-acetyl-*D*-glucosamine and *L*-fucose. Once bound to the receptors, enzyme is rapidly internalized and transported to the lysosomes. The discoveries of specific recognition of phosphorylated mannose residues, mannose receptor-mediated uptake of lysosomal enzymes, and the presence of these receptors on macrophages, demonstrated that a lysosomal enzyme needs to be specifically recognized by its target cells. These insights led on to the first successful enzyme replacement therapy (ERT) for type I Gaucher disease that occurs with a frequency of 1 in 75,000 births worldwide, making it the most prevalent of the sphingolipid storage disorders. Gaucher disease is the result of an inborn error of metabolism due to a deficiency of the lysosomal acid β-glucosidase glycoprotein, β-glucocerebrosidase (glucosylceramidase; β-glucosyl-*N*-acylsphingosine glucohydrolase) which cleaves β-*D*-glucosylceramide (glucocerebroside) into glucose and ceramide. The enzyme's substrate is a widely distributed cell membrane component and, in the absence of β-glucocerebrosidase, glucocerebroside, and other glycolipids accumulate by as much as 20–100-fold in the lysosomes of cells, particularly macrophages and other cells of the reticuloendothelial system. With this background, glucocerebrosidase prepared from human placenta and marketed as Ceredase® was used to reverse the clinical manifestations of type I Gaucher disease by targeting the patients' macrophages after sequential deglycosylation to expose mannose residues. Imiglucerase, a recombinant, deglycosylated glucocerebrosidase prepared in Chinese hamster ovary cells by DNA technology, was soon introduced and because it proved at least as clinically effective as Ceredase® and could provide a pathogen-free preparation in almost unlimited amounts, it (as Cerezyme®) soon replaced its predecessor.

Approved Enzymes as Replacement Therapy for Lysosomal Storage Diseases

In an effort to encourage the development of products for the diagnosis and treatment of rare or the so-called orphan diseases, countries, beginning with the USA, have introduced legislation to provide incentives for developers and manufacturers who would otherwise have been unable to cover the costs involved in bringing a drug to a very small market. For example, the Food and Drug Administration (FDA) Office of Orphan Products Development (OOPD) provides incentives in cases involving less than 200,000 people in the USA where, since 1983, more than 400 drugs and biologic products for rare diseases have been brought to market under the Orphan Drug Designation programs. This compares to a total of ten products developed in the previous decade. The recombinant enzyme biologics covered in this review are regulated and approved by the FDA Center for Drug Evaluation and Research (CDER) rather than the Center for Biologics Evaluation and Research which retains regulatory responsibility for bacterial and human cellular products, gene therapy products, vaccines, allergenic extracts, antitoxins, antivenoms, blood, blood components, and plasma-derived products.

The emphasis in this review is on the recombinant proteins involved in ERT for lysosomal storage diseases as well as other recombinant/purified approved orphan drug enzymes for the treatment of other rare diseases. The relatively poorly defined nature of pancrelipase (Creon®), an extract of porcine pancreas glands containing multiple enzymes, including lipases, proteases, and amylases and indicated for the treatment of pancreatic insufficiency due to cystic fibrosis and chronic pancreatitis, is therefore not covered. The serine protease and transglutaminase coagulation factors involved in the coagulation pathways are examined in Chap. 10.

Utilizing the scientific insights summarized above and the still expanding manufacturing expertize in biologics, orphan drug research, and development programs have made remarkable progress in introducing efficacious and safe enzyme-based therapies for a range of rare disorders. Central amongst these disorders, and as a direct result of the knowledge and experience gained in the development of ERT for Gaucher disease, are successful ERTs for a small but growing number of other lysosomal storage diseases. A list of these diseases together with the responsible deficient enzymes, and the accumulating products that cause the signs and symptoms of the diseases currently being treated with approved recombinant enzyme preparations are set out in Table 9.1. Depending on the particular MPS, the accumulated products produce clinical manifestations such as short stature, coarse features, motor dysfunction, cardiomyopathy, ocular abnormalities, skeletal dysplasia, hepatosplenomegaly, and mental retardation. Approved enzyme replacement treatments are now administered for a second lipid storage disorder, Fabry disease, for the glycogen storage disorder, Pompe disease, and for MPS types I, II, IVA, and VI. In addition to the individual deficient enzymes and resultant accumulated products (Table 9.1), the specificities, properties, and mechanisms of action of the ten approved recombinant enzymes currently used as ERT for lysosomal storage diseases (alglucosidase alfa, algalsidase beta, sebelipase alfa, imiglucerase, taliglucerase alfa, velaglucerase alfa, laronidase, idursulfase, elosulfase alfa, and galsulfase) are set out in Table 9.2. Three of these recombinant enzymes, the human β-glucocerebrosidases **imiglucerase**, **taliglucerase alfa**, and **velaglucerase alfa**, are each approved for the treatment of **Gaucher disease** but one of these, taliglucerase alfa, differs from the others in two significant ways. Taliglucerase alfa, produced in carrot root cells, is the first recombinant enzyme used for therapy that has been derived from a plant cell expression system. Taliglucerase alfa also shows differences in its glycosylation, a result of its core α(1–2)-*D*-xylose and α(1–3)-*L*-fucose which is unique to plant-derived proteins.

Table 9.1 Approved enzymes[a] used as replacement therapy for lysosomal storage diseases: syndromes, deficient enzymes, and accumulated products

Approved recombinant enzyme[b] used for ERT	Type of disorder	Name of syndrome	Deficient enzyme	Accumulated substrate/product(s)
Agalsidase beta[c] (Fabrazyme®)	Lipid storage (Sphingolipidosis)	Fabry disease	α-*D*-Galactosidase A	Ceramide trihexoside[d]
Alglucosidase alfa (Lumizyme®; Myozyme®)	Glycogen storage	Pompe disease[e]	Acid α-glucosidase	Glycogen
Imiglucerase[f] (Cerezyme®)	Lipid storage (Sphingolipidosis)	Gaucher disease	β-Glucocerebrosidase[g]	Glucocerebroside (Glucosylceramide)
Taliglucerase alfa[f] (Elelyso®)				
Velaglucerase alfa[f] (VPRIV®)				
Laronidase[h] (Aldurazyme®)	Mucopolysaccharidosis type I (MPS I)	Hurler, Scheie and Hurler-Scheie syndromes	α-*L*-Iduronidase	Dermatan sulfate Heparan sulfate
Idursulfase[h] (Elaprase®)	Mucopolysaccharidosis type II (MPS II)	Hunter syndrome	Iduronate-2-sulfatase	Dermatan sulfate Heparan sulfate
Elosulfase alfa[h] (Vimizim®)	Mucopolysaccharidosis type IVA (MPS IVA)	Morquio A syndrome	*N*-Acetylgalactosamine-6-sulfatase	Keratan sulfate Chondroitin 6-sulfate
Galsulfase[h] (Naglazyme®)	Mucopolysaccharidosis type VI (MPS VI)	Maroteaux-Lamy syndrome	*N*-Acetylgalactosamine-4-sulfatase	Dermatan sulfate Chondroitin 4-sulfate
Sebelipase alfa (Kanuma®)	Lipid storage Dyslipidemia: impaired degradation of lysosomal lipid[i]	Lysosomal acid lipase deficiency[j]	Lysosomal acid lipase	Cholesteryl esters Triglycerides LDL-C

CDER FDA Center for Drug Evaluation and Research, *EMA* European Medicines Agency, *ERT* enzyme-replacement therapy, *FDA* US Food and Drug Administration, *HDL-C* high density lipoprotein-cholesterol, *LDL-C* low density lipoprotein-cholesterol, *MPS* mucopolysaccharidosis

[a]Approved by FDA CDER or EMA or both

[b]For information on individual enzymes refer to Table 9.2

[c]Agalsidase alfa (Replagal®), another recombinant form of the enzyme, is also used (see Table 9.2)
[d]Globotriaosylceramide (Gb3, GL-3)
[e]Also known as glycogen storage disease type II and acid maltase deficiency
[f]For details and differences of these 3 enzymes, see Table 9.2
[g]Also called acid β-glucosidase, β-glucosidase, *D*-glucosyl-*N*-acylsphingosine glucohydrolase or GCase
[h]Enzymes that catabolize or cleave glycosaminoglycans dermatan, heparan, keratan, and chondroitin sulfates, preventing their accumulation which causes the clinical manifestations seen in the four disorders
[i]Cholesteryl esters and triglycerides accumulate within lysosomes. Leads to elevated total cholesterol and LDL-C and reduced HDL-C
[j]A heterogeneous disease showing a clinical continuum from Wolman disease (rapidly progressing) to cholesteryl ester storage disease (CESD)
From Baldo BA. Enzymes approved for human therapy: Indications, mechanisms and adverse effects. BioDrugs 2015;29:31–55. Adapted and reproduced with permission from Springer Science+Business Media

Table 9.2 Enzymes approved for human therapy[a]: properties, approved indications[a], mechanisms and adverse effects

Generic and trade names	Enzyme specificity and properties	Approved indications[a]	Mechanism(s) of action relevant to indications	Adverse effects, serious and common
Agalsidase beta[b] (Fabrazyme®)	A recombinant α-*D*-galactosidase A; glycosylated, 2 subunits MW ~ 100 kDa 398 amino acids	Fabry disease (α-*D*-galactosidase A deficiency)	Hydrolyzes globotriaosylceramide (GL-3, ceramide trihexoside) and lipids[c] with terminal α-*D*-galactose thus reducing GL-3 deposition in capillary endothelia	Infusion reactions; immunogenicity and allergic reactions (anaphylaxis, hives, pruritus, rash); stroke; ataxia; pain; cardiac abnormalities; nephrotic syndrome
Alglucosidase alfa (Lumizyme®; Myozyme®)	A recombinant human glycogen-specific acid alfa-glucosidase (GAA). Glycoprotein, MW 109 kDa, 883 residues	Pompe disease (GAA deficiency)	Binds cell surface mannose-6-PO_4, transported to lysosomes where it degrades glycogen (and maltose and isomaltose) by hydrolyzing α-1,4- and α-1,6-glycosidic linkages	*Boxed warning*: anaphylaxis, severe allergic reactions; cardiorespiratory failure. Also: infusion reactions; respiratory problems; pneumonia; fever; rash; ↑BP; nausea; rash; urticaria
Alteplase (Activase®; Actilyse®; Cathflo® Activase®)	Recombinant tissue plasminogen activator (tPA), a serine protease; single glycosylated polypeptide chain 527 amino acids MW ~59 kDa	Myocardial infarction with ST elevation; acute ischemic stroke; pulmonary embolism	Enzyme binds fibrin in thrombus, cleaves Arg561-Val562 bond in plasminogen to → plasmin which causes local fibrinolysis	Bleeding (especially GI); sepsis; venous thrombosis; allergy including anaphylaxis

Generic and trade names	Enzyme specificity and properties	Approved indications[a]	Mechanism(s) of action relevant to indications	Adverse effects, serious and common
Asfotase alfa (Strensiq®)	Recombinant nonspecific alkaline phosphatase; two identical chains of 726 amino acids MW ~161 kDa linked by two disulfide bonds	Treatment of perinatal/infantile- and juvenile-onset hypophosphatasia	Enzyme reduces elevated substrate levels including inorganic pyrophosphate which inhibit bone mineralization leading to rickets, bone deformation and osteomalacia	*Warnings*: Hypersensitivity reactions; lipodystrophy; ectopic calcifications. Also: ISR; ectopic calcifications (eye and kidney); hypersensitivity
L-Asparaginase (Elspar®; Erwinase®)	From *E. coli* (Elspar®) or *Erwinia chrysanthemi* (Erwinaze®). *L*-Asparagine specific; 4 subunits MW35kDa *E. coli*	Acute lymphocytic leukemia	Hydrolyzes *L*-asparagine → aspartic acid and ammonia leading to inhibition of protein synthesis in some leukemia cells which cannot synthesize *L*-asparagine	Allergic reactions including anaphylaxis; thrombotic events; pancreatitis; glucose intolerance; hepatotoxicity; PRES; coagulopathy
Pegaspargase[d] (Oncaspar®)	enzyme linked to monomethylpolyethylene glycol (PEG)	Erwinaze® administered to patients hypersensitive to *E. coli* *L*-asparaginase		
Collagenase[e] (Xiaflex®; Xiapex®; Santyl®)	Proteases that hydrolyze collagen; 2 enzymes, collagenases AUX-I (MW 114 kDa) and AUX-II (MW 113 kDa), each ~ 1000 amino acids[f]	Dupuytren's contracture with palpable cord (Xiaflex®) Wound cleaning and healing (Santyl® ointment)	Breaks peptide bonds of collagen in its native triple helical conformation	Peripheral edema; allergic reactions; ISR; contusion; pain in treated tissue; tendon rupture; ligament damage; pruritus
Dornase alfa (Pulmozyme®)	Recombinant human deoxyribonuclease I (rhDNase I); 260 amino acids, identical to natural enzyme.	Cystic fibrosis	Hydrolyzes extracellular DNA released from neutrophils into sputum reducing viscosity and aiding clearance from lungs[g]	Voice alteration; sore throat; rash; pharyngitis, laryngitis; dyspnea; dyspepsia; chest pain; urticaria

(continued)

Table 9.2 (continued)

Generic and trade names	Enzyme specificity and properties	Approved indications[a]	Mechanism(s) of action relevant to indications	Adverse effects, serious and common
Elosulfase alfa (Vimizim®)	Recombinant *N*-acetylgalactosamine –6-sulfatase (rhGALNS); monomer 496 amino acids, MW 55.4 kDa, 2 glycosylation sites. Cys53 (in active site) modified to formylglycine[h]	Enzyme replacement therapy for mucopolysaccharidosis IV type A (MPS IVA; Morquio A syndrome)	Enzyme increases catabolism of glycosaminoglycans[i] that otherwise remain incompletely catabolized causing skeletal dysplasia and other abnormalities	*Boxed warning*: anaphylaxis. Also: urticaria, dyspnea, flushing; pyrexia; immunogenicity; vomiting; nausea; headache; chills; abdominal pain
Galsulfase (Naglazyme®)	Recombinant *N*-acetyl-galactosamine –4-sulfatase (rhASB), MW ~56 kDa 495 amino acids; 6 glycosylation sites, 4 with mannose-6-phosphate. Cys53 modified to formylglycine[h]	Enzyme replacement therapy for mucopolysaccharidosis VI (Maroteaux-Lamy syndrome)	Cleaves SO_4 of sulfate ester from terminal *N*-acetylgalactosamine of glycosaminoglycans reducing disease manifestations[j]	Infusion reactions, anaphylaxis, rash; urticaria; dyspnea; pyrexia; vomiting; nausea; headache; abdominal pain
Glucarpidase (Voraxaze®)	Recombinant pseudomonas sp. carboxypeptidase G2 produced in *E. coli*	Treatment of toxic concentrations of methotrexate due to impaired renal clearance	Methotrexate hydrolyzed to glutamate and less toxic 2,4-diamino-N^{10}-methyl-pteroic acid largely excreted by the liver	Allergic reactions including anaphylaxis; flushing; paresthesia; headache; nausea; vomiting; hypotension
Hyaluronidase (Hylenex® recombinant; Amphadase® k; Hydase™ [k]; Vitrase® [l])	Recombinant human hyaluronidase (rhuPH20); glycoprotein 447 amino acids, MW ~ 61 kDa produced in CHO cells; bovine[k] and ovine[l] purified preparations	Dispersion and absorption of injected drugs; sc fluid administration for hydration; sc urography for resorption of radiopaque agents	Decreases cellular cement viscosity by hydrolyzing glucosaminidic bond of hyaluronic acid between C1 of *N*-acetyl-*D*-glucosamine and C4 of *D*-glucuronic acid[m]	*Warnings*: spread of localized infection[n]; not to be applied to cornea; inactivated iv; Also: edema in association with hypodermoclysis; ISR; allergic reactions

Generic and trade names	Enzyme specificity and properties	Approved indications[a]	Mechanism(s) of action relevant to indications	Adverse effects, serious and common
Idursulfase (Elaprase®)	Recombinant human iduronate-2-sulfatase; 525 amino acids; MW 76k Da; 8 Asp-linked glycosylation sites. Cys modified to formylglycine[h]	Enzyme replacement therapy for mucopolysaccharidosis II (Hunter syndrome)	Reduces disease manifestations by hydrolyzing 2-SO_4 esters of terminal iduronate SO_4 residues from glycosaminoglycans	*Boxed warning*: anaphylaxis. Also: hypersensitivity including rash, urticaria, flushing; pyrexia; diarrhea; pruritus; headache; vomiting; cough
Imiglucerase (Cerezyme®)	Recombinant monomeric glycoprotein human β-glucocerebrosidase[o] M_r 60430, 497 amino acids; His not Arg at pos 495 of placental enzyme[p] Oligosaccharide chains modified to expose terminal mannose residues	Long-term enzyme replacement therapy for type 1 Gaucher disease[q]	Terminal mannose recognized by receptors on macrophages. Catalyzes hydrolysis of glucocerebroside to glucose and ceramide preventing secondary hematologic, spleen, liver, and skeletal complications	Hypersensitivity[r] (hypotension, dyspnea, pruritus, urticaria, angioedema, anaphylaxis); nausea; vomiting; headache; chills; tachycardia; diarrhea; rash[s]
Laronidase (Aldurazyme®)	Recombinant variant of human α-*L*-iduronidase; 628 amino acids; MW 83 kDa; 6 glycosylation sites with at least 2 mannose-6-phosphorylated	Enzyme replacement for mucopolysaccharidosis I (Hurler, Hurler-Scheie, Scheie forms)	Enzyme cleaves α-*L*-iduronic acid from nonreducing end of dermatan and heparan sulfates that remain stored in lysosomes	*Boxed warning*: anaphylaxis. Also: ISR; rash; upper respiratory infections; hyperreflexia; paresthesia; pyrexia
Ocriplasmin (Jetrea®)	A recombinant truncated human plasmin, 2 polypeptide chains of 19 and 230 residues linked by 2 disulfide bonds, MW 27.24 kDa	Treatment of symptomatic vitreomacular adhesion	Active against fibronectin and laminin, components of the vitreomacular interface. Enzyme dissolves the protein that links the vitreous to the macular	Vitreous floaters; macular hole; eye pain; conjunctival hemorrhage; photopsia; vision blurring and impairment;

(continued)

Table 9.2 (continued)

Generic and trade names	Enzyme specificity and properties	Approved indications[a]	Mechanism(s) of action relevant to indications	Adverse effects, serious and common
Pegademase bovine (Adagen®)	Extensively pegylated bovine adenosine deaminase (ADA) [(monomethoxypolyethyleneglycol succinimidyl) 11–17-adenosine deaminase]	Enzyme replacement for ADA deficiency in patients with SCID[t]	Enzyme deaminates toxic[u] adenosine and 2′-deoxyadenosine to less toxic inosine and 2′-deoxyinosine	Hemolytic anemia; autoimmune hemolytic anemia; thrombocytopenia; ISR; urticaria[v]
Pegloticase[w] (Krystexxa®)	A pegylated recombinant porcine-like uricase (urate oxidase)[x]; 4 identical noncovalently linked chains each 300 amino acids, MW 136.8 kDa.	Treatment of chronic gout refractory to conventional therapy	Lowers uric acid levels and prevents deposition of crystals by conversion of uric acid to allantoin excreted by kidneys	*Boxed warning*: anaphylaxis; infusion reaction. Also: gout flares; nausea; vomiting; contusion or ecchymosis; chest pain[y]
Rasburicase (Elitek®; Fasturtec®)	Recombinant *Aspergillus*-derived urate oxidase[z] expressed in yeast; tetrameric protein, identical subunits 301 amino acids, MW 34 kDa	For management of plasma uric acid levels during anti-cancer therapy[aa]	Converts uric acid to allantoin in patients with hyperuricemia. Soluble allantoin excreted via kidneys	*Boxed warnings*: Anaphylaxis; hemolysis[w]; methemoglobinemia; inhibits uric acid *in vitro*. Also: vomiting, headache; abdominal pain; pyrexia; peripheral edema; rash; anxiety
Reteplase[ab] (Retavase®; Rapilysin®)	Recombinant non-glycosylated form of human tPA; contains 355[ab] of 527 amino acids MW 39.571 kDa	Acute myocardial infarction	Thrombolytic. As for alteplase but reteplase is fibrin selective and works faster and longer than tPA[ab]	Bleeding; allergic reactions; fever; ISR; nausea; vomiting; cardiac effects[ac]

Generic and trade names	Enzyme specificity and properties	Approved indications[a]	Mechanism(s) of action relevant to indications	Adverse effects, serious and common
Sebelipase alfa (Kanuma®)	Recombinant human lysosomal acid lipase with same amino acid sequence; MW 55 kDa; produced in chicken egg white; has 6 *N*-linked glycosylation sites	Treatment of patients with lysosomal acid lipase deficiency	Enzyme binds to cell surface receptors via its glycan structures; internalized into lysosomes where it hydrolyses cholesteryl esters and triglycerides to free cholesterol, glycerol, and free fatty acids	*Warnings*: Hypersensitivity including to eggs or egg products. Also: 1st 6 months of life: vomiting; diarrhea; cough; nausea; anemia; nasopharyngitis; fever. Older children and adults; fever; nausea; nasopharyngitis; headache; constipation
Streptokinase (Streptase®)	From *Streptococcus equisimilis*, MW 47 kDa 414 amino acids; activates plasminogen[ad] to plasmin responsible for dissolution of fibrin clot	Acute myocardial infarction within 12 h onset with ST elevation[ae] or recent left bundle-branch block	Binds to plasminogen producing conformational[af] expression of an active catalytic site on plasminogen that cleaves the Arg561-Val562 bond → plasmin and fibrinolysis[ag]	Bleeding; immune disorders especially antibodies and anaphylaxis[ah]; GI[ai] and CV disorders[aj]; headache; fever; chills; back pain; myalgia; asthenia; malaise
Taliglucerase alfa (Elelyso®)	Recombinant[ak] human lysosomal β-glucocerebrosidase, MW 60.8 kDa; differs from natural enzyme by 2 amino acids at *N*-terminal and up to 7 at C-terminal	Long-term enzyme replacement therapy for adults with type I Gaucher disease	Enzyme taken up by target cells via mannose receptors[ak]. Catalyzes hydrolysis of glucocerebroside to glucose and ceramide preventing secondary hematologic, spleen, liver, and skeletal complications	*Warnings*: anaphylaxis, allergic and infusion reactions. Also: urti; headache; pharyngitis, arthralgia; back and extremity pain; urinary tract infection; flu-like symptoms

(continued)

Table 9.2 (continued)

Generic and trade names	Enzyme specificity and properties	Approved indications[a]	Mechanism(s) of action relevant to indications	Adverse effects, serious and common
Tenecteplase (TNKase®, Metalyse®)	Recombinant tPA of 527 amino acids with Asn for Thr at 103, Gln for Asn at 117 and tetra Ala at 296–299	Acute myocardial infarction with fewer bleeding complications	As for alteplase but more specific for fibrin and more resistant to inactivation by PAI-1[al]	Bleeding; allergic reactions; nausea; vomiting; fever; hypotension; cardiac effects[ac]
Velaglucerase alfa (VPRIV®)	Recombinant human β-glucocerebrosidase, MW 63kDa, 497 amino acids; same sequence as natural enzyme but has added mannose *N*-linked glycans for target cell recognition	Long-term enzyme replacement therapy for adults with type I Gaucher disease	As for Taliglucerase alfa	*Warnings*: hypersensitivity including anaphylaxis; infusion rate. Also: infusion reactions; headache; pyrexia; arthralgia; prolonged aPTT; nasopharyngitis; dizziness; bone pain

ADA adenosine deaminase, *ALL* acute lymphocytic leukemia, *aPTT* activated partial thromboplastin time, *BP* blood pressure, *CDER* FDA Center for Drug Evaluation and Research, *CHO* Chinese hamster ovary cells, *EMA* European Medicines Agency, *ERT* enzyme-replacement therapy, *FDA* US Food and Drug Administration, *GI* gastrointestinal, *IM* intramuscular, *ISR* injection site reaction, *iv* intravenous, *MMP* matrix metalloproteinase, *MW* molecular weight, *PAI* plasminogen activator inhibitor, *PRES* posterior reversible encephalopathy syndrome, *sc* subcutaneous, *SCID* severe combined immunodeficiency disease, *tPA* tissue plasminogen activator, *URTI* upper respiratory tract infection

[a]Approved by FDA CDER or EMA or both

[b]Agalsidase alfa (Replagal®) is another recombinant form of the enzyme. Beta and alfa forms differ only in the quantitative levels of the sialic acid and mannose-6-phosphate in the oligosaccharide side chains

[c]E.g., Galabiosylceramide and blood group B substance

[d]Allows IM injection of smaller doses and with less frequent administration than native *E. coli L*-asparaginase

[e]Collagenase from *Clostridium histolyticum*

[f]AUX-I and AUX-II show some sequence homology with human MMPs and antibodies to these collagenase proteins may have the potential to cross-react with the MMPs

[g]DNA, a viscous polyanion, may also reduce effectiveness of aminoglycoside antibiotics, e.g., tobramycin, by binding to the antibiotic

[h]Required for sulfatase enzyme activity

[i]Glycosylaminoglycans keratan sulfate and chondroitin sulfate catabolized in lysosomes
[j]Uptake of enzyme by lysosomes thought to be mediated by binding of terminal mannose-6-phosphate residues on oligosaccharide chains of galsulfase to receptors complementary to mannose phosphate
[k]Amphadase® and Hydase™ are preparations of purified bovine testicular hyaluronidase
[l]Vitrase® is a preparation of purified ovine testicular hyaluronidase
[m]Also hydrolyzes some other acid mucopolysaccharides of connective tissue to a variable degree
[n]Not to be injected into or around infected or inflamed areas
[o]β-*D*-glucosyl-*N*-acyl-sphingosine glucohydrolase
[p]Alglucerase (Ceredase®), a placentally derived glucocerebrosidase, was the first enzyme to receive approval for Gaucher disease. Ceredase® was replaced by imiglucerase (Cerezyme®) in which oligosaccharide chains are modified to terminate in mannose sugars recognized by macrophages that accumulate lipid in Gaucher disease
[q]In children and adults with one or more of anemia, thrombocytopenia, bone disease, hepatomegaly, or splenomegaly
[r]6.6 % of patients
[s]13.8 % of patients experienced adverse events related to the enzyme
[t]To be used in patients who are not suitable candidates for, or have failed, bone marrow transplantation
[u]Toxic to lymphocytes
[v]Enzyme is an orphan drug and because of small population and voluntary reporting in post-marketing period, clinical experience is still not sufficient to be confident that the full range and extent of adverse events have been seen/reported
[w]Contraindicated in patients with glucose-6-phosphate dehydrogenase deficiency since hydrogen peroxide is produced during the conversion of uric acid to allantoin and severe oxidative hemolytic anemia and methemoglobinemia may result
[x]Pegylation increases half-life from ~8 h to 10–12 days
[y]May exacerbate congestive heart failure but not yet formally studied
[z]Enzyme does not occur in humans
[aa]Used in pediatric and adult patients with leukemia, lymphoma, and solid tumors receiving anti-cancer therapy expected to result in tumor lysis syndrome with subsequent elevation of plasma uric acid
[ab]Compare tPA and alteplase, each composed of 527 amino acids; longer half-life (13–16 min) than alteplase
[ac]Frequent sequelae of patients' underlying disease; symptoms may not be due to the drug
[ad]Both conformational and proteolytic activations contribute to streptokinase-induced activation of plasminogen
[ae]Patients without ECG changes should not be given fibrinolytic therapy
[af]Streptokinase-induced conformational activation of plasminogen leading to fibrin clot dissolution
[ag]Plasmin also generated by proteolytic cleavage of Arg561-Val562 with streptokinase-plasmin complex leading to plasminogen activation via a substrate recognition mechanism

(continued)

Table 9.2 (continued)

[ah]Including serum sickness and vasculitis
[ai]Nausea, vomiting, diarrhea, epigastric pain
[aj]Including hypotension, heart rate and rhythm disorders, angina pectoris
[ak]Produced in genetically modified carrot root cells since plant-derived enzyme contains terminal mannose residues necessary for binding to target phagocytic cell receptors
[al]PAI-1, plasminogen activator inhibitor-1 (also called endothelial plasminogen activator inhibitor or serpin E1). A serine protease inhibitor (serpin) and inhibitor of tPA and urokinase, the activators of plasminogen and ultimately fibrinolysis

From Baldo BA. Enzymes approved for human therapy: Indications, mechanisms and adverse effects. BioDrugs 2015;29:31–55. Adapted and reproduced with permission from Springer Science+Business Media

Besides Gaucher disease, **Fabry disease** is a second lipid storage disorder (sphingolipidosis) but in this case caused by a deficiency of an **α-*D*-galactosidase** which results in a buildup of globotriaosylceramide, more commonly known as Gb3 or ceramide trihexoside. Gb3 accumulates in capillary endothelia affecting in particular the kidneys, heart, eyes, brain, nervous tissue, and skin. Progress in determining the efficacy of enzyme replacement therapy for Fabry disease was greatly aided by ensuring an ongoing supply of recombinant human enzyme prepared in Chinese hamster ovary (CHO) cells and development of a mouse model. Both **algalsidase alfa** (Replagal®) and **algalsidase beta** (Fabrazyme®) (Table 9.2) have received regulatory approval for the treatment of Fabry disease although the two preparations appear to be functionally indistinguishable.

Another type of lipid storage disorder is caused by a marked decrease, or complete loss of activity, of the enzyme **lysosomal acid lipase** (**LAL**) involved in the breakdown in the lysosome of lipids including low density lipoprotein-cholesterol (LDL-C). This lysosomal storage disorder, known as **lysosomal acid lipase or LAL deficiency**, is an autosomal recessive disease characterized by lysosomal accumulation of cholesteryl esters and triglycerides in organs including liver, intestine, and blood vessel walls. This results in increased liver fat and liver disease, lipid accumulation in intestinal walls leading to malabsorption, and growth failure, and dyslipidemia with elevated)LDL-C and triglycerides together with low high density lipoprotein-cholesterol (HDL-C). **Sebelipase alfa** (Kanuma®) (Table 9.1), a recombinant human LAL produced in egg white of genetically engineered chickens, has the same amino acid sequence as natural human LAL. In the lysosome, the enzyme catalyzes the hydolysis of cholesteryl esters and triglycerides to free cholesterol, glycerol, and free fatty acids (Table 9.2). Said to be under-recognized, LAL deficiency is a heterogeneous disease presenting as a clinical continuum from its most rapid progression in infants (Wolman disease) to a later onset form called cholesteryl ester storage disease (CESD). Both forms of the disease share the same underlying pathology resulting from mutations in the *LIPA* gene which encodes LAL. The varying presentation and rates of progression of the disease appear to be due to the degree of retained enzyme activity but there may be other factors involved.

A deficiency of lysosomal acid **α-glucosidase** in **Pompe disease** leads to an accumulation of glycogen which causes myopathy throughout the body but especially in skeletal muscles, heart, liver, and the nervous system. In the early ERT studies, recombinant human α-glucosidase, or **alglucosidase alfa**, was produced in large scale in CHO cells and in milk of transgenic rabbits. Both enzyme preparations reduced lysosomal glycogen storage in animals and had a prominent effect on cardiac hypertrophy, cardiac function, and survival in human infants. Alglucosidase alfa currently approved by both the FDA and EMA is produced by CHO cells (Table 9.2). Two commercial preparations are available, Lumizyme® and Myozyme.® Despite their apparent identity, the two products are seen as different by the FDA due to their different manufacturing process.

MPS types I, II, IVA, and VI are due to deficiencies of **α-*L*-iduronidase**, **iduronate-2-sulfatase**, ***N*-acetylgalactosamine-6-sulfatase**, and ***N*-acetylgalactosamine-4-sulfatase**, respectively. Each produce accumulations of glycosaminoglycans, in particular dermatan sulfate, heparan sulfate, and chondroitin sulfate. **Laronidase**, a recombinant human α-*L*-iduronidase enzyme secreted by overexpressing CHO cells and modified to contain complex oligosaccharides, is used in ERT for **MPS I** (also called **Hurler syndrome**) and clinically milder variants (Tables 9.1 and 9.2). MPS I is a progressive multisystem disorder with mild to severe clinical features including short stature, course facial features with prominent forehead, enlarged liver and spleen, corneal clouding, and dysostosis multiplex (Figs. 9.1 and 9.2). **Idursulfase**, a recombinant form of iduronate-2-sulfatase (Tables 9.1 and 9.2), has been approved by both the FDA and EMA as a safe and effective treatment for **MPS II** or **Hunter syndrome**. However, in common with each of the enzymes used for ERT in lysosomal storage diseases and which are too large to cross the blood–brain barrier, idursulfase is probably not an effective treatment for the CNS manifestations of the disease. The enzyme removes the sulfate group from the 2-position of dermatan and heparan sulfates and while the cDNA sequence predicts a 550 amino acids precursor, the mature protein is secreted as a 525 amino acid protein after cleavage of the 25 amino acid signal sequence; the following eight amino acids are also removed from the proprotein. **Elosulfase alfa**, a CHO-cell-derived recombinant form of *N*-acetylgalactosamine-6-sulfatase (Tables 9.1 and 9.2), was granted marketing approval by the FDA and EMA in 2014 for the treatment of patients with **MPS IVA**, also known as **Morquio A syndrome**, an autosomal recessive disorder caused by a deficiency of the enzyme. In 2005, **galsulfase**, a recombinant form of *N*-acetylgalactosamine-4-sulfatase

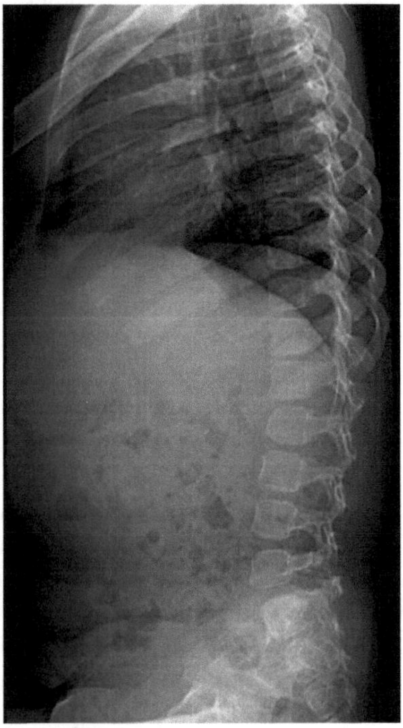

Fig. 9.1 Radiograph of the spine and ribs of a female child with mucopolysaccharidosis I showing anterior notching in thoracolumbar vertebral bodies and oar-shaped ribs. Reproduced from Anand R, Bhatia D, Yadav DS. Radiol Case Rep. 2011;7:641. doi:10.2484/rcr.v7i2.641, an open-access article distributed under the terms of the Creative Commons Attribution License

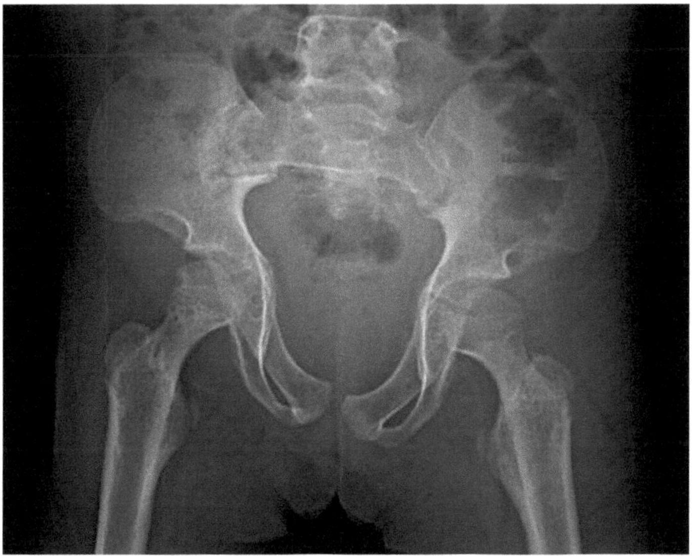

Fig. 9.2 Radiograph of the lower spine and pelvic area of a female child with mucopolysaccharidosis I showing a narrow pelvis with flared iliac wings. Reproduced from Anand R, Bhatia D, Yadav DS. Radiol Case Rep. 2011;7:641. doi:10.2484/rcr.v7i2.641, an open-access article distributed under the terms of the Creative Commons Attribution License

(Tables 9.1 and 9.2), was approved by the FDA for the treatment of patients with **MPS VI** (**Maroteaux-Lamy syndrome**), an autosomal recessive disease due to deficiency of *N*-acetylgalactosamine-4-sulfatase leading to an accumulation of dermatin sulfate and chondroitin sulfate. The disease has a wide range of symptoms that may progress slowly or rapidly. There is characteristic skeletal disease with dysplasia (Figs. 9.3, 9.4, and 9.5) including short stature, degenerative joint disease, and dysostosis multiplex. Other clinical abnormalities may include pulmonary impairment, enlarged liver and spleen, heart valve disease, sinusitis and otitis media, sleep apnea, corneal clouding (Fig. 9.6), cervical spinal instability, bony stenosis, and meningeal thickening. Although ERT with galsulfase has been relatively successful in treating some aspects of MPS VI disease such as growth, puberty, and pulmonary function, it has been disappointing in alleviating joint, ophthalmic, and CNS symptoms. Starting treatment at an earlier age may be beneficial for growth and some aspects of the syndrome may be modified by longer periods of treatment.

Other Enzymes Approved for Therapy

Of the remaining enzymes in the current CDER Therapeutic Biologic Products list, three are tissue plasminogen activators (tPA) while the other eight cover a range of diseases including cancers, immunodeficiency, heart disorders, cystic fibrosis, macular adhesion, and disorders involving collagen.

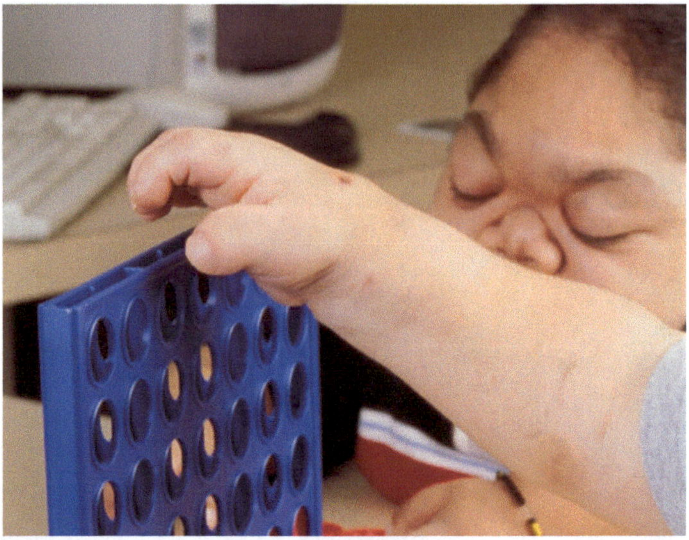

Fig. 9.3 Photograph of a claw hand deformity in a 16-year-old male with rapidly progressing mucopolysaccharidosis VI. Claw hand is often secondary to flexion contractures and carpal tunnel disease. Reproduced from Valayannopoulos V, Nicely H, Harmatz P, et al. Mucopolysaccharidosis VI. Orphanet J Rare Dis. 2010;5:5. doi:10.1186/1750-1172-5-5, an open-access article distributed under the terms of the Creative Commons Attribution License

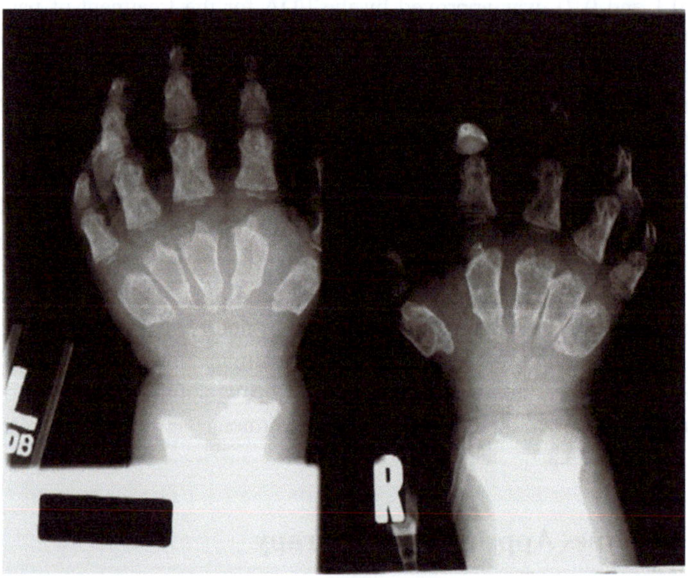

Fig. 9.4 Radiograph of the hands of a 3-year-old male with rapidly progressing mucopolysaccharidosis VI showing short, thickened metacarpal bones with proximal pointing and thin cortices and irregular carpal bones. Reproduced from Valayannopoulos V, Nicely H, Harmatz P, et al. Mucopolysaccharidosis VI. Orphanet J Rare Dis. 2010;5:5. doi:10.1186/1750-1172-5-5, an open-access article distributed under the terms of the Creative Commons Attribution License

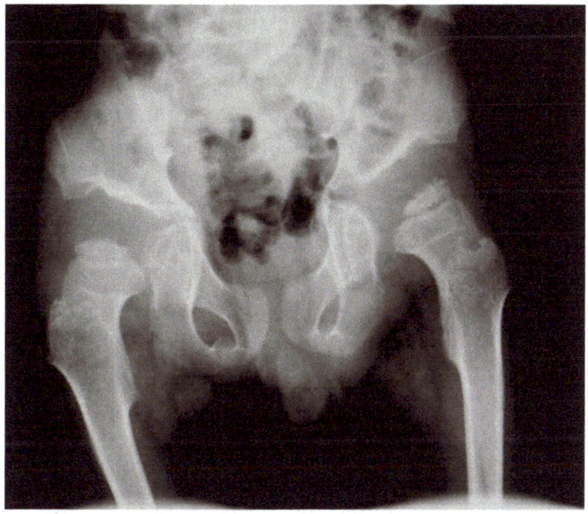

Fig. 9.5 Radiograph of the hip of a 9-year-old female with rapidly progressing mucopolysaccharidosis VI showing a dysplastic femoral head and severe hip dysplasia. Reproduced from Valayannopoulos V, Nicely H, Harmatz P, et al. Mucopolysaccharidosis VI. Orphanet J Rare Dis. 2010;5:5. doi:10.1186/1750-1172-5-5, an open-access article distributed under the terms of the Creative Commons Attribution License

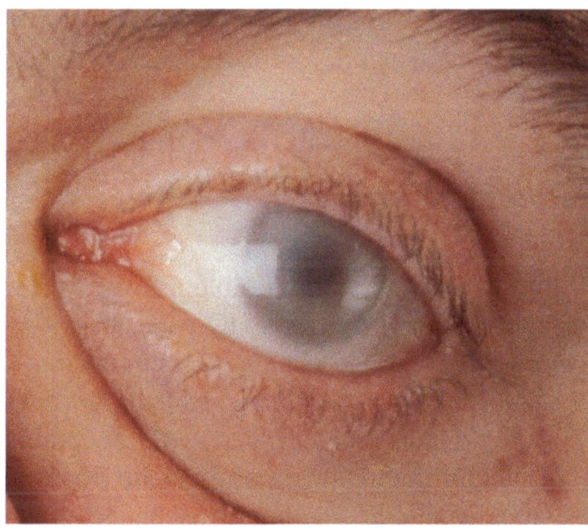

Fig. 9.6 Photograph of an eye of a 30-year-old male with rapidly progressing mucopolysaccharidosis VI showing clouding of the cornea. Reproduced from Valayannopoulos V, Nicely H, Harmatz P, et al. Mucopolysaccharidosis VI. Orphanet J Rare Dis. 2010;5:5. doi:10.1186/1750-1172-5-5, an open-access article distributed under the terms of the Creative Commons Attribution License

Tissue Plasminogen Activators

Natural human tPA is a serine protease synthesized by endothelial cells. It is a single polypeptide chain of 527 amino acids, MW ~ 70 kDa but in the presence of plasmin it is cleaved into two chains linked by one interchain disulfide bond. For the correct folding of tPA, correct pairing of the 17 disulfide bonds in the molecule is required. tPA is made up of five structural domains—a looped "finger" domain near the *N*-terminal, a growth factor domain, the kringle 1 and kringle 2 domains and, next to the latter, the serine protease domain. The finger and kringle 2 domains bind fibrin clots while the protease domain with its catalytic site at the C-terminus catalyzes the conversion of plasminogen to plasmin. Since tPA does not cause side effects such as systemic hemorrhaging and depletion of fibrinogen, investigations were undertaken to provide a suitable high yielding, reliable source of the protein. This led to transfection of CHO cells with the tPA gene, extraction of the recombinant product from the culture medium and attempts to produce the protein from *E. coli*. Currently approved tPAs are alteplase, reteplase, and tenecteplase (Table 9.2). Whereas alteplase is glycosylated and has 527 amino acids, the so-called third generation recombinant tPAs reteplase and tenecteplase are variants of tPA engineered for a longer half-life and resistance to inhibition. Deletion mutant reteplase is not glycosylated and has only 355 amino acids due to deletion of three of the tPA domains, kringle 1, finger, and epidermal growth factor domains. The kringle 2 and protease domains are retained, representing amino acids 1–3 and 176–527. The aim in developing reteplase was to produce a faster thrombolytic agent and one with a longer effective half-life without increasing the risk of thrombosis. Tenecteplase is the most fibrin specific of the tPAs. The 527 amino acid recombinant protein has modifications at three sites, positions 103, 117, and 296–9 (Table 9.2), that result in an increase in half-life, resistance to PAI-1 and thrombolytic potency against platelet-rich thrombi. Like natural human tPA, the recombinant forms bind fibrin in clots via the fibronectin finger (except reteplase) and kringle 2 domains before protease domain-mediated cleavage of a plasminogen Arg-Val bond to form plasmin which, in turn, exerts a thrombolytic action by degrading fibrin.

Asfotase Alfa

Asfotase alfa (Strensiq®) (Table 9.2) is a recombinant tissue-nonspecific alkaline phosphatase produced in CHO cells. It is a metallo-enzyme glycoprotein of two identical polypeptide chains each containing 726 amino acids and linked by two disulfide bonds. Each chain is made up of the enzyme catalytic domain, the human IgG1 Fc domain, and an aspartate decapeptide used to target bone. Asfotase alfa is indicated for the treatment of hypophosphatasia, an inherited bone disorder caused by mutations in the alkaline phosphatase gene that causes malformed bones, bone fractures, and loss of teeth. In the absence of alkaline phosphatase, the enzyme's substrates including inorganic pyrophosphate, accumulate, blocking hydroxyapatite crystal growth and inhibiting bone mineralization. This results in rickets, bone deformation, osteomalacia, and muscle weakness.

Asparaginase

Specific for the nonessential amino acid *L*-asparagine, the amidohydrolase *L*-asparaginase (Fig. 9.7) is prepared from *E. coli* and *Dickeya dadantii*, formerly named *Erwinia chrysanthemi*. Three asparaginase preparations are currently

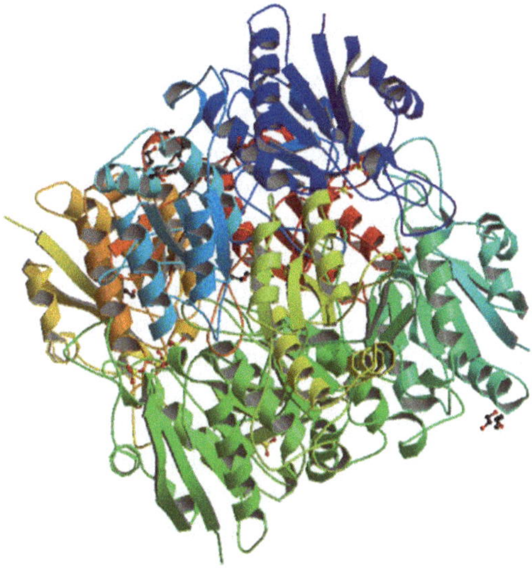

Fig. 9.7 X-ray structure of *L*-asparaginase from *Dickeya dadantii* (formerly *Erwinia chrysanthemi*) at 1Å resolution showing a crystal with four molecules of the enzyme, each of 327 amino acids. *L*-Asparaginase catalyzes the hydrolysis of *L*-asparagine to aspartic acid. Details of the enzymatic reaction and substrate specificity have not yet been fully worked out. Substrates for *L*-asparaginase include *L*-glutamine, *D*-asparagine, and succinic acid as well as *L*-asparagine. Image from RCSB Protein Databank PDB file 1O7J (Lubkowski J, Dauter M, Aghaiypour K, et al. Acta Crystallogr D Biol Crystallogr. 2003;59(Pt 1):84–92)

approved, one each from *E. coli* (Elspar®) and *Dickeya dadantii* (Erwinase®) and a pegylated preparation, pegaspargase (Oncaspar®), from *E. coli* (Table 9.2). A recombinant *E. coli* form has been developed and is being tested. Asparaginase catalyzes the hydrolysis of *L*-asparagine to *L*-aspartic acid and ammonia. Depletion of the supply of asparagine leads to cell cycle arrest in the G1 phase, inhibition of protein synthesis and apoptosis of lymphocytic leukemia cells. Use of the enzyme in acute lymphocytic leukemia (ALL) is based on the fact that susceptible leukemia cells cannot synthesize asparagine due to lack of the enzyme asparagine synthase and depend on an endogenous source of the amino acid for survival. Pegaspargase is a monomethoxypolyethylene glycol succinimidyl conjugate of *E. coli* *L*-asparaginase. It shows increased half-life and decreased immunogenicity and appears to be associated with improved outcomes when administered for ALL.

Collagenase

Collagenases are proteinases that hydrolyze collagen, and it is this property that has seen the enzyme from *Clostridium histolyticum* become the first approved nonsurgical treatment for Dupuytren's contracture (Table 9.2). Collagenase for this purpose was approved by the FDA in 2010 and by the EMA in 2011. Injection of collagenase into a Dupuytren's cord, which is mainly interstitial collagen, results in enzymatic cleavage and disruption of the cord. Collagenase as Xiaflex® (US) and Xiapex® (Europe) is a mixture of class I (AUX-I) and class II (AUX-II) *C. histolyticum* collagenases in a required ratio. Each class hydrolyzes collagen at different sites but act in a complementary manner to degrade the protein. Class I collagenases (α, β, γ, and η), products of the *colG* gene, hydrolyze collagen near the amino and carboxy termini generating large proteolytic fragments. Class II collagenases (δ, ε, and ζ), products of the *colH* gene, cleave interior sites of the molecule generating smaller fragments. Together, this different, but complementary, substrate specificity leads to effective degradation of the entire collagen molecule. All seven collagenases, α, β, γ, δ, ε, ζ, and η, are zinc proteinases functionally related to matrix metalloproteinases (MMPs) which, amongst other activities, degrade the extracellular matrix. Although it has been suggested that the sequence homology shared between AUX-I, AUX-II, and human MMPs indicates potential immunological cross-reactivity between the proteins with resultant antibody-induced inhibition of MMPs, studies so far show no evidence of any clinical adverse events associated with MMP inhibition in patients treated with collagenase.

In December 2013, the FDA approved collagenase (Xiaflex®) through a restricted treatment program called the Xiaflex® REMS Program for Peyronie's disease with a palpable plaque and curvature deformity of at least 30 degrees. Xiapex® was approved in Europe in 2015.

Dornase Alfa

Dornase alfa is a 260 amino acid recombinant human deoxyribonuclease I (rhDNase I) identical in composition to the natural enzyme (Table 9.2, Fig. 9.8). Viscous extracellular DNA, released mainly by disintegrating neutrophils during infection, accumulates in sputum of patients with cystic fibrosis contributing to reduced pulmonary function and frequent pulmonary infection. Mucus containing significant amounts of extracellular DNA from degenerating leukocytes is also a problem in bronchiectasis patients, patients with respiratory syncytial virus bronchiolitis, and in non-cystic fibrosis pediatric patients with atelectasis. Dornase alfa (Pulmozyme®) (Table 9.2) converts extracellular DNA to 5′-phosphonucleotide end products reducing both sputum viscosity in the airways and adhesiveness of lung secretions without affecting intracellular DNA and has proven to be an effective treatment in cystic fibrosis and the other conditions mentioned.

Glucarpidase

Glucarpidase (Voraxaze®) (Table 9.2), a recombinant pseudomonas carboxypeptidase G2 produced in *E. coli*, is used clinically to hydrolyze methotrexate and other antifolates. Methotrexate, administered for various cancers, is eliminated in the

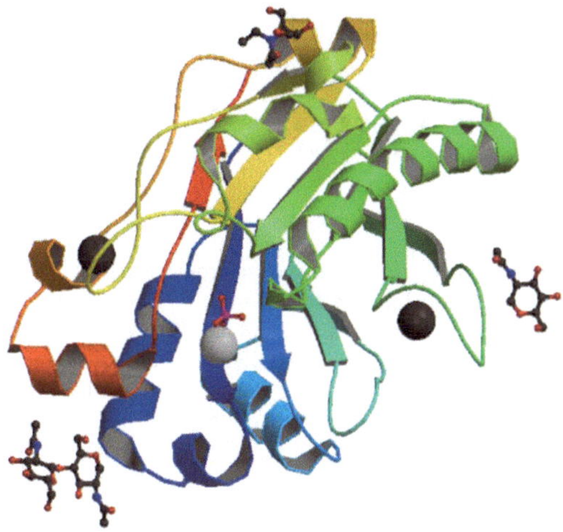

Fig. 9.8 Crystal structure at 1.95Å resolution of recombinant human Dnase (dornase alfa, Pulmozyme®) complexed with magnesium and phosphate ions, both of which are bound in the active site. Mutagenesis studies and structural analyses indicate a key catalytic role for Asn7. Dornase alfa hydrolyzes high concentrations of DNA making it a useful treatment for the lungs of cystic fibrosis patients. Image from RCSB Protein Databank PDB file 4AWN (Parsiegla G, Noguere C, Santell L, et al. Biochemistry. 2012;51:10250–8)

urine so patients with renal impairment given the drug may experience high plasma concentrations. Glucarpidase is indicated in such patients ensuring that methotrexate is eliminated enzymically, mainly by hepatic mechanisms, and not by the kidneys. Toxic blood levels of the drug can be rapidly decreased by intravenous administration of glucarpidase. Leucovorin, a reduced folate and potential substrate for glucarpidase, should not be given with the enzyme which degrades it. In addition, the drug competes with methotrexate for the enzyme.

Hyaluronidase

Hyaluronidases (also called hyals) are enzymes that degrade hyaluronan (also called hyaluronic acid) although, except for the enzymes from a few bacteria, substrate specificity generally extends to chondroitin and chondroitin sulfates. Found in the extracellular matrix of most types of connective tissue, hyaluronan is a high molecular mass glycosaminoglycan that may be as large as 6–8 MDa. It is a linear polysaccharide polymer of disaccharides of *D*-glucuronic acid and *N*-acetyl-*D*-glucosamine linked by alternating β 1,3 and β 1,4 glycosidic bonds (Fig. 9.9). Connective tissue, which acts as a barrier to the flow of fluid through the interstitial matrix, can become permeable after treatment with hyaluronidase which breaks down the hyaluronan component of the matrix. This increased permeability lasts for up to about 48 h after which collagen remains unchanged, and there are no signs of inflammation.

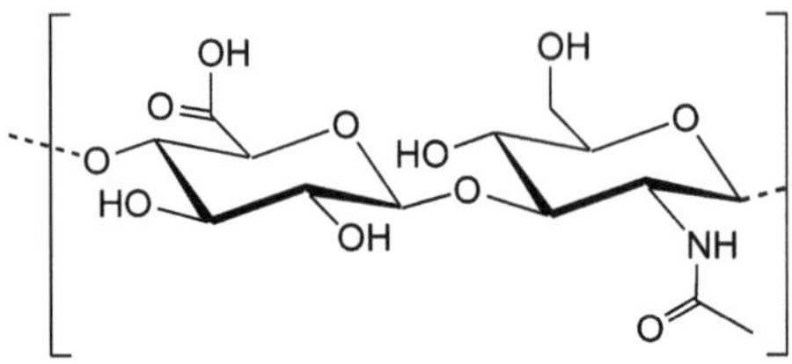

Fig. 9.9 Structure of hyaluronan (also known as hyaluronic acid), a high molecular mass glycosaminoglycan present in the extracellular matrix of most types of connective tissue. It is a linear polymer of disaccharides of *D*-glucuronic acid and *N*-acetyl-*D*-glucosamine linked by alternating β 1,3 and β 1,4 glycosidic bonds. Hyaluronidases (hyals) including the recombinant human hyaluronidase (rhuPH20) hydrolyze the glycosidic bonds between the β 1,4 linked *N*-acetyl-*D*-glucosamine and *D*-glucuronic acid sugars of hyaluronan

Hyaluronidases have been classified into different categories: prokaryotic enzymes, for example from bacteria, that act as β-endoglycosidases cleaving the β-(1→4) linkage in hyaluronan or as β-exoglycosidases that degrade the polymer by removing single monosaccharides; and two classes of eukaryotic endoglycosidase hydrolases, β-endoglycosidases (endo-β-*n*-acetylhexosaminidases and β-endoglucuronidases), and β-exoglycosidases (β-exoglucuronidase and exo-β-*N*-acetylglucosaminidase).

There are six endo-β-*n*-acetylhexosaminidase hyaluronidase-like sequences in the human genome but only three of the gene products, Hyal1, Hyal2, and PH20, are known to demonstrate hyaluronidase enzyme activity and, of these, only PH20 degrades glycosylaminoglycans under physiological conditions. PH20 (sperm adhesion molecule 1[SPAM1]; zona pellucida binding) is the predominant hyaluronidase in mammalian testes and is enzymically active at neutral pH. The human enzyme (Table 9.2) is a 509 amino acid glycoprotein attached to the outer cell surface membrane via a glycosylphosphatidylinositol linkage that facilitates sperm penetration and fertilization. For many years, hyaluronidase extracts used medicinally have been derived from animal testes, usually bovine or ovine, but these preparations have a number of undesirable features including low enzyme concentrations (typically < 1 % enzyme per unit weight of protein), the presence of contaminants (e.g., proteases, immunoglobulins, anticoagulants, and vasopermeability factors), and immunogenicity. To overcome the shortcomings of non-human, impure, animal-derived extracts, a recombinant form of PH20, rhuPH20 (Hylenex® recombinant), was generated in Chinese hamster ovary cells. This preparation was approved by the FDA in 2005. The recombinant enzyme which lacks the glycosylphosphatidylinositol linkage is a soluble single chain polypeptide with *N*-linked glycans. Like the animal-derived hyaluronidase preparations, it hydrolyzes the glycosidic bonds between the β 1,4 linked *N*-acetyl-*D*-glucosamine and *D*-glucuronic acid sugars of hyaluronan but appears to be well tolerated, eliciting neither inflammatory nor immunogenic responses following repeated subcutaneous injections. In a revealing comparison of the recombinant preparation with animal-derived enzyme preparations, five pharmacy-compounded formulations of the latter were found to have a low specific activity in the range 500–700 U/mg of protein, whereas an animal-derived preparation from a pharmaceutical company had an activity of 18,000 U/mg and rhuPH20 an activity of 120,000 U/mg. In a comparison of the recombinant with the compounded and pharmaceutical animal-derived forms of the enzyme, these figures represent increases in activity of ~240- to 7-fold. Electrophoretic analyses revealed that these differences could largely be accounted for by the presence of impurities in the animal-derived preparations.

For many years, hyaluronidase given subcutaneously has been used to facilitate the distribution and absorption of medications, for rehydration therapy and to administer urographic contrast media when the intravenous route is precluded. One important, and specialized, application of the enzyme is in ophthalmology, a use dating back 65 years when it was first included in retrobulbar blocks. Nowadays, it is used routinely for that purpose but also in peribulbar and sub-Tenon's blocks.

Questions such as lack of solubility, tissue irritation, and injection site pain, the need for rapid action, bioavailability, and tissue distortion have led to the administration of many drugs being restricted to the intravenous route. To help avoid this in some cases and overcome problems of drug absorption and dispersion, the so-called spreading agents have been utilized. Co-injection of animal-derived hyaluronidase preparations has been used for this purpose but usually in a restricted manner, for example, limited to emergency situations like the need for fluid hydration or during periocular anesthesia. Co-injection of the highly purified rhuPH20 in nanogram doses showed an absence of toxicity while permitting the administration of a volume up to five times greater than that normally given for drugs injected subcutaneously. Findings indicate that some drugs presently restricted to intravenous administration may now be accessible by a subcutaneous co-injection with the enzyme, thus making self-administration a possibility.

In October 2014, the FDA granted Orphan Drug designation for PEGylated recombinant human hyaluronidase (PEGrhuPH20) for the treatment of pancreatic cancer. The EMA granted the same designation in December 2014. The PEGylated form of the enzyme is under development for the systemic treatment of tumors that accumulate hyaluronan. PEGrhuPH20 is currently being investigated in a phase 2 study in combination with gemcitabine and nab-paclitaxel (Abraxane®) in metastatic pancreatic cancer. Nab-paclitaxel is albumin-bound paclitaxel, the albumin being in nanoparticle form to overcome the toxicity associated with the solvent Cremophor EL normally used to dissolve paclitaxel.

Ocriplasmin

The use of plasmin to degrade fibrin polymers has been a priority in human therapy for more than 60 years. In the recombinant protein age, it was found that intracellular plasminogen activators made the production of plasminogen, the precursor of plasmin, difficult. This led to the production of short forms of plasmin that retained fibrinolytic activity. Ocriplasmin (Jetrea®; also known as des-kringle 1–5 plasmin or microplasmin) (Table 9.2), a proteolytic enzyme and truncated form of human plasmin produced in a yeast *Pichia pastoris* expression system, lacks all five kringle domains and is composed of 249 amino acids in the form of two polypeptide chains linked by two disulfide bonds joining residues 6 and 124 and 16 and 24. Four intra-chain disulfide bonds stabilize the larger polypeptide chain of 230 amino acids. Ocriplasmin has activity against the clinically relevant plasmin receptors fibronectin and laminin, components of the vitreoretinal interface. Vitreomacular adhesion may lead to traction and macular hole but clinical trials have shown that intravitreal injection of ocriplasmin can induce separation of the vitreous and macular surfaces thereby resolving vitreomacular traction and closing macular holes.

Pegademase Bovine

Pegademase bovine, the first enzyme approved by the FDA as an orphan drug, is adenosine deaminase of bovine intestine origin extensively pegylated (Table 9.2) to increase half-life. Pegademase bovine as Adagen®, used as therapy for severe combined immunodeficiency disease (SCID), also provided the first successful enzyme treatment for an inherited disease. SCID is a primary (i.e., inherited) immune deficiency caused by several different genetic mutations affecting the immune system with at least 12 genes implicated. The disease is referred to as "combined" because both cell-mediated (T lymphocytes) and humoral (B lymphocytes) immunity are affected. The most common genetic defect in SCID is an X-linked mutation or the so-called gamma chain defect leading to decreased amounts of immunoglobulin G. This form represents about 50 % of cases while defects involving the gene for adenosine deaminase account for ~15 %. A deficiency of adenosine deaminase leads to the accumulation of adenosine, 2′-deoxyadenosine, and their metabolites which are toxic to lymphocytes. A deficiency of T lymphocytes, particularly functional helper T cells and B lymphocytes, results in a markedly decreased production of antibodies and impairment of both arms of the adaptive immune response. Therapy with pegademase bovine has been shown to diminish the frequency of opportunistic infections and relieve symptoms of diarrhea, failure to thrive, and dermatitis. Treatment with the enzyme does not preclude a subsequent bone marrow transplant from an HLA-identical donor.

Pegloticase

Rasburicase (see below), developed for treating tumor lysis syndrome (TLS) in pediatric cancer patients, was also found to lower serum urate levels in patients with gout but the use of rasburicase for this purpose was found to be limited by its immunogenicity and relatively short half-life. Pegloticase (Table 9.2) was developed to overcome this. Pegylated to reduce its potential for immunogenicity and to increase its circulatory half-life, pegloticase (Krystexxa®), a recombinant porcine/baboon variant uricase (urate oxidase) produced in *E. coli*, is used for therapy of gout previously refractory to conventional therapy with uricostatic (e.g., allopurinol) and uricosuric drugs (e.g., probenecid, colchicine, sulfinpyrazone). Nine of the 30 lysines of this tetrameric peptide are pegylated giving a final MW of 545 kDa. The enzyme urate oxidase, which is absent in humans, catalyzes the oxidation of uric acid to 5-hydroxyisourate and hydrogen peroxide. The former compound is unstable and breaks down to racemic allantoin which is much more soluble than uric acid and readily excreted. Pegloticase has been evaluated for efficacy in at least eight clinical studies.

Rasburicase

Rasburicase (Table 9.2), a recombinant *Aspergillus flavus*-derived urate oxidase expressed in *Saccharomyces cerevisiae*, is almost identical to the natural *Aspergillus* enzyme, differing only by having a higher specific activity and by a modified reactive cysteine. Rasburicase (Elitek®; Fasturtec®) has proved effective in managing TLS which usually occurs 48–72 h after initiation of cancer therapy when large numbers of tumor cells undergo apoptosis in a short time, releasing their intracellular contents into the circulation. This causes an ionic imbalance involving hyperkalemia and hyperphosphatemia, secondary hypercalcemia and hyperuricemia, and possibly acute kidney injury and death. In pediatric patients with acute leukemia and lymphoma, rasburicase has proved to be the treatment of choice and superior to allopurinol but results with adults have been less clear. However, in a study of the efficacy of the preparation for the prevention and treatment of hyperuricemia during induction of chemotherapy for aggressive non-Hodgkin lymphoma in adults, results showed the enzyme to be a highly effective, fast-acting, and reliable therapy.

Streptokinase

Streptokinase (Table 9.2) is a potent plasminogen activator but, unlike tPA, streptokinase is not a protease. Streptokinase (Streptase®; Kabikinase®) has been used for more than 35 years as a thrombolytic agent to treat blood vessel blockages such as acute myocardial infarction. Despite this, its mechanism of action remains incompletely understood. Streptokinase forms complexes with human plasminogen, a single chain protein of 791 amino acids, and plasmin and these complexes can hydrolytically activate other plasminogen molecules. Neither streptokinase nor human plasminogen has enzymatic activity but when streptokinase binds plasminogen it induces within it an active site by a non-proteolytic mechanism. Combined, the streptokinase-plasminogen complex assumes amidolytic activity and is rapidly converted to streptokinase-plasmin even though the activating peptide Arg560-Val561 remains intact. After formation of streptokinase-plasmin, both entities are able to act as plasminogen activators to catalyze the hydrolysis of the Arg560-Val561 bond of further plasminogen molecules (Fig. 9.10). Binding of streptokinase to plasmin also decreases the inactivation of the complex by α_2-antiplasmin and α_2-macroglobulin. The mechanisms of the conformational activation of plasminogen and the change in substrate specificity shown by plasmin after streptokinase

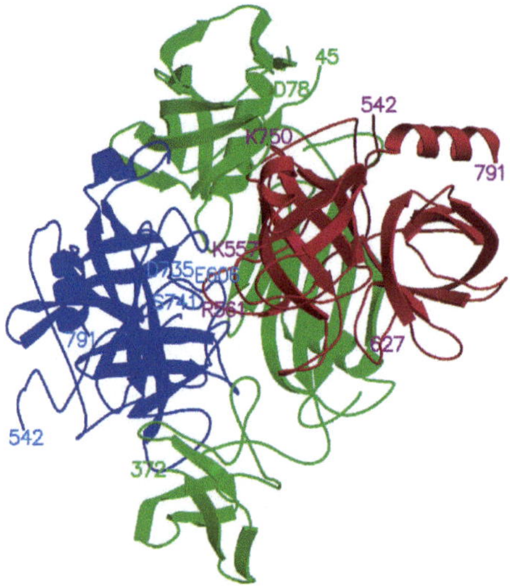

Fig. 9.10 X-ray structure at 2.9Å resolution of the streptokinase-plasmin catalytic domain complex docking with the catalytic domain (residues 542–791) of plasminogen. Streptokinase is shown in green, the plasmin catalytic domain is in blue, and the substrate plasminogen catalytic domain is in red. The activation bond Arg561-Val562 of plasminogen is in the catalytic site of the streptokinase-plasmin catalytic complex. The *N*-terminal Val562 forms a critical salt linkage to Asp740 but in the absence of the valine, a lysine at position 698 can form the salt bridge. Besides salt linkages, interactions include hydrogen bonds involving residues 625–629 of plasminogen. Active site residues Arg561from the substrate and Ser741from plasmin are labeled. Reproduced from Wang X, Lin X, Loy JA, et al. Science. 1998;281:1662–5 with permission from The American Association for the Advancement of Science

binding remain unknown. The proteolytic conversion of plasminogen to plasmin is also poorly understood. Streptokinase is composed of three domains designated α (amino acid residues 1–150), β (residues 151–287), and γ (residues 288–414). Although each of these domains bind plasminogen, no single domain can, by itself, activate plasminogen but residues 1–59 of the streptokinase α domain appear to be particularly important in the activation of plasminogen.

Fibrinolytic therapy is given to dissolve thrombus and restore blood flow, and this is usually effected by administration of streptokinase or tPAs (section "Tissue Plasminogen Activators"). Indications for fibrinolytic therapy are symptoms of myocardial infarction of less than 12 h duration with ECG changes of ST elevation or left bundle block. Without the ECG changes, patients should not be given fibrinolytic therapy. Bleeding, especially intracerebral bleeding, associated with fibrinolytic therapy is a concern and tPAs produce more of both than streptokinase resulting in two to three extra strokes with one death per 1000 treated patients. This higher incidence of strokes and the significantly increased cost of tPAs compared to streptokinase may be offset by the former's net clinical benefit reflected in the reported 10 extra survivors for every 1000 treated patients.

Safety of Approved Enzymes Used as Therapy for Lysosomal Storage Diseases

Two general points need emphasis in relation to the safety of enzymes administered as enzyme replacement therapies. Since the enzyme preparations are, in the main, similar to the enzymes produced naturally in humans, adverse events, especially severe ones, are generally not expected. They do occur, of course, but the symptom range is often relatively small and reactions tend to be mild. Note, however, that since enzymes are proteins there is always the *potential* for immediate allergic reactions, including anaphylaxis, and boxed warnings mentioning anaphylaxis have been issued for six of the enzymes listed in Table 9.2. Secondly, many of the enzymes have been developed and administered under orphan drug programs and have therefore so far been given to relatively small numbers of patients meaning that collected safety data constitutes a much smaller body of results than is usually the case with newly approved medical agents.

Agalsidase Beta for Fabry Disease

An infusion reaction, occurring in ~10–14 % of patients, is the most common adverse event seen during administration of agalsidase for Fabry disease. Reactions, usually mild with no sign of respiratory symptoms, urticaria or changes in vital signs, occur with a higher frequency (~18 %) in children. Symptoms tend to be easily controllable with antihistamines and corticosteroids. The overall general absence of severe adverse events is reflected in each of the reported adverse event categories (Table 9.2), for example, cardiac disorders, when they occur, are commonly tachycardia and/or palpitations and gastrointestinal reactions tend to be nausea, vomiting, and diarrhea.

Alglucosidase Alfa for Pompe Disease

Now more than a decade old, the Genzyme Corporation MA, USA-sponsored Pompe Registry, the world's largest repository of worldwide data on Pompe disease (section "Living with Enzyme Replacement Therapy"), was set up to further understand the natural history and clinical symptomatology of, and treatment outcomes for, the disorder.

The two registered commercial preparations of alglucosidase alfa (Table 9.2), Lumizyme® and Myozyme®, are seen by the FDA as being biologically different preparations due to a different manufacturing process. Lumizyme® is approved by the FDA as ERT for late-onset, i.e., non-infantile, Pompe disease without evidence of cardiac hypertrophy in patients more than 8 years old; Myozyme® is approved for infantile-onset Pompe disease. Despite the apparent enzyme identity of the two preparations, the listed adverse reactions of each show some significant differences.

Whereas both carry warnings for severe allergic reactions including life-threatening anaphylaxis and cardiorespiratory failure, Lumizyme® is responsible for a wide range of reactions including pyrexia, flushing, hyperhidrosis, headache, hypertension, dizziness, rash, and urticaria while for Myozyme®, infusion reactions, infections such as pneumonia and respiratory syncytial virus, respiratory distress/failure, and gastrointestinal problems predominate. This difference is probably related as much to the age of patients and the administered dose of enzyme as to the alleged "biological difference" between the preparations with Lumizyme® given to older patients with late-onset disease and Myozyme® used to treat infants. In a randomized study of alglucosidase alfa for late-onset Pompe disease in 60 patients, most enzyme-induced adverse events were not serious except for three cases of anaphylaxis, two with respiratory and cutaneous reactions and the third with severe tongue swelling. Serum IgE antibodies to alglucosidase alfa were detected in two of the patients. Myozyme® treatment of 18 patients with infantile-onset Pompe disease caused mild to moderate infusion reactions in 11 patients within 2 h of the infusion. The most common adverse manifestations, urticaria, fever, and decreased oxygen saturation, occurred in the patients given 40 mg/kg.

Recombinant Enzymes Used to Treat Gaucher Disease: Imiglucerase, Taliglucerase Alfa, and Velaglucerase Alfa

Imiglucerase, taliglucerase alfa, and velaglucerase alfa, each administered for long-term enzyme replacement therapy for Gaucher disease, show a similar spectrum of adverse events with immediate type I hypersensitivity reactions being the most prominent (Table 9.2). As with all orphan drugs, limited safety information was available at the time of marketing authorization. A long-term international post-marketing safety surveillance of imiglucerase carried out on more than 4500 patients treated with the drug over an 8-year period (1997–2004) found that the most common and consistently reported adverse events could be classified into three categories: general disorders and administration site reactions (i.e., pyrexia, chills, chest discomfort); skin and subcutaneous tissue disorders (pruritus, rash, urticaria); respiratory, thoracic, and mediastinal disorders (dyspnea, cough, throat irritation). Adverse events related to imiglucerase were most frequently associated with infusion of enzyme, they occurred with a low incidence, generally <1 %, and were predominantly self-limiting. Post-marketing analysis of adverse events in children (2–12 years) given imiglucerase identified dyspnea, fever, nausea, flushing, vomiting, and coughing as the most common events. The most common events for adolescents and adults were headache, pruritus, and rash. The affect on safety of the frequency of administration of imiglucerase was assessed in a comparison of intravenous infusions given to adults once every 4 or every 2 weeks with the same total monthly dose. The frequency of adverse events (which were generally mild) was comparable between the treatment groups.

A successful protocol for rapid desensitization to imiglucerase in a hypersensitive adult female patient with Gaucher disease was recently published. Imiglucerase therapy was terminated at the age of 19 because of intolerance (headache, flushing, tachycardia) to the drug. However, nearly 3 years of substrate reduction therapy with miglustat (N-butyldeoxynojirimycin, Zavesca®) resulted in deterioration of the patient's condition and a return to imiglucerase infusion therapy every 2 weeks. After tolerating the first three infusions, the patient experienced flushing, periorbital angioedema and pruritus of the wrists, ankles, and at the infusion site, ten minutes after the beginning of the fourth infusion. At a subsequent attempt, the patient was premedicated with intravenous dimethindene maleate before an attempt was made to deliver the required drug dose in four consecutive steps of increasing administration rate but, once again, infusion had to be stopped 5 min into the final step due to the patient's drug intolerance. Intradermal skin testing undertaken with imiglucerase solution proved positive prompting efforts to develop a 12-step rapid desensitization protocol beginning 20 min after premedication with intravenous dimethindene maleate. Stepwise increases of drug from an initial dose of 0.03 U were given at 15 min intervals except for the final large dose which was prolonged to deliver 92 % of the target dose of 1500 U. The whole procedure was successfully completed in less than 6 h and the patient remained free of adverse effects after two infusions per week over a (so far) 2-month period.

A summary of the adverse events following taliglucerase alfa, including a warning for anaphylaxis, allergy, and infusion reactions, recorded by the FDA is shown in Table 9.2. In a phase III multinational double-blind, 9-month, 20-infusion clinical trial, taliglucerase alfa-related adverse events were mild/moderate and transient. No serious drug-related events occurred although two patients (6 %) developed hypersensitivity reactions. Twenty adverse events related to taliglucerase alfa treatment were reported in eight patients with Gaucher disease studied for bone marrow responses; one experienced a hypersensitivity reaction during infusion, one had infusion-related dizziness, chills, and nausea and one developed a fixed drug reaction. Some clinicians believe that taliglucerase alfa seems to have the poorest safety profile of the three recombinant enzymes used to treat Gaucher disease and while usage may reflect this, there is currently an absence of figures to decide the question.

As with imiglucerase and taliglucerase alfa used for ERT, velaglucerase alfa is generally well tolerated with drug-induced adverse events usually mild to moderate. The list of adverse events (Table 9.2) for velaglucerase alfa is similar to the lists for the other two β-glucocerebrosidases although there are reports of severe prolonged activated partial thromboplastin time and allergic dermatitis occurring with velaglucerase alfa. The safety of the enzyme was assessed in 12 patients with type I Gaucher disease in a 9-month phase I/II and extension study. Adverse events with an incidence of >10 % were infusion reactions, dizziness, headache, nausea, back pain, bone pain, and pyrexia. No serious events related to therapy were reported. Events with an incidence of >10 % and related to treatment during the extension study were tremor, epistaxis, abdominal pain, pain in extremity, and fatigue. Results from an FDA-approved Early Access Program using pre-licensed velaglucerase alfa revealed no enzyme-related serious adverse events and no withdrawals. One patient out of the 71 treated who had previously had an allergic reaction to taliglucerase, experienced an allergic reaction during the first infusion of velaglucerase and a second switch-over patient had a fixed drug reaction at the first infusion.

Sebelipase Alfa for Lysosomal Acid Lipase Deficiency

The possibility of a hypersensitivity reaction, including anaphylaxis, is prominent amongst the issued precautions to be aware of during the use of this enzyme. In clinical trials, ~20 % of 106 infused patients experienced symptoms of hypersensitivity (fever, chills, pallor, rash, laryngeal edema, nausea, rash, pruritus, vomiting) while in ~3 % of patients the reaction was judged to be anaphylactic. Anaphylaxis has been reported as early as the sixth infusion and even after a year of treatment. Sebelipase alfa (Kanuma®) is produced in chicken egg white and the risk and warning of the possibility of anaphylaxis has therefore been extended to eggs and egg products. In clinical trials so far, the most common adverse reactions to the enzyme in patients with rapidly progressing LAL deficiency presenting in the first 6 months of life are diarrhea, vomiting, fever, rhinitis, anemia, cough, and nasopharyngitis. In older infants/children and adults, the list of adverse reactions includes headaches, fever, oropharyngeal pain, nasopharyngitis, asthenia, constipation, and nausea (Table 9.2).

Laronidase for MPS I

Results of trials designed to examine the efficacy and safety of laronidase (α-*L*-iduronidase) in the treatment of MPS I showed the enzyme to be generally well tolerated with few treatment-related events and few, if any, serious adverse reactions. Apart from an FDA boxed warning issued for the potential of anaphylaxis and in common with other enzymes used in ERT, the list of side effects is not extensive and lacks serious events (Table 9.2). In a long-term trial, laronidase infusions provoked mostly mild and easily managed reactions in 53 % of 40 patients. One patient experienced an anaphylactic reaction but, in the main, reactions diminished markedly after 6 months.

Idursulfase for MPS II

A phase II/III idursulfase ERT study of 32 Hunter syndrome patients (ages 5–31) showed that the enzyme was well tolerated over the 1-year treatment period. Again, the most common adverse events were infusion-based reactions. The incidence of these reactions reached a maximum between weeks 4 and 12 and declined thereafter. Adverse events, occurring with at least a 9 % greater frequency than in the placebo-treated group, were headache, nasopharyngitis, abdominal pain, arthralgia, pruritus, pruritic rash, swelling at the infusion site, urticaria, dyspepsia, anxiety, and chest wall pain. A follow-up long-term, open-label extension study showed that 50 of the 94 patients (53 %) experienced at least one infusion-related reaction with headache, urticaria, and pyrexia the most common symptoms. In a Japanese idursulfase ERT study of ten adult patients with MPS II, five of ten patients experienced drug-related infusion reactions occurring within 24 h of the infusion. Skin reactions, namely urticaria and erythema, resulted in patients who showed the highest incidences. Two patients experienced serious reactions, one involving flushing, diffuse urticaria, and numbness of the tongue after commencement of the fifth infusion. A study designed to evaluate the occurrence of infusion-related reactions in patients with MPS II receiving idursulfase reported 65 reactions in 33 patients (31.7 %) in the first year of ERT with almost all of the initial reactions occurring within the first 3 months. Most reactions were mild to moderate and could be managed by slowing the rate of infusion or by giving antihistamines or antipyretics. Idursulfase is yet another enzyme used for ERT that carries an FDA black box warning of the risk of anaphylaxis after/during infusion of the drug. In a recent Korean study, anti-idursulfase IgE antibodies were detected in 34 patients taking the drug for MPS II. IgE antibodies were detected in three patients (8.8 %) by skin prick testing, an enzyme-linked immunosorbent assay, and Western blotting. All three patients with anaphylaxis proved to be skin test positive. The authors stated that they were unable to identify any risk factors for the development of infusion-related immediate allergic reactions to idursulfase. A list of the adverse events to idursulfase assembled by the FDA is shown in Table 9.2.

Elosulfase Alfa for MPS IVA

Collected results from six different clinical trials involving 235 patients revealed that ~19 % of patients infused with elosulfase alfa experienced a hypersensitivity reaction with 8 % classified as anaphylactic. This finding is reflected in an FDA box warning (Table 9.2). Enzyme efficacy did not appear to be impaired in the hypersensitive patients. Safety data collected in a phase III controlled study of ERT with elosulfase alfa for MPS IVA (176 patients aged ≥ 5 years) showed that 22.4 % of patients had an adverse reaction leading to an extended interruption or discontinuation of infusions. A list of adverse events to the enzyme reveals that most are mild such as nausea, vomiting, chills, headache, and abdominal pain.

Galsulfase for MPS VI

Galsulfase is generally considered to be a relatively safe drug when used for ERT with infusion-related reactions being the most commonly occurring, and predominant, adverse event. This is not unusual with the enzymic preparations used for ERT or, in fact, for intravenously administered protein therapeutics in general. Other reported adverse events following galsulfase include headache, pyrexia, limb, chest and ear pain, visual abnormalities, anxiety, dyspepsia, upper respiratory infections, and cough (Table 9.2). In their extended examination of galsulfase therapy for MPS

VI in phase I, II, III, and long-term follow-up studies, Harmatz and coworkers observed an incidence of infusion reactions of 20–75 %. Serious drug-related adverse events were uncommon with a total of only 1.8 % of patients experiencing such reactions in the published studies; three of the more serious ones were asthma, apnea, and urticaria. An assessment of the efficacy and safety of galsulfase for MPS VI in 34 children less than 5 years of age showed no serious infusion-related reactions although eight children (24 %) experienced some treatment-related adverse events. Infusions were continued in each case with the aid of slower infusion rates and antipyretics if needed. Skin rash and edema occurred in one patient who had been on ERT for 2 years and single episodes of tachycardia, skin rash, nausea, elevated blood pressure, and pyrexia were recorded.

Slowing the infusion rate and pretreatment with antihistamines, steroids, and/or antipyretics has often been used in attempts to manage infusion reactions during ERT. For example, in one report of the successful management of a reaction in a young patient that precluded further therapy, a significant reduction in the rate of infusion was slowly reversed over a period of months while at the same time steroid premedication was introduced. The patient ultimately tolerated infusions with no adverse effects.

There has been a report of a single case of thrombocytopenia after the third ERT infusion of galsulfase. A decrease in dose of the enzyme followed by a subsequent return to the normal dose led to a reversal of the condition. The authors speculated that antibodies to galsulfase may have been the cause.

With ocular pathology such as corneal clouding, ocular hypertension, and optic nerve swelling (papilledema) common in MPS VI patients, the report of reversed papilledema and improved visual acuity in an 11-year-old MPS VI patient receiving galsulfase is both curious and encouraging.

Adverse Events Caused by Other Approved Enzymes

Tissue Plasminogen Activators

The most frequently seen and potentially serious adverse event associated with thrombolytic agents is bleeding, especially at intracranial, gastrointestinal, retroperitoneal, and pericardial sites. In an international randomized trial (GUSTO) conducted in 1081 hospitals in 15 countries and involving 41,021 patients with evolving myocardial infarction, tPA was compared with streptokinase and heparin given subcutaneously (sc) and intravenously (iv). Mortality rates in the four treatment groups were streptokinase and heparin sc, 7.2 %; streptokinase and heparin iv, 7.5 %; tPA and iv heparin, 6.3 %; and the combination of both thrombolytics with iv heparin, 7 %. These figures reveal a 14 % reduction (95 % confidence interval 5.9–21.3 %) in mortality compared to tPA. Corresponding rates for hemorrhagic stroke were 0.49 %, 0.54 %, 0.72 % and 0.94 %, respectively, a significantly higher incidence for tPA (p=0.03). For death and disabling stroke combined, the incidence was lower in the tPA group (6.9 %) than the streptokinase groups (7.8 %; $p=0.006$). Combined results from trials conducted in the late 1980s showed an incidence of stroke of 1.2 % for patients receiving alteplase and 0.9 % for patients receiving placebo. Intracerebral hemorrhage is a severe adverse event for tPA therapy for acute myocardial infarction but for alteplase, the frequency of intracerebral hemorrhage combined with cerebral infarction and subdural hematoma is comparable to incidences of adverse events seen with other thrombolytic agents in myocardial infarction. Use of alteplase to salvage dysfunctional central venous access devices due to thrombosis was undertaken to analyze the efficacy and safety of the drug after administration of up to 2 mg instilled into the lumen of the central venous catheter and allowed to remain for up to 2 h. If the device was still occluded, the procedure was repeated. A total of 1064 patients with dysfunctional catheters were treated. Serious adverse events occurring within 30 days of treatment were gastrointestinal bleeding (0.3 %), thrombosis (0.3 %), and sepsis (0.4 %). No cases of intracranial hemorrhage or embolic events were seen. Intrapleural installation has also produced adverse reactions to alteplase. For example, in one study where the enzyme was used in the management of complicated pleural effusion or empyema, chest pain, and bleeding at the chest tube site were seen in 6 % and 2 %, respectively of 120 patients. At least 41 cases of orolingual angioedema after alteplase administration have been recorded. Although rare, the condition, characterized by swelling of the upper lip and tongue, develops within minutes of injection causing airway obstruction and breathing problems. These characteristics of the reaction show a clear resemblance to a type I immediate hypersensitivity anaphylactic-like response mediated by IgE antibodies. A number of studies have reported that the development of orolingual angioedema is associated with the use of angiotensin-converting enzyme (ACE) inhibitors. Sepsis, venous thrombosis, and allergic reactions are other well-known adverse events resulting from alteplase therapy.

As with other thrombolytics, bleeding is the most common adverse event seen with reteplase and tenecteplase (Table 9.2). There appears to be no significant difference in the risk of hemorrhage and stroke between reteplase and alteplase and the risk of stroke (1.2 % for reteplase in 3288 patients) is similar to other thrombolytic agents. The incidence of intracranial hemorrhage for reteplase is ~0.8 %, and this increases with age and elevated blood pressure. Any bleeding, regardless of severity, is said to occur with an incidence of ~20 %. In a comparison of thrombolysis achieved with reteplase and alteplase in patients with acute myocardial infarction, there was no significant difference in bleedings requiring a transfusion or in the incidence of hemorrhagic stroke. Chest pain is a common side effect of reteplase occurring in more than 10 % of patients. Cardiac events such as arrhythmias, circulatory collapse, and another heart attack are seen in 1–10 % and heart, or heart valve damage, a blood clot in the lungs, and hypersensitivity occur in 0.1–1 % of treated patients. To evaluate the safety of tenecteplase, 3235 patients with acute myocardial infarction were given the enzyme as a single bolus (30–50 mg). Total stroke rate at 30 days was 1.5 % and intracranial hemorrhage occurred in 25 patients (0.77 %). Death, death or nonfatal stroke, or severe bleeding occurred in 6.4 %, 7.4 %, and 2.8 % of patients, respectively. A comparison of single-bolus tenecteplase with front-loaded alteplase in acute myocardial infarction revealed an almost identical 30-day mortality rate (6.18 % tenecteplase vs. 6.15 % alteplase), similar rates of intracranial hemorrhage (0.93 % vs. 0.94 %), and a similar rate of death or nonfatal stroke at 30 days but there were fewer non-cerebral bleeding complications and less need for blood transfusions with tenecteplase. The similarity in side effects between alteplase, reteplase, and tenecteplase extends to the possibility of allergic reactions including anaphylaxis; the incidence of such reactions to tenecteplase is <1 %. Other recorded adverse events to tenecteplase include nausea, vomiting, fever, hypotension, and a range of cardiac abnormalities. Expanded lists of side effects for the three tPAs are set out in the regulatory literature.

Asfotase Alfa

Warnings and precautions for this recombinant nonspecific alkaline phosphatase formulation cover hypersensitivity reactions including anaphylaxis, localized lipodystrophy including lipoatrophy and lipohypertrophy at injection sites, and ectopic calcifications particularly of the eye (including the cornea and conjunctiva), and the kidneys. These adverse events constitute the most common safety considerations during asfotase alfa therapy. Less commonly occurring adverse reactions are hypocalcemia, renal stones, chronic hepatitis, and decreased vitamin B6 levels.

L-*Asparaginase*

Allergic reactions (including anaphylaxis), pancreatitis, hyperglycemia, hepatotoxicity and abnormal liver function, posterior reversible encephalopathy syndrome (PRES), CNS dysfunction, and thrombosis are listed as the most serious adverse events amongst multiple toxic effects induced by *L*-asparaginase therapy (Table 9.2). More than 30 years ago, *L*-asparaginase-induced coagulopathy involving intracranial hemorrhage with thrombosis of the extremities, immune hemolytic anemia, and abnormal collagen-stimulated platelet aggregation was described and discussed. This profound effect on the coagulation and fibrinolytic systems appears to be the result of drug-impaired synthesis of proteins including fibrinogen, antithrombin III, protein C, and plasminogen (Chap. 10). The incidence of thrombotic complications is in the range 3–7 %, usually occurring within days of *L*-asparaginase administration. Hypersensitivity reactions to the enzyme, occasionally fatal, are well known especially in children where allergic sensitivity can sometimes be avoided by switching from the *E. coli* to the *Dickeya* (*Erwinia*) enzyme. The impact of *L*-asparaginase-induced hypersensitivity on the duration of survival was assessed in adults with ALL. Leukemia-free survival did not correlate with the occurrence of hypersensitivity reactions and, in general, *E. coli L*-asparaginase was well tolerated. The authors concluded that the possibility of *L*-asparaginase becoming inactivated by an immune response becomes less important when secondary treatment with

Dickeya enzyme is available. A comparison of *E. coli* *L*-asparaginase (half-life 26 h) and polyethylene glycol (PEG)-conjugated enzyme (half-life 5.5 days) for the treatment of children with ALL showed no correlation between *L*-asparaginase activity and serum levels. This trial, involving 118 children, ultimately led on to FDA approval of pegaspargase for the first-line treatment of ALL. A recent successful pediatric treatment regimen of multiple doses of intravenous pegaspargase for adult leukemia patients showed hyperbilirubinemia and transaminitis as the most common grade 3/4 toxicities. An interesting combination of two complications of *L*-asparaginase therapy was recently reported in a patient who developed transient diabetes mellitus with ketoacidosis and acute pancreatitis.

Collagenase

Many of the adverse events recorded for collagenase (Table 9.2) are associated with the use of the enzyme to treat Dupuytren's contracture and Peyronie's disease. Reactions occurring in ≥5 % of patients with Dupuytren's contracture include peripheral edema, contusion, injection site reactions, pain in extremities, and pruritus. In Peyronie's disease, the most frequently reported events are penile hematoma, swelling, pain and ecchymoses, contusion, injection site hemorrhage, erectile dysfunction, and genital pruritus. The FDA issued a boxed warning for corporal rupture or penile fracture when collagenase is used for the treatment of Peyronie's disease and stated that cases of severe penile hematoma have been reported. Precautions are necessary to avoid the risk of injecting into the urethra, nerves, blood vessels, and corpora cavernosa. Other issued precautions relate to the possibility of hypersensitivity reactions including anaphylaxis and avoidance of the enzyme in patients with coagulation disorders, including those receiving concomitant anticoagulants (except low-dose aspirin).

Dornase Alfa

Treatment of 968 adult and child cystic fibrosis patients with aerosoled recombinant human DNase for 24 weeks proved to be well tolerated with only voice alteration and laryngitis observed more often in the treatment group than the placebo group. Even these reactions were mild and resolved within 21 days. No signs of anaphylaxis were seen. Adverse events occurring in a 12-week trial of dornase alfa administered to patients with advanced cystic fibrosis predominately involved the respiratory system but no overall difference between the test and placebo groups were seen. Cystic fibrosis-related adverse events such as chest pain, dyspnea, and hemoptysis occurred with similar frequency in both groups. No serious adverse events were recorded. Non-respiratory events that occurred more frequently (≥3 %) in the dornase alfa recipients included fever, rhinitis, pharyngitis, a decrease in forced vital capacity (FVC), voice alteration, and dyspepsia. Adverse events involving sputum changes (consistency, color, increase or decrease) and its mobilization (e.g., inability to bring it up), and deaths during the treatment period, were similar in the treated and control groups. A controlled trial of dornase alfa conducted before and after physiotherapy, produced 34 adverse events in 26 of 52 cystic fibrosis patients. Upper respiratory tract infections caused by viruses and suppurative lung disease exacerbated by sputum pathogens accounted for the majority of the events. Rash, urticaria, abdominal pain, nausea, vomiting, pyrexia, and headache have also been reported following dornase alfa therapy.

Glucarpidase

High-dose methotrexate therapy used to treat a number of malignancies can be problematic, leading to toxicities such as liver dysfunction and renal failure that delay clearance of the drug, thus significantly increasing the risk of even greater toxic effects. Even though implementation of a number of measures specifically designed to reduce the incidence of life-threatening methotrexate-induced toxicities have proved largely effective, renal failure may still occur resulting in the drug's accumulation. A number of studies show that glucarpidase is generally well tolerated with no major adverse events requiring intervention. The most common adverse events reported from clinical trials and compassionate use experience are flushing, paresthesia, headache, shaking, hypotension, nausea and vomiting, tingling fingers, burning of face and extremities, and a feeling of warmth. Some drugs such as penicillins, sulfonamides, nonsteroidal antiinflammatory drugs, and proton pump inhibitors may interfere with methotrexate elimination and should not be given with the enzyme. In cases of the presence or likelihood of ascites or pleural effusions, the accumulation of high concentrations of methotrexate may lead to delayed clearance of the drug. The high cost of glucarpidase has led to the fairly often expressed opinion that it should be reserved for those patients who do not respond to other standard therapeutic and supportive measures.

Hyaluronidase

In addition to the warnings and precautions issued by the regulatory agencies for hyaluronidase set out in Table 9.2, the enzyme should not be used to reduce the swelling of bites and stings. In relation to its use with other drugs, hyaluronidase has been found to be incompatible with furosemide, benzodiazepines, and phenytoin and it should not be used to enhance absorption or dispersion of dopamine and/or alfa agonists. The main recorded adverse events are injection site reactions, edema in association with hypodermoclysis, and allergic reactions including urticaria, angioedema, and rarely anaphylaxis. Not surprisingly, unpurified preparations of bovine and ovine hyaluronidase are much more likely to elicit immediate IgE-mediated allergic reactions, especially after repeat administrations. Some animal-derived preparations were reported to provoke allergic responses in up to 10 % of patients.

Adverse reactions to animal-derived hyaluronidase formulations have been reported in the literature following the enzyme's use in local anesthetic ophthalmic blocks. Between January 1998 and August 2011, 98 cases were reported to the FDA. Symptoms included swelling, tenderness, inflammation, reduced visual acuity, increased intraocular pressure, disc hemorrhage, exophthalmos, eye pain, orbital and eyelid edema, hypersensitivity, and blindness. Disability, hospitalization, and/or medical intervention was often necessary. Hypersensitivity reactions after retrobulbar or peribulbar blocks, seen almost immediately in some patients and up to a few days later in others, manifested as periorbital edema with erythema, itch, pain, conjunctival chemosis, proptosis, restricted eye movement, and angioedema. Many of the affected patients had received animal-derived hyaluronidase before. There are a number of reports related to the use of hyaluronidase in cataract surgery. These involve cases of both acute and delayed onsets of severe edema and exophthalmos, the former simulating choroidal hemorrhage, retrobulbar hemorrhage, or orbital cellulitis and the latter, pseudotumor. In three cases reported from the Mayo Clinic, angioedema after cataract surgery occurred following the use of pharmacy-compounded hyaluronidase included in an anesthesia injection of lidocaine. Skin prick tests to the hyaluronidase preparation (150 U/mL) were positive in all three patients, indicating probable IgE antibody-mediated immediate allergic reactions.

Ocriplasmin

Two double-blind, phase III clinical trials comparing a single intravitreal injection of ocriplasmin with placebo injection of patients with symptomatic vitreomacular adhesion, found a similar incidence of ocular events in the two groups. Events associated with vitreous detachment, namely, vitreous floaters, photopsia, injection site pain, and conjunctival hemorrhage occurred in 68.4 % of eyes injected with the recombinant enzyme and in 53.5 % of placebo-injected eyes ($p<0.001$). Vitreous floaters occurred in 16.8 % of the treated group and 7.5 % of the placebo group. There was no difference in the incidence of serious ocular events such as macular hole, retinal detachment, or visual acuity. The most commonly reported adverse reactions (≥5 %) in patients treated with ocriplasmin listed by the FDA are vitreous floaters, conjunctival hemorrhage, eye pain, photopsia, blurred vision, macular hole, reduced visual acuity, visual impairment, and retinal edema (Table 9.2). A recently published post-marketing safety survey of 2465 retinal physicians in the USA seeking responses to questions on the frequency of use of ocriplasmin and incidences of ocular events, received 270 responses concerning 1056 eyes treated with the enzyme. Reported incidences of adverse events were: acute decline in visual acuity 179 (17 %); development of submacular fluid or serous retinal detachment 108 (10.2 %); dyschromatopsia 96 (9.1 %); progression of vitreomacular traction to macular hole 92 (8.7 %); development of retinal detachment 28 (2.7 %); development of retinal tear 21 (2 %); development of afferent pupillary defect 19 (1.8 %); electroretinographic abnormalities 6 (0.6 %); crystalline lens instability 4 (0.4 %); and vasculitis 3 (0.3 %). Interestingly, only 15.9 % of the physicians who observed an adverse event reported the event to the FDA. There have been a few other important reports in the post-marketing period on side effects following intravitreal ocriplasmin including vision loss that correlated with outer retinal disruption and darkened vision even though there were improvements in vitreomacular adhesion and visual acuity. A somewhat more disturbing case of acute panretinal structural and functional abnormalities involving visual acuity loss, pupillary abnormality, and visual field constriction after ocriplasmin suggested to the authors that the many adverse effects were due to the widespread occurrence of laminin in the eye and the consequent diffuse protease effect of ocriplasmin throughout the retina.

Pegademase Bovine

Because pegademase bovine is an orphan drug, the relatively small pool of patients and voluntary nature of reporting adverse events in the post-marketing period means that clinical experience with the enzyme is still less than what is considered desirable (Table 9.2). Hypersensitivity reactions have so far not been reported and a fairly small range of adverse events, including injection site reactions, headache, and pain, were noted in early clinical trials. In the first months of a developing restoration of the immune response, a transient immune dysregulation may occur. This has seen the development of cases of thrombocytopenia and hemolytic anemia, sometimes associated with virus infections or sepsis. Transient thrombocytosis has also been reported. The long-term effects of pegademase therapy are not yet known but it is already clear that, in some patients, full T cell immune function is not achieved. Low T cell numbers have led to cases of malignancy, likely reflecting reduced immune surveillance.

Pegloticase

The FDA has issued a black box warning for anaphylaxis and infusion reactions during and after administration of pegloticase and listed warnings and precautions for gout flares and the possibility of exacerbation of congestive heart failure. Other adverse reactions listed by regulatory agencies are chest pain, contusion or ecchymosis, nausea, vomiting, nasopharyngitis, and constipation (Table 9.2). A recently published retrospective analysis of the FDA Adverse Event Reporting System database identified 118 cases of adverse events involving pegloticase in the USA. Fourteen were pegloticase-associated cardiovascular events; 35 were infusion reactions; 26 were related to gout; and 11 were cases of anaphylaxis induced by the enzyme. Cardiovascular events, infusion-related reactions, gout flares, and anaphylaxis were found to occur more frequently than statistically expected. An evaluation of the long-term (up to 3 years) safety of pegloticase in the treatment of refractory chronic gout revealed that infusion reactions were the most frequent adverse event. Reactions were less often seen in patients receiving biweekly treatment and those who showed a sustained low urate response. Note that pegloticase is contraindicated in patients with glucose-6-phosphate dehydrogenase deficiency (see following section).

Rasburicase

Anaphylaxis to rasburicase and hemolysis and methemoglobinemia are the subjects of an FDA black box warning for the enzyme. The latter two events are relevant to patients with glucose-6-phosphate dehydrogenase deficiency whose erythrocytes are subject to the oxidative stress of hydrogen peroxide produced during the rasburicase-catalyzed conversion of uric acid to allantoin. Deficiency of glucose-6-phosphate dehydrogenase, most often seen in males from Africa, South East Asia, and the Mediterranean region, is one of the most common enzyme deficiencies. Methemoglobinemia and/or hemolysis results from the production of peroxide which oxidizes the Fe^{2+} of hemoglobin to Fe^{3+}. In the ferric state, methemoglobin cannot bind oxygen, increasing the risk of tissue hypoxia and ischemia. Consequently, rasburicase is contraindicated in patients with known glucose-6-phosphate dehydrogenase deficiency. There are now a number of reports in the literature of methemoglobinemia and/or hemolysis in adults and children leading to a recommendation to screen patients thought to be at risk. Rasburicase degrades uric acid in blood and plasma at room temperature causing interference with the measurement of uric acid. This has resulted in a fourth contribution to the FDA boxed warning for the enzyme with the directive to immediately chill collected blood samples and perform the assay within 4 h of collection. In an assessment of the safety of rasburicase and a comparison with allopurinol in adults at risk of tumor lysis syndrome, adverse events, mainly allergic in nature, were more common in the rasburicase groups. Most reactions were grade 1 or 2; no anaphylaxis or grade 4 hypersensitivity was reported. The most common adverse reactions (incidence $\geq$20 %) occurring in rasburicase-treated patients with hematologic malignancies are vomiting, nausea, pyrexia, peripheral edema, anxiety, headache, abdominal pain, constipation, and diarrhea (Table 9.2).

Streptokinase

Hypotension commonly results if streptokinase is given too quickly but this is usually overcome by slowing the infusion rate and giving fluids. Streptokinase may induce a number of different immune disorders (Table 9.2). A serum sickness-like reaction involving arthralgia, fever, rash, and immune complexes may occur despite the dose or route of administration and leukocytoclastic vasculitis, another type III hypersensitivity response, has been seen after intravenous streptokinase. There are reports of acute renal failure complicating a serum sickness-like reaction following prolonged infusion of streptokinase but also a few cases of serum sickness with renal impairment following a short infusion of the enzyme. A renal biopsy undertaken in one patient showed acute tubular necrosis with structurally normal glomeruli and mesangial deposits of IgA, IgM, and complement C3. Vasculitis resembling Schönlein–Henoch purpura associated with streptokinase therapy has been described and acute renal failure developed in one patient 10 days after only a single dose of streptokinase. A renal biopsy revealed pronounced granular staining of glomeruli with prominent deposits of IgA, segmental mesangial hypercellularity, and mesangial expansion. An unusual case of crescentic glomerulonephritis, judged to be a hypersensitivity reaction and apparently associated with a streptococcal infection and streptokinase therapy, has been reported. The reaction developed 2 weeks after a β-hemolytic streptococcal throat infection and 33 days after the administration of streptokinase for myocardial infarction. The authors speculated that the prior exposure to the streptococcus-derived enzyme had sensitized the patient for the subsequent hypersensitive response.

A possible type II or mixed type II and type III hypersensitivity in the form of thrombocytopenia together with acute renal failure after streptokinase administration was found to be improved by corticosteroid therapy.

Being of bacterial origin, antibody responses, including immediate IgE responses, to streptokinase are not unexpected. In practice, antibodies to the enzyme generally appear in treated patients after only a few days and persist for some years. This has often led to a consensus that streptokinase should be used only once in a patient and that the patient should be provided with some sort of record in case of a future infarction. Type I allergic hypersensitivities recorded after streptokinase administration cover the spectrum of immediate reactions namely, flushing, itching, rash, dyspnea, bronchospasm, hypotension, angioedema, urticaria, and anaphylaxis.

Retroperitoneal, lingual, and uvula hematomas, and a hematoma of the rectus muscle after streptokinase for acute myocardial infarction have been described.

Antibody Responses to Enzymes

In theory at least, there is always the possibility of an immune response to enzyme therapy, and this carries with it the potential to adversely affect both the efficacy and the safety and perhaps to also influence subsequent treatments. Immunologically based adverse events may manifest as anaphylaxis in the form of cardiovascular collapse, bronchospasm, angioedema, urticaria, and erythema; as an infusion reaction; cytokine release syndrome; an autoimmune reaction; a cytotoxic type II and immune complex type III hypersensitivities; or as delayed cell-mediated type IV cutaneous hypersensitivities. On the other hand, as often reported, the presence of antibodies to a particular protein may have no clinical consequences. Recently, the FDA issued a draft guidance, "Immunogenicity assessment for therapeutic protein products," an approach to evaluating and mitigating immune responses to a therapeutic proteins that might affect its therapeutic action. Attention is drawn to the fact that immunogenicity is influenced by both product-specific and patient-specific factors, conclusions borne out by experience obtained with ERT, especially for patients with lysosomal storage diseases. In 2003, Brooks and coworkers drew attention to studies that collectively showed that antibody responses to enzymes used in ERT are variable.

Percentages of patients with an antibody response to therapy for Gaucher disease, MPS I, MPS II, MPS IV, MPS VI, Fabry disease, and Pompe disease were 15 %, 91 %, 11 %, 100 %, 89 %, 55 %, and 66 %, respectively. Most of the antibodies were of the IgG and not the IgE class. It was further pointed out that the relationship of enzyme pharmacokinetics to antibody production was not clear and that over the long term, some patients develop immune tolerance with the adverse effects of antibodies becoming of less concern. The authors concluded that most patients with lysosomal storage diseases receiving ERT and who mounted an immune response, continued therapy, sometimes with premedication and/or a reduced infusion rate. The production of antibodies inhibiting the therapeutic effect appeared to be a rare occurrence but it was suggested that the occasional high titer, high affinity antibodies might be important. Later, enzyme therapy with either **agalsidase alfa or beta** for Fabry disease revealed antibodies in males occurred more often to agalsidase beta and. that unlike males, females did not develop detectable antibodies to the enzymes. In another study, a long-term one, 40 % of adult females developed antibodies only to agalsidase beta. Agalsidase antibodies that developed in male Fabry disease patients were found to frequently interfere with the urinary excretion of Gl3 (deacylated Gb3 or globotriaosylceramide), a glycosphingolipid more elevated in plasma than Gb3. However, a retrospective analysis of data from 134 patients of the potential impact of IgG antibodies to agalsidase beta on the efficacy of ERT for Fabry disease failed to show a correlation between antibody titers to the enzyme and the onset of clinical events. In addition, no significant association was found between IgG titers and elevated levels of Gl3 but there was a suggestion that clearance of Gl3 may be impaired in patients with high antibody titers. The questions of the presence of antibodies and the reduction of Gl3 in plasma and urine following infusion of agalsidase (alfa and beta), and resultant treatment outcomes, were further examined in a long-term study. Anti-agalsidase antibodies remained for up to 10 years of ERT, and their presence was associated with a subdued decrease in plasma Gl3 and a negative effect on urinary Gb3 reduction. The authors concluded that these results "may reflect worse treatment outcome." Women with Fabry disease proved more tolerant to agalsidase beta than males with the disease (58 % vs. 11 %); most men developed antibodies to the enzyme (73 %) whereas most women did not (12 %). Men who developed anti-agalsidase antibodies appear to be more likely to experience infusion reactions (26 %) than women (11 %) and those who are seronegative.

Results from clinical trials on infantile-onset Pompe disease showed 34 of 38 (89 %) patients had IgG antibodies to **alglucosidase alfa,** and there were indications that those with high titers may experience reduced. clinical efficacy. In late-onset disease, a high proportion of patients, if not all, developed antibodies and did so within the first 3 months. There appeared to be no associated inhibition of enzyme activity or adverse events. Approximately 15 % of patients with Gaucher disease developed IgG antibodies to **imiglucerase** in the first year of therapy but antibodies rarely developed after a year of therapy. Almost half of the patients with antibodies to imiglucerase experienced symptoms of hypersensitivity. In two studies, IgG antibodies. were detected in 53 % and 14 % of patients treated with **taliglucerase alfa** but the relevance of these antibodies to the enzyme's efficacy and to adverse events such as infusion reactions remains unclear. Only about 4 % of patients receiving **velaglucerase** developed antibodies to the enzyme. These developed late in a 1-year trial, were of the IgG class, and neutralizing in action.

The development of anti-drug antibodies to **sebelipase alfa** has so far received little attention. Four of seven infants with rapidly progressing LAL deficiency presenting. in the first 6 months of life developed antibodies to the enzyme. Two of these patients had neutralizing antibodies that inhibited the enzyme in vitro and in cell uptake studies. Antibodies developed in three of the patients within the first 2 months of treatment, antibodies persisted in one patient and declined to undetectable levels in three patients. Hypersensitivity reactions occurred in the four antibody-positive patients and one of the antibody-negative patients.

Most patients with MPS I who received **laronidase** developed antibodies by week 12 (mean ~ 53 days). During weeks 1–12, plasma clearance of enzyme increased in proportion to antibody titer but by week 26 clearance was back to week one levels in spite of sometimes elevated antibody titers. MPS II patients aged 7 years and younger with gene deletion, rearrangement or different mutations, experienced a higher incidence of **anti-idursulfase** antibody formation than patients with missense mutations. Half of patients 5 years and older given the enzyme developed anti-idursulfase antibodies and the incidence of hypersensitivity reactions was higher in patients with the antibodies. Approximately 40 % of the antibody-positive patients had antibodies that neutralized enzyme activity or cell uptake of enzyme. The relationship between anti-idursulfase antibody status and safety in MPS II pediatric patients treated with idursulfase was examined in the pivotal phase II/III trial involving 63 treatment-naïve patients and designed to examine the relationship between antibody status and outcomes 32 patients (51 %) proved positive for anti-idursalfase IgG antibodies and in 23 of these patients (37 %) the antibodies persisted. However, the presence of antibodies did not significantly impair patient assessment tests and the presence of antibodies was not associated with a higher adverse event rate. On the other hand, antibody-positive patients were 2.3 times more likely to have an infusion-related reaction than patients without serum antibodies. A genotype analysis found that patients with nonsense or frameshift mutations may be more likely to develop antibodies, infusion reactions, and show higher urinary concentrations of glycosaminoglycans than those with missense mutations.

Trials have shown that by week 4 of treatment of MPS IVA patients with **elosulfase alfa**, antibodies developed in all patients and antibody titers are maintained or increased thereafter during therapy. All patients had antibodies that inhibited binding of the enzyme to the mannose-6-phosphate receptor but titers were not assessed, thus not allowing an examination of any association between neutralizing antibody and treatment. Clinical studies showed that 98 % of MPS VI patients treated with **galsulfase** developed complementary antibodies within 4–8 weeks of treatment. Evaluation of sera for potential relationships of antibodies to clinical outcomes revealed no clear and consistent relationship between antibody titers on the one hand and neutralizing antibody, infusion reactions, IgE antibodies, and urinary glycosaminglycan levels on the other.

With regard to tissue plasminogen activators, no readministration studies or any other studies for immunogenicity or tolerance have been undertaken with **alteplase**, **reteplase,** and **tenecteplase**. Three of 487 patients tested for anti-tenecteplase antibodies had a positive titer at 30 days.

In immunogenicity studies, 76 of 98 hypophosphatasia patients (76 %) tested positive for anti-asfotase alfa antibodies at some point during treatment. Of these 76 patients, 34 (45 %) had neutralizing antibodies but no correlation was seen between antibody titers and neutralizing antibody.

Approximately one quarter of patients treated with ***L*-asparaginase** from E. coli develop antibodies to the bacterial enzyme. The incidence of antibodies is higher after the second exposure, hypersensitivity reactions occur more often in patients with antibodies and reactions may be associated with increased clearance of enzyme. There is inadequate data on neutralizing antibodies but higher levels appear to correlate with decreased enzymic activity. In a study of children with ALL, high titers of antibody were associated with low *E. coli L*-asparaginase activity but not with pegasparaginase activity.

Thirty days after the commencement of administration of **collagenase** to patients with Dupuytren's contracture, high incidences of an antibody response were seen to AUX-I (92 %) and AUX-II (86 %). High titers to both proteins were recorded after the fourth injection. Neutralizing antibodies to AUX-1 and AUX-2 were detected in 10 % and 21 % of patients, respectively, but no correlations of antibody frequency, titers or neutralizing capacity to clinical response or adverse events were seen. After the eighth injection of Xiaflex, > 99 % of Peyronie's disease patients developed high titers of antibodies to AUX-I and AUX-II. Antibodies proved to be neutralizing in 60 % and 51.8 % of patients, respectively.

Less than 5 % of patients treated with **dornase alfa** have developed complementary antibodies, no patients have developed IgE antibodies to the enzyme and there have been no reports of serious allergic reactions to dornase alfa. Importantly, improvements in pulmonary function still occurred even after the appearance of antibodies.

In clinical studies reported by the FDA, 16 of 96 patients (17 %) developed **anti-glucarpidase** antibodies following administration of the recombinant carboxypeptidase. Twelve patients developed antibodies after a single injection while the remaining four patients received two doses.

There appears to be no reports of hypersensitivity reactions to **pegademase bovine** injection although antibodies are produced to the agent in some patients. Antibodies were detected three to eight months after the beginning of therapy and a small percentage of the patients with antibodies displayed enhanced enzyme clearance. After five months of therapy, 1 of 12 patients demonstrated enhanced clearance of plasma pegademase bovine that correlated with the appearance of antibodies reactive with the bovine enzyme and with adenosine deaminase. In what was essentially an attempt to induce tolerance, the patient was given an increased dose of enzyme intramuscularly twice a week instead of weekly. No adverse effects resulted, plasma levels of adenosine deaminase were restored, and the patient was returned to weekly doses after 4 weeks.

Anti-pegloticase antibodies were detected in a high proportion of patients (89–92 %) given the enzyme. High antibody titers were detected 3 weeks after the beginning of treatment, indicating the presence of IgM. In some studies, high titer antibodies were associated with failure to maintain pegloticase-induced normal uric acid levels and showed some correlation with a higher incidence of infusion reactions. In a study of the relationship between efficacy and antibody development in patients treated for chronic refractory gout, anti-pegloticase antibodies were determined and antibody titers were examined for possible relationships with serum pegloticase, uric acid concentrations, and risk of infusion reactions. Patients' diminished responsiveness to pegloticase was found to be associated with high titers of antibody that increase clearance of the enzyme with the consequent failure to decrease uric acid levels. Lack of response to the enzyme also saw an increased risk of infusion reactions.

Anti-rasburicase antibodies may develop and inhibit enzyme activity in patients given the drug. Eleven percent of pediatric patients with a hematologic malignancy developed antibodies in clinical trials within a month of first administration while trials with adults with hematologic malignancies showed antibodies in 18 %, enzyme neutralizing IgG antibodies in 8 %, and anti-rasburicase IgE antibodies in 6 % of patients from day 14 to 24 months following five daily doses of rasburicase. However, in a phase III study comparing rasburicase to allopurinol in controlling plasma uric acid in pediatric patients with hematologic malignancies, the enzyme did not prove to be highly immunogenic. Overall, the incidence of anti-rasburicase antibodies was low with only 2 % positive for enzyme neutralizing antibodies and no patients positive for IgE antibodies to rasburicase.

Streptokinase is an enzyme of bacterial origin, the protein is highly immunogenic and antibodies to streptokinase tend to form in most patients receiving the drug. Such antibodies are likely to increase resistance to successful therapy so repeat treatment with streptokinase is to be avoided if administration has been given for more than 5 days, and particularly if continued for 5 days to 12 months after the initiation of treatment. A reduced therapeutic benefit may also be seen if patients have recently experienced a streptococcal infection such as streptococcal pharyngitis, acute rheumatic fever, or acute glomerulonephritis.

Immunogenicity has not yet been evaluated for **ocriplasmin**.

Although it seems to be assumed that PEG is not immunogenic and non-antigenic, some pegylated agents elicit anti-PEG antibodies in animal studies and anti-PEG may limit therapeutic efficacy of pegasparaginase in ALL and of pegloticase in patients with gout. It has therefore been suggested that the immunogenicity and antigenicity of approved pegylated compounds be carefully examined in humans.

Other Therapies for Lysosomal Storage Diseases

While in the past, treatment for lysosomal storage diseases was essentially palliative, direct enzyme replacement therapy is now available for some mucopolysaccharidoses and although this has led to substantial improvements in some somatic symptoms for patients with MPS I, II, IVA, and VI, neurological and other symptoms often remain a problem at least, in part, because the enzymes do not cross the blood–brain barrier. Apart from surgical interventions, other therapies are therefore being sought and a number of approaches are showing promise as potentially successful treatments.

Stem Cell Transplantation

In this procedure, the patient's hematopoietic stem cells are replaced by cells of the selected compatible donor (or perhaps cells from cord blood) which then proliferate and produce sufficient enzyme to overcome the original enzyme deficiency. Since the first bone marrow transplant about 35 years ago, several hundred patients with a lysosomal storage disease have received a hematopoietic stem cell transplant with quite good results in children with MPS I under the age of 2 years. After a successful transplant, glycosaminoglycan levels in urine decline, joint mobility, vision, hearing, cardiopulmonary function and airways symptoms improve and organ sizes are reduced but skeletal abnormalities and intellectual capacity generally show far less improvement. Clinical experience with hematopoietic stem cell transplants remains limited for other MPSs although there has been some success with MPS VI. Doubts surround its efficacy in MPS II and MPS III. Interestingly, ERT does not appear to increase the incidence of graft rejection or graft-versus-host disease. For MPS I, Australian guidelines recommend a maximum of 12 weeks of ERT before transplantation and a maximum of 15–17 weeks after the transplant.

Substrate Reduction Therapy

Rather than providing the absent or deficient degrading enzyme, substrate reduction therapy aims to deplete the amount of storage material to compensate for the defective enzyme activity. For example, in Gaucher disease, the imino sugar *N*-butyldeoxynojirimycin (miglustat, Zavesca®) is used to inhibit ceramide glucosyltransferase, the enzyme that synthesizes the storage compound, glucocerebroside. Also unlike ERT, substrate reduction therapy is undertaken with "small" molecules that are able to pass the blood–brain barrier and hence offer the chance of alleviating neuronal as well as peripheral symptoms. The phytoestrogen genistein (4′,5,7-trihydroxyisoflavone) reduces glycosaminoglycan synthesis in cell culture and some mouse models and in an open label clinical trial of ten patients affected by MPS IIIA or MPS IIIB, reduced urinary glycosaminoglycan levels were noted together with improvements in hair morphology and behavior. However, these promising results have not always been confirmed leading to the suggestion that higher doses might be necessary to obtain clinical improvements. The fluorescent dye rhodamine B has been shown to inhibit glycosaminoglycan synthesis in cell culture and an animal model of MPS IIIA but more information is needed on its mechanism of action and safety.

Chaperones

Chaperones such as heat shock proteins and calnexin aid the correct folding of proteins that would otherwise be degraded by proteasomes. In lymphoblasts of Fabry disease patients, 1-deoxygalactonojirimycin has been used as a competitive inhibitor and chaperone of the mutant enzyme α-galactosidase by inducing the correct conformation and restoring the catalytic activity of the enzyme. This concept of employing inhibitors as chaperones has been extended to in vitro investigations of cells from patients with Gaucher disease and Pompe disease. Being small molecules, chemical chaperones can pass the blood–brain barrier. A major disadvantage, however, is their restriction to patients with missense mutations, bearing in mind that with Pompe disease for example, only 10–15 % of patients might benefit from the therapy.

Gene Therapy

In gene therapy in vivo, a correctly functioning gene inserted into a recombinant vector is deposited into organs such as the liver and lung where it can be a continuous source of enzyme for correction of the metabolic disorder. The efficacy of the strategy has been demonstrated in a number of animal models. The procedure also has the potential to provide a stable source of enzyme for bone and the brain and thus reverse, or slow, central as well as peripheral symptoms. Vector delivery systems in animal models have been examined for gene transfer into the central nervous system, and this was extended to a clinical study on ten children with the neurodegenerative lysosomal storage disorder, late infantile neuronal ceroid lipofuscinosis. A reduced rate of neurological decline was seen 18 months after the surgery. In the so-called ex vivo gene therapy, genetically modified autologous hematopoietic stem cells or cells from a healthy donor are transplanted with the aim of cross-correcting an enzyme defect. Since hematopoietic cells can cross the blood–brain barrier, defects in neuronal and glial cells may be corrected. The efficacy of the therapy has been confirmed in animal studies, in particular in arylsulfatase deficient mice. Graft-versus-host disease remains a significant risk if autologous stem cells are not used.

Despite the undoubted progress being made in gene therapy research, studies are still largely investigational and there is still much to do and learn before the strategy becomes a realistic option for patients with lysosomal storage disorders. As well as efficacy of the therapies, issues of safety are still a long way from being defined and thoroughly investigated.

Living with Enzyme Replacement Therapy

As can be readily imagined, living with ERT for a lysosomal storage disease and the medical interventions it necessitates, make up a major part of patients', and their families', lives since ERT is time-consuming, costly, disruptive due to required hospital visits, and distressing because of many uncertainties associated with the disorders. With these points in mind, a recent exploration of the experiences of young patients and their families in Australia examined the impact of receiving ERT for a lysosomal storage disorder on the health-related quality of life of patients with Pompe disease, Gaucher disease, or mucopolysaccharidosis types I or II. Findings highlighted the challenges and coping strategies of living with the disorders and the lifetime treatments required. Communication with family members and professionals were deemed especially important and intervention to deal with the disease was given a high priority. On the positive side, some improvements in physical and psychosocial well-being were noted and the importance of positive thinking and ways to manage uncertainty were described.

Data on age at symptom onset, methods of diagnosis, common mutations, symptoms, and clinical manifestations of infantile-onset versus adult-onset disease are contained in the Pompe Registry set up to further understand the natural history, clinical symptomatology, and treatment outcomes for the disorder. Said to be the largest repository of worldwide data on the disease and with more than 28 countries enrolled in the Registry, it is hoped that continuing analyses will improve recognition as well as understanding of Pompe disease. Analyses so far reveal that 70 % of patients have symptom onset at more than 12 months of age, 23 % have symptom onset at less than 12 months of age and most patients in the latter group (~70 %) have cardiomyopathy. Patients with cardiomyopathy have more respiratory, neuromuscular, and gastrointestinal symptoms and delays in development compared to infant patients with no cardiac involvement.

In a survey published in 2014 by the European Gaucher Alliance, an umbrella group established in 1994 to support patient organizations for the disease, valuable information was obtained in response to important questions such as numbers affected, specific treatments, availability of and access to treatments, support for patients, patient organizations, and funding sources. Inequalities in access to treatment in different countries were revealed, for example, 6 % of patients in 20 countries were untreated because of lack of funding and 3 of 27 countries relied entirely on humanitarian aid. Main concerns expressed were difficulties associated with access to treatment, reimbursements, specialist treatment centers, home infusions, and finding doctors with expertize in Gaucher disease. The survey revealed that funding was almost always limited to one or a very few sources and two member organizations had no external funding source at all. Of great concern is the fact that awareness of, and expertise in, the medical community is low, sometimes resulting in long delays or gaps in receiving a correct diagnosis and appropriate medical treatment. Other support organizations for patients with Gaucher disease, their families, helpers, and physicians are the International Collaborative Gaucher Group, Gaucher Registry, Genzyme Corporation, USA; Gaucer Registry, Cambridge, USA; Gaucher Association, London, UK; Children's Gaucher Research Fund, Granite Bay, CA, USA; and International Gaucher Disease Associations, http://www.gaucherdisease.org.uk/overseas.htm

Amongst rare diseases, Gaucher disease at least has the relative good fortune of having a number of fairly effective and specific non-neuronopathic treatments available. Imiglucerase is the most widely employed treatment, being used in 64 % of patients, followed by velaglucerase alfa (23 %), and taliglucerase alfa and miglustat (each 4 %.). Of great concern for now and the future is the high cost of ERT for the lysosomal storage diseases. Current treatments for Gaucher disease for example, depending on the patient's weight and dosage, is estimated to be US$200,000–$380,000 or about 154,000–292,000 Euros per patient per year. Often with deteriorating economic conditions, it is hard to see how patients, and most countries, can afford such seemingly ever-increasing, and already high, if not unaffordable, medical costs.

Summary

- The discoveries of specific recognition of phosphorylated mannose residues, mannose receptor-mediated uptake of lysosomal enzymes and the presence of these receptors on macrophages, demonstrated that a lysosomal enzyme needs to be specifically recognized by its target cells. These insights led on to the first successful enzyme replacement therapy (ERT) for type I Gaucher disease which occurs with a frequency of 1 in 75,000 births worldwide, making it the most prevalent of the sphingolipid storage disorders.
- Gaucher disease is the result of an inborn error of metabolism due to a deficiency of the lysosomal acid β-glucosidase glycoprotein, β-glucocerebrosidase (glucosylceramidase; β-glucosyl-*N*-acylsphingosine glucohydrolase) which cleaves β-*D*-glucosylceramide (glucocerebroside) into glucose and ceramide. The enzyme's substrate is a widely distributed cell membrane component and, in the absence of β-glucocerebrosidase, glucocerebroside, and other glycolipids accumulate by as much as 20–100-fold in the lysosomes of cells, particularly macrophages and other cells of the reticuloendothelial system.
- As a direct result of the knowledge and experience gained in the development of ERT for Gaucher disease, successful ERTs have been developed for a small but growing number of other lysosomal storage diseases otherwise known as mucopolysaccharide storage diseases or mucopolysaccharidoses (MPSs). Approved enzyme replacement treatments are now administered for a second lipid storage disorder, Fabry disease, for the glycogen storage disorder, Pompe disease, and for types I, II, IVA, and VI MPS.
- Depending on the particular MPS, the accumulated products produce clinical manifestations such as short stature, coarse features, motor dysfunction, cardiomyopathy, ocular abnormalities, skeletal dysplasia, hepatosplenomegaly, and mental retardation.
- Three recombinant enzymes, the human β-glucocerebrosidases imiglucerase, taliglucerase alfa, and velaglucerase alfa are each approved for the treatment of Gaucher disease.
- Fabry disease is a second lipid storage disorder (sphingolipidosis) but in this case caused by a deficiency of an α-*D*-galactosidase which results in a buildup of globotriaosylceramide, more commonly known as Gb3 or ceramide trihexoside. Gb3 accumulates in capillary endothelia affecting in particular the kidneys,

heart, eyes, brain, nervous tissue, and skin. Both algalsidase alfa (Replagal®) and algalsidase beta (Fabrazyme®) have received regulatory approval for the treatment of Fabry disease.

- Another type of lipid storage disorder, lysosomal acid lipase deficiency, is caused by a decrease of the enzyme lysosomal acid lipase (LAL) involved in the breakdown of lipids in the lysosome. This disorder, characterized by lysosomal accumulation of cholesteryl esters and triglycerides in organs including liver, intestine, and blood vessel walls, results in increased liver fat and liver disease, lipid accumulation in intestinal walls leading to malabsorption and growth failure, and dyslipidemia with elevated LDL-C and triglycerides together with low high HDL-C.
- Sebelipase alfa (Kanuma®), a recombinant human LAL produced in egg white of genetically engineered chickens, has the same amino acid sequence as natural human LAL. In the lysosome, the enzyme catalyzes the hydolysis of cholesteryl esters and triglycerides to free cholesterol, glycerol, and free fatty acids.
- A deficiency of lysosomal acid α-glucosidase in Pompe disease leads to an accumulation of glycogen which causes myopathy throughout the body but especially in skeletal muscles, heart, liver, and the nervous system. A recombinant alglucosidase alfa used as ERT for Pompe disease is currently approved by both the FDA and EMA.
- MPS types I, II, IVA, and VI are due to deficiencies of α-*L*-iduronidase, iduronate-2-sulfatase, *N*-acetylgalactosamine-6-sulfatase, and *N*-acetylgalactosamine-4-sulfatase, respectively. Each produces accumulations of glycosaminoglycans, in particular dermatan sulfate, heparan sulfate, and chondroitin sulfate.
- Laronidase is a recombinant human α-*L*-iduronidase enzyme used in ERT for MPS I (also called Hurler syndrome) and clinically milder variants. Idursulfase, a recombinant form of iduronate-2-sulfatase, has been approved by both the FDA and EMA as a safe and effective treatment for MPS II or Hunter syndrome. The enzyme removes the sulfate group from the 2-position of dermatan and heparan sulfates.
- Elosulfase alfa, a recombinant form of *N*-acetylgalactosamine-6-sulfatase, was granted marketing approval for the treatment of patients with MPS IVA, also known as Morquio A syndrome, an autosomal recessive disorder caused by a deficiency of the enzyme. In 2005, galsulfase, a recombinant form of *N*-acetylgalactosamine-4-sulfatase, was approved by the FDA for the treatment of patients with MPS VI (Maroteaux-Lamy syndrome), an autosomal recessive disease due to deficiency of *N*-acetylgalactosamine-4-sulfatase.
- Natural human tPA is a serine protease synthesized by endothelial cells. tPA is made up of five structural domains—a looped "finger" domain near the *N*-terminal, a growth factor domain, the kringle 1 and kringle 2 domains and, next to the latter, the serine protease domain. The finger and kringle 2 domains bind fibrin clots while the protease domain with its catalytic site at the C-terminus catalyzes the conversion of plasminogen to plasmin. Currently approved tPAs are alteplase, reteplase, and tenecteplase.
- Asfotase alfa (Strensiq®), a recombinant tissue-nonspecific alkaline phosphatase, is indicated for the treatment of hypophosphatasia, an inherited bone disorder caused by mutations in the alkaline phosphatase gene. The enzyme's substrates,

including inorganic pyrophosphate, accumulate, blocking hydroxyapatite crystal growth and inhibiting bone mineralization. This results in rickets, bone deformation, osteomalacia, and muscle weakness.

- Specific for the nonessential amino acid *L*-asparagine, the amidohydrolase *L*-asparaginase is prepared from *E. coli* and *Dickeya dadantii*. Three asparaginase preparations are currently approved, one each from *E. coli* and *Dickeya dadantii* and a pegylated preparation, pegaspargase, from *E. coli*. Use of the enzyme in acute lymphocytic leukemia is based on the fact that susceptible leukemia cells cannot synthesize asparagine due to lack of the enzyme asparagine synthase and depend on an endogenous source of the amino acid for survival.
- Collagenases are proteinases that hydrolyze collagen, and it is this property that has seen the enzyme from *Clostridium histolyticum* become the first approved nonsurgical treatment for Dupuytren's contracture and Peyronie's disease. Collagenase as Xiaflex® and Xiapex® is a mixture of class I (AUX-I) and class II (AUX-II) *C. histolyticum* collagenases in a required ratio. Each class hydrolyzes collagen at different sites but act in a complementary manner to degrade the protein.
- Dornase alfa is a recombinant human deoxyribonuclease I (rhDNase I) identical in composition to the natural enzyme. Viscous extracellular DNA, released mainly by disintegrating neutrophils during infection, accumulates in sputum of patients with cystic fibrosis contributing to reduced pulmonary function and frequent pulmonary infection. Dornase alfa converts extracellular DNA to 5′-phosphonucleotide end products reducing both sputum viscosity in the airways and adhesiveness of lung secretions.
- Glucarpidase, a recombinant pseudomonas carboxypeptidase G2 produced in *E. coli*, is used clinically to hydrolyze methotrexate and other antifolates. Methotrexate, administered for various cancers, is eliminated in the urine so patients with renal impairment given the drug may experience high plasma concentrations. Glucarpidase is indicated in such patients ensuring that methotrexate is eliminated enzymically.
- Hyaluronidases are enzymes that degrade hyaluronan (hyaluronic acid). Found in the extracellular matrix of most types of connective tissue, hyaluronan is a high molecular mass linear glycosaminoglycan polymer of disaccharides of *D*-glucuronic acid and *N*-acetyl-*D*-glucosamine linked by alternating β 1,3 and β 1,4 glycosidic bonds.
- To overcome the shortcomings of non-human, impure, animal-derived extracts, a recombinant form of hyaluronidase PH20, rhuPH20, was prepared. Like the animal-derived hyaluronidase preparations, rhuPH20 hydrolyzes the glycosidic bonds between the β 1,4 linked *N*-acetyl-*D*-glucosamine and *D*-glucuronic acid sugars of hyaluronan but appears to be well tolerated, eliciting neither inflammatory nor immunogenic responses following repeated subcutaneous injections.
- Ocriplasmin is a proteolytic enzyme and truncated form of human plasmin produced in a yeast. Ocriplasmin has activity against the clinically relevant plasmin receptors fibronectin and laminin, components of the vitreoretinal interface.

Intravitreal injection of ocriplasmin can induce separation of the vitreous and macular surfaces thereby resolving vitreomacular traction and closing macular holes.

- Pegademase bovine is adenosine deaminase of bovine intestine origin extensively pegylated to increase half-life. It is used as therapy for severe combined immunodeficiency disease (SCID). A deficiency of adenosine deaminase leads to the accumulation of adenosine, 2′-deoxyadenosine, and their metabolites which are toxic to lymphocytes. A deficiency of T lymphocytes, particularly functional helper T cells and B lymphocytes, results in a markedly decreased production of antibodies and impairment of both arms of the adaptive immune response.
- Pegylated to reduce its potential for immunogenicity and to increase its circulatory half-life, pegloticase, a recombinant porcine/baboon variant uricase (urate oxidase) produced in *E. coli*, is used for therapy of gout previously refractory to conventional therapy with uricostatic (e.g., allopurinol) and uricosuric drugs (e.g., probenecid). The enzyme urate oxidase which is absent in humans, catalyzes the oxidation of uric acid to 5-hydroxyisourate and hydrogen peroxide.
- Rasburicase a recombinant urate oxidase is effective in managing tumor lysis syndrome which causes an ionic imbalance involving hyperkalemia and hyperphosphatemia, secondary hypercalcemia, and hyperuricemia and possibly acute kidney injury and death. In pediatric patients with acute leukemia and lymphoma, rasburicase has proved to be the treatment of choice and superior to allopurinol.
- Streptokinase is a potent plasminogen activator but unlike tPA it is not a protease. Used for more than 35 years as a thrombolytic agent to treat blood vessel blockages such as acute myocardial infarction, the mechanism of action of streptokinase remains incompletely understood. It forms complexes with human plasminogen and plasmin, and these complexes can hydrolytically activate other plasminogen molecules. The streptokinase-plasminogen complex is rapidly converted to streptokinase-plasmin. After formation of streptokinase-plasmin, both entities are able to act as plasminogen activators to catalyze the hydrolysis of further plasminogen molecules.
- An infusion reaction, occurring in ~10–14 % of patients, is the most common adverse event seen when agalsidase beta is given for Fabry Disease. Reactions are usually mild with no sign of respiratory symptoms, urticaria or changes in vital signs. Symptoms tend to be easily controllable with antihistamines and corticosteroids.
- Despite the apparent enzyme identity of the two preparations of alglucosidase alfa for Pompe Disease, viz., Lumizyme®, and Myozyme®, the listed adverse reactions of each show some significant differences. Both carry warnings for severe allergic reactions including life-threatening anaphylaxis and cardiorespiratory failure. Reactions to Lumizyme® include pyrexia flushing, hyperhidrosis, headache, hypertension, dizziness, rash, and urticaria while for Myozyme® infusion reactions, infections such as pneumonia and respiratory syncytial virus, respiratory distress/failure, and gastrointestinal problems predominate. The different adverse reactions are probably related as much to the age of patients and the administered dose of enzyme as to the alleged "biological difference."

Lumizyme® is given to older patients with late-onset disease while Myozyme® is mainly used to treat infants.

- Imiglucerase, taliglucerase alfa, and velaglucerase alfa, each administered for long-term enzyme replacement therapy for Gaucher disease, show a similar spectrum of adverse events with immediate type I hypersensitivity reactions being the most prominent. Most common adverse events for imiglucerase are general disorders and administration site reactions (i.e., pyrexia, chills, chest discomfort); skin and subcutaneous tissue disorders (pruritus, rash, urticaria); respiratory, thoracic, and mediastinal disorders (dyspnea, cough, throat irritation). Adverse events following taliglucerase alfa include allergic and infusion reactions while for velaglucerase alfa there are reports of severe prolonged activated partial thromboplastin time and allergic dermatitis.
- The possibility of a hypersensitivity reaction, including anaphylaxis, is prominent amongst the issued warnings and precautions for sebelipase alfa. Sebelipase alfa is produced in chicken egg white and the risk and warning of the possibility of anaphylaxis has therefore been extended to eggs and egg products
- Apart from an FDA boxed warning issued for the potential of anaphylaxis, laronidase (α-*L*-iduronidase) treatment of MPS I is generally well tolerated with few treatment-related events and few, if any, serious adverse reactions.
- Idursulfase carries an FDA black box warning of the risk of anaphylaxis after/during infusion. Given for MPS II, the enzyme is generally well tolerated with the most common adverse events being infusion-based reactions. The incidence of these reactions reaches a maximum between weeks 4 and 12 and declines thereafter. Adverse events include headache, nasopharyngitis, abdominal pain, arthralgia, pruritus, pruritic rash, swelling at the infusion site, erythema, urticaria, dyspepsia, anxiety, and chest wall pain.
- Approximately 19 % of patients infused with elosulfase alfa for MPS IVA experience a hypersensitivity reaction with 8 % classified as anaphylactic. This finding is reflected in an FDA boxed warning. Enzyme efficacy does not appear to be impaired in the hypersensitive patients.
- Galsulfase given for MPS VI is generally considered to be a relatively safe drug with infusion-related reactions being the most commonly occurring, and predominant, adverse event. Other reported events include headache, pyrexia, limb, chest and ear pain, visual abnormalities, anxiety, dyspepsia, upper respiratory infections, and cough.
- The most frequently seen and potentially serious adverse event associated with thrombolytic agents is bleeding, especially at intracranial, gastrointestinal, retroperitoneal, and pericardial sites. Results from trials showed an incidence of stroke of 1.2 % for patients receiving alteplase. Orolingual angioedema after alteplase administration has been recorded. The reactions show a clear resemblance to a type I immediate hypersensitivity anaphylactic-like responses. Sepsis, venous thrombosis, and allergic reactions are other well-known adverse events after alteplase therapy.
- As with other thrombolytics, bleeding is the most common adverse event seen with reteplase and tenecteplase. There appears to be no significant difference in

the risk of hemorrhage and stroke between reteplase and alteplase, and the risk of stroke is similar to other thrombolytic agents. Chest pain is a common side effect of reteplase occurring in more than 10 % of patients. Cardiac events such as arrhythmias, circulatory collapse, and another heart attack are seen in 1–10 % and heart or heart valve damage, a blood clot in the lungs and hypersensitivity occur in 0.1–1 % of treated patients.

- A comparison of single-bolus tenecteplase with front-loaded alteplase in acute myocardial infarction revealed an almost identical 30-day mortality rate (6.18 % tenecteplase vs. 6.15 % alteplase), similar rates of intracranial hemorrhage (0.93 % vs. 0.94 %) and a similar rate of death or nonfatal stroke at 30 days but there were fewer non-cerebral bleeding complications and less need for blood transfusions with tenecteplase. The similarity in side effects between alteplase, reteplase, and tenecteplase extends to the possibility of allergic reactions; the incidence of such reactions to tenecteplase is <1 %. Other recorded adverse events to tenecteplase include nausea, vomiting, fever, hypotension, and a range of cardiac abnormalities.
- Warnings and precautions for the recombinant nonspecific alkaline phosphatase asfotase alfa cover hypersensitivity reactions including anaphylaxis; localized lipodystrophy including lipoatrophy and lipohypertrophy at injection sites; and ectopic calcifications particularly of the eye (cornea and conjunctiva), and the kidneys.
- Allergic reactions (including anaphylaxis), pancreatitis, hyperglycemia, hepatotoxicity, and abnormal liver function, posterior reversible encephalopathy syndrome, and CNS dysfunction and thrombosis are the most serious adverse events induced by *L*-asparaginase therapy. *L*-asparaginase induces coagulopathy involving intracranial hemorrhage with thrombosis of the extremities, immune hemolytic anemia, and abnormal collagen-stimulated platelet aggregation. A recent successful pediatric treatment regimen of multiple doses of intravenous pegaspargase for adult leukemia patients showed hyperbilirubinemia and transaminitis as the most common grade 3/4 toxicities.
- Reactions occurring in patients with Dupuytren's contracture include peripheral edema, contusion, injection site reactions, pain in extremities, and pruritus. In Peyronie's disease, the most frequently reported events are penile hematoma, swelling, pain and ecchymoses, erectile dysfunction, and genital pruritus. A boxed warning has been issued for the possibility of corporal rupture or penile fracture when collagenase is injected for Peyronie's disease.
- Treatment of cystic fibrosis patients with aerosoled recombinant human DNase (dornase alfa) is generally well tolerated with only voice alteration and laryngitis sometimes seen. Even these reactions tend to be mild and usually resolve within 21 days.
- Glucarpidase is generally well tolerated. The most common adverse events are flushing, paresthesia, headache, shaking, hypotension, nausea and vomiting, tingling fingers, burning of face and extremities, and a feeling of warmth.
- The main adverse events following hyaluronidase are injection site reactions, edema in association with hypodermoclysis, and allergic reactions including

urticaria, angioedema, and rarely anaphylaxis. Unpurified preparations of bovine and ovine hyaluronidase are much more likely to elicit immediate IgE-mediated allergic reactions, especially after repeat administrations. Adverse reactions, including hypersensitivities, to animal-derived hyaluronidase formulations have been reported following the enzyme's use in local anesthetic ophthalmic blocks.

- The most commonly reported adverse reactions in patients treated with ocriplasmin are vitreous floaters, conjunctival hemorrhage, eye pain, photopsia, blurred vision, macular hole, reduced visual acuity, visual impairment, and retinal edema. The main adverse events reported in a recent post-marketing safety survey of US retinal physicians were acute decline in visual acuity, development of submacular fluid or serous retinal detachment, dyschromatopsia, progression of vitreomacular traction to macular hole, development of retinal detachment, and development of retinal tear.
- A fairly small range of adverse events to pegademase bovine, including injection site reactions, headache, and pain have been reported. So far, hypersensitivity reactions have not proved to be a significant problem. In the first months of a developing restoration of the immune response, a transient immune dysregulation may occur. This has seen the development of cases of thrombocytopenia, transient thrombocytosis, hemolytic anemia, and virus infections or sepsis. Low T cell numbers have led to cases of malignancy, likely reflecting reduced immune surveillance.
- The FDA has issued a black box warning for anaphylaxis and infusion reactions during and after administration of pegloticase and listed warnings and precautions for gout flares and the possibility of exacerbation of congestive heart failure. Other adverse reactions are chest pain, contusion or ecchymosis, nausea, vomiting, nasopharyngitis, and constipation. Pegloticase is contraindicated in patients with glucose-6-phosphate dehydrogenase deficiency.
- Anaphylaxis to rasburicase and hemolysis and methemoglobinemia are the subjects of an FDA black box warning for the enzyme. Rasburicase, contraindicated in patients with known glucose-6-phosphate dehydrogenase deficiency, degrades uric acid in blood and plasma at room temperature. This has resulted in a fourth contribution to the FDA boxed warning with the directive to immediately chill collected blood samples and perform the assay within 4 h of collection.
- Streptokinase may induce a number of different immune disorders including a serum sickness-like reaction involving arthralgia, fever, rash, and immune complexes; serum sickness with renal impairment; leukocytoclastic vasculitis (another type III hypersensitivity response); vasculitis resembling Schönlein–Henoch purpura; and crescentic glomerulonephritis. A possible type II or mixed type II and type III hypersensitivity in the form of thrombocytopenia together with acute renal failure after streptokinase administration was found to be improved by corticosteroid therapy. Type I allergic hypersensitivities recorded after streptokinase administration has often led to a consensus that streptokinase should be used only once in a patient.

- Percentages of patients with an antibody response to therapy for Gaucher disease, MPS I, MPS II, MPSIVA, MPS VI, Fabry disease, and Pompe disease were 15 %, 91 %, 11 %, 100 %, 89 %, 55 %, and 66 %, respectively. Most of the antibodies are of the IgG and not the IgE class. Most patients who receive ERT and mount an immune response continue therapy, sometimes with premedication and/or a reduced infusion rate.
- In Fabry disease patients on agalsidase therapy, clearance of Gl3 may be impaired in patients with high antibody titers. Women with Fabry disease proved more tolerant to agalsidase beta than males with the disease; most men developed antibodies to the enzyme whereas most women did not. Men who developed anti-agalsidase antibodies appear to be more likely to experience infusion reactions than women and those who are seronegative.
- Infants with Pompe disease had IgG antibodies to alglucosidase alfa, and there were indications that those with high titers may experience reduced clinical efficacy. In late-onset disease, a high proportion of patients, if not all, developed antibodies and did so within the first 3 months. There appeared to be no associated inhibition of enzyme activity or adverse events.
- Approximately 15% of patients with Gaucher disease developed IgG antibodies to imiglucerase in the first year of therapy but antibodies rarely developed after a year of therapy. Almost half of the patients with antibodies to imiglucerase experienced symptoms of hypersensitivity. The relevance of anti-taliglucerase alfa and anti-velaglucerase antibodies to the enzymes' efficacy and to adverse events remains unclear.
- Most patients with MPS I who received laronidase developed antibodies by week 12. During weeks 1–12, plasma clearance of enzyme increased in proportion to antibody titer but by week 26 clearance was back to week one levels in spite of sometimes elevated antibody titers.
- MPS II patients aged seven years and younger with gene deletion, rearrangement or different mutations, experienced a higher incidence of anti-idursulfase antibody formation than patients with missense mutations. Half of patients five years and older given the enzyme developed anti-idursulfase antibodies and the incidence of hypersensitivity reactions was higher in patients with the antibodies.
- By week four of treatment of MPS IVA patients with elosulfase alfa, antibodies developed in all recipients and antibody titers were maintained or increased thereafter during therapy. All patients had antibodies that inhibited binding of the enzyme to the mannose-6-phosphate receptor but titers were not assessed, thus not allowing an examination of any association between neutralizing antibody and treatment.
- Approximately one quarter of patients treated with *L*-asparaginase from E. coli develop antibodies to the bacterial enzyme. The incidence of antibodies is higher after the second exposure, hypersensitivity reactions occur more often in patients with antibodies, and reactions may be associated with increased clearance of enzyme. There is inadequate data on neutralizing antibodies but higher levels appear to correlate with decreased enzymic activity.

- Thirty days after the commencement of administration of collagenase to patients with Dupuytren's contracture, high incidences of an antibody response were seen to AUX-I and AUX-II. No correlations of antibody frequency, titers or neutralizing capacity to clinical response or adverse events were seen. Antibodies to AUX-I and AUX-II, some neutralizing, have been found in a high proportion of patients given collagenase for Peyronie's disease.
- Less than 5 % of patients treated with dornase alfa developed complementary antibodies, no patients developed IgE antibodies to the enzyme, and there were no reports of serious allergic reactions to dornase alfa. Improvements in pulmonary function occurred even after the appearance of antibodies.
- Antibodies have been detected 3–8 months after the beginning of pegademase bovine therapy. A small percentage of the patients with antibodies display enhanced enzyme clearance.
- Patients' diminished responsiveness to pegloticase was found to be associated with high titers of antibody that increase clearance of the enzyme with the consequent failure to decrease uric acid levels and an increased risk of infusion reactions.
- In a phase III study comparing rasburicase to allopurinol in controlling plasma uric acid in pediatric patients with hematologic malignancies, the enzyme did not prove to be highly immunogenic. Overall, the incidence of anti-rasburicase antibodies was low with only 2 % positive for enzyme-neutralizing antibodies.
- Antibodies to streptokinase tend to form in most patients receiving the drug. Such antibodies are likely to increase resistance to successful therapy so repeat treatment with streptokinase is to be avoided if administration has been given for more than 5 days, and particularly if continued for 5 days to 12 months after the initiation of treatment.
- Apart from ERT and surgical interventions, other therapies are being sought and a number of approaches are showing promise as potentially successful treatments. The most promising are stem cell transplantation, substrate reduction therapy, the use of chaperones, and gene therapy.

Further Reading

Baldo BA. Enzymes approved for human therapy: indications, mechanisms and adverse effects. BioDrugs. 2015;29:31–55.

Barton NW, Brady RO, Dambrosia JM, et al. Replacement therapy for inherited enzyme deficiency—macrophage-targeted glucocerebrosidase for Gaucher disease. N Engl J Med. 1991;324:1464–70.

Becker MA, Baraf HSB, Yood RA, et al. Long-term safety of pegloticase in chronic gout refractory to conventional treatment. Ann Rheum Dis. 2013;72:1469–74.

Bookbinder LH, Hofer A, Haller MF, et al. A recombinant human enzyme for enhanced interstitial transport of therapeutics. J Control Release. 2006;114:230–41.

Booth C, Gaspar HB. Pegademase bovine (PEG-ADA) for the treatment of infants and children with severe combined immunodeficiency (SCID). Biol Targets Ther. 2009;3:349–58.

Cortes J, Moore JO, Maziarz RT, et al. Control of plasma uric acid in adults at risk for tumor lysis syndrome: efficacy and safety of rasburicase alone and rasburicase followed by allopurinol compared with allopurinol alone—results of a multicenter phase III study. J Clin Oncol. 2010;28:4207–13.

Deegan PB, Cox TM. Imiglucerase in the treatment of Gaucher disease: a history and perspective. Drug Des Devel Ther. 2012;6:81–106.

Hershfield MS, Mitchell BS. Immunodeficiency diseases caused by adenosine deaminase deficiency and purine nucleoside phosphorylase deficiency. In: Scriver CR, Beaudet al, Sly WS, et al., editors. The metabolic and molecular basis of inherited disease. 7th ed. New York: McGraw-Hill; 1995. p. 1725–68.

Hurst LC, Badalamente MA, Hentz VT, et al. Injectable collagenase clostridium histolyticum for Dupuytren's contracture. N Engl J Med. 2009;361:968–79.

Jones AP, Wallis C. Dornase alfa for cystic fibrosis. Cochrane Database Syst Rev. 2010;March 17(3):CD001127. doi:10.1002/14651858.CD001127.pub2.

Mundada LV, Prorok M, DeFord ME, et al. Structure-function analysis of the streptokinase amino terminus (residues 1–59). J Biol Chem. 2003;278:24421–7.

Neufeld EF, Muenzer J. The mucopolysaccharidoses. In: Scriver CR, Beaudet al, Sly WS, et al., editors. The metabolic and molecular basis of inherited disease. 8th ed. New York: McGraw-Hill; 2001. p. 3421–52.

Rattu MA, Shah N, Lee JM, et al. Glucarpidase (voraxaze), a carboxypeptidase enzyme for methotrexate toxicity. PT. 2013;38:732–44.

Reiner Z, Guardamagna O, Nair D, et al. Lysosomal acid lipase deficiency—an under-recognized cause of dyslipidemia and liver dysfunction. Artherosclerosis. 2014;235:21–30.

Stalmans P, Benz MS, Gandorfer A, et al. Enzymatic vitreolysis with ocriplasmin for vitreomacular traction with macular holes. N Engl J Med. 2012;367:606–15.

The GUSTO Angiographic Investigators. The effects of tissue plasminogen activator, streptokinase, or both on coronary-artery patency, ventricular function, and survival after acute myocardial infarction. N Engl J Med. 1993;329:1615–22.

van der Ploeg AT, Reuser AJJ. Lysosomal storage disease 2. Pompe's disease. Lancet. 2008;372:1342–63.

Chapter 10
Blood Coagulation

Essential for survival, the biologically highly conserved process of coagulation or clotting results in homeostasis, that is, the stemming of blood loss and ultimately repair of the damaged vessels and other tissues. The components and steps involved in the retention of homeostasis are complex with the participation of cells, the kallikrein-kinin system, and a cascade of primary and activated protein coagulation or clotting factors together with a number of cofactors and regulatory proteins. These participate in a series of predominately enzyme activations and catalyzed reactions, ultimately leading to the formation of insoluble cross-linked fibrin and cessation of blood loss. In keeping with this book's emphasis on the safety aspects of the current list of the major biologic agents approved for human therapy, no attempt will be made to provide an all-embracing coverage of the history, physiology, pharmacology, and biochemical aspects of blood coagulation; rather, sufficient background will be provided for nonspecialists to become aware of the range of available blood products approved by regulatory agencies, their nature, indications, mechanisms of action, the warnings and precautions associated with their prescribing, and adverse events resulting from their use.

Platelet Activation and von Willebrand Factor

Collagen, activates and aids the spreading of platelets via platelet surface glycoprotein (GP) Ia/IIa receptors, a process that is activated upon contact of circulating platelets with exposed collagen following damage to endothelium and some other tissues. Adhesion of platelets to wound sites is aided by von Willebrand factor**,** a large multimeric glycoprotein produced by, and stored in, platelet alfa granules. This complex macromolecule carries blood group ABO antigens and is composed of monomers each of 2050 amino acids. Isolation and characterization of a collagen-binding domain has shown that von Willebrand factor interacts with collagen type I alfa 1. Importantly, von Willebrand factor acts as a bridge connecting collagen

B.A. Baldo, *Safety of Biologics Therapy*, DOI 10.1007/978-3-319-30472-4_10

fibrils to the glycoprotein complex GPIb alfa-V–IX (CD42a-d complex) (Fig. 10.1), each encoded by distinct genes, on the platelet surface while interaction between GPIb alfa and thrombin (Fig. 10.2) is required for platelet aggregation at sites of vascular injury. Defects in proteins of the GPIb alfa-V–IX complex leads to a condition with a range of molecular defects, the autosomal recessive disorder known as Bernard–Soulier syndrome (also called hemorrhagiparous thrombocytic dystrophy or giant platelet syndrome), with symptoms of mucocutaneous bleeding, low platelet count, purpura, gingival bleeding, menorrhagia in women, and gastrointestinal hemorrhage. A deficiency of von Willebrand factor is responsible for a usually, but not exclusively, hereditary coagulation abnormality known as von Willebrand disease, often with no, or mild, symptoms such as epistaxis. Depending on the disease type, desmopressin (Chap. 7, section "Desmopressin"), which stimulates the release of von Willebrand factor, may be given intranasally or intravenously. For more severe cases, human factor VIII preparations that have not been highly purified such as Wilate® can be used as a source of von Willebrand factor (Table 10.1). Purified factor VIII, including recombinant preparations, contains an insignificant amount of, or no, von Willebrand factor so are therefore not clinically useful. A recombinant preparation of von Willebrand factor (vonicog alfa; Vonvendi®) expressed in CHO

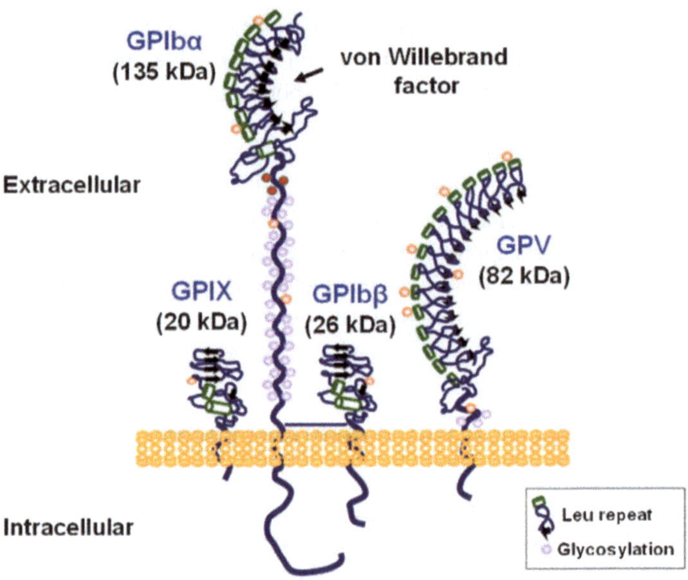

Fig. 10.1 Diagrammatic representation of the structure of the glycoprotein (GP) complex GPIb alfa-V–IX (CD42a-d complex) on the platelet surface. The complex, whose main function is to ensure normal primary homeostasis by initiating platelet adhesion at sites of vascular injury, is composed of four distinct transmembrane proteins, GPIbα, GPIX, GPV, and GPIbβ. The GPIbα chain provides the von Willebrand factor-binding site, an interaction that facilitates platelet adhesion and signaling events that lead to platelet activation, thrombosis, and hemostasis. Reproduced from Lanza F. Bernard–Soulier syndrome (Hemorrhagiparous thrombocytic dystrophy). Orphanet J Rare Dis 2006;1:46. doi:10.1186/1750-1172-1-46. An open access article distributed under the terms of the Creative Commons Attribution License

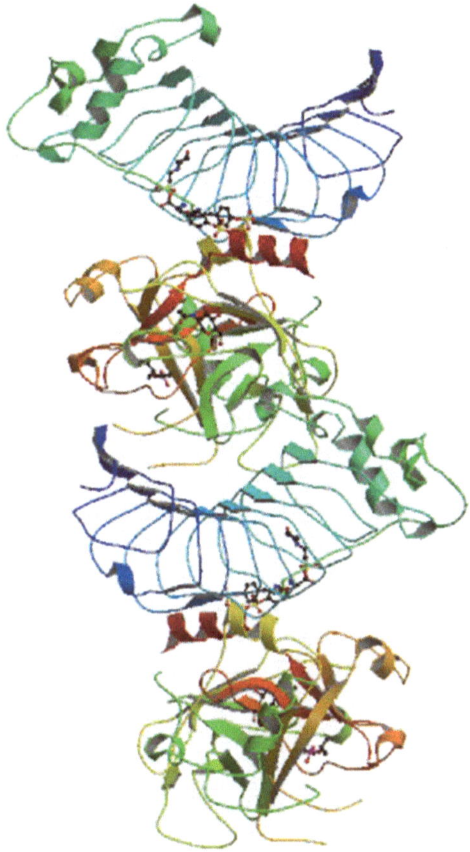

Fig. 10.2 Crystal structure of the platelet receptor glycoprotein GPIb alfa-thrombin complex at 2.6 Å resolution showing interactions of the platelet glycoprotein with exocite (secondary binding site distinct from the active site) I of one thrombin molecule and exocite II of a second thrombin molecule. The structure is from Protein Data Bank RCSB PDB file 1P8V (Dumas JJ, Kumar R, Seehra J, et al. Science 2003;301:222–6)

cells and indicated for on-demand treatment and control of bleeding in adults with von Willebrand disease, was recently approved by the FDA (Table 10.1).

In the absence of platelet aggregation and without the involvement of platelet surface phospholipids, the blood coagulation cascade does not proceed. Activation of platelets leading to a signal transduction cascade is initiated by thrombin interacting with surface G protein-coupled receptors. The activation of protein kinase C (PKC) and, in particular, the release of intracellular Ca^{2+} stores together with the adherence of platelets to collagen results in activation of phospholipase A_2 (PL A_2), hydrolysis of membrane phospholipids liberating arachidonic acid, and release of thromboxane A_2 (TXA_2), a potent inducer of platelet activation. Platelet activation is further stimulated by PKC-induced activation of a 47 kDa platelet protein that, in turn, induces the release of granule contents, most importantly, ADP which further stimulates platelets and leads to the exposure of the glycoprotein receptor complex GPIIb-GPIIIa (also called integrin αIIb-β3), a receptor for von Willebrand factor and fibrinogen. Interaction between this receptor and fibrinogen also induces platelet aggregation. An inherited bleeding disorder, Glanzmann thrombasthenia, results from a defect in the GPIIb protein of the GPIIb-GPIIIa receptor complex. The monoclonal antibody, abciximab (ReoPro®) (Chap. 4, section "Abciximab"), targeted to GPIIb-GPIIIa, the

Table 10.1 Coagulation factors and related agents approved for human therapy[a]: descriptions, indications[a], and adverse effects as at June 2016

Generic and trade names	Description	Approved indications and usage	Warnings and precautions	Adverse effects, serious and common
Factor VIIa[b] (NovoSeven®)	Recombinant human coagulation factor VIIa. A vitamin K-dependent serine protease glycoprotein of 406 amino acids (MW 50 kDa)	Treatment of bleeding in hemophilia A or B patients[c] and in patients with congenital factor VII deficiency; prevention of bleeding in surgical/invasive procedures in hemophilia A or B patients[c] and patients with congenital factor VII deficiency	Increased risk of thrombotic events in patients with circulating TF or predisposing coagulopathies[d]; monitor for prothrombin time and factor VII before and after administration	Serious: Thromboembolic events[e]; anaphylaxis Common: Pyrexia; hemorrhage; ISR; arthralgia; headache; nausea; vomiting; hypertension; hypotension; pain; edema; rash; urticaria
Factor VIII[f] Full length—Octocog alfa (Kogenate® FS; Advate®[g]; Kovaltry®; Recombinate®)	Recombinant glycoprotein of multiple peptides (incl. 80 kDa and extensions of 90 subunit); produced in BHK cells; associates in inactive form with von Willebrand factor	Bleeding in adults and children with hemophilia A (congenital factor VIII deficiency); perioperative management of patients with hemophilia A; prophylaxis to reduce frequency of bleeding and joint damage	Anaphylaxis/hypersensitivity[h]; development of neutralizing antibodies; possible hypersensitivity to mouse or hamster protein[i]; monitor factor VIII activity[j] and inhibitors[k]	Serious: Systemic hypersensitivity; high titer inhibitors Common: Antibody inhibitor formation; pruritus; rash; ISR; CVAD line-associated infections
Factor VIII modified — Moroctocog alfa (Xyntha®; ReFacto AF®)	Recombinant glycoprotein human BDDrFVIII produced in CHO cells; 1438 amino acids, MW 170 kDa (90+80 kDa forms)	Control and prevention of bleeding episodes and for surgical prophylaxis in patients with hemophilia A	Allergic hypersensitivity; neutralizing antibodies; antibodies to murine and hamster proteins; catheter-related complications; reports of lack of effect—record batch number	Factor VIII inhibitors especially in PUPs; vomiting; nausea; headache; arthralgia; vascular access complications; asthenia; headache; hemorrhage/hematoma; pyrexia
Factor VIII modified—Turoctocog alfa (NovoEight®)	Human BDDrFVIII made in CHO cells; consists of heavy and light chains (87 kDa and 79 kDa resp.) and 21 residue truncated B-domain	Bleeding in adults and children with hemophilia A; perioperative management of patients with hemophilia A; routine prophylaxis to reduce frequency of bleeding episodes	Hypersensitivity reactions; neutralizing antibodies; monitor and test for factor VIII activity levels and development of factor VIII inhibitor antibodies	ISR; increased hepatic enzymes; pyrexia

Factor VIII human recombinant—Simoctocog alfa (Nuwiq®)	Human BDDrFVIII derived from human HEK 293 F cells, 170 kDa. B domain (H chain ~ 90 kDa, L chain ~ 80 kDa) replaced by 16 amino acid linker	In adults and children with hemophilia A for on-demand treatment and control of bleeding; routine prophylaxis to reduce bleeding frequency; perioperative management of bleeding	Hypersensitivity reactions; neutralizing antibodies; monitor and test for factor VIII activity levels and development of factor VIII inhibitor antibodies	Paresthesia; headache; ISR; back pain; dry mouth; vertigo; formation of anti-FVIII antibodies
Factor VIII porcine analog recombinant – Susoctocog alfa (Obizur®)	Porcine BDDrFVIII replaced by 24 amino acid linker. Produced in BHK cells; 1448 amino acids, MW ~170 kDa (~90 kDa H chain, ~80 kDa L chain)	Bleeding episodes in adults with acquired hemophilia A. Note: Safety and efficacy not established in patients with anti-porcine FVIII inhibitor titer > 2 BU; not indicated for congenital hemophilia or von Willebrand disease	Hypersensitivity reactions; neutralizing antibodies; test for porcine FVIII inhibitor concentration if expected FVIII levels not attained or bleeding is not controlled	Development of inhibitors to porcine FVIII
Factor VIII Fc fusion protein[l] (Eloctate®; Elocta®)	Recombinant glycoprotein of B-domain deleted analog of human factor VIII linked to human IgG1 Fc; 1890 amino acids, MW 330 kDa[m]. Produced in HEK cells	Control and prevention of bleeding episodes in adults and children with hemophilia A; for routine prophylaxis and surgical prophylaxis (perioperative management) to prevent or reduce the frequency of bleeding episodes	Hypersensitivity reactions[h]; neutralizing antibodies; monitor factor VIII activity[j] and inhibitors[k]	Arthralgia; malaise; rash[n]
Factor IX[o] Complex (Profilnine® SD[p])	Concentrate of non-activated factors IX, X, II (prothrombin) and low levels of factor VII derived from pooled human plasma[q]	Prevention and control of bleeding in patients with factor IX deficiency due to hemophilia B (congenital factor IX deficiency)	From pooled human plasma, therefore possible risk of transmitting infectious agents[r]; thrombosis or DIC potentially fatal reactions in patients undergoing surgery or with liver disease	Thrombosis; DIC; urticaria; fever; chills; nausea; vomiting; headache; somnolence; lethargy; flushing; tingling[s]
Factor IX Fc Fusion Protein (Alprolix®)	Recombinant factor IX expressed in HEK cells covalently linked to Fc domain of human IgG1; fusion protein has 867 amino acids, MW ~98 kDa	Indicated in adults and children with hemophilia B for control and prevention of bleeding; perioperative management; routine prophylaxis to prevent or reduce frequency of bleeding episodes	Hypersensitivity including anaphylaxis; development of neutralizing antibodies; thromboembolic complications	Common (incidence ≥1 %): Headache; oral paresthesia

(continued)

Table 10.1 (continued)

Generic and trade names	Description	Approved indications and usage	Warnings and precautions	Adverse effects, serious and common
Factor IX albumin fusion protein (rIX-FP) (Idelvion®)	Recombinant factor IX genetically fused to recombinant albumin via a cleavable linkage. Glycoprotein of 1018 amino acids. Secreted by CHO cell line. Factor IX identical to Thr148 allelic form in plasma	Indicated for adults and children with hemophilia B for on-demand control of bleeding; perioperative bleeding; and routine prophylaxis to prevent/reduce bleeding	Hypersensitivity including anaphylaxis; development of neutralizing antibodies; thromboembolic complications; nephrotic syndrome; need to monitor plasma levels of factor IX and development of inhibitors	Headache; dizziness; hypersensitivity; rash; eczema
Factor XIII A-Subunit (Tretten®)	Recombinant human factor XIII-A_2 homodimer of two factor 731 amino acid XIII A-subunits with an *N*-acetylated terminal Ser[t]	Routine prophylaxis for bleeding in patients with congenital factor XIII A-subunit deficiency	Hypersensitivity reactions; thromboembolic risk; inhibitory antibodies	Headache; extremity pain; ISR (pain); increase in fibrin D dimer levels[u]; immunogenicity
Factor XIII Concentrate (Human) (Corifact®)	Concentrate of factor XIII from pooled human plasma consisting of two A-subunits and two B-subunits	Routine prophylactic treatment and perioperative management of surgical bleeding in patients with congenital factor XIII deficiency	Hypersensitivity; thromboembolic risk; inhibitory antibodies; transmission of infectious agents; monitor patient's trough factor XIII levels and inhibitory antibody levels[v]	Hypersensitivity; joint inflammation; arthralgia; headache; elevated thrombin-antithrombin levels; increased blood lactate dehydrogenase; rash; pruritus; erythema; hematoma
von Willebrand Factor/Coagulation Factor VIII Complex (Wilate®)	Derived from pooled human plasma; manufactured from cryoprecipitate by $Al(OH)_3$ treatment, chromatography, ultra-, dia-, and sterile filtration[w]	Treatment of spontaneous and trauma-induced bleeding episodes in patients with severe von Willebrand disease and patients with mild-moderate disease in whom use of desmopressin is known or suspected to be ineffective or contraindicated	Hypersensitivity; thromboembolic risk; inhibitory antibodies; transmission of infectious agents; monitor von Willebrand factor and factor VIII activity levels and inhibitory antibody levels	Serious: Hypersensitivity reactions Common: Urticaria; dizziness; dyspnea; nausea; vomiting; cough

Antihemophilic Factor/von Willebrand Factor Complex (Humate-P®)	Purified, dried, pasteurized concentrate. Purified from cold insoluble fraction of pooled human fresh-frozen plasma with a claimed higher factor potency than from cryoprecipitates[x]	Hemophilia A (classical hemophilia): treatment and prevention of bleeding. von Willebrand disease: treatment of spontaneous and trauma-induced bleeding; prevention of excessive bleeding during and after surgery[y]	Hypersensitivity; thromboembolic risk; transmission of infectious agents[z]; monitor von Willebrand factor and factor VIII activity levels and for hemolysis in blood group A, B and AB patients[aa]	Serious: Anaphylaxis; thromboembolic events Common: Hypersensitivities including urticaria, chest tightness, rash, pruritus, edema, shock; post-operative hemorrhage, nausea, pain[ab]
Recombinant von Willebrand Factor – Vonicog alfa (Vonvendi®)	Expressed in CHO cells. Contains large multimers in addition to multimers found in plasma. Cleaved by proteolytic enzyme ADAMTS13 in blood	On-demand treatment and control of bleeding in adults with von Willebrand disease	Thromboembolc reactions; inhibitors of von Willebrand factor; hypersensitivity; monitor for VWF:RCo and factor VIII and antibody inhibitors	Generalized pruritus
Anti-Inhibitor Coagulant Complex (FEIBA[ac]NF®)[ad]	Human plasma fraction containing mainly non-activated factors II, IX and X and activated factor VII plus 1–6 units of factor VIII coagulant antigen per ml[ae]	Control and prevention of bleeding, perioperative management and routine prophylaxis to prevent or reduce bleeding in hemophilia A and B patients with inhibitors	*Boxed warning* for thromboembolic events. Other warnings/precautions: hypersensitivity including anaphylaxis; transmission of infectious agents	Chills; chest pain; dizziness; dysgeusia; dyspnea; hypoesthesia; nausea; pyrexia, anemia; somnolence; increase in inhibitory antibody titer
Prothrombin Complex Concentrate (Human) (Kcentra®)	Non-activated 4-factor concentrate of factors II, VII, IX and X plus anti-thrombotic proteins C and S. Factor IX is the lead factor for potency	Urgent reversal of acquired coagulation factor deficiency induced by vitamin K antagonist therapy (e.g., warfarin) in adults with acute major bleeding[af]	*Boxed warning* for thromboembolic complications. Other warnings and precautions: hypersensitivity including anaphylaxis; transmission of infectious agents	Serious: thromboembolic events (stroke, pulmonary embolism, deep vein thrombosis). Common: nausea/vomiting; headache; arthralgia, hypotension

(continued)

Table 10.1 (continued)

Generic and trade names	Description	Approved indications and usage	Warnings and precautions	Adverse effects, serious and common
Fibrinogen concentrate (Human) (RiaSTAP®)	Heat-treated, lyophilized fibrinogen (factor I) from pooled human plasma; a glycoprotein MW ~340 kDa; dimer of three pairs of polypeptide chains Aα, Bβ and γ	Acute bleeding in patients with congenital fibrinogen deficiency, including afibrinogenemia and hypofibrinogenemia[ag]	Hypersensitivity reactions; thrombotic events; transmission of infectious agents	Serious: thromboembolic events (deep vein thrombosis, pulmonary embolism, myocardial infarction, arterial thrombosis); allergy/ anaphylaxis. Common: Chills; fever, nausea; vomiting
Fibrin Sealant (Tisseel®)[ah]	Two component sealant from pooled human plasma: sealer protein (fibrinogen plus aprotinin[ai]) and thrombin	An adjunct to hemostasis in patients undergoing surgery when conventional control of bleeding is ineffective or impractical; adjunct to surgery to prevent leakage from colonic anastomoses	Aprotinin known to be associated with anaphylaxis; transmission of infectious agents; excess application impairs wound healing; risk of intravascular application; denatured by heavy metals alcohol, iodine; air or gas embolism[aj]	Hypersensitivity including anaphylaxis. Post-marketing: Bradycardia, tachycardia; vascular disorders; nausea; wheezing; bronchospasm; dyspnea; angioedema; flushing; erythema; urticaria; pruritus
Fibrin Sealant Patch (Evarrest™)	Human plasma-derived fibrinogen and thrombin embedded in a patch of oxidized regenerated cellulose and polygalactin fibers	An adjunct to hemostasis for soft tissue bleeding during open retroperitoneal, intra-abdominal, pelvic and non-cardiac thoracic surgery when control of bleeding by standard surgery is ineffective or impractical	Thrombosis; hypersensitivity; risk of transmission of infectious agents; infection; inadvertent adhesions;	Abdominal distension; increased blood fibrinogen; post procedural and intra-abdominal hemorrhage; pulmonary embolism
Protein C Concentrate (Human) (Ceprotin®)	Vitamin K-dependent glycoprotein; converted to activated protein C,[ak] an anticoagulant[al] serine protease. Inactivates factors Va and VIIIa[am]	Replacement therapy for severe congenital protein C deficiency and prevention and treatment of venous thrombosis and purpura fulminans	Hypersensitivity reactions; bleeding episodes; transmission of infectious agents; renal impairment on low Na^+ diet[an]; heparin-induced thrombocytopenia[ao]	Rash; itching; lightheadedness; ISR; hemothorax; hypotension

Generic and trade names	Description	Approved indications and usage	Warnings and precautions	Adverse effects, serious and common
Antithrombin Recombinant) (ATryn®)	Recombinant human antithrombin, 432 amino acid glycoprotein,[ap] MW ~57 kDa; produced in scrapie-free goats,[aq] protein expressed into the milk[ar]; three disulfide bridges	Prevention of perioperative and peripartum thrombo-embolic events in hereditary antithrombin deficient patients[as]	Anaphylaxis and severe hypersensitivity reactions[at]; may alter anticoagulant effect of drugs that cause antithrombin-mediated anticoagulation.	Hemorrhage; ISR; enhances anticoagulant effect of heparin

AF albumin free, *BDDrFVIII* B-domain deleted recombinant factor VIII, *BHK* baby hamster kidney, *CHO* Chinese hamster ovary, *CVAD* central venous access device, *DIC* disseminated intravascular coagulation, *HEK* human embryonic kidney, *ISR* injection site reactions, *PUPs* previously untreated patients, *TF* tissue factor (factor III), VWF:RCo von Willebrand factor ristocetin cofactor

[a]Approved by FDA or EMA or both

[b]Sometimes called proconvertin

[c]Patients with inhibitors to factor VIII or factor IX and patients with acquired hemophilia

[d]Patients with disseminated intravascular coagulation, advanced artherosclerotic disease, crush injury, septicemia or concomitant treatment with prothrombin complex concentrates

[e]Including myocardial infarction, myocardial ischemia, cerebral infarction, and/or ischemia, thrombophlebitis, arterial thrombosis, deep vein thrombosis, pulmonary embolism

[f]Anti-hemophilic factor A. Kogenate FS is produced by BHK cells into which the human factor VIII gene has been introduced

[g]Advate®, a glycoprotein of 2332 amino acids, is a full length recombinant (expressed in CHO cells) anti-hemophilic factor A (octocog alfa) with no human or animal-derived plasma proteins. Its most commonly recorded adverse reactions are pyrexia, headache, cough, nasopharyngitis, vomiting, arthralgia, limb injury

[h]Including anaphylaxis, pruritus, rash, urticaria, facial swelling, dizziness, hypertension, nausea, chest discomfort, cough, dyspnea, wheezing, flushing

[i]Produced by recombinant DNA technology in BHK cells; preparation contains trace amounts of mouse immunoglobulin and BHK proteins

[j]By the one stage clotting assay

[k]Use Bethesda units (BU) to titer antibody inhibitors

[l]B-domain deleted recombinant factor VIII Fc fusion protein (BDD-rFVIIIFc). Also known as efraloctocog alfa

[m]The B-domain portion of the factor VIII portion of the fusion protein is covalently attached to the Fc fragment by a 14 amino acid linkage

[n]Surprisingly few adverse events. Results from a trial with only 164 patients; more experience needed with more patients

[o]Anti-hemophilic factor B or Christmas factor

[p]Solvent detergent-treated to inactivate viruses

[q]Complex temporarily increases the plasma level of factor IX, minimizing the hazards of hemorrhage

(continued)

Table 10.1 (continued)

[r]Risk of infectious agents cannot be totally eliminated; hepatitis A and B vaccinations encouraged
[s]Slower infusion rates generally relieve symptoms. Highly reactive patients may tolerate product from a different lot number
[t]Activation of factor XIII by thrombin cleaves a 37 amino acid peptide from the A subunit N-terminus
[u]Without evidence of thrombolic events
[v]If breakthrough bleeding occurs or if expected peak plasma factor XIII levels are not attained
[w]Labeled with von Willebrand factor (von Willebrand factor: Ristocetin Cofactor [VWF:RCo]) and factor VII activities in IU/vial determined with reference to WHO International Standards
[x]Claimed to be a highly purified and concentrated human anti-hemophilic factor/von Willebrand factor complex with low amounts of non-factor proteins. Contains labeled amounts of factor VIII in IU and von Willebrand factor: ristocetin cofactor activity expressed in IU
[y]Patients with severe von Willebrand disease and patients with mild-moderate disease where desmopressin is known or suspected to be inadequate
[z]Parvoviris B19 and hepatitis A are particularly difficult to remove. Physicians should therefore be alert for symptoms of these infections, particularly in pregnant women and immune-compromised individuals
[aa]Humate-P® contains anti-A and anti-B isoagglutinins
[ab]In patients undergoing surgery
[ac]FEIBA, factor VIII inhibitor bypassing activity, anti-inhibitor coagulant complex
[ad]Nanofiltered and vapor heated preparation. Contains approximately equal unitages of factor VIII bypassing activity and prothrombin complex factors
[ae]Plus traces of factors of the kinin generating system
[af]Not indicated for such patients without acute major bleeding
[ag]Not indicated for disfibrinogenemia
[ah]For topical use only
[ai]Synthetic fibrinolysis inhibitor that delays fibrinolysis
[aj]Embolism may occur when sealant is administered using pressurized gas
[ak]Converted by thrombin-thrombomodulin complex
[al]In the presence of its cofactor, protein S
[am]Leads to a decrease in thrombin formation
[an]Quantity of sodium in the maximum daily dose of cepotin exceeds 200 mg
[ao]Ceprotin contains trace amounts of heparin which may lead to heparin-induced thrombocytopenia
[ap]Glycosylation profile is different from plasma-derived antithrombin which results in an increased heparin affinity
[aq]Genetically engineered goats
[ar]Antithrombin DNA coding sequence and a mammary gland-specific DNA sequence introduced to direct expression of antithrombin into the milk
[as]Not indicated for *treatment* of thromboembolic events in hereditary antithrombin deficient patients
[at]Contraindicated in patients with hypersensitivity to goat and goat milk proteins

tyrosine derivative tirofiban (Aggrastat®) (Fig. 10.3), and the cyclic heptapeptide eptifibatide (Integrelin®) (Fig. 10.4) from pygmy rattlesnake venom, create a thrombasthenia-like state, blocking the interaction between fibrinogen and the receptor thus inhibiting platelet aggregation. This property makes the drugs useful for the treatment of patients undergoing angioplasty and unstable angina.

Factor VIII (see below, section "Factor VIII"), which degrades rapidly in the circulation, also binds to von Willebrand factor where it remains inactive, but stabilized, until released by the action of thrombin.

The Kallikrein-Kinin System and Coagulation

The plasma kallikrein-kinin system leads to the release of vasoactive kinins from inactive precursor kininogen macromolecules as a result of the activation of kallikrein from its prekallikrein inactive form. The released kinins produce a wide range of actions such as vasodilation, increased vascular permeability, nitric oxide production, release of tissue plasminogen activator (tPA), and arachidonic acid-mediated processes, particularly the production of prostacyclin (PGI_2). As a consequence of these various activities, kinins participate in numerous physiological and pathological processes including the regulation of blood pressure via the renin-angiotensin system, angiogenesis, inflammation, cell growth and proliferation, apoptosis, and the one we are considering here, blood coagulation.

The plasma kallikrein-kinin system consists of factor XII (also called Hageman factor), factor XI, high molecular weight kininogen (HMWK; Fitzgerald factor) and prekallikrein (PK; Fletcher factor). Both factor XI and factor XII are zymogen (inactive) forms of the corresponding factor XIa and factor XIIa active serine proteases. HMWK occurs in the plasma as a single chain polypeptide, MW 88–120 kDa depending on its extent of glycosylation. It is seen as comprising six domains of different compositions and/or functions such as platelet and endothelial cell-binding activity, anti-angiogenic properties, the presence of a low affinity Ca^{2+}-binding site, the bradykinin sequence, and

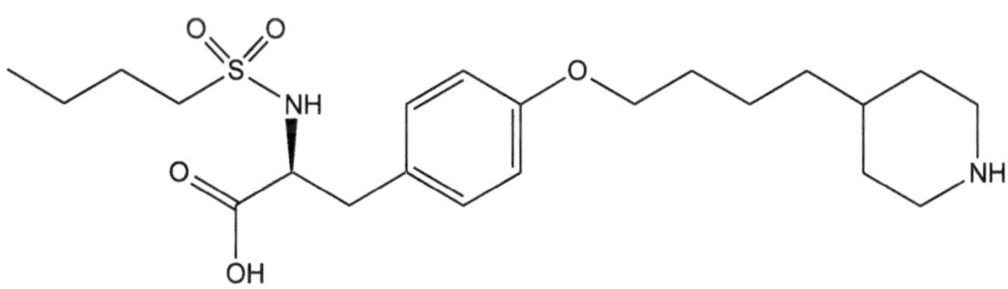

Tirofiban

Fig. 10.3 Structure of the synthetic, non-peptide anticoagulant tirofiban (Aggrastat®) which inhibits platelet aggregation by blocking the binding of fibrinogen and von Willebrand factor to the GPIIb-GPIIIa receptor on platelet surfaces. Tirofiban, a small molecule mimic of the arginine-glycine-aspartate (RGD) sequence and reversible antagonist of the platelet receptor, has a rapid onset and short duration of action. Its structure, based on an anticoagulant isolated from the saw-scaled viper, *Echis carinatus*, contains terminal 4-(4-piperidinyl)butyl and (S)-butylsulfonylamino linkages at the molecule's two termini

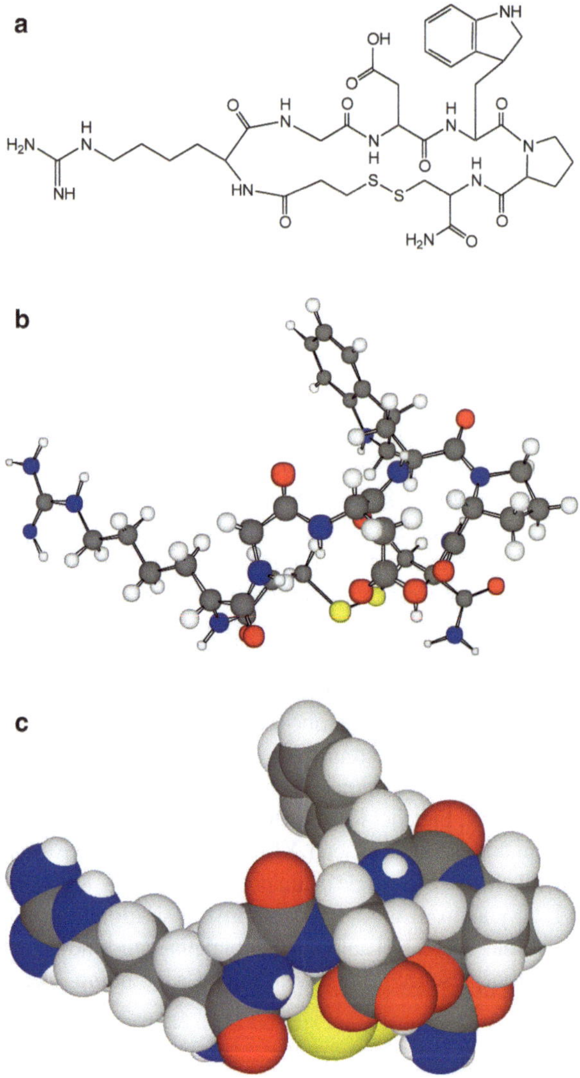

Fig. 10.4 Two-dimensional (**a**), three-dimensional ball-and-stick (**b**) and CPK (**c**) models of the cyclic heptapeptide antiplatelet drug eptifibatide (Integrelin®) derived from a protein found in the venom of the southeastern pygmy rattlesnake (*Sistrurus miliarius barbouri*) and classed as an arginine-glycine-aspartat mimetic. Eptifibatide, which has a short half-life, blocks the binding of fibrinogen and von Willebrand factor to the GPIIb-GPIIIa receptor on the platelet surface. The molecule contains six amino acids and one mercaptopropionyl group. Note the interchain disulfide bridge between the cysteine amide and the mercaptopropionyl moieties. Five of the amino acids are chiral, existing in their natural L-configuration. There are therefore 32 potential optical isomers. Specifications and tests for enantiomers control the optical purity of eptifibatide

sequences that inhibit cysteine proteases or bind heparin. HMWK binds to negatively charged surfaces via domain 5 while at the same time, the adjacent domain 6 containing the prekallikrein and factor XI-binding sites, binds these two proteins.

The Clotting Cascade

The formation of fibrin, necessary for clot formation, can be initiated by two distinct pathways, the extrinsic, or tissue factor, pathway triggered by damage to blood vessel walls or the so-called intrinsic pathway, more accurately termed the contact activation pathway. Formation of factor XIIa via activation of its zymogen precursor factor XII by negatively charged surfaces (e.g., glass) is the initiation point for the latter system. Factor XIIa activates prekallikrein to kallikrein which, in turn, cleaves HMWK releasing bradykinin but kallikrein also initiates a feedback amplification process by reciprocally activating factor XII (Fig. 10.5). Released bradykinin, via interaction with the kinin B2 receptor (B2R), activates proinflammatory signaling pathways that result in dilatation of blood vessels, increased vascular permeability and chemotaxis of neutrophils. In other words, events initiated by factor XII may manifest as inflammatory responses mediated by the kallikrein-kinin system and procoagulant responses resulting from the contact coagulation pathway. Subsequent events in this pathway proceed by the activation of factor IX (also known as Christmas factor) to factor IXa by factor XIa in the presence of Ca^{2+}. Factor IX, like factors II, VII, and X (Stuart–Prower factor), is a serine protease zymogen

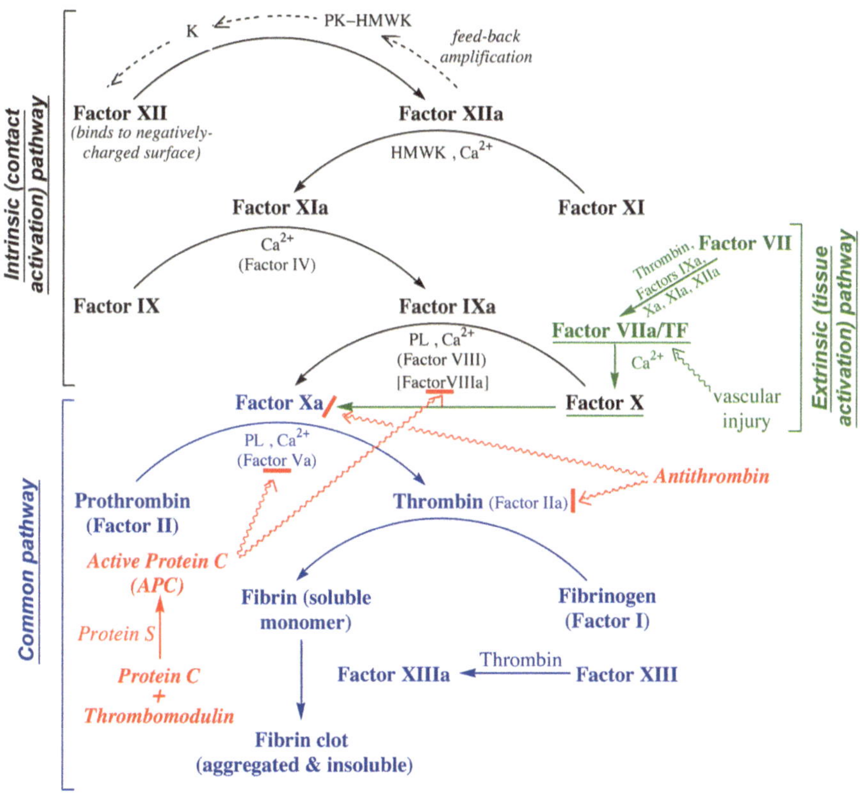

Fig. 10.5 Scheme of the coagulation or clotting cascade showing the relationships between the contact activation (intrinsic), tissue activation (extrinsic), and common pathways together with a summarized representation of the roles of protein C, leading to a decrease in thrombin formation, and antithrombin, the main inhibitor of thrombin and factor Xa. HMWK, high molecular weight kininogen

containing vitamin K-dependent γ-carboxyglutamate residues and activation is effected by the binding of Ca^{2+} to these residues. In the next step in the pathway, factor X is activated to factor Xa by factor IXa-induced cleavage in a reaction that requires Ca^{2+}, factors VIIIa, IXa, and X on the surface of activated platelets. This complex of molecules, referred to as the tenase complex, forms with the involvement of the phospholipids phosphatidylserine and phosphatidylinositol on the platelet surface. Activation of factor VIII to factor VIIIa is induced by a small concentration of thrombin (too much thrombin causes inactivation) and, in its activated form, factor VIII acts as a receptor for factors IXa and X.

Factor Xa can be viewed as the junction or meeting place of the contact activation and extrinsic (tissue factor) coagulation pathways (Fig. 10.5). Following vascular injury, released tissue factor (factor III) initiates the extrinsic clotting cascade by functioning as a cofactor in the Ca^{2+}-dependent activation of factor X by factor VIIa. Activation of factor VII requires the involvement of thrombin or factor Xa, the latter creating the link between the intrinsic and extrinsic pathways.

The shared sections of both pathways are the activation of factor X to factor Xa and events subsequent to that leading to formation of the fibrin clot (Fig. 10.5). Upon the production of factor Xa, prothrombin (factor II) is activated to thrombin (factor IIa) on the surface of activated platelets in interactions that require the formation of the so-called prothrombinase complex, an assemblage of platelet phospholipids, Ca^{2+}, factor Xa, factor Va (a cofactor in the complex and formed by activation of factor V), and, of course, prothrombin. Like factor VIII, factor V is activated to factor Va by traces of thrombin but inactivated by increased levels of the serine protease. Thrombin cleaves fibrinogen (factor I) at four Arg-Gly bonds generating soluble fibrin monomers that spontaneously aggregate to form a weak clot but, more importantly, thrombin activates factor XIII to form the transglutaminase factor XIIIa that forms an aggregated and insoluble clot by cross-linking the amide nitrogen of glutamines to the ε-amino group of lysins in the fibrin monomers.

The low molecular weight heparin anticoagulant **enoxaparin** (Lovenox®, Clexane®, Xaparin®) (Fig. 10.6), given subcutaneously as the sodium salt for prophylaxis and treatment of deep vein thrombosis after surgery, unstable angina, and myocardial infarction, acts by inhibiting factor Xa. This results in a decrease in thrombin and the prevention of clots. Enoxaparin carries an FDA black box warning for the risk of epidural or spinal hematomas, severe events that may result in paralysis in patients treated with low molecular weight heparins. Other warnings relate to an increased risk of hemorrhage, percutaneous coronary revascularization, induction of thrombocytopenia, and the need for caution when used in patients with a history of heparin-induced thrombocytopenia; bleeding diathesis; diabetic retinopathy; renal dysfunction; uncontrolled arterial hypertension; gastrointestinal ulceration; or hemorrhage. The most common adverse reactions to enoxaparin include bleeding and bruising, anemia, thrombocytopenia, diarrhea, nausea, and elevated levels of serum aminotransferase. Post-marketing experience has produced reports of the formation of epidural or spinal hematomas, many causing neurologic injury; injection site reactions; hyperkalemia; thrombocytosis and thrombocytopenia with thrombosis; allergic reactions (including anaphylaxis/anaphylactoid reactions, urticaria, and pruritus);

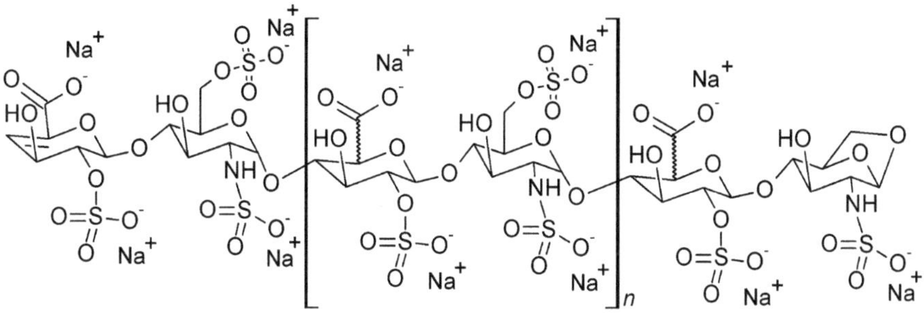

Fig. 10.6 The low molecular weight heparin (average MW ~ 4500) enoxaparin, used prophylactically as an anticoagulant and to treat deep vein thrombosis after surgery, is produced by depolymerization of heparin benzyl ester. The molecule, produced as the sodium salt, contains a 2-*O*-sulfo-4-enepyranosuronic group at the nonreducing end and a 2-*N*,6-*O*-disulfo-*D*-glucosamine at the reducing end. About 20 % of the structure has a 1,6-anhydro derivative at the reducing end

hypersensitivity cutaneous vasculitis; vesiculobullous rash; purpura; and skin necrosis, and drug-induced liver injury. The latter may occur and appear within 1 week of the commencement of enoxaparin therapy. The FDA Adverse Events Reporting System (FAERS) database reveals that ~4 % of all enoxaparin-related adverse events involve hepatic injury. Patients are generally asymptomatic, displaying no features of an immune-mediated reaction including hypersensitivity. The absence of signs of an immune mechanism and other evidence point to a direct toxic effect of the drug on the liver. High dosage and long duration of therapy may lead to hospitalization and/or death. These outcomes are likely the result of hemorrhage and other severe adverse events generally linked to patient comorbidities requiring anticoagulant therapy. Liver injury is generally rapidly reversible upon drug withdrawal.

The vitamin K-dependent glycoprotein **protein C**, also known as autoprothrombin IIA and coagulation factor XIV, is a zymogen for its activated form, the serine protease activated protein C (APC). Activation of protein C is promoted by thrombomodulin and endothelial protein C receptor found on cells of blood vessels. This reaction process proceeds with the participation of cofactors, protein S, factor V, high-density lipoprotein, glycosphingolipids, and phospholipids. APC inactivates the highly procoagulant factor Va and factor VIIIa (Fig. 10.5, Table 10.1), leading to a decrease in thrombin formation. Factor Va and factor VIIIa are important components in blood clotting and the protein C pathway is an important control mechanism preventing excessive procoagulant responses. A deficiency of protein C may manifest as a significantly increased risk of venous thrombosis, purpura fulminans, severe disseminated intravascular coagulation, and simultaneous venous thromboembolism in the womb. Resistance to APC also occurs, most commonly due to a mutation at the cleavage site of factor V, producing factor V Leiden which prevents inactivation of both factor Va and factor VIIIa. Thrombin generation and a marked increase risk of thrombosis results. FDA-approved human protein C concentrate (Ceprotin®) is indicated for severe congenital protein C deficiency and thus the prevention and treatment of venous thrombosis and purpura fulminans. Safety aspects of human protein C concentrate are summarized in Table 10.1.

Antithrombin has an important role in the regulation of homeostasis being the main inhibitor of the serine proteases thrombin and factor Xa and, to a lesser extent, factors IXa, XIa, XIIa, trypsin, plasmin, and kallikrein (Fig. 10.5). Neutralization of the serine proteases is effected by the formation of a complex that is rapidly removed from the circulation. In the presence of heparin, complex formation is accelerated and the capacity of antithrombin to inhibit thrombin is increased a 1000-fold. Atryn® is a recombinant human antithrombin produced in the milk of transgenic goats. Indications and safety considerations for this approved preparation are summarized in Table 10.1.

Factor XII and Coagulation

The role of factor XII in contact-mediated coagulation in vitro is well known if only for the activated partial thromboplastin time (aPTT) test, extensively used for monitoring anticoagulation therapy, screening for deficiencies of coagulation factors, and detection of inhibitors of coagulation. Despite its obvious role and importance in vitro clotting via contact-driven fibrin formation, factor XII-initiated coagulation in vivo was not considered significant and, in fact, was thought to be redundant for normal hemostasis. Furthermore, although deficiencies in a number of other participants in the coagulation cascade lead to bleeding disorders, for example, factors VII, VIII, and IX, this is not the case with factor XII. The conclusion that fibrin clot formation in vivo proceeded largely, if not exclusively, via the extrinsic tissue activation coagulation pathway was therefore not surprising. To more intensively study the in vivo effect of factor XII on the plasma coagulation system, a murine knockout factor $XII^{-/-}$ model was developed. Factor $XII^{-/-}$ mice were found to have a normal hemostatic capacity but the formation of clots was impaired. Intriguingly, however, attenuation of arterial thrombus formation appears to be linked to protection from experimental cerebral ischemia and pulmonary embolism. These findings have led to a new interest in factor XII, raising the possibility of anticoagulation therapy that targets thrombosis without influencing hemostasis. The observation that mice lacking factor XI are protected from occlusive clots further suggests that factor XII influences pathologic clotting via the intrinsic pathway.

The discovery that factor XII appears to be necessary for occlusive thrombus formation has led to a search for endogenous activators that might initiate factor XII-dependent clotting in vivo. For many years, platelets have been linked to the intrinsic pathway with suggestions that factor XII is activated on procoagulant platelets but universal acceptance of proposed mechanisms has not eventuated. Recently, platelet polyphosphate, an inorganic, linear polymer of orthophosphate units linked by phosphoanhydride bonds, has been claimed to be the long sought-after surface that triggers fibrin formation by activated platelets. Enriched in platelet dense granules, polyphosphate is released upon platelet activation and has been shown to modulate plasma coagulation in a factor XII-dependent manner indicating that it is the endogenous, platelet-derived in vivo activator of factor XII. Identification of polyphosphate as a platelet-derived procoagulant has been welcomed by many as the previously elusive link between primary and secondary hemostasis.

Factor VIII

The human factor VIII gene, first cloned in 1984, encodes a single chain polypeptide of 2332 amino acids made up of three different types of domains, A, B, and C arranged as A1-A2-B-A3-C1-C2 (Fig. 10.7). The A domains, each approximately 330 residues, show ~40% homology with each other and cupredoxin-type domains of ceruloplasmin. The highly glycosylated central B domain has a variable sequence in factor VIII proteins from different species and the C domains, each approximately 160 residues, are distantly related to discoidin proteins such as galactose oxidase. In post-translational processing, glycosylation and sulfation of some tyrosines occur and the polypeptide is converted into a heterodimer of a heavy chain made up of the A1-A2-B domains and a light chain of the A3-C1-C2 domains held together by copper and zinc ions. While the B domain has no known structural homologs and appears to be unnecessary for procoagulant activity, it regulates the cellular expression and secretion of factor VIII. The carboxy terminal C2 domain of 159 amino acids is involved in binding von Willebrand factor and binding to platelet membrane surfaces. Thrombin-induced activation of factor VIII leads to the production of the A1, A2, and A3-C1-C2 fragments of factor VIIIa, generation of the tenase complex (factor VIIIa/factor IX) and factor Xa, the production of thrombin, and ultimately a stable fibrin clot (Fig. 10.5). The recombinant form of full length factor VIII is known as octocog alfa.

B-Domain Deleted Recombinant Factor VIII

Factor VIII is synthesized as a large single-chain protein of 2332 amino acids (MW ~ 300 kDa) with several distinct domains, A1-A2-B-A3-C1-C2 (Fig. 10.7). Recombinant, human, full-length factor VIII has the International Nonproprietary Name (INN), octocog alfa. The large B domain which has no known function or homology to any other protein and which has 19 of the 25 glycosylation sites is processed to generate a heavy chain (A1-A2-B; 90–200 kDa in size) and a light chain (A3-C1-C2; 80 kDa). It was found that deletion of a large section of the B domain from Ser743 to Gln1638 causing a 38% reduction in protein size largely to a 90 kDa heavy chain and a 80 kDa light chain had no effect on the activity of factor VIII but its expression was enhanced. This enabled the construction of a functional truncated

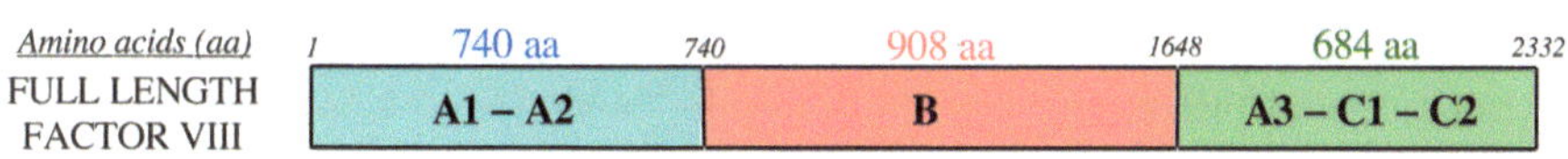

Fig. 10.7 Representation of the full-length factor VIII single polypeptide chain of 2332 amino acids showing the arrangement of the domains A1, A2, B, A3, C1, and C2. The polypeptide is a heterodimer of a heavy chain made up of the A1-A2-B domains and a light chain of the A3-C1-C2 domains held together by metal ions. The carboxy terminal C2 domain of 159 amino acids is involved in binding von Willebrand factor and binding to platelet membrane surfaces. Thrombin-induced activation of factor VIII leads to the production of the A1, A2, and A3-C1-C2 fragments of factor VIIIa

enzyme in the form of the so-called B-domain deleted recombinant factor VIII protein with its five domains (Fig. 10.8) and which retains six potential *N*-linked glycosylation sites at Asp residues 41, 239, 582, 1685, 1810, and 2118. This preparation, designated here BDDrFVIII, proved comparable to factor VIII with retained B-domain when compared for specific activity in chromogenic substrate and clotting assays.

As well as full length recombinant factor VIII preparations such as Kogenate® FS, Kovaltry®, and Advate®, B-domain deleted recombinant factor VIII, or BDDrFVIII preparations, given the international nonproprietary name moroctocog alfa, are marketed as ReFacto® and Xyntha® (Table 10.1). The first BDDrFVIII, ReFacto®, was licensed by the FDA in 2000. This was replaced in the US market in 2008 by an albumin-free preparation Xyntha® (marketed as Refacto AF® in Europe). For ReFacto®, the deleted B-domain was replaced by a short 14 amino acid linker. Post-translational modifications and in vitro functional characteristics of the moroctocog alfa preparations are comparable to those of endogenous factor VIII. The full-length recombinant factor VIII preparations Kogenate® FS and Kovaltry® have the same amino acid sequence but glycosylation, in particular branching and sialylation capping of the latter, is said to produce a clinically superior product.

As well as the full-length preparation octocog alfa and the BDD preparation moroctocog alfa, four other recombinant BDD-modified factor VIII preparations are used as antihemophilic factors to treat hemophilia A. These are turoctocog alfa, simoctocog alfa, susoctocog alfa, and factor VIII Fc fusion protein, also known as efmoroctocog alfa.

Turoctocog Alfa

Turoctocog alfa (Table 10.1), produced in CHO cells and licensed by the FDA in 2013, is a recombinant factor VIII preparation with a truncated B-domain of only 21 amino acids linking the A1-A2 domains to the A3-C1-C2 domains. The linking

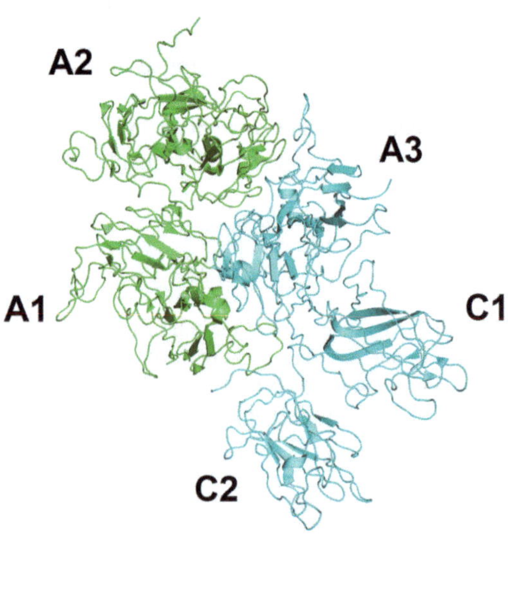

Fig. 10.8 Ribbon diagram of B-domain deleted factor VIII showing the relative locations of the A1, A2, A3, C1, and C2 domains. Own work of, and uploaded by, Mattkosloski (Wikimedia Commons, file Fviii 2R7E.png with Pymol (http://www.pymol.org/)) from PDB 2R7E in the Protein Data Bank (ref. Shen BW, Spiegel PC, Chang CH, et al. Blood 2008;111:1240–7)

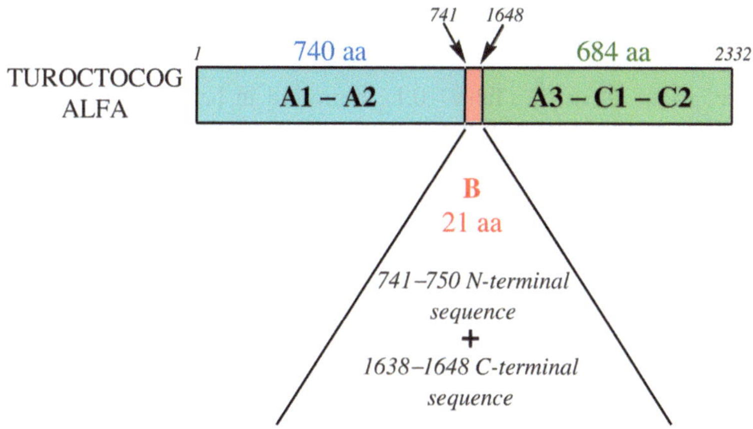

Fig. 10.9 Representation of the structural arrangement of the domains A1, A2, A3, C1, and C2 of the 1445 amino acid polypeptide turoctocog alfa which retains a heavy chain containing a highly truncated B-domain of only 21 amino acids. Compare with Fig. 10.7 which depicts the full-length factor VIII single polypeptide chain of 2332 amino acids showing the arrangement of the domains A1, A2, B, A3, C1, and C2. The thrombin cleavage site upon activation of turoctocog alfa is between residues 740 and 741

sequence is made up of 10 amino acids (residues 741–750) from the *N*-terminus and 11 amino acids (residues 1638–1648) from the C-terminus of the original B-domain (Fig. 10.9). Turoctocog alfa contains four *N*-linked glycosylations and six tyrosine sulfation sites, two of which, Tyr346 and Tyr1664, are necessary for optimal interaction with thrombin. The mean half-life of turoctocog alfa in humans is 10.8 h (c.f., Advate® 11.2 h) and the clearance rate is 4.1 mL/h/kg (Advate® 4.2 mL/h/kg).

Simoctocog Alfa

Simoctocog alfa (Table 10.1), produced in human embryonic kidney cells is the latest BDDrFVIII. It is licensed by the EMA (2014), FDA (2015), and in Australia, Canada, and a number of other countries. Marketed as Nuwiq®, simoctocog alfa comprises the FVIII domains A1-A2 plus A3-C1-C2 (Fig. 10.7) with a 16 amino acid linker between the A2 and A3 domains replacing the deleted B-domain. The first eight amino acids of the linker sequence and the N-terminal B-domain are the same; the remaining octapeptide is arginine rich to ensure similar proteolytic processing as the full length FVIII molecule. Tyrosine 1680 of simoctocog alfa is sulfated (important for binding to von Willebrand factor) as occurs in plasma-derived FVIII.

Susoctocog Alfa

Susoctocog alfa (Obizur®) (Table 10.1), expressed in hamster kidney cells, is a BDDrFVIII analog of porcine factor VIII (pFVIII) in which the B-domain is replaced with a 24 amino acid linker. Like muroctocog alfa, turoctocog alfa, and simoctocog alfa, susoctocog alfa has a MW of ~ 170 kDa with heavy (H) and light (L) chains of MWs 90 kDa and 80 kDa, respectively. When activated, rpFVIII shows comparable activity to endogenous human FVIII.

Clotting Factor Fusion Proteins

For a detailed discussion of approved fusion proteins used for human therapy, see Chap. 6.

Regular infusions of factor VIII or factor IX products have long been the basis of hemophilia care but the inconvenience of repeated intravenous injections, especially in children, has led to the desire for products that reduce the number of infusions and the need for venous access devices. Efforts have consequently been directed at producing preparations of factor VIII and factor IX that have an increased in vivo half-life. In 2014, the FDA approved the first antihemophilic fusion proteins, a recombinant factor VIII Fc fusion protein (Eloctate®), indicated for the treatment of adults and children with hemophilia A or congenital factor VIII deficiency, and a recombinant factor IX Fc fusion protein (Alprolix®), indicated for the treatment of adults and children with hemophilia B or congenital factor IX deficiency (Table 10.1). Attention is drawn to both fusion proteins as examples of the conversion of first-generation therapies with recombinant enzymes, each limited by their short in vivo half-life, into a second-generation therapy of extended half-life and with consequent improved clinical value.

Factor VIII Fc Fusion Protein

The FDA-approved factor VIII Fc fusion protein Eloctate® (also known as efmoroctocog alfa and designated here as rFVIIIFc; see also Chap. 6, section "Factor VIII Fc Fusion Protein") is composed of BDDrFVIII (section "B-Domain Deleted Recombinant Factor VIII") covalently linked to the Fc fragment of human immunoglobulin G1 (IgG1) (Table 10.1, Fig. 10.10), equipping it to bind to the neonatal Fc receptor FcRn. FcRn prolongs immunoglobulin half-life by delaying degradation in lysosomes and recirculating the protein back into the plasma (Chap. 2, section "IgG Antibody Subclasses" and Chap. 6, section "Fc Fusion Proteins"). In several preclinical animal models and in humans, the half-life of rFVIIIFc was found to be increased 1.5–2-fold to ~19.7 h. BDDrFVIII is produced by recombinant DNA technology in human embryonic kidney (HEK) cells. A plasmid for BDDrFVIII (Ser743 to Gln1638 fusion) fused to human IgG1 Fc fragment with no linker sequence was expressed in HEK293 cells. Although an rFVIIIFc dimer was expected to be produced by

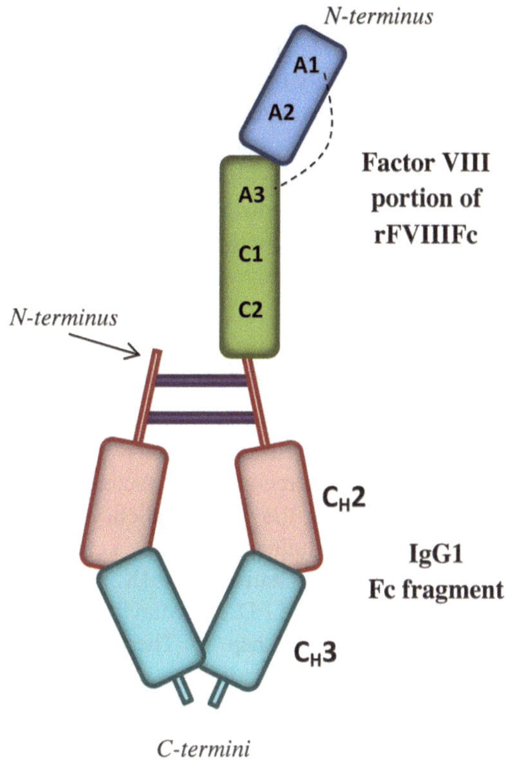

Fig. 10.10 Diagrammatic representation of the structure of the monomeric factor VIII human Fc fusion protein (rFVIIIFc) produced by covalently linking B-domain deleted recombinant factor VIII to the Fc fragment of human IgG1. Effector molecules may be linked to one or both of the Fc chains creating a monomeric or dimeric fusion protein, respectively. rFVIIIFc is expressed as a monomer, a configuration shown to have enhanced pharmacokinetic and pharmacodynamic properties in vivo with some other effector molecules in fusion proteins. The dashed line between A1 and A3 represents a metal ion-mediated noncovalent bond

transfected cells, only an Fc fusion monomer (Fig.10.10) could be extracted and purified. Perhaps surprisingly, some Fc fusion monomers show enhanced pharmacodynamic and pharmacokinetic properties to their dimeric counterparts in, for example, improved transport, longer half-life, and enhanced biologic activity. This has been demonstrated with erythropoietin, factor IX, and interferon beta fusion proteins (see below). Factor VIII Fc fusion protein was found to contain the expected sequence based on the DNA coding sequence as well as the expected *N*-linked glycosylation and sulfation post-translational modifications. Analyses also confirmed that the attached Fc protein did not impair in vitro activity in comparison to BDDrFVIII; specific activity by chromogenic data proved similar on a molar basis, and similar results were obtained in coagulation activity assays, importantly when activated human platelets were used. Both rFVIIIFc and BDDrFVIII bound von Willebrand factor with high affinity, with KD values of 0.34 and 0.26 nM, respectively.

Factor IX Fc Fusion Protein

Factor IX, a ~55 kDa, vitamin K-dependent serine protease has a relatively short half-life of 14–34 h. This prompted the development of the approved monovalent recombinant factor IX Fc fusion protein Alprolix® (rFIXFc; see also Chap. 6, section

"Factor IX Fc Fusion Protein") with single ligand specificity and a half-life extended to ~3.5 days (Table 10.1). rFIXFc is composed of the human coagulation factor IX sequence covalently linked to the Fc fragment of human IgG1. The recombinant factor IX has an amino acid sequence identical to the Thr148 allelic form of factor IX derived from plasma and structural and functional properties similar to natural human factor IX. As with rFVIIIFc, the monomeric factor IX-Fc construct proved pharmacodynamically and pharmacokinetically superior to a dimeric factor IX-Fc fusion construct. Production of the recombinant form in human embryonic kidney cells allowed all the necessary post-translational modifications to be incorporated.

Factor IX Albumin Fusion Protein

Factor IX albumin fusion protein is generated by genetic fusion of recombinant human albumin to recombinant factor IX, producing a single protein designated rIX-FP containing the Thr148 allelic form of human plasma-derived factor IX. Factor IX and albumin are joined by a cleavable linker derived from the endogenous activation peptide in native factor IX. Factor IX albumin fusion protein as Idelvion® was approved by the FDA in March 2016. Its approved indications are set out in Table 10.1 (see also Chap. 6). The fusion protein is claimed to be effective for an extended period with dosing schedules of up to 14 days. Pharmacokinetic parameters revealed a half-life of 87 - 93 hours.

Safety of Approved Blood Coagulation Preparations

Not-for-profit professional bodies such as the Hemostasis and Thrombosis Research Society (HTRS), a North American medical society focused on disorders of hemostasis and thrombosis, the European Thrombosis Research Organization (ETRO), and the International Society on Thrombosis and Haemostasis (ISTH), are organizations that aim to advance the understanding, prevention, diagnosis, and treatment of all aspects of thrombosis and hemostasis. Through, for example, post-market surveillance programs, these societies can play a valuable role in collecting and recording data and expanding the experience base being built up by the use of the various coagulation factors and related agents. For example, the HTRS Registry surveillance program is designed to collect data on all uses of factor VIIa (NovoSeven®) to expand the experience base for this agent.

Inhibitors in Hemophilia and Bypass Therapy

Fifty years ago, life expectancy for hemophiliacs lacking appropriate treatment was about 16 years with hemarthroses commonly leading to the development of chronic arthropathy. Prophylactic administration of factor VIII and factor IX concentrates sometimes reduced the number of spontaneous hemarthroses but the development of inhibitory antibodies often made the treatment less effective, or even ineffective. Patients with antibodies to factor VIII or other coagulation factors may have severe bleeding that is difficult to control and, according to Lollar: "The development of inhibitory antibodies in hemophilia A and B frequently transforms these disorders from treatable to refractory and is now considered to be the most significant complication of treatment." DiMichele has pointed out that inhibitory antibodies complicate treatment, neutralizing even large doses of the factors in ~20–33 % of patients with severe hemophilia A and 1–6 % of patients with severe hemophilia B. Four percent of patients without inhibitors were found to be hospitalized for orthopedic procedures whereas rates of hospitalization for patients with inhibitors were 16 % for 14–35 year olds and 27 % for 36–65 year olds. For factor VIII and IX treatments, the most successful procedure for eradication of the inhibitor is the induction of immune tolerance, otherwise known as desensitization or immunotherapy when used by allergists to dampen or eliminate the allergic response to inhalant, ingested, injected, or contact allergens. The procedure is an attempt to induce tolerance to the agent in question so that treatment can continue at optimal dosage. Increasing amounts of the agent are administered in an incremental and stepwise manner while avoiding or minimizing life threatening, or even lesser adverse symptoms. Induction of immune tolerance to factor VIII and factor IX involves regular (daily or several times weekly) infusions of increasing doses of the agents which are administered for periods of weeks to years. Although the procedure is time-consuming, costly and potentially hazardous, it has, over a 30-year period, proved effective in 70–85 % of patients with factor VIII inhibitors but effective in only about 30 % of patients with antibodies to factor IX.

Inhibitors of factor VIII are the most commonly seen pathogenic antibodies to blood coagulation factors, developing in response to infusions, usually within the first year of treatment. Inhibitors have been found in approximately 20 % of normal healthy donors and anti-factor VIII IgG antibodies appear to occur in all normal sera. Inhibitors of factor VIII almost always consist of polyclonal IgG with the IgG4 subclass being the major component even though IgG4 represents only ~5 % of the total IgG in plasma. IgG1 and IgG2, but not IgG3, are also present. For both hereditary and acquired hemophilia A, antibodies are primarily complementary to the A2 and C2 domains of factor VIII (Figs. 10.7 and 10.8).

Factor IX has a structure made up of five domains, Gla-EGF1-EGF2-AP-SP where Gla is γ-carboxyglutamic acid, EFG refers to epidermal growth factor like, AP stands for activation peptide, and SP is serine protease. Inhibitors of factor IX have been found to be polyclonal IgG antibodies, mainly IgG1 and IgG4. These antibodies mainly recognize the Gla and SP domains with no recognition of the EGF and AP domains.

Nowadays, the choice of treatment for hemophilia A and B is the so-called bypass approach avoiding the specific requirement for factor VIII or IX. For both hemophilia A and B, the preparations of choice for patients with inhibitors are the prothrombin complex concentrates, activated prothrombin concentrate (see sections "Factor Eight Inhibitor Bypassing Activity (FEIBA),

Anti-inhibitor Coagulant Complex" and "Prothrombin Complex Concentrate" below), and recombinant activated factor VII (rFVIIa; eptacog alfa) (Table 10.1). Prothrombin complex concentrates and activated prothrombin concentrates, both prepared from human plasma, contain prothrombin (factor II), factors VII, IX, X, and small quantities of factor VIII. Activated prothrombin concentrates contain activated factor VII and mainly nonactivated factors II, IX, and X. Currently, prothrombin complex concentrates are seldom used to treat inhibitor patients, being less effective than the activated preparations and showing a higher rate of adverse reactions. Two bypassing agents, the activated prothrombin concentrate, anti-inhibitor coagulant complex (FEIBA NF®) and recombinant activated factor VII (NovoSeven®) are currently available for use in patients with inhibitors (Table 10.1). Note that 1–6 units of factor VIII coagulant antigen per mL are included in FEIBA NF®. Efficacy rates for activated prothrombin concentrate and activated factor VII treatments of bleeding episodes in hemophiliacs with inhibitors range from 57–79 % to 87–100 %, respectively.

Despite the success of vapor treatment and nanofiltration in apparently overcoming the possibility of transmission of infectious agents (in particular viruses and prions), these preparations are not without their own drawbacks. The bypass therapy is, by its very nature, short-acting and, if given too often, may itself increase bleeding or lead to excess clotting. Also, these preparations contain small amounts of factor VIII and some factor IX which may provoke antibodies to factor VIII in patients with hemophilia A and antibodies to factor IX in hemophilia B.

Table 10.1 summarizes the adverse effects together with warnings, precautions, and approved indications of the registered blood coagulation preparations in current usage. Across the 22 different preparations, including the recombinant proteins, a number of recurring features are seen, in particular, the possibility of the development of neutralizing antibodies, hypersensitivity reactions including severe allergic manifestations such as anaphylaxis, the risk of thrombotic events, and the ever-present, if small, chance of the transmission of infectious agents. In addition to the information shown in Table 10.1, the following safety-related data relevant to some of the individual preparations are noted.

Factor VIIa

Occasional patients treated with factor VII developed serum antibodies to the protein and, in some cases, the antibodies showed an in vitro inhibitory effect. Results from clinical trials indicate that the use of factor VIIa is associated with a low incidence of thrombotic events (0.2 %). Furthermore, information from the FDA Adverse Event Reporting System Database reveals that most of the serious thromboembolic events reported for factor VIIa occurred in its off-label usage in non-hemophilic patients. Of 185 thromboembolic events, 17 (9.2 %) occurred in hemophiliacs while 151 (81.6 %) occurred in patients with bleeding due to other causes. The thromboembolic events most frequently reported included stroke, acute myocardial infarction, and pulmonary embolism. Note, however, the absence of controls in such a case series does not permit an immediate conclusion of a causal link between factor VIIa treatment and thromboembolic adverse events.

Factor VIII Full Length Preparations

Serious hypersensitivity reactions to unmodified (full length) factor VIII recombinant preparations have been identified during its post-approval use. Reactions reported include facial swelling, a fall in blood pressure, flushing, chest tightness, shortness of breath, tachycardia, vomiting, and urticaria as well as nausea, rash, restlessness, and tingling. Immunogenicity investigations of 73 previously treated patients revealed only one with a preexisting inhibitor; no inhibitors were detected in the other 72 patients over a 4-year period. Inhibitors developed in 9 of 60 previously untreated and minimally treated pediatric patients over a period ranging from 2 to 16 days of exposure. Six of the patients developed high titer antibody inhibitors. In another study, 8 of 64 patients with negative baseline values developed antibodies; two of the patients had to withdraw from the study because of high titer inhibitors. Antibodies were detected between 5 and 151 days after exposure with a median time of 44 days. The 15 most frequently reported serious adverse events associated with the use of Advate® in worldwide reports amongst all ages between December 2011 and June 2013 are, in descending order of frequency: factor VIII inhibition (30 reports); central line infection (six reports); bleeding and condition aggravated (both five reports); joint bleeding and lack of drug effect (both four reports); cyanosis (three reports); and (all two reports), cerebral hemorrhage, chronic hepatitis C, hemoarthrosis, headache, hemophilia A, hemophilic arthropathy, pyrexia, and pain.

Moroctocog Alfa: B-Domain Depleted Factor VIII

The incidence of factor VIII inhibitors (usually low titer), in patients previously treated with B-domain depleted factor VIII (Refacto®AF) was found to be ~2 % with, depending on the study, median exposure days of at least 50 and as high as 169 days. In a clinical trial with previously untreated patients, 32 of 101 patients developed inhibitors and of the 32, half were classified as high titer ($\geq$5 BU) and half as low titer (<5 BU). The median number of exposure days before the appearance of inhibitors in the 32 patients was 12 days. For adverse reactions in general, age is an important consideration. Children aged 7–16 years show a higher frequency of adverse reactions compared to adults. In cases of suspected hypersensitivity, stopping the infusion or slowing the rate may be required. For both inhibitors and hypersensitivity responses, the induction of immune tolerance or desensitization may be required.

Other B-Domain-Depleted Factor VIII Preparations: Turoctocog Alfa, Simoctocog Alfa and Susoctocog Alfa

As reported by the EMA, in clinical trials up to November 2011, 503 adverse events occurred in 154 patients during the prevention and treatment of bleeds with turoctocog alfa, giving an overall rate of 2.45 events per patient year of exposure. Adverse events judged to be possibly or probably related to turoctocog alfa administration were injection site reactions, pyrexia, and increased hepatic enzymes (Table 10.1). No allergic hypersensitivities or thromboembolic events occurred during all of the trials with turoctocog alfa and no differences in the safety profile of the agent were observed between adults and children. Results from a recent large multinational clinical trial using prophylactic treatment with turoctocog in 150 previously treated adult and adolescent patients and 63 previously treated child patients with severe hemophilia A showed a median annualized bleeding rate of 3.7 bleeds per patient per year in adults and adolescents while in children the figure was 3 bleeds per patient per year. The large majority (89–95 %) of bleeding episodes were controlled within 1–2 infusions of turoctocog alfa. None of the patients developed inhibitors to factor VIII and no other safety concerns were identified.

As for turoctocog alfa, important precautions when using simoctocog alfa include the possibility of hypersensitivity reactions and the development of neutralizing anti-FVIII antibodies. Consequently, patients should be monitored for FVIII activity levels and the development of inhibitory antibodies. Adverse events revealed so far in clinical studies include anti-FVIII antibody formation, paresthesia, headache, injection site inflammation and pain, vertigo, back pain, and dry mouth (Table 10.1).

Hypersensitivity reactions, inhibitory antibodies, and the need to monitor the development of antibodies, constitute the issued warnings and precautions for susoctocog alfa. The most common adverse reaction to the glycoprotein seen so far is the development of antibody inhibitors to pFVIII (Table 10.1).

Factor VIII Fc Fusion Protein

A recent phase 3 study evaluating the safety, efficacy, and pharmacokinetics of rFVIIIFc for prophylaxis, treatment of acute bleeding, and perioperative hemostatic control in 165 previously treated males aged ≥12 years with severe hemophilia, found that five subjects tested positive for non-neutralizing antibodies at baseline. In each case, antibodies were low titer, directed against factor VIII not Fc, and declined over the course of the study. Six patients, negative at baseline, developed non-neutralizing antibodies during the study but in four of the subjects antibodies were no longer present at the final testing. Overall, the inhibitor incidence was 0 %. Of 164 patients exposed to the fusion protein, 108 reported ≥1 adverse event, most commonly, nasopharyngitis, arthralgia, headache, and upper respiratory infections. Adverse events related to rFVIIIFc treatment occurred in 10 (6.1 %) patients while malaise and arthralgia occurred in more than one patient. No serious adverse events, including hypersensitivity or vascular thrombotic events, were judged to be related to rFVIIIFc treatment. FDA warnings for the approved factor VIII fusion protein cover the possibilities of hypersensitivity reactions and the formation of neutralizing antibodies plus the need to monitor inhibitors and plasma factor VIII activity to confirm that adequate levels of the factor have been achieved and maintained. The relatively small list of adverse effects indicates that more experience with more patients is needed with the fusion protein (Table 10.1). Refer also to Chap. 6, section "Factor VIII Fc Fusion Protein".

Factor IX Fusion Proteins

For the two factor IX fusion proteins, refer also to Chap. 6, section "Factor IX Fc Fusion Protein" and "Factor IX Albumin Fusion Protein". Formation of neutralizing antibody inhibitors to factor IX and an association between the occurrence of inhibitors and allergic reactions have been reported. In a multicenter, prospective, open-label clinical trial with the Fc fusion protein, 115 patients previously exposed to a factor IX-containing product were treated with rFIXFc for at least 26 weeks and 56 patients were treated for at least 52 weeks. Adverse reactions were seen in 8.4 % of patients given rFIXFc for routine prophylaxis or on-demand therapy, inhibitors were not detected, and no anaphylactic events occurred. The adverse events seen were headache, dizziness, dysgeusia, paresthesia oral, breath odor, fatigue, infusion site pain, palpitations, obstructive uropathy, and hypotension. Headaches occurred in two patients; all the other reactions occurred in only one patient. As with the approved factor VIII fusion protein, the FDA has warnings for hypersensitivity, the development of neutralizing antibodies, and the need to check factor IX concentrations. A further warning draws attention to a known association of factor IX products with the development of thromboembolic complications (Table 10.1). Also in common with rFVIIIFc, the list of adverse effects so far compiled for rFIXFc is relatively few in number, probably due to its limited usage so far.

Table 10.1 sets out the warnings, precautions, and adverse events presented in the prescribing information for the factor IX albumin fusion protein Idelvion® (rIX-FP). In immunogenicity investigations undertaken so far, no subjects developed antibodies to factor IX or albumin after sample periods of two to four weeks, 12 weeks, and three or six months thereafter.

Factor XIII A-Subunit

In a study involving 50 healthy subjects given two doses of the recombinant factor XIII A-subunit preparation Tretten®, one subject developed low titer, non-neutralizing, non-inhibitory antibodies that were not detected 6 months later. In

77 patients all under 18 years and with congenital factor XIII A-subunit deficiency, low titer, non-neutralizing antibodies were found in four patients after receiving at least two doses The antibodies proved to be not clinically significant. The adverse reactions to factor XIII A-subunit shown in Table 10.1 were seen in a study involving 68 patients with congenital factor XIII A-subunit deficiency given a total of 1979 doses of the coagulation factor preparation for routine prophylaxis of bleeding.

Von Willebrand Factor/Coagulation Factor VIII Complex

Patients with von Willebrand disease may potentially develop inhibitory antibodies to von Willebrand factor or factor VIII. This may result in failure to obtain the desired plasma levels of von Willebrand factor and/or the inability to control bleeding with normally adequate doses of the complex. Monitoring for the development of antibody inhibitors is therefore recommended as is the monitoring of plasma levels for von Willebrand factor ristocetin cofactor (VWF:RCo) and factor VIII activities to avoid excessive levels of these agents with possible consequent thrombotic events. Being prepared from human plasma, von Willebrand Factor/Coagulation Factor VIII complex must always be considered as a possible carrier of infectious agents such as viruses, variant Creutzfeldt–Jakob disease, and as yet unknown agents. Although the risk of transmission of viruses has been reduced by the screening of plasma donors, vaccination against hepatitis A and B virus should be considered and the batch number of the product should be recorded for every administration. Relevant to this was the finding of four von Willebrand disease patients given the complex who showed seroconversion for antibodies to parvovirus B19, although clinical signs of the disease were not seen.

Antihemophilic Factor/Von Willebrand Factor Complex

As with other human plasma-derived preparations, the possibility of viral transmission such as parvovirus B19 and hepatitis A infections needs to be kept in mind when using antihemophilic factor/von Willebrand factor complex (Humate-P®). The presence of isoagglutinins anti-A and anti-B in this preparation means that blood group A, B, and AB patients need to be monitored for signs of intravascular hemolysis and decreasing hematocrit values when large or frequently repeated doses are administered. Monitoring VWF:RCo in von Willebrand disease patients undergoing surgery should be considered to avoid excessive bleeding. There are reports of thromboembolic events following coagulation factor replacement therapy in von Willebrand disease patients and, for the moment at least, there appears to be a higher incidence in females. Warnings, precautions, and adverse effects associated with the preparation are summarized in Table 10.1. Formation of inhibitors has not been observed in any of the clinical trials.

Recombinant Von Willebrand Factor

Approved as Vovendi® by the FDA in December 2015, recombinant von Willebrand factor (rVWF; vonicog alfa; BAX 111) is subject to a warning for thromboembolic reactions including disseminated intravascular coagulation, venous thrombosis, pulmonary embolism, myocardial infarction, and stroke. Patients should be monitored for VWF:RCo and factor VIII to avoid the risk of thrombotic events. Other issued warnings and precautions concern hypersensitivity reactions including anaphylaxis, and the development of neutralizing antibody inhibitors to von Willebrand factor and factor VIII. The most common adverse reaction to Vonvendi® seen in clinical trials was generalized pruritus (Table 10.1).

Factor Eight Inhibitor Bypassing Activity (FEIBA), Anti-inhibitor Coagulant Complex

Thromboembolic events, for which there is an FDA boxed warning, may occur particularly following the administration of high doses of anti-inhibitor coagulant complex, that is >200 units per kg per day. Other associated adverse events are shown in Table 10.1. Patients with disseminated intravascular coagulation, advanced atherosclerotic disease, crush injury, septicemia, or concomitant treatment with factor VIIa have an increased risk of developing thrombotic events.

Prothrombin Complex Concentrate

The approved preparation of this concentrate (Table 10.1) is subject to an FDA black box warning for arterial and venous thromboembolic complications in relation to patients being treated with vitamin K antagonists who have underlying diseases that predispose them to thromboembolic events. Clinical trials and post-marketing surveillance have revealed fatal and nonfatal arterial and venous thromboembolic complications following prothrombin complex concentrate. The FDA further warns that the concentrate may not be suitable for patients who experienced a thromboembolic event in the prior 3 months.

Fibrinogen Preparations to Control Bleeding

Table 10.1 provides brief descriptions and summarizes the approved indications, usages, warnings, precautions, and adverse effects of three preparations containing fibrinogen approved for the control of bleeding: human fibrinogen concentrate (RiaSTAP®) for intravenous use, fibrin sealant (Tisseel®), and a fibrin sealant patch (Evarrest®).

Summary

- Essential for survival, the coagulation or clotting system results in homeostasis and ultimately repair of the damaged vessels and other tissues. Cells, the kallikrein-kinin system, a cascade of primary and activated protein coagulation or clotting factors, and a number of cofactors and regulatory proteins, participate in a series of predominately enzyme activations and catalyzed reactions that lead to the formation of insoluble cross-linked fibrin and cessation of blood loss.
- Adhesion of platelets to wound sites is aided by von Willebrand factor, a large multimeric glycoprotein produced by, and stored in, platelet alfa granules. Von Willebrand factor interacts with collagen type I alfa 1 and acts as a bridge connecting collagen fibrils to the glycoprotein complex GPIb alfa-V–IX (CD42a-d complex). Interaction between GPIb alfa and thrombin is required for platelet aggregation at sites of vascular injury.
- Defects in proteins of the GPIb alfa-V–IX complex leads to a condition with a range of molecular defects, the autosomal recessive disorder known as Bernard–Soulier syndrome, also called hemorrhagiparous thrombocytic dystrophy or giant platelet syndrome. Symptoms include mucocutaneous bleeding, low platelet count, purpura, gingival bleeding, menorrhagia in women, and gastrointestinal hemorrhage.
- A deficiency of von Willebrand factor is responsible for an abnormality known as von Willebrand disease, often with no, or mild, symptoms such as epistaxis.
- A recombinant preparation of von Willebrand factor (vonicog alfa; Vonvendi®) indicated for on-demand treatment and control of bleeding in adults with von Willebrand disease, was recently approved by the FDA.
- In the absence of platelet aggregation and without the involvement of platelet surface phospholipids, the blood coagulation cascade does not proceed. Activation of platelets leading to a signal transduction cascade is initiated by thrombin interacting with surface G protein-coupled receptors.
- The glycoprotein receptor complex GPIIb-GPIIIa (also called integrin αIIb-β3) is a receptor for von Willebrand factor and fibrinogen. Interaction between this receptor and fibrinogen also induces platelet aggregation. An inherited bleeding disorder, Glanzmann thrombasthenia, results from a defect in the GPIIb protein of the GPIIb-GPIIIa receptor complex.
- The monoclonal antibody, abciximab (ReoPro®), targeted to GPIIb-GPIIIa, the tyrosine derivative tirofiban (Aggrastat®), and the cyclic heptapeptide eptifibatide (Integrelin®) create a thrombasthenia-like state blocking the interaction between fibrinogen and the receptor thus inhibiting platelet aggregation. This property makes the drugs useful for the treatment of patients undergoing angioplasty and unstable angina.
- The plasma kallikrein-kinin system leads to the release of vasoactive kinins from inactive precursor kininogen macromolecules as a result of the activation of kallikrein from its prekallikrein inactive form. The released kinins produce a wide range of actions such as vasodilation, increased vascular permeability, nitric

oxide production, and release of tissue plasminogen activator. As a consequence of these various activities, kinins participate in numerous physiological and pathological processes including blood coagulation.

- The plasma kallikrein-kinin system consists of factor XII, factor XI, high molecular weight kininogen (HMWK), and prekallikrein. Both factor XI and factor XII are zymogen (inactive) forms of the corresponding factor XIa and factor XIIa active serine proteases.
- The formation of fibrin, necessary for clot formation, can be initiated by two distinct pathways, the extrinsic, or tissue factor pathway triggered by damage to blood vessel walls or the so-called intrinsic pathway, more accurately termed the contact activation pathway.
- Formation of factor XIIa via activation of factor XII by negatively charged surfaces initiates the contact activation pathway. Factor XIIa activates prekallikrein to kallikrein which cleaves HMWK releasing bradykinin. Released bradykinin, via interaction with the kinin B2 receptor (B2R), activates proinflammatory signaling pathways that result in dilatation of blood vessels, increased vascular permeability and chemotaxis of neutrophils.
- Subsequent events in the pathway proceed by the activation of factor IX to factor IXa by factor XIa in the presence of Ca^{2+}. Factor X is activated to factor Xa by factor IXa-induced cleavage in a reaction that requires Ca^{2+}, factors VIIIA, IXa, and X on the surface of activated platelets. This complex of molecules, the tenase complex, forms with the involvement of the phospholipids phosphatidylserine and phosphatidylinositol on the platelet surface. Activation of factor VIII to factor VIIIa is induced by a small concentration of thrombin. In its activated form, factor VIII acts as a receptor for factors IXa and X.
- Following vascular injury, released tissue factor (factor III) initiates the extrinsic clotting cascade by functioning as a cofactor in the Ca^{2+}-dependent activation of factor X by factor VIIa. Activation of factor VII requires the involvement of thrombin or factor Xa, the latter creating the link between the intrinsic and extrinsic pathways.
- The shared sections of both pathways are the activation of factor X to factor Xa and the events subsequent to that leading to the formation of the fibrin clot. Upon the production of factor Xa, prothrombin (factor II) is activated to thrombin (factor IIa) on the surface of activated platelets in interactions that require the formation of the prothrombinase complex—platelet phospholipids, Ca^{2+}, factor Xa, factor Va, and prothrombin. Thrombin cleaves fibrinogen (factor I) generating soluble fibrin monomers that spontaneously aggregate to form a weak clot but it also activates factor XIII which results in an insoluble clot by cross-linking fibrin monomers.
- The low molecular weight heparin anticoagulant enoxaparin, given for prophylaxis and treatment of deep vein thrombosis after surgery, unstable angina, and myocardial infarction, acts by inhibiting factor Xa. Enoxaparin carries an FDA boxed warning for the risk of epidural or spinal hematomas and warnings for an increased risk of hemorrhage, percutaneous coronary revascularization, and induction of thrombocytopenia. The most common adverse reactions to enoxaparin include bleeding and bruising, anemia, and thrombocytopenia.

- The vitamin K-dependent glycoprotein protein C is a zymogen for its activated form, the serine protease activated protein C (APC). Activation of protein C is promoted by thrombomodulin and endothelial protein C receptor found on cells of blood vessels. APC inactivates the highly procoagulant factor Va and factor VIIIa leading to a decrease in thrombin formation. The protein C pathway is an important control mechanism preventing excessive procoagulant responses.
- A deficiency of protein C may manifest as a significantly increased risk of venous thrombosis, purpura fulminans, severe disseminated intravascular coagulation, and simultaneous venous thromboembolism in the womb. Resistance to APC also occurs producing factor V Leiden which prevents inactivation of both factor Va and factor VIIIa. Thrombin generation and a marked increased risk of thrombosis results.
- Antithrombin has an important role in the regulation of homeostasis being the main inhibitor of the serine proteases thrombin and factor Xa.
- Attenuation of arterial thrombus formation appears to be linked to protection from experimental cerebral ischemia and pulmonary embolism. These findings raise the possibility of anticoagulation therapy that targets thrombosis without influencing hemostasis. Factor XII may influence pathologic clotting via the intrinsic pathway.
- Platelets have been linked to the intrinsic pathway suggesting that factor XII is activated on procoagulant platelets. Recently, platelet polyphosphate, an inorganic, linear polymer of orthophosphate units linked by phosphoanhydride bonds, has been claimed to be the long sought-after surface that triggers fibrin formation by activated platelets.
- Polyphosphate as a platelet-derived procoagulant may be the previously elusive link between primary and secondary hemostasis.
- Human antihemophilic factor VIII is made up of three different types of domains, A, B, and C arranged as A1-A2-B-A3-C1-C2. Thrombin-induced activation of factor VIII leads to the production of the A1, A2, and A3-C1-C2 fragments of factor VIIIa, generation of the tenase complex and factor Xa, the production of thrombin and ultimately a stable fibrin clot.
- The large B domain of factor VIII which has no known function or homology to any other protein is processed to generate a heavy chain (A1-A2-B) and a light chain (A3-C1-C2). Deletion of a large section of the B domain causing a reduction in protein size of the heavy chain has no affect on the activity of factor VIII but its expression is enhanced. This enables the construction of a functional truncated enzyme in the form of the B-domain deleted recombinant factor VIII protein (BDDrFVIII) also called moroctocog alfa.
- Turoctocog alfa, used to treat hemophilia A, is a recombinant factor VIII preparation with a truncated B-domain of only 21 amino acids linking the A1-A2 domains to the A3-C1-C2 domains. The linking sequence is made up of 10 amino acids from the *N*-terminus and 11 amino acids from the C-terminus of the original B-domain.
- Simoctocog alfa, produced in human embryonic kidney cells is the latest BDDrFVIII. Marketed as Nuwiq®, simoctocog alfa comprises the FVIII domains A1-A2 plus A3-C1-C2 with a 16 amino acid linker between the A2 and A3 domains replacing the deleted B-domain. The first eight amino acids of the linker sequence and the N-terminal B-domain are the same.

- Susoctocog alfa (Obizur®), is a recombinant BDDrFVIII analog of porcine factor VIII (pFVIII) in which the domain is replaced with a 24 amino acid linker. Like muroctocog alfa, turoctocog alfa, and simoctocog alfa, susoctocog alfa shows camparable activity to endogenous human factor VIII.
- The FDA-approved factor VIII Fc fusion protein Eloctate® (rFVIIIFc), extracted as an Fc fusion monomer and composed of BDDrFVIII covalently linked to the Fc fragment of human IgG1, increases the half-life of factor VIII by 1.5–2-fold.
- Factor IX, a ~55 kDa, vitamin K-dependent serine protease has a relatively short half-life of 14–34 h. This prompted the development of the monovalent recombinant factor IX Fc fusion protein (rFIXFc) with single ligand specificity. rFIXFc is composed of the human coagulation factor IX sequence covalently linked to the Fc fragment of human IgG1.
- Factor IX albumin fusion protein is generated by genetic fusion of recombinant human albumin to recombinant factor IX, producing a single protein designated rIX-FP approved by the FDA in March 2016 as Idelvion®. rIX-FP's half-life of 87–93 hours enables dosing schedules of up to 14 days.
- Across the 22 different approved blood coagulation preparations, including the recombinant proteins, a number of recurring adverse events are seen, in particular, the possibility of the development of neutralizing antibodies, hypersensitivity reactions including severe allergic manifestations such as anaphylaxis, the risk of thrombotic events and the small possibility of transmission of infectious agents.
- Inhibitory antibodies complicate treatment, neutralizing even large doses of the clotting factors in patients with severe hemophilias A and B.
- The current choice of treatment for hemophilia A and B is the so-called bypass approach avoiding the specific requirement for factor VIII or IX. For both hemophilia A and B, the favored preparations for patients with inhibitors are the prothrombin complex concentrates, activated prothrombin concentrate and recombinant activated factor VII.
- Prothrombin complex concentrates and activated prothrombin concentrates, both prepared from human plasma, contain prothrombin (factor II), factors VII, IX, X, and small quantities of factor VIII. Activated prothrombin concentrates contain activated factor VII and mainly nonactivated factors II, IX, and X. Currently, prothrombin complex concentrates are seldom used to treat inhibitor patients, being less effective than the activated preparations and showing a higher rate of adverse reactions. Two bypassing agents, the activated prothrombin concentrate, anti-inhibitor coagulant complex (FEIBA NF®) and recombinant activated factor VII (NovoSeven®), are currently available for use in patients with inhibitors.
- Occasional patients treated with factor VII develop serum antibodies to the protein and, in some cases, the antibodies show an in vitro inhibitory effect.
- Serious hypersensitivity reactions to unmodified (full length) factor VIII recombinant preparations have been identified during its post-approval use. Reactions reported include facial swelling, a fall in blood pressure, flushing, chest tightness, shortness of breath, tachycardia, vomiting, and urticaria.
- Immunogenicity investigations of recombinant factor VIII have revealed antibody inhibitors in a proportion of previously untreated and minimally treated pediatric patients after only 2–16 days of exposure.

- As for turoctocog alfa, important precautions when using simoctocog alfa include the possibility of hypersensitivity reactions and the development of neutralizing anti-FVIII antibodies. Consequently, patients should be monitored for FVIII activity levels and the development of inhibitory antibodies.
- Hypersensitivity reactions, inhibitory antibodies, and the need to monitor the development of the antibodies, make up the issued warnings and precautions for susoctocog alfa.
- Non-neutralizing low titer antibodies to factor VIII Fc fusion protein were demonstrated at baseline in a small number of males aged ≥12 years with severe hemophilia. Antibodies declined over the course of the study. Overall, the inhibitor incidence of antibodies to the fusion protein appears to be 0 %.
- Formation of neutralizing antibody inhibitors to factor IX and an association between the occurrence of inhibitors and allergic reactions have been reported.
- In patients given von Willebrand factor/coagulation factor VIII complex, monitoring for the development of antibody inhibitors is recommended as is the monitoring of plasma levels for von Willebrand factor ristocetin cofactor (VWF:RCo) and factor VIII activities to avoid excessive levels of these agents and possible consequent thrombotic events.
- The presence of isoagglutinins anti-A and anti-B in antihemophilic factor/von Willebrand factor complex means that blood group A, B, and AB patients need to be monitored for signs of intravascular hemolysis and decreasing hematocrit values when large or frequently repeated doses are administered.
- Thromboembolic events, the subject of an FDA boxed warning, may occur particularly following the administration of high doses of anti-inhibitor coagulant complex.
- Prothrombin complex concentrate is subject to a boxed warning for arterial and venous thromboembolic complications in relation to patients being treated with vitamin K antagonists who have underlying diseases that predispose them to thromboembolic events. Fatal and nonfatal arterial and venous thromboembolic complications have occurred following prothrombin complex concentrate.
- Three preparations containing fibrinogen are approved for the control of bleeding: human fibrinogen concentrate (RiaSTAP®) for intravenous use, fibrin sealant (Tisseel®), and a fibrin sealant patch (Evarrest®). Thrombosis, hypersensitivity, and transmission of infectious agents are three possible adverse effects.

Further Reading

Alving B. Transfusion of the patient with acquired coagulation defects. In Hillyer CD, Silberstein LE, Ness PM, et al, editors. Blood banking and transfusion medicine; basic principles and practice. 2nd ed. Philadelphia: Churchill Livingstone Elsevier; 2007. p. 482–95.

Ezban M, Vad K, Kjalke M. Turoctocog alfa (NovoEight®)—from design to clinical proof of concept. Eur J Haematol. 2014;93:369–76.

Gill JC, Castaman G, Windyga J, et al. Hemostatic efficacy, safety and pharmacokinetics of a recombinant von Willebrand factor in severe von Willebrand disease. Blood. 2015;126: 2038–46.

Hilgartner M, Aledort L, Andes A, et al. Efficacy and safety of vapor-heated anti-inhibitor coagulant complex in hemophilia patients. FEIBA Study Group. Transfusion. 1990;30:626–30.
Lanza F. Bernard-Soulier syndrome (Hemorrhagiparous thrombocytic dystrophy). Orphinet J Rare Dis. 2006;1:46. doi:10.1186/1750-1172-1-46.
Lee C, Berntorp E, Hoots K, editors. Textbook of hemophilia. 2nd ed. Wiley, Chichester; 2011.
Lin PC, Liao YM, Tsai SP, et al. Immune tolerance induction therapy for patients with hemophilia A and FVIII inhibitors particularly using low-dose regimens. Pediatr Blood Cancer. 2011;57:1029–33.
Lind P, Larsson K, Spira J, et al. Novel forms of B-domain-deleted recombinant factor VIII molecules. Construction and biochemical characterization. Eur J Biochem. 1995;232:19–27.
Lollar P. Pathogenic antibodies to coagulation factors. Part one: Factor VIII and factor IX. J Thromb Haemost. 2004;2:1082–95.
Lollar P. Pathogenic antibodies to coagulation factors. Part II. Fibrinogen, prothrombin, thrombin, factor V, factor XI, factor XII, factor XIII, the protein C system and von Willebrand factor. J Thromb Haemost. 2005;3:1385–91.
Luu H, Ewenstein B. FEIBA safety profile in multiple modes of clinical and home-therapy application. Haemophilia. 2004;10(Suppl 2):10–6.
Mahlangu J, Powell JS, Ragni MV, et al. Phase 3 study of recombinant factor VIII Fc fusion protein in severe hemophilia A. Blood. 2014;123:317–25.
Müller F, Mutch NJ, Schenk WA, et al. Platelet polyphosphates are proinflammatory and procoagulant mediators in vivo. Cell. 2009;139:1143–56.
Peters RT, Toby G, Lu Q, et al. Biochemical and functional characterization of a recombinant monomeric factor VIII-Fc fusion protein. J Thromb Haemost. 2013;11:132–41.
Powell JS, Pasi KJ, Ragni MV, et al. Phase 3 study of recombinant factor IX Fc fusion protein in hemophilia B. N Engl J Med. 2013;369:2313–23.
Renné T, Schmaier AH, Nickel KF, et al. In Vivo roles of factor XII. Blood 2012;120:4296–303.
Sandberg H, Almstedt A, Brandt J, et al. Structural and functional characteristics of the B-domain-deleted recombinant factor VIII protein, r-VIII SQ. Thromb Haemost. 2001;85:93–100.
Santagostino E, Martinowitz U, Lissitchkov T, et al. Long-acting recombinant coagulation factor IX albumin fusion protein (rIX-FP) in hemophilia B: results of a phase 3 trial. Blood. 2016;127:1761–9.
Shapiro AD. Long-lasting recombinant factor VIII proteins for hemophilia A. Hematology Am Soc Hematol Educ Program. 2013;2013:37–43.
Shapiro AD, Hedner U. Advances in bypassing agent therapy for hemophilia patients with inhibitors to close care gaps and improve outcomes. Ther Adv Drug Saf. 2011;2:213–25.
Shusterman S, Manno CS. Transfusion of the patient with congenital coagulation defects. In Hillyer CD, Silberstein LE, Ness PM, et al, editors. Blood banking and transfusion medicine; basic principles and practice. 2nd ed. Philadelphia: Churchill Livingstone Elsevier; 2007. p. 467–81.

Chapter 11
Vaccines

Growth of the science of microbiology; the vital contribution of sanitation in terms of public health measures such as a clean water supply and sewage disposal; widening awareness through education of the need for personal hygiene; and greater insights into the pathogenic mechanisms involved in host–parasite relationships, infection, and the acquired immune response have each contributed to the central role vaccines now have in humans' continuing campaign to better understand and eradicate infectious diseases. Vaccination has been described as the greatest triumph of modern immunology and the most successful exploitation of our knowledge of the workings of the immune response. Despite the impressive, if not undreamt of, progress in our knowledge and many outstanding clinical applications of immunobiological insights over the last two decades, few might argue against this opinion.

Vaccines: Definition, Attenuation, and Subunit, Acellular, Carbohydrate, Conjugate and DNA Vaccines

A vaccine may be defined as an antigenic biological preparation of a disease causative agent(s) obtained by extraction, modification, synthesis, or by being left unchanged, and used to induce protection by challenging the immune system to stimulate an acquired memory response in the form of long-lasting protective immunity. The protection provided by most vaccines is in the form of neutralizing antibodies but for some organisms, in particular, malaria, tuberculosis, and HIV, antibodies alone are not sufficient for protection and ongoing cell-mediated immunity is required for long-term protection and pathogen elimination. So far, achieving the insights and technology necessary to produce effective vaccines for these three high-profile diseases has remained insurmountable. Vaccines are generally given prophylactically although what are often described as cancer vaccines are being developed and tested as therapeutic agents to treat various cancers and antitoxins such as those prepared from bacteria causing tetanus, diphtheria, and botulism are likewise used as passively

B.A. Baldo, *Safety of Biologics Therapy*, DOI 10.1007/978-3-319-30472-4_11

administered antibodies to directly treat existing conditions. Largely because of the risk of types I and III hypersensitivity reactions, and loss of protection due to antigen binding and protein catabolism, passive immunization is now largely restricted to the administration of antivenoms and some viral infections such as rabies and cytomegalovirus. In some circumstances, however, passive immunization is the best available and/or preferred treatment method, for example, the humanized monoclonal antibody (mAb) palivizumab (Synagis®) given for the prevention of lower respiratory tract disease caused by respiratory syncytial virus (RSV) in infants (Chap. 4, section "Palivisumab") and the fully human raxibacumab (Abthrax®) and chimeric obiltoxaximab (Anthim®) mAbs developed to treat inhalational anthrax in response to bioterrorism threats (Chap. 4, sections "Raxibacumab" and "Recent Approvals: Obiltoxaximab, Ikexizumab, Reslizumab"). For circumstances such as these, plus the mounting problems with antibiotic-resistant bacteria, and the effectiveness and utility of carefully targeted mAbs, it seems likely that these highly specific agents will be increasingly developed and used for passive immunization therapies.

From the early days of vaccine development, efforts to overcome microbial virulency and disease induction while achieving protective immunity have been based on two empirical approaches, *attenuation* of organisms to reduce pathogenicity and production of vaccines using *killed organisms*. Attenuation has been achieved in a number of different ways including altering in vitro growth conditions (e.g., temperature and anaerobic conditions) of bacteria and viruses, particularly polio, measles, mumps, rubella, and varicella, and passaging the pathogen through a normally foreign host such as embryonated eggs, tissue culture, and live animals. Utilizing a foreign host in this way may lead to mutations (generally several point mutations) allowing growth of the pathogen in the new host and the ultimate formation of a population of organisms that is significantly different to the originally introduced population, especially with respect to no longer being able to grow, or to only grow poorly, in the original host. Thus, the overall desired effect is to produce immunity but not disease. *Live attenuated vaccines* are generally more potent then *killed vaccines* ("killed" for viruses meaning loss of the capacity to replicate) since live organisms usually elicit more effector mechanisms including the activation of CD4 and CD8 cytotoxic T cells. A potential advantage of killed vaccines lies in the possibility of a live vaccine causing a lethal systemic infection in immunosuppressed or immunoglobulin-deficient individuals. If the pathogen cannot be cleared in such individuals, an increased chance of mutation remains as the organism continues to reside in the host and one or more mutations may lead on to potentially fatal disease. On the other hand, killed vaccines cannot generate killer T-cell responses and are not a realistic option for some diseases. To overcome some of the problems and risks of live vaccines where there are many other antigens in addition to the desired protective antigen(s), vaccination is often carried out with isolated protective antigens, a procedure called *subunit vaccination*. Subunit vaccines generally consist of well-defined proteins such as tetanus and diphtheria exotoxins that are chemically inactivated to produce toxoids. Recombinant DNA technology may also be utilized to produce a purified protein component from a pathogen to be used as a subunit vaccine, for example, the hepatitis B surface antigen of hepatitis B virus which is associated with hepatic cancer. Subunit vaccines are able to generate helper T cell but not killer T-cell responses.

Findings over the years, especially with *Bordetella pertussis* (the bacterium that causes whooping cough), showed that *acellular vaccines* are generally safer than vaccines formulated from whole organisms. Side effects of erythema, pain, and swelling at the injection site after inoculation with *B. pertussis* vaccines were relatively common and this was sometimes accompanied by an unacceptably high incidence of persistent crying in infants and elevated temperature leading to fits and more severe symptoms. Research showed that the antibody response to *B. pertussis* was largely directed to four components of the bacterium—the pertussis toxin, the membrane protein virulence factor pertactin, fimbrial antigens, and filamentous hemagglutinin (see Table 11.1). These insights led to the development of acellular pertussis vaccines which contain the pertussis toxoid prepared from the toxin by treatment with hydrogen peroxide or formaldehyde, sometimes with one or more of the other three antigenic components. Acellular pertussis vaccines are proving as effective as the whole-cell vaccines but without the side effects of the latter preparations.

Many bacteria, including *Neisseria meningitidis*, *Streptococcus pneumoniae*, and *Haemophilus influenzae*, each capable of causing serious and life-threatening infections, have a dense surface distribution of glycans often in the form of an outer capsular polysaccharide that is species- and type-specific. These complex polysaccharide structures have a number of properties that present special problems when attempting to develop *carbohydrate vaccines* that induce good protective antibody responses. In the first place, carbohydrates often tend to be poorly immunogenic and need to be coupled to a carrier protein to participate in the immune events associated with CD4+ T helper cells. Anticarbohydrate antibodies when they do form, typically show low affinity and when this is linked to the extensive heterogeneity shown by many complex glycan structures, the effective antibody response often tends to be less avid and less concentrated target wise than desired. Since opsonization of the polysaccharide coating of these bacteria is usually an effective immunologic defense and carbohydrates tend to be T-cell independent antigens, capsular polysaccharides are sometimes isolated and used alone to prepare vaccines. However, this strategy is not effective for infants and young children under 2 years old who cannot mount a good T-cell independent antibody response and hence cannot be effectively protected with a carbohydrate vaccine. This problem can be overcome by forming so-called *conjugate vaccines* whereby the bacterial glycan structures are chemically coupled to suitable protein carriers that can act as a source of peptides recognizable by T cells. Such conjugated vaccines have been developed for a number of important vaccines including *H. influenzae* type b (Hib) and some *N. meningitidis* serotypes (Table 11.1) by coupling their carbohydrate antigens to tetanus toxoid. The nontoxic variant of diphtheria toxin, CRM19, has also been coupled in this way to provide helper T-cell epitopes in the preparation of meningococcal and pneumococcal vaccines (Table 11.1).

DNA vaccines, sometimes referred to as third generation vaccines, contain DNA encoding a pathogen antigen that, after injection, results in production of the antigen by the transfected cells of the host before transfer to dendritic cells for presentation to T cells. The process of DNA vaccination is usually initiated by intramuscular injection of the selected DNA sequence with a polyA terminator in a plasmid with a

Table 11.1 Licensed vaccines[a] for immunization: indications, warnings, and adverse events

Vaccine	Description	Indications and usage	Warnings and precautions	Adverse events
Adenovirus types 4 and 7	Live, nonattenuated, oral vaccine given as two enteric-coated tablets Virus replicates in intestine	Prevention of febrile acute respiratory disease caused by adenovirus types 4 and 7[b]	Live virus shed in stools—may infect children, pregnant women[c] and immunocompromised individuals	*Serious*: Hematuria; gastroenteritis; febrile gastroenteritis; gastritis; hematochezia; pneumonia. *Common*: URTI; headache; cough; nasal congestion; abdominal pain; diarrhea
Anthrax vaccine adsorbed (Biothrax®)	Cell-free cultural filtrate of avirulent nonencapsulated strain of *Bacillus anthracis*; contains PA protein[d]	Prevention of disease caused by *B. anthracis*	Hypersensitivity; fetal harm; may not protect all individuals; history of disease may lead to severe local reactions	isr; muscle aches; fatigue; headaches; allergic reactions; malaise; cutaneous reactions[e]
BCG Live	Attenuated, live culture preparation of BCG (strain of *Mycobacterium bovis*), TICE[f] strain	Prevention of tuberculosis in persons not previously infected with *M. tuberculosis*	Administer by percutaneous route only; disseminated BCG infection	Local reactions[g]; flu-like syndrome; disseminated BCG disease
BCG Live (Tice® BCG)	Attenuated, live culture preparation of BCG strain of *M. bovis* for intravesical use[h]	Treatment and prophylaxis of carcinoma in situ of bladder and recurrence of papillary tumors[h]	*Boxed warning*: Risk of transmission of live, infectious bacteria. *Other*: May cause tuberculin sensitivity; handle aseptically; not an anticancer vaccine	Flu-like syndrome; systemic BCG infection; bladder irritability; dysuria; urinary frequency; hematuria; fever; avoid use in immunosuppressed and tuberculosis patients

Vaccine	Description	Indications and usage	Warnings and precautions	Adverse events
Diphtheria and tetanus[i] toxoids adsorbed	USP preparation for pediatric use Alum-precipitated toxoid for IM use	Active immunization of children up to 7 years old. Used when pertussis is contraindicated	Not to be used in patients over 7[j]; isr; anaphylaxis; vial stopper is natural latex	Local reaction; malaise; fever; pain; hypotension; arthralgia; urticaria; rash; Arthus-type reaction; neurological symptoms[k]
Diphtheria and tetanus toxoids and acellular pertussis adsorbed (Infanrix®; Tripedia®; Daptacel®)	Diphtheria and tetanus toxoids and pertussis antigenic preparations[l] are each $Al(OH)_3$-adsorbed	Active immunization against the three bacteria	GBS; syncope; pertussis-associated events[m]; apnea in premature infants; allergic reaction[n]	isr; fever; drowsiness; irritability; loss of appetite; hypersensitivity[o]; URTI; lymphadenopathy; cough; headache
Diphtheria and tetanus toxoids, acellular pertussis adsorbed, hepatitis B[p] and inactivated poliovirus[q] combined (Pediarix®)	Diphtheria and tetanus toxoids and pertussis antigenic preparations[l] are each $Al(OH)_3$-adsorbed. Hepatitis B surface antigen adsorbed to $AlPO_4$	Active immunization against the three bacteria and both viruses	Fever; GBS; latex allergy; syncope; pertussis-associated events[m]; apnea in premature infants; children at risk of seizures	isr; fever; drowsiness; irritability; loss of appetite; URTI; hypersensitivity[o]; syncope; cough; fatigue
Diphtheria and tetanus toxoids, acellular pertussis adsorbed and inactivated poliovirus (Quadracel®; Kinrix®)	Diphtheria and tetanus toxoids and pertussis antigenic preparations[l] are each $Al(OH)_3$-adsorbed; 3 strains of polio virus	Active immunization against the three bacteria and poliovirus	GBS; latex allergy; syncope; pertussis-associated events[m]; children at risk of seizures; preventing/managing allergic reaction	isr; fever; drowsiness; irritability; loss of appetite; syncope; injection site vesicles; pruritus

(continued)

Table 11.1 (continued)

Vaccine	Description	Indications and usage	Warnings and precautions	Adverse events
Diphtheria and tetanus toxoids, acellular pertussis adsorbed, inactivated poliovirus and haemophilus b, tetanus toxoid conjugate (Pentacel®)	Diphtheria and tetanus toxoids and pertussis antigens[l] AlPO4-adsorbed; poliovirus types 1,2,3; *Haemophilus influenzae* capsular antigen[r]	Active immunization against the three bacteria, poliovirus and *H. influenzae* type b	GBS and brachial neuritis; pertussis-associated events[m]; children at risk of seizures; apnea in premature infants; hypersensitivity	isr; fever; irritability; GI; hypersensitivity; viral infection; loss of appetite; apnea, cough; pallor; somnolence; cyanosis
Haemophilus b conjugate vaccine (tetanus toxoid conjugate) (ActHIB®; Hiberix®)	*H. influenzae* capsular antigen[r] covalently bound to tetanus toxoid For IM use only	Active immunization of children 2 months to 5 years against *H. influenzae*[s]	Hypersensitivity; latex allergy; syncope; risk analysis for GBS	isr; fever; irritability; loss of appetite; allergic reactions; convulsions; somnolence; syncope; apnea; lymphadenopathy; febrile seizures; rash
Haemophilus b conjugate vaccine (meningococcal protein conjugate) (PedvaxHIB®)	Purified capsular polysaccharide of *H. influenzae* covalently bound to membrane protein complex of *Neisseria meningitidis* serotype B[t]	For vaccination against *H. influenzae* type b in infants and children 2 to 71 months of age	Hib disease[u]; care that injection not enter blood vessel; latex exposure	Irritability; somnolence; isr; GI; URTI; otitis media; rash
Hepatitis A vaccine, inactivated (Havrix®; Vaqta®)	Inactivated virus strain HM175, $Al(OH)_3$-adsorbed. For IM use	Active immunization against hepatitis A in persons ≥12 months old	Latex exposure; syncope; preventing/managing allergic reaction	isr; irritability; loss of appetite; hypersensitivity; somnolence; GBS; headache; thrombocytopenia; encephalitis; rhinitis; rash; EM

Vaccine	Description	Indications and usage	Warnings and precautions	Adverse events
Hepatitis B vaccine (recombinant) (Engerix-B®)	HbsAg expressed in *Saccharomyces cerevisiae*, $Al(OH)_3$-adsorbed	Immunization against all known subtypes of hepatitis B virus	Latex; syncope; apnea in premature infants; defer for infants <2 kg	Dizziness; headache; isr; URTI; hypersensitivity[v]; encephalopathy; GBS; seizures; skin reactions[w]
Human papillomavirus quadrivalent (types 6,11,16, 18) vaccine, recombinant (Gardasil®; Cervarix®[x])	Capsid (L1) proteins of the 4 types of HPV expressed in *S. cerevisiae* Alum-adsorbed[y]	For females and males 9–26 years for prevention of HPV disease caused by types 6, 11, 16, 18[z]	Syncope; managing allergic reaction (avoid in yeast-allergic patients; yeast used in vaccine production)	Syncope; headache; fever; nausea; dizziness; isr; anaphylaxis; arthralgia; GBS; GI; seizures; lymphadenopathy; asthenia; chills
Human papillomavirus 9-valent (types 6,11,16,18,31,33,45,52,58) vaccine, recombinant (Gardasil 9®)	Capsid (L1) proteins of the 9 types of HPV expressed in *S. cerevisiae* Alum-adsorbed	For females 9–26 and males 9–15 years for prevention of HPV disease[aa] caused by 9 HPV types	Syncope; managing allergic reaction (avoid in yeast-allergic patients; yeast used in vaccine production)	Syncope; headache; isr; AHI; GI; hypersensitivity; GBS; asthenia; chills; malaise; lymphadenopathy; dizziness; seizures
Influenza A (H1N1) monovalent vaccine	Influenza virus (H1N1) grown in chicken eggs, inactivated and disrupted to produce "split virion"	Active immunization against influenza caused by pandemic H1N1 2009 virus in persons ≥6 months old	GBS; managing allergic reaction	isr; headache, malaise; muscle aches. *Children*: irritability; rhinitis; fever; cough; loss of appetite; GI; headache; muscle aches; sore throat. *PM*: Hypersensitivity; neuralgia; vasculitis; rash

(continued)

Table 11.1 (continued)

Vaccine	Description	Indications and usage	Warnings and precautions	Adverse events
Influenza A virus monovalent (H5N1) vaccine[ab]	Influenza virus (H5N1) grown in chicken eggs, inactivated and disrupted to produce split virus	Active immunization against H5 N1 virus in persons 18–64 years	Hypersensitivity; GBS; managing allergic reaction; neurological disorders; vasculitis	isr; headache, malaise; muscle aches; anaphylaxis; neurological disorders; vasculitis
Influenza A (H5N1) virus monovalent vaccine, adjuvanted emulsion	Supplied as 2 vials: 1. Influenza virus (H5N1) grown in chicken eggs, inactivated "split virion". 2. Vial of AS03 adjuvant emulsion[ac]	Active immunization against H5 N1 virus in persons 18 and older	Hypersensitivity; GBS; syncope	Muscle aches; headache; fatigue; joint pain; isr; shivering; sweating; hypersensitivity; narcolepsy; somnolence; febrile convulsions
Influenza virus vaccine, quadrivalent types A and B (Fluarix quadrivalent®; FluLaval quadrivalent®; Fluzone quadrivalent®) for IM injection[ad]	Influenza virus grown in embryonated chicken eggs, inactivated and disrupted to produce split virus. No preservatives	Active immunization against influenza types A and B viruses for persons 3 years and older	GBS; syncope; managing allergic reaction	isr; muscle aches; headache; fatigue. *Children*: isr; drowsiness; irritability; loss of appetite; headache; muscle aches; fatigue; GI; arthralgia. *PM*: Asthenia; lymphadenopathy; vertigo; eye disorders; GBS; hypersensitivity; cutaneous reactions[ae]; vasculitis
Influenza virus vaccine, trivalent, types A and B (Afluria®; Flublok®; Flucelvax®; and multiple others[af])	Influenza virus grown in embryonated chicken eggs, inactivated and disrupted to produce split virus or recombinant preparations[af]	Active immunization against influenza types A and B viruses[af]	Fever and febrile seizures; GBS; managing allergic reaction	isr; myalgia; malaise; headache; irritability; hypersensitivity; transient thrombocytopenia; neuralgia; vasculitis; pruritus; urticaria; rash
Japanese encephalitis vaccine, inactivated, adsorbed (Ixiaro®)	Inactivated, $Al(OH)_3$-adsorbed JEV No preservatives	Active immunization against JEV. Approved for patients ≥2 months of age	Hypersensitivity	isr; fever; irritability; diarrhea; headache; myalgia; paresthesia; neuritis

Vaccine	Description	Indications and usage	Warnings and precautions	Adverse events
Japanese encephalitis virus vaccine, inactivated (JE-Vax®)	JEV Nakayama-NIH strain inactivated. For sc use	Active immunization against JEV for persons 1 year of age and older	Generalized urticaria and angioedema[ag]; administer other vaccines concurrently;↑alcohol intake associated with hypersensitivity	isr; fever; headache; malaise; rash; chills; myalgia; dizziness; GI; urticaria; angioedema; wheezing; itching; neurologic events
Measles, mumps, and rubella virus vaccine live (M–M–R®II)	Attenuated measles and mumps[ah] viruses grown in chick embryo; rubella virus Wistar DA 27/3 strain[ai]	Simultaneous vaccination against measles, mumps, and rubella in infants 1 year or older	Hypersensitivity (including to egg); thrombocytopenia	isr; fever; syncope; headache; dizziness; malaise; irritability; vasculitis; arthritis; encephalitis; skin reactions[aj]; panniculitis; pneumonitis
Measles, mumps, rubella, and varicella virus vaccine live (ProQuad®)[ak]	Measles, mumps, and rubella[al] plus Oka/Merck strain of varicella zoster grown in MRC-5 cells. Contains small amounts of neomycin	Active immunization against the four viruses in children 1–12 years of age	Hypersensitivity to eggs; fever and febrile seizures; history of cerebral injury or seizures; thrombocytopenia; contact hypersensitivity to neomycin[am]	isr; fever; irritability; measles-like rash; arthritis; hypersensitivity; ADEM; convulsions/seizures; bronchospasm; GI; dizziness; headache; infections; thrombocytopenia; cutaneous reactions[an]

(continued)

Table 11.1 (continued)

Vaccine	Description	Indications and usage	Warnings and precautions	Adverse events
Meningococcal (groups A, C, Y, and W-135) oligosaccharide diphtheria CRM197 conjugate vaccine (Menveo®)	A,C,Y, and W-135 oligosaccharides conjugated to *C. diphtheriae* CRM 197 protein	Active immunization to prevent *Neisseria meningitidis* (A,C,Y,W-135)-induced meningococcal disease in persons 2 months to 55 years	Managing allergic reaction; syncope; GBS; apnea in premature infants	isr; irritability; persistent crying; sleepiness; eating habit changes; GI; malaise; headache; myalgia; hypersensitivity; dizziness; arthralgia; syncope
Meningococcal group B vaccine (Bexsero®; Trumenba®)	Recombinant Neisserial adhesin A, NHBA and fHbp plus OMV from *N. meningitidis* NZ98/254 inactivated and adsorbed to $Al(OH)_3$	Active immunization to prevent *N. meningitidis serotype B*-induced disease in persons 10–25 years	Managing allergic reaction; syncope; latex exposure	isr; myalgia; fatigue; headache; induration; nausea; arthralgia; hypersensitivity; syncope; vasovagal responses; rash
Meningococcal groups C and Y and haemophilus b tetanus toxoid conjugate vaccine (MenHibrix®)	C and Y capsular polysaccharides and *Haemophilus* b capsular polysaccharide bound to inactivated tetanus toxoid	Active immunization to prevent *N. meningitidis* C and Y and *H. influenzae* b-induced disease in children 6 weeks to 18 months	GBS; syncope; apnea in premature infants; managing allergic reaction	isr; irritability; drowsiness; loss of appetite; fever; hypersensitivity; convulsions; apnea; syncope or vasovagal responses; urticaria; rash

Vaccine	Description	Indications and usage	Warnings and precautions	Adverse events
Meningococcal (groups A, C, Y, and W-135) polysaccharide diphtheria toxoid conjugate vaccine (Menactra®)	A,C,Y, and W-135 polysaccharides conjugated to diphtheria toxoid protein	Active immunization to prevent *N. meningitidis* (A,C,Y,W-135)-induced meningococcal disease in persons 9 months to 55 years	GBS; syncope; managing allergic reaction	isr; irritability; drowsiness; loss of appetite; fever; abnormal crying; vomiting; headache; fatigue; malaise; arthralgia; hypersensitivity; myalgia; GBS; paresthesia; vasovagal syncope; dizziness
Meningococcal polysaccharide vaccine, groups A,C,Y, and W-135 combined (Menomune-A/C/Y/W-135®)	A,C,Y, and W-135 polysaccharides combined in aqueous vehicle	Active immunization to prevent *N. meningitidis* (A,C,Y,W-135)-induced meningococcal disease in persons 2 years and older	Latex; managing allergic reaction; postpone immunization in cases of severe acute illness	isr; irritability; drowsiness; diarrhea; headache; fatigue; malaise; arthralgia; vasovagal syncope; paresthesia; GBS; hypersensitivity

(continued)

Table 11.1 (continued)

Vaccine	Description	Indications and usage	Warnings and precautions	Adverse events
Pneumococcal vaccine polyvalent (Pneumovax 23®)	Mixture of purified capsular polysaccharides from *Streptococcus pneumoniae* types 1,2,3,4,5,6B,7F,8,9N,9V,10A,11A,12F,14,15B,17F,18C,19F,19A,20,22F,23F,33F	Active immunization to prevent pneumococcal disease caused by the 23 serotypes in the vaccine in persons ≥50 years or those ≥2 years at increased risk	Defer in persons with moderate–severe illness; use with caution in patients with compromised CV/pulmonary function; does not replace antibiotic therapy; may be ineffective in CCSFL[ao]	isr; fatigue; headache, asthenia; myalgia; arthritis; lymphadenitis; hypersensitivity; paresthesia; GBS; febrile convulsions; GI; rash; urticaria; EM
Pneumococcal 7-valent conjugate vaccine (diphtheria CRM197 protein) (Prevnar®)	Mixture of purified capsular polysaccharides from *S. pneumoniae* types 4,6B,9V,14,18C,19F,23F conjugated to diphtheria CRM197 protein; contains $AlPO_4$ adjuvant	Active immunization to prevent pneumococcal disease caused by the 7 serotypes in infants; routine schedule is 2,4,6, and 12–15 months of age	Not to be given to those with thrombocytopenia or coagulation disorders; IM only, never IV; managing allergic reaction; does not replace use of 23-valent vaccine in SCD, asplenia; HIV infection and the immunocompromised	Irritability; isr; decreased appetite; fever; restless sleep; drowsiness; GI; rash; hypersensitivity; EM; angioneurotic edema

Vaccine	Description	Indications and usage	Warnings and precautions	Adverse events
Pneumococcal 13-valent conjugate vaccine (diphtheria CRM197 protein) (Prevnar 13®)	Mixture of purified capsular polysaccharides from *S. pneumoniae* types 1,3,4,5,6A,6B,7F,9V,14,18C, 19A,19F,23F linked to diphtheria CRM197 protein; $AlPO_4$ adjuvant	Active immunization to prevent disease caused by the 13 sero-types in children 6 weeks to 5 years, 6–17 years and adults ≥50; to prevent otitis media[ap]	Apnea in premature infants; managing allergic reaction; reduced antibody response in cases of altered immunocompetence	Irritability; isr; decreased appetite; fever; restless sleep; increased sleep; fatigue; headache; muscle and joint pain; chills; rash; hypersensitivity; apnea; hypotonia; pallor
Poliovirus vaccine inactivated (monkey kidney cell) (IPOL®)	Purified inactivated poliovirus types 1(Mahoney), 2(MEF-1), 3 (Saukett) grown in monkey kidney cells	Active immunization of infants (as young as 6 weeks), children and adults for polio types 1,2 and 3	Possible poor response in cases of altered immunocompetence; traces of neomycin, streptomycin, polymyxin B	Irritability; isr; anorexia; tiredness; fever; vomiting; lymphadenopathy; hypersensitivity; arthralgia; myalgia; headache; convulsion; somnolence; rash

(continued)

Table 11.1 (continued)

Vaccine	Description	Indications and usage	Warnings and precautions	Adverse events
Rabies vaccine (Imovax®; RabAvert®)	Imovax—strain PM-1503-3M harvested from human diploid cells MRC-5 strain; RabAvert—virus strain Flury LEP grown in cultures of chicken fibroblasts	Indicated for pre- and postexposure to rabies in all age groups	Not to be injected into gluteal area[aq]; serum sickness[ar]; neurologic illness resembling GBS; traces of antibiotics are a small risk of hypersensitivity	isr; headache; nausea; abdominal pain; dizziness; muscle aches; lymphadenopathy; hypersensitivity; paresthesia; neuropathy; convulsion, encephalitis; arthralgia; GI; asthenia; malaise; chills; wheezing
Rotavirus vaccine, live, oral (Rotarix®)	Live attenuated human rotavirus strain 89–12 propagated in Vero cells[as]	Prevention of rotavirus gastroenteritis caused by rotavirus G1 and non-G1 types (G3,G4 ,G9) in infants 6–24 weeks of age	Delay vaccine in cases of GI disorders; infants with SCID should not receive vaccine; altered immunocompetence; intussusception; latex exposure	Irritability; loss of appetite; cough; runny nose; fever; vomiting; gastroenteritis; hematochezia; viral shedding in SCID patients; Kawasaki disease; thrombocytopenic purpura
Rotavirus vaccine, live, oral pentavalent (RotaTeq®)	Contains 5 live reassortant rotaviruses G1,G2, G3, G4, P1A(8) from human and bovine hosts Propagated in Vero cells[as]	Prevention of rotavirus gastroenteritis caused by G1,G2,G3, and G4 serotypes	Managing allergic reaction; use with caution with GI and febrile illnesses; intussusception; safety and efficacy not known in immunocompromised infants; viral shedding/ transmission	Diarrhea; vomiting; irritability; otitis media; nasopharyngitis; bronchospasm; anaphylaxis; hematochezia; gastroenteritis with viral shedding in SCID patients; Kawasaki disease; angioedema; urticaria

Vaccine	Description	Indications and usage	Warnings and precautions	Adverse events
Smallpox (Vaccinia) vaccine, live (ACAM2000®)	Live vaccinia virus from cloning from Dryvax®[at] grown in Vero cells	Active immunization against smallpox for persons at high risk	*Boxed warning*: Risk of serious systemic and cutaneous adverse events[au]; accidental eye infection; risk of systemic vaccinia infection in immune deficiency; risk of eczema vaccinatum; management of smallpox vaccine complications	Inoculation site reactions; inadvertent inoculation of other sites (mouth, face, nose, lips, genitalia, anus); lymphadenopathy; malaise; fatigue; fever; myalgia; headache; GI; musculoskeletal and connective tissue disorders; skin and sc disorders including urticaria and folliculitis
Tetanus and diphtheria toxoids, adsorbed	Tetanus and diphtheria toxoids detoxified with formaldehyde adsorbed onto $AlPO_4$	Active immunization to prevent tetanus and diphtheria for persons 7 years of age and older	Persons with previous Arthus reaction and GBS should not receive vaccine	isr; dizziness; malaise; headache; convulsions; rash; musculoskeletal pain; arthralgia
Tetanus and diphtheria toxoids, adsorbed for adult use (Decavac®; Tenivac®)	Tetanus and diphtheria toxoids detoxified with formaldehyde adsorbed onto alum	Active immunization to prevent tetanus and diphtheria for persons 7 years of age and older	Risk for persons with Arthus reaction or GBS; latex exposure; managing allergic reaction	isr; headache; myalgia; muscle weakness; chills; tiredness; diarrhea; nausea; sore joints; syncope; paresthesia; dizziness; convulsions; asthenia; fatigue hypersensitivity; lymphadenopathy

(continued)

Table 11.1 (continued)

Vaccine	Description	Indications and usage	Warnings and precautions	Adverse events
Diphtheria and tetanus toxoids adsorbed STN 103944	Tetanus and diphtheria toxoids detoxified with formaldehyde adsorbed onto $AlPO_4$	Active immunization to prevent tetanus and diphtheria for children 6 weeks to 6 years of age	Managing allergic reaction; apnea in premature infants; syncope; altered immunocompetence; GBS	isr; crying; fever; loss of appetite; somnolence; syncope; headache; convulsion; nausea; pallor; lymphadenopathy
Tetanus toxoid, reduced diphtheriae toxoid, and acellular pertussis vaccine adsorbed (Adacel®; Boostrix®)	Diphtheria and tetanus toxoids and pertussis antigenic preparations[l] are each $Al(OH)_3$-adsorbed[av]	Booster immunization against tetanus, diphtheria, pertussis. As a single dose for persons ≥10 years of age	GBS and brachial neuritis; syncope; latex; neurologic disorders[aw]; Arthus-type reaction; managing allergic reaction; altered immunocompetence	isr; headache; fatigue; GI; arthralgia; lymphadenitis; back pain; hypersensitivity; myocarditis; myalgia; convulsions; encephalitis; facial palsy; paresthesia; syncope; angioedema; Henoch–Schönlein purpura
Typhoid vaccine live oral Ty21a (Vivotif®)	Live attenuated *Salmonella typhi* Ty21a, stabilized, lyophilized, and put into enteric-coated capsules	Immunization of adults and children >6 years old. Selective rather than routine vaccination is recommended	Not to be given during acute GI illness or to those given antibiotics and sulfonamides; some antimalarials interfere with immunogenicity	Diarrhea; abdominal pain; nausea; fever; vomiting; headache; rash; urticaria; anaphylaxis

Vaccine	Description	Indications and usage	Warnings and precautions	Adverse events
Typhoid Vi polysaccharide vaccine (Typhim Vi®)	Cell surface Vi polysaccharide from *Salmonella enterica* serovar *Typhi*, *S typhi* Ty2 strain	Active immunization to prevent *S. typhi*-induced typhoid fever in children ≥2 years. Selective vaccination recommended	Allergic reactions; safety in children <2 years old unproven; protects against *S. typhi* but not other species; delay use in febrile illness or infection; syncope	isr; malaise; headache; myalgia; nausea; fever; GI; hypersensitivity; arthralgia; syncope; tremor; asthma lymphadenopathy; flu-like episode
Varicella virus vaccine, live (Varivax®)[ax]	Oka/Merck strain of live attenuated varicella virus [ay] passaged in human diploid cell cultures MRC-5	Active immunization to prevent varicella in individuals 12 months of age and older	Managing allergic reaction; family history of immunodeficiency[az]; risk of virus transmission; immunoglobulins[ba] salicylate therapy[bb]	Fever; isr; rash; URTI; irritability; fatigue; cough; loss of appetite; disturbed sleep; GI; headache; chills; malaise; allergic reactions; arthralgia; eye complaints; encephalitis; GBS; paresthesia; ataxia; pharyngitis; varicella; cutaneous reactions[bc]

(continued)

Table 11.1 (continued)

Vaccine	Description	Indications and usage	Warnings and precautions	Adverse events
Yellow fever vaccine (YF-Vax®)	Yellow fever virus strain 17D-204 cultured in avian virus-free chicken embryos. Vaccine is stabilized, lyophilized, and sealed under N	Active immunization of persons ≥9 months traveling in endemic areas, for international travel and exposed laboratory personnel	Contraindicated in infants <9 months; postpone in cases of acute febrile disease; immunosuppressed patients[bd]; latex; rare risks of viscerotropic and neurotropic diseases; hypersensitivity	Headache; myalgia; fever; isr; hypersensitivity; neurotropic disease[be]; viscerotropic disease[bf]
Zoster vaccine, live (Zostavax®)	Oka/Merck strain of live attenuated varicella-zoster virus	Prevention of herpes zoster in persons 50 years and older	Hypersensitivity; transmission of vaccine virus; defer immunization in cases of concurrent illness	Headache; isr; fever; flu syndrome; diarrhea; rash; rhinitis; respiratory disorders; asthenia; arthralgia; myalgia; herpes zoster (vaccine strain); hypersensitivity

ADEM acute disseminated encephalomyelitis, *AHI* autoimmune hemolytic anemia, *BCG* bacillus Calmette–Guerin, *CCFL* chronic cerebrospinal fluid leakage, *CV* cardiovascular, *EM* erythema multiforme, *FDA* US Food and Drug Administration, *fHbp* factor H binding protein of *N. meningitidis*, *GBS* Guillain-Barré syndrome, *GI* gastrointestinal disorders, nausea, vomiting, diarrhea etc, *HBsAg* hepatitis B surface antigen, *Hib Haemophilus influenza* type b, *HPV* human papillomavirus, *IM* intramuscular, *isr* injection site reaction, *IV* intravenous, *JEV* Japanese encephalitis virus, *NHBA* Neisserial heparin binding antigen, *OMV* outer membrane vesicles of *N. meningitidis*, *PA* protective antigen, *PM* postmarketing, *sc* subcutaneous, *SCD* sickle cell disease, *SCID* severe combined immunodeficiency disease, *SJS* Stevens-Johnson syndrome, *URTI* upper respiratory tract infection

[a]Licensed by FDA

[b]Approved for use in military populations 17–50 years old

[c]Infection of pregnant women during period of virus shedding may cause fetal harm

[d]An 83 kDa protective antigen released during growth period

[e]Rash, pruritus, and urticaria. Cases of EM and SJS also seen

[f]TICE® strain, University of Illinois. From original Pasteur Institute strain

[g]Axillary or cervical lymphadenopathy, induration, and pustule formation at injection site

[h]Used as prophylaxis and as therapy for recurrent tumors in patients with carcinoma in situ of the urinary bladder and to prevent recurrence of stage Ta and/or T1 papillary tumors following transurethral resection. Not indicated for papillary tumors of stages higher than T1

[i]*Corynebacterium diphtheriae* and *Clostridium tetani*

[j]Adverse reactions to diphtheria toxoid may be more severe in older patients

[k]Usually temporary: brachial plexus neuropathies; paralysis of radial and recurrent nerves; accommodation paresis; GBS; EEG disturbances with encephalopathy

[l]*Bordetella pertussis* antigen preparation is made up of inactivated *B. pertussis* toxin and formaldehyde-treated filamentous hemagglutinin and pertactin

[m]Temperature ≥40 °C, collapse or shock state, persistent crying lasting >3 h (all within 48 h), seizures without fever within 3 days. An appropriate antipyretic may be administered to children at risk of seizures

[n]Vaccine hypersensitivity and possible allergic reaction to latex in prefilled syringes

[o]Angioedema, urticaria, rash, and pruritus

[p]Recombinant preparation

[q]Three strains of polio virus

[r]*H. influenzae* type b capsular polysaccharide (polyribosylribitol-phosphate) covalently bound to tetanus toxoid to provide helper T-cell epitopes for the carbohydrate structures

[s]*H. influenzae* type b also known as Hib

[t]Covalently bound to *N. meningitidis* to enhance immunogenicity of the *H. influenzae* polysaccharide

[u]Hib disease may occur in the week after vaccination prior to protection developing

[v]Including an IgE-mediated allergic local reaction and a serum sickness-like reaction of delayed onset with arthralgia, fever, and dermatologic reactions

[w]Including urticaria, ecchymosis, lichen planus, EM, and SJS

[x]Cervarix®: Human papillomavirus bivalent (types 16 and 18) vaccine, recombinant

[y]Adsorbed on amorphous aluminium hydroxyphosphate sulfate

[z]Females: Cervical, vulvar, vaginal, and anal cancers (types 16 and 18); genital warts (types 6 and 11); precancerous and dysplastic lesions caused by HPV (types 6,11,16, and 18). Males: Anal cancer (types 16 and 18); genital warts (types 6 and 11); anal intraepithelial neoplasia (types 6,11,16 and 18)

[aa]Females: Cervical, vulvar, vaginal, and anal cancers caused by HPV types 16,18,31,33,45,52, and 58; genital warts, types 6 and 11; precancerous and dysplastic lesions caused by HPV types 6,11,16,18, 31,33,45,52, and 58. Males: Anal cancer caused by HPV types 16,18,31,33,45,52, and 58; genital warts, types 6 and 11; anal intraepithelial neoplasia caused by types 6,11,16,18,31,33,45,52, and 58

[ab]For US national stockpile

[ac]AS03 adjuvant: an homogenized sterile emulsion of squalene *DL*-α-tocopherol and polysorbate 80

[ad]Intranasal vaccine also available: FluMist quadrivalent® for persons 2–49 years old

[ae]EM, SJS, urticaria, facial swelling, and pruritus have been reported

(continued)

Table 11.1 (continued)

[af]Other trade names: Agriflu®, FluLaval®, Fluarix®, Flublok®, Flucelvax®, Fluvirin® and FluMist® intranasal. Target ages: Afluria® 5 years and older; FluLaval®, Fluarix® ≥ 3 years; and Fluvirin® ≥ 4 years; others ≥ 18 years. Flublok® and Flucelvax® are recombinant preparations; neither are grown in egg embryo and therefore contain no egg proteins (see text, section "Allergic Reactions to Egg Proteins in Vaccines")

[ag]May occur within minutes or as late as 17 days after injection. Vaccinees should be observed for 30 min after vaccination and remain in areas of medical facilities for 10 days after injection. Increased risk of hypersensitivity if other vaccines given within 7 days prior to JEV vaccine

[ah]Jeryl Lynn™ (B level) strain

[ai]Grown in human lung fibroblasts

[aj]Including SJS, EM, urticaria, rash, measles-like rash, and pruritus

[ak]Available as refrigerator-stable and frozen formulations

[al]For details of measles, mumps, and rubella viruses see "Measles, mumps, and rubella virus vaccine (M–M–R®II)" this table. Measles and mumps viruses propagated in chick embryo cell culture; rubella propagated in human diploid lung fibroblasts

[am]Transmission of varicella virus may occur with contacts susceptible to varicella, viz., immunocompromised individuals, pregnant women, and newborn infants. Defer vaccination for at least 3 months following transfusions to avoid antivirus-interfering antibodies

[an]Including EM, panniculitis, impetigo, pruritus, SJS, and sunburn

[ao]May be ineffective in pneumococcal meningitis in patients who have CCSFL

[ap]For the prevention of otitis media caused by serotypes 4,6B,9V,14,18C,19F, and 23F

[aq] Gluteal injection may result in lower antibody titers

[ar]In persons receiving booster doses. Reactions (generalized urticaria, arthralgia, arthritis, angioedema, nausea, vomiting, fever, and malaise) occur 2–21 days after booster injection. Reported in up to 7 % of those receiving booster dose

[as]Porcine-derived material used in manufacture. Vaccine contains porcine circovirus type 1 which is not known to cause disease in humans

[at]Live virus preparation prepared from calf lymph. Made by Wyeth; world's oldest smallpox vaccine. Manufacture ceased after smallpox declared eradicated in 1980s

[au]Myocarditis and pericarditis, encephalomyelitis, encephalopathy, progressive vaccinia, generalized vaccinia, severe permanent sequelae or death, ocular complications, blindness, and fetal death have occurred following primary or revaccination. Severe vaccinial skin infections, EM, and SJS have been reported

[av]$AlPO_4$ for Adacel®

[aw]For progressive or unstable neurologic conditions, defer vaccination with pertussis-containing vaccines

[ax]Refrigerator-stable and frozen preparations available

[ay]Originally isolated from a child with wild-type varicella

[az]Defer vaccination in patients with family history of immunodeficiency until immune status has been evaluated and found to be immunocompetent

[ba]Immunoglobulin should not be given concomitantly and for 2 months after. Defer vaccination for at least 5 months following transfusions

[bb]Avoid salicylates in children and adolescents 1–17 years old for 6 weeks following vaccination because of association of Reye syndrome with aspirin therapy and varicella infection

[bc]Including EM, SDS, Henoch–Schönlein purpura, impetigo, and cellulitis

[bd]Risk of encephalitis in immunosuppressed patients

[be]Also known as postvaccinal encephalitis

[bf]Previously described as multiple organ system failure

promotor and a CpG bacterial sequence as adjuvant. Although DNA vaccination has not yet been as successful in humans as in mice, the approach has many potential advantages including the simplicity of the methodology and the promising strategies it appears to open up, in particular, for the eventual protection of malaria, influenza, and HIV infections.

Currently Approved Vaccines: Description, Indications, Warnings, Precautions, and Adverse Events

Immunization is now considered to be so effective, safe and such an important public health measure that most advanced countries require all infants/children to be immunized against at least a core of important life-threatening diseases such as tetanus, diphtheria, polio, and whooping cough. In the United States, the required list generally includes not only these four diseases but also measles, mumps, rubella, and chicken pox (varicella). In recent years, vaccines have become available for *H. influenzae*, childhood diarrhea caused by rotavirus, meningococcal infections caused by *N. meningitidis*, pneumococcal infections, and human papillomavirus. To simplify immunization schedules, decrease the number of injections required for separate administrations, extend the range of disease protection coverage, and improve patient compliance, combinations of vaccines are often used as so-called polyvalent preparations. Commonly used examples of such vaccines include diphtheria–tetanus–pertussis (DTP), DTP with inactivated poliovirus (and sometimes also hepatitis B or haemophilus b capsular antigen), measles–mumps–rubella, and measles–mumps–rubella–varicella. Combination vaccines offer some obvious advantages, but their use may occasionally lead to some unwanted outcomes. For example, while there is an increased risk of febrile convulsions during the first 2 weeks after administration of measles–mumps–rubella vaccine (Table 11.1), parents tend to proceed with vaccination, but varicella vaccination is often not taken up, apparently because of doubts about its benefits. Although the risk of febrile convulsions after measles–mumps–rubella–varicella vaccination is about twice as high as separate administrations of measles–mumps–rubella and varicella vaccines, the number of hospitalization days are claimed to be substantially reduced in children who received varicella vaccination in addition to measles–mumps–rubella immune protection. Such trade-offs need to be considered in devising immunization programs and it is up to clinicians to make parents aware of such subtleties when recommending which vaccinations should be included.

Table 11.1 lists 46 vaccine preparations currently licensed for therapy by the FDA together with a description of each vaccine's composition, approved indications and usage, associated warnings and precautions, and main recorded adverse events. Some vaccines are offered by a number of different manufacturers. For example, influenza A H1N1 monovalent vaccine is available from five different manufacturers, there are two preparations of influenza A H5N1 monovalent vaccine, and nine different manufacturers offer licensed preparations of influenza trivalent types A and B vaccine. Under "Description" of the vaccines in Table 11.1, details include the species of organisms together with serotypes; whether pathogens are live, killed, attenuated, acellular, or presented in subunit(s) form; how organisms were grown, for example, influenza, measles, and mumps viruses grown in chicken eggs; those vaccines formulated as unconjugated or conjugated toxoids, oligosaccharides, or polysaccharides; the route of administration, generally parenteral but sometimes oral and, if oral, whether the antigenic material is enteric coated (as with, e.g., live oral typhoid vaccine); presence and type of adjuvant included; use of recombinant protein (as with hepatitis B virus vaccine); and if viruses are included in split virion form.

Despite the fact that live vaccines are generally more potent and elicit more effective immune responses than killed vaccines (section "Vaccines: Definition, Attenuation, and Subunit, Acellular, Carbohydrate, Conjugate and DNA Vaccines"), killed bacterial and inactivated viral vaccines have the inherent advantage of being generally safer since live organisms may retain both the capacity to mutate and the risk of causing infection in immune compromised individuals. This risk is reflected in the proportions of live and killed vaccines amongst the FDA approved products with approximately three times as many killed vaccines registered as live vaccines (Table 11.2). Under "Warnings and precautions," some adverse events and potential risks are common to a number of different vaccines and are recurring entries throughout the table (Table 11.1). The most commonly recurring events listed are injection site reactions and hypersensitivities/allergic reactions with particular emphasis on the possibility of potentially life-threatening anaphylaxis. Injection site reactions, often incorrectly labeled as allergic, occur often enough and produce sufficient discomfort to have a significant impact on

Table 11.2 Live and killed vaccines approved for therapy by the FDA

Live vaccines	Killed vaccines
Adenovirus types 4 and 7	Anthrax vaccine adsorbed
BCG Live (2 preparations)	Diphtheria and tetanus toxoids adsorbed
Measles, mumps, and rubella virus vaccine live	Diphtheria and tetanus toxoids and acellular pertussis adsorbed
Measles, mumps, rubella, and varicella virus vaccine live	Diphtheria and tetanus toxoids, acellular pertussis adsorbed, hepatitis B, and inactivated poliovirus combined
Rotavirus vaccine, live, oral	Diphtheria and tetanus toxoids, acellular pertussis adsorbed and inactivated poliovirus
Rotavirus vaccine, live, oral pentavalent	Diphtheria and tetanus toxoids, acellular pertussis adsorbed, inactivated poliovirus and haemophilus b, tetanus toxoid conjugate
Smallpox (Vaccinia) vaccine, live	Haemophilus b conjugate vaccine (tetanus toxoid conjugate)
Typhoid vaccine live, oral Ty21a	Haemophilus b conjugate vaccine (meningococcal protein conjugate)
Varicella virus vaccine, live	Hepatitis A vaccine, inactivated
Yellow fever vaccine	Hepatitis B vaccine (recombinant)

(continued)

Table 11.2 (continued)

Live vaccines	Killed vaccines
Zoster vaccine, live	Human papillomavirus quadrivalent (types 6,11,16, and 18) vaccine, recombinant
	Human papillomavirus 9-valent (types 6,11,16,18,31,33,45,52, and 58) vaccine, recombinant
	Influenza A (H1N1) monovalent vaccine
	Influenza A virus monovalent (H5N1) vaccine
	Influenza A (H5N1) virus monovalent vaccine, adjuvanted emulsion
	Influenza virus vaccine, quadrivalent types A and B
	Influenza virus vaccine, trivalent, types A and B
	Japanese encephalitis vaccine, inactivated, adsorbed
	Japanese encephalitis virus vaccine, inactivated
	Meningococcal (groups A, C, Y, and W-135) oligosaccharide diphtheria
	CRM197 conjugate vaccine
	Meningococcal group B vaccine
	Meningococcal groups C and Y and haemophilus b tetanus toxoid conjugate vaccine
	Meningococcal (groups A, C, Y, and W-135) polysaccharide diphtheria toxoid conjugate vaccine
	Meningococcal polysaccharide vaccine, groups A,C,Y, and W-135 combined
	Pneumococcal vaccine polyvalent
	Pneumococcal 7-valent conjugate vaccine (diphtheria CRM197 protein)
	Pneumococcal 13-valent conjugate vaccine (diphtheria CRM197 protein)
	Poliovirus vaccine inactivated (monkey kidney cell)
	Rabies vaccine
	Tetanus and diphtheria toxoids, adsorbed
	Tetanus and diphtheria toxoids, adsorbed for adult use
	Diphtheria and tetanus toxoids adsorbed STN 103944
	Tetanus toxoid, reduced diphtheriae toxoid and acellular pertussis vaccine adsorbed
	Typhoid Vi polysaccharide vaccine

clinical practice. These local reactions may involve pain, local swelling, redness, edema, limb swelling, subcutaneous nodules, local eczematous lesions, and nevi associated with hypertrichosis (e.g., after BCG, tetanus, and smallpox vaccines). With the possibility of severe allergic reactions in mind, particularly type I immediate hypersensitivity manifesting as anaphylaxis, many of the approved vaccines carry a warning for clinicians to review patients' immunization histories for possible allergic sensitivities and previous adverse events. As well as this reminder to aid prevention, adequate preparations in the form of ready availability of epinephrine and other necessary drugs and facilities should be made to manage a severe allergic reaction should it occur. Connected to the frequent warning of possible hypersensitivity is the risk of severe reactions provoked by latex when it is used as latex stoppers and in prefilled syringes. Other often-repeated warnings/precautions highlight the frequent occurrence of syncope following inoculation; a reduced or poor response in patients with reduced immunocompetence; the risk of severe infection with the immunizing agent in immunosuppressed/immunodeficient patients given live vaccines; possible induction of Guillain–Barré syndrome (see below); the need for extra care with children at risk of seizures; the need to defer immunization in cases of concurrent illness; and a risk of apnea in premature infants. Pregnant women should not be given live vaccines, but inactivated influenza virus vaccine, tetanus, and hepatitis B vaccine may be given. Two vaccines, BCG live for intravesical use and smallpox (vaccinia) live, carry FDA boxed warnings, the former for the risk of transmission of live, infectious bacteria and vaccinia for risks of cardiac, central nervous system (CNS), and severe cutaneous events. Myocarditis and/or pericarditis have been observed at an incidence of 5.7 per 1000 healthy adult primary smallpox vaccinees. Serious CNS adverse events include encephalitis, encephalomyelitis, and encephalopathy. Severe cutaneous reactions seen are severe vaccinial skin infection, erythema multiforme, and Stevens–Johnson syndrome.

Guillain-Barré syndrome, a rare disorder causing muscle weakness, sometimes flaccid paralysis, and infrequently death, is thought to be immune-mediated but its underlying aetiology and pathophysiology are not understood. A recent analysis of 23 million vaccinated people in the US revealed that the 2009 influenza A H1N1 vaccines were associated with a small risk of Guillain-Barré syndrome (1–6 excess cases/1 million). Similarly, results in Spain showed a small, statistically significant association between influenza vaccines and Guillain-Barré syndrome and in Germany, an increased risk of the syndrome in temporal association with influenza H1N1 vaccination was identified. It has been suggested that some vaccines other than influenza might also be associated with Guillain-Barré syndrome.

In addition to the adverse effects already mentioned, Table 11.1 collectively reveals an extensive list of adverse events, again with some events recorded for many of the 46 different vaccines. The list of commonly recurring events includes syncope, irritability, malaise, loss of appetite, fever, arthralgia, drowsiness, headache, dizziness, somnolence, and rash. Although some vaccines in particular, and even vaccination in general, have been claimed to confer long-term adverse effects such as autism, atopy, and multiple sclerosis, no evidence has been forthcoming to support such associations.

Combination vaccines offering protection against six diseases, diphtheria, tetanus, pertussis, hepatitis B, poliomyelitis, and *H. influenzae* are available in the European Union. These preparations have the potential to simplify pediatric immunization schedules, improve patient compliance, and reduce costs. Hexyon®/Hexacima®, a fully liquid, ready-to-use, thimerosal-free polyvalent pediatric vaccine containing acellular pertussis, is indicated for primary and booster vaccination of infants 6 weeks to 24 months of age. Another hexavalent preparation, Infanrix Hexa®, approved in the European Union and marketed widely, contains diphtheria and tetanus toxoids, inactivated polio virus types 1, 2, and 3, recombinant hepatitis B surface antigen, *H. influenzae* type b polysaccharide conjugated to tetanus toxoid, and pertussis antigens pertactin, filamentous hemagglutinin, and pertussis toxoid as well as aluminium hydroxide and aluminium phosphate adjuvants. To overcome instability of the *H. influenzae* type b polysaccharide, it is supplied as a lyophilized powder while the remaining components are presented as a turbid fine suspension. Attention is drawn to the possible presence of traces of formaldehyde, neomycin, and polymyxin used during the manufacturing process. Vaccine-related adverse reactions recorded from studies on more than 16,000 subjects included loss of appetite, abnormal crying/irritability/restlessness, diarrhea, vomiting, and fever ($\geq$38 °C) and, as observed with other DTP combination vaccines, local reactions and fever may occur. Postmarketing reports suggest a potential risk of convulsions, fever, and hypotonic hyporesponsive episode.

"Allergy"/Adverse Reactions to Vaccines and Added Components

Vaccines are administered to millions of individuals worldwide, particularly children, and apart from the occasional adverse events they provoke and periodic resultant extensive media attention, vaccination and its management is generally routine and uneventful, occupying a small but significant proportion of everyday clinical practice. Adverse events after vaccination, usually mild, are fairly commonly seen and although side effects ranging from mild local reactions to rare fatal outcomes constitute a significant health issue, sometimes with accompanying exaggerated claims and fears, many adverse reactions attributed to vaccines are often without justification. As mentioned above, the most frequently reported adverse events following immunization are local reactions manifesting simply as redness, transient pain, subcutaneous nodules, eczematous lesions, or swollen limbs. Local reactions may lead on to delayed skin eruptions such as urticaria and maculopapular skin rash or, in rare instances, an immediate type I allergic response. That most reactions are not true hypersensitivity responses is borne out by recorded data over many years. For example, in one revaccination study of 421 children with a past history of adverse vaccine reactions, only ~10 % of children with a suspected allergic reaction to a vaccine produced a response on re-exposure, but responses were not truly allergic. True hypersensitivities following vaccination are known and may be seen as rare immediate reactions occurring within 1 h of administration and manifesting as urticaria, angioedema, wheezing, rhinitis, hypotension, and in severe instances, bronchospasm and cardiovascular collapse. Anaphylaxis to toxoids is well known although the incidence has fallen to less than 1 case per 10,000 administrations with the introduction of highly purified preparations. Confirmation that reactions are true type I IgE antibody-mediated responses has been provided in some cases by the demonstrations of positive skin tests and specific IgE antibodies. Following an unusually high incidence of anaphylaxis to influenza vaccine in Japan in the 2011–2012 immunization season, 19 of the affected patients (average age 62 months) and age-matched controls, including 10 with egg allergy, were tested for IgE antibodies to trivalent vaccines and influenza hemagglutinin proteins derived from egg and cell cultures. IgE antibodies to influenza vaccine antigens, whole vaccines from different manufacturers, and hemagglutinin proteins A (H1N1 and H3N2) and B were significantly increased in the anaphylactic patients and influenza vaccine-induced CD203c expression in basophil activation tests was also enhanced in the patients but not in controls. Neither egg proteins nor vaccine excipients 2-phenoxy-ethanol and thimerosal could be implicated as allergic sensitizing agents. Urticaria, angioedema, and anaphylaxis have been attributed to *B. pertussis* antigens but although IgE antibodies to the bacteria have been detected following immunization and IgE levels correlate with IgG responses, this has been interpreted as a reflection of the immunogenicity of the organism rather than its allergenicity. Anaphylaxis, a positive skin test, and specific IgE antibodies have been found in a small number of patients given *S. pneumoniae* vaccine. Delayed reactions, occurring some hours after vaccination, are relatively common but are generally not well understood; data confirming these reactions as one or more of the four types of true hypersensitivities are often absent.

The risk of anaphylaxis after vaccination in children and adults was recently assessed using data from the Vaccine Safety Datalink, a collaboration between the Centers for Disease Control and Prevention's Immunization Safety Office and a number of integrated health care systems in The United States. Rates of anaphylaxis after vaccination were determined for the period January 2009 to December 2011 using the Brighton Collaboration definition to confirm anaphylaxis. Thirty three cases of anaphylaxis induced by vaccines were confirmed out of a total of 25,173,965 vaccine administrations (rate 1.31 per million vaccine doses). There were ten cases of anaphylaxis to inactivated trivalent influenza vaccine (1.35 per million doses) and two cases to inactivated monovalent influenza vaccine (1.83 per million doses). The rate of anaphylaxis that did not involve trivalent influenza vaccine was 19 cases from 14,394,021 vaccinations (1.32 per million doses). These results confirmed the rarity of postvaccination anaphylaxis found in previous surveys.

As well as the microbial components of vaccines, other additives, namely preservatives, bacteriostatics, stabilizers, and adjuvants have been suspected and occasionally implicated in adverse effects. In a few cases, these adverse effects were shown to have an allergic basis. Gelatin and egg proteins are the most commonly implicated nonmicrobial additives, but other suspected added components have included yeast proteins, dextrans, thimerosal, formaldehyde, phenoxyethanol, antibiotics, and adjuvants, often aluminium hydroxide or aluminium phosphate.

Allergic Reactions to Egg Proteins in Vaccines

The presence of egg proteins, principally ovalbumin, in some vaccines such as influenza, measles, mumps and rubella, and yellow fever, and the administration of these vaccines to patients allergic to eggs, have been ongoing concerns. Even so, studies on the safety of influenza vaccines have not revealed any cases of anaphylaxis to egg proteins despite finding a few individuals with mild generalized urticaria. In addition, the risk of a reaction to egg protein was similar in patients with positive or negative skin tests leading to the conclusion that skin testing with influenza vaccine would be of no value in assessing egg-allergic patients for vaccine sensitivity. On the basis of the ovalbumin content of influenza vaccines, some have concluded that vaccines using recombinant antigens contain no egg protein and are therefore safe to administer; vaccines produced in chicken egg embryo cells containing very small amounts of ovalbumin (e.g., <1 μg/0.5 mL) can be administered although perhaps with some caution in severely egg-allergic patients; and vaccines containing higher amounts may pose a risk for egg-allergic individuals and should be avoided. It is open to question, however, if this classification of risk is as clear cut as it is presented since patients allergic to eggs have been found to safely tolerate influenza vaccines with ovalbumin contents clearly in the supposedly high-risk range. A more up-to-date and likely assessment of the situation began in 2012 with the preparation and publication of "Adverse reactions to vaccines practice parameters 2012 update" developed by the Joint Task Force on Practice Parameters (JTFPP) representing the American Academy of Allergy, Asthma and Immunology, The American College of Allergy, Asthma and Immunology and the Joint Council of Allergy, Asthma and Immunology. Consistent with recommendations from the Centers for Disease Control's Advisory Committee on Immunization Practices (ACIP), the 2012 practice parameters recommended that egg-allergic patients receive inactivated influenza vaccine as a single dose without prior skin testing and patients then be observed for 30 min for the appearance of any allergic signs. An additional recommendation was that if a reaction following the ingestion of eggs consisted of hives only, the vaccine could be administered in a primary care setting but if the reaction was more severe, vaccine administration should be undertaken in an allergist's office. Subsequent to the publication of the 2012 update, the Vaccine Adverse Events Reporting System showed no disproportionate reporting of allergy and anaphylaxis to influenza vaccines and further published studies on the administration of inactivated influenza virus to egg-allergic patients revealed no vaccine-related reactions in 4172 patients receiving 4729 doses of influenza vaccine or in 143 individuals with severe egg allergy. These findings indicated that even for patients with an allergy to egg, the risk of an adverse reaction to inactivated influenza virus vaccine is so low that vaccination by an allergist is not necessary. The relatively recent availability of recombinant influenza preparations has also had a major bearing on recommendations for egg-allergic patients receiving influenza vaccination. Two new preparations of trivalent influenza virus vaccine types A and B, Flublok® and Flucelvax®, are composed of recombinant proteins not grown in eggs and both have been approved for patients 18 years and older. Flublok® contains recombinant hemagglutinin proteins from three influenza viruses formulated for intramuscular injection. The recombinant proteins are produced in insect cell lines derived from Sf9 cells of the fall armyworm *Spodoptera frugiperda*. Each hemagglutinin is expressed using a baculovirus vector. Flucelvax® is a subunit influenza vaccine prepared from virus propagated in Madin Darby Canine Kidney cells. The virus is inactivated by β-propiolactone. In the 2013 update issued by the JTFPP, the prior recommendation that egg-allergic patients can receive influenza vaccine as a single dose without prior vaccine skin testing was repeated. It was declared that either egg-based or egg-free vaccine can be used and special precautions after administration of the vaccine are not warranted. In line with the ACIP General Recommendations on Immunization, it was stated: "Although anaphylactic reactions are rare after vaccination, their immediate onset and life-threatening nature require that all personnel and facilities providing vaccinations have procedures in place for anaphylaxis management." Recommendations were also extended to remove from guidelines and product labeling, all special precautions and any language that describes egg-allergic recipients as being at risk compared with nonegg-allergic recipients.

Allergic Reactions to Gelatin in Vaccines

Allergic reactions to gelatin are well known, being reported after eating flavored fruit gums and condiments, after infusion of plasma expanders containing gelatin, and following injection of vaccines containing gelatin as a heat stabilizer. Anaphylaxis to measles, mumps, rubella (MMR) vaccines is known to occur and after an initial report of antigelatin IgE antibodies demonstrated by immunoblotting in a patient who experienced an anaphylactic reaction in response to MMR vaccine, Japanese investigators confirmed the allergenic role of this protein in a number of vaccine reactions including anaphylaxis and cutaneous reactions such as urticaria. In early investigations, 10 of 11 children with systemic immediate-type reactions to vaccines had antigelatin IgE antibodies and 24 of 26 children who had immediate reactions to MMR vaccines also had IgE antibodies to gelatin; seven of the 24 subjects experienced an allergic reaction after ingesting gelatin and of these, two had reactions before vaccination and five had reactions after vaccination. Results therefore demonstrated that some children who had anaphylaxis to MMR vaccine have food allergy to gelatin. The authors further showed that such IgE-mediated reactions to vaccines were boosted by gelatin. It was concluded that the patients were probably sensitized to gelatin present in foods and/or previous vaccine injections. In a study designed to investigate gelatin-specific humoral and cellular immune responses in children with immediate- and nonimmediate-type reactions to live MMR and varicella vaccines, six patients with immediate reactions had IgE antibodies to gelatin, whereas no IgE antibodies to gelatin were found in 21 patients who experienced nonimmediate reactions to the protein. All six immediate reactors and 17 of the 21 nonimmediate reactors showed gelatin-specific T-cell responses. These results suggest that cell-mediate immunity to gelatin may have a significant role in the pathogenesis of nonimmediate reactions in some adverse events induced by some vaccines. Gelatin-induced immediate reactions in the form of both IgE antibody-mediated systemic reactions and generalized urticaria have also been reported after varicella vaccinations. Speculation that Japanese children may have been sensitized to gelatin as a result of exposure to diphtheria, tetanus, and acellular pertussis (DTAP) vaccines before the introduction of live viral vaccination appears to have been supported by the finding of what was described as a strong causal relationship between DTAP vaccination, the production of antigelatin IgE antibodies, and the risk of anaphylaxis following subsequent immunization with live vaccines containing relatively larger amounts of gelatin. Systemic immediate reactions to gelatin have also been reported following immunization with Japanese encephalitis vaccines. In one study, three children who had immediate systemic reactions to Japanese encephalitis vaccine had antigelatin IgE antibodies in their sera. Two of the patients also experienced systemic allergic reactions to foods containing gelatin before vaccination.

Aluminium Adjuvants

Used as an immunologic adjuvant in some vaccines since 1927, aluminium salts have become the standard adjuvant in vaccines such as diphtheria, tetanus, pertussis, *H. influenzae* type b, pneumococcus, and hepatitis A and B. Aluminium salts are added to vaccines as alum (potassium aluminium sulfate), aluminium sulfate, aluminium phosphate, or aluminium hydroxide. Itching nodules or granulomas at the injection site after vaccination with aluminium-adsorbed vaccines have been known for over 50 years but remain rare when viewed against the large numbers of immunizations administered worldwide. Reactions to aluminium adjuvants seem to be seen most often after injection of aluminium hydroxide as adjuvant in tetanus, diphtheria–tetanus, diphtheria–tetanus–pertussis (whole cell), diphtheria–polio, DTP (acellular) –polio, hepatitis B, and influenza vaccines. Other adverse reactions to aluminium salts, although rare, are also well recognized as contact allergies with persistent itching nodules, and after desensitization of allergic patients with aluminium precipitated allergen extracts. In one well-known trial in Sweden of aluminium hydroxide-adsorbed DTP (acellular) vaccine from a single manufacturer, persistent itching nodules occurred at the injection site after subcutaneous and intramuscular injections in 645 children out of ~76,000 vaccines, an incidence of ~0.8 %. Itching was intense and long-lasting with about three quarters of the subjects still experiencing symptoms 4 years after immunization. Seventy seven percent of the children with itching nodules were found to have contact hypersensitivity to aluminium; however, the reason for the high incidence of nodules remained unexplained. Commenting on the relatively high incidence of reactions to their vaccine in Sweden, the Danish manufacturer drew attention to the contrastingly low incidences of reactions in Denmark of similar vaccines containing the same diphtheria, tetanus, and pertussis toxoids and the same aluminium hydroxide gel from the same producer. In suggesting that the injection technique was a likely explanation for the reactions seen in Sweden, the vaccine manufacturer stated that the incidence of itching granulomas is low following the correct intramuscular administration of aluminium-adsorbed vaccines. The authors of the Swedish trial rejected the suggestion that the high incidence of itching nodules was due to subcutaneous injection technique and in turn suggested that in Denmark a different product, DTP (acellular)–polio vaccine was used, and this product contained more aluminium hydroxide and sodium dihydrogen phosphate as buffer instead of sodium hydroxide. It was further suggested that this difference might affect the solubility of the final product and, in turn, the occurrence of itching nodules. Pointing out that there are many different diphtheria–tetanus and DTP (acellular)–polio vaccines in the world, the Swedish authors also stated (rather vaguely) that minor differences between these vaccines may cause differences in adverse events. With the above results in mind, a further study from Sweden undertook a prospective surveillance of 22,365 vaccinations of children aged 10 years old comparing a new booster diphtheria–tetanus preparation containing 6.25 Lf (limit of flocculation) units of diphtheria toxoid, 6.25 Lf tetanus toxoid and 0.5 mg Al^{3+} as aluminium hydroxide per 0.5 mL with a previously released preparation containing 7.5 Lf diphtheria toxoid, 1.8 Lf tetanus toxoid, and 0.25 mg Al^{3+} as aluminium phosphate per 0.25 mL dose. Vaccines were injected intramuscularly in the deltoid muscle using a Luer needle number 16 size 0.6×25 (23G× 1 in.) at an angle of 90° against the skin surface. Injections were performed using a muscle mass grasp between the thumb and fingers. Before withdrawing the vaccine, the vial stopper was swabbed with 70 % isopropyl alcohol. Three to six children per 10,000 injected were found to have local itching persisting for at least 2 months and no significant differences were detected between the two vaccine groups. In contrast to the previous Swedish study, no positive epicutaneous tests to aluminium were seen and while the authors concluded that either vaccine could safely be used in the Swedish vaccination program, mild, late-occurring itching nodules with no sign of aluminium hypersensitivity were seen after the fourth diphtheria–tetanus dose at much higher rates than previously recognized from passive surveillance information. Interestingly, attention was also drawn to the pertussis component of the vaccine used in the prior Swedish study where a relatively high dose of 40 μg of pertussis toxin was administered compared to the 3.5–24 μg dose often employed in many vaccines. Pertussis extracts are themselves effective adjuvants boosting both IgE and IgG responses and the possibility that pertussis toxin may have potentiated the development of delayed hypersensitivity to aluminium was suggested.

In a systematic review with meta-analysis, vaccines against diphtheria, tetanus, and pertussis, alone or in combination, were compared with the same vaccines either without or with aluminium. In young children up to 18 months, vaccines containing aluminium hydroxide caused significantly more erythema and induration but significantly fewer reactions of all types than vaccines without aluminium. In older children aged 10–16 years, no associations between exposure to aluminium and onset of local induration, swelling, or raised temperature were seen, but there was an association with local pain lasting up to 14 days. Overall, the authors found no evidence that aluminium salts cause any serious or long-term adverse events and came to what might be seen as the rather dogmatic recommendation that there need be no further research on this topic. As to suggestions by some that aluminium salts in currently licensed vaccines be replaced, the reviewers gave the timely reminder that the introduction a new adjuvant(s) would require thorough investigation before approvals could be given and, at present, there are no obvious replacement adjuvants.

Macrophagic myofasciitis (MMF), sometimes categorized as an autoimmune/inflammatory syndrome induced by adjuvants, is a rare systemic inflammatory disease characterized by chronic fatigue, diffuse myalgias, cognitive dysfunction, and postvaccination granulomas. Associated symptoms include muscle and generalized weakness, joint pain, and depression. In ~20 % of patients there is an association with autoimmune disease, for example, multiple sclerosis and autoimmune necrotizing myopathy. The syndrome appears to be triggered in genetically predisposed patients, in particular in subjects with the HLA-DRB1*01 allele, by parenteral exposure to aluminium (usually aluminium hydroxide) as adjuvant. Manifestation of the disease may occur up to 10 years after vaccination. Histology of biopsies from muscle injection sites shows granulomas with Schiff-positive infiltration of macrophages with inclusions containing alum crystals. The World Health Organization Vaccine Advisory Committee has defined MMF as "a predisposed subset of individuals with an impaired ability to clear aluminium from the deltoid muscle after vaccination;" however, the use of alternative adjuvants for vaccines was deemed to be unwarranted.

Although there has been interest and speculation on the likely association between the onset of dermatomyositis/polymyositis and vaccine immunization in some subjects with a certain genetic predisposition, retrospective and epidemiological studies have so far failed to identify such an association and no increase in the incidence of these disorders has been observed after vaccination programs involving large numbers. This subject can be followed in greater detail in "Further Reading" at the end of this chapter.

Overall, it can be said that while identification and purification of antigens has seen improvements in the efficacy and safety of vaccines, research aimed at improving vaccine adjuvants has marked time with few improved introductions and changes over half a century. This may be at the point of change with recent initiatives by the National Institutes of Health to support adjuvant research and development. Although existing adjuvants have a good long-term safety record, improved immunogenicity, fewer adverse vaccine reactions, and understanding the relationship between adjuvants and rare reactions, such as MMF and narcolepsy, are aims worth pursuing.

Reactions to Other Vaccine Additives

Human papillomavirus and hepatitis B vaccines may contain residual traces of *yeast proteins* left over from cell cultures, but yeast-associated anaphylaxis after vaccination appears to be a rare event. There is at least one report of a type I hypersensitivity reaction together with a pronounced late phase reaction following a first dose of yeast-derived (recombinant) hepatitis B vaccine in an adult who claimed to be intolerant to yeast and who had a positive skin prick test reaction to baker's yeast *Saccharomyces cerevisiae.* A higher than expected rate of apparent anaphylaxis, 2.6 cases per 100,000 doses, was seen in Australian children following vaccination with the quadrivalent human papillomavirus vaccine. In the United States, 15 cases of anaphylaxis/anaphylactoid reactions following human papillomavirus vaccination were reported to the Vaccine Adverse Events Reporting System in 2007 after distribution of over 13 million doses by the end of that year. This rate of about one case per one million vaccinations is consistent with the incidences of anaphylaxis recorded with other vaccines (Sect. "'Allergy'/Adverse Reactions to Vaccines and Added Components"). Hypersensitivity reactions to human papillomavirus vaccine were the fifth most common adverse event reported to the Vaccine Adverse Events Reporting System in 2007. Skin rashes appearing within 48 h made up about one-quarter of the reports of adverse events elicited by the vaccine in 2007 in the state of New South Wales, Australia. Being a recombinant yeast-associated preparation, yeast proteins were considered as a possible allergenic source of the observed anaphylaxis in the Australian Gardasil® immunization program, but only four of the anaphylactic patients were skin tested and all proved skin test-negative to baker's yeast, Gardasil®, Cervarix®, and polysorbate 80. The authors' conclusion, namely, "reasons for the increased rate of anaphylaxis after HPV vaccine....are not clear" is not surprising since the inadequate investigations carried out on the patients could hardly be expected to yield likely clues let alone definitive answers. However, an important consequence of the national immunization program and subsequent verification of the cases of anaphylaxis led to the package insert and product information for the quadrivalent papillomavirus vaccine being updated to reflect reports of anaphylaxis and general practitioners were alerted to the possibility of anaphylaxis following papillomavirus vaccination. In summary, the warning that human papillomavirus vaccine should be avoided in yeast-allergic patients is probably prudent and, for the moment at least, should be observed, but there is at present no clear evidence that yeast proteins have, or will, provoke anaphylaxis in immunized individuals or that the vaccine poses a significant risk for yeast-sensitive subjects. In the JTFPP's 2012 practice parameters for adverse reactions to vaccines, allergy testing is recommended for individuals suspected of being allergic to yeast. *S. cerevisiae* is available commercially as a skin test solution and as a solid phase for in vitro detection of specific IgE antibodies.

Dextrans, polysaccharides of varying chain length, are composed of *D*-glucose units linked α-(1–6) with branches linked α-(1-3). Rare dextran-induced anaphylactic reactions are known to occur and range in severity from mild erythema to death. Antidextran IgG antibodies cross the placenta and the occurrence of neurologic impairment and death following antidextran administration means that dextran administration should normally be avoided in pregnant women. The most likely mechanism of severe anaphylaxis-like reactions to dextrans is considered to be immune complex formation with IgG antibodies, complement activation, and release of anaphylatoxin. Dextran-specific IgG antibody hypersensitivity responses to MMR and BCG vaccines have been described leading to the withdrawal of some products from the market. Dextran is also occasionally found in some other vaccine preparations such as rotavirus vaccine.

Some vaccines such as MMR formulations, influenza, and polio may contain traces of the antibiotics *neomycin*, gentamycin, polymyxin B, or streptomycin added to counteract microbial contamination during production. The aminoglycoside antibiotics neomycin and streptomycin are known to occasionally induce hypersensitivities although the two compounds share no structural similarities. Reports of anaphylaxis to streptomycin go back over 50 years and with its decline in therapeutic usage, there are now only rare instances of reactions to the drug when it is present in foods, culture media, and during vitro fertilization and immunotherapy procedures. By contrast, neomycin, present in several vaccines, has consistently ranked in the top 10 % of contact allergens causing delayed eczematous contact dermatitis and the antibiotic is also known to provoke generalized reactions such as exfoliative dermatitis and erythroderma. There appears to be, however, only a single report of a hypersensitivity reaction caused by neomycin present in a vaccine. The reaction was of the immediate type in a patient who had previously experienced a maculopapular rash and local erythema to the topical application of the drug on two separate occasions. Within 5 min of the subcutaneous administration of 0.05 mL of MMR vaccine, the 6-year-old patient experienced shortness of breath, wheezing, and tachycardia. No investigations for the presence of IgE antibodies to neomycin or the vaccine were undertaken. There appears to be no definitive reports of any adverse reactions to streptomycin or any other antibiotic present in any vaccine.

Thimerosal (thiomersal; merthiolate), an organomercurial bacteriostatic/bactericide, has long been added as a preservative to vaccines although its use was controversial and it has been phased out of vaccines (with the exception of inactivated influenza vaccine) used for young children in a number of countries including states of the European Union and the United States. There is doubt that the fears surrounding thimerosal in vaccines are justified. Thimerosal is, however, a well-known contact allergen with a relatively high incidence of sensitivity reported in some counties, an incidence often attributed to its use in vaccines. In an early study of thimerosal sensitivity and vaccination reactions in the United Kingdom, delayed hypersensitivity to the agent was found in 1 % of individuals. In further studies of 50 patients with positive patch tests, no association was found between thimerosal hypersensitivity and an increased risk of vaccination reactions. A high incidence of thimerosal sensitivity in Austria demonstrated by patch test results was attributed to recent immunization with vaccines containing the antimicrobial but in a prospective study, no adverse events were seen when 12 thimerosal-sensitized individuals received a follow-up injection. The investigators concluded that although sensitization had probably occurred via vaccine exposure, thimerosal given intramuscularly as a follow-up dose did not elicit clinical symptoms and a positive patch test to thimerosal is not a contraindication to intramuscular immunization with vaccines containing the preservative. In another example of a relatively small number of documented sensitivities to thimerosal, patch testing confirmed a T-cell-mediated reaction to the bacteriostatic in a patient who developed a generalized maculopapular eruption following immunization with an influenza vaccine containing thimerosal.

Phenoxyethanol and formaldehyde have been increasingly used as preservatives in vaccines and both have been shown to trigger adverse reactions, generally nonimmediate in their occurrence and resembling delayed hypersensitivity responses in appearance. *Phenoxyethanol* is recorded as provoking a generalized eczematous reactions in a young child following immunization with diphtheria, tetanus, pertussis vaccine, and *formaldehyde*, claimed by some to be a major contact allergen affecting 3.2–6.3 % of people, has been implicated in cases of contact dermatitis and maculopapular rash following vaccinations. In one case, an adult female developed itchy, erythematous, and vesicular eruptions of the hands 1 day after a hepatitis B injection and this reoccurred 3–4 h after the second injection in the course.

Although not an additive to vaccine formulations, natural rubber or *latex* in vaccine containers either as stoppers or syringe plungers theoretically pose a risk for patients allergic to latex. A review a decade ago of the Vaccine Adverse Events Reporting System, which contains more than 160,000 reports of adverse events following vaccine administration, revealed only 28 cases of possible immediate-type hypersensitivity reactions in immunized patients who had a history of latex allergy. The investigators concluded that given the large numbers of vaccinations carried out each year in the US, the risk of an allergic reaction to a natural rubber latex contaminant in a vaccine appears to be very small.

Cutaneous Reactions to Vaccinations

Local nonspecific reactions with erythema, soreness, and sometimes swelling and itching are the most common cutaneous reactions seen after vaccination. Some rare, generalized hypersensitivity reactions and dermatologic conditions may also occur.

Injection Site Reactions

Livedoid dermatits with necrosis, or Nicolau syndrome, sometimes called embolia cutis medicamentosa, is a rare iatrogenic cutaneous reaction that begins with immediate intense pain and swelling followed by bluish skin discoloration (Fig. 11.1), progressing to livedoid erythema with hemorrhagic areas and finally necrosis that may leave scarring. The reaction usually occurs after intramuscular injections of drugs including vaccines such as diphtheria, tetanus, pertussis, polio, influenza, *H. influenzae*, and hepatitis B. The mechanism is not understood but vasospasm resulting from needle prick, the injected drug itself, embolization of the injected material, or local pressure around the vessel have been suggested.

The best-known injection site reaction is that caused intentionally by introducing the smallpox vaccine into superficial scratches in the skin, a process originally termed "variolation." This is done with the aid of a bifurcated needle previously dipped into the vaccine solution. Within the next few days, a local inflammatory reaction is clearly visible and, as immunity develops over the next few weeks, the reaction progresses (Fig. 11.2). Examples of injection site reactions to different vaccines are nodules and granulomas seen with hepatitis B vaccine, vesicles and ery-

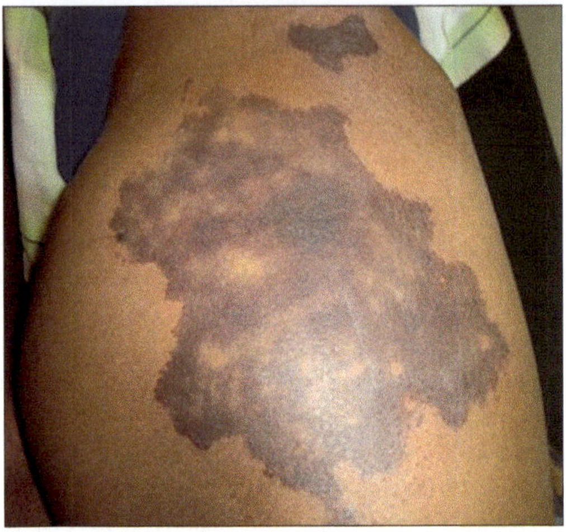

Fig. 11.1 An example of Nicolau syndrome showing bluish discoloration of the skin on a patient's hip after an intramuscular injection. The photograph shows a well-defined large patch, and a smaller patch, each with reticulate pigmentation. Reproduced from Nischal KC, et al. Nicolau syndrome: An iatrogenic cutaneous necrosis. J Cutan Aesthet Surg. 2009;2:92-5, an Open Access article distributed under the terms of the Creative Commons Attribution License

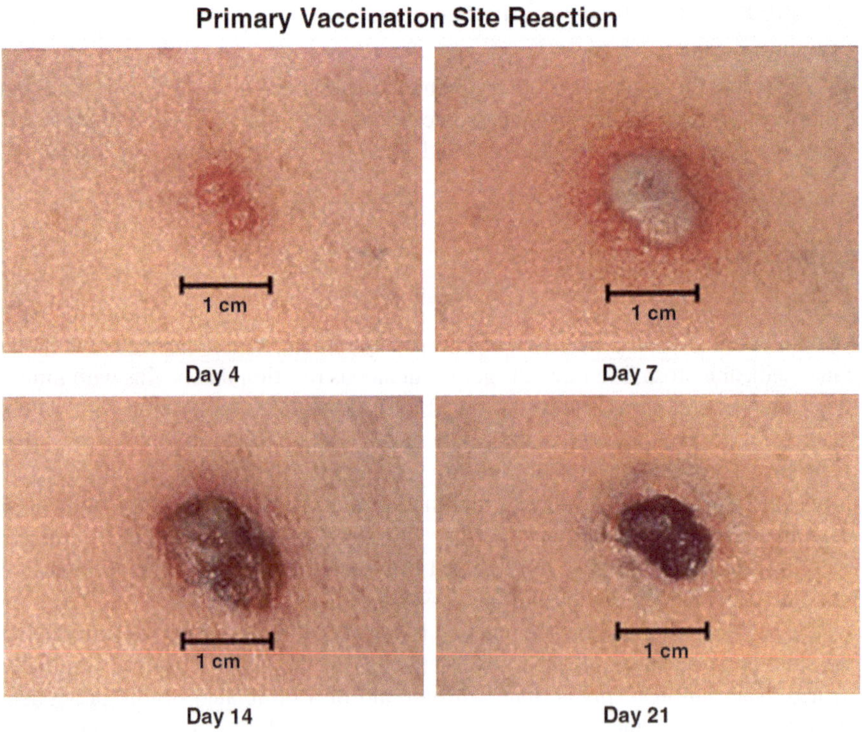

Fig. 11.2 Primary vaccination site reaction and reaction progression from papule to scab following smallpox vaccination. Sometimes still referred to as variolation, smallpox inoculation is carried out by introducing the vaccine into superficial scratches in the skin with the aid of a bifurcated needle previously dipped into the vaccine solution. Image from the Public Health Image Library, Centers for Disease Control and Prevention

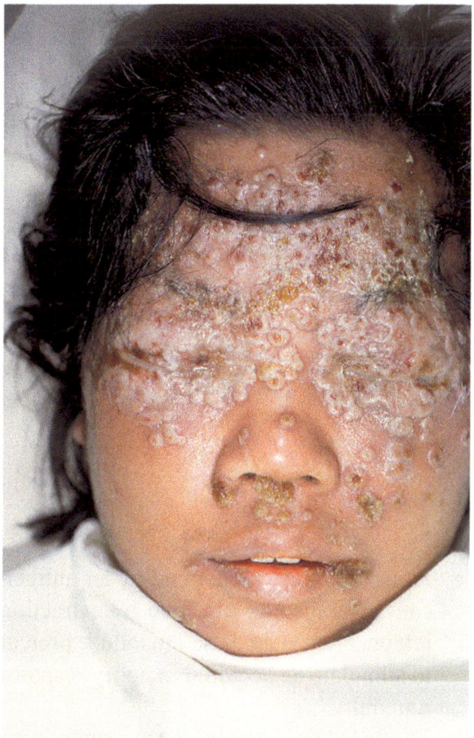

Fig. 11.3 Adult female with eczema vaccinatum contacted from an immunized child. The patient had a history of atopic dermatitis. Image from the Public Health Image Library, Centers for Disease Control and Prevention. Contact providers A.W. Mathies and J. Leedom; Photograph credit A. W. Mathies

thema mimicking chickenpox after varicella vaccine, and three different reactions after smallpox inoculation: progressive vaccinia, generalized vaccinia, and eczema vaccinatum (Fig. 11.3). Eczema vaccinatum, typically occurring in a patient with a history of atopic dermatitis, is characterized by disseminated painful vesicles while patients with generalized vaccinia develop vesicular, pustular, or papular eruptions at distant sites. Progressive vaccinia, characterized by failure to heal within a fortnight and an ulcer with central necrosis and eschar formation, is a rare adverse effect usually seen in immunocompromised individuals.

Cutaneous Hypersensitivity Reactions to Vaccines

The type IV cutaneous hypersensitivity, erythema multiforme has been reported in association with smallpox, human papilloma virus, varicella, hepatitis B, pneumococcal, meningococcal, influenza, diphtheria–pertussis–tetanus, *H. influenzae*, and measles–mumps–rubella vaccines. The more severe blistering toxidermia, Stevens–Johnson syndrome, has developed after measles–mumps–rubella vaccine and there are numerous reports of urticaria, sometimes with angioedema, after the same vaccine. Urticaria, angioedema, and morbilliform eruptions have been associated with human papilloma virus vaccine.

Other Dermatologic Reactions

There are a number of reports of adverse dermatologic events apparently triggered by vaccination, but it often remains unclear if a direct association exists between the vaccine and the skin reaction or if coincidence alone was responsible. Some of the dermatologic conditions include bullous pemphigoid and Gianotti–Crosti syndrome (infantile papular acrodermatitis) seen after DTP, influenza, and measles–mumps–rubella vaccines; granuloma annulare after DTP and hepatitis B vaccines; lichenoid-like eruptions after hepatitis B; and folliculitis associated with smallpox vaccine; Sweet's syndrome; and, rarely, some immune-based diseases such as cutaneous leukocytoclastic vasculitis, Henoch–Schönlein purpura, and systemic lupus erythematosus following influenza vaccination.

Summary

- A vaccine may be defined as an antigenic biological preparation of a disease causative agent(s) obtained by extraction, modification, synthesis, or by being left unchanged, and used to induce protection by challenging the immune system to stimulate an acquired memory response in the form of long-lasting protective immunity.
- Vaccines are generally given prophylactically although antitoxins such as those prepared from bacteria causing tetanus, diphtheria, and botulism are used as passively administered antibodies to directly treat existing conditions.
- Efforts to overcome microbial virulency and disease induction while achieving protective immunity are based on two empirical approaches: attenuation of organisms to reduce pathogenicity and production of vaccines using killed organisms.
- Attenuation has been achieved in a number of different ways including altering in vitro growth conditions (e.g., temperature and anaerobic conditions) of bacteria and viruses, particularly polio, measles, mumps, rubella, and varicella and passaging the pathogen through a normally foreign host such as embryonated eggs, tissue culture, and live animals.
- Live attenuated vaccines are generally more potent then killed vaccines since live organisms usually elicit more effector mechanisms including the activation of CD4 and CD8 cytotoxic T cells.
- A potential advantage of killed vaccines resides in the possibility of a live vaccine causing a lethal systemic infection in immunosuppressed or immunoglobulin-deficient individuals. On the other hand, killed vaccines cannot generate killer T-cell responses and are not a realistic option for some diseases.
- To overcome some of the problems and risks of live vaccines where there are many other antigens in addition to the desired protective antigen(s), vaccination is often carried out with isolated protective antigens, a procedure called subunit

vaccination. Subunit vaccines generally consist of well-defined proteins such as tetanus and diphtheria exotoxins that are chemically inactivated to produce toxoids.

- Findings, especially with *Bordetella pertussis*, showed that acellular vaccines are generally safer than vaccines formulated from whole organisms. Acellular pertussis vaccines are proving as effective as the whole-cell vaccines but without the side effects of the latter preparations.
- Many bacteria, including *Neisseria meningitidis*, *Streptococcus pneumoniae*, and *Haemophilus influenzae*, have a dense surface distribution of glycans often in the form of an outer capsular polysaccharide that is species- and type-specific. Carbohydrates often tend to be poorly immunogenic and anticarbohydrate antibodies typically show low affinity. These problems can be overcome by forming so-called conjugate vaccines whereby the bacterial glycan structures are chemically coupled to suitable protein carriers that can act as a source of peptides recognizable by T cells. Such conjugated vaccines have been developed for a number of important vaccines including *H. influenzae* type b and some *N. meningitidis* serotypes by coupling their carbohydrate antigens to tetanus toxoid or the nontoxic variant if diphtheria toxin, CRM19.
- DNA vaccines, sometimes referred to as third generation vaccines, contain DNA encoding a pathogen antigen that, after injection, results in production of the antigen by the transfected cells of the host before transfer to dendritic cells for presentation to T cells.
- In the United States, the list of vaccines required for infants and children generally includes not only tetanus, diphtheria, polio, and whooping cough but also measles, mumps, rubella, and chicken pox (varicella). In recent years, vaccines have become available for *H. influenzae*, childhood diarrhea caused by rotavirus, *N. meningitidis*, pneumococcal infections, and human papillomavirus.
- Forty six vaccine preparations are currently licensed for therapy by the FDA.
- Combination vaccines offering protection against six diseases, diphtheria, tetanus, pertussis, hepatitis B, poliomyelitis, and *H. influenzae* are available in the European Union.
- The most commonly recurring adverse events listed for vaccines are injection site reactions and hypersensitivities/allergic reactions with particular emphasis on the possibility of potentially life-threatening anaphylaxis. Injection site reactions, often incorrectly labeled as allergic, may involve pain, local swelling, redness, edema, limb swelling, subcutaneous nodules, local eczematous lesions, and nevi associated with hypertrichosis.
- Other warnings/precautions highlight the frequent occurrence of syncope following inoculation; a reduced or poor response in patients with reduced immunocompetence; the risk of severe infection with the immunizing agent in immunosuppressed/immunodeficient patients given live vaccines; possible induction of Guillain–Barré syndrome; the need for extra care with children at risk of seizures; the need to defer immunization in cases of concurrent illness; and a risk of apnea in premature infants. Pregnant women should not be given live vaccines but inactivated influenza virus, tetanus, and hepatitis B vaccines may be given.

- A recent analysis of 23 million vaccinated people in the US revealed that the 2009 influenza H1N1 vaccines were associated with a small risk of Guillain-Barré syndrome.
- Two vaccines, BCG live for intravesical use and smallpox (vaccinia) live, carry FDA boxed warnings, the former for the risk of transmission of live, infectious bacteria and vaccinia for risks of cardiac (myocarditis and/or pericarditis), CNS (encephalitis and encephalomyelitis), and severe cutaneous events (erythema multiforme, and Stevens–Johnson syndrome).
- The list of commonly recurring events for vaccines includes syncope, irritability, malaise, loss of appetite, fever, arthralgia, drowsiness, headache, dizziness, somnolence, and rash.
- Although some vaccines in particular, and even vaccination in general, have been claimed to confer long-term adverse effects such as autism, atopy, and multiple sclerosis, no evidence has been forthcoming to support such associations.
- True hypersensitivities following vaccination are known and may be seen as rare immediate reactions occurring within 1 h of administration. Reactions manifest as urticaria, angioedema, wheezing, rhinitis, hypotension, and in severe instances, bronchospasm and cardiovascular collapse. Anaphylaxis to toxoids is well known although the incidence has fallen to less than 1 case per 10,000 administrations with the introduction of highly purified preparations.
- Urticaria, angioedema, and anaphylaxis have been attributed to *B. pertussis* antigens but although IgE antibodies to the bacteria have been detected following immunization and IgE levels correlate with IgG responses, this has been interpreted as a reflection of the immunogenicity of the organism rather than its allergenicity.
- As well as the microbial components of vaccines, other additives, namely preservatives, bacteriostatics, stabilizers, and adjuvants have been suspected and occasionally implicated in adverse effects. Gelatin and egg proteins are the most commonly implicated nonmicrobial additives, but other suspected added components have included yeast proteins, dextrans, thimerosal, formaldehyde, phenoxyethanol, antibiotics, and adjuvants, often aluminium hydroxide or aluminium phosphate.
- The risk of a reaction to egg protein in influenza vaccine is similar in patients with positive or negative skin tests leading to the conclusion that skin testing with influenza vaccine is of no value in assessing egg-allergic patients for vaccine sensitivity.
- For patients with an allergy to egg, the risk of an adverse reaction to inactivated influenza virus vaccine is so low that vaccination in an allergist's surgery is not necessary.
- Recombinant influenza preparations have had a major bearing on recommendations for egg-allergic patients receiving influenza vaccination. Two new preparations of trivalent influenza virus vaccine types A and B, Flublok® and Flucelvax®, are composed of recombinant proteins not grown in eggs and both have been approved for patients 18 years and older.

- Allergic reactions to gelatin have occurred following injection of vaccines containing gelatin as a heat stabilizer. Anaphylaxis with antigelatin IgE antibodies to measles–mumps–rubella vaccines has been found in patients who experienced an anaphylactic reaction in response to the vaccine. Gelatin-induced immediate reactions in the form of both IgE antibody-mediated systemic reactions, generalized urticaria after varicella vaccines, and systemic immediate reactions to gelatin following immunization with Japanese encephalitis vaccines are known.
- Used as an immunologic adjuvant in some vaccines since 1927, aluminium salts have become the standard adjuvant in vaccines such as diphtheria, tetanus, pertussis, *H. influenzae* type b, pneumococcus, and hepatitis A and B. Aluminium salts are added to vaccines as alum (potassium aluminium sulfate), aluminium sulfate, aluminium phosphate, or aluminium hydroxide.
- Itching nodules or granulomas at the injection site after vaccination with aluminium-adsorbed vaccines have been known for over 50 years but remain rare events given the large numbers of immunizations administered worldwide. Reactions to aluminium adjuvants seem to be seen most often after injection of aluminium hydroxide as adjuvant in tetanus, diphtheria–tetanus, diphtheria–tetanus–pertussis (whole cell), diphtheria–polio, DTP (acellular)–polio, hepatitis B, and influenza vaccines.
- In a systematic review with meta-analysis, vaccines against diphtheria, tetanus, and pertussis, alone or in combination, were compared with the same vaccines either without or with aluminium. In young children up to 18 months, vaccines containing aluminium hydroxide caused significantly more erythema and induration but significantly fewer reactions of all types than vaccines without aluminium. In older children aged 10–16 years, no associations between exposure to aluminium and onset of local induration, swelling, or raised temperature were seen, but there was an association with local pain lasting up to 14 days. Overall, no evidence was found that aluminium salts cause any serious or long-term adverse events.
- Macrophagic myofasciitis (MMF), a rare systemic inflammatory disease characterized by chronic fatigue, diffuse myalgias, cognitive dysfunction, and postvaccination granulomas, is sometimes categorized as an autoimmune/inflammatory syndrome induced by adjuvants. The syndrome appears to be triggered in genetically predisposed patients, in particular in subjects with the HLA-DRB1*01 allele, by parenteral exposure to aluminium (usually aluminium hydroxide) as an adjuvant. Histology of biopsies from muscle injection sites shows granulomas with Schiff-positive infiltration of macrophages with inclusions containing alum crystals.
- The warning that human papillomavirus vaccine should be avoided in yeast-allergic patients is probably prudent and, for the moment at least, should be observed, but there is at present no clear evidence that yeast proteins have, or will, provoke anaphylaxis in immunized individuals. The vaccine poses a minimal risk for yeast-sensitive subjects.
- Dextran-specific IgG antibody hypersensitivity responses to MMR and BCG vaccines have been described, leading to the withdrawal of some products from

the market. Dextran is occasionally found in some other vaccine preparations such as rotavirus vaccine.

- Vaccines, such as MMR formulations, influenza, and polio, may contain traces of the antibiotics neomycin, gentamycin, polymyxin B, or streptomycin. Neomycin, present in several vaccines, has consistently ranked in the top 10 % of contact allergens causing delayed eczematous contact dermatitis. Neomycin is also known to provoke generalized reactions such as exfoliative dermatitis and erythroderma. However, there appears, to be only a single report of a hypersensitivity reaction caused by neomycin present in a vaccine.
- Thimerosal, an organomercurial bacteriostatic/bactericide, has long been added as a preservative to vaccines although its use was controversial and it has been phased out of vaccines (with the exception of inactivated influenza vaccine) used for young children in a number of countries including states of the European Union and the United States. There is doubt that the fears surrounding thimerosal in vaccines are justified.
- Phenoxyethanol and formaldehyde have been increasingly used as preservatives in vaccines and both have rarely been shown to trigger adverse reactions, generally nonimmediate in their occurrence and resembling delayed hypersensitivities in appearance.
- Natural rubber or latex in vaccine containers, either as stoppers or syringe plungers, theoretically pose a risk for patients allergic to latex, but the risk of an allergic reaction to a natural rubber latex contaminant in a vaccine appears to be very small.
- Injection site reactions with erythema and pain are the most common cutaneous reactions seen after vaccination. Some rare, generalized hypersensitivity reactions and dermatologic conditions may also occur.
- A range of different vaccines may induce type IV hypersensitivities including erythema multiforme. The toxidermia Stevens–Johnson syndrome, and urticaria, sometimes with angioedema, have been seen after measles–mumps–rubella vaccine. Urticaria, angioedema, and morbilliform eruptions have been associated with human papilloma virus vaccine.
- Dermatologic conditions following vaccine injection include bullous pemphigoid, Gianotti–Crosti syndrome (infantile papular acrodermatitis), granuloma annulare, folliculitis, Sweet's syndrome, cutaneous leukocytoclastic vasculitis, and lichen planus-like eruptions.

Further Reading

Centers for Disease Control and Prevention. Recommended adult immunization schedule—United States—2015. Available at: http://www.cdc.gov/vaccines/schedules/downloads/adult/adult-schedule.pdf. Accessed 1 Sept 2015.

Centers for Disease Control and Prevention. Updated recommendations for use of tetanus toxoid, reduced diphtheria toxoid and acellular pertussis (Tdap) vaccine from the Advisory Committee on Immunization Practices, 2010. Morb Mortal Wkly Rep (MMWR). 2011;60:13–5.

Centers for Disease Control and Prevention. Vaccines and immunizations. Birth-18 years & "catch-up" immunization schedules. United States 2015. Available at: http://www.cdc.gov/vaccines/schedules/hcp/child-adolescent.html. Accessed 1 Sept 2015.

Food and Drug Administration. Thimerosal in vaccines. Available at: http://www.fda.gov/BiologicsBloodVaccines/SafetyAvailability/VaccineSafety/UCM096228. Accessed 24 Nov 2015

Goulleret N, Mauvisseau E, Essevaz-Roulet M, et al. Safety profile of live varicella virus vaccine (Oka/Merck): five-year results of the European Varicella Zoster Virus Identification Program (EU VZVIP). Vaccine. 2010;28:5878–82.

Grohskopf LA, Sokolow LZ, Olsen SJ, et al. Prevention and control of influenza with vaccines: recommendations of the Advisory Committee on Immunization Practices, United States, 2015-16 season. Centers for Disease Control and Prevention. Morb Mortal Wkly Rep (MMWR). 2015;64:818–25.

Hamborsky J, Kroger A, Wolfe S, editors. Centers for Disease Control and Prevention. Epidemiology and prevention of vaccine-preventable diseases. 13th ed. Washington, DC: Public Health Foundation; 2015.

Kelso JM, Greenhawt MJ, Li JT. Adverse reactions to vaccines practice parameter 2012 update. J Allergy Clin Immunol. 2012;130:25–43.

Madsen KM, Hviid A, Vestergaard M, et al. A population-based study of measles, mumps, and rubella vaccination and autism. N Engl J Med. 2002;347:1477–82.

McNeil MM, Weintraub ES, Duffy JD, et al. Risk of anaphylaxis after vaccination in children and adults. J Allergy Clin Immunol. 2016;137:868–78. doi:/10.1016/j.jaci.2015.07.048.

Nagao M, Fujisawa T, Ihara T, et al. Highly increased levels of IgE antibodies to vaccine components in children with influenza-associated anaphylaxis. J Allergy Clin Immunol. 2016;137:861–7. doi:10.1016/j.jaci.2015.08.001.

Plotkin S, Orenstein W, Offit P. Vaccines. 6th ed. Philadelphia: Saunders; 2012.

Salmon DA, Proschan M, Forshee R, et al. Association between Guillain-Barré syndrome and influenza A (H1N1) 2009 monovalent inactivated vaccines in the USA; a meta-analysis. Lancet 2013;381:1461–8.

Shimabukuro TT, Nguyen M, Martin D, et al. Safety monitoring in the vaccine adverse event reporting system (VAERS). Vaccine 2015;33:4398–405.

Sivadon-Tardy V, Orlikowski D, Porcher R, et al. Guillain-Barre syndrome and influenza virus infection. Clin Infect Dis. 2009;48:48–56.

Stübgen J-P. A review on the association between inflammatory myopathies and vaccination. Autoimmun Rev. 2014;13:31–9.

Vaccine Adverse Event Reporting System (VAERS). Available at: vaers.hhs.gov. Accessed 1 Sept 2015.

Chapter 12
Botulinum Neurotoxins

Botulinum Neurotoxin Serotypes

Said to be the deadliest toxin known, botulinum toxin, the cause of the human paralytic disease botulism, is a neurotoxin produced by heat-resistant spores of the obligate anaerobe *Clostridium botulinum*, a Gram-positive, motile bacillus commonly found in soil, water, and the intestinal tracts of animals. Botulism, a serious but fortunately rare, disease mainly results from ingestion of contaminated food, chiefly meat and fish, but also vegetables. Cases may also occur from intestinal colonization, particularly in infants, or from an infected wound. Cases of botulism in the USA from food contamination, chiefly home-canned vegetables, number about 20–30 per year. Infant botulism accounts for more than half of the cases of botulism while only ~/10 % are due to wound contamination.

C. botulinum demonstrates heterogeneity within the species and, in 1910, toxin from two different strains of the organism was shown to differ antigenically. By the 1970s, seven antigenically distinguishable botulinum neurotoxin (BoNT) serotypes designated A, B, C, D, E, F, and G (BoNT/A-G) had been identified. Even more diversity is seen with at least 6 of the 7 serotypes existing as a number of subtypes. For example, 7 subtypes are recognized for BoNT/A, 5 for BoNT/B, 8 for BoNT/E, and 7 for BoNT/F. These differences, reflected at the amino acid level, can affect toxin binding to different antisera and specific recognition of BoNTs in vivo. BoNT/A is the most potent serotype followed by the types B and F toxins; types A, B, E, and rarely F cause human botulism, types C and D are toxic in a number of animals and type G has not been associated with clinical symptoms. Strains of bacteria that produce two BoNT serotypes, termed bivalent strains, have also been identified. In 2013 in the laboratory of Stephen S. Arnon, California Department of Public Health, two BoNT serotypes, type B, present in greater amount, and a new serotype present in lesser amount and termed type H, were identified in a novel bivalent, proteolytic strain of *C. botulinum* recovered from an infant with botulism. Strains that produce toxins in different amounts are indicted by upper case letters for the larger amount (in this case B) and lower case letters for the smaller (in this case h). Serotype H was

B.A. Baldo, *Safety of Biologics Therapy*, DOI 10.1007/978-3-319-30472-4_12

distinguished by the long-established criterion of the mouse bioassay employing seven different monovalent antisera to the seven different BoNT serotypes A to G. Subsequently, studies involving gene sequencing, comparative genomics, and phylogenetic analyses were used to identify and characterize the BoNT/H gene (*bont/H*) and to demonstrate that it differs from the other *bont* genes. These genes (~3880 base pairs) show 34–97 % sequence similarity for the seven BoNT serotypes. In addition to its clear gene sequence differences to the known genes *bont/A-G*, BoNT/H was found to have BoNT/A-like, /B-like, and F-like epitopes.

In a curious twist to the study identifying and characterizing the eighth BoNT serotype, the decision to publish the findings was considered from the perspective of "dual use research of concern" (DURC), a view that led to the withholding from publication of the gene sequence for the toxin until the development of an antitoxin to the novel toxin. Prior to publication, the authors had detailed consultations with a number of US government agencies as well as the editors of the selected journal, the Journal of Infectious Diseases. The question of trying to balance the free flow of information and realization of all possible benefits from the discovery, against a possible public health vulnerability involving the probability of misuse, presented a difficult dilemma. Agencies involved in the deliberations on publication included appropriate divisions of the National Institute of Allergy and Infectious Diseases and National Institutes of Health; the Department of Homeland Security; Office of the Director of National Intelligence; US Army Medical Research Institute of Infectious Diseases; Centers for Disease Control and Prevention; and significantly the Biological Countermeasures Unit, FBI Weapons of Mass Destruction Directorate. As summed up by David A. Relman, Departments of Medicine, Microbiology and Immunology, Center for International Security and Cooperation, Stanford University, and a member of the National Science Advisory Board for Biosecurity: "Until anti-BoNT/H can be created, shown to be effective, and deployed, both the strain itself and the sequence of this toxin (with which recombinant protein can be easily made) pose serious risks to public health because of the unusually severe, widespread harm that could result from misuse of either."

Toxin Structure and Mechanisms of Action

The term "botulinum toxin" has been used in the literature to designate different forms of the *C. botulinum* toxic proteins including BoNT serotypes and progenitor protein complexes of toxin with other nontoxic proteins. Toxic preparations employed in therapy range in size from 150 kDa to complexed forms with molecular masses up to 900 kDa.

Structure of Botulinum Neurotoxin BoNT

Each BoNT serotype is synthesized as a single polypeptide chain progenitor toxin of molecular mass ~150 kDa. In this precursor form the protein displays relatively low neurotoxic activity. Subsequent cleavage, e.g., of BoNT/A, by clostridial or tissue

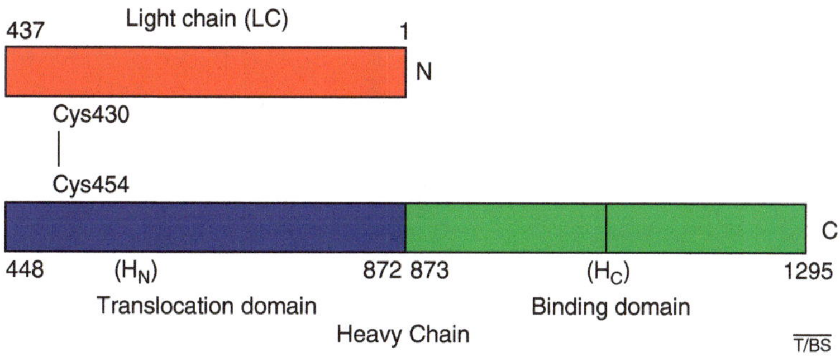

Fig. 12.1 Synthesized as a single polypeptide chain MW ~150 kDa, the botulinum neurotoxin BoNT/A is posttranslationally proteolyzed to the activated di-chain form made up of a zinc-dependent light chain (MW ~50 kDa) endopeptidase or catalytic domain (shown in *red*), and a heavy chain (MW ~100 kDa) made up of a translocation domain (H_N) (*blue*) and a binding domain (H_C) (*green*) at the C-terminal end. Compare with Fig. 12.2. The light and heavy chains are linked by a disulfide bond, Cys430–Cys454. Reproduced from Turton K, Chaddock JA, Ravi Acharya K. Botulinum and tetanus neurotoxins: structure, function and therapeutic utility. Trends Biochem Sci. 2002;27:552-8 with permission from Elsevier Limited

proteases produces a two-chain, single disulfide bond-linked (Cys430–Cys454) structure of three functionally different and structurally distinct domains: an *N*-terminal light chain (~50 kDa, amino acids 1–447) zinc-dependent endopeptidase termed the catalytic domain; a heavy chain (~100 kDa, amino acids 448–1295) consisting of the *N*-terminal ~50 kDa (amino acids 448–872) translocation domain (H_N); and the C-terminal ~50 kDa (amino acids 873–1295) receptor-binding domain (H_C) involved in binding to the target cell membrane and internalizing the toxin (Figs 12.1 and 12.2). Note that the proportion of single to di-chain toxin, that is, the amount of bacterial protease-induced cleavage, depends on the presence in the organism of the appropriate protease and the toxin serotype. BoNT/A, for example, is 95 % cleaved while BoNT/B is less cleaved and BoNT/E is not cleaved at all, leaving only the single polypeptide chain. In the latter case, toxin activity can still be released by host proteases.

Structure and Absorption of Botulinum Neurotoxin Complex

BoNT/A-H, the agents that cause botulism, are denatured by the acid environment and proteases of the gastrointestinal tract but, in practice, toxicity is retained by secretion of the neurotoxin together with up to four associated nontoxic neurotoxin-associated proteins (NAPs) (Sect. "Structure and Absorption of Botulinum Neurotoxin Complex") in the form of a progenitor toxin complex (PTC) which may be as much as 1600 times more toxic orally than the free BoNT. The size of the toxin complex released depends on the bacterial strain and hence the BoNT serotype. BoNT/A is released as 300, 500, or 900 kDa complexes; BoNT/B and C as 500 and 700 kDa complexes; BoNT/D as 300 and 500 kDa complexes; and

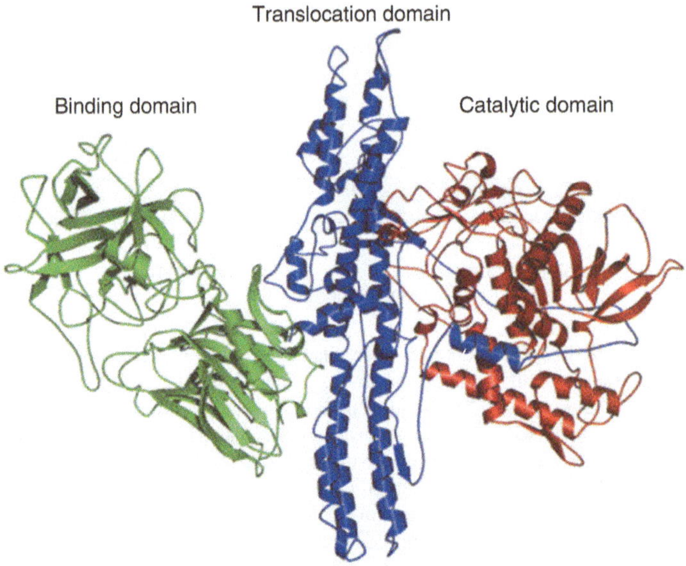

Fig. 12.2 Crystal structure of botulinum neurotoxin BoNT/A showing the three structurally distinct functional domains such that there is no contact between the catalytic (*red*) and the binding (*green*) domains which are separated by the translocation (*blue*) domain. The catalytic zinc atom is in the catalytic domain. Compare Fig. 12.1. Reproduced from Turton K, Chaddock JA, Ravi Acharya K. Botulinum and tetanus neurotoxins: structure, function and therapeutic utility. Trends Biochem Sci. 2002;27:552-8 with permission from Elsevier Limited

BONT/E and F as a ~300 kDa complex. The associated nontoxic proteins are encoded with the *bont* gene in a gene cluster that encodes the so-called nontoxic, non-hemagglutinin (NTNHA) protein as well as 3 hemagglutinins, HA17, HA33, and HA70. Together, the BoNT, NTNHA, and HA components constitute what has been termed the large PTC (L-PTC) (Fig. 12.3). Note that a second gene cluster, the orfX cluster, also occurs but the orfX proteins have yet to be characterized. The functions of the associated nontoxic proteins remain to be fully defined. The NTNHA protein protects BoNT against low pH and proteolysis in the gastrointestinal tract but it is not clear whether or not the HAs also have a protective role. In pursuing studies to better define the L-PTC structure and understand the molecular mechanisms involved in toxin shielding, delivery, and action, X-ray crystallography, single-particle electron microscopy (EM), and 3D reconstruction were recently applied to define the structure of a ~760 kDa L-PTC/A. L-PTC/A was shown to be composed of a minimally functional PTC (M-PTC) positioned above the HA complex (Fig. 12.4) in what has been described as an extended 3-blade architectural form. Recent EM studies have demonstrated the same stoichiometry and architectural form for L-PTC/A, B, and D.

Following previous work on intestinal absorption of L-PTC in Japan which demonstrated binding to cell surface glycans, binding studies with selected monosaccharides and oligosaccharides helped to confirm that absorption of BoNT/A is

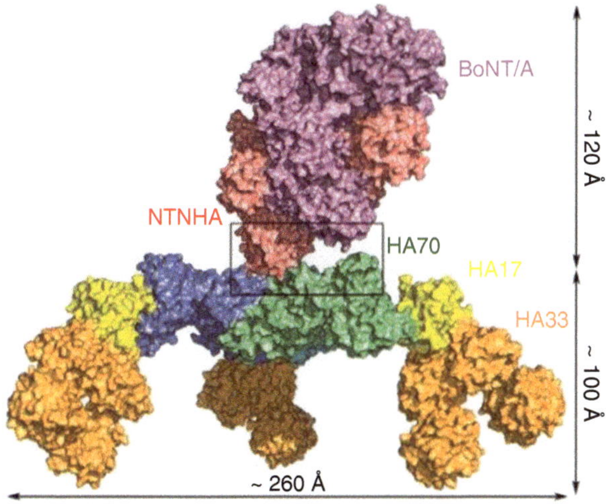

Fig. 12.3 Structural model of the botulinum neurotoxin BoNT/A large progenitor toxin complex (L-PTC) assembled from X-ray crystallographic data, single particle electron microscopy, and three-dimensional reconstruction. The L-PTC is composed of the protective minimally functional PTC (M-PTC) which sits on top of the nontoxic hemagglutinin complex (HA70, HA17, HA33). NTNHA, nontoxic, non-hemagglutinin protein. Reproduced from Lee K, Gu S, Jin L, et al. Structure of a bimodular botulinum neurotoxin complex provides insights into its oral toxicity. PloS Pathog. 9(10):e1003690. doi:10.1371/journal.ppat.1003690, an open-access article distributed under the terms of the Creative Commons Attribution Licence

effected by nine glycan-binding sites (Fig. 12.4) on the HA complex interacting with carbohydrate receptors on intestinal epithelial cells. HA33 bound D-galactose, lactose, *N*-acetyllactosamine, and isopropyl-β-D-1-thiogalactopyranoside (IPTG) with fairly high affinity while HA70 bound α-2-3-sialyllactose and α-2-6-sialyllactose with high specificity but showed low affinity for *N*-acetylneuraminic acid. No overlap between the carbohydrate recognition selectivity of HA33 and HA70 was seen. Overall, results suggested that the binding of HAs to epithelial glycans containing derivatives of D-galactose and *N*-acetylneuraminic acid is necessary for transport of BoNT across the intestinal wall. Inhibition of BONT/A oral toxicity in mice by similar carbohydrate binding with D-galactose in β-D-glycosidic linkage (as IPTG) has been suggested as a possible preventive treatment for BoNT poisoning.

Mechanism of Action at the Neuromuscular Junction

All BoNT serotypes exert their neurotoxic effect by interfering with neurotransmission. This results from neuromuscular blockade following presynaptic binding of toxin via the H_C domain which blocks motor neuron release of acetylcholine, the principal neurotransmitter at the neuromuscular junction. Once bound, BoNT

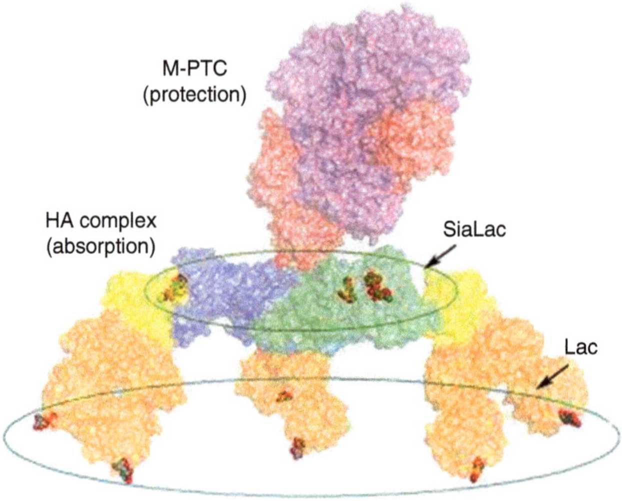

Fig. 12.4 Three-dimensional model of the BoNT/A L-PTC showing the M-PTC above the HA complex and the nine glycan binding sites containing derivatives of D-galactose (e.g., lactose) and *N*-acetylneuraminic acid, necessary for interaction with carbohydrate receptors and transport of BoNT/A across the intestinal cell wall. See also Fig. 12.3. Reproduced from Lee K, Gu S, Jin L, et al. Structure of a bimodular botulinum neurotoxin complex provides insights into its oral toxicity. PloS Pathog. 9(10):e1003690. doi:10.1371/journal.ppat.1003690, an open-access article distributed under the terms of the Creative Commons Attribution Licence

internalization is effected by insertion of the heavy chain into the synaptic vesicle membrane. This creates a transmembrane channel that aids entry to the cytosol of the light chain released by disruption of the disulfide bond. Once in the cell, light chains cleave the SNARE (acronym derived from SNAP [Soluble *N*-ethylmaleimide-sensitive factor Attachment protein] REceptor) complex required for synaptic vesicle fusion. This leads to inhibition of synaptic activity. The light chain of BoNT/A, for example, specifically cleaves the 25 kDa synaptosome-associated protein (SNAP25), blocking release of neurotransmitter while for BoNT/B, D, and F, the secretory vesicle-associated protein synaptobrevin, part of the vesicle-associated membrane protein family (VAMP), is involved as well as cellubrevin or VAMP3, a non-neuronal isoform of the VAMP/synaptobrevin family. For BoNT/C, both SNAP25 and the SNARE syntaxin are cleaved. Cellular substrates for BoNT/E and BoNT/G are SNAP25 and VAMP/synaptobrevin, respectively. BoNT/A induces long-term (months) inhibition of neurotransmitter release whereas recovery after BoNT/E is more rapid (weeks). However, inhibition of synaptic transmission by BoNT is temporary and reversible. Following synaptic blockade by BoNT, nerve sprouting at motor terminals begins and this leads to reestablishment of synaptic contact. Perisynaptic Schwann cells and some growth factors appear to be involved in this regeneration process.

Therapeutic Applications of Botulinum Neurotoxin

Although at first site the toxicity and fairly limited number of initially approved indications appear to indicate that BoNTs may not have wide therapeutic application, a moment's thought on the neuromuscular action of the toxin brings the realization of its potential application to a wide range of muscular, neurologic, gastrointestinal, urologic, ophthalmic, and oropharyngeal disorders. In fact, since the first product approval in 1989, applications to a myriad of conditions have skyrocketed and probably now exceed the most optimistic early estimates.

Nomenclature and Equivalence of Different Botulinum Toxin Preparations

Recognizing the differences between botulinum toxin products on the market, and in an attempt to reduce the potential for errors of product selection, potency, and prescription mix-ups, the FDA in August 2009 issued a Note to Correspondents advising name changes by adding what some have called "meaningless prefixes," ona-, abo-, inco-, and rima-, to each of the established USAN botulinum toxin names. The new nomenclature, former names, and trade names are set out in Table 12.1 (Note the possible relevance of this nomenclature approach to nomenclature difficulties associated with biosimilars; see Chap. 13). In announcing the changes, the FDA stated: "Changes to the established drug names to reinforce individual potencies and prevent medication errors. The potency units are specific to each botulinum toxin product, and the doses or units of biological activity cannot be compared or converted from one product to any other botulinum toxin product. The new established names reinforce these differences and the lack of interchangeability among products." The changes do help to avoid mixing up the products but, as might be expected, changes to the established nomenclature have not been universally welcomed. Considering all of these factors, clinicians need to thoroughly familiarize themselves with, and distinguish between, the available preparations.

From the above summaries, it is already clear that preparations containing BoNT may differ in a number of ways and are not interchangeable. Differences may relate to amino acid composition, serotype, the presence of associated proteins, protein and toxin concentrations, specific neurotoxin potency, mechanism of action, onset of action, duration of effect, immunogenicity, local diffusion of toxin, additives to formulations, adverse effects and, importantly, units used to measure activity and potency of different formulations cannot be easily converted one to the other (Table 12.2). There are three main products containing BoNT/A under various trade names on the market: Botox®/Vistabel® (Allergan Inc., California), Dysport®/Azzalure® (Ipsen, Slough, UK/Galderma, Paris), and Xeomin®/Bocouture (Merz Pharmaceuticals, Germany). Botulinum

Table 12.1 FDA changes[a] to established drug names of botulinum neurotoxins

Trade name	Former drug name	New generic name
Botox®	Botulinum toxin type A	OnabotulinumtoxinA
Botox® Cosmetic	Botulinum toxin type A	OnabotulinumtoxinA
Dysport®	Botulinum toxin type A	AbobotulinumtoxinA
MYOBLOC®	Botulinum toxin type B	RimabotulinumtoxinB
Xeomin®[b]	Botulinum toxin type A	IncobotulinumtoxinA

[a]Announced August 2009 for Botox®, Botox® Cosmetic, Dysport®, and MYOBLOC®
[b]Approved August 2010 for cervical dystonia

neurotoxin from the Hall strain of *C. botulinum* type A is used in each. Whereas Dysport® was originally reported to contain a mixture of the M-PTC (300 kDa) and L-PTC (600 kDa) complexes and Botox® is a 900 kDa complex with one molecule (150 kDA) of BoNT/A, Xeomin® contains only the pure BoNT/A neurotoxin. Note that there is a lack of equivalence of unit potency between BoNT serotypes, but differences exist not only between different formulations of the same serotype but also in the claims made for the relative potencies of the three FDA-approved products. For all BoNT preparations, dosage concentrations are expressed in "mouse units," a unit being the amount of toxin protein lethal by intraperitoneal injection for 50 % of female Swiss-Webster mice. While a 1:1 ratio between Botox® and Xeomin® has been claimed, 1 unit of BoNT/A as Botox is said to be approximately equivalent to 2–5 units of Dysport® and ratios between the two of up to 1:6 have been claimed. Since both the pure neurotoxin and the accompanying nontoxic proteins have the capacity to induce an immune response (see Sect. "Immunogenicity and Clinical Relevance of Botulinum Neurotoxin"), it is important to know the relative concentrations and specific potencies of the different preparations. A study designed to obtain this information was undertaken by J. Frevert in Germany using rabbit and guinea pig antisera against the 150 kDa BoNT/A neurotoxin in a sensitive sandwich ELISA. Mean concentrations of BoNT/A per 100 unit vial were found to be 0.73 ng for Botox®, 0.65 ng for Dysport®, and 0.44 ng for Xeomin®. Corresponding specific potencies of neurotoxin for the three preparations (Table 12.1) were 137, 154, and 227 units/ng, respectively (Table 12.2). The high specific activity of Xeomin® might be due to the presence of pure neurotoxin without any accompanying proteins and to retention of activity during the manufacturing process whereas with Dysport®, free neurotoxin is accompanied by approximately 30 % complexing proteins and patients given Botox® receive about 60 % more protein made up of nontoxic proteins and denatured or inactive neurotoxin. Higher antigen loads may stimulate a more pronounced immune response of neutralizing antibodies leading to treatment failure (Sect. "Immunogenicity and Clinical Relevance of Botulinum Neurotoxin").

Table 12.2 Comparison of botulinum toxin products and formulations

Points of comparison	OnabotulinumtoxinA	AbobotulinumtoxinA	IncobotulinumtoxinA	RimabotulinumtoxinA	Chinese botulinum toxin A
Trade name	Botox®; Botox® Cosmetic	Dysport®	Xeomin®	MYOBLOC®	BTXA; Prosigne
Manufacturer	Allergan Inc., Ireland	Ipsen Ltd, UK	Merz Pharmaceuticals GmbH, Germany	Solstice Neurosciences Inc., USA	Lanzhou Institute of Biologic Products, China
Clostridial strain	Hall A	Hall A	Hall A	"Bean" strain	Hall A
Toxin serotype	A	A	A	B	A
Dose (U per vial)	100 or 200	500	50 or 100	2500, 5000, or 10,000	50 or 100
Presentation	Vacuum-dried powder for reconstitution	Lyophilized powder for reconstitution	Lyophilized powder for reconstitution	Liquid	Lyophilized powder for reconstitution
Molecular composition	150 kDa neurotoxin plus complexing proteins	150 kDa neurotoxin plus complexing proteins	150 kaD neurotoxin	150 kDa neurotoxin plus complexing proteins	150 kDa neurotoxin plus complexing proteins
Total protein per 100 U	5.0 ng	0.87 ng	0.44 ng	Not known	Not known
Total neurotoxin per 100 U (proportion of total protein)	0.73 ng (15 %)	0.65 ng (75 %)	0.44 ng (100 %)	Not known	Not known
Specific neurotoxin potency	137 U/ng	154 U/ng	227 U/ng	Not known	Not known
Clinical conversion ratio to onabotulinumtoxinA	–	3:1	1:1	Not known	Not known
Protein stabilizer	0.5 mg HSA[a]	0.125 mg HSA	1 mg HSA[b]	0.5 mg/mL HSA[c]	5 mg gelatin[b]

(continued)

Table 12.2 (continued)

Points of comparison	OnabotulinumtoxinA	AbobotulinumtoxinA	IncobotulinumtoxinA	RimabotulinumtoxinA	Chinese botulinum toxin A
Other excipients	0.9 mg NaCl[a]	2.5 mg lactose	4.7 mg sucrose[b]	0.1 M NaCl, 0.01 M disodium succinate, water for injection[c,d]	25 mg dextran, 25 mg sucrose[b]

Adapted from, Benecke R. Clinical relevance of botulinum toxin immunogenicity. BioDrugs 2012;26:e1-9 with permission from Springer Science + Business Media

[a]In 100 U vial; 1.0 mg HSA (human serum albumin) and 1.8 mg NaCl in 200 U vial

[b]In both 50 and 100 U vials

[c]In 5000 U vial

[d]pH adjusted with HCl

Approved Indications of FDA Registered Botulinum Neurotoxin Preparations

Protein preparations of botulinum toxin are being increasingly used to inhibit excessive muscle spasms and in the treatment of movement disorders. Typically, however, effects of the toxin last no more than several months and patients must therefore receive repeated injections for control of symptoms. Table 12.3 lists the FDA-approved indications for the four registered botulinum toxin preparations, onabotulinumtoxinA, abobotulinumtoxinA, incobotulinumtoxinA, and rimabotulinumtoxinB. The history of regulatory approvals for botulinum toxin preparations began 25 years ago with FDA approval for what was then called BoNT/A (Botox®) for the treatment of strabismus, blepharospasm, and hemifacial spasm in children ≥12 years of age. Eleven years later, the agency extended approval of Botox® and added MYOBLOC® (Solstice Neurosciences; BoNT/B; rimabotulinumtoxinB), and then in 2009, Dysport® (abobotulinumtoxinA), and in 2010, Xeomin® (incobotulinumtoxinA), for the treatment of cervical dystonia to reduce the severity of abnormal head position and neck pain. In 2001, Botox® was approved in the United Kingdom and Canada for axillary hyperhidrosis, 3 years before the FDA did likewise. In 2001, the Canadian agency also issued approval for focal muscle spasticity and cosmetic use for the treatment of wrinkles and the following year, Botox® Cosmetic

Table 12.3 Approved indications of botulinum toxins[a]

Generic and trade names	Approved indications[a]
OnabotulinumtoxinA Botox®; Vistabel® Botox® Cosmetic[e]	• Urinary incontinence[b] • Upper limb spasticity • Cervical dystonia in adults • Headaches in migraine patients[c] • Severe hyperhidrosis • Blepharospasm associated with dystonia[d] • Strabismus[d] • Hemifacial spasm • Temporary improvement in glabellar and lateral canthal lines
AbobotulinumtoxinA Dysport®; Azzalure®	• Cervical dystonia • Glabellar lines
IncobotulinumtoxinA Xeomin®; Bocouture®	• Cervical dystonia • Blepharospasm[f] • Glabellar lines
RimabotulinumtoxinB MYOBLOC®	• Cervical dystonia

[a]Approved for human therapy by FDA or EMA or both
[b]Due to detrusor overactivity associated with neurologic conditions, e.g., multiple sclerosis, spinal cord injury
[c]Chronic migraine ≥15 days/month with headache lasting 4 h/day or longer
[d]In patients ≥12 years of age
[e]Reconstituted to give 4 U/0.1 mL
[f]In patients previously treated with onabotulinumtoxinA

received FDA approval to treat glabellar lines (Fig. 12.5). In 2010, after the renaming of the toxins, the FDA approved Botox® for treatment of spasticity of flexor muscles of the arm of adults affected by stroke, brain injury, or multiple sclerosis and for chronic migraine, and Xeomin® for blepharospasm in adults previously given Botox®. Incontinence due to detrusor overactivity in patients with a neurological condition such as spinal cord injury, multiple sclerosis, cerebral palsy, meningomyelocele, and stroke affects approximately 340,000 people in the USA. Although anticholinergic therapy is used, troublesome side effects at the dosages necessary can lead to intolerance and patient compliance becomes an issue. In phase III clinical trials, Botox® (200 units) injected into the bladder muscle of patients reduced episodes of urinary incontinence by approximately 20 episodes per week and in August 2011 the FDA granted approval for the drug to treat this condition. More recently, September 2013 saw FDA approval of Botox® for severe lateral canthal lines (crow's feet). Figure 12.6 shows a grading scale for assessing

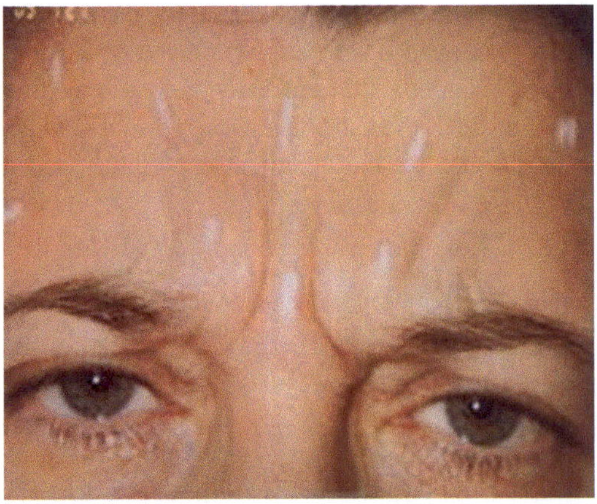

Fig. 12.5 Example of glabellar lines in a patient also assessed for brow elevation. Reproduced from Lowe NJ, Lowe P. Botulinum toxins for facial lines: A concise review. Dermatol Ther. 2012;2:14. Published with open access at Springerlink.com

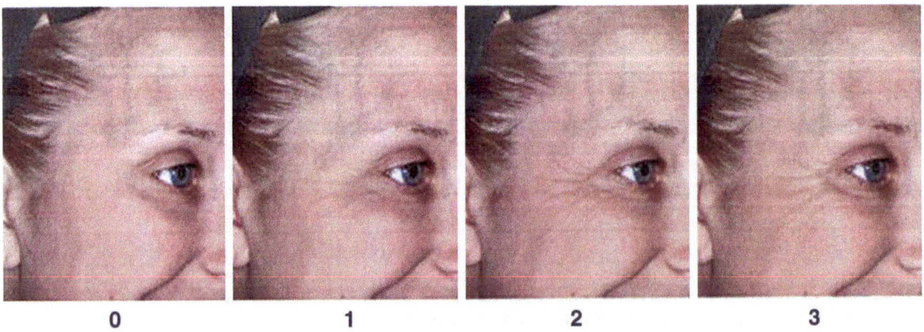

Fig. 12.6 Photonumeric grading scale for assessing wrinkle severity in the lateral canthus area. (0) no wrinkles; (1) mild wrinkles; (2) moderate wrinkles; (3) severe wrinkles. Adapted from Curruthers A, Curruthers CJ, Hardas B, et al. A validated grading scale for crow's feet. Dermatol Surg 2008;34(Suppl 2):S173-8 with permission from Wolters Kluwer Health Inc

wrinkle severity in the lateral canthus area. Figure 12.7 shows a typical example of the appearance of crow's feet when the subject is smiling and the improvement 2 weeks later showing a reduction of lateral canthal lines and a small elevation of the lateral eyebrow following injection of BoNTA.

In 2008, in an assessment of an evidence-based review of botulinum neurotoxin for the treatment of spasticity, the Therapeutics and Technology Assessment Subcommittee of the American Academy of Neurology recommended that the neurotoxin should be offered as a treatment option for spasticity in adults and children (level A). In a further evidence-based review, the same subcommittee assessed the neurotoxin for the treatment of movement disorders and concluded that it: (1) should be offered as a treatment option for cervical dystonia (level A); (2) may be offered for blepharospasm, focal upper extremity dystonia, adductor laryngeal dystonia, and upper extremity essential tremor (level B); and (3) may be considered for hemifacial spasm, focal lower limb dystonia, and motor tics (level C). In 2010, in an international consensus statement from Australia, it was concluded that dystonia of the neck can be safely and effectively reduced by BoNT/A and B, that evidence for efficacy and safety in patients with secondary dystonia in the neck is unclear, and more research is needed to answer questions about safety and efficacy in secondary spastic neck dystonia, effective adjunctive therapy, dosing, and favorable injection techniques. An updated version of its guidelines on diagnosis and treatment of primary dystonias published in 2011 by the European Federation of Neurological Societies concluded that BoNT/A is the first-line treatment for primary cranial (excluding oromandibular) or cervical dystonia, that BoNT/A is also effective in righting dystonia, and BoNT/B is not inferior to BoNT/A in cervical dystonia.

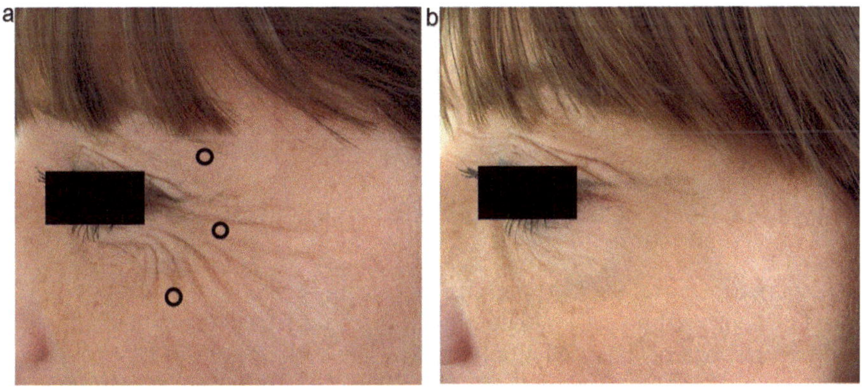

Fig. 12.7 (**a**) Example of lateral canthal lines or crow's feet when patient smiles. Small circles show points for BoNTA injections. (**b**) Same patient 2 weeks later after injection with the toxin showing a clear reduction of lines and a slight elevation of the lateral eye brow. Reproduced from Jaspers GWC, Pijpe J, Jansma J. The use of botulinum toxin type A in cosmetic facial procedures. Int J Oral Maxillofac Surg 2011;40:127-33 with permission from Elsevier Limited

Other Possible Indications for Botulinum Neurotoxins

The list of approved indications for BoNTs is already large as is the list of potential indications. The FDA does not have the authority to regulate medical practice by controlling the prescribing of registered medical practitioners and off-label use of some licensed drugs is not unusual. Given the wide utility and effectiveness of BoNT therapy, it is not surprising that these agents are being evaluated outside controlled clinical trials in a large number of off-label treatments for a range of possible new indications. Some of these indications not currently approved by the FDA but which are being investigated and reported, are set out in Table 12.4 and include pain of various type and location, cerebral palsy, extracervical spasticities, piriformis syndrome, arthrofibrosis, stuttering, bruxism, keloid, scarring, Raynaud's phenomena, and many more as well as cosmetic applications and a wide variety of gastrointestinal, neurologic, exocrine gland, urologic, ocular, ENT, oropharyngeal, and dermatologic disorders. Since the approval of BoNT/A for the treatment of glabellar and lateral canthal lines (Sect. "Approved Indications of FDA Registered Botulinum Neurotoxin Preparations"; Figs 12.5, 12.6, and 12.7), cosmetic use of the toxin has exceeded even what seemed to be the most inflated prediction of its likely use, testimony perhaps to the preoccupation with appearance, vanity, and affluence of so many. Off-label indications for botulinum toxin A have now been extended to a variety of intended cosmetic improvements (Table 12.4) including, for example, browlift, upper lip rhytides, gummy smile, masseter hypertrophy, naso-labial fold, bunny nose, and different types of dimples on the human body such as mentalis dimples or so-called pebbly chin (Fig. 12.8). In summing up the popularity and heavy cosmetic usage of Botox®, P. K. Nigam and A. Nigam (see Further Reading) stated: "its application ranges from correction of lines, creases and wrinkling all over the face, chin (see for example, neck, and chest, depressor anguli oris, nasolabial folds, mentalis, medial and lateral brow lifts, to lessen shadows on one's face and maintain a smooth outline of the jaw and cheeks from all directions…"

Adverse Effects of Botulinum Neurotoxin

As mentioned above, the large and still increasing number of therapeutic applications of registered potent botulinum toxin products has surprised and exceeded the expectations of many. Similarly, the variety and severity of adverse events induced by these preparations is also somewhat surprising to many in that the numbers involved appear to be smaller than expected.

Table 12.4 Possible new indications investigated but yet to be approved for BoNT therapy

Muscle spasms and pain	*ENT and oropharyngeal*
Neuropathic pain	Oromandibular disorders
Headaches, tension and other types	Oromandibular dystonia
Lower back pain	Bruxism
Piriformis syndrome	Masseter hypertrophy
Radicular pain	Temporomandibular joint dysfunction
Myofacial pain syndrome	Pharyngeal disorders
Refractory knee pain	Cricopharyngeal dysphagia
Tennis elbow	Larynx closure in chronic aspiration
Diabetic neuropathic pain	Laryngeal disorders
Application to trigeminal neuralgia	Vocal fold granuloma
trigger-zone to control pain	Ventricular dysphonia
Fibromyalgia	Mutational dysphonia
Gastrointestinal	Spasmodic dysphonia
Esophageal achalasia	Palatal myoclonus
Chagasic achylasia	Stuttering with glottal block
Hirschsprung disease	Oesophageal diverticulosis
Sphincter of Oddi dysfunction	*Dermatologic*
Anal fissures	Lichen simplex
Anismus	Hailey-Hailey disease
Hemorrhoids	*Cosmetic applications*
Zenker's diverticulum	Browlift
Hyperreactivity of exocrine glands	Frontalis frown
Sialorrhoea	Upper lip rhytides
Palmar and plantar hyperhidrosis	Hypertrophic platysma muscle bands
Frey's syndrome	Horizontal neck rhytides
Gustatory hyperlacrimation	Naso-labial fold
Neurologic	Bunny nose
Cerebral palsy	Pebbly chin
Parkinson's disease	*Cancer—measures to increase efficacy of therapy*
Benign cramp fasciculation	Used in treatment of:
Neurogenic tibialis anterior	Severe radiation-induced proctosis
hypertrophy	Spastic leg contractures associated with vincristine
Bladder (detrusor) sphincter	Ocular motility disturbances (diplopia) induced by
dyssynergia	plaque brachytherapy for uveal melanoma
Myokymia	BoNT/A to compensate diplopia-associated tumor
Synkinesis	resulting from sixth nerve palsy
Urologic	*Miscellaneous*
Bladder pain	Tics
Benign prostatic disease	Tourette syndrome
Pelvic pain syndrome	Keloid
Urinary retention	Scarring after surgery
Eyes	Raynaud's phenomena
Nystagmus	Arthrofibrosis
Oscillopsia	Tardive dystonia
Protective ptosis in corneal disease	Vaginismus
Sixth nerve palsy	Depression

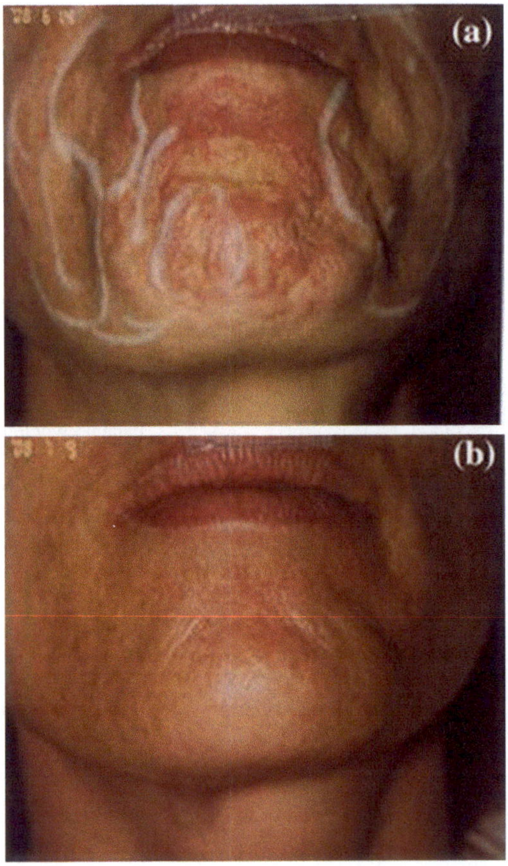

Fig. 12.8 Botulinum neurotoxin A is also administered for lines of the lower face, for example, for mentalis dimples or so-called "pebbly chin." (**a**) AbobotulinumtoxinA was administered to the mid mentalis. (**b**) Patient 4 weeks post-treatment. Reproduced from Lowe NJ, Lowe P. Botulinum toxins for facial lines: A concise review. Dermatol Ther. 2012;2:14. Published with open access at Springerlink.com

Warnings and Precautions

The FDA has issued a boxed warning for all BoNT products concerning the possibility of symptoms of botulinum toxic effects resulting from the spread of toxin from the injection site, including swallowing and breathing difficulties leading to death. Symptoms may occur in hours to weeks after injection and may be more pronounced in children treated for spasticity. Other warnings and precautions issued by regulatory agencies for Botox® (onabotulinumtoxinA) are the possibilities of (1) exacerbation of clinical effects by concomitant neuromuscular disorders; (2) corneal ulceration due to reduced blinking when administered for blepharospasm; (3) retrobulbar hemorrhage and impaired retinal circulation when given for strabismus; and (4) bronchitis and upper respiratory tract infection in patients treated for upper limb spasticity. Attention has also been drawn to the need for caution when Botox® is administered to patients with compromised respiratory function and the possibility of urinary retention in patients treated for detrusor overactivity, particularly multiple sclerosis patients. A warning of the possibility of exacerbation of clinical effects by concomitant neuromuscular disorders has also been issued for Xeomin® (incobotulinumtoxinA), Dysport® (abobotulinumtoxinA), and MYOBLOC® (rimabotulinumtoxinB). As for Botox®, caution is recommended for Xeomin®-treated patients with compromised respiratory function or dysphagia and reduced blinking in patients treated for blepharospasm may lead to corneal exposure and ulceration. The need for caution has been emphasized when Dysport® is administered to patients with surgical alterations to facial anatomy and asymmetry, injection site inflammation, ptosis, excessive dermatochalasis, dermal scarring, or thick sebaceous skin. Dysport® and MYOBLOC® contain human albumin and therefore carry what the FDA calls "an extremely remote risk" for transmission of viral diseases and Creutzfeldt–Jakob disease. Note that transmission of the latter has never been linked with albumin.

Adverse Events Following Therapeutic and Cosmetic Use of Botulinum Neurotoxin

Injected preparations of BoNT are usually well tolerated with adverse effects, when they do occur, generally mild and transient. The more common side effects include mild injection pain, transient numbness, erythema, local edema, headache, and malaise. As already pointed out, the most serious and potentially dangerous adverse event is a weakening of the muscles, especially those involved in breathing and swallowing, in the area of the injection, and occasionally at distant sites. Aspiration may result from toxin-induced dysphagia, and deaths as a consequence of severe dysphagia have been reported. Some post-marketing reports of respiratory failure following BoNT in cervical dystonia patients with respiratory disorders appear to result from weakening of neck muscles that serve as accessory muscles for ventilation. Muscle weakness usually resolves within a few weeks or months. Diffusion of toxin to the levator palpebrae superioris muscle causes temporary upper eyelid ptosis (Fig. 12.9) lasting up to about 6 weeks in approximately 1–3 % of patients. The range of adverse reactions induced by the approved preparations and listed by the FDA in its prescribing information on its access data sites is summarized in Table 12.5.

Adverse events to Botox® reported to the FDA and detailed in 1437 and 1031 reports following therapeutic and cosmetic use, respectively, were reviewed in the wake of the wide spectrum of post-marketing off-label usage of the agent and with the aim of determining the seriousness of the reported events. On- and off-label usage, indications for use, and incidences of adverse events associated with each of the indications are summarized in Table 12.6. Adverse events occurred predominately in females, with a median age of 50 years. In the therapeutic use group, there were 217 serious reports and 189 non-serious; in the cosmetic group, 36 reports covered serious events and 995 non-serious events. The 217 serious reports in the therapeutic group involved a wide range of events including 28 (12.9 %) deaths. On-label usage contributed 75 (34.1 %) events for treatment of 54 (73 %) cases of

Fig. 12.9 Daguerreotype of man with ptosis. Taken by William Bell, 1852. https://www.flickr.com/photos/22719239@N04/2260115819 (Wikimedia Commons, file 1852 ptosis patient.jpg. Uploaded by Blurpeace)

Table 12.5 Range of side effects of botulinum toxin preparations approved for human therapy[a]

Boxed warning: Distant spread of toxin effect[b]
Hypersensitivity[c]
Immunogenicity[d]
Cardiovascular[e]
Cutaneous[f]
Side effects when used for:
Detrusor overactivity: urinary retention and urinary tract infection
Spasticity: pain in extremities
Cervical dystonia: dysphagia; upper respiratory tract infection; flu syndrome; cough; rhinitis; neck and back pain; musculoskeletal pain; headache; dysphonia; dry mouth; injection site discomfort; eye disorders
Migraine: headache; neck pain
Hyperhidrosis: injection site pain and hemorrhage; non-axillary sweating; pharyngitis; flu syndrome
Blepharospasm: eyelid ptosis; dry eyes; dry mouth; diarrhea; headache; visual impairment; dyspnea; nasopharyngitis; respiratory tract infection
Glabellar and lateral canthal lines: eyelid edema and ptosis; nausea; headache; respiratory infection; flu syndrome

[a]Approved by FDA or EMA or both

[b]FDA Boxed Warning applies to all four registered botulinum toxin preparations. Symptoms are those of botulism, viz., loss of strength and muscle weakness; double vision; blurred vision; drooping eyelids; dysphonia; dysarthria; incontinence; dyspnea; dysphagia; pneumonia

[c]Includes anaphylaxis, urticaria; serum sickness, dyspnea

[d]Potential for the formation of neutralizing antibodies to the toxin, thus reducing the effectiveness of treatment

[e]Arrhythmia and myocardial infarction, sometimes fatal

[f]Skin rash including erythema multiforme and psoriasiform eruption identified during post approval use

Table 12.6 Incidences of serious[a] and non-serious adverse events[b] following administration of botulinum toxin[c] for therapeutic and cosmetic purposes

Indications for use[d]	Therapeutic use		Cosmetic use	
	Serious (*N*=217)	Non-serious (*N*=189)	Serious (*N*=36)	Non-serious (*N*=995)
	N (%)	*N* (%)	*N* (%)	*N* (%)
On-label	74(34.1)	87(46)	15(41.7)	532(53.5)
Cervical dystonia	54(73)	48(55.2)		
Blepharospasm	19(25.7)	37(42.5)		
Strabismus	2(2.7)	2(2.3)		
Facial wrinkle: glabellar			15	532
Possibly on-label	42(19.3)	24(12.7)	11(30.6)	369(37.1)
Spasticity	38(90.5)	20(83.3)		
Neck pain	3(7.1)	4(16.7)		
Facial wrinkle[e]			11	369
Off-label	98(45)	83(43.9)	14(38.9)	322(32.4)
Spasticity[f]	37(37.8)	22(26.5)		
Pain[f]	24(24.5)	14(16.9)		
Achalasia	10(10.2)	2(2.4)		
Migraine	9(9.2)	30(36.1)		
Spasmodic dysphonia	6(6.1)	4(4.8)		
Other	15(15.3)	17(20.5)		
Facial wrinkle[g]			14	322

Adapted from, Coté TR, Mohan AK, Polder JA, et al. Botulinum toxin type A injections: Adverse events reported to the US Food and Drug Administration in therapeutic and cosmetic cases. J Am Acad Dermatol 2005;53:407-15 with permission from Elsevier

[a]According to US Code of Federal Regulations 600.80 regulatory definition of serious

[b]2 patients with both therapeutic and cosmetic indications in the serious group and 6 patients with both indications in the non-serious group

[c]OnabotulinumtoxinA (Botox®)

[d]More than one indication for use for some patients

[e]Site unknown

[f]Other than neck

[g]Not glabellar

cervical dystonia, 19 (25.7 %) of blepharospasm, and 2 (2.7 %) cases of strabismus. Off-label usage representing 98 (45.2 %) of the serious reports was made up of treatments for spasticity (other than neck) 37 (37.8 %), pain (other than neck) 24 (24.5 %), achalasia 10 (10.2 %), migraine 9 (9.2 %), spasmodic dysphonia 6 (6.1 %), plus small numbers of a few other indications. Usage judged as possibly on-label accounted for 42 (19.4 %) reports with 38 (90.5 %) being for spasticity and 3 (7.1 %) neck pain. In general, non-serious reactions associated with the different indications paralleled the percentages seen with the serious events except for adverse events associated with migraine which were approximately four times higher in the non-serious reaction group. A breakdown of the 217 serious events revealed that the highest numbers of adverse responses were contributed by the respiratory (15.2 %), cardiovascular (10.1 %), nervous (23.5 %), and gastrointestinal (18.9 %) systems. Individual adverse events showing incidences above 3 % were respiratory compromise (3.7 %), pneumonia (4.1 %), flu-like syndrome (4.6 %), arrhythmia (4.1 %), muscle weakness/spasm (6 %), seizure (7.8 %), dysphagia (12 %), injection trauma (4.1 %), allergic reaction (5.1 %), and syncope (3.2 %). The most obvious differences seen in the list of non-serious adverse events were the relatively high incidences of reactions involving the eye, muscle weakness/spasm, skin and injection site reactions, and pain at the injection site. The highest incidences of both serious and non-serious reactions following cosmetic treatments were those involving the nervous system, gastrointestinal system, skin and injection site reactions, and the eye. The most common individual adverse events in the cosmetic group were focal facial paralysis, muscle weakness/spasm, dysphagia, rash, edema and pain at the injection site, ptosis, and headache. Overall conclusions were that serious adverse events were more common in the therapeutic group who received higher doses and frequently had underlying diseases; all the deaths occurred in this group; and serious events were seen less often in the cosmetic cases. The authors noted "numerous departures from the drug dose, dilution, administration, handling, storage, and site of injection advised in the approved labeling."

In relation to the cosmetic applications of anabotulinumtoxinA and with its associated warnings, precautions, and adverse events in mind, the extensive worldwide use of the agent reveals a remarkable record of safety. The large numbers of patients involved and procedures undertaken confirm the tolerability and utility of the treatments and show that when adverse reactions do occur, they are generally mild. Botox® Cosmetic was approved in 2002 after two placebo-controlled, randomized identical trials in which results obtained from 537 patients were pooled. The most frequently occurring treatment-related adverse events were headache and blepharoptosis but a striking feature of the results was the drop in incidences of reactions after the second and third injections. For example, the incidence of headache fell from 8.2 % after the first injection to 0.6 % after the second and 0.8 % after the third injection. Cases of blepharoptosis went from an incidence of 3 % after injection 1–2.2 % after injection 2 and 0.8 % after injection 3. Corresponding figures for face pain, the third most common adverse event, were 1.8, 0 and 0 %. Three double-blind, randomized trials with abobotulinumtoxinA in 2009 involving 398 treated patients found low rates of generally mild to moderately severe eye disorders, including a 2 % incidence of ptosis, which did not increase with treatment. The development of neutralizing antibodies was also low. Data from long-term safety and repeated treatments with onabotulinumtoxinA for facial aesthetic applications also showed acceptable safety profiles with low incidences of mild and transient brow and eyelid ptosis and dysphagia. A retrospective review of 945 patients administered abobotulinumtoxinA for glabellar lines over three treatment cycles, found no adverse events in 91 % of the patients. Nineteen instances of brow or eyelid ptosis were seen in 16 patients.

Immunogenicity and Clinical Relevance of Botulinum Neurotoxin

Although an immune response to botulinum toxin therapy is not normally considered to be detrimental or a significant risk factor for the patient, for example, any humoral and cellular responses to the toxin do not cross-react with endogenous antigens, there is always the possibility that therapy may be impaired and clinical improvement stalled.

Anti-Neurotoxin Antibodies

BoNTs in pure or complexed forms are proteins and this fact, plus the need for repeated injections required for many long-term therapies, means that an antibody response may occur. Such antibodies may, or may not, affect the biological activity of the toxin. Each of the commercially available botulinum toxin products except incobotulinumtoxinA (Xeomin®) contain NAPs meaning that antibodies may form to both the toxin and nontoxic proteins but no convincing studies have compared the immunogenicity of the different marketed products. Proteins involved in complexing may have an adjuvant effect, a conclusion supported by studies demonstrating that toxin complexes are more immunogenic than purified neurotoxin, so employment of purified toxin may potentially reduce the rate of secondary treatment failure. Other factors that may affect the immunogenicity of neurotoxin include the toxin source, the manufacturing process and, importantly, previous exposure of the patient. Antibodies to the toxin may be neutralizing or non-neutralizing. The former presumably inhibit the biological activity of the toxin by inhibiting interaction with its receptor. Higher protein loads, that is, antigen content, dosing frequency, and number of injections, appear to produce more neutralizing antibodies. Some studies have shown that rates of neutralizing antibodies are higher in pediatric patients. Neutralizing antibodies that recognize the heavy chain of BoNT have been recorded but antibodies to other determinants on the BoNT structure are also known. Neutralizing antibodies are thought to persist for several years, so avoidance of their formation in the first place is the best strategy. While most of the available information on antibody formation relates to BoNT/A, antibodies, including those with neutralizing capacity, to BoNT/B are known to occur and these can be found in some patients within 2 years of the commencement of treatment. In attempts to overcome treatment failure, different toxin serotypes have been employed. BoNT/B, for example, has demonstrated efficacy in the treatment of type A-resistant cervical dystonia but improvements may only be temporary with the development of antibodies to the substituted toxin. There is also the possibility that the substituted toxin may cross-react with the one originally administered meaning a rapid interaction between the second toxin and existing antibodies in the already primed patient's serum.

Detection and Measurement of Neutralizing Antibodies

ELISAs, Western blotting, fluorescent immunoassays, radioimmunoprecipitation assays, and a number of other in vitro test procedures have been utilized for the examination of antibodies to botulinum toxin. However, a common drawback with these procedures is the failure to distinguish neutralizing from non-neutralizing antibodies. Clearly therefore, it is necessary to include a clinically relevant test when examining the sera of patients with suspected antibody-induced treatment failure. A number of assays have been utilized to detect and measure neutralizing antibodies.

Mouse Protection Assay

Patient's serum is preincubated with a lethal dose of neurotoxin. If the proportion of mice dying in the test population is less than the proportion of deaths in the control population (mice injected with toxin after preincubation with normal serum), neutralizing antibodies are presumed to be present. In other words, mice survival indicates the presence of neutralizing antibodies. Expense and the length of time needed to obtain the results are disadvantages of the test.

Mouse Hemidiaphragm Assay

The assay employs a mouse hemidiaphragm with its attached phrenic nerve. After placing the tissue in an organ bath, the phrenic nerve is stimulated and the contractions and muscle twitches are recorded with a force transducer. BoNT is added and the time required for a reduction in the twitch amplitude by 50 %, called the paralysis half-time, is measured as a function of the dose of BoNT. When neutralizing antibodies to the toxin are present, paralysis is inversely proportional to the amount of antibodies in the serum being tested. A recent comparison of the mouse hemidiaphragm and protection assays showed a detection limit of 0.17 mU/mL for the former and 1 mU/mL for the latter assay, a sixfold difference in sensitivity. In fact, the clinical relevance of the hemidiaphragm assay is in question since it may be too sensitive, producing false positives. This conclusion is supported by the detection of apparently neutralizing antibodies in patients without therapy failure.

Unilateral Brow Injection Test

Although there are a number of minor variations of this test, the procedure is essentially as follows. The botulinum toxin preparation (e.g., Botox® 15–20 U or MYOBLOC® 1000 U) is injected into a corrugator supercilii (frowning muscle) of the patient with suspected toxin-neutralizing antibodies and after about 2 weeks, the patient is examined for the ability to furrow the brow. If the patient is unable to frown, the toxin has remained active, paralyzing the brow and indicating the absence of toxin-neutralizing antibodies. Retention of the ability to frown shows absence of paralysis and the presence of antibodies blocking the neuromuscular action of the toxin.

Extensor Digitorum Brevis Assay

A recent case report by Santamato and coworkers (see Further Reading for more details) is presented to illustrate the principle of the extensor digitorum brevis (EDB) test and as a good example of the occurrence of toxin-neutralizing antibodies and their relationship to different BoNT preparations with or without complexing proteins. Because of a suspicion of secondary immune resistance to BoNT/A, an adult male patient with spasticity after ischaemic stroke was assessed using the EDB assay. Amplitudes of EDB compound muscle action potential (CMAP) of both sides were recorded by transcutaneous supramaximal electrical nerve stimulation of the peroneal nerve before and after the injection of what was described as "a complexing protein-containing product" into the EDB muscle. Four weeks later, no decrease in CMAP amplitude was seen in the injected muscle and the patient was judged to be a secondary nonresponder. Months later, after treatment with Xeomin® (a preparation containing pure BoNT/A in the absence of complexing nontoxic proteins and considered to be of low immunogenicity), the injection procedures were repeated using the same dosage. At this stage, a reduction of approximately 2 points was seen in the Modified Ashworth Scale for grading spasticity. This result was supported by the EDB test which showed a reduction in CMAP (13.6 to 2.9 mV) in the injected muscle but was unchanged in the non-injected muscle. These findings were interpreted as an initial failure in treatment due to the administration of a toxin preparation containing complexing proteins and the consequent formation of neutralizing antibodies. Subsequent switching of the therapy to a purified toxin preparation with low immunogenicity led to a clinical improvement in the patient's condition with concurrent reduction in, or absence of, neutralizing antibodies confirmed by the EDB test.

Sudomotor Sweat Test

Sudomotor innervation is the stimulation and production of sweat via activation of muscarinic acetylcholine receptors. The sweat-inhibiting property of BoNT/A, an action used to test the efficacy of the toxin, may also be applied to discriminate between patients with and without antibodies to BoNT/A. In the sudomotor sweat test, toxin is injected subcutaneously into (for example) patients with a spasticity and into control subjects. Sweating is detected using iodine-starch staining and quantitation may be achieved with capacitance hygrometry. Inhibition of BoNT/A--induced sweating at the injection site occurs in the absence of neutralizing antibodies while continued sweating is seen in the presence of antitoxin antibodies.

Summary

- Botulinum toxin, the cause of the human paralytic disease botulism, is a neurotoxin produced by heat-resistant spores of the obligate anaerobe *C. botulinum*. Botulism, a rare disease, results mainly from ingestion of contaminated food, chiefly meat and fish, but also vegetables. Cases may occur from intestinal colonization, particularly in infants, and from an infected wound.
- *C. botulinum* demonstrates heterogeneity within the species. By the 1970s, seven antigenically distinguishable botulinum neurotoxin (BoNT) serotypes designated A, B, C, D, E, F, and G (BoNT/A-G) had been identified. Even more diversity is seen with at least 6 of the 7 serotypes existing as a number of subtypes. BoNT/A is the most potent serotype followed by the types B and F toxins; types A, B, E, and rarely F cause human botulism, types C and D are toxic in a number of animals, and type G has not been associated with clinical symptoms.
- In 2013, a new serotype termed type H was identified in a novel bivalent, proteolytic strain of *C. botulinum* recovered from an infant with botulism. Subsequent studies identified and characterized the BoNT/H gene (*bont/H*) and demonstrated that it differs from the other *bont* genes.
- Each BoNT serotype is synthesized as a single polypeptide chain progenitor toxin of molecular mass ~150 kDa. In this precursor form the protein displays relatively low neurotoxic activity. Subsequent cleavage by clostridial or tissue proteases produces a two-chain, single disulfide bond-linked (Cys430–Cys454) structure of three functionally different and structurally distinct domains: an *N*-terminal light chain termed the catalytic domain; a heavy chain consisting of the *N*-terminal translocation domain (H_N); and the C-terminal receptor-binding domain (H_C) involved in binding to the target cell membrane and internalizing the toxin.
- BoNT/A-H are denatured by the acid environment and proteases of the gastrointestinal tract but toxicity is retained by secretion of the neurotoxin together with up to four associated nontoxic neurotoxin-associated proteins (NAPs) in the form of a progenitor toxin complex (PTC). The size of the toxin complex released depends on the BoNT serotype. BoNT/A is released as 300, 500, or 900 kDa complexes.
- The associated nontoxic proteins are encoded with the *bont* gene in a gene cluster that encodes the nontoxic, non-hemagglutinin (NTNHA) protein as well as three hemagglutinins, HA17, HA33, and HA70. Together, the BoNT, NTNHA, and HA components constitute what has been termed the large PTC (L-PTC). The functions of the associated nontoxic proteins remain to be fully defined. The NTNHA protein protects BoNT against low pH and proteolysis in the gastrointestinal tract but it is not clear whether or not the HAs also have a protective role.
- The 3D structure of a ~760 kDa L-PTC/A was recently defined and shown to be composed of a minimally functional PTC (M-PTC) positioned above the HA complex in what has been described as an extended 3-blade architectural form.

- Absorption of BoNT/A is effected by nine glycan-binding sites on the HA complex interacting with carbohydrate receptors on intestinal epithelial cells. HA33 binds D-galactose, lactose, *N*-acetyllactosamine, and isopropyl-β-D-1-thiogalactopyranoside (IPTG) with fairly high affinity while HA70 binds α-2-3-sialyllactose and α-2-6-sialyllactose with high specificity but shows low affinity for *N*-acetylneuraminic acid. Overall, results suggest that the binding of HAs to epithelial glycans containing derivatives of D-galactose and *N*-acetylneuraminic acid is necessary for transport of BoNT across the intestinal wall.
- All BoNT serotypes exert their neurotoxic effect by neuromuscular blockade following presynaptic binding of toxin via the H_C domain which blocks motor neuron release of acetylcholine. Once bound, BoNT internalization is effected by insertion of the heavy chain into the synaptic vesicle membrane. This creates a transmembrane channel that aids entry to the cytosol of the light chain released by disruption of the disulfide bond. Once in the cell, light chains cleave the SNARE complex required for synaptic vesicle fusion. This leads to inhibition of synaptic activity. The light chain of BoNT/A, for example, specifically cleaves the 25 kDa synaptosome-associated protein (SNAP25), blocking release of neurotransmitter.
- Inhibition of synaptic transmission by BoNT is temporary and reversible. Following synaptic blockade by BoNT, nerve sprouting at motor terminals begins and this leads to reestablishment of synaptic contact.
- Preparations containing BoNT may differ in a number of ways and are not interchangeable; potency of different formulations cannot be easily converted one to the other. There are three main products containing BoNT/A under various trade names on the market: Botox®/Vistabel® (Allergan Inc., California), Dysport®/Azzalure® (Ipsen, Slough, UK/Galderma, Paris), and Xeomin®/Bocouture® (Merz Pharmaceuticals, Germany). Whereas Dysport® was originally reported to contain a mixture of the M-PTC (300 kDa) and L-PTC (600 kDa) complexes and Botox® is a 900 kDa complex with one molecule (150 kDA) of BoNT/A, Xeomin® contains only the pure BoNT/A neurotoxin.
- For all BoNT preparations, dosage concentrations are expressed in "mouse units," a unit being the amount of toxin protein lethal by intraperitoneal injection for 50 % of female Swiss-Webster mice. Mean concentrations of BoNT/A per 100 unit vial were found to be 0.73 ng for Botox®, 0.65 ng for Dysport®, and 0.44 ng for Xeomin®. Corresponding specific potencies of neurotoxin for the three preparations were 137, 154, and 227 units/ng, respectively.
- Recognizing the differences between botulinum toxin products on the market, and in an attempt to reduce the potential for errors of product selection, potency, and prescription mix-ups, the FDA in 2009 issued name changes by adding the prefixes, ona-, abo-, inco-, and rima-, to each of the former botulinum toxin names. For example, botulinum toxin type A became onabotulinumtoxinA.
- Protein preparations of botulinum toxin are being increasingly used to inhibit excessive muscle spasms and in the treatment of movement disorders. Typically, however, effects of the toxin last no more than several months and patients must therefore receive repeated injections for control of symptoms.

- The list of approved indications for botulinum toxin preparations includes urinary incontinence, upper limb spasticity, cervical dystonia, headaches in migraine patients, severe hyperhidrosis, blepharospasm associated with dystonia, strabismus, hemifacial spasm, and temporary improvement in glabellar and lateral canthal lines.
- Incontinence due to detrusor overactivity in patients with a neurological condition such as spinal cord injury, multiple sclerosis, cerebral palsy, meningomyelocele, and stroke affects approximately 340,000 people in the USA. In 2011 the FDA granted approval for Botox® for the treatment of this condition.
- BoNT therapy is being evaluated outside controlled clinical trials in a large number of off-label treatments for a range of possible new indications. Some of these indications not currently approved by the FDA include pain of various type and location, cerebral palsy, extracervical spasticities, piriformis syndrome, arthrofibrosis, stuttering, bruxism, keloid, scarring, Raynaud's phenomena, and many more as well as cosmetic applications and a wide variety of gastrointestinal, neurologic, exocrine gland, urologic, ocular, ENT, oropharyngeal, and dermatologic disorders.
- The FDA has issued a boxed warning for all BoNT products concerning the possibility of symptoms of botulinum toxic effects resulting from the spread of toxin from the injection site, including swallowing and breathing difficulties leading to death. Other warnings and precautions include (1) exacerbation of clinical effects by concomitant neuromuscular disorders; (2) corneal ulceration due to reduced blinking when administered for blepharospasm; (3) retrobulbar hemorrhage and impaired retinal circulation when given for strabismus; and (4) bronchitis and upper respiratory tract infection in patients treated for upper limb spasticity. Attention has also been drawn to the need for caution when Botox® is administered to patients with compromised respiratory function and the possibility of urinary retention in patients treated for detrusor overactivity, particularly multiple sclerosis patients.
- Adverse effects of botulinum toxins when used for: *Detrusor overactivity*: urinary retention and urinary tract infection; *Spasticity*: pain in extremities; *Cervical dystonia*: dysphagia, upper respiratory tract infection, flu syndrome, cough, rhinitis, neck and back pain, musculoskeletal pain, headache, dysphonia, dry mouth, injection site discomfort, eye disorders; *Migraine*: headache; neck pain; *Hyperhidrosis*: injection site pain and hemorrhage, non-axillary sweating, pharyngitis, flu syndrome; *Blepharospasm*: eyelid ptosis, dry eyes, dry mouth, diarrhea, headache, visual impairment, dyspnea, nasopharyngitis, respiratory tract infection; *Glabellar and lateral canthal lines*: eyelid edema and ptosis, nausea, headache, respiratory infection, flu syndrome.
- The protein nature of botulinum toxin, plus the need for repeated injections for many long-term therapies, means that an antibody response may occur. Such antibodies may, or may not, affect the biological activity of the toxin. Proteins involved in complexing may have an adjuvant effect and toxin complexes are more immunogenic than purified neurotoxin, so employment of purified toxin may potentially reduce the rate of secondary treatment failure.

- ELISAs, Western blotting, fluorescent immunoassays, radioimmunoprecipitation assays, and a number of other in vitro test procedures have been utilized for the examination of antibodies to botulinum toxin. A common drawback with these procedures is the failure to distinguish neutralizing from non-neutralizing antibodies. It is necessary to include a clinically relevant test when examining the sera of patients with suspected antibody-induced treatment failure. Such assays utilized to detect and measure neutralizing antibodies include the mouse protection assay; mouse hemidiaphragm assay; unilateral brow injection test; extensor digitorum brevis assay; and sudomotor sweat test.

Further Reading

Barash JR, Arnon SS. A novel strain of Clostridium botulinum that produces type B and Type H botulinum toxins. J Infect Dis. 2014;209:183–91.
De Boulle K, Fagien S, Sommer B, et al. Treating glabellar lines with botulinum toxin type A-hemagglutinin complex: a review of the science, the clinical data, and patient satisfaction. Clin Interv Aging. 2010;5:101–18.
Curruthers J, Curruthers A, editors. Botulinum toxin: procedures in cosmetic dermatology. 3rd ed. Amsterdam: Elsevier Health Sciences; 2013.
Dolly JO, Aoki KR. The structure and mode of action of different botulinum toxins. Eur J Neurol. 2006;13 Suppl 4:1–9.
Dover ND, Barash JR, Hill KK, et al. Molecular characterization of a novel botulinum neurotoxin type H gene. J Infect Dis. 2014;209:192–202.
Frevert J. Content of botulinum neurotoxin in Botox®/Vistabel®, Dysport®/Azzalure® and Xeomin®/Bocouture®. Drugs R D. 2010;10:67–73.
Hill KK, Smith TJ. Genetic diversity within Clostridium botulinum serotypes, botulinum neurotoxin gene clusters and toxin subtypes. Curr Top Microbiol Immunol. 2013;364:1–20.
Jancovik J, Brin MF. Botulinum toxin: historical perspective and potential new indication. Muscle Nerve Suppl. 1997;6:S129–45.
Lacy DB, Stevens RC. Sequence homology and structural analysis of the Clostridial neurotoxins. J Mol Biol. 1999;291:1091–104.
Lee K, Gu S, Jin L, et al. Structure of a bimodular botulinum neurotoxin complex provides insights into its oral toxicity. PLoS Pathog. 2013;9(10), e1003690. doi:10.1371/journal.ppat.1003690.
Nigam PK, Nigam A. Botulinum toxin. Indian J Dermatol. 2010;55:8–14.
Novak I, Campbell L, Boyce M, et al. Botulinum toxin assessment, intervention and aftercare for cervical dystonia and other causes of hypertonia of the neck: international consensus statement. Eur J Neurol. 2010;17 Suppl 2:94–108.
Oguma K, Fujinaga Y, Inoue K. Structure and function of *Clostridium botulinum* toxins. Microbiol Immunol. 1995;39:161–8.
Raffestin S, Couesnon A, Pereira Y, et al. Botulinum and tetanus neurotoxins: molecular biology, toxin gene regulation and mode of action. In: Bruggermann H, Gottschalk G, editors. Clostridia: molecular biology in the post-genomic era. Norfolk: Caister Academic Press; 2009. p. 1–28.
Santamoto A, Ranieri M, Panza F, et al. Effectiveness of switching therapy from complexing protein-containing botulinum toxin type A to a formulation with low immunogenicity in spasticity after stroke: a case report. J Rehabil Med. 2012;44:795–7.
Turton K, Chaddock JA, Ravi Acharya K. Botulinum and tetanus neurotoxins: structure, function and therapeutic utility. Trends Biochem Sci. 2002;27:552–8.
Walker TJ, Dayan SH. Comparison and overview of currently available neurotoxins. J Clin Aesthet Dermatol. 2014;7:31–9.

Chapter 13
Biosimilars

The Continuing Evolution of Biosimilars

In the words of the EMA, "a biosimilar medicine is a biological medicine that is developed to be similar to an existing biological medicine (the 'reference medicine'). Biosimilars are not the same as generics, which have simpler chemical structures and are considered to be identical to their reference medicines." The definition preferred here is that of McCamish and Woollett: "a follow-on biologic that meets extremely high standards for comparability or similarity to the originator biologic drug that is approved for use in the same indication." Biosimilars have been given a number of different names, including follow-on biologics in the USA and subsequent entry biologics in Canada, but they should not be confused with "biobetters," "biosuperiors," and "next-generation biologics" which are categories of biologic therapies that show physicochemical differences to existing marketed products in, for example, primary structure, glycosylation patterns, or added molecules such as a drug or toxin. Biosimilars may be authorized as high-quality, lower cost biologics when patents expire, that is several years after the initial approval of the reference medicine which benefits from a period of exclusivity during which biosimilars cannot be authorized. Now that patent protection of a number of recombinant biologics is drawing to a close, it is estimated that by 2020, more than $100 billion of biologic sales will be for off-patent products. It has been said that all, or most, biologics are potential reference products for biosimilars in the long term. The therapeutically active material of a biosimilar is developed to be the same as its reference medicine but, given the complexities of most biologic agents and the often complex production methods, in practice there will be differences. However, approval will only follow if any variability and differences are shown to not affect efficacy and safety. As well as being authorized to treat the same condition as the reference medicine, a biosimilar will generally be used at the same dose.

In 2004, enabling legislation was enacted in Europe for the approval of biosimilar products. Based on a comparability approach, the EMA was enabled to approve biologic products that are similar to a previously approved biologic. In 2006, the

B.A. Baldo, *Safety of Biologics Therapy*, DOI 10.1007/978-3-319-30472-4_13

EMA approved the recombinant somatropin (somatropin; human growth hormone [Chap. 7, section "Human Growth Hormone"]) preparation Omnitrope®, making it the first licensed biosimilar. In the USA, the Biologics Price Competition and Innovation Act (BPCIA) of 2009 allowed the FDA to develop procedures that lead to regulatory approval of biosimilars and in early 2012 the agency released drafts outlining the pathway that needed to be followed for sponsors to obtain biosimilar approval. As in the European Union, biosimilars for the US market are subjected to strict comparability requirements with "comparability" being invoked when the manufacturing process of a biologic product is altered. In such situations, sponsors must demonstrate that the original reference product and the biosimilar are "comparable." Note, however, that comparability of the *manufacturing processes* as distinct from the reference and biosimilar products themselves is not mandated by either the FDA or the EMA. Structural and functional comparisons with the reference product underlie the development of biosimilars, meaning that the two preparations should be highly similar and have an identical primary amino acid sequence but, because of post-translational modifications of glycosylated preparations, different isoforms with the same amino acid sequence may result and different batches of a preparation may exhibit heterogeneity. The expected difficulty of achieving complete identity remains problematic and a controversial regulatory question. Although variation in amino acid sequence from the reference disqualifies a preparation as a biosimilar in Europe and the USA and does not accord with the WHO definition, there is no common definition of a "biosimilar" between regulatory agencies throughout the world (e.g., Europe compared to China and India). A study in which two commercially available biosimilars were characterized and compared to their reference product etanercept (Enbrel®) helps to illustrate the problem of heterogeneity and the extent to which a biosimilar should demonstrate similarity to its reference product. Two commercially available intended biosimilar TNF receptor protein products were characterized and compared with the etanercept reference for potency and by carrying out analyses for peptide mapping, charge variations, tandem mass spectroscopy, affinity, and N-glycosylation. One of the products showed high similarity to the reference Enbrel® whereas the second preparation had a two amino acid variance in the Fc region, a significantly reduced content of sialylated *N*-oligosaccharides, and a significant difference in charge but its affinity and bioactivity were similar to the reference preparation. Therefore, despite differences in primary structure and glycosylation which would disqualify the second preparation from being called a biosimilar in Europe, its comparable bioactivity with the reference prompted the authors to draw attention to different definitions of a biosimilar by regulatory agencies and to ask whether more comparability studies should sometimes be undertaken to verify safety and efficacy of such products.

The above discussion leads on to the question of whether exceptions should or might be made in the requirements for at least some biosimilars in poorer countries where affordability of treatments is an even bigger consideration than in more affluent countries. It has been estimated that the cost of development of a biosimilar in the USA or Europe will be in the region of $75–250 million, and this will restrict the number of enterprises able to bring an approved product to market. Although the development of biosimilars in highly regulated markets such as the USA, Europe, Australia, Canada, and Japan will demand a high degree of similarity to the reference product plus purity, efficacy, and safety, in some less regulated jurisdictions where economic considerations and other factors lead to lower standards, the so-called alternative biologics may be increasingly approved without demonstrating biosimilarity. In both analytical and clinical terms, such products are not biosimilars but copycat "follow-on" biologics.

By mid 2015 the EMA had approved 22 biosimilars (20 still approved after one preparation each of filgrastim and somatropin were withdrawn in 2011 and 2012, respectively) including eight for filgrastim (recombinant human granulocyte colony-stimulating factor [G-CSF]; reference or originator product Neupogen®; Chap. 5, section "Colony-Stimulating Factors: Filgrastim, Sargramostim and Tbo-Filgrastim"), five for erythropoietin (Epogen®; Chap. 5, section "Epoetins"), and three for the mAb infliximab (Remicade®; Chap. 4, section "Infliximab") (Table 13.1). By April 2016, 45 biosimilars have been approved worldwide with, again, filgrastim (reference Neupogen®; 16 approvals) heading the list followed by infliximab (Remicade®; nine approvals), erythropoietin (Epogen®; seven approvals), somatropin (Genotropin®; Chap. 7, section "Indications and Usage of Somatropin"; 4 approvals), and trastuzumab (Herceptin; Chap. 3, section "Trastuzumab"; two approvals). Other biosimilars have been approved for follitropin-alfa (Gonal-f®; Chap. 8, section "Follicle-Stimulating Hormone"), insulin glargine (Lantus®; Chap. 7, section "Different Available Insulin Preparations"), adalimumab (Humira®; Chap. 4, section "Adalimumab"), rituximab (Rituxan®; Chap. 3, section "Rituximab"), and etanercept (Enbrel®; Chap. 6, section "Etanercept") (Table 13.1). In March 2015, the FDA approved its first biosimilar, Zarxio® (reference Neupogen®), approved in Europe in 2009 as Zarzio®. Some commentators have predicted that five or more biosimiolars may be approved in the USA by the end of 2015 but by April 2016, only one more, the Remicade biosimilar Inflectra (infliximab-dyyb), had received FDA approval. In accordance with the FDA's recently released (March 2016) draft guidance for the labeling of biosimilar products, the common name of the product, infliximab-dyyb, contains a suffix which, unlike Sandoz's Zarxio® (filgrastim-sndz), does not identify the manufacturer and has no meaning (see "Naming of Biosimilars and Pharmacovigilance Considerations" below). Other biosimilars under review by the EMA include the antithrombotic enoxaparin sodium (reference product Lovenox®; Chap. 10, section "The Clotting Cascade"), etanercept (Enbrel®) used to treat rheumatoid arthritis, and human insulin (Insuman®). Another biosimilar candidate for infliximab has also been accepted for review by the EMA.

Naming of Biosimilars and Pharmacovigilance Considerations

The importance placed on rigorous pharmacovigilance for biologics means that biosimilars will need to be accurately identified in safety reports and health records, and it is therefore essential that adverse events associated with biosimilars and

Table 13.1 Approved biosimilars worldwide

INN	Trade name (™/®)	Reference product (®)	Market	Year of approval
Somatropin	Omnitrope	Genotropin	Australia	2004[a]
Somatropin	Omnitrope	Genotropin	EU	2006
Somatropin	Somatropin BS Sandoz KK	Genotropin	Japan	2009
Somatropin	Omnitrope	Genotropin	New Zealand	2013
EPO-alfa	Binocrit	Epogen	EU	2007
EPO-alfa	EPO-alfa Hexal	Epogen	EU	2007
EPO-alfa	Abseamed	Epogen	EU	2007
EPO-zeta	Silapo	Epogen	EU	2007
EPO-zeta	Retacrit	Epogen	EU	2007
EPO-kappa	EPO-kappa BS JCR Pharmaceuticals	Epogen	Japan	2010
EPO-alfa	Binocrit	Epogen	New Zealand	2012
Filgrastim	Ratiograstim	Neupogen	EU	2008
Filgrastim	TevaGrastim[b]	Neupogen	EU	2008
Filgrastim	Biograstim	Neupogen	EU	2008
Filgrastim	Filgrastim Hexal	Neupogen	EU	2009
Filgrastim	Zarzio	Neupogen	EU	2009
Filgrastim	Nivestim	Neupogen	EU	2010
Filgrastim	Nivestim	Neupogen	Australia	2010
Filgrastim	Nivestim	Neupogen	New Zealand	2012
Filgrastim	Filgrastim BS Injection Syringe Mochida	Neupogen	Japan	2012
Filgrastim	Filgrastim BS Injection Syringe Nippon Kayaku	Neupogen	Japan	2013
Filgrastim	Zarzio	Neupogen	Australia	2013
Filgrastim	Grastofil	Neupogen	EU	2013
Filgrastim	Filgrastim BS Injection Syringe Sandoz	Neupogen	Japan	2014
Filgrastim	Zarzio	Neupogen	New Zealand	2014
Filgrastim	Accofil	Neupogen	EU	2014
Filgrastim-sndz	Zarxio	Neupogen	USA	2015
Follitropin-alfa	Ovaleap	Gonal-f	EU	2013
Follitropin-alfa	Bemfola	Gonal-f	EU	2014
Insulin glargine	Abasria	Lantus	EU	2014
Insulin glargine	Insulin glargine BS Lilly	Lantus	Japan	2015
Infliximab	Remsima	Remicade	Korea	2012
Infliximab	Remsima	Remicade	EU	2013
Infliximab	Inflectra	Remicade	EU	2013

(continued)

Table 13.1 (continued)

INN	Trade name (™/®)	Reference product (®)	Market	Year of approval
Infliximab	Flammegis	Remicade	EU	2013
Infliximab	Remsima	Remicade	Canada	2014
Infliximab	Inflectra	Remicade	Canada	2014
Infliximab	Infliximab BS for IV infusion 100 mg NK	Remicade	Japan	2014
Infliximab	Infimab	Remicade	India	2014
Infliximab	Inflectra	Remicade	USA	2016
Trastuzumab	CANMab	Herceptin	India	2013
Trastuzumab	Herzuma	Herceptin	Korea	2014
Adalimumab	Exemptia	Humira	India	2014
Rituximab	Acellbia	Rituxan	Russia	2014
Etanercept	Davictrel[c]	Enbrel	Korea	2014

Benepali, an etanercept biosimilar referencing Enbrel, was recently approved in the EU (2016) and launched in the UK
INN International nonproprietary name
[a]Not approved as a biosimilar; approved before biosimilar guidelines were issued
[b]Approved under a Biologics License Application in the USA where it is known as Granix®
[c]Technology to be transferred from Hanwha, South Korea, to Merck

reference products are detected and distinguished. In the post-approval phase, monitoring of safety of biological products relies on active surveillance and spontaneous reporting and both may be compromised when there is more than one manufacturer of a biosimilar based on a common reference product. Workable measures to ensure traceability for biologics are therefore essential and, at the least, such measures should include some form of coding and distinguishable nomenclature. It is clear, however, that nomenclature for biosimilars remains an outstanding and potentially controversial issue. There is an obvious need for names that clearly distinguish each biosimilar product from its reference product and from other biosimilars. The need for this is reinforced in the USA by the BPCIA which allows interchangeability, stating that the biosimilar product may be substituted (e.g., by the pharmacist) for the reference product without consultation with the prescriber of the reference product. There is no regulation in Europe for interchangeability of biosimilars where regulations for substitution of biologics differ between European Union member states. However, in some countries, some steps have already been taken to tackle the naming problem. In Australia, a biosimilar is given the Australian Biologic Name of the reference product plus the term "sim" and a three-letter code issued by the WHO INN Committee. In Japan, the suffix "BS" to denote a biosimilar is added to the reference name.

Naming actions already adopted by the FDA may give an indication of its future naming policy for biosimilars. Several biologics, for example, botulinum toxins onabotulinumtoxinA and abobotulinumtoxinA (Chap. 12), the mAb ado-trastuzumab emtansine, tbo-filgrastim, and the fusion protein ziv-aflibercept have been named by adding a prefix to the nonproprietary name, and it has been suggested that this practice of the addition of a distinguishable prefix or suffix should be followed to distinguish each biosimilar from its reference product and from other biosimilars. For the newly approved Zarxio®, reference Neupogen® and generic name filgrastim, the FDA called the drug filgrastim-sndz, the suffix representing the name of the manufacturer, Sandoz. The FDA stated that this naming did not necessarily indicate the agency's future naming policy for biosimilars.

In August 2015, the FDA released nonlegally enforceable draft guidance entitled "Nonproprietary Naming of Biological Products; Draft Guidance for Industry; Availability"(https://www.federalregister.gov/articles/2015/08/28/2015-21383/nonproprietary-naming-of-biological-products-draft-guidance-for-industry-availability) along with a proposed rule related to the nonproprietary names for six FDA-approved products that are either biosimilars, reference products for biosimilars, or a related biologic ("Designation of Official Names and Proper Names for Certain Biological Products"; https://www.federalregister.gov/articles/2015/08/28/2015-21382/designation-of-official-names-and-proper-names-for-certain-biological-products). The use of nonproprietary names was proposed for reference and biosimilar products with a common drug substance name, and this is to be followed by a random combination of four letters to uniquely identify each product. For example, for Epogen®, marketed by Amgen, the proposed proper name plus four letters, epoetin alfa-cgkn, was suggested. If the suffix is to be manufacturer derived, the assigned proposed proper name would be epoetin alfa-amgn. For Amgen's Neupogen®, the suggested corresponding proposed names were filgrastim-jcwp and filgrastim-amgn, and for the biosimilar Zarxio® (Sandoz), filgrastim-bflm, and filgrastim-sndz, respectively. In attempting to ensure adequate pharmacovigilance for biosimilars and to enable clinicians to distinguish individual biologicals, particularly biosimilars, the FDA may have recognized that prescribers and patients are more likely to remember trade than nonproprietary names. Nonproprietary names without a suffix are also likely to be more easily remembered. Some critics believe that the use of nonproprietary names for biosimilars may create confusion for providers, patients, and billing and contribute to limiting the uptake of biosimilars. In inviting comments on the draft guidance, the FDA has directed attention to possible potential approaches for designating and incorporating suffixes retrospectively and prospectively into the nonproprietary names of *all* biological products and has also solicited comments on ways to improve pharmacovigilance systems for monitoring safety.

In its recently released draft guidance, "Labeling for Biosimilar Products" mentioned above, the FDA recommends that a biosimilar product label should contain the relevant information from the reference product's label plus any "product-specific modifications". It is already clear that interchangeability between products such as infliximab-dyyb and Remicade is not universally acceptable and responses to defend intellectual property rights are likely to follow. In the meantime, discussion remains open and is sure to be taken up by a variety of interested parties.

Biobetters

From the above discussion, it is clear that while biosimilars offer the promise of fostering competition, allowing the treatment of more patients at lower cost, and helping to lower ever-increasing health costs for governments, their development costs, and the many difficulties involved in achieving the required comparability or similarity and eventual regulatory approval, are considerable. With all of these difficulties, it seems likely that some companies will choose to pursue the so-called biobetters, that is improved next-generation versions of originator biologics, rather than biosimilar products. At the end of 2014, 452 biobetters were said to be in development compared to 655 biosimilars and the relative development numbers of biobetters to biosimilars for the most popular reference products were Epogen® (epoetin alfa) 25 biobetters, 82 biosimilars; Neupogen® (filgrastim) 15, 56; Enbrel® (etanercept) 11, 26; Remicade® (infliximab) 8, 13; Humira® (adalimumab) 7, 20.

A number of strategies may be employed to produce a biobetter including modification of an existing biologic agent and improving a manufacturing process. Such changes may enhance clinical activity, increase half-life, reduce immunogenicity, and/or lessen or eliminate a particular adverse reaction. PEGylation, for example, may prolong half-life; production of a humanized protein can remove unwanted foreign antigens; manipulation of the attached glycan structures on antibody Fab and Fc components may produce higher affinity for antigen and mediate normal effector functions via Fc receptors and the C1q component of complement; and addition of a carefully selected small molecule drug or toxin can improve clinical efficacy. An example of the latter modification is the linkage of the cytotoxic agent DM1 to the mAb trastuzumab (Chap. 3, section "Ado-Trastuzumab Emtansine") which, in unconjugated form, prevents the growth of cancer cells by binding the HER2 receptor. After addition of the toxin to form ado-trastuzumab emtansine, cell killing is increased by the mAb-drug conjugate entering the tumor cells where it inhibits the assembly of microtubules leading to cell cycle arrest and ultimately cell death.

Biosimilars in the Immediate Future

Globally, the World Health Organization has generated guidelines for the consideration of biosimilars in the less regulated markets. These guidelines have been mainly derived from the European Union, Japanese, Australian, and Canadian regulated markets. Pointing out that the USA is already the leading market for biotechnology-based products and the FDA led the world in 1996 with the development of comparability guidance, McCamish and Woollett believe that the FDA's "support of a scientifically-justified and appropriate reduction in the data burdens for biosimilars through the FDA's recognition of quality characterization and support of abbreviated clinical development programs would create great confidence amongst sponsors, and no changes in risks for consumers."

From the times of the earliest usage of biologic therapies and the introduction of biologic drugs in recombinant form in the 1980s, it seemed unlikely, if not impossible, that generic versions of the products could be made and licensed. Despite all the perceived complexities in achieving such an outcome, a number of factors have combined to demonstrate that the production of biologic generic-like "similars" suitable for approval as therapies by regulatory agencies is entirely feasible. These factors are the stimulus of expiring patents; the potential and need of large financial rewards; application of strategies based on comparability to a well-defined reference; scientific advances in physicochemical and analytical technologies; and impressive developments in biopharmaceutical manufacturing processes. Although development costs will be high and problems related to naming and the substitution of biosimilars for brand-name drugs remain to be resolved, the introduction of these drugs looks set to make cheaper lifesaving drugs available to millions and save billions of dollars in drug costs over the next decade.

Summary

- A biosimilar medicine is a biological medicine developed to be similar to an existing biological medicine (the "reference medicine"). Biosimilars may be authorized as high-quality, lower cost biologics when patents expire. A biosimilar may be defined as a follow-on biologic that meets extremely high standards for comparability or similarity to the originator biologic drug that is approved for use in the same indication. They are not the same as generics which are considered to be identical to their reference medicines.
- A biosimilar is developed to be the same as its reference medicine but in practice there will be differences. Marketing approval is only given if any variability and differences are shown to not affect efficacy and safety.
- In the European Union and the USA, biosimilars are subjected to strict comparability requirements with "comparability" being used when the manufacturing process of a biologic product is altered. Sponsors must demonstrate that the original reference product and the biosimilar are "comparable."
- In 2006, the EMA approved the recombinant somatropin preparation, Omnitrope®, making it the first approved biosimilar.
- The reference product and biosimilars should be highly similar and have an identical primary amino acid sequence but, because of post-translational modifications of glycosylated preparations, different isoforms may result and different batches of a preparation may exhibit heterogeneity.
- There is no common definition of a "biosimilar" between regulatory agencies throughout the world. In some less regulated jurisdictions where economic con-

siderations and other factors lead to lower standards, the so-called alternative biologics may be increasingly approved without demonstrating biosimilarity.

- By mid 2015, the EMA had approved 20 biosimilars including filgrastim (originator product Neupogen®), erythropoietin (Epogen®), and the mAb infliximab (Remicade®). By April 2016, 45 biosimilars have been approved worldwide with, again, filgrastim heading the list followed by infliximab, erythropoietin, somatropin, and trastuzumab (Herceptin®). Other biosimilars have been approved for follitropin-alfa (Gonal-f®), insulin glargine (Lantus®), adalimumab (Humira®), rituximab (Rituxan®), and etanercept (Enbrel®). In March 2015, the FDA approved its first biosimilar, Zarxio® (filgrastim, reference Neupogen®), approved in Europe in 2009 as Zarzio®. Inflectra® (infliximab-dyyb; reference product Remicade®) became the second product approved by the FDA in April 2016.
- The importance placed on rigorous pharmacovigilance for biologics means that biosimilars will need to be accurately identified in safety reports and health records. It is therefore essential that adverse events associated with biosimilars and reference products are detected and distinguished. In the post-approval phase, monitoring of safety of biological products relies on active surveillance and spontaneous reporting and both may be compromised when there is more than one manufacturer of a biosimilar based on a common reference product.
- There is an obvious need for names that clearly distinguish each biosimilar product from its reference product and from other biosimilars. The need for this is reinforced in the USA by the Biologics Price Competition and Innovation Act which allows interchangeability. It has been suggested that a prefix or suffix should be added to the root nonproprietary name and this has been done for the mAb biosimilar Inflectra, recently approved by the FDA.
- Because of development costs, and the many difficulties involved in achieving the required comparability, some companies will choose to pursue the so-called biobetters, that is improved next-generation versions of originator biologics, rather than biosimilar products.
- A number of strategies may be employed to produce a biobetter including modification of an existing biologic agent and improving a manufacturing process. Such changes may enhance clinical activity, increase half-life, reduce immunogenicity, and/or lessen or eliminate a particular adverse reaction.
- For the global situation, the World Health Organization has generated guidelines for the consideration of biosimilars in the less regulated markets. These guidelines have been mainly derived from the European Union, Japanese, Australian, and Canadian regulated markets.
- Despite all of the perceived complexities and obvious difficulties, the production of biologic generic-like "similars" suitable for approval as therapies by regulatory agencies has been shown to be feasible.
- Although development costs will be high and problems related to naming and the substitution of biosimilars for brand-name drugs remain to be resolved, the introduction of these drugs looks set to make cheaper lifesaving drugs available to millions and save billions of dollars in drug costs over the next decade.

Further Reading

Beck A, Diemer H, Ayoub D, et al. Analytical characterization of biosimilar antibodies and Fc-fusion proteins. Trends Anal Chem. 2013;48:81–95.

Beck A, Sanglier-Cianférani S, Van Dorsselaer A. Biosimilar, biobetter, and next generation antibody characterization by mass spectroscopy. Anal Chem. 2012;84:4637–46.

Berkowitz SA, Engen JR, Mazzeo JR, et al. Analytical tools for characterizing biopharmaceuticals and the implications for biosimilars. Nat Rev Drug Discov. 2012;11:527–40.

Casadevall N, Felix T, Strober BE, et al. Similar names for similar biologics. BioDrugs. 2014;28:439–44.

Fang J, Doneanu C, Alley WR Jr, et al. Advanced assessment of the physicochemical characteristics of Remicade® and Inflectra® by sensitive LC/MS techniques. mAbs. 2016;8:1021–34.

Grampp G, Felix T. Pharmacovigilance considerations for biosimilars in the USA. BioDrugs. 2015;29:309–21.

Kálmán-Szekeres Z, Olajos M, Ganzler K. Analytical aspects of biosimilarity issues of protein drugs. J Pharm Biomed Anal. 2012;69:185–95.

Macdonald JC, Hartman H, Jacobs IA. Regulatory considerations in oncologic biosimilar drug development. mAbs. 2015;7:653–61.

McCamish M, Woollett G. Worldwide experience with biosimilar development. mAbs. 2011;3:209–17.

Tan Q, Guo Q, Fang C, et al. Characterization and comparison of commercially available TNF receptor 2-Fc fusion protein products. mAbs. 2011;3:209–17.

van Aerts LA, de Smet K, Reichmann G, et al. Biosimilars entering the clinic without animal studies. mAbs. 2014;6:1155–62.

Weise M, Bielsky M-C, de Smet K, et al. Biosimilars: what clinicians should know. Blood. 2012;120:5111–7.

Index

B.A. Baldo, *Safety of Biologics Therapy*, DOI 10.1007/978-3-319-30472-4

N

O

P

MIX
Papier aus verantwortungsvollen Quellen
Paper from responsible sources
FSC® C105338

If you have any concerns about our products,
you can contact us on
ProductSafety@springernature.com

In case Publisher is established outside the EU,
the EU authorized representative is:
Springer Nature Customer Service Center GmbH
Europaplatz 3, 69115 Heidelberg, Germany

Printed by Libri Plureos GmbH
in Hamburg, Germany